ANNUAL REVIEW OF NEUROSCIENCE

ANNUAL REVIEW OF NEUROSCIENCE

VOLUME 8, 1985

W. MAXWELL COWAN, *Editor*
Salk Institute for Biological Studies

ERIC M. SHOOTER, *Associate Editor*
Stanford University School of Medicine

CHARLES F. STEVENS, *Associate Editor*
Yale University School of Medicine

RICHARD F. THOMPSON, *Associate Editor*
Stanford University

ANNUAL REVIEWS INC. 4139 EL CAMINO WAY PALO ALTO, CALIFORNIA 94306 USA

ANNUAL REVIEWS INC.
Palo Alto, California, USA

International Standard Serial Number: 0147-006X
International Standard Book Number: 0-8243-2408-0

Typesetting and Composition by Graphic Typesetting Service, Los Angeles, California
Typesetting Production Coordinator, Barbara Menkes

PRINTED AND BOUND IN THE UNITED STATES OF AMERICA

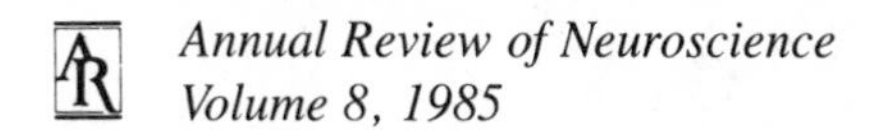
Annual Review of Neuroscience
Volume 8, 1985

CONTENTS

(Note: Titles of chapters in Volumes 4–8 are arranged by category on pages 600–3.)

(*Continued*)

SOME RELATED ARTICLES IN OTHER *ANNUAL REVIEWS*

From the *Annual Review of Biochemistry*, Volume 54 (1985)

Cell Adhesion and the Molecular Processes of Morphogenesis, G. M. Edelman

Growth Hormone Releasing Factors, R. Guillemin, N. Ling, P. Bohlen, F. Esch, F. Zeytin, A. Baird, and W. Wehrenberg

Protein Kinases in the Brain, A. C. Nairn, P. Greengard, and H. C. Hemmings, Jr.

From the *Annual Review of Biophysics and Biophysical Chemistry*, Volume 14 (1985)

Adventures in Molecular Biology, F. O. Schmitt

Trigger and Amplification Mechanisms in Visual Phototransduction, M. Chabre

From the *Annual Review of Genetics*, Volume 18 (1984)

The Genetic Control of Cell Lineage During Nematode Development, P. W. Sternberg and H. R. Horvitz

From the *Annual Review of Medicine*, Volume 36 (1985)

Beta-Adrenergic Blocking Drugs, J. G. Gerber and A. S. Nies

Alzheimer's Disease, D. L. Price, P. J. Whitehouse, and R. G. Struble

From the *Annual Review of Psychology*, Volume 36 (1985)

Animal Behavior Genetics: A Search for the Biological Foundations of Behavior, R. E. Wimer and C. C. Wimer

Cellular Mechanisms of Learning, Memory, and Information Storage, J. Farley and D. L. Alkon

From the *Annual Review of Pharmacology and Toxicology*, Volume 25 (1985)

Physiological Studies with Somatocrinin, A Growth Hormone–Releasing Factor, W. B. Wehrenberg, A. Baird, F. Zeytin, F. Esch, P. Böhlen, N. Ling, S. Y. Ying, R. Guillemin

The Physiology and Pharmacology of Spinal Opiates, T. L. Yaksh and R. Noueihed

Alterations in the Release of Norephinephrine at the Vascular Neuroeffector Junction in Hypertension, T. C. Westfall and M. J. Meldrum

From the *Annual Review of Physiology*, Volume 47 (1985)

Physiology and Pharmacology of Gap Junctions, D. C. Spray and M. V. L. Bennett

Antibody Probes in the Study of Gap Junctional Communication, E. L. Hertzberg

The Role of Gap Junctions in Development, S. Caveney

Physiological Roles of Permeable Junctions: Some Possibilities, J. D. Sheridan and M. M. Atkinson

Mechanism of Calcium Transport, G. Inesi

Ion Movements Through the Sodium Pump, J. H. Kaplan

Oscillating Neural Networks, A. I. Selverston

Circadian Neural Rhythms in Mammals, M. Moulins and F. W. Turek

ANNUAL REVIEWS INC. is a nonprofit scientific publisher established to promote the advancement of the sciences. Beginning in 1932 with the *Annual Review of Biochemistry,* the Company has pursued as its principal function the publication of high quality, reasonably priced *Annual Review* volumes. The volumes are organized by Editors and Editorial Committees who invite qualified authors to contribute critical articles reviewing significant developments within each major discipline. The Editor-in-Chief invites those interested in serving as future Editorial Committee members to communicate directly with him. Annual Reviews Inc. is administered by a Board of Directors, whose members serve without compensation.

Ann. Rev. Neurosci. 1985. 8:1–19

THE PRIMATE PREMOTOR CORTEX: Past, Present, and Preparatory†

Steven P. Wise

Laboratory of Neurophysiology, National Institute of Mental Health, Bethesda, Maryland 20205

THE CONCEPT AND HISTORY OF THE PREMOTOR COMPLEX

Campbell and the Invocation of J. Hughlings Jackson

The concept of a premotor cortex was first proposed by Campbell (1905), who called it the "intermediate precentral cortex." Campbell linked the precentral motor cortex to the middle of J. Hughlings Jackson's three levels of motor organization and postulated that the rostrally adjacent region, the intermediate precentral cortex, might be the site of Jackson's highest motor level. Brodmann (1905, 1909) agreed with Campbell's cytoarchitectonic definitions and emphasized the idea that together the precentral cortex (area 4) and the rostrally adjacent field (area 6) constituted a group of functionally related, presumably motor, areas.

Other investigators also thought that they could identify architectonic subdivisions of the primate frontal cortex (Vogt & Vogt 1919, von Economo 1929, Bucy 1933, 1935, von Bonin & Bailey 1947, von Bonin 1949, Bailey & von Bonin 1951, Jones et al 1978), although their schemes differed markedly from one another. In the years since the beginning of the century, several of these subdivisions have been identified as the "premotor cortex," but that term has been employed with an inconsistency and vagueness that has hindered the study of this cortical region. This review is submitted in an attempt to clarify some of the issues involved in the definition of the premotor cortex and the study of its function. Transcending the narrow questions about the definition of premotor cortex, however, are problems representative of those widely encountered in cortical areas not linked in a straightforward fashion with either a receptor system or effector organ. In this sense, then, study of the premotor cortex is relevant to the more general problem of nonprimary, or "association," cortical areas.

Terminology of the Premotor Cortex

A large part of the frontal lobe of primates is thought to play an important role in the cerebral control of movement. Those regions commonly designated as motor fields lack a clearly defined internal granular layer, layer IV, and thus often are referred to as the *agranular frontal cortex*. When used in this way, the term "agranular frontal cortex" excludes not only the granular prefrontal cortex, but also medial and orbital frontal regions of cortex. A three-part division of the agranular frontal cortex (Figure 1) is possible on the basis of a constellation of electrophysiological, connectional, architectonic, and behavioral findings that are outlined below. Two subdivisions, the *primary* (or *precentral) motor cortex* (MI) and the *supplementary motor cortex* (MII), are located approximately as described by Woolsey and his colleagues (1952).

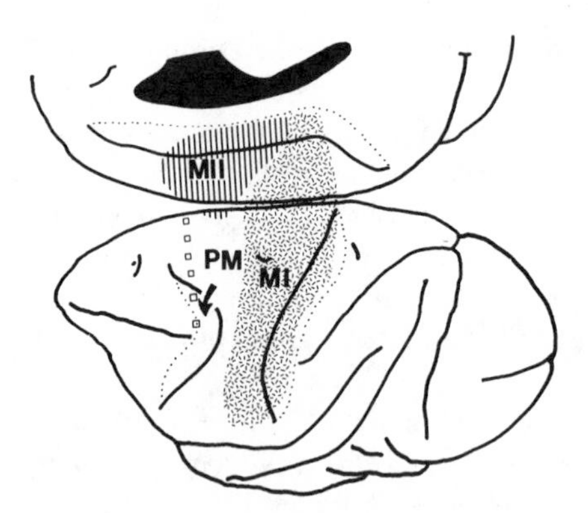

Figure 1 Surface view of the left cerebral hemisphere of a macaque monkey. The medial surface of the hemisphere is shown, inverted, at top. The approximate locations of the primary motor cortex (MI), the supplementary motor cortex (MII), and the premotor cortex (PM) are indicated. Each *dotted line* represents the fundus of a sulcus. The *open squares* show a rough estimation of the rostral boundary of PM.

The third subdivision, the focus of this review, is the part of the agranular frontal cortex lateral to MII and rostral to MI: the *premotor cortex* (PM).

Hines (1929) appears to have been the first to use the term "premotor cortex," which was adopted later by Fulton and his collaborators (Fulton et al 1932, Bucy & Fulton 1933, Fulton 1934, 1935, Jacobsen 1934, Kennard et al 1934). However, their "premotor area" contained the supplementary motor cortex as it was later defined by physiological methods (Woolsey et al 1952, Penfield & Welch 1951). It seems advisable, therefore, to restrict the name "premotor cortex" to the more laterally situated part of the agranular frontal cortex, the region outside of the boundaries of both the primary motor cortex (MI) and the supplementary motor cortex (MII) as defined physiologically. Although the terminology used here is not identical to all of the historical uses of the terms "premotor cortex" or "premotor area," this system of nomenclature is consistent with that of several contemporary investigators (Brown et al 1979, Brinkman & Porter 1983, Roland et al 1980a, b, Wiesendanger 1981, Weinrich & Wise 1982, Freund & Hummelsheim 1984). Together, MII and PM are called the "nonprimary motor cortex" in this review in order to contrast them with the primary motor cortex (Wise 1984), and the term "motor cortex" is used in its general sense to refer to all motor cortical fields.

The relation of MI, MII, and PM to cortical cytoarchitectonics remains uncertain. In one view, MI is taken to be identical to Brodmann's area 4, and PM and MII together occupy area 6 (see e.g. Hartmann-von Monakow 1979, Asanuma et al 1983). There are many alternative views, however, most of which depend upon the architectonic descriptions of Vogt & Vogt (1919). Based on a comparison of MI as defined by physiological methods and the cortical map of Vogt & Vogt, several investigators (e.g. Woolsey et al 1952, Humphrey 1979, Wiesendanger 1981) have concluded that MI corresponds to area 4 plus a part of area 6. It is often assumed, invariably without architectonic documentation, that the part of area 6 to be included within MI corresponds to a cortical field termed 6aα by Vogt & Vogt (1919), but no one since the Vogts has been able to distinguish area 6aα from the rostrally adjacent field, 6aβ. Accordingly, and in view of the uncertainty about the exact location of the area 4/6 boundary within the precentral gyrus, it seems most useful to abandon these architectonic terms and instead use those derived from neurophysiological studies.

History of Premotor Cortex Study

FULTON'S FORMULATION Fulton's studies of the nonprimary motor cortex marked the starting point of the experimental investigation of this region. One should not be too surprised to find that these early studies now appear primitive and inconclusive, or that they engendered controversy when other investigators, with different views of the nebulous cortical boundaries involved, approached similar questions. However, many of the concepts entertained by Fulton and his colleagues in early studies of the premotor cortex appear more accurate, in the main, than many of the more popular views that dominated the intervening years.

The primary contribution of Fulton and his school was to solidify the concept of a nonprimary motor cortex with distinct motor functions. The lengthy and confusing arguments about the site necessary for and types of spasticity and abnormal grasping reflexes are not reviewed here. Several treatments of this subject are available (e.g. Humphrey 1979, Wiesendanger 1981), but such observations have failed to elucidate the problem of motor cortex function and specialization. In brief, Fulton and his colleagues asserted that ablations of the nonprimary motor cortex in monkeys, apes, and man led to spasticity, forced grasping, and vasomotor disturbances (Fulton et al 1932, Kennard et al 1934, Jacobson 1934, Kennard 1935, Fulton 1934, 1935), but the most interesting effect they claimed to observe was a specific deficit in the execution of skilled movements.

Of course, there are serious problems with Fulton's experiments. In none of the studies purporting to describe a premotor syndrome was the extent and location of the cortical damage accurately assessed. Indeed, in view of the lack of agreement concerning the architectonic correlates of frontal cortical fields, it is easy to understand the problems these investigators had in the localization of ablations and tumors. Even accurate localization of the damaged sites would not have entirely solved the problem, since the cortical fields had not been well defined. In the context of these difficulties and imprecise formulations, controversy and disagreement would seem inevitable, and they developed rather early in the study of the premotor cortex. Foerster (1936), for example, stated that in his experience with 40 patients with surgical damage to the premotor cortex, he observed none of the symptoms that Fulton described. Walshe (1935, p. 61) went further to deny almost all of Fulton's fundamental observations as "so paradoxical that their interpretation has so far proved impossible." Reviewing these controversies in detail would not be fruitful, but it should be noted that it was in the context of such uncertainty that subsequent investigations, most notably Woolsey's, took place.

WOOLSEY'S WORK Woolsey, along with Penfield & Welch (1951), definitively established the concept of a nonprimary motor cortex by discovering the supplementary motor cortex, a motor representation on the medial aspect of the agranular frontal cortex. However, the studies of Woolsey and his colleagues (Woolsey et al 1952, Travis 1955) led them to the conclusion that the premotor cortex was not a part of the motor cortex and played no role in the cerebral control of movement. Two sets of observations led Woolsey and Travis to these conclusions: (*a*) electrical shocks applied to the premotor cortex of monkeys deeply anesthetized with barbiturates did not evoke movements, and (*b*) ablations of premotor cortex that were added to previous lesions of either the supplementary motor cortex or the precentral motor cortex did not cause additional gross abnormalities in locomotion or muscle tone. Certain problems with these studies are described below, but even if one were to accept these negative results, it is difficult to understand the strength and confidence with which these investigators eliminated the premotor cortex from a role in the cerebral control of movement. Woolsey et al (1952, p. 259) concluded that it " . . . should be evident that our findings are incompatible with the idea of a premotor cortex as that concept has been understood for the past two decades [1931–1950]." And Travis (1955, p. 186) concluded, in the same vein, that " . . . lesions of the dorsal and lateral 'area 6' not included in the precentral motor area do not contribute observable motor deficits in the limbs. The only functional portions of 'area 6' are those included in either precentral or supplementary motor areas."

However, aside from their negative nature, the observations that formed the basis of these conclusions were further weakened by methodological problems. It had already been reported that the level of anesthesia was a critical variable in experiments involving premotor cortical stimulation in monkeys (Bucy 1933, Bucy & Fulton 1933). It is therefore not surprising that stimulation of the premotor cortex failed to evoke movements in the deeply anesthetized preparations used by Woolsey. Moreover, the aphysiological nature of electrical stimulation should lead to rather cautious interpretation of such results; certainly the absence of "electrical excitability" of a cortical field should not be construed as demonstrating its lack of participation in the cerebral control of movements. As for the study of Travis, an assessment of the effects of premotor cortex ablations that includes mainly gross observation of locomotion, muscle tone, etc. without quantitative measures, should be viewed with skepticism and interpreted with caution. In any event, the lack of chronic gross motor deficits after ablation of the premotor cortex should have been expected if, as had already been suggested at the time of Travis's study, it is involved in "higher-order" aspects of motor control (for further discussion see Wise 1984).

In conclusion, neither the work of Woolsey nor that of his collaborators yielded any convincing evidence against the concept of a premotor cortex asa distinct cortical field with a specialized role in the cerebral control of movement. However, although a few neuroanatomists and clinical neurologists continued to have an interest in the premotor cortex, neurophysiologists and neuropsychologists rarely studied this region for the next quarter of a century.

Definition of the Premotor Cortex

GENERAL CONSIDERATIONS Cortical field definition typically requires synthesis of data obtained by many methods from a large number of different laboratories. The criteria that can be used to define a cortical field include architectonics, connectivity, effects of electrical stimulation, activity of neurons or groups of neurons in the field, and the behavioral effects of damage to the cortical field. Rarely does any one of these features clearly and unambiguously serve to define a cortical field, and even more rarely can such a criterion be applied uniformly to a large number of species or to more than one boundary of a field. Usually, and this is clearly the case for the premotor cortex, many bits of information must be collated and compared to arrive at a constellation of properties that serve to define a cortical field. Complicating this process further is the unfortunate but unavoidable fact that not all the published reports are equally thorough and reliable in their description and documentation of results. Additional problems have been caused by the practice of using several terms and concepts in vague, ambiguous, and often confusing ways. In the discussion below, the premotor cortex of macaque monkeys is treated as a single cortical field located roughly as indicated in Figure 1. The term "premotor cortex" is used regardless of the terminology of the cited reports, and it is taken to include the arcuate premotor area (Schell & Strick 1984), peri-arcuate cortex (Godschalk et al 1981), post-arcuate cortex (Rizzolatti et al 1983), and area 6 of others.

DEFINITION BY EFFERENT CONNECTIVITY In this review, I discuss only those aspects of cortical connectivity that may help to distinguish premotor cortex from adjacent cortical fields. Two aspects of corticofugal organization are most relevant. First, it has been reported that PM (along with MII) projects only to the parvocellular part of the red nucleus, whereas MI projects to both its parvocellular and magnocellular parts (Kuypers & Lawrence 1967, Catsman-Berrevoets et al 1979, Hartmann-von Monakow et al 1979). Second, according to Sessle & Wiesendanger (1982), whereas MI and MII project directly to the spinal cord, PM does not. This view is consistent with older reports that PM sends its most prominent corticofugal projection to the medullary reticular formation (Kuypers & Brinkman 1970, Catsman-Berrevoets & Kuypers 1976). It should be noted, however, that other reports indicate that some corticospinal neurons may be located within PM (Murray & Coulter 1981, see also Toyoshima & Sakai 1982). Nevertheless, it seems likely that the lack of a prominent, direct corticospinal projection may serve to distinguish PM from both MI and MII. Other corticofugal projections (see Künzle 1978) are not very useful for the definition of PM because, like the corticopontine and corticostriatal projections, they arise from most of cerebral cortex (Kemp

& Powell 1970, Brodal 1978). Perhaps future studies will reveal specific corticofugal connections, e.g. corticopretectal or corticosubthalamic projections, that will serve to improve the definition of PM. Descriptions of corticocortical projections from the premotor cortex to the primary motor cortex (Pandya & Vignolo 1971, Muakassa & Strick 1979) have also helped to define the field, and more detailed analysis promises a further contribution in this regard.

DEFINITION BY AFFERENT CONNECTIVITY Schell & Strick (1984) have recently concluded that the main input to at least a part of premotor cortex arises from the caudal parts of the deep cerebellar nuclei and relays via nucleus X of the ventrolateral complex of thalamic nuclei (see also Rizzolatti et al 1983). The thalamic input from this part of the ventrolateral thalamus distinguishes it from MI and MII, which receive their main thalamic input from other parts of the thalamus. Further, Avendaño et al (1983) reported recently that the amygdaloid complex projects to PM, but not to MI.

Corticocortical inputs also help to distinguish PM from other motor cortical fields. Lateral parts of area 5, but more particularly areas 1 and 2, have been reported to project to MII and MI, but not to PM (Jones et al 1978). In addition, a number of frontal, visual, and auditory areas (Pandya & Kuypers 1969, Chavis & Pandya 1976, Godschalk et al 1983), including area 7, are thought to project to PM and not to MI.

DEFINITION BY NEUROPHYSIOLOGICAL METHODS Weinrich & Wise (1982) and Sessle & Wiesendanger (1982), supporting an earlier observation of Sakai (1978), have reported that intracortical electrical stimulation in the cortex rostral to MI rarely evokes movements, whereas within MI such movements are quite commonly elicited (Asanuma & Rosén 1972, Kwan et al 1978). This is not to say that movements cannot be evoked from premotor cortex, especially if stimulation points are narrowly spaced and higher currents are used, but rather that the frequency with which such movements can be evoked by relatively low current levels serves as a useful and convenient artifice for distinguishing PM from MI.

In addition to the features described above that may serve to distinguish PM from other cortical fields, the patterns of neuronal activity observed during the performance of a motor task are quantitatively different in MI and PM. The proportion of neurons apparently responding to visual stimuli that instruct a monkey about forthcoming movements (signal-related cells), or that show sustained changes in discharge rate during periods of motor preparation (set-related cells), is much larger in PM than in MI (Weinrich & Wise 1982, Weinrich et al 1984, Godschalk et al 1981, see also Lamarre et al 1983, but cf Kwan et al 1981). Rizzolatti et al (1981b) also report visual inputs to a region just lateral to the main recording site used by Weinrich & Wise. It seems most likely that the region studied by Rizzolatti and his colleagues is also within the premotor cortex. Brinkman & Porter (1979, 1983, Porter 1983) have reported that PM neurons rarely respond to somatic sensory input, in contrast to neurons in MI, and, if true, this feature may help to distinguish these cortical fields. However, Rizzolatti et al (1981a) found substantial somatic sensory inputs to PM, so further study is needed to determine whether the relative sensitivity to somatic sensory inputs can be used to distinguish PM and MI.

DEFINITION BY CYTOARCHITECTONICS Since the differences between the structure of PM and MI are much less pronounced than their similarities, it is at best difficult to describe and document reliably the cytoarchitectonic differences between these fields. This difference is often stated to be the presence of giant layer V pyramidal cells (of Betz) in MI and their absence in PM, a criterion that dates from Brodmann (1905, 1909). While there is an obvious difference between the rostral and caudal parts of the agranular frontal cortex in the density of giant pyramidal cells, the "Betz-cell criterion" suffers from two major difficulties. First, the distribution of cell-body diameters in the agranular frontal cortex shows no obvious bimodality, and therefore the distinction between Betz and non-Betz cells on the basis of size is, by necessity, arbitrary. Second, there are some very large cells, Betz cells by any published definition, located so far rostral in the agranular frontal cortex that it is inconceivable that they are all confined to the primary motor cortex (or area 4 by anyone's definition). Thus, the attempt at a simple distinction between MI and PM on the basis of the existence of Betz cells alone (e.g. Brodmann 1909, Bucy 1933, von Bonin & Bailey 1947, Denny-Brown & Botterell 1948) is bound to be unreliable.

In spite of these difficulties, it has been possible to show that the change in microstimulation threshold described above for the PM/MI boundary corresponds, at least within 2 mm, with a marked change in the density of giant layer V pyramidal cells. As electrode penetrations move from the lower-threshold (MI) to higher-threshold (PM) microstimulation zones, the density of large layer V pyramidal cells abruptly drops (Weinrich & Wise 1982, Wise 1984, see also Sessle & Wiesendanger 1982). Thus, the classical distinction drawn between PM and MI is generally correct, but the differences are not absolute. Other cytoarchitectonic differences are apparently subtle or inconsistent, although the possibility remains that future quantitative architectonic investigations, in conjunction with axonal fiber tracing techniques, neurophysiology, or other methods, may yield a clearer structural definition of PM. Regarding the boundary between PM and MII, no architectonic difference has ever been reported.

Most of this discussion focuses on an attempt to distinguish PM from MI. The distinction between premotor cortex and the "prefrontal" cortex, i.e. the frontal granular cortex, is presumably possible on purely cytoarchitectonic grounds, but very little work has been directed toward this question. It is known, however, that the distribution of contralaterally directed corticotectal cells respects this boundary. These cells are found almost exclusively within the rostral part of the agranular frontal cortex (Distel & Fries 1982). Whether this part of the frontal cortex is a separate cortical field or part of the premotor cortex is not yet known.

CONCLUSIONS On the basis of a constellation of physiological, connectional, and architectonic properties it is possible to draw the conclusion that the premotor cortex constitutes a distinct cortical field within the frontal agranular cortex. MI can be distinguished from PM on the basis of the following characteristics: its lower threshold of microstimulation; its complete representation of the body as revealed by electrical stimulation; its lower proportion of certain neuronal activity patterns, e.g. those indicating visual inputs; its higher density of giant layer V pyramidal cells; its projection to the magnocellular red nucleus; its lack of input from the amygdala; its input from a different main relay nucleus in the thalamus; and its major projection to the spinal cord.

MII, in turn, can be distinguished from PM by virtue of its complete body representation, its pronounced corticospinal projection, and its input from a distinct thalamic main-relay nucleus. The prefrontal cortex and PM can, presumably, be distinguished on cytoarchitectonic grounds. The former has a clearly observable internal granular layer, whereas the latter is agranular. The behavioral effects of damage to the premotor cortex (Halsband & Passingham 1982), discussed below, provide further evidence that it is a distinct cortical field.

That the premotor cortex should be considered part of the motor cortex is supported by the pattern of its connections with structures involved in motor control: its input from the cerebellum via the thalamus (Schell & Strick 1984); its output to the red nucleus, striatum, basilar pontine nuclei, subthalamic nucleus, and medullary reticular formation; and the high proportion of neurons closely related to voluntary limb movements (Kubota & Hamada 1978, Weinrich & Wise 1982, Brinkman & Porter 1983).

PREMOTOR CORTEX AND ITS ROLE IN THE CEREBRAL CONTROL OF MOVEMENT: PREPARATION, PROGRAMMING, AND OTHER PROPOSALS

Complex Movements

As mentioned above, Fulton (1935, 1938) and, later, Luria (1980) speculated that the primate premotor cortex is especially important in the synthesis of skilled motor sequences. The fundamental observation in this area was that of Jacobsen (1934), who reported that a chimpanzee, after ablation of the premotor cortex, was deficient on a task involving the movement of a series of latches in order to open a box. However, because Jacobsen included the supplementary motor cortex within his "premotor cortex," it is possible to attribute any observed deficit to MII damage. Recently, Deuel (1977) showed some deficits in latch-box opening in macaque monkeys after frontal cortex damage, but she also ablated a large amount of cortex outside the premotor cortex. In any event, correct performance of the latch-box and related tasks requires complicated combinations of sensorially guided movements, independent movements of muscle groups that usually work in concert, and memory of motor sequences. In addition, possible motivational and attentional variables make these experiments difficult to interpret. Still, the idea that the premotor cortex is necessary for complex sequencing of movement remains an intriguing possibility and one that might be important in understanding human speech mechanisms (Passingham 1981). However, Halsband & Passingham (1982) reported no deficit in the performance of a fixed sequence of forelimb movements after ablation of the macaque premotor cortex. Thus, if one accepts that their monkeys' lesions were as intended, the premotor cortex is not necessary for the execution of a *fixed* motor sequence. Of course, this finding does not imply that the premotor cortex is uninvolved in the performance of such tasks or less fixed sequences.

Roland and his collaborators (1980a,b) have shown, in human subjects, that significant increases in regional cerebral blood flow occur in the premotor (and the supplementary motor) cortex when subjects perform a complicated sequence of digit movements, but not when they perform a single repetitive digit movement or a sustained isometric contraction. This finding supports the idea that the premotor cortex may play an important role, along with other parts of the nonprimary motor cortex, in the execution of relatively complex movements. However, this view requires further direct experimental analysis and should be considered speculative at this time (see also Freund & Hummelsheim 1984).

Sensory Guidance of Movement

A number of studies support the idea that the premotor cortex may play an important role in the sensory guidance of movement. Neuroanatomical studies (e.g. Pandya & Kuypers 1969, Chavis & Pandya 1976) have been taken to support this view insofar as they show putatively visual (and other sensory) cortical areas projecting to the general region of the premotor cortex. In addition, some specific experimental evidence supports this hypothesis. Attempts to disconnect visual areas from premotor regions in macaque monkeys lead to marked deficits in visually guided movements, especially those requiring independent digit movements (Keating 1973, Haaxma & Kuypers 1975).

Several other behavioral reports could be interpreted in terms of deficits in sensorial guidance of movements. For example, Halsband & Passingham (1982) have reported that monkeys with ablations of the premotor cortex are deficient in an abstractly guided visuomotor task (see also Petrides 1982). Specifically, the monkeys were impaired in performing a task that required them to turn a handle by pronation in response to a signal of one color, and to pull the same handle in response to a signal of another color. Rizzolatti et al (1983) have also found deficits in sensorially guided movements after more restricted premotor cortex lesions. Many effects of cortical ablations directed toward the arcuate sulcus or prefrontal cortex, but probably involving the premotor cortex, can also be interpreted in terms of deficits in certain types of sensorially guided movements (see Wise 1984).

In man, Roland et al (1980b) have shown that the premotor cortex (along with MII) increases its blood flow during the execution of a sensorially guided movement. In one of these tasks, the "maze task" in their terminology, a subject is verbally instructed to make a limb movement of a certain magnitude and direction. There are, of course, many areas that show blood flow increases during such a task, but increases in the premotor cortex are consistent with the idea that this region is important for sensorially guided movements, especially those of an abstract nature. In addition, Roland & Larsen (1976) report blood flow increases in the premotor cortex during stereognostic testing, which may also be considered to involve the sensory guidance of movement.

Single-unit recording methods, as applied to unanesthetized monkeys, have also been used to support the argument that the premotor cortex plays a role in sensorially guided movements. For example, Rizzolatti et al (1981b) have reported visual inputs to premotor cortex. The receptive fields of premotor cortex cells may be independent of gaze direction (Gentilucci et al 1983), an important property for visual control of movement. Other investigators have reported similar neuronal activity without explicitly studying visual receptive fields (Godschalk et al 1981, Weinrich & Wise 1982, Godschalk & Lemon 1983, Wise et al 1983, Brinkman & Porter 1983). The role of this visual input in visuomotor mechanisms is unknown, however. Regarding the motor aspects of sensorially guided movements, the vast majority of premotor cortex neurons have discharge modulations temporally correlated with and preceding the onset of movement (e.g. the "movement-related cells" of Weinrich & Wise 1982, Brinkman & Porter 1983). The magnitude of modulation in some cells is correlated with movement parameters such as acceleration or velocity (Kubota & Hamada 1978, Weinrich et al 1984), but the number of such cells appears to be small. At present, the role of these patterns of activity in the control of sensorially guided or any other movements is unknown.

Preparation for Motor Action

Most commonly, the concept of motor preparation is thought of in terms of motor programming, i.e. the assembly of motor subroutines for later execution

(Sternberg et al 1978) or the assembly of force and time specifications for all of the muscles involved in an individual movement (Brooks 1979, Paillard 1983), including prime mover and postural support muscles. But there are several aspects of the preparation for movement—motor set (which includes the concepts of programming, postural stabilization, and reflex suppression), selective attention, arousal, goal selection, and strategy formulation—and all of these concepts may be important in understanding premotor cortex function.

MOTOR SET A general aspect of motor preparation is *motor set,* the state of readiness to make a particular movement (Evarts et al 1984). It is possible to test the idea that the premotor cortex plays a role in motor set by examining the activity of premotor cortex cells during an *instructed delay period,* i.e. after an animal has received an instruction for a limb movement, but before it has received a signal allowing it to execute the movement. Neuronal activity during periods of time when the monkey can reasonably be inferred to be set to move his limb can then be compared with conditions in which such preparation is unlikely.

One class of cells in premotor cortex exhibits directional specificity; that is, these cells show sustained excitation after an instruction for movement in one direction and show inhibition or unchanged activity after an instruction for movement in the opposite direction (Figure 2). However, for most of this class of premotor cortex neuron, when the visuospatial stimulus signals the monkey to withhold movement during a given behavioral trial, there is little or no change in neuronal activity (Wise et al 1983, Weinrich et al 1984). Thus, the activity of these cells seems to be specific for situations in which the monkey is preparing to move the limb. This class of premotor cortex neuron also typically shows directional specificity regardless of whether the instruction is given in the auditory or visual modality (Weinrich & Wise 1982)—a feature to be expected of a cell involved in the preparation for movement. Further, if activity during an instructed delay period reflects motor set, then (*a*) removing the instruction stimulus should have no long-lasting effect on that activity and (*b*) changing the instruction should overtly change the pattern of neuronal activity. Both of these predictions have been confirmed (Mauritz & Wise 1983). For example, as the instruction is changed from one requiring leftward limb movement to one requiring rightward movement, the pattern of activity rapidly changes to reflect the new motor set (Figure 2). These data are consistent with the view that the premotor cortex plays a role in the preparation for specific voluntary movements, i.e. motor set.

MOTOR PROGRAMMING Substantial interest in premotor cortex has focused on motor programming, one aspect of motor set. Roland et al (1980b, pp. 137, 146) have speculated on the basis of their regional cerebral blood flow

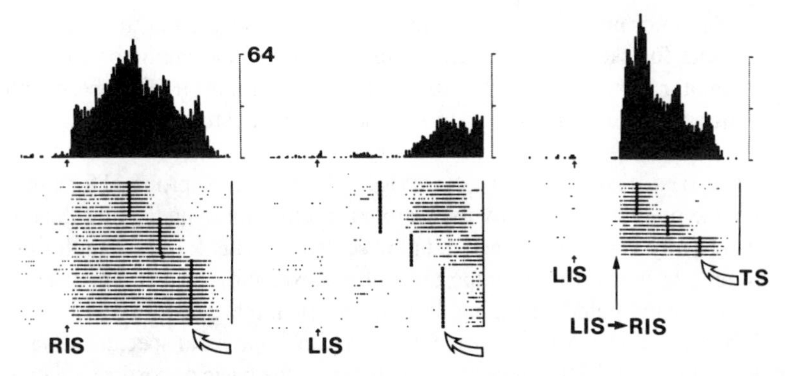

Figure 2 Three peri-event rasters and histograms for a premotor cortex neuron. Each line of each raster represents the neuronal discharge during one behavioral trial, centered on the presentation of a visuospatial instruction stimulus (IS). The IS informed the animal about the next movement to be made with the forelimb. An IS for a limb movement to the right (RIS) or left (LIS) on any given trial was followed, at a variable time, by another stimulus that signaled the monkey to begin the movement. The time of this later signal, the trigger stimulus (TS, *open arrow*), is indicated by the *heavy square mark* on each raster line. In the right panel, the instruction is changed at the time indicated by the *second vertical arrow* (LIS → RIS), 1 s after the first instruction for those trials. The three conditions illustrated separately here were randomly intermixed during recording of the cell's activity. The unit activity appears to represent the animal's motor set for future movements to the right, and the unit also shows modulation in association with the movement of the limb to the right. Histogram scale, impulses/s; binwidth, 40 ms. The time from the IS to the vertical bar bordering each raster on the right is 4 s.

studies in man that the "premotor areas are activated when a new motor program is established or a previously learned motor program is modulated," i.e. during nonrepetitive voluntary movements, or when a program is "changed on the basis of sensory information, as will be the case during exploratory manipulation of objects." Rizzolatti et al (1981a–c, 1983), based on their studies of macaque monkeys, also have suggested a role for premotor cortex in programming ("praxic") functions for complex movements, with subregions of the premotor cortex involved in the organization of specific motor acts, such as bringing food to the mouth by hand.

Although these suggestions are interesting, the hypothesis that the premotor cortex is necessary for motor programming or for the storage of learned motor programs still lacks clear experimental support. These speculations need to be elaborated and pursued with experiments employing controlled motor tasks. It seems likely, though, on the basis of presently available information, that the premotor cortex is not necessary for all types of programming. For example, monkeys with premotor cortex ablations remain capable for performing a variety of motor tasks, including motor sequences and movements involving finely controlled digit movements (Halsband & Passingham 1982). Thus, the aspect of motor programming that involves the specification of time and force parameters for skeletal movements does not require the premotor cortex. If the premotor cortex is necessary for motor programming, it might be specific for those programs guided by external environmental stimuli.

AXIAL AND PROXIMAL LIMB STABILIZATION One common proposal for a function of the premotor cortex involves a role for it in stabilizing the limb, mainly or exclusively via proximal and axial musculature. This aspect of motor programming is thought to be important for the execution of distal movements such as visually guided grasping of an object (Humphrey 1979).

Pribram et al (1955/1956) proposed a proximal and axial specialization for the rostral part of the agranular frontal cortex on the basic of cortical ablations in macaque monkeys. However, their lesions included the supplementary motor cortex, and those cases with the clearest proximal motor dysfunction also involved the rostral part of the primary motor cortex. Accordingly, it is difficult to attribute the deficits they observed to the premotor cortex. The historical development of the idea that the premotor cortex controls exclusively proximal muscles also might have depended, in part, on confusion over the term "area 6." The most rostral part of the primary motor cortex, sometimes held to overlap into area 6, is usually thought to contain the proximal and axial parts of the motor representation (Woolsey et al 1952, Kwan et al 1978, Fetz et al 1980). Accordingly, the image of axial and proximal motor control has sometimes been unjustifiably extended to area 6 as a whole, including the premotor cortex (and the supplementary motor cortex). A recent clinical study has indicated that humans with damage to the premotor cortex have weakness of the proximal musculature and difficulty in motor tasks that require proximal support of the arm (Freund & Hummelsheim 1984). This finding implies a significant role for the premotor cortex in the control of proximal limb muscle under certain circumstances, but should not be construed to rule out an additional role in the control of distal limb movements (see also Brinkman & Porter 1983).

The hypothesis that the premotor cortex participates in a limb-stabilization function may find support in the older observations of cortical electrical stimulation effects, which are suggestive of an axial or proximal specialization (e.g. Bucy 1933, Bucy & Fulton 1933). However, these findings and similar ones in humans are difficult to interpret for a variety of reasons, mainly the aphysiological nature of the technique itself and the effects of anesthesia.

As for neuroanatomical data, Künzle (1978) states that the projections from area 6 to precentral motor cortex are directed to the proximal parts of its representation. This finding might also suggest a proximal specialization, but Künzle presents no documentation for his contention. Muakassa & Strick (1979), on the other hand, report a strong input from the premotor cortex to the "wrist representation" of the primary motor cortex. Of course, an input to a distal part of the primary motor representation does not, in itself, provide evidence that premotor cortex is important in distal motor control, but the converse proposition is unsupported by current anatomical data.

Behavioral experiments are also relevant to this problem. Halsband & Passingham (1982) compared the effects of premotor ablations on a number of behaviors. The requirement for postural upper arm support was not a determining factor in the pattern of deficits they observed; their monkeys could perform a fixed sequence of movements that required comparable postural support to their "conditional motor task," described above, on which the animals were deficient. At the present time, a cautious estimate would be that whereas the premotor cortex may be very important in the control of proximal and axial muscles, it is unlikely to be exclusively concerned with such muscles. In the present context, it should be noted that proximal fixation of the limb for distal action can be considered an aspect of the preparation or programming for such movements.

SUPPRESSION OF "LOWER" MOTOR ACTS One of the roles of "higher motor centers" might be to suppress or facilitate "lower" motor control circuits. This concept, although not always explicitly stated in these terms, has been influential in thinking about the function of the premotor cortex. Denny-Brown (1966, Denny-Brown & Botterell 1948) thought that the premotor cortex functioned to suppress certain reflexes and that its removal caused "release" of these reflexes. Although this view cannot be ruled out, it reflects a curious pattern of thinking about the effects of brain lesions. In this view, ablations reveal a negative image of the functional specialization of the region in question. In the analysis of complex central nervous system circuits, such interpretations are unlikely to be of any enduring conceptual value.

Nevertheless, it may be realistic to view one important role of premotor cortex as suppression of relatively automatic responses to certain sensory stimuli. Moll & Kuypers (1977) interpreted the effects of large nonprimary motor cortex ablations (including the premotor cortex and several additional frontal fields) in this context. In their experiment, monkeys with such brain damage continued to reach directly toward a desired object, even when the pathway was obstructed by a transparent barrier and an alternative trajectory would have been successful. It was suggested that the monkeys were deficient in this task because they could not suppress direct and immediate reaching for the object. Such suppression might be another aspect of the preparation for movements, especially for delayed movements or those of a complex nature.

VASOMOTOR EFFECTS Very little is known about possible autonomic functions of the cerebral cortex in primates, but vasomotor regulation might be important in the preparation for movement. Vasomotor changes were reported after premotor cortex damage in both humans (Kennard et al 1934) and monkeys (Kennard 1935), and Woolsey et al (1952) speculated about an autonomic role for this part of the cortex. Of course, these effects may well have been due to involvement of cortical fields other than the premotor cortex.

ATTENTION Rizzolatti and his colleagues (1983) have recently proposed that the premotor cortex of monkeys is necessary for selective attention, especially to stimuli near the animal. It is difficult, however, to distinguish attentional deficits from those involving the sensory guidance of movements. Since the animals' motor orientation toward a stimulus is the method used to determine an attention deficit, it remains possible that such deficits reflect difficulty in sensorially guided movements. If this is the case, then the results of Rizzolatti et al suggest that somewhat different mechanisms guide movements in relation to near and distant stimuli. Nevertheless, mechanisms of selective attention are undoubtedly important in the preparation for motor action and have many features in common with motor set (see Evarts et al 1984).

SUMMARY

The concept of a separate premotor cortical field involved in the cerebral control of movement went out of favor among neurophysiologists during the quarter century from 1952 to 1977, but recent studies have led to its rehabilitation. The premotor cortex appears to be one of at least three fields within the motor cortex, two others being the primary motor cortex and the supplementary motor cortex. Several proposals have been presented concerning the functional specializations of the premotor cortex. Although no specific hypotheses have very strong support at present, the best evidence favors a role for premotor cortex in the preparation for and the sensory guidance of movement.

Literature Cited

Asanuma, C., Thach, W. T., Jones, E. G. 1983. Distribution of cerebellar terminations and their relation to other afferent terminations in the ventral lateral thalamic region of the monkey. *Brain Res. Rev.* 5:237–65

Asanuma, H., Rosén, I. 1972. Topographical organization of cortical efferent zones projecting to distal forelimb muscles in the monkey. *Exp. Brain Res.* 14:243–56

Avendaño, C. J., Price, L., Amaral, D. G. 1983. Evidence for an amygdaloid projection to premotor cortex but not to motor cortex in the monkey. *Brain Res.* 264:111–17

Bailey, P., von Bonin, G. 1951. *The Isocortex of Man,* Vol. 6. Urbana: Univ. Ill. Press

Brinkman, J., Porter, R. 1979. 'Premotor' area of the monkey's cerebral cortex: Activity of neurons during the performance of a learned motor task. *Proc. Aust. Physiol. Pharmacol. Soc.* 10:198

Brinkman, J., Porter, R. 1983. Supplementary motor area and premotor area of monkey cerebral cortex: Functional organization and activities of single neurons during performance of a learned movement. *Adv. Neurol.* 39:393–420

Brodal, P. 1978. The corticopontine projection in the rhesus monkey. *Brain* 101:251–83

Brodmann, K. 1905. Beiträge zur histologischen Lokalisation der Grosshirnrinde III. Die Rindenfelder der niederen Affen. *J. Psychol. Neurol. Leipzig* 4:177–226

Brodmann, K. 1909. *Vergleichende Lokalizationlehre der grosshirnrinde in ihren Prin-*

zipien dargestellt auf Grund des Zellenbaues. Leipzig: Barth

Brooks, V. B. 1979. Motor programs revisited. In *Posture and Movement,* ed. R. E. Talbot, D. R. Humphrey, pp. 13–49. New York: Raven

Brown, R. M., Crane, A. M., Goldman, P. S. 1979. Regional distribution of monoamines in the cerebral cortex and subcortical structures of the rhesus monkey: Concentrations and in vivo synthesis rates. *Brain Res*. 168:133–50

Bucy, P. C. 1933. Electrical excitability and cyto-architecture of the premotor cortex in monkeys. *Arch. Neurol. Psychiatr.* 30:1205–24

Bucy, P. C. 1935. A comparative cytoarchitectonic study of the motor and premotor areas in the primate *J. Comp. Neurol*. 62:293–331

Bucy, P. C., Fulton, J. F. 1933. Ipsilateral representation in the motor and premotor cortex of monkeys. *Brain* 56:318–42

Campbell, A. W. 1905. *Histological studies on the Localization of Cerebral Function*. New York: Cambridge Univ. Press

Catsman-Berrevoets, C. E., Kuypers, H. G. J. M. 1976. Cells of origin of cortical projections to dorsal column nuclei, spinal cord and bulbar medial reticular formation in the rhesus monkey. *Neurosci. Lett*. 3:245–52

Catsman-Berrevoets, C. E., Kuypers, H. G. J. M., Lemon, R. N. 1979. Cells of origin of the frontal projections to magnocellular and parvocellular red nucleus and superior colliculus in cynomolgus monkey. An HRP study. *Neurosci. Lett*. 12:41–46

Chavis, D. A., Pandya, D. N. 1976. Further observations on corticofrontal connections in the rhesus monkey. *Brain Res*. 117:369–86

Denny-Brown, D. 1966. *The Cerebral Control of Movement*. Liverpool: Liverpool Univ. Press

Denny-Brown, D., Botterell, E. H. 1948. The motor functions of the agranular frontal cortex. *Res. Publ. Assoc. Res. Nerv. Ment. Dis*. 27:235–345

Deuel, R. K. 1977. Loss of motor habits after cortical lesions. *Neuropsychologia* 15:205–15

Distel, H., Fries, W. 1982. Contralateral cortical projections to the superior colliculus in the macaque monkey. *Exp. Brain Res*. 48:157–62

Evarts, E. V., Shinoda, Y., Wise, S. P. 1984. *Neurophysiological Approaches to Higher Brain Functions*. New York: Wiley

Fetz, E. E., Finocchio, D. V., Baker, M. A., Soso, M. J. 1980. Sensory and motor responses of precentral cortex cells during comparable passive and active joint movements. *J. Neurophysiol*. 43:1070–89

Foerster, O. 1936. The motor cortex in man in light of Hughlings Jackson's observations. *Brain* 59:135–59

Freund, H.-J., Hummelsheim, H. 1984. The premotor syndrome in man: Evidence for innervation of proximal limb muscles. *Exp. Brain Res*. 53:479–82

Fulton, J. F. 1934. Forced grasping in relation to the syndrome of the premotor area. *Arch. Neurol. Psychiatr.* 31:221–35

Fulton, J. F. 1935. Definition of the motor and premotor areas. *Brain Res*. 58:311–16

Fulton, J. F. 1938. *Physiology of the Nervous System*. New York: Oxford Univ. Press. 1st ed.

Fulton, J. F., Jacobsen, C. F., Kennard, M. A. 1932. A note concerning the relation of the frontal lobes to posture and forced grasping in monkeys. *Brain* 55:524–36

Gentilucci, M., Scandolara, C., Pigarev, I. N., Rizzolatti, G. 1983. Visual responses in the postarcuate cortex (area 6) of the monkey that are independent of eye position. *Exp. Brain Res*. 50:464–68

Godschalk, M., Lemon, R. N. 1983. Involvement of monkey premotor cortex in the preparation of arm movements. In *Neuronal Coding of Motor Performance,* ed. J. Massion, J. Paillard, W. Schultz, M. Wiesendanger, pp. 114–19. New York: Springer-Verlag

Godschalk, M., Lemon, R. N., Nijs, H. G. T., Kuypers, H. G. J. M. 1981. Behavior of neurons in monkey peri-arcuate and precentral cortex before and during visually guided arm and hand movements. *Exp. Brain Res*. 44:113–16

Godschalk, M., Lemon, R. N., Kuypers, H. G. J. M. 1983. Afferent and efferent connections of the postarcuate region of the monkey cerebral cortex. *Soc. Neurosci. Abstr.* 9:490

Haaxma, R., Kuypers, H. G. J. M. 1975. Intrahemispheric cortical connections and visual guidance of hand and finger movements in the rhesus monkey. *Brain* 98:239–60

Halsband, U., Passingham, R. 1982. The role of premotor and parietal cortex in the direction of action. *Brain Res*. 240:368–72

Hartmann-von Monakow, K., Akert, K., Künzle, H. 1979. Projections of precentral and premotor cortex to the red nucleus and other midbrain areas in *Macaca fascicularis*. *Exp. Brain Res*. 34:91–105

Hines, M. 1929. On cerebral localization. *Physiol. Rev.* 9:462–574

Humphrey, D. R. 1979. On the cortical control of visually directed reaching: Contributions by nonprecentral motor areas. See Brooks 1979, pp. 51–112

Jacobsen, C. F. 1934. Influence of motor and premotor area lesions upon the retention of skilled movements in monkeys and chimpanzees. *Res. Publ. Assoc. Res. Nerv. Ment. Dis.* 13:225–47

Jones, E. G., Coulter, J. D., Hendry, S. H. C. 1978. Intracortical connectivity of architectonic fields in the somatic sensory, motor and parietal cortex of monkeys. *J. Comp. Neurol.* 181:291–348

Keating, E. G. 1973. Loss of visual control of the forelimb after interruption of cortical pathways. *Exp. Neurol.* 41:635–48

Kemp, J. M., Powell, T. P. S. 1970. The cortico-striate projection in the monkey. *Brain* 93:525–46

Kennard, M. A. 1935. Vasomotor disturbances resulting from cortical lesions. *Arch. Neurol. Psychiatr.* 33:537–45

Kennard, M. A., Viets, H. R., Fulton, J. F. 1934. The syndrome of the premotor cortex in man: Impairment in skilled movements, forced grasping, spasticity and vasomotor disturbance. *Brain* 57:69–84

Kubota, K., Hamada, I. 1978. Visual tracking and neuron activity in the postarcuate area in monkeys. *J. Physiol. Paris* 74:297–312

Künzle, H. 1978. An autoradiographic analysis of the efferent connections from premotor and adjacent prefrontal regions (areas 6 and 9) in *Macaca fascicularis. Brain Behav. Evol.* 15:185–234

Kuypers, H. G. J. M., Brinkman, J. 1970. Precentral projections to different parts of the spinal intermediate zone in the rhesus monkey. *Brain Res.* 24:29–48

Kuypers, H. G. J. M., Lawrence, D. G. 1967. Cortical projections to the red nucleus and the brain stem in the rhesus monkey. *Brain Res.* 4:151–88

Kwan, H. C., MacKay, W. A., Murphy, J. T., Wong, Y. C. 1978. Spatial organization of precentral cortex in awake primates. II. Motor outputs. *J. Neurophysiol.* 41:1120–31

Kwan, H. C., MacKay, W. A., Murphy, J. T., Wong, Y. C. 1981. Distribution of responses to visual cues for movement in precentral cortex of awake primates. *Neurosci. Lett.* 24:123–28

Lamarre, Y., Busby, L., Spidalieri, G. 1983. Fast ballistic arm movements triggered by visual, auditory, and somesthetic stimuli in the monkey. I. Activity of precentral cortical neurons. *J. Neurophysiol.* 50:1343–58

Luria, A. R. 1980. *Higher Cortical Functions in Man*. New York: Basic Books

Mauritz, K.-H., Wise, S. P. 1983. Motor programming and the premotor cortex of the rhesus monkey. *Soc. Neurosci. Abstr.* 9:308

Moll, L., Kuypers, H. G. J. M. 1977. Premotor cortical ablations in monkeys: Contralateral changes in visually guided reaching behavior. *Science* 198:317–19

Muakassa, K. F., Strick, P. L. 1979. Frontal lobe inputs to primate motor cortex: Evidence for four somatotopically organized 'premotor' areas. *Brain Res.* 177:176–82

Murray, E. A., Coulter J. D. 1981. Organization of corticospinal neurons in the monkey. *J. Comp. Neurol.* 195:339–65

Paillard, J. 1983. The functional labelling of neural codes. See Godschalk & Lemon 1983, pp. 1–19

Pandya, D. N., Kuypers, H. G. J. M. 1969. Cortico-cortical connections in the rhesus monkey. *Brain Res.* 13:13–36

Pandya, D. N., Vignolo, L. A. 1971. Intra- and interhemispheric projections of the precentral, premotor and arcuate areas in the rhesus monkey. *Brain Res.* 26:217–33

Passingham, R. 1981. Broca's area and the origin of human vocal skills. *Philos. Trans. R. Soc. London Ser. B* 292:167–75

Penfield, W., Welch, K. 1951. The supplementary motor area in the cerebral cortex. *AMA Arch. Neurol. Psychiatr.* 66:289–317

Petrides, M. 1982. Motor conditional associative-learning after selective prefrontal lesions in the monkey. *Behav. Brain Res.* 5:407–13

Porter, R. 1983. Neuronal activities in primary motor area and premotor regions. See Godschalk & Lemon 1983, pp. 23–29

Pribram, K., Kruger, L., Robinson, F., Berman, A. 1955/1956. The effects of precentral lesions on the behavior of monkeys. *Yale J. Biol. Med.* 28:428–43

Rizzolatti, G., Scandolara, C., Matelli, M., Gentilucci, M. 1981a. Afferent properties of periarcuate neurons in macaque monkeys. 1. Somatosensory responses. *Behav. Brain Res.* 2:125–46

Rizzolatti, G., Scandolara, C., Matelli, M., Gentilucci, M. 1981b. Afferent properties of periarcuate neurons in macaque monkeys. 2. Visual responses. *Behav. Brain Res.* 2:147–63

Rizzolatti, G., Scandolara, C., Matelli, M., Gentilucci, M., Camarda, R. 1981c. Response properties and behavioral modulation of 'mouth' neurons of the postarcuate cortex (area 6) in macaque monkeys. *Brain Res.* 225:421–24

Rizzolatti, G., Matelli, M., Pavesi, G. 1983. Deficits in attention and movement following the removal of postarcuate (Area 6) and prearcuate (Area 8) cortex in macaque monkeys. *Brain* 106:655–73

Roland, P. E., Larsen, B. 1976. Focal increase of cerebral blood flow during stereognostic testing in man. *Arch. Neurol.* 33:551–58

Roland, P. E., Larsen, B., Lassen, N. A., Skinhoj, E. 1980a. Supplementary motor area and other cortical areas in organization of voluntary movements in man. *J. Neurophysiol.* 43:118–36

Roland, P. E., Skinhoj, E., Lassen, N. A., Larsen, B., 1980b. Different cortical areas in man in organization of voluntary movements in extrapersonal space. *J. Neurophysiol.* 43:137–50

Sakai, M. 1978. Single unit activity in a border area between the dorsal prefrontal and premotor regions in the visually conditioned motor task of monkeys. *Brain Res.* 147:377–83

Schell, G. R., Strick, P. L. 1984. The origin of thalamic inputs to the arcuate premotor and supplementary motor areas. *J. Neurosci.* 4:539–60

Sessle, B. J., Wiesendanger, M. 1982. Structural and functional definition of the motor cortex in the monkey. *J. Physiol. London* 323:245–65

Sternberg, S., Monsell, S., Knoll, R. L., Wright, C. E. 1978. The latency and duration of rapid movement sequences: Comparisons of speech and typewriting. In *Information Processing in Motor Control and Learning,* ed. G. E. Stelmach, pp. 117–52. New York: Academic

Toyoshima, K., Sakai, H. 1982. Exact cortical extent of the origin of the corticospinal tract (CST) and the quantitative contribution to the CST in different cytoarchitectonic areas. A study with horseradish peroxidase in the monkey. *J. Hirnforsch.* 23:257–69

Travis, A. M. 1955. Neurological deficiencies following supplementary motor area lesions in *Macaca mulatta. Brain* 78:174–98

Vogt, O., Vogt, C. 1919. Ergebisse unserer Hirnforschung. *J. Psychol. Neurol. Leipzig* 25:277–462

Von Bonin, G. 1949. Architecture of the precentral motor cortex and some adjacent areas. *The Precentral Motor Cortex,* ed. P. C. Bucy, pp. 7–82. Urbana: Univ. Ill. Press

Von Bonin, G., Bailey, P. 1947. *The Neocortex of Macaca Mulatta.* Urbana: Univ. Ill. Press

Von Economo, C. 1929. *The Cytoarchitecture of the Human Cerebral Cortex,* trans. S. Parker. Oxford: Oxford Univ. Press

Walshe, F. M. R. 1935. On the "syndrome of the premotor cortex" (Fulton) and the definition of the terms "premotor" and "motor": With consideration of Jackson's views on the cortical representation of movements. *Brain* 58:49–80

Weinrich, M., Wise, S. P., Mauritz, K.-H. 1984. A neurophysiological analysis of the premotor cortex of the monkey. *Brain* 107:385–414

Weinrich M., Wise, S. P. 1982. The premotor cortex of the monkey. *J. Neurosci.* 2:1329–45

Wiesendanger, M. 1981. Organization of secondary motor areas of cerebral cortex. pp. 1121–47. *Hand. Physiol. (Sect. 1)2(1):*1121–47

Wise, S. P. 1984. Non-primary motor cortex and its role in the cerebral control of movement. In *Dynamic Aspects of Neocortical Function,* ed. G. Edelman, W. M. Cowan, E. Gall. New York: Wiley

Wise, S. P., Weinrich, M., Mauritz, K.-H. 1983. Motor aspects of cue-related neuronal activity in premotor cortex of the rhesus monkey. *Brain Res.* 260:301–5

Woolsey, C. N., Settlage, P. H., Meyer, D. R., Sencer, W., Pinto Hamuy, T., Travis, A. M. 1952. Patterns of localization in precentral and "supplementary" motor areas and their relation to the concept of a premotor area. *Res. Publ. Assoc. Res. Nerv. Ment. Dis.* 30:238–64

Ann. Rev. Neurosci. 1985. 8:21–44

THE GABA-ERGIC SYSTEM: A LOCUS OF BENZODIAZEPINE ACTION

John F. Tallman and Dorothy W. Gallager

Department of Psychiatry, Connecticut Mental Health Center, Yale University School of Medicine, New Haven, Connecticut 06508

INTRODUCTION

γ-Aminobutyric acid is regarded as one of the major inhibitory amino acid transmitters in the mammalian brain. Over 30 years have elapsed since its presence in the brain was demonstrated (Roberts & Frankel 1950, Udenfriend 1950). Since that time, an enormous amount of effort has been devoted to implicating GABA in the etiology of a host of neurological and psychiatric disorders. Widely, although unequally, distributed throughout the mammalian brain, GABA is said to be a transmitter at approximately 30% of the synapses in the brain and its levels are measured in micro- rather than nanomoles per gram, typical of monoamines (Fahn & Cote 1968). In most regions of the brain, GABA is associated with local inhibitory neurons and only in two regions is GABA associated with longer projections (Fonnum & Storm-Mathisen 1978). GABA mediates its actions through a complex of proteins localized both on cell bodies and nerve endings; postsynaptic responses to GABA are mediated through alterations in chloride conductance that generally, although not invariably, lead to hyperpolarization of the cell (Curtis & Johnston 1974). Recent investigations have indicated that the complex of proteins associated with postsynaptic GABA responses is a major site of action for a number of structurally unrelated compounds capable of modifying postsynaptic responses to GABA. Depending on the mode of interaction, these compounds are capable of producing a spectrum of activities (either sedative, anxiolytic, and anticonvulsant, or wakefulness, seizures, and anxiety). The present article describes how interaction of binding sites for GABA, benzodiazepines (BDZ), and

0147-006X/85/0301-0021$02.00

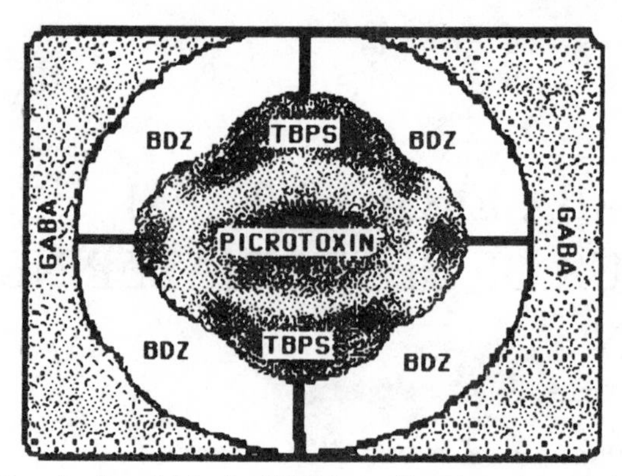

Figure 1 A model for the GABA/benzodiazepine anion complex.

chloride channel blockers (picrotoxin, TBPS) results in the pharmacological activities of these compounds (Figure 1).

Benzodiazepines Alter GABA Responses

1,4-Benzodiazepines continue to be among the most widely used drugs in the world. In spite of some decrease in the use of these compounds in the last few years, it can be estimated that close to 10% of the population will admit to the use of a benzodiazepine in the past year. Principal among the benzodiazepines marketed are chlordiazepoxide, diazepam, flurazepam, and triazolam. These compounds are widely used as anxiolytics, sedative-hypnotics, muscle relaxants, and anticonvulsants. A number of these compounds are extremely potent drugs; such potency indicates a site of action with a high affinity and specificity for individual structures.

Early electrophysiological studies indicated that a major action of benzodiazepines was to enhance GABAergic inhibition. The benzodiazepines were capable of enhancing presynaptic inhibition of a monosynaptic ventral root reflex (Schmidt et al 1967), a GABA-mediated event. All subsequent electrophysiological studies (reviewed in Tallman et al 1980, Haefley et al 1981) have generally confirmed this finding, and by the mid-1970s, there was a general consensus among electrophysiologists that the benzodiazepines could enhance the actions of GABA. This view was supported by the indirect biochemical evidence that pretreatment with benzodiazepines would antagonize the rise in cGMP due to the depletion of GABA in cerebellum caused by the inhibition of glutamic acid dehydrogenase with isoniazid (Costa et al 1975).

At that time, a number of other neurotransmitter systems (glycine, serotonin, adenosine, etc) were thought to be the primary sites of action for the benzodiazepines (see Costa & Greengard 1975 for review). With the discovery of the "receptor" for the benzodiazepines and the subsequent definition of the nature of the interaction between GABA and the benzodiazepines, it appears that the behaviorally important interactions of the benzodiazepines with different neurotransmitter systems are due in large part to the enhanced ability of GABA itself to modify these systems. Each modified system, in turn, may be associated with the expression of a behavior.

Benzodiazepine Receptors

Studies on the mechanistic nature of these interactions depended on the demonstration of a high-affinity benzodiazepine binding site (receptor). This receptor is present in the CNS of all vertebrates phylogenetically newer than the boney fishes (Squires & Braestrup 1977, Mohler & Okada 1977). Originally, by using titrated diazepam, and now with a wide variety of compounds, it was was possible to demonstrate that these benzodiazepine binding sites fulfilled many of the criteria of pharmacological receptors; binding to these sites in vitro is rapid, reversible, stereospecific, and saturable. More importantly, it was possible to show highly significant correlations between the ability of benzodiazepines to displace diazepam from its binding site and activity in a number of animal behavioral tests predictive of benzodiazepine potency (Braestrup & Squires 1978, Mohler & Okada 1978). The average therapeutic doses of these drugs in man also correlates with receptor potency (Tallman et al 1980).

A number of studies have demonstrated that these central benzodiazepine receptors (high affinity for clonazepam) exist on neurons (Braestrup et al 1979b). A second type of binding site (high affinity for Ro 5-4864, *p*-chlorodiazepam) exists on nonneuronal cells (Taniguchi et al 1980, Wang et al 1980), on a number of continuous cell lines (Syapin & Skolnick 1979, McCarthy & Harden 1981), and on glial elements in brain (Gallager et al 1981, Schoemaker et al 1981). The pharmacology of these sites is not consistent with the central actions of benzodiazepines and may be related to other actions of the benzodiazepines [induction of differentiation in cells or antiproliferative effects (Matthew et al 1981)]. A third benzodiazepine site of much lower affinity has been described in the brain (Bowling & DeLorenzo 1982). It appears to be associated with the calcium-calmodulin protein kinase system in the brain (DeLorenzo et al 1981). The localization and function of these sites is obscure.

In contrast to the other sites, it now is clear that many of the central actions of the benzodiazepines are related to their ability to interact with the neuronal clonazepam sensitive sites. The rest of this article deals only with these sites.

GABA MODIFIES BENZODIAZEPINE BINDING In 1978, it became clear that GABA and related analogs could interact at a low affinity (~ 1 μM) GABA binding site to enhance the binding of benzodiazepines to the clonazepam-sensitive site (Tallman et al 1978). The enhancement was caused by an increase in the affinity of the benzodiazepine binding site due to occupancy of the GABA site. The data were interpreted to mean that both GABA and benzodiazepine sites were allosterically linked in the membrane as part of a complex of proteins. For a number of GABA analogs, the ability to enhance diazepam binding by 50% of maximum and the ability to inhibit the binding of GABA to brain membranes by 50% could be directly correlated. However, the concentrations required to enhance diazepam binding are 100 times higher than those required to inhibit GABA binding. From these data, it appears that the structural requirements of the binding and enhancement are identical, although the absolute affinities are different. Partial GABA agonists can also be determined by correlating enhancement of diazepam binding, inhibition of GABA binding, and pharmacological activity (Braestrup et al 1979a). Thus several (at least two) GABA receptors seem to exist by these criteria and also by direct binding studies with GABA (Enna & Snyder 1977); it now appears that these different GABA receptors are multiple affinity states within the same complex (see below). It remains to be shown that these affinity states are interconvertible. These sites are different from a recently defined GABA-B site (Bowery 1983).

The enhancement of benzodiazepine binding by GABA agonists is blocked by the GABA receptor antagonist (+)bicuculline; the stereoisomer (−)bicuculline is much less active (Tallman et al 1978). By utilizing (+)(^{3}H) bicuculline methochloride binding in the presence of thiocyanate, it has been possible to label these lower affinity GABA sites. These sites are similar to those involved in the enhancement of benzodiazepine receptor binding (Mohler & Okada 1977, Olsen & Snowman 1983). Along with being a requirement for the binding of (+)bicuculline, thiocyanate has been shown to obliterate reversible binding to the high affinity GABA binding site without altering the lower affinity, GABA-receptor-mediated enhancement of diazepam binding (Browner et al 1981).

Our current interpretation of these results is that there are several potential affinity states of the GABA receptor. The highest affinity state (10 nM) is the most stable thermodynamically and results from treatment of the membrane with detergent or solubilization. Probably two lower affinity states of the GABA receptor exist; at least one of these is involved in the activation of benzodiazepine binding. The lower affinity states are also far more likely candidates for the functional form of the GABA receptor, since the levels of GABA in brain are high enough to offer significant saturation of a low affinity site and the total saturation of the high affinity site. In addition, electrophysiological experiments indicate that micromolar concentrations of GABA are required for the activation of chloride conductances (Krespan et al 1984). It is not clear whether the highest affinity state of the GABA receptor exists in normal circumstances or has functional significance. Such a high affinity state would dissociate GABA much more slowly. This would be of great importance if the high affinity state were a desensitized form of the receptor; such a state has been proposed for the acetylcholine receptor (Changeux 1981).

CHANNEL FUNCTION SITE INTERACTS WITH BENZODIAZEPINE SITE

Anion site The binding of diazepam to benzodiazepine receptors is enhanced by a number of anions, including iodide, bromide, nitrate, thiocyanate, and chloride (Costa et al 1979). The selectivity of the effective anions has been correlated with their selectivity for penetrating the activated postsynaptic membrane of cat motorneurons and was initially interpreted to indicate the proximity of this site to chloride gating function. This view of the data was challenged (Candy & Martin 1979); however, a rather large number of reports have now demonstrated a chloride dependence for a number of effects that seem to be mediated by a site close to the anion channel, and the potential importance of the anion binding site seems to be considerable for these funtional interactions.

Picrotoxin site It was known for a number of years that picrotoxin, found in various plants of the *Menispermaceae* family, was an inhibitor of GABA-mediated inhibitory responses. In addition, picrotoxin (composed of a 50:50 mixture of two compounds, the more active picrotoxin and the less active picrotin) was a potent convulsant when administered to animals (Olsen 1982). In contrast to bicuculline, the convulsant activity of picrotoxin was not mediated by inhibition of GABA receptor binding; the site at which picrotoxin seems to work is the anion channel. The overall effect is to block changes in chloride conductance; hence, it appears to be a GABA antagonist (Haefley et al 1981, Simmonds 1980).

It was initially possible to study binding to the picrotoxin sites by using (^{3}H)-dihydropicrotoxinin (DHP); although not suitable for careful investigation of interactive effects (because of the high nonspecific binding), this ligand allowed early investigation of a site close to the channel to be carried out (Ticku et al 1978). Binding to this site was inhibited by a large number of excitatory and inhibitory drugs that affect the central nervous system, many of which are thought to interact with GABA-ergic transmission (Olsen 1982). Among these compounds are the barbiturates, which interact with DHP binding in a manner roughly consistent with their sedative-hypnotic properties (Ticku & Olsen 1978, Ticku 1981, Leeb-Lundberg et al 1981).

Another interesting set of compounds showing interaction with this site are the "cage convulsants" (Ticku & Olsen 1979), a set of bicyclo-phosphates and carboxylates that are related to anticholinesterase insecticides, and are thought to be postsynaptic GABAergic inhibitors (Bowery et al 1976, 1977). These compounds do not interact directly with the GABA binding sites and seem to interfere with the change in chloride conductance induced by GABA. By using a ^{35}S cage convulsant (*t*-butylbicyclophosphorothionate, TBPS), Squires and colleagues (1982) have developed a much better assay for this anion-dependent, channel-related site. This site recognizes a number of the barbiturates, and binding to it is potentially antagonized by picrotoxin. It shows functional interaction with the GABA receptor, and GABA agonists antagonize the anion requirements for binding to this site (in the presence of bromide, GABA inhibits the binding of TBPS to membranes). Thus, this site seems to overlap with the picrotoxin site (perhaps not identical to it) and shows interactions with a number of compounds that are involved either with anion channel opening or closing. It is difficult to envision this diverse group of compounds as simply blocking a channel with a high affinity. It is also hard to envision these compounds as interacting identically with a single site. Rather, these compounds can be thought to interact with overlapping topologies and a number of different amino acids that compose the anion channel or pore. Such amino acids could interact with these drugs to enhance or block the conformational change required for the opening of the chloride channel in response to GABA.

The anion requirements of this site are consistent with the presence of charged amino acids at the mouth of the channel or inside the channel itself. Such a general structure is consistent with the current views of nicotinic cholinergic receptors, in which the channel is envisioned as a funnel-shaped water-filled pore, with a number of charged groups present (Stroud 1983).

In addition to the direct binding studies described above, the channel-related site has been studied by examining its ability to interact with the benzodiazepine site. A number of compounds with GABA-like activity interact with the picrotoxin site and enhance benzodiazepine binding. The sedative barbiturates fall into this category and possess similar affinities for picrotoxin binding displacement and benzodiazepine enhancement (Leeb-Lundberg et al 1980). This enhancement is anion dependent (Skolnick et al 1980, 1981). In addition, these sites are capable of interaction with GABA binding sites to make GABA more potent in activating benzodiazepine binding (Supavilai et al 1982). A number of interesting anxiolytics, pyrazolopyridines, enhance benzodiazepine binding by interacting within this site, and their properties seem to be similar to those of the barbiturates (Williams & Risely 1979, Supavilai & Karobath 1979, 1980, 1981, Meiners & Salama 1982). Finally, a new, potentially interesting, antihelmintic agent, avermectin B1a, which potentiates invertebrate GABA receptors (Campbell et al 1983), also enhances benzodiazepine receptor binding in a similar manner (Williams & Yarbrough 1979, Pong et al 1982).

MULTIPLE BENZODIAZEPINE RECEPTORS Soon after the discovery of high affinity binding sites for the benzodiazepines, it was discovered that a triazolopyridazine, CL218,872, could interact with benzodiazepine receptors in a number of regions of the brain in a manner consistent with receptor heterogeneity or negative cooperativity. In these studies, Hill coefficients significantly less than one were observed in a number of brain regions, including cortex, hippocampus, and striatum. In cerebellum, CL218,872 interacted with the benzodiazepine sites with a Hill coefficient of 1 (Squires et al 1979, Klepner et al 1979). Thus, multiple benzodiazepine receptors were predicted in the cortex, hippocampus, striatum, but not in the cerebellum. At that time, it was also predicted that the triazolopyridazines increased the ability to respond in the face of punishment in a test situation (conflict test) and protected against pentylenetetrazole-induced seizures, but did not interact with GABAergic systems or serotonin (Lippa et al 1979). The conclusion drawn from this study was that the triazolopyridazines were nonsedating anxiolytics.

Based on these studies, extensive receptor autoradiographic localization studies were carried out at a light microscopic level. Although receptor heterogeneity has been demonstrated (Young & Kuhar 1980, Young et al 1981, Niehoff et al 1982), no simple correlation between localization of receptor subtypes and the behaviors associated with the region has emerged. In addition, in the cerebellum, where one receptor was predicted from binding studies, autoradiography revealed heterogeneity of receptors (Niehoff et al 1982).

Furthermore, it now appears that the triazolopyridazines possess proconvulsant effects under certain conditions (Melchior et al 1983), so the initial reports of the behavioral properties may not have been complete. It also appears now that these compounds possess a GABA shift (Braestrup et al 1982) slightly lower than the benzodiazepines, and these compounds may be classified more appropriately as partial agonists than full benzodiazepine agonists (Gee et al 1982). The triazolopyridazines also possess a smaller photoshift (see below), which is consistent with this view.

One of the pieces of evidence for the neuronal localization of benzodiazepine receptors is the elegant electron microscopic study of Mohler et al (1981b) that demonstrated the localization of benzodiazepine receptors in GABAergic synapses. These synapses were identified by immunocytochemical staining for glutamic acid dehydrogenase, the committed step in GABA synthesis. Such studies do not provide evidence for a presynaptic or postsynaptic localization of the receptors because of the limitation in resolution of the autoradiographic method at the electron-microscope level. Using lesioning techniques in combination with light microscopic autoradiography, Lo et al (1983) have demonstrated the preferential presynaptic localization of benzodiazepine receptors with a high affinity for the triazolopyridazines in substantia nigra. Thus, it is possible that the regional differences in receptor type may be due to altered proportions of pre- and postsynaptically localized receptors.

A physical basis for the differences in drug specificity for the two apparent subtypes of benzodiazepine sites has been demonstrated by Sieghart & Karobath (1980). Using gel electrophoresis in the presence of sodium dodecyl sulfate, the presence of several molecular weight receptors for the benzodiazepines has been reported. The receptors were identified by the covalent incorporation of radioactive flunitrazepam (see below). The major labeled bands have molecular weights of 50,000, 53,000, 55,000, and 57,000 and the triazolopyridazines inhibit labeling of the slightly higher molecular weight forms (53,000, 55,000, 57,000) (Sieghart et al 1983). In these studies, no attempt was made to inhibit proteolysis, and endogenous proteases are apparently present in the membrane fractions and could contribute to the observed heterogeneity (Klotz et al 1984). Another hypothesis is that heterogeneity may result from different processing of a precursor form of the receptor at pre- and postsynaptic sites, respectively, altered glycosylation, or removal of a leader sequence after insertion of newly synthesized receptors into the membrane. Altered processing could also account for changes in the cellular localization of the receptor or microheterogeneity due to incomplete carbohydrate incorporation; the presence of variable carbohydrate sequences is already demonstrated by differential interactions with the lectin, concanavalin A (Lo et al 1982). Altered processing could also account for the changes in molecular weight seen during development; alternatively, pre- and postsynaptic receptors may develop at different rates or be susceptible to differences in neuronal sprouting (Mallorga et al 1980, Lippa et al 1981, Regan et al 1980). These changes may also be correlated to changes in responsiveness to GABA. Neonatal animals show a much larger effect of GABA on benzodiazepine binding than adult animals (Mallorga et al 1980).

Another, more tantalizing, possibility is that the multiple forms of the receptor represent "isoreceptors" or multiple allelic forms of the receptor. Although common for enzymes, genetically distinct forms for receptors have not generally been described. As we begin to study receptors using specific radioactive probes and electrophoretic techniques, it is almost certain that isoreceptors will emerge as important in investigations of the etiology of psychiatric disorders in people.

Endogenous Benzodiazepine Ligands

Soon after the initial description of the high affinity binding site for the benzodiazepines, a number of investigators began a search for the "endogenous Valium." This search was spurred by the description of endogenous opiates and led to the demonstration of weak interactions of a number of compounds with benzodiazepine receptors. These included hypoxanthine, inosine, nicotinamide, and thromboxanes (reviewed extensively in Tallman et al 1980); the present status of these compounds as putative endogenous ligands for the benzodiazepine receptor is weak. The one interaction that has overshadowed the endogenous ligand concept is the GABAergic interaction. The distinct possibility exists that the benzodiazepines may function purely as allosteric modifiers of this site.

Some of the initial reports identified possible peptide ligands for the benzodiazepine receptor and for the GABA receptor (Tallman et al 1980). Only one group has actively pursued the purification of these peptides to yield material suitable for sequencing (Guidotti et al 1978, 1982). The first is a noncompetitive inhibitor of high affinity GABA binding and has a fully characterized structure. This peptide is a very basic protein of 16,000 molecular weight; it has over 90% homology with a small myelin basic protein of rat brain but is not identical to it (Vaccarino et al 1983). Because of its properties this compound has been termed *GABA-modulin*. The second protein(s) is a competitive inhibitor of benzodiazepine receptor binding and has been extensively purified; its molecular weight is 11,000. By itself, inhibition of ligand binding in vitro is a poor criterion for an endogenous ligand, as many substances, including serum albumin, could be classified as pseudocompetitive inhibitors by binding the radioactive trace and altering the amount of free ligand in an incubation. However, the 11,000 molecular weight peptide isolated by this group has been shown to be behaviorally active and to possess apparent benzodiazepine antagonist activity (Guidotti et al 1983).

Both these materials are of a high enough molecular weight to make their roles as classical transmitters suspect. In consideration of this point, it has been proposed that the peptides are much less active, that they might serve as precursors to smaller sized molecules, or that they could serve as cotransmitters with GABA (Guidotti et al 1982, 1983). Further work is needed to support this hypothesis. The actual storage and release of these peptides or the replication of the behavioral properties of these compounds with synthetic peptides needs to be demonstrated. It had been previously reported that the actions of GABA-modulin could be attributed to the presence of GABA tightly bound to proteins during purification (Napias et al 1980).

The most interesting of the ligands is the series of compounds isolated by Braestrup and colleagues from human urine (Nielsen et al 1979, Braestrup et al 1980). These compounds are beta-carbolines formed from the cyclization of l-tryptophan; a series of esters, depending on the extraction procedure, were isolated. It is now thought that these compounds were formed during the extraction and probably are not endogenous ligands (Braestrup & Nielsen 1983). However, they are the first compounds that have actions opposite to that of the benzodiazepines and which would be considered antagonists at the benzodiazepine receptor.

Agonist and Antagonist Domains of Benzodiazepine Receptors

The beta-carbolines were first isolated based upon their ability to inhibit competitively the binding of diazepam to its binding site (Nielsen et al 1979). As mentioned above, such an assay is not instructive about biological activity; agonists, partial agonists, and antagonists can inhibit binding. When the beta-carboline structure was determined, it was possible to synthesize a number of analogs and test these compounds behaviorally. It was immediately realized that the beta-carbolines could antagonize the actions of diazepam behaviorally (Tenen & Hirsch 1980) and could electrophysiologically antagonize both GABA and the benzodiazepines (Polc et al 1981). In addition to this antagonism, the beta-carbolines possess intrinsic activity of their own opposite to that of the benzodiazepines. This activity is described in greater detail below.

Radioactive beta-carboline esters were synthesized and used as ligands in in vitro binding assays. In spite of some difficulty with brain esterases and hydrolysis of the beta-carbolines, it was possible to show that they bound to brain membranes with a high affinity and to the same number of sites as the benzodiazepines (Nielsen et al 1981). Beta-carboline binding was competitively inhibited by benzodiazepines. Some of the beta-carboline esters (e.g. the propyl ester) were capable of distinguishing receptor subtypes; these were similar but not identical to the subtypes described above (Braestrup & Nielsen 1981a). In addition to inhibiting benzodiazepine binding, the beta-carbolines were able to distinguish differences in the benzodiazepine receptor structure (Nielsen & Braestrup 1980). In contrast to the benzodiazepines, binding to the site labeled by beta-carbolines is inhibited by GABA rather than enhanced. This is consistent with the behavioral and electrophysiological antagonism described above (Braestrup & Nielsen 1981).

In addition to the beta-carbolines, a number of other specific antagonists of the benzodiazepine receptor were developed based on their ability to inhibit the binding of benzodiazepines. The best studied of these compounds is an imidazodiazepine, Ro15-1788 (Hunkeler et al 1981). This compound is a high affinity competitive inhibitor of benzodiazepine and beta-carboline binding and is capable of blocking the pharmacological actions of both these classes of compounds. By itself, it possesses little intrinsic pharmacological activity in animals and humans (Hunkeler et al 1981, Moh et al 1981, Darragh et al 1983). When a radiolabeled form of this compound was studied (Mohler & Richards 1981, Mohler et al 1981a), it was demonstrated that this compound would interact with the same number of sites as the benzodiazepines and beta-carbolines, and that the interactions of these compounds were purely competitive. Another synthetic antagonist of extremely high activity has also been prepared (Yokoyama et al 1982, Czernick et al 1982). This compound, an arylpyrazoloquinoline, has some activity opposite to that of the benzodiazepines (File & Lister 1983) and is the parent compound to a set of other partial benzodiazepine agonists.

The study of the interactions of a wide variety of compounds similar to the above has led to the renaming of these compounds. At present, those compounds possessing activity similar to the benzodiazepines are called *benzodiazepine agonists*. Those compounds, primarily beta-carbolines, possessing activity opposite to the benzodiazepines are called *inverse agonists,* and the compounds, primarily imidazodiazepines, blocking both types of activity have been termed *antagonists* (Braestrup & Nielsen 1983). This nomenclature has been developed to emphasize the fact that a wide variety of compounds can produce a spectrum of pharmacological effects, to indicate that compounds can interact at the same receptor to produce opposite effects, and to indicate that beta-carbolines and antagonists with intrinsic anxiogenic effect are not synonymous. Beta-carbolines with agonist effects have been recently prepared (C. Braestrup, personal communication).

A biochemical test for the pharmacological and behavioral properties of compounds that interact with the benzodiazepine receptor continues to emphasize the interaction with the GABAergic system. In contrast to the benzodiazepines, which show an increase in their affinity due to GABA (Tallman et al 1978, 1980), compounds with antagonist properties show little GABA shift (i.e. change in receptor affinity due to GABA) (Mohler & Richards 1981), and the inverse agonists actually show a decrease in affinity due to GABA (Braestrup & Nielsen 1981b). Thus the GABA shift predicts generally (although not perfectly) the expected behavioral properties of the compounds. Partial agonists are also predicted by the GABA shift, and this may be a useful prebehavioral screen in pharmaceutical development.

Behavioral Effects of Benzodiazepine Antagonists and Inverse Agonists

Although it has not been demonstrated that endogenous ligands interact with benzodiazepine receptors, it is clear from the discussion above that a diverse set of chemicals can elicit behavioral actions by interacting with this site. Many of the beta-carbolines, for example, possess activity opposite to that of benzodiazepines, and initial reports of this activity in rats noted proconvulsant or convulsant effects (Tenen & Hirsch 1980, Oakley & Jones 1980, Robertson 1980, Braestrup et al 1982, Nutt et al 1982). It has been hypothesized that these activities are directly related to the ability of the beta-carbolines to interact with GABA-ergic systems (Braestrup & Nielsen 1983). The structure activity relationships of beta-carbolines and their synthesis have been reviewed elsewhere (Cain et al 1982). In experiments where the effect on sleep in rats was observed, one of the beta-carbolines produced a state of watchful alertness, opposite to the effect of benzodiazepines (Mendelson et al 1982), as well as blocking benzodiazepine-induced sedation (Cowen et al 1981).

The anxiolytic effects of benzodiazepines have always been difficult to assess experimentally in nonprimates (Petersen et al 1982); thus, anecdotal reports of the potent anxiogenic activity of a beta-carboline in healthy human subjects is of great interest (Braestrup & Nielsen 1982). When a metabolically stable beta-carboline was given to humans, two of the subjects who had detectable blood levels of the compound indicated states of extreme agitation and panic. These experiences were reported to be among the worst felt by these experienced drug users. The subjective actions were rapidly terminated by intravenous diazepam. In an analogous series of experiments, the intravenous administration of a beta-carboline to monkeys resulted in intense arousal reactions (Ninan et al 1982); these reactions were prevented by pretreatment with a benzodiazepine or by a benzodiazepine antagonist. Concomitant with the behavioral signs, an activation of the noradrenergic system and release of corticosteroids occurred. The behaviors elicited are similar to those elicited by stimulation of the noradrenergic system with drugs or stimulation of the locus coeruleus (Redmond & Huang 1979). If the proper drug and administration route can be determined, similar pharmacological challenges could be used to study anxiety and panic disorders in human subjects. Current methods of drug challenge are indirect (eg. sodium lactate infusion) or directed only to the examination of the noradrenergic system (Charney et al 1983).

Chronic Benzodiazepine Administration

A reduction in drug response, or tolerance, has been demonstrated following chronic benzodiazepine administration. A large clinical (Greenblatt & Shader 1978, Browne & Penry 1973, Wilder & Bruni 1981, Grundstrom et al 1978) and experimental (Goldberg et al 1967, Sepinwall et al 1978, Hironaka et al 1983) literature has documented decrease in sedative and anticonvulsant effects; the development of tolerance to the anxiolytic effects is somewhat more controversial. A number of behavioral experiments have demonstrated (Vellucci & File 1979) or failed to find (Margules & Stein 1968, Sepinwall et al 1978) tolerance to chronic benzodiazepine effects in experimental paradigms reported to measure anxiety; however, little evidence is found in the clinical literature for development of tolerance to antianxiety effects, even after prolonged continuous administration (Rickels et al 1983, Lader & Petursson 1983). This discrepancy may be explained by differences in ability to predict anxiolytic behavior by animal models (File 1980); an alternative hypothesis might be that in contrast to antiepileptic studies, clinical studies of anxiety have never employed threshold anxiolytic doses of benzodiazepines. Anxiolytic studies in humans have used relatively large doses of benzodiazepines, which do not produce marked side effects. Because the benzodiazepines are so efficacious (high therapeutic index) in the treatment of anxiety (Greenblatt & Shader 1974), the high doses employed in such studies may far exceed the ability to develop demonstrable tolerance even following prolonged treatment.

As with other sedative-hypnotic drugs, tolerance to benzodiazepines has been associated with development of physical dependence (McNicholas et al

1983, Ryan & Boisse 1983). Withdrawal reactions to chronic benzodiazepine administration can be demonstrated both clinically and with experimental animals (Hollister et al 1961, Rickels 1983). The benzodiazepine antagonist, Ro15-1788, has been reported to induce severe withdrawal responses in a number of species (including primates) following chronic benzodiazepine administration (Lukas & Griffiths 1982, McNicholas & Martin 1982, Rosenberg & Chiu 1982).

However, despite our increasing knowledge about the various components of the GABAergic complex and the ability to predict some of the behavioral properties of drugs in this system, mechanisms by which chronic occupation of the benzodiazepine sites could lead to decreased drug efficacy remain elusive (Braestrup & Nielsen 1983).

Based on the involvement of GABA with the actions of the benzodiazepines, one possible site of alteration in the complex would be at the GABA receptor. Electrophysiological studies on cerebellar Purkinje cells (Waterhouse et al 1984) and in spinal cord cultures (Sher et al 1983) have shown that prolonged exposure to benzodiazepines reduces their ability to enhance GABA-mediated inhibition. Recently, electrophysiological evidence for decreased postsynaptic sensitivity to GABA has been reported following chronic benzodiazepine administration (Gallager et al 1984a). The microionotophoretic sensitivity to the neurotransmitters GABA and serotonin were tested in serotoninergic cells of the dorsal raphe nucleus. These cells are known to receive GABAergic input (Gallager 1978, Belin et al 1979). In animals exposed chronically (≥ 3 wk) to a moderate dose of diazepam, cells were found to be significantly less sensitive to GABA (Figure 2, Gallager et al 1984a). In contrast, a single acute exposure to diazepam results in a marked increase in GABA sensitivity, which is postulated to be the mode of action of the benzodiazepines (see second section, above). Thus, neurons in the dorsal raphe nucleus were found to respond differently to a normal neurotransmitter, GABA, depending upon their prior history of exposure to diazepam. The electrophysiologically measurable changes in GABA sensitivity following chronic administration of benzodiazepine were found to be selective since serotonin responses were not altered by either chronic or acute exposure to diazepam. These chronic responses are indicative of a measurable functional subsensitivity to GABA. However, in these studies, drug treatments were discontinued 16 to 24 hours prior to the electrophysiological recording sessions. Thus, the subsensitivity to GABA observed in chronic benzodiazepine-treated rats could have been related either to development of tolerance or to withdrawal from the drug. To address this issue the specific benzodiazepine antagonist, Ro15-1788, was administered. Ro15-1788 administration rapidly reversed GABA subsensitivity in chronic benzodiazepine-treated rats. This indicates that the decreased sensitivity to GABA is not a withdrawal response (a response that would be predicted to be

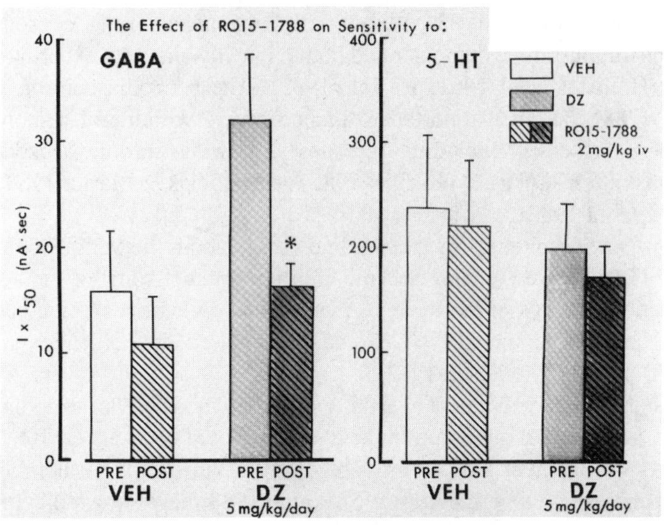

Figure 2 Sensitivity of spontaneously firing serotonin-containing cells in the dorsal raphe nucleus to microinotophoretically applied GABA and serotonin (5HT). Mean $I \times T_{50}$ values as defined by the ejection current I (in nA) applied multiplied by the time T_{50} (s) required to depress cell firing by 50% from the basal firing rate. Data shows that prior to Ro15-1788, sensitivity in chronic diazepam-treated animals to GABA was reduced (indicated by greater $I \times T_{50}$, DZ: 32 ± 6 vs VEH: $= 15 \pm 6$) without significantly altering 5HT sensitivity. Immediately following an intravenous injection of 2 mg/kg of the benzodiazepine antagonist, Ro15-1788, sensitivity to GABA in chronic diazepam-treated animals returned to control $I \times T_{50}$ values (16 ± 3); Ro15-1788 had no effect on microiontophoretic sensitivities to GABA or 5HT in other treatment groups.

enhanced rather than reversed by Ro15-1788). Since the anticonvulsant activity of benzodiazepines is mediated by the facilitation of GABAergic transmission, the demonstration of a decrease in sensitivity to GABA following chronic exposure to diazepam is consistent with the development of tolerance to anticonvulsant effects after prolonged use.

In order to investigate whether receptor alterations might accompany these electrophysiological changes in GABA sensitivity, various binding-site components of the GABA/benzodiazepine receptor complex were examined (Gallager et al 1984a). The same chronic treatment regimen as in the electrophysiological experiments was utilized and no significant alteration in the number or affinity of benzodiazepine binding sites was noted. This indicated that chronic exposure to diazepam modified some element in the GABA/benzodiazepine macromolecular complex other than the benzodiazepine binding site. However, a significant decrease in the ability of GABA to enhance benzodiazepine binding was observed in cortical membranes from animals chronically treated with diazepam. This decrease is consistent with the electrophysiological evidence for subsensitivity. Decreases in the ability of GABA to enhance diazepam binding have also been observed in developing rats following prenatal exposure to benzodiazepines (Massotti et al 1980, Braestrup & Nielsen 1981c). A number of studies have suggested that GABA enhancement of benzodiazepine binding is associated with a low affinity component of the GABA recognition site (see above). On the basis of this information, low and high affinity components of the GABA recognition portion of the GABA/benzodiazepine complex were investigated (Gallager et al 1984b). In these studies, no significant alteration in the high-affinity component of the GABA recognition site was found. However, using ^{3}H-bicuculline combined with thiocyanate treatment to measure a lower affinity component of the GABA recognition site, significant alterations were observed. Increases in the affinity of GABA for the ^{3}H-bicuculline binding site and increases in ^{3}H-bicuculline binding were noted; these increases may be indicative of a conversion of the GABA recognition site to a higher affinity form in the chronically treated animal, which may be a desensitized form (see above). Such a mechanism would be consistent with the physiological data showing reduced sensitivity to GABA following chronic benzodiazepine administration.

The conclusion that can be drawn from these data is that chronic diazepam exposure leads to a functionally significant decrease in GABA sensitivity. Results from the binding studies suggest that chronic diazepam exposure may induce a change in the GABA recognition site of the GABA/benzodiazepine receptor complex. The electrophysiological data suggest that this altered GABA recognition site is functionally less active. Such an effect might be one mechanism for the development of tolerance to anticonvulsant effects after prolonged benzodiazepine use. In pathological states in which the GABAergic system may already be compromised, as is speculated for some seizure disorders (Worms & Lloyd 1981), chronic administration of benzodiazepines might actually lead to a more seizure-prone nervous system.

What is most interesting about such a postulated mechanism is that the chronic occupancy of the benzodiazepine site appears to affect not the site itself but the GABA recognition site. Such an interaction would provide an example of how alteration of a modulatory component in the GABAergic system can change other elements of the receptor complex. In fact, alterations of the GABA receptor site might be expected to affect responses in the GABA system more profoundly than a change at the benzodiazepine modulatory site. This type of adaptive change might be representative of changes due to occupancy of other modulatory sites in different transmitter systems. Such adaptive changes may explain unexpected consequences of chronic occupation of drug receptors.

Molecular Characterization of Benzodiazepine Receptors

In addition to the behavioral properties of the different ligands for the benzodiazepine receptor, the way in which they interact with the binding site is under study at the molecular level. The ability of GABA to alter the binding of benzodiazepine agonists and antagonists differentially means that the loci to which agonists and antagonists bind cannot be identical. Probably several amino acids are involved in the binding of ligands to the receptor protein and contribute to the driving force for this reaction. The binding of benzodiazepines is highly temperature dependent and both the agonist and antagonist binding appear to be enthalpy driven (reviewed in Thomas & Tallman 1984). There are only minor entropic contributions; however, in the presence of GABA, the increase in affinity at 4°C appears to be due primarily to an entropic change, whereas at 37°C an enthalpic component change is also involved. Taken together, these studies would indicate that the topology of the benzodiazepine site is altered by GABA and this altered topology is reflected in different interactions between amino acids and drugs. It is likely that although benzodiazepines and other compounds interact with each others' binding sites, both in membrane-bound and solubilized receptors, they interact at different domains of the binding site.

Another way of studying these separate domains is through photolabeling. During the studies described above using photolabeling with (^{3}H)-flunitrazepam, it was discovered that ultraviolet illumination could cause the ligand to form a covalent crosslink with the receptor (Mohler et al 1980). This crosslink would occur only with the neuronal receptor and was inhibited by compounds that compete for binding sites. In addition to the membrane receptor, it was possible to photolabel solubilized benzodiazepine receptors, and differences have been demonstrated between membrane and soluble preparations (Thomas & Tallman 1981). In solubilized preparations, the same number of sites are photolabeled with (^{3}H)-flunitrazepam as are inactivated when nonradioactive flunitrazepam is used to photolabel; the remaining sites are counted in equilibrium-binding studies. In contrast, analysis of binding data on membranes indicated that the number of sites photolabeled with (^{3}H)-flunitrazepam is about one-quarter the number of sites inactivated when nonradioactive flunitrazepam is used (Mohler et al 1980, Thomas & Tallman 1981). The antagonist domain of the benzodiazepine site is not affected by photolabeling; however, the ability of benzodiazepines to inhibit binding to the antagonist domain is decreased proportional to the extent of photolabeling (Hirsch et al 1982, Mohler 1983, Thomas & Tallman 1983). It is possible that photolabeling could be used to study the life-cycle and turnover of the benzodiazepine receptor (Chan et al 1983).

Solubilization and Purification of the Complex of Proteins

A number of research groups have been attempting to solubilize and purify the membrane proteins involved in the actions of the benzodiazepines. The early studies in this area have been extensively reviewed (Thomas & Tallman 1984). In brief, a number of research groups have solubilized these proteins.

Most studies have been able to maintain high-affinity binding for GABA and all benzodiazepine ligands. In one report, the picrotoxinin site was solubilized along with these other activities and separated from the others by gel filtration. Although the allosteric nature of the interactions indicates a close proximity, it is not known whether the binding sites for the drugs described in this review all exist on the same polypeptide chain. Estimates of the molecular weight of the detergent solubilized material are greater than 200,000, while the molecular weight of the photolabeled benzodiazepine subunit is 50,000. Some reports of heterogeneity in the solubilized material have also appeared, and a differential extraction with detergent and salt has been reported (Lo et al 1982). The model of a multimeric complex containing more than just a single benzodiazepine subunit is currently favored. The exact stoichiometry of the sites remains to be determined and will require purification of the complex to homogeneity.

Affinity chromatographic methods have been used to purify the benzodiazepine binding sites. The most successful methods used to date involve the use of a column with a long hydrophilic spacer and incorporation of a charge moiety, along with a potent benzodiazepine (Martini et al 1982, Sigel et al 1983). These methods allow a successful competitive ligand-based elution and

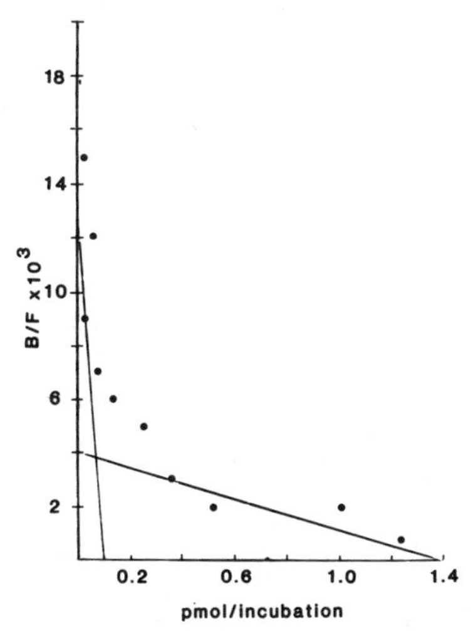

Figure 3 Binding of ^{3}H-GABA to rat brain GABA receptors purified by affinity chromatography on an agarose aminoclonazepam column. Detergent solubilized receptors were adsorbed to the affinity column and eluted with 100 μM flurazepam (J. F. Tallman, unpublished). Flurazepam was removed from the preparation by dialysis and filtration and the supernatant assayed for GABA binding activity (Yousufi et al 1979). Binding data were resolved into two curves with affinity of 8.3 and 350 nM, respectively. The same preparation contained high affinity binding sites for muscimol, flunitrazepam, beta-carbolines, and Ro15-1788.

a high degree of purification. Using variants of these methods, the presence of benzodiazepine receptors, high and low affinity GABA binding sites (Figure 3), and GABA activation of benzodiazepine binding have already been demonstrated in preparations more than a thousand-fold purified from crude homogenate. It should be possible to use these purified preparations for reconstitution studies and the preparation of anti-GABA receptor complex antibodies. In turn, the domains of the complex important for particular functions can be identified. It will then be possible to examine the function of this complex at a molecular level and understand how the drugs are capable of inducing conformational changes necessary for chloride conductance changes.

THEORETICAL PROPERTIES
OF THE GABA/BENZODIAZEPINE COMPLEX

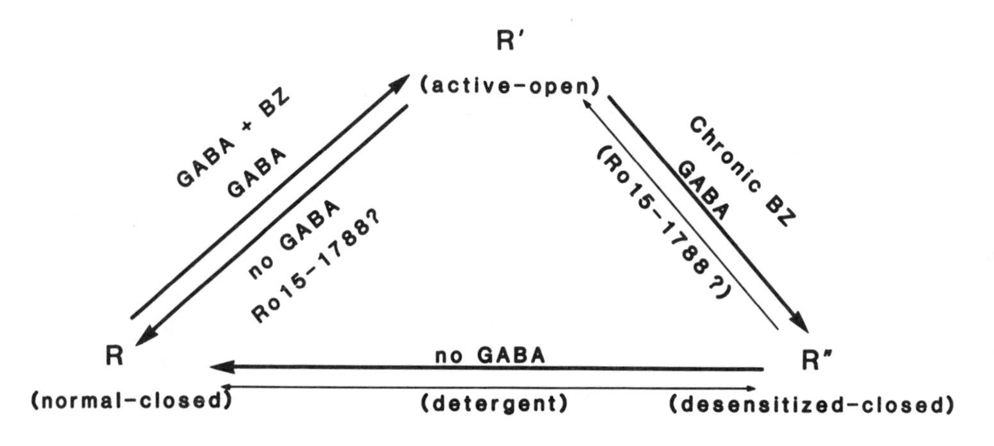

AFFINITY FOR GABA	CONFORMATION OF ACTIVE SITE	METHOD OF STUDY	STABILITY	ANION CHANNEL CONFORMATION
Intermediate	R	GABA-BZ interaction GABA-BIC	metastable	closed
Low	R'	?Fast kinetics ?Patch clamp	unstable	open
High	R"	Agonist desensitized Chronic BZ	stable	closed

Figure 4 Theoretical properties of the GABA/benzodiazepine complex.

Conclusions

One of the major directions of research is toward the definition of the molecular interactions that occur at the various binding sites in the GABA-ergic complex. A full understanding of the function of the complex will require a more integrated view of the functional interactions between the sites. Thus, while most of the reviews concerning the pharmacological and molecular actions of the benzodiazepines have focused on the benzodiazepine binding site itself, it seems more appropriate to emphasize the modulatory role of benzodiazepines in interacting with the GABAergic system. Each drug would be considered in light of its effects on the GABAergic system and its modification of the actions of this important inhibitory transmitter. We have summarized the current view of the properties of the various components of the GABAergic complex (Figure 4). According to our model, the probability that the active site will be in the R′ conformation is increased following occupation by GABA and is even higher after GABA in combination with benzodiazepines. From the R′ conformation, the active site can decay to its native state R or proceed to R″, depending upon the amount of GABA available. Since benzodiazepines increase the probability of attaining the R′ conformation, chronic benzodiazepines would increase the likelihood of conversion to R″. Once in R″, the receptor will slowly return to R, dependent on the removal of GABA from the site. Such a model is suggested to account for the rapid physiological desensitization following agonist (GABA) activation and the ability of benzodiazepines to modify GABA actions acutely and chronically. As more information becomes available in this rapidly progressing area, a much clearer picture of the molecular interactions occurring at the complex should lead to a more complete knowledge of functional importance of GABAergic interactions in brain.

Literature Cited

Belin, M., Augera, M., Tappaz, M., McRae-Degueurce, A., Bobillier, P., Pujol, J. 1979. GABA-accumulating neurons in the nucleus raphe dorsalis and periaqueductal gray in the rat: A biochemical and autoradiographic study. *Brain Res*. 170:279–97

Bowery, N. G. 1983. Classification of GABA receptors. In *The GABA Receptors*, ed. S. Enna, pp. 178–205. Clifton, N.J.: Humana.

Bowery, N. G., Collins, J. F., Hill, R. G., Pearson, S. 1976. GABA antagonism as a possible basis for the convulsant action of a series of bicyclic phosphorus esters. *Br. J. Pharmacol*. 57:435–36

Bowery, N. G., Collins, J. F., Hill, R. G., Pearson, S. 1977. *t*-Butyl bicyclo phosphate: A convulsant and GABA antagonist more potent than bicuculline. *Br. J. Pharmacol*. 60:175–76

Bowling, A. C., DeLorenzo, R. J. 1982. Micromolar affinity benzodiazepine receptors: Identification and characterization in central nervous systems. *Science* 216(4551):1247–50

Braestrup, C., Nielsen, M. 1981b. GABA reduces binding of H-methyl beta-carboxylate to brain benzodiazepine receptors. *Nature* 294(5840):472–74

Braestrup, C., Nielsen, M. 1981c. Modulation of benzodiazepine receptors. *Adv. Biosci*. 31:221–27

Braestrup, C., Nielsen, M. 1982. Anxiety. *Lancet* 2(8306):1030–34

Braestrup, C., Nielsen, M. 1983. Benzodiazepine receptors. *Handb. Psychopharmacol*. 17:285–384

Braestrup, C., Nielsen, M., Krogsgaard-Larsen, P., Falch, E. 1979a. Partial agonists for brain GABA/benzodiazepine receptor complex. *Nature* 280(5729):331–33

Braestrup, C., Nielsen, M., Biggio, G., Squires, R. F. 1979b. Neuronal localization of benzodiazepine receptors in cerebellum. *Neurosci. Lett.* 13(3):219–24

Braestrup, C., Nielsen, M., Olsen, C. E. 1980. Urinary and brain beta-carboline-3-carboxylates as potent inhibitors of brain benzodiazepine receptors. *Proc. Natl. Acad. Sci. USA* 77(4):2288–92

Braestrup, C., Schmiechen, R., Neef, G., Nielsen, M., Petersen, E. N. 1982. Interaction of convulsive ligands with benzodiazepine receptors. *Science* 216:1241–43

Braestrup, C., Squires, R. F. 1978. Brain specific benzodiazepine receptors. *Br. J. Psychiatry* 133:249–60

Browne, T. R., Penry, J. K. 1973. Benzodiazepines in the treatment of epilepsy. A review. *Epilepsia* 14:277–310

Browner, M., Ferkany, J. W., Enna, S. J. 1981. Biochemical identification of pharmacologically and functionally distinct GABA receptors in rat brain. *J. Neuroscience* 1:514–18

Cain, M., Weber, R. W., Guzman, F., Cook, J. M., Barker, S. A., Rice, K. C., Crawley, J. N., Paul, S. M., Skolnick, P. 1982. Beta-carbolines: Synthesis and neurochemical and pharmacological actions on brain benzodiazepine receptors. *J. Med. Chem.* 25(9):1081–91

Campbell, W. C., Fisher, M. H., Stapley, E. O., Albers-Schonberg, G., Jacob, T. A. 1983. Ivermectin: A potent new antiparasitic agent. *Science* 221:823–28

Candy, J. M., Martin, I. L. 1979. Is the benzodiazepine receptor coupled to a chloride anion channel? (letter.) *Nature* 280 (5718):172–74

Chan, C. Y., Gibbs, T. T., Borden, L. A., Farb, D. H. 1983. Multiple embryonic benzodiazepine binding sites: Evidence for functionality. *Life Sci.* 33:2061–69

Changeux, J. P. 1981. The acetylcholine receptor: An allosteric membrane protein. *Harvey Lectures* 75:85–254

Charney, D. S., Henninger, G. R., Hafstad, K., Capelli, S., Redmond, D. E. 1983. Neurobiological mechanisms in human anxiety: Recent clinical studies. *Psychopharm. Bull.* 19:470–74

Costa, E., Greengard, P., eds. 1975. *Mechanism of Action of Benzodiazepines*. New York: Raven

Costa, E., Guidotti, A., Mao, C. C., Suria, A. 1975. New concepts on the mechanism of action of the benzodiazepines. *Life Sci.* 17:167–86

Costa, T., Rodbard, D., Pert, C. B. 1979. Is the benzodiazepine receptor coupled to a chloride anion channel? *Nature* 177 (5694):315–17

Cowen, P. J., Green, A. R., Nutt, D. J. 1981. Ethyl beta-carboline carboxylate lowers seizure threshold and antagonizes flurazepam-induced sedation in rats. *Nature* 290(5801):54–55

Curtis, D. R., Johnston, G. A. R. 1974. Amino acid transmitters in the mammalian central nervous system. *Ergeb. Physiol. Biol. Chem. Exp. Pharamakol.* 69:97–188

Czernick, A. J., Petrack, B., Kalinsky, J., Psychoyos, S., Cash, W. D., et al. 1982. Receptor binding characteristics of a potent benzodiazepine antagonist. *Life Sci.* 30:363–72

Darragh, A., Lambe, R., Kenny, M., Brick, J. 1983. Tolerance of health volunteers to intravenous administration of the clenzodiazepine antagonist Ro15-1788. *Eur. J. Clin. Pharmacol.* 14:569–70

DeLorenzo, R. J., Burdette, S., Holderness, J. 1981. Benzodiazepine inhibition of the calcium-calmodulin protein kinase system in brain membrane. *Science* 213(4507): 546–49

Enna, S. J., Snyder, S. H. 1977. Influences of ions, enzyme and detergents on gamma-aminobutyric acid receptor binding in synaptic membranes of rat brain. *Mol. Pharmacol.* 13:442–53

Fahn, S., Cote, L. J. 1968. Regional distribution of gamma-amino butyric acid (GABA) in the brain of the Rhesus monkey. *J. Neurochem.* 15:209–13

File, S. A. 1980. The use of social interaction as a method for detecting anxiolytic activity of chlordiazepoxide-like drugs. *J. Neurosci. Meth.* 2:219–38

File, S., Lister, R. 1983. Quinolines and anxiety: Anxiogenic effects of CGS-8216 and partial anxiolytic profile of PK 9084. *Pharmacol. Biochem. Behav.* 18:185–88

Fonnum, F., Storm-Mathisen, J. 1978. Localization of GABAergic neurons in the CNS. *Handb. Psychopharmacol.* 9:357–401

Gallager, D. W. 1978. Benzodiazepines: Potentiation of a GABA inhibitory response in the dorsal raphe nucleus. *Eur. J. Pharmacol.* 49:133–43

Gallager, D., Lakoski, J., Gonsalves, S., Rauch, S. 1984a. Chronic benzodiazepine treatment decreases postsynaptic GABA sensitivity. *Nature* 308:74–77

Gallager, D., Rauch, S., Malcolm, A. 1984b. Alterations in a low affinity GABA recognition site following chronic benzodiazepine. *Eur. J. Pharmac.* 98:159–60

Gallager, D. W., Mallorga, P., Oertel, W., Henneberry, R., Tallman, J. F. 1981. ^{3}H Diazepam binding in a mammalian central nervous system: A pharmacological characterization. *J. Neurosci.* 1(2):218–25

Gee, K. W., Morelli, M., Yamamura, H. I.

1982. The effect of temperature on CL218872 and propyl beta-carboline-3-carboxylate inhibition of ^{3}H-flunitrazepam binding in rat brain. *Biochem. Biophys. Res. Commun.* 105(4):1532–37

Goldberg, M. E., Manian, A. A., Efron, D. H. 1967. A comparative study of certain pharmacologic responses following acute and chronic administrations of chlordiazepoxide. *Life Sci.* 6:481–99

Greenblatt, D. T., Shader, R. I. 1974. *Benzodiazepines in Clinical Practice*. New York: Raven

Greenblatt, D. J., Shader, R. I. 1978. Dependence, tolerance, and addiction to benzodiazepines: Clinical and pharmacokinetic considerations. *Drug Metab.* 8:13–28

Grundstrom, R., Holmberg, G., Hansen, T. 1978. Degree of sedation obtained with various doses of diazepam and nitrazepam. *Acta Pharmacol. Toxicol.* 43:13–18

Guidotti, A., Konkel, D. R., Epstein, B., Corda, M. G., Wise, B. C., Krutzsch, H., Meek, J. L., Costa, E. 1982. Isolation, characterization and purification to homogeneity of a rat brain protein (GABA-modulin). *Proc. Natl. Acad. Sci. USA* 79:6084-88

Guidotti, A., Toffano, G., Costa, E. 1978. An endogenous protein modulates the affinity of GABA and benzodiazepine receptors in rat brain. *Nature* 157:553–55

Haefely, W., Pieri, L., Polc, P., Schaffner, R. 1981. General pharmacology and neuropharmacology of benzodiazepine derivatives. *Handb. Exptl. Pharmacol.* 55:1–262

Hironaka, T., Fuchino, K., Fujii, T. 1983. The benzodiazepine receptor and receptor tolerance produced by chronic treatment of diazepam. *Jpn. J. Pharmacol.* 33:95–102

Hirsch, J. D., Kochman, R. L., Sumner, P. R. 1982. Heterogeneity of brain benzodiazepine receptors demonstrated by ^{3}H-propyl B-carboline-3-carboxylate binding. *Mol. Pharmacol.* 21:618–28

Hollister, L. E., Motzenbacker, F. P., Degan, R. O. 1961. Withdrawal drug reactions from chlordiazepoxide (Librium). *Psychopharmacologia* 2:63–68

Hunkeler, W., Mohler, H., Pieri, L., Polc, P., Bonetti, E. P., Cumin, R., Schaffner, R., Haefely, W. 1981. Selective antagonists of benzodiazepines. *Nature* 290:514–16

Klepner, C. A., Lippa, A. S., Benson, D. I., Sano, M. C., Beer, B. 1979. Resolution of two biochemically and pharmacologically distinct benzodiazepine receptors. *Pharmacol. Biochem. Behav.* 11:457–62

Klotz, K. L., Bocchetta, A., Neale, J., Thomas, J. W., Tallman, J. F. 1984. Proteolytic degradation of neuronal benzodiazepine binding sites. *Life Sci.* 34:293–99

Krespan, B., Springfield, S. A., Haas, H., Geller, H. M. 1984. Electrophysiological studies on benzodiazepine antagonists. *Brain Res.* In press

Lader, M., Petursson, H. 1983. Long-term effects of benzodiazepines. *Neuropharmacology* 22:527–33

Leeb-Lundberg, F., Napias, C., Olsen, R. W. 1981. Dihydropicrotoxinin binding sites in mammalian brain: Interaction with convulsant and depressant benzodiazepines. *Brain Res.* 216:339–408

Leeb-Lundberg, F., Snowman, A., Olsen, R. W. 1980. Barbiturate receptor sites are coupled to benzodiazepine receptors. *Proc. Natl. Acad. Sci. USA* 77:7468–72

Lippa, A. S., Beer, B., Sano, M. C., Vogel, R. A., Meyerson, L. R. 1981. Differential ontogeny of type 1 and type 2 benzodiazepine receptors. *Life Sci.* 28:2343–47

Lippa, A. S., Coupet, J., Greenblatt, E. N., Klepner, C. A., Beer, B. 1979. A synthetic non-benzodiazepine ligand for benzodiazepine receptors: A probe for investigating neuronal substrates of anxiety. *Pharmacol. Biochem. Behav.* 11:99–106

Lo, M., Strittmatter, S. M., Snyder, S. H. 1982. Physical separation and characterization of two types of benzodiazepine receptors. *Proc. Natl. Acad. Sci. USA* 79:680–84

Lo, M., Niehoff, D., Kuhar, M. J., Snyder, S. H. 1983. Differential localization for type 1 and type 2 benzodiazepine binding sites in substantia nigra. *Nature* 306:57–60

Lukas, S. E., Griffiths, R. R. 1982. Precipitated withdrawal by a benzodiazepine receptor antagonist (Ro15-1788) after 7 days of diazepam. *Science* 217:1161–63

Mallorga, P., Hamburg, M., Tallman, J. F., Gallager, D. W. 1980. Ontogenetic changes in GABA modulation of brain benzodiazepine binding. *Neuropharmacology* 19:405–8

Margules, D., Stein, L. 1968. Increase of "antianxiety" activity and tolerance of behavioral depression during chronic administration of oxazepam. *Psychopharmacologia* 13:74–80

Martini, C., Lucacchini, A., Ronca, G., Hrelis, S., Rossi, C. 1982. Isolation of putative benzodiazepine receptors from rat brain membranes by affinity chromatography. *J. Neurochem.* 38:15–19

Massotti, M., Allera, F., Balazs, T., Guidotti, A. 1980. GABA and benzodiazepine receptors in the offspring of dams receiving diazepam: ontogenetic studies. *Neuropharmacology* 19:951–56

Matthew, E., Laskin, J. D., Zimmerman, E. A., Weinstein, I. B., Hsu, K. C., Engelhardt, D. L. 1981. Benzodiazepines have high-affinity binding sites and induce melanogenesis in B16-C3 melanoma cells. *Proc. Natl. Acad. Sci. USA* 78:3935–39

McCarthy, K. D., Harden, T. K. 1981. Identification of two benzodiazepine binding sites on cells cultured from rat cerebral cortex. *J. Pharmacol. Exp. Ther.* 216:183–91

McNicholas, L. F., Martin, W. R. 1982. The effect of a benzodiazepine antagonist, Ro15-1788, in diazepam dependent rats. *Life Sci.* 31:731–37

NcNicholas, L. F., Martin, W. R., Cherian, S. 1983. Physical dependence on diazepam and lorazepam in the dog. *J. Pharm. Exp. Ther.* 226:783–89

Meiners, B. A., Salama, A. I. 1982. Enhancement of benzodiazepine and GABA binding by the novel anxiolytic, tracazolate. *Eur. J. Pharmacol.* 78:315–22

Melchior, C. L., Garrett, K., Tabakoff, B. 1983. Proconvulsant effects of the benzodiazepine agonist CL 218,872. *Soc. Neurosci. Abstr.* 9:129

Mendelson, W. B., Cain, M., Cook, J. M., Paul, S. M., Skolnick, P. 1982. Do benzodiazepine receptors play a role in sleep regulation? Studies with the benzodiazepine antagonist, 3-hydroxymethyl-beta-carboline (3-HMC). *Prog. Clin. Biol. Res.* 90:253–61

Mohler, H. 1982. Benzodiazepine receptors: Differential interaction of benzodiazepine agonists and antagonists after photoaffinity labeling with flunitrazepam. *Eur. J. Pharmacol.* 80:435–36

Mohler, H., Okada, T. 1977. Gamma-aminobutyric acid receptor binding with (+) bicuculline methiodide in rat CNS. *Nature* 267:65–67

Mohler, H., Okada, T. 1977. Benzodiazepine receptor: Demonstration in the central nervous system. *Science* 198:845–51

Mohler, H., Okada, T. 1978. The benzodiazepine receptor in normal and pathological human brain. *Br. J. Psychiatry* 133:261–68

Mohler, H., Richards, J. G. 1981. Agonist and antagonist benzodiazepine receptor interaction in vitro. *Nature* 294:763–65

Mohler, H., Battersby, M. K., Richards, J. G. 1980. Benzodiazepine receptor protein identified and visualized in brain tissue by a photoaffinity label. *Proc. Natl. Acad. Sci. USA* 77:1166–1670

Mohler, H., Burkard, W. P., Keller, H. H., Richards, J. G., Haefely, W. 1981a. Benzodiazepine antagonist Ro15-1788: Binding characteristics and interaction with drug-induced changes in dopamine turnover and cerebellar cGMP levels. *J. Neurochem.* 37:714–22

Mohler, H., Richards, J. G., Wu, J. Y. 1981b. Autoradiographic localization of benzodiazepine receptors in immunocytochemically identified gamma-aminobutyrergic synapses. *Proc. Natl. Acad. Sci. USA* 78:1935–38

Napias, C., Bergman, M. O., VanNess, P. C., Greenlee, D. V., Olsen, R. W. 1980. GABA binding in mammalian brain inhibition by endogenous GABA. *Life Sci.* 27:1001–11

Niehoff, D. L., Mashal, R. D., Horst, W. D., O'Brien, R. A., Palacios, J. M., Kuhar, M. J. 1982. Binding of a radiolabeled triazolopyridazine to a subtype of benzodiazepine receptor in the rat cerebellum. *J. Pharmacol. Exp. Ther.* 221:670–75

Ninan, P., Insel, T. R., Cohen, R. M., Skolnick, P., Paul, S. M. 1982. A benzodiazepine receptor mediated model of anxiety. *Science* 218:1332

Nielsen, M., Braestrup, C. 1980. Ethyl beta-carboline-3-carboxylate shows differential benzodiazepine receptor interaction. *Nature* 286:606–7

Nielsen, M., Gredal, O., Braestrup, C. 1979. Some properties of ^{3}H-diazepam displacing activity from human urine. *Life Sci.* 25:679–86

Nielsen, M., Schou, H., Braestrup, C. 1981. ^{3}H propyl beta-carboline-3-carboxylate binds specifically to brain benzodiazepine receptors. *J. Neurochem.* 36:276–85

Nutt, D. J., Cowen, P. J., Little, H. J. 1982. Unusual interactions of benzodiazepine receptor antagonists. *Nature* 295:436–38

Oakley, N. R., Jones, B. J. 1980. The proconvulsant and diazepam-reversing effects of ethyl-beta-carboline-3-carboxylate. *Eur. J. Pharmacol.* 68:381–82

Olsen, R. W. 1982. Drug interactions at the GABA receptor-ionophore complex. *Ann. Rev. Pharmacol. Toxicol.* 22:245–77

Olsen, R. W., Snowman, A. M. 1983. ^{3}H-bicuculline methochloride binding to low-affinity gamma-amino butyric acid receptor sites. *J. Neurochem.* 41:1653–63

Petersen, E. N., Paschelke, G., Kehr, W., Nielsen, M., Braestrup, C. 1982. Does the reversal of the anticonflict effect of phenobarbital by beta-CCE and FG 7142 indicate benzodiazepine receptor-mediated anxiogenic properties? *Eur. J. Pharmacol.* 82:217–21

Polc, P., Ropert, N., Wright, D. M. 1981. Ethyl beta-carboline-3-carboxylate antagonizes the action of GABA and benzodiazepines in the hippocampus. *Brain Res.* 217:216–20

Pong, S. S., DeHaven, R., Wang, C. C. 1982. A comparative study of avermectin B1a and other modulators of the gamma-aminobutyric acid receptor—chloride ion channel complex. *J. Neurosci.* 2:966–71

Redmond, D. E., Huang, Y. H. 1979. New evidence for a locus coeruleus-norepinephrine connection with anxiety. *Life Sci.* 25:2149

Regan, J. W., Roeske, W. R., Yamamura, H. I. 1980. The benzodiazepine receptor: Its development and its modulation by gamma-aminobutyric acid. *J. Pharmacol. Exp. Ther.* 212:137–43

Rickels, K., Case, W., Downing, R., Winokur, A. 1983. Long-term diazepam therapy and clinical outcome. *JAMA* 250:767–71

Roberts, E., Frankel, S. 1950. Gamma-amino butyric acid in brain: Its formation from glutamic acid. *J. Biol. Chem.* 187:55–63

Robertson, H. A. 1980. Harmaline-induced tremor: The benzodiazepine receptor as a site of action. *Eur. J. Pharmacol.* 67:129–32

Rosenberg, H. C., Chiu, T. H. 1982. An antagonist-induced benzodiazepine abstinence syndrome. *Eur. J. Pharmacol.* 81:153–57

Ryan, G., Boisse, N. 1983. Experimental induction of benzodiazepine tolerance and physical dependence. *J. Pharmacol. Exp. Ther.* 226:100–7

Schmidt, R. F., Vogel, M. E., Zimmerman, M. 1967. Die Wirkung von Diazepam auf die prasynaptische Hemmung und andre Ruckenmarksreflexe. *Arch. Exp. Path. Pharmakol.* 258:69–82

Schoemaker, H., Bliss, M., Yamamura, H. I. 1981. Specific high-affinity saturable binding of ^{3}H RO5-4864 to benzodiazepine binding sites in the rat cerebral cortex. *Eur. J. Pharmacol.* 71:173–75

Sepinwall, J., Grodsky, F. S., Cook, L. 1978. Conflict behavior in the squirrel monkey: Effects of chlordiazepoxide, diazepam and *N*-desmethyldiazepam. *J. Pharmacol. Exp. Ther.* 204:88–102

Sher, P. K., Study, R. E., Mazzetta, J., Barker, J. L., Nelson, P. G. 1983. Depression of benzodiazepine binding and diazepam potentiation of GABA-mediated inhibition after chronic exposure of spinal cord cultures to diazepam. *Brain Res.* 268:1711–76

Sigel, E., Stephenson, A., Mamalaki, C., Barnard, E. 1983. A gamma-amino butyric acid/benzodiazepine receptor complex of bovine cerebral cortex. *J. Biol. Chem.* 258:6965–71

Skolnick, P., Moncada, V., Barker, J. L., Paul, S. M. 1981. Pentobarbital: Dual actions to increase brain benzodiazepine receptor affinity. *Science* 211:1448–50

Skolnick, P., Paul, S. M., Barker, J. L. 1980. Pentobarbital potentiates GABA-enhanced ^{3}H-diazepam binding to benzodiazepine receptors. *Eur. J. Pharmacol.* 65:125–27

Sieghart, W., Karobath, M. 1980. Molecular heterogeneity of benzodiazepine receptors. *Nature* 286:285–87

Sieghart, W., Mayer, A., Drexler, G. 1983. Properties of flunitrazepam binding to different benzodiazepine binding proteins, *Eur. J. Pharmacol.* 88:291–99

Simmonds, M. A. 1980. Evidence that bicuculline and picrotoxin act at separate sites to antagonize gamma-aminobutyric acid in rat cuneate nucleus. *Neuropharmacology* 19:39–45

Squires, R. F., Braestrup, C. 1977. Benzodiazepine receptors in rat brain. *Nature* 166:732–34

Squires, R. F., Benson, D. I., Braestrup, C., Coupet, J., Klepner, C. A., Myers, V., Beer. B. 1979. Some properties of brain specific benzodiazepine receptors: New evidence for multiple receptors. *Pharmacol. Biochem. Behav.* 10:825–30

Squires, R. F., Casida, J., Richardson, M., Saederup, E. 1982. (^{35}S) *t*-butylbicyclophosphorothionate binds with high affinity to brain specific sites coupled to gamma-amino butyric acid-A and ion recognition sites. *Mol. Pharmacol.* 23:326–36

Stroud, R. M. 1983. Acetylcholine receptor structure. *Neurosci. Comment.* 1:124–38

Supavilai, P., Karobath, M. 1979. Stimulation of benzodiazepine receptor binding by SQ 20009 is chloride-dependent and picrotoxin-sensitive. *Eur. J. Pharmacol.* 60:111–13

Supavilai, P., Karobath, M. 1980. Interaction of SQ 20009 and GABA-like drugs as modulators of benzodiazepine receptor binding. *Eur. J. Pharmacol.* 62:229–33

Supavilai, P., Karobath, M. 1981. Action of pyrazolopyridines as modulators of ^{3}H-flunitrazepam binding to the GABA/benzodiazepine receptor complex of the cerebellum. *Eur. J. Pharmacol.* 70:183–93

Supavilai, P., Mannonen, A., Collins, J. F., Karobath, M. 1982. Anion-dependent modulation of ^{3}H-muscimol binding and of GABA-stimulated ^{3}H-flunitrazepmam binding by picrotoxin and related CNS consultants. *Eur. J. Pharmacol.* 81:687–91

Syapin, P. J., Skolnick, P. 1979. Characterization of benzodiazepine binding sites in cultured cells of neural origin. *J. Neurochem.* 32:1047–51

Tallman, J. F., Paul, S. M., Skolnick, P., Gallager, D. W. 1980. Receptors for the age of anxiety: Pharmacology of the benzodiazepines. *Science* 207:274–81

Tallman, J. F., Thomas, J. W., Gallager, D. W. 1978. GABAergic modulation of benzodiazepine binding site sensitivity. *Nature* 274:383–85

Taniguchi, T., Wang, J. K., Spector, S. 1980. Properties of ^{3}H diazepam binding to rat peritoneal mast cells. *Life Sci.* 27:171–78

Tenen, S. S., Hirsch, J. D. 1980. Beta-carboline-3-carboxylic acid ethyl ester antagonizes diazepam activity. *Nature* 288:609–10

Thomas, J. W., Tallman, J. F. 1981. Characterization of photoaffinity labeling of benzodiazepine binding sites. *J. Biol. Chem.* 156:9838–42

Thomas, J. W., Tallman, J. F. 1983. Photoaffinity labeling of benzodiazepine receptors causes altered agonist-antagonist interactions. *J. Neurosci.* 3:433–40

Thomas, J., Tallman, J. F. 1984. Solubilization and characterization of brain benzodiazepine binding sites. In *Brain Neurotransmitter and Neuromodulator Receptor Methodology,* ed. P. Marangos et al. New York: Academic. In press

Ticku, M. K. 1981. Interaction of stereoisomers of barbiturates with ^{3}H alpha-dihydropicrotoxin binding sites. *Brain Res.* 211:127–33

Ticku, M. K., Olsen, R. W. 1978. Interaction of barbiturates with dihydropicro-toxinin binding sites to the GABA receptor-ionophore system. *Life Sci.* 22:1643–52

Ticku, M. K., Olsen, R. W. 1979. Cage convulsants inhibit picrotoxinin binding. *Neuropharmacology* 18:315–18

Ticku, M. K., Ban, M., Olsen, R. W. 1978. Binding of ^{3}H-alpha-dihydropicrotoxinin, a gamma-aminobutyric acid synaptic antagonist, to rat brain membranes. *Mol. Pharmacol.* 14:391–402

Udenfriend, S. 1950. Identification of gamma-aminobutyric acid in brain by the isotope derivative method. *J. Biol. Chem.* 187:65–69

Vaccarino, F., Costa, E., Guidotti, A. 1983. Synaptosomal basic proteins: Differences from myelin basic proteins. *Soc. Neurosci. Abstr.* 9(2):1042

Vellucci, S., File, S. 1979. Chlordiazepoxide loses its anxiolytic action with long-term treatment. *Psychopharmacologia* 21:1–7

Wang, J. K., Taniguchi, T., Spector, S. 1980. Properties of ^{3}H diazepam binding sites on rat blood platelets. *Life Sci.* 27:1881–88

Waterhouse, B. D., Moises, H. C., Yeh, H. H., Geller, H. M., Woodward, D. J. 1984. Comparison of norepinephrine- and benzodiazepine-induced augmentation of Purkinje cell responses to GABA. *J. Pharm. Exp. Ther.* 228:257–67

Wilder, B. J., Bruni, J. 1981. Benzodiazepines: Diazepam, clonazepam, nitrazepam, chlordiazepoxide, chlorazepate and lorazepam. In *Seizure Disorders: A Pharmacological Approach to Treatment,* pp. 105–113. New York: Raven

Williams, M., Risely, E. A. 1979. Enhancement of the binding of ^{3}H-diazepam to rat brain membrane in vitro by SQ 20009, a novel anxiolytic, GABA, and muscimol. *Life Sci.* 24:833–42

Williams, M., Yarbrough, G. G. 1979. Enhancement of in vitro binding and some of the pharmacological properties of diazepam by a novel anthelmintic agent, Avermectin B1a. *Eur. J. Pharmacol.* 56:273–76

Worms, P., Lloyd, K. G. 1981. Functional alterations of GABA synapses in relation to seizures. In *Neurotransmitters, Seizures and Epilepsy,* eds. P. L. Morselli, K. G. Lloyd, W. Loscher, B. Meldrum, E. H. Reynolds, pp. 37–46. New York: Raven

Yokoyama, N., Ritter, B., Neubert, A. D. 1982. 2-Arylpyrazolo (4,3-c)quinolin-3-ones: Novel agonist, partial agonist, and antagonist of benzodiazepines. *J. Med. Chem.* 25:337–39

Young, W. S. III, Kuhar, M. J. 1980. Radiohistochemical localization of benzodiazepine receptors in rat brain. *J. Pharmacol. Exp. Ther.* 212:337–46

Young, W. S. III, Niehoff, D., Kuhar, M. J., Beer, B., Lippa, A. S. 1981. Multiple benzodiazepine receptor localization by light microscopic radiohistochemistry. *J. Pharmacol. Exp. Ther.* 216:425–30

Yousufi, M. A. K., Thomas, J. N., Tallman, J. F. 1970. Solubilization of benzodiazepine binding site from rat cortex. *Life Sci.* 25:463–70

Ann. Rev. Neurosci. 1985. 8:45–70

CELL LINEAGE IN THE DEVELOPMENT OF INVERTEBRATE NERVOUS SYSTEMS

Gunther S. Stent and David A. Weisblat

Departments of Molecular Biology and Zoology, University of California, Berkeley, California 94720

Historical Background

Studies of developmental cell lineage, i.e. of the fate of individual blastomeres that arise in an embryo, were begun in the 1870s, in the context of the controversy then raging about the "biogenetic," or "recapitulation," law promulgated by Ernst Haeckel. The biogenetic law seemed to imply that the early, pre-gastrula stages of metazoan embryogenesis recapitulate the nondifferentiated condition of a remote colonial ancestor. Hence, prior to gastrulation, all blastomeres should be of equivalent developmental potency. Only after gastrulation would particular domains of the embryo become committed to the differentiated tissues characteristic of more recent metazoan ancestors. This implication was tested by a group of American biologists, led by Charles O. Whitman (1878, 1887). By observing the cleavage pattern of early leech embryos, Whitman traced the fate of individual blastomeres, from the uncleaved egg to the gastrular germ layer stage and concluded that, contrary to the simplest interpretation of the biogenetic law, even the earliest blastomeres are developmentally distinct and that each identified blastomere, and the clone of its descendant cells, plays a specific role in later development.

Whitman set the pattern for all subsequent lineage studies. His disciples, including such future leaders of American cell biology as E. B. Wilson, E. G. Conklin, and F. R. Lillie, studied the embryos of other annelids, ascidians, and molluscs. Comparisons of their data revealed significant cross-phyletic similarities, as well as differences, in developmental cell lineage relations. Hence they concluded that there must be some relation between ontogeny and

0147-006X/85/0301-0045$02.00

phylogeny, but that that relation cannot be one of simple recapitulation. [A full account of these origins of cell lineage studies has been provided by Maienschein (1978).] The study of developmental cell lineage went into decline during the subsequent half century. In fact, cell lineage analyses are still not mentioned in most contemporary textbooks of embryology. One notable exception to the eclipse of cell lineage studies occurred in the 1920s, when A. H. Sturtevant (1929) devised a genetic method for mapping the developmental fate of cells of the *Drosophila* embryo. But it was only in the 1960s and 1970s that there occurred a revival of interest in the role of cell lineage in development, accompanied by the invention of more precise and far-reaching analytical techniques.

Conceptual Background

FATE MAPS AND CELL LINEAGE Before reviewing the results of recent cell lineage analyses, we attempt to clarify a few developmental concepts and problems to which these studies pertain. [An excellent explication of the concepts of experimental embryology is given in a recent monograph by Slack (1983), on which part of the following discussion is based.] A *fate map* is a diagram that shows what becomes of each region of an embryo in the course of subsequent normal development to some arbitrarily chosen endpoint of maturity. *Cell lineage* analysis is a form of fate mapping in which a single cell and its clone of progeny is followed. Two principal methods are available for cell lineage analysis. One consists of continuous observation of the entire course of development, following the blastomeres and their progeny visually all the way to the endpoint tissues. This method can be used only as long as the embryo remains transparent to the chosen endpoint and consists of a relatively small number of cells. The other principal method consists of the labeling of a specific blastomere at an early developmental stage and examining the location and distribution of the label at the endpoint. This is the method that must be used when the embryo is insufficiently transparent or the number of cells is too large to permit tracing of their fate by direct visual observation.

ABLATION In addition to these two principal methods of lineage analysis there is a third method, which consists of ablating a specific cell of the embryo and noting which particular organs and tissues are missing at the endpoint. The missing parts might then be inferred to represent the normal fate of the ablated cell. Strictly speaking, this method does not provide a real fate map, since ablation precludes normal development. On the one hand, an organ or tissue might be missing at the endpoint, not because its precursor cell had been ablated, but because an interaction with the cell that was ablated, or its normal progeny, is needed for the precursor cell to express its normal fate. On the other hand, an organ or tissue might be present at the endpoint, even though its normal precursor cell had been ablated, because it arose from another cell among whose progeny it is not normally included. Nevertheless, even though the ablation method cannot yield definitive information regarding the normal fate of embryonic cells, it may provide suggestive data. And in case a fate map has been established by either of the other methods, the ablation method can be used to probe the possible role of interactive or regulative processes in the determination of normal fate.

COMMITMENT A central focus of interest in the study of fate maps, especially at the cell lineage level, is the process by which a cell is *committed,* or commits its descendants, to express some trait *A* rather than another trait *B*. The differential commitment to *A* is said to be *clonal* if a group of cells expresses it that comprises all the descendants of a single ancestor cell. The concept of differential commitment implies that the cell has taken on a state R_A, which persists and which at some later time is bound to lead to another state S_A in some (or all) of its descendants, sufficient for expression of *A*. Suppose that under the conditions of normal development all of the descendants of the cell express *A* and none *B*. Does this mean that the cell had entered state R_A and was thus differentially committed? This question has no empirical answer, unless a set of *abnormal* developmental conditions, such as tissue transplantations or explantations, ablations of neighbors, or perfusion, is specified under which R_A still persists and leads to state S_A. If the cell responds to such abnormal conditions by giving rise to descendants that express *B* rather than *A*, the cell is judged to have been in a reversible state, and hence differentially uncommitted with respect to traits *A* and *B*. But if, despite these interventions, the descendants of the cell still express only *A*, then the cell can be said to be committed under those experimental conditions which did not result in the expression of *B* rather than *A*. (Of course, it is always possible that a new set of conditions of abnormal development can be found under which the cell would not be committed to the expression of *A*.) In this latter case the expression of *A* can be said to be *autonomous,* in the sense that the persistence of state R_A and the path leading from it to state S_A does not require the entire set of conditions to which the cell is exposed in normal development.

MODELS The significance for normal development of empirical tests for commitment under abnormal conditions lies in their use for distinguishing between different models of commitment, of which the two most common are the following. One model envisages that taking on state R_A (differential commitment to *A*) requires an *intracellular determinant, a,* whereas taking on state R_B requires another determinant *b*. A pluripotent cell contains and passes on to its descendants both determinants *a* and *b,* and a differential commitment to *A* (and a restriction of potency for *B*) occurs at an asymmetric cell division in which at least one of the daughter cells received only *a*. These intracellular determinants could be cytoplasmic structures or molecules that are distributed anisotropically in the egg. The determinants could also be nuclear structures, especially parts of the DNA, that are differentially modified in successive cell divisions and passed on in that modified form. Under this model cell lineage would play a crucial role in differential commitment, because the line of descent of any cell would govern which particular subset of intracellular determinants has been passed on to it.

The other model envisages that taking on states R_A or R_B depends on the anisotropic distribution of *intercellular inductive signals* α and β over the volume of the embryo. A pluripotent cell is capable of responding to either inductive signal, and once having responded to α at some crucial stage of development it has taken on state R_A. These intercellular inductive signals could be electrical potentials, diffusible molecules, or nondiffusible surface structures that signal by direct contact. Here cell lineage would play a role in differential cell commitment, because the line of descent of a cell would govern its position in a determinant field, and hence the set of inductive signals to which it is exposed at the critical stage of normal development. As formulated here, the intracellular determinant model equates differential commitment with restriction of potency. If both models are combined, however, i.e. if taking on state R_B (differential commitment to *B*) requires an interaction of determinant *b* with signal β, the potency for expression of *B* can be independent of the differential commitment to state R_A.

MODES OF CELL DIVISION Another important focus of interest in cell lineage analyses is the mode of division by which a cell gives rise to its clone of descendants. There are three principal modes:

1. The *proliferative* mode, under which a cell divides symmetrically to produce two equal daughter cells, both of which also divide symmetrically.
2. The *stem cell mode,* under which a cell of type A divides asymmetrically to give rise to two unequal daughters, of which one is of type A and the other is of type B. The (regenerative) daughter of type A divides again, as did its mother cell, to yield one daughter of type A and one of type B, and division can be said to proceed according to a *parental reiteration pattern*. Under one variant of the stem cell mode, the regenerative daughter cell is of type A′, different from A but dividing asymmetrically to yield one daughter of type A and another of a fourth type, C. This variant of the stem cell mode is referred to as a *grandparental reiteration pattern* (Chalfie et al 1981).

3. The *diversification mode,* under which a cell of type A divides to yield two unequal daughters of types B and C, neither of which ever gives rise again to a cell of type A. The diversification mode is a characteristic feature of early embryogenesis in invertebrates. Usually it terminates upon the generation of daughter cell types which, if they divide at all, do so according to either the proliferation or stem cell modes. In embryos of annelids and mollusks, where the stem cell division mode plays a prominent role in development, blastomeres that divide according to that mode are designated *teloblasts.*

Nematodes

The entire cell lineage is now known for the nematode *Caenorhabditis elegans,* thus completing a project that was begun late in the last century by another group of pioneering students of cell lineage (Boveri 1887, 1892, zur Strassen 1896; for review cf von Ehrenstein & Schierenberg 1980). Despite being built of only 810 nongonadal (i.e. somatic) cells, *C. elegans* contains the principal metazoan tissue types, such as nerve, muscle, epidermis, and intestine. The number of cells is constant in all somatic tissues, and in the case of the nervous system, the organization of its 302 cells is known also in its ultrastructural details (White et al 1976, 1983). The nervous system includes a cephalic part, with sensory sensilla and their nerves, a circumpharyngeal nerve ring, as well as a dorsal and ventral nerve cord, and a variety of sensory organs and ganglia.

The fertilized nematode egg cleaves asymmetrically in the diversification mode to generate a set of blastomeres designated as *founder cells* AB, MS, E, C, D, and P_4, of which AB is removed by one division, MS, E, and C by three divisions, and D and P_4 by four divisions from the uncleaved egg, respectively. Embryonic development culminates in the hatching of a larva, designated as L1, comprised of 550 nongonadal cells. Postembryonic development continues through three more larval stages (L2–L4) to the sexually mature, adult worm with its 810 nongonadal cells. In postembryonic development, the 260 additional somatic cells, including 61 neurons, are generated according to a stereotyped lineage pattern, as descendants of 55 blast cells carried over from the embryo to the larva.

The complete description of developmental cell lineage in the nematode was accomplished by continuous observations of living embryos and larvae, using time-lapse video recording and Normarski differential interference contrast optics (Deppe et al 1978, Sulston & Horvitz 1977, Kimble & Hirsch 1979, Sulston et al 1983). Of the founder cells, AB is the largest single contributor of somatic cells. During embryogenesis, 214 of the 222 neurons of the newly hatched L1 larva derive from AB, as well as a substantial fraction of the cells of the hypodermis and of the pharyngeal and trunk muscles. Moreover, all of the neurons formed postembryonically are derived from blast cells descended from the AB founder cell. The next largest single contributor to somatic cells is MS, which gives rise mainly to tissues regarded as mesodermal, including muscles, glands, and coelomocytes, and had been designated as the mesodermal founder cell. However, just as the mainly ectodermal founder AB includes muscles among its descendants, so does the mainly mesodermal founder MS include six neurons among its descendants. A single blast cell descended from MS accounts for all of the mesoderm produced postembryonically. Of the remaining founder cells, C gives rise to muscles and hypodermis, as well as to the remaining two prelarval neurons; D gives rise exclusively to muscles, E exclusively to intestine (i.e. endoderm), and P_4 to the germ line (whose cell lineage is not included in this review).

In *C. elegans,* the sequence of events leading from each founder cell to the differentiated, postmitotic cells of larva and adult is highly invariant with respect to timing and equality or inequality of cell divisions, as well as relative cell positions and movements. This invariant sequence also includes the death of identifiable cells at exactly defined stages of development. There are some exceptional groups of cells, however, whose fates are not invariant. Each such group is called an *equivalence group,* whose members resemble each other closely in structure and function and are usually of similar origin (Kimble et al 1979). Some equivalence groups consist of a bilaterally symmetric pair of cells, which move to the midline and meet; subsequently, one cell (sometimes from the left and sometimes from the right) takes on one particular fate while the other takes on another fate, suggesting the intervention of an element of chance in the alternative commitment of two equally pluripotent cells.

To learn at which stage of the invariant lineage pathway there occurs commitment to the normal fate, ablation experiments have been carried out, in which various identified cells were killed by irradiation with a laser microbeam (Sulston & White 1980, Kimble 1981, Sulston et al 1983). The result of the majority of these ablation experiments was that those cells, and only those cells, failed to develop in the lesioned embryo which, on the basis of the fate map, are known to be the normal descendants of the ablated cell. Thus, commitment to developmental fate appears to proceed autonomously in most cell lineages, there being neither regulative restoration of the ablated cell line from an as yet uncommitted, abnormal source, nor a need for an inductive interaction of another cell line with the ablated line to become committed to its normal fate. However, in a minority of the experiments a different result was obtained. These cases were used to define equivalence groups, in which another cell of the group may abandon its normal fate and take on the fate of the missing cell (designated in this case as the *primary* fate of one group). This result shows that commitment to a particular fate among the initially equally pluripotent members of an equivalence group is the result of intercellular interactions. The alternative of autonomous versus interactive commitment applies also when the fate in question is cell death: in some cases an identified cell normally destined to die will do so at the normal time, regardless of ablation of any neighboring cell ("suicide"), whereas in other cases the normally moribund cell survives if a particular neighboring cell has been ablated ("murder").

Comparison of the cell lineages of *C. elegans* with lineages that have been partially elucidated in other nematode species, such as *Turbatrix aceti, Panagrellus redivivus,* and *Aphelencoides blastophorus* (Sulston et al 1983, Sternberg & Horvitz 1982), reveals striking similarities in developmental pattern. Moreover, morphological differences by which one species is distinguished taxonomically from another have been traced in several instances to focal differences in otherwise homologous cell lineages, such as the death or terminal differentiation in one species of an identified cell, whose homolog undergoes further divisions in another species.

To apply genetic techniques to the study of cell lineages in *C. elegans,* mutants were isolated in which normal development is disrupted (Sulston & Horvitz 1981, Chalfie et al 1981, Greenwald et al 1983). Many of these mutant strains display widespread abnormalities, which are the likely consequence of some general disturbance of cell function. But some mutant strains were found, in which the mutation causes specific alterations of particular cell lineages, while leaving other lines of descent apparently unaffected. One such group of mutations, which occur at a genetic locus designated *lin*-4, induces three different kinds of supernumerary divisions, or lineage reiterations, in various postembryonic blast cell lines. The first, and simplest type of reiteration, equivalent to the proliferative division mode, consists of supernumerary equal divisions by an ordinarily postmitotic cell, leading to a geometric increase in the number of (presumably) equivalent supernumerary cells. The second and third types of lineage reiterations arise from the conversion of an ordinarily postmitotic cell into a stem cell, dividing either in the parental or grandparental reiteration pattern, respectively. Another group of lineage-specific mutations, mapped to the *unc*-86 locus, induces abnormal reiterative cell lineages in several postembryonic blast cell lineages. The mutations at both the *lin*-4 and the *unc*-86 loci are recessive; thus their mutant phenotype is likely to be the result of a reduction in activity, or loss of a gene product. This finding suggests that the capacity for lineage reiteration is latently present in the normal wild type strain, where its expression is specifically suppressed by the products of the normal alleles of the mutant loci.

Another genetic locus, designated *lin*-12, has also been found relevant for the cell lineage pattern. The effect of mutations within *lin*-12 is analogous at the cellular level to the effect of homeotic mutations at the tissue level, as known in insects, especially in *Drosophila* (Morata & Lawrence 1977, Ouweneel 1976). Homeotic mutations cause one group of cells to adopt the fate normally associated with another group, resulting in the transformation of one structure into another. Mutations at the *lin*-12 locus effect a number of such transformations throughout the embryo, especially between members of equivalence groups, both within neural and nonneural cell lineages and between neural and nonneural lineages. By shifting a homozygous mutant embryo carrying a temperature-sensitive mutation at the *lin*-12 locus from the permissive to the restrictive temperature at various developmental stages, it was found that the time at which the mutant gene acts to induce the transformation of cells in an equivalence group corresponds just to the stage at which the pluripotent cells of the group become committed to a particular fate, as defined by their lack of response to laser ablation of a fellow-member of their group.

Although most of the cells affected by mutations at the *lin*-12 locus are members of equivalence groups, some of them are not. In fact, one example of the effect of mutations at the *lin*-12 locus on the development of the nematode nervous system is provided by two bilaterally homologous descendants of founder cell AB, whose commitment appears to be autonomous, as judged by laser ablation experiments. Normally, the right homolog becomes a motor neuron, designated PDA, and the left homolog becomes a different motor neuron, designated DA9. But in animals homozygous for a semidominant mutation at *lin*-12, both cells take on the PDA fate. By contrast, in animals homozygous for a recessive null mutation at *lin*-12 (i.e. one that eliminates the gene product altogether), both cells take on the DA9 fate. Such findings led Greenwald et al (1983) to suggest that, in analogy with the proposals made for the function of homeotic mutant loci in *Drosophila* (Morata & Lawrence 1977), "*lin*-12 functions as a binary switch to control decisions between alternative cell fates during *C. elegans* development."

Leeches

The total number of somatic cells in the leech is several orders of magnitude greater than in *C. elegans;* so the kind of total cell lineage analysis now available for the nematode seems out of reach for the leech. Indeed, such an undertaking might not even be meaningful for animals in which, unlike in the nematode, the number of cells making up a particular organ or tissue, such as muscle or epidermis, is variable from specimen to specimen. Nevertheless, the prospect of establishing a very extensive genealogy of the approximately 15,000 cells of the leech CNS is by no means out of sight. Each of the 32 segmental ganglia of the leech ventral nerve cord is composed of about 400 bilaterally paired neurons, eight paired giant glial cells, as well as a few unpaired neurons (Muller et al 1981). Since most of these neurons are serially homologous, not only with respect to their traits, but also, insofar as is presently known with respect to their embryonic lines of descent, the cell lineage analyst of the leech CNS seeks to account for the origins of about 200 cell types, a task comparable in magnitude to that already achieved for the nematode nervous system. (Reviews of leech development can be found in Weisblat 1981, Stent et al 1982, Fernandez & Olean 1982.)

The initial cleavages divide the leech egg in the diversification mode into four large cells, macromeres A, B, C, and D. Each macromere buds off a micromere at the animal pole, designated by the corresponding lower case letter a, b, c, or d. Cell D continues to divide in the diversification mode to yield five bilateral blastomere pairs, namely M, N, Q, and two sister pairs, both designated O/P. Separation of the embryo into the three germinal layers has now been accomplished: A, B, and C give rise to endoderm, N, Q, and the O/Ps to ectoderm, and M to mesoderm (Whitman 1887). The paired M, N, O/P, O/P, and Q blastomeres divide in the stem cell mode, and hence are designated as teloblasts. Each of them carries out a series of 40–100 highly asymmetric divisions, producing a bandlet of *primary blast cells*. The bandlets merge on either side of the midline to form the right and left *germinal bands*. In either band, the mesodermal bandlet lies under the four ectodermal bandlets. Gastrulation is represented by a movement of the right and left germinal bands over the surface of the embryo and their coalescence on the ventral midline to form the *germinal plate,* in which the superficial ectodermal blast cell bandlets lie in mediolateral order n, o, p, q. (In the case of the M, N, and Q teloblasts, their primary blast cells and bandlets are designated by lower case letters corresponding to their own upper case letter; in the case of the two sister O/P teloblasts, their primary blast cells and bandlets are designated o and p, according to which bandlet lies proximal and which distal, respectively, to the n bandlet.) The germinal plate is partitioned along its length into a series of tissue blocks, each separated from its neighbors by transverse septa. Each block corresponds to a future body segment, including a globular ganglion containing about the same number of cell bodies as an adult ganglion.

To establish the line of descent of identified cells of the leech CNS, Whitman's century-old cell lineage studies were refined and extended, by means of a novel tracer technique (Stent et al 1982). This technique consists of injecting a tracer molecule, such as horseradish peroxidase (HRP) or a fluorescent dye conjugated to a large carrier molecule, into an identified cell of the early embryo, allowing embryonic development to proceed to the endpoint, and then observing the distribution pattern of the tracer within the tissues (Weisblat et al 1978, 1980a, 1980b).

Use of the lineage tracer technique has shown that the leech CNS is derived from all five teloblast pairs. The progeny of the N teloblast are found almost exclusively within the ganglia of the ventral nerve cord. Descendants of the O/P and Q teloblasts, however, give rise to characteristic patterns of cell clusters, both in the segmental ganglia and in the epidermis. The dorsal aspect of the segmental epidermis is derived from the ipsilateral Q teloblast and the ventral aspect mainly from the ipsilateral O/P sister teloblasts, with the N teloblast providing a few epidermal cells on the ventral midline (Weisblat et al 1984). In addition, three or four paired neurons of the ganglion are derived from the mesodermal teloblast M, whose main contribution is made to tissues and organs to which a mesodermal origin is generally assigned, namely connective tissue, nephridia, and muscle.

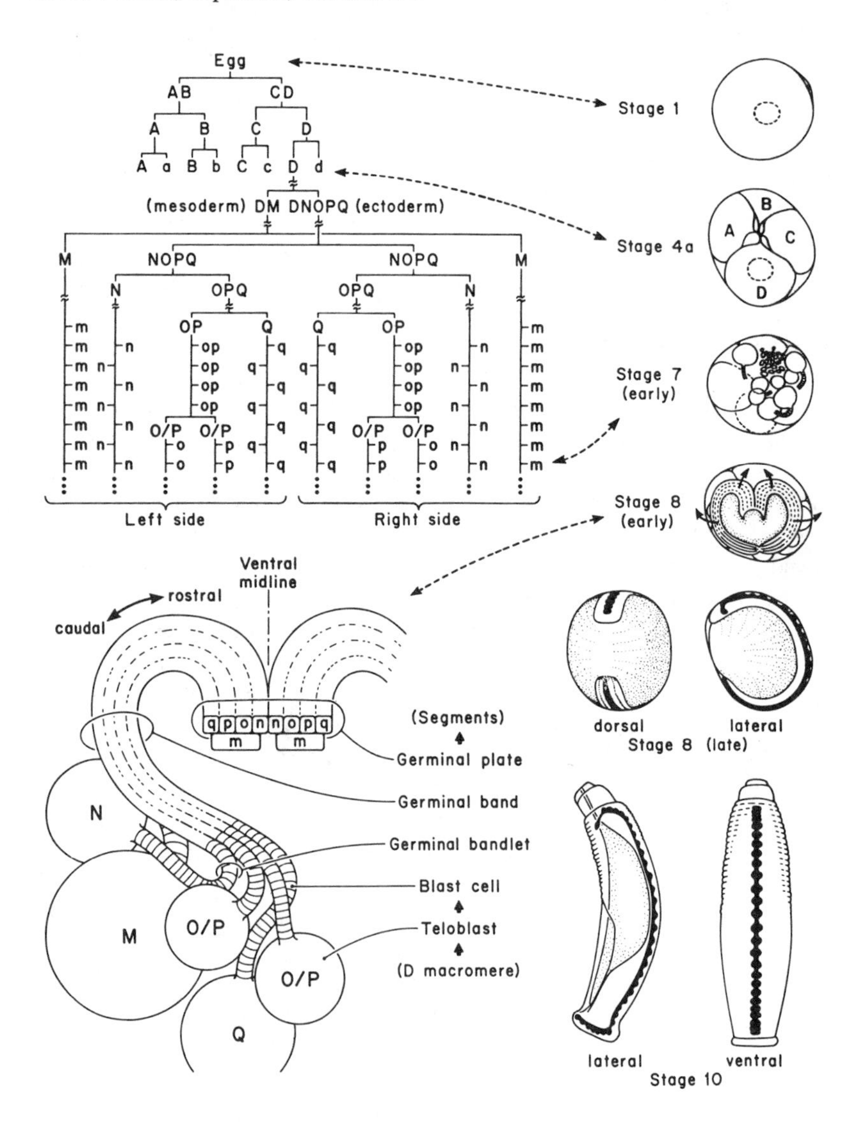

The tracer technique showed that individual segments arise as cell clones, each segment being founded by a fixed number of primary blast cells: one primary m, o, and p blast cell and two primary n and q blast cells per half segment on each side. Of the two primary n and q founder blast cells, one has a *mixed* fate, giving rise to descendants in both the CNS and periphery, while the other has a *pure* fate: purely CNS for n and purely peripheral for q (Zackson 1982, 1984, Weisblat & Shankland 1984). To give rise to its segmental complement of descendants, each primary blast cell divides according to a stereotyped and lineage-specific manner. In the underlying m bandlet pair, blast cell divisions occur in rapid succession in three dimensions, so that the clones of m-derived progeny form a chain of isomorphic, protosegmental clusters. In the overlying ectodermal bandlets, each of the four bandlet pairs is characterized by its own unique cell division pattern. In the n and q bandlets, moreover, there occur two alternating types of primary blast cell divisions, corresponding to the two different types of founder blast cells per segment. Later blast cell divisions show a diversity of characteristic patterns as well, including both proliferative and stem cell division modes (Zackson 1984). These findings suggest that the primary blast cells of different bandlets are endowed by their parent teloblasts with different states of commitment that cause them to follow lineage-specific pathways of cell division, and hence to take on different lineage-specific fates. In the case of the N and Q teloblasts, moreover, it would appear that each of them passes on in alternating sequence two different states of commitment to its daughter blast cells, and hence divides according to the grandparental reiteration pattern of the stem cell mode.

Figure 1 Schematic summary of the development of the leech *Helobdella triserialis*. *Upper left:* Cell pedigree leading from the uncleaved egg to the macromeres A, B, and C; the micromeres a, b, c, and d; the teloblast pairs M, N, O/P, O/P, and Q; and the paired primary blast cell bandlets. Breaks in the lineage indicate points where additional micromeres may be produced. The number of op blast cells produced prior to cleavage of proteloblast OP varies from four to seven. *Lower left:* Hemilateral disposition of the teloblasts and their primary blast cell bandlets within the germinal band and germinal plate. *Right margin:* Diagrammatic views of the embryo at various stages. The *dashed circle* in the uncleaved egg (stage 1) signifies the right M teloblast (which is invisible from the dorsal aspect), and the many small, closed contours in the upper midportion indicate the micromere cap. In the stage 8 (early) embryo, the heart-shaped germinal bands migrate over the surface of the embryo in the directions indicated by the *arrows*. The incipient larval integument is shown as a *stippled area* lying in between. In the stage 8 (late) embryo the germinal plate is shown to be on the ventral midline, with the nascent ventral nerve cord and its ganglia and ganglionic primordia indicated in black. The stippled larval integument covers the entire embryo, from one edge of the germinal plate to the other. In the stage 10 embryo shown, body closure is nearly complete. Here, the *stippled areas* signify the yolky remnant of the macromeres and teloblasts, now enclosed in the gut of the embryo. The chain of ganglia linked via connectives, shown in black, already closely resembles the adult nerve cord (from Weisblat et al 1984).

The ganglionic subpopulations of neurons arising from each teloblast pair form five identifiable neuronal *kinship groups:* M, N, O, P, and Q (Stent et al 1982, Weisblat et al 1984). (The kinship groups designated as O and P refer to those descendants of the O/P sister teloblasts which are respectively derived from the o and p bandlets.) Combined use of the fluorescent lineage tracers with electrophysiological, anatomical, and histochemical identification techniques showed that each kinship group invariably contains a particular set of identified neurons and glial cells, along with a number of as yet unidentified neurons (Kramer & Weisblat 1984). Thus far, no unique set of traits has come to light that separates the members of one kinship group from another, such as functional category (glia or sensory, motor or interneuron) or type of neurotransmitter synthesized, except that all serotonergic neurons belong to group N and all dopaminergic neurons to groups O, P, and Q.

In addition to its 32 segmental ganglia, the leech CNS has a supraesophageal ganglion at its rostral end. The origin of that ganglion has long been the subject of controversy (Whitman 1892), but it has now been shown that these frontmost cells of the CNS arise from the a, b, c, and d micromeres, rather than from the blast cells of the germinal bands (Weisblat et al 1980a, 1984). This finding indicates that the supraesophageal ganglion of leeches is developmentally homologous to the much more elaborate supraesophageal ganglion of polychaetes, which is known to arise as the neural tissue of a nonsegmented larva entirely derived from the micromeres (Dawydoff 1959).

Cell lineage analyses thus indicate that in the leech, neurogenesis is as highly determinate as it is in the nematode. Ablation experiments show here, too, that commitment to cell fate is largely, but not wholly, autonomous. Ablation of a teloblast of the early embryo by intracellular injection of a toxic enzyme results in an embryo whose nervous system lacks all those identified neurons which are normally derived from that teloblast (Weisblat et al 1980b, Blair & Weisblat 1982, Blair 1983). (The special cases of the O/P sister teloblasts and of unpaired neurons are considered below.) But in the defective ganglia of such embryos, the positions of the neurons that are present may be highly abnormal, and in some cases even the neurons normally descended from a nonablated teloblast may be missing (Blair & Weisblat 1982, Weisblat et al 1980b). These findings indicate the role of a morphogenetic interaction during gangliogenesis between the blast cells derived from different bandlets. In contrast to the lack of restoration of identified neurons following ablation of the teleoblast from which these neurons are normally derived, some kind of regulative restoration does occur from blast cell progeny derived from other bandlets for the portion of the epidermis normally derived from that very same teleoblast (Blair & Weisblat 1984). Hence the developmental commitment of a given primary blast cell is more autonomous in regard to its neural than to its epidermal fate.

The consequence of direct ablation of individual blast cells rather than of their teloblast progenitors can be studied by focal photoablation of cells labeled with fluorescent lineage tracers (Shankland 1984). This method can provide information regarding cell lineage relations among the descendants of primary blast cells (subject to the usual limitations placed on the inference of fate maps from ablation data). The method can be used also to induce rearward slippage of one bandlet relative to the other bandlets. By means of such slippage the cause of death of the last, supernumerary blast cells produced by each teloblast in excess of the number needed to found the 32 body segments (Fernandez & Stent 1982, Zackson 1982) has been clarified: blast cells that would, on the basis of their birth rank, have survived in a normal embryo and participated in segment formation, degenerate along with the other supernumeraries if they slip rearward into positions behind the caudal end of the germinal band. Hence the birth rank of a primary blast cell does not commit it autonomously to survival or death; rather its fate is decided by an inductive signal received after entrance into or exclusion from the germinal band (Shankland 1984).

Notwithstanding the highly determinate fate of each teloblast in the course of neuronal development, the primary blast cells derived from the O/P sister teloblasts can interchange their fate and thus form an equivalence group. In the leech such interchange has been designated "transfating" (Weisblat & Blair 1984). The underlying cause of O/P transfating is that the bandlets of primary blast cells generated by either of the two sister teloblasts may take either position in the germinal band, and that the blast cells are committed to their specific fates, designated as O fate and P fate, only after the bandlet has come to lie in either the o or p position (Weisblat & Blair 1984, Shankland & Weisblat 1984). But upon ablating either of the O/P sister teloblasts, progeny of the surviving teloblast take on the P fate (Weisblat & Blair 1984). Thus, in the parlance of nematode lineage analysis, the P fate is the primary fate of the O/P equivalence group. Cytological examination of the bandlets has shown that the commitment to these alternative fates occurs within the first few divisions of the blast cell, once it lies in the germinal band (Zackson 1984, Shankland & Weisblat 1984).

The unpaired interneurons of the segmental ganglion provide another example of an equivalence group. Initially, an unpaired interneuron is present as a bilateral pair of cells, of which one member later dies. Lineage tracers show that the teloblast of origin, right or left, of the surviving member of the pair varies randomly from ganglion to ganglion, and from specimen to specimen. Moreover, upon ablating one of the parent teloblasts, none of the unpaired interneurons is missing (Blair & Stuart 1982, Blair 1983). Hence, in the formation of the ganglionic primordia, both right and left primary blast cells give rise to an interneuronal precursor cell, and by random outcome of a competitive process one interneuronal precursor cell is committed to survival and the other to death.

Insects

The late developmental stages and adult forms of insects and leeches demonstrate homologous segmental body plans. This homology is particularly obvious in the case of the CNS. In insects, as in leeches, the CNS consists of a ventral nerve cord composed of a chain of bilaterally symmetric, metameric ganglia, joined via longitudinal connective nerves. The dorsally situated brain seems homologous to, but very much more elaborate than, the supraesophageal ganglion of leeches. In embryogenesis each ganglion arises as a distinct pair of primordia—one neuromere per segment—but as development proceeds some ganglia fuse to give rise to the rostrocaudally differentiated adult CNS. However, the initial stages of embryogenesis in insects are radically different from those of leeches. The insect egg begins its development with a series of synchronous mitotic divisions of the zygote nucleus without cell cleavage. This process gives rise to an embryonic syncytium containing thousands of nuclei. Eventually most of these syncytial nuclei migrate to the periphery of the egg, where each nucleus becomes enclosed by an infolding of the egg cell membrane. Thus the insect embryo comes to consist of a uniform sheet of several thousand cells, the *blastoderm,* on its outer surface, which encloses an acellular yolky interior. The cells of the blastoderm continue the process of proliferation and differentiation. Along the length of the ventral aspect of the embryo there eventually forms a multilayered cell structure designated as the *germ band,* of which the outer and inner layers correspond to ectoderm and mesoderm, respectively. The germ band shows clear signs of segmentation and already reflects the later body plan. In its general structure and role in subsequent embryogenesis the insect germ band is evidently homologous to the germinal plate of leeches, even though the two structures are generated by radically different processes in early embryogenesis.

GRASSHOPPER Cell lineage analyses have been carried out by visual observation for the final stages of development of the segmental ganglia in the grasshopper, onwards from the germ band stage (for review cf Goodman 1982). As in the case of the germinal plate of the leech, the ectodermal cell layer of the insect germ band that gives rise to the CNS extends longitudinally, flanking the body midline. Within that cell layer, some cells become recognizable as neuronal precursors: they round up and enlarge relative to other ectodermal cells. In the grasshopper embryo two types of such neuronal precursors can be identified: one type, designated *neuroblast,* or NB (Bate 1976), is a stem cell, and the other type, which is not a stem cell, is designated *midline precursor,* or MP (Bate & Grunewald 1981). The number of both types of neuronal precursors per segment is fixed. There are two bilaterally paired sets of 30 unique NBs each, arranged in seven transverse rows, plus one unpaired NB, designated MNB, lying on the midline at the posterior margin of the segment. And there are seven MPs, lying in a stereotyped arrangement along the midline, of which two, designated MP2L and MP2R, are bilaterally paired, and the remainder, numbered in rostrocaudal order MP1, MP3, MP4, MP5, and MP6, are unpaired. Upon subsequent development each MP divides only once to produce two daughter cells straddling the midline (in the case of the unpaired cells) or a pair of dorsoventral two-cell stacks (in the case of the paired MP2 cells). By contrast, each NB undergoes a series of stem cell divisions, to give rise to a chain of smaller cells designated *ganglion mother cells,* or GMCs, each of which, in turn, divides once to produce a pair of daughter cells designated *ganglion cells,* or GCs, which eventually differentiate into neurons. Thus each NB contributes a clone of prospective neurons to the CNS; ultimately, all NBs die and degenerate, with some of them having contributed as few as 10 and others as many as 100 neuronal progeny to their ganglion. According to recent experiments, each NB_x arises from an equivalence group of a few neuroepithelial cells, of which any one may become NB_x, but no other NB. By the time that NB_x has made its first division, however, other members of its equivalence group can no longer replace it (Taghert et al 1984).

The paired descendants of the MPs are the first to extend neuronal processes from their cell bodies. Each pioneers a characteristic and segmentally stereotyped, central or peripheral, hemilateral or bilateral, axonal pathway (Bate & Grunewald 1981, Goodman et al 1981). Of the MP3 pair, one sister differentiates into a neuron designated as cell H, identifiable on the basis of its characteristic bilateral axonal branching pattern, whereas the other sister projects only hemilaterally (Goodman & Spitzer 1979, Goodman & Bate 1981). As the MNB stem cell carries out its iterated divisions, the string of GMC progeny cells advances anteriorly. Each GMC divides once to produce a GC pair that straddles the midline.

The first six MNB-derived GC pairs differentiated into 12 identified neurons, designated collectively as *dorsal unpaired medial neurons,* or DUM. They project their axons bilaterally, with the axons of each pair initially following the axonal pathway previously laid down by one or another of the MP cell pairs. Any two sister DUM cells initially project their axons along the pathway laid down by the same MP pair, but eventually their axonal branching patterns diverge, resulting in the generation of two identifiably different neurons. However, which of the two DUM sisters, right or left, develops which of the two different branching patterns depends on which sister happens to have been the first to extend an axonal growth cone from its cell body. Thus they form an equivalence group. The DUM neurons share one striking biochemical characteristic: they all contain the neurotransmitter, octopamine (Evans & O'Shea 1977, Goodman et al 1979), and it is likely that they are the only neurons in the segmental ganglia that do.

Cell lineage analyses have also illuminated the problem of how the segmentally iterated sets of neuronal precursor cells in the germ band give rise to adult ganglia that are specialized to function in the regionally differentiated body segments of the adult insect. For example, whereas the ganglia of the three thoracic segments (T1, T2, and T3) each contain about 3000 neurons, the ganglia of the 11 abdominal segments (A1–A11) each contain only about 500 neurons. Much of this difference in cell number per ganglion is attributable to cell death during embryogenesis (Bate 1982, Bate et al 1981): after the NBs have already produced their crop of descendants, the degeneration of hundreds of cells can be observed visually in each embryonic abdominal ganglion. Specifically, in the adjacent segments T3 and A1, the homologous MNB stem cells give rise to about 100 and 90 descendants, respectively. But in segment T3 all 100 descendants survive, while as many as 45 descendants die in segment A1, among them several of the identifiable DUM cells. Death of the abdominal DUM cells occurs only after they have already begun to project axons into their characteristic pathways within the CNS and are about to enter the periphery. Similarly, some of the identified descendants of MPs, such as cell H, die in segments A3 through A6.

Other regional differences in segmental ganglion structure are attributable to specific differences in the pattern of differentiation of the neurons that do survive (Bate et al 1981). For instance, homologous H cells form the characteristic H-shaped axon branching pattern (from which the cell's name is derived) only in segments T1, T2, and T3; in segments A1 and A2 they develop only part of the axonal H pattern.

These neurodevelopmental studies in the grasshopper have thus shown that here, too, as in the nematode and in the leech, specific identified neurons arise by a specific sequence of cell divisions from an identifiable embryonic precursor cell and that serially homologous neurons have corresponding cell pedigrees. This work has revealed, moreover, that the sixfold higher neuron number per ganglion in the adult thoracic segments is attributable mainly to the specific death of particular cells in the abdominal segments, after the cells had already begun to differentiate. However, these studies have not as yet provided much direct information regarding the mechanism—partition of intracellular determinants or positioning in a determinant field—by which the line of descent of a neuron governs its commitment to the expression of one trait rather than another. In the case of the MNB descendants, their common neurotransmitter trait, whose expression depends on the presence (and function) of just a few specific enzymes, seems more plausibly explained by partition of an intracellular determinant, whereas their individual axonal projection patterns seem more likely to be the consequence of their positions relative to the set of MP descendants to which their rank order of birth from the stem cell has consigned them.

DROSOPHILA Because of the syncytial character of the early insect embryo, it is impossible, indeed meaningless, to trace back the cellular pedigree of any neuron of the insect nervous system to the egg. (Indirect evidence indicates, moreover, that the nuclei of the syncytium have no definable fates.) What has been done, however, is to establish fate maps for various regions of the blastoderm, particularly in the case of the embryo of the *Drosophila*. One such fate map was established by direct histological observation by Poulson (1950). This map shows that the rostrocaudal segmental sequence manifest in the ectodermal and mesodermal layer of the germ band is already presaged in the blastoderm, and that the ventral nerve cord arises from the bilateral bands of cells extending longitudinally on the ventral aspect of the blastoderm, spearated by a band of mesodermal presursor cells lying on the future ventral midline. The brain, by contrast, arises from two paired blastoderm patches that lie front-and dorsalward to the nerve cord precursor bands. The finding of a separate origin for the insect brain is in agreement with the results of the cell lineage analyses of the leech, which assigned to the supraesophageal ganglion a line of descent separate from that of the ventral nerve cord ganglia.

Another fate map of the *Drosophila* blastoderm was established upon revival of Sturtevant's (1929) genetic mapping method (Garcia-Bellido & Meriam 1969, Hotta & Benzer 1972, Janning 1978). This method is based on the experimental generation of gynanders, or flies whose tissues form a mosaic of male and female cells. The gynander map of the embryonic origin of the insect nervous system showed that in accord with Poulson's direct fate map, the precursors of the segmental nerve cord ganglia lie bilaterally on the ventral aspect of the blastoderm in their eventual rostrocaudal sequence, with the precursors of the brain and its optic lobe being more dorsally disposed (Kankel & Hall 1976). It was possible also to estimate that only a few blastoderm cells (from three to ten) are the precursors of each ganglion on the right or left side of the body. If these findings apply also to the grasshopper, then it would follow that the set of NBs and MPs that make up the ganglionic primordium in the germ band arise by multiplication of a much smaller number of blastodermal founder cells.

A further method is available for producing insects with genetically mosaic bodies, which can likewise be used for developmental cell lineage analyses. This method is based on the discovery by Stern (1936, 1968) of genetic recombination between homologous chromosomes during the mitotic nuclear divisions in the somatic tissue of *Drosophila*. The somatic recombination method has been used for cell lineage analysis in the arthropod compound eye. The regular array of ommatidia, each with a fixed number of regularly arranged photoreceptor cells, had led to the suggestion that each ommatidium arises as a clone, i.e. that its set of photoreceptor cells is descended from a single ommatidial founder cell (Bernard 1937). Later radiological and genetic experiments seemed to support this view of the development of the 700 to 800 ommatidia of the *Drosophila* compound eye (Becker 1957). But more recent findings made with the somatic recombination method, argue against this view (Hofbauer & Campos-Ortega 1976, Ready et al 1976). The eight photoreceptors within a single ommatidium are not all descendants of a single founder cell, and the commitment of eight photoreceptor cells to form a given ommatidium is not determined by their lineage. Moreover, even the possibility that *Drosophila* photoreceptors do arise as ommatidium-sized clones but that the members of a clone are not constrained to take part in the formation of the same ommatidium (Campos-Ortega & Hofbauer 1977, Campos-Ortega et al 1978) was eliminated by statistical analysis of the size distribution of identified clones (Lawrence & Green 1979). Thus, fixed cell lineage does not seem to play a determinative role in ommatidial development.

Ascidians

One of Whitman's disciples, E. G. Conklin (1905), had studied cell lineage in the embryos of ascidians. In their development, these sessile marine animals pass through a free-living tadpole stage whose morphology is very similar to that of the vertebrates: it has a notochord, segmented tail muscles, and a CNS consisting of a brain, brainstem, and a spinal cord. The ascidian egg cleaves meridionally, to yield a bilateral cell pair designated AB2. The second cleavage, also meridional, is orthogonal to the first and results on either side in an anterior blastomere pair designated A3 and a posterior blastomere pair designated B3. The third cleavage is equatorial and results in two cell pairs, a4.2 and b4.2, in the animal hemisphere and two cell pairs, A4.1 and B4.1, in the vegetal hemisphere. A series of further, highly regular cleavages follows, leading to a 64-cell blastula composed of individually identifiable blastomeres. Gastrulation now begins, leading to the formation of neural ectoderm and an underlying mesodermal layer on the dorsal aspect of the embryo. Conklin (1905) managed to establish a fate map for the 64-cell ascidian blastula, which was later refined by Ortolani (1955). On that map the prospective region of the CNS is located near the future dorsal midline in the anterior hemisphere, with the future rostrocaudal array of brain, brainstem, and spinal cord oriented in the animal-vegetal direction.

Although on this map the prospective CNS regions are contiguous, the differential commitment to nervous vs nonnervous tissue is not clonal: at the 16-cell stage, eight cells each give rise to some part of the CNS, as well as to non-CNS tissues, such as notochord, gut, and epidermis. This means that the boundaries between prospective CNS and non-CNS regions do not correspond to cellular boundaries, making it unlikely that differential commitment by segregation of nuclear determinants occurs at this early stage of development. Nishida & Satoh (1983) have recently applied the intracellular cell lineage tracer technique to the ascidian embryo, injecting HRP into identified blastomeres at various early stages of development. They obtained a fate map that generally confirmed Conklin's classical map, except that the HRP label showed that muscles are derived also from blastomeres A4.1 and b4.2, and not only from blastomere B4.1, previously identified as their sole source.

The ascidian embryo has provided one of the few convincing demonstrations of the existence of intracellular determinants that are distributed anisotropically in the egg and later partitioned unequally over daughter cells in the course of asymmetric cell divisions. This demonstration derives from the work of Whittaker (1973, 1979) on the cellular commitment for expression of acetylcholinesterase present in the tail muscles of the tadpole. This enzyme normally makes its first appearance in the tail muscle cell line at the time of formation of the neural tube. Upon inhibiting further cell division in the embryo at various stages of early development by exposing the embryo to cytochalasin B, Whittaker found that acetylcholinesterase still makes its appearance in the arrested embryo after the normal lapse of time in, and only in, those blastomeres which, according to the classical fate map, are precursors of muscle cells. Thus if cleavage is inhibited at either the one- or two-cell stage, acetylcholinesterase eventually appears throughout the arrested embryo. But if cleavage is inhibited at the four- or eight-cell stage, the enzyme appears only in the B3 or B4.1 blastomere pairs, respectively. Moreover, commitment to expression of acetylcholinesterase in the B4.1 cell line, as well to its nonexpression in other cell lines, is *autonomous:* the enzyme will appear after the normal lapse of time in a B4.1 blastomere pair surgically removed from the eight-cell embryo and cultured in isolation, while the remainder of the embryo lacking these blastomeres does not produce the enzyme (Whittaker et al 1977).

To demonstrate that the commitment to differential expression of acetylcholinesterase is, in fact, attributable to the partition of a cytoplasmic determinant, Whittaker (1980) compressed the embryo just prior to its third cleavage. This operation causes transmission to the b4.2 blastomere pair of some of the cytoplasm ordinarily passed on only to the B4.1 pair. Upon inhibition of further cleavage by cytochalasin B in such manipulated embryos, acetylcholinesterase is now expressed in both the B4.1 and the b4.2 blastomere pairs. Whittaker (1982) also apportioned cytoplasm destined for the B4.1 blastomere pair to the b4.2 pair microsurgically. He found upon culturing the cytoplasmically enriched b4.2 blastomeres in isolation that their abnormal expression of acetylcholinesterase develops autonomously, just as does the normal expression in the B4.1 cell line. Unfortunately, the force of these impressive results is slightly weakened by Nishida & Satoh's later findings by use of the HRP lineage tracer method that blastomere B4.1 is not, in fact, the sole precursor of muscle cells, which are derived also from blastomeres b4.2 and A4.1. Hence, according to Whittaker's argumentation on behalf of the role of cytoplasmic determinants in commitment, acetylcholinesterase should have been expressed not only in the B4.1 blastomere but also in the b4.2 and A4.1 blastomeres upon inhibition of further cell division at the eight-cell stage.

Satoh (1979) and Satoh & Ikegami (1981a,b) have also used another inhibitor of cell division, namely aphidicolin. In contrast to cytochalasin B, which acts by blocking the cytokinesis phase of cell division while permitting the indefinite continuation of successive rounds of DNA replication, aphidicolin stops cell division in ascidian embryos by blocking DNA replication. The effect of blocking cell division by arrest of DNA replication turns out to be dramatically different from that found after blocking cytokinesis; if the embryo is exposed to aphidicolin at any time prior to gastrulation (or about the seventh division), no acetylcholinesterase appears in the cells of the lineage normally destined to express it. However, if aphidicolin is added at about the 76-cell stage (by which time the cell division rhythm has become asynchronous), the enzyme is eventually expressed in some, but not all, of the cells belonging to the known muscle lineage. At this stage there have been seven to nine rounds of DNA replication in the line of ancestry of different muscle precursor cells. Satoh & Ikegami (1981a) were able to show that only those cells whose DNA had undergone eight or nine rounds of replication eventually express the enzyme in the arrested embryo. Hence, they suggest that the rounds of DNA replication provide a developmental clock, and that it is only after the eighth round of its replication that the genome becomes competent to interact with the cytoplasmic determinant to effect the commitment for eventual expression of the enzyme. It is worthy of note that it is also after the eighth round of replication that the muscle cell lineage has finally become clonal, i.e. that there are precursor cells that give rise only to muscle cells and to no other cell types.

Amphibia

Although the focus of this review is on the analysis of cell lineage in the development of the nervous system of invertebrates, we consider finally but briefly one such analysis performed on a vertebrate nervous system. We present this case—M. Jacobson's (1982) application to the frog embryo of the methodology of cell lineage analysis by injection of tracers—only because it has sown confusion among developmental neurobiologists.

Fate maps of the amphibian embryo, obtained by labeling its various regions with externally applied vital stains, became available in the 1920s (Vogt 1929) and were later refined by a succession of workers (Pasteels 1942, Keller 1975, 1976). The fate maps, whose regularity revealed that there is very little random mixing of surface cells at any stage of amphibian development, indicated that the precursors of the CNS lie in the dorsal quadrant of the animal hemisphere of the amphibian blastula, symmetrically disposed on either side of the meridian of the future body midline. Moreover, the prospective regions of the rostrocaudally sequential subdivisions of the CNS—forebrain, midbrain, hindbrain, spinal cord—lie in that same order, from the animal pole toward the equator of the blastula. To bring the classical amphibian fate map down to the single blastomere, i.e. cell lineage, level, individual blastomeres of frog embryos at various early developmental stages, from the two-cell to the 1024-cell stage, were HRP-injected, and the distribution of label in the CNS was observed in the resulting larva (Jacobson & Hirose 1978, 1981, Hirose & Jacobson 1979). The results of these cell lineage analyses confirmed the classical fate map, and, in addition, showed at a higher level of resolution that deep cells in clones labeled prior to the 512 cell stage mix extensively, and that surface cells mix somewhat also, in the course of development. Nevertheless, it was found that the later the developmental stage at which an individual blastomere is labeled, the smaller is the domain of the larval CNS that contains labeled cells.

These results were interpreted to mean (Jacobson 1980) that, by the 512-cell stage, blastomeres and their descendant clones are already committed to express a neural phenotype. This interpretation was contrary to the accepted view that the descendants of the dorsal quadrant cells identified on the fate map as prospective neural tissue must receive an inductive signal from the underlying dorsal mesoderm at, or after, gastrulation to become differentially committed to develop as neurons rather than epidermis (Nieuwkoop 1952). Moreover, the commitment of blastomeres at the 512-cell stage would be difficult to reconcile with the finding by Spemann & Mangold (1924) that grafting a second dorsal blastoporal lip on the ventral aspect of an amphibian gastrula, containing by then more than 10,000 cells, results in the development of a second, ventrally situated CNS in the host tissue. It was therefore proposed (Jacobson 1982) that, contrary to the usual interpretation, in this experiment the second CNS was not "induced" from previously uncommitted cells on the ventral aspect, whose normal fate is belly and tail epidermis, but that it arose by an abnormal, ventral-ward *migration* from the dorsal aspect of cells committed long ago to a neural fate or by a self-differentiation of the graft.

This radical reinterpretation of one of the classical experiments in the history of embryology was shown to be incorrect by Gimlich & Cooke (1983), by use of a modification (Gimlich & Braun 1984) of one of the fluorescent cell lineage tracers devised for use with leech embryos. In a series of frog morulas at the 32-cell stage, Gimlich & Cooke injected a fluorescent lineage tracer into either of two identified blastomeres, D2 or V3, lying in the prospective CNS or ventral epidermis regions, respectively. At early gastrula, a second blastoporal lip, from synchronous but unlabeled donor embryos, was implanted on the ventral aspect of the labeled embryos, to induce formation of a second, ventral CNS. The result of this experiment was completely unambiguous. In embryos with a labeled D2 blastomere, only the normal, dorsal CNS contained any labeled cells, whereas the second, ventral CNS was free of label. But in embryos with a labeled V3 blastomere, only the second, ventral CNS contained any labeled cells, whereas the normal, dorsal CNS was free of label. Hence there can be no doubt that Spemann & Mangold (1924) had, in fact, interpreted the result of their experiment correctly and that the second, ventral CNS does arise from cells that would normally have become belly and tail epidermis.

Conclusion

The examples of cell lineages presented here show that in metazoan development, the line of descent of a cell plays a critical role in determining its fate. The strongest indicator of this fact is the finding that in the embryos of nematodes, leeches, and insects, most rostrocaudally homologous neurons arise on rostrocaudally homologous branches of the cell lineage tree. But just how that determinative role is played has been elucidated so far in only a very few cases, despite the fact that for one animal, the nematode *C. elegans,* the exact line of descent of every somatic cell, neuronal and non-neuronal, is now known. One of the most surprising facts to emerge from this superbly detailed pedigree is the bewildering diversity of ontogenetic processes that are at work even at this comparatively modest level of metazoan complexity. For the nematode provides examples of almost any developmental mechanism that can be reasonably put forward to explain how a nerve cell becomes differentially committed to express the set of traits that make it uniquely identifiable. Moreover, the data also provide a counterexample for almost any nontrivial generalization that might be proposed regarding the mechanism of commitment. For instance, although most similar cell types expressing the same trait arise via corresponding branches of homologous sublineages, some arise via proliferative divisions of a common precursor cell, and yet others via seemingly chance interactive recruitment of genealogically unrelated cells that happen to lie in appropriate parts of a morphogenetic field. Or, by way of another example, most bilaterally homologous cells take on the same fate, while some have two different fates. And of those homologous cell pairs that do take on different fates under normal conditions, some pairs belong to an equivalence group, so that whenever either cell is ablated, the survivor takes on the primary fate, while in other pairs each cell is autonomously committed to its fate, even if the precursor cell of another group member has been ablated several generations previously. Finally there are a few equivalence groups whose members, as judged by their lines of descent, are not even homologs.

The much less complete cell lineage analysis of leech development leads to similar conclusions. The leech, more complex than the nematode, yet still far removed from the complexity of the vertebrates, resorts for its development to the same variety of developmental processes. This patchwork of mechanisms, which achieves what appear to be essentially similar ends by a great diversity of means, supports the notion set forth by Francois Jacob (1982) that ontogeny is related to philogeny by "tinkering," i.e. that evolution changed the course of embryogenesis by resort to any tool or trick that may happen to have been handy when it was needed. These findings suggest that by the time evolution had put the pseudoceolomate nematode worm on the scene, it had already tried most of the items in its bag of tools and tricks for determining cell fate. Thus it does not seem very probable that in the relatively brief period of subsequent metazoan evolution there have emerged many novel developmental mechanisms at the cellular level. Rather, what does seem likely is that the vertebrate nervous system arose by opportunistic variations in the timing, in the number of iterations, and in the spatial localization of the determinative processes that were already at work in the embryos of invertebrates.

ACKNOWLEDGMENT

Our research has been supported by NIH grants NS 12818 and HD17088, NSF grant BNS79–12400, and by grants from the March of Dimes and Rowland Foundations.

Literature Cited

Bate, C. M. 1976. Embryogenesis of an insect nervous system. I. A map of the thoracic and abdominal neuroblasts in *Locusta migratoria*. *J. Embryol. Exp. Morphol.* 35:107–23

Bate, C. M. 1982. Proliferation and pattern formation in the embryonic nervous system of the grasshopper. *NRP Bull.* 20:803–13

Bate, C. M., Goodman, C. S., Spitzer, N. C. 1981. Embryonic development of identified neurons: Segment specific differences in H cell homologues. *J. Neurosci.* 1:103–6

Bate, C. M., Grunewald, E. B. 1981. Embryogenesis of an insect nervous system. II. A second class of neuron precursor cells and the origin of the intersegment connectives. *J. Embryol. Exp. Morphol.* 61:317–30

Becker, H. J. 1957. Über Röntgenmosaikflecken und Defektmutationen am Auge von *Drosophila melanogaster* und die Entwicklungsphysiologie des Auges. *Z. Indukt. Abstamm. Vererbungsl.* 88:333–73

Bernard, F. 1937. Recherches sur la morphogènese des yeux composés d'arthropodes. *Bull. Biol. Fr. Belg.* 23:1–162 (Suppl.)

Blair, S. S. 1983. Blastomere ablation and the developmental origin of identified monoamine-containing neurons in the leech. *Dev. Biol.* 95:65–72

Blair, S. S., Stuart, D. K. 1982. Monoamine containing neurons of the leech and their teloblast of origin. *Soc. Neurosci. Abstr.* 8:16

Blair, S. S., Weisblat, D. A. 1982. Ectodermal interactions during neurogenesis in the Glossiphoniid leech *Helobdella triserialis*. *Dev. Biol.* 91:74–82

Blair, S. S., Weisblat, D. A. 1984. Cell interactions in the developing epidermis of the leech Helobdella triserialis. *Dev. Biol.* 101:318–25

Boveri, T. 1887. Über die Differenzierung der Zellkerne während der Furchung des Eies von Ascaris megalocephela. *Anat. Anz.* 2:668–93

Boveri, T. Über die Entstehung des Gegensatzes zwischen den Geschlechts Zellen und den somatischen Zellen bei Ascaris megalocephela. *Sitzungsber. Ges. Morphol. Physiol.* 8:114–25

Campos-Ortega, J. A., Hofbauer, A. 1977. Cell clones and pattern formation in the lineage of photoreceptor cells in the compound eye

of *Drosophila*. *Wilhelm Roux's Arch. Dev. Biol*. 181:227–45
Campos-Ortega, J. A., Jurgens, G., Hofbauer, A. 1978. Clonal segregation and positional information in late ommatidial development in *Drosophila*. *Nature* 274:584–86
Chalfie, M., Horwitz, H. R., Sulston, J. E. 1981. Mutations that lead to reiterations of cell lineages of *C. elegans*. *Cell* 24:59–69
Conklin, E. G. 1905. The organization and cell lineage of the ascidian egg. *J. Acad. Nat. Sci. Philadelphia: 13:1–119*
Dawydoff, C. 1959. Ontogenèse des Annelides. In *Traité de Zoologie*, ed. P. P. Grassé, 5:594–686. Paris: Masson
Deppe, V., Schierenberg, E., Cole, T., Krieg, C., Schmitt, D., Yoder, B., von Ehrenstein, G. 1978. Cell lineages of the embryo of the nematode *Caenorhabditis elegans*. *Proc. Natl Acad. Sci. USA* 75:376–80
Evans, P., O'Shea, M. 1977. The identification of an octopaminergic neuron which modulates neuromuscular transmission in the locust. *Nature* 270:275–79
Fernandez, J., Olea, N. 1982. Embryonic development of gloosiphoniid leeches. In *Developmental Biology of Freshwater Invertebrates,* ed. F. W. Harrison, R. R. Cowden, pp. 317–61. New York: Alan Liss
Fernandez, J., Stent, G. S. 1983. Embryonic development of the hirudiniid leech *Hirudo medicinalis*. Structure, development and segmentation of the germinal plate. *J. Embryol. Exp. Morphol*. 72:71–96
Garcia-Bellido, A., Merriam, J. R. 1969. Cell lineage of the imaginal disk in *Drosophila* gynandromorphs. *J. Exp. Zool*. 170:61–76
Gimlich, R. L., Cooke, J. L. 1983. Cell lineage and the induction of second nervous systems in amphibian development. *Nature* 306: 471–73
Gimlich, R. L., Braun, J. 1984. Bright, fixable cell-lineage tracers. *Dev. Biol*. Submitted
Goodman, C. 1982. Embryonic development of identified neurons in the grasshopper. In *Neuronal Development,* ed. N. C. Spitzer, pp. 171–212. New York/London: Plenum
Goodman, C. S., Bate, C. M. 1981. Neuronal development in the grasshopper. *Trends Neurosci*. July:163–69
Goodman, C. S., Bate, C. M., Spitzer, N. C. 1981. Embryonic development of identified neurons: Origin and transformation of the H cell. *J. Neurosci*. 1:94–102
Goodman, C. S., O'Shea, M., McCaman, R., Spitzer, N. C. 1979. Embryonic development of identified neurons: Temporal pattern of morphological and biochemical differentiation. *Science* 204:1219–22
Goodman, C. S., Spitzer, N. C. 1979. Embryonic development of identified neurones: Differentiation from neuroblast to neurone. *Nature* 280:208–14
Greenwald, I. S., Sternberg, P. W., Horvitz, H. R. 1983. The *lin*-12 locus specifies cell fates in *Caenorhabditis elegans*. *Cell* 34: 435–44
Hirose, G., Jacobson, M. 1979. Clonal organization of the central nervous system of the frog. I. Clones stemming from individual blastomeres of the 16 cell and earlier stages. *Dev. Biol*. 71:191–202
Hofbauer, A., Campos-Ortega, J. A. 1976. Cell clones and pattern formation: Genetic eye mosaics in *Drosophila melanogaster. Wilhelm Roux's Arch. Dev. Biol*. 179:275–89
Hotta, J., Benzer, S. 1972. Mapping of behavior in *Drosophila* mosaics. *Nature* 240: 527–35
Jacob, F. 1982. *The Possible and the Actual,* p. 71. Seattle/London: Univ. Washington Press
Jacobson, M. 1980. Clones and compartments in the vertebrate central nervous system. *Trends Neurosci*. Jan: 3–5
Jacobson, M. 1982. Origins of the nervous system in amphibians. In *Neuronal Development,* ed. N. C. Spitzer, pp. 45–99. New York/London: Plenum
Jacobson, M., Hirose, G. 1978. Origin of the retina from both sides of the embryonic brain: A contribution to the problem of crossing at the optic chiasma. *Science* 202:637–39
Jacobson, M., Hirose, G. 1981. Clonal organization of the central nervous system of the frog. II. Clones stemming from individual blastomeres of the 32- and 64-cell stages. *J. Neurosci*. 1:271–84
Janning, W. 1978. Gynandromorph fate maps in *Drosophila*. In *Genetic Mosaics and Cell Differentiation,* ed. W. J. Gehring, pp. 1–28. Berlin/Heidelberg: Springer Verlag
Kankel, D. R., Hall, J. C. 1976. Fate mapping of nervous system and other internal tissues in genetic mosaics of *Drosophila melanogaster. Dev. Biol*. 48:1–24
Keller, R. E. 1975. Vital dye mapping of the gastrula and neurula of *Xenopus laevis*. I. Prospective areas and morphogenetic movements of the superficial layer. *Dev. Biol*. 42:222–41
Keller, R. E. 1976. Vital dye mapping of the gastrula and neurula of *Xenopus laevis*. II. Prospective areas and morphogenetic movements of the deep layer. *Dev. Biol*. 51:118–37
Kimble, J. 1981. Alterations in cell lineage following laser ablation of cells in the somatic gonad of *Caenorhabditis elegans*. *Dev. Biol*. 87:286–300
Kimble, J., Hirsch, D. 1979. The postembryonic cell lineages of the hermaphrodite and male gonads in *Caenorhabditis elegans*. *Dev. Biol*. 70:396–417
Kimble, J., Sulston, J., White, J. 1979. In *Cell Lineage, Stem Cells and Cell Determination*.

INSERM Symp. No. 10, ed. N. le Douarin, pp. 59–68. Amsterdam: Elsevier

Kramer, A. P., Weisblat, D. A. 1984. Developmental neural kinship group in the leech. *Dev. Biol.* In press

Lawrence, P. A., Green, S. M. 1979. Cell lineage in the developing retina of *Drosophila. Dev. Biol.* 71:142–52

Maienschein, J. 1978. Cell lineage, ancestral reminiscence, and the biogenetic law. *J. History Biol.* 11:129–58

Morata, G., Lawrence, P. A. 1977. Homeotic genes, compartments and cell determination in *Drosophila. Nature* 265:211–16

Muller, K. J., Nicholls, J. G., Stent, G. S., eds. 1981. *Neurobiology of the Leech.* New York: Cold Spring Harbor Lab. 320 pp.

Nieuwkoop, P. D. 1952. Activation and organization of the amphibian central nervous system. *J. Exp. Zool.* 120:1–130

Nishida, H., Satoh, N. 1983. Cell lineage analysis in ascidian embryos by intracellular injection of a tracer enzyme. I. Up to the eight cell stage. *Dev. Biol.* 99:382–94

Ortolani, G. 1955. The presumptive territory of the mesoderm in the ascidian germ. *Experientia* 11:445–46

Ouweneel, W. J. 1976. Developmental genetics of homeosis. *Adv. Genet.* 18:179–248

Pasteels, J. 1942. New observations concerning the maps of presumptive areas of the young amphibian gastrula (*Ambystoma* and *Discoglossus*). *J. Exp. Zool.* 89:255–81

Poulson, D. F. 1950. Histogenesis, organogenesis, and differentiation in the embryo of *Drosophila melanogaster.* In *The Biology of Drosophila,* ed. M. Demerec, pp. 168–274. New York: Wiley

Ready, D. F., Hanson, T. E., Benzer, S. 1976. Development of the *Drosophila* retina, a neurocrystalline lattice. *Dev. Biol.* 53:217–40

Satoh, N. 1979. On the clock mechanism determining the time of tissue-specific enzyme development during ascidian embryogenesis. I. Acetylcholinesterase development in cleavage arrested embryos. *J. Embryol. Exp. Morphol.* 54:131–39

Satoh, N., Ikegami, S. 1981a. A definite number of aphidicolin sensitive cell cycle events are required for acetylcholinesterase development in the presumptive muscle cells of the ascidian embryo, *J. Embryol. Exp. Morphol.* 61:1–13

Satoh, N., Ikegami, S. 1981b. On the "clock" mechanism determining the time of tissue specific enzyme development during ascidian embryogenesis. II. Evidence for association of the clock with the cycle of DNA replication. *J. Embryol. Exp. Morphol.* 64:61–71

Shankland, M. 1984. Positional control of supernumerary blast cell death in the leech embryo. *Nature* 307:541–43

Shankland, M., Weisblat, D. A. 1984. Stepwise loss of neighbor cell interactions during positional specification of blast cell fates in the leech embryo. *Dev. Biol.* In press

Slack, J. M. W. 1983. *From Egg to Embryo.* London/New York: Cambridge Univ. Press. 241 pp.

Spemann, H., Mangold, H. 1924. Über Induktion von Embryonalanlagen durch Implantation artfremder Organisatoren. *Arch. Mikrosk. Anat. Entwmech.* 100:599–638

Stent, G. S., Weisblat, D. A., Blair, S. S., Zackson, S. L. 1982. Cell lineage in the development of the leech nervous system. *Neuronal Development,* ed. N. Spitzer, pp. 1–44. New York: Plenum

Stern, C. 1936. Somatic crossing over and segregation in *Drosophila melanogaster. Genetics* 21:625–730

Stern, C. 1968. *Genetic Mosaics and Other Essays.* Boston: Harvard Univ. Press

Sternberg, P. W., Horwitz, H. R. 1982. Postembryonic non-gonadal cell lineages of the nematode *Panagrellus redivivus:* Description and comparison with those of *Caenorhabditis elegans. Dev. Biol.* 93:181–205

Sturtevant, A. H. 1929. The claret mutant type of *Drosophila simulans:* A study of chromosome elimination and cell lineage. *Z. Wiss. Zool.* 135:325–56

Sulston, J. E., Horwitz, H. R. 1977. Postembryonic cell lineages of the nematode *Caenorhabditis elegans. Dev. Biol.* 56:110–56

Sulston, J. E., Horwitz, H. R. 1981. Abnormal cell lineages in mutants of the nematode *Caenorhabditis elegans. Dev. Biol.* 82:41–55

Sulston, J. E., Schierenberg, E., White, J. G., Thomson, J. N. 1983. The embryonic cell lineage of the nematode *Caenorhabditis elegans. Dev. Biol.* 100:64–119

Sulston, J. E., White, J. G. 1980. Regulation of cell autonomy during postembryonic development of *Caenorhabditis elegans. Dev. Biol.* 78:577–97

Taghert, P. H., Doe, C. Q., Goodman, C. S. 1984. Cell determination and regulation during development of neuroblasts and neurons in grasshopper embryo. *Nature* 307:163–65

von Ehrenstein, G., Schierenberg, E. 1980. Cell lineages and development of *Caenorhabditis elegans* and other nematodes. In *Nematodes as Biological Models,* ed. B. Zuckerman, 1:1–71. New York: Academic

Vogt, W. 1929. Gestaltungsanalyse am Amphibienkern mit orlicher Vitalfarbung. II. Gastrulation und Mesodermbildung bei Urodelen und Anvren *Wilhelm Roux' Arch. Entwicklungsmech. Org.* 120:384–706

Weisblat, D. A. 1981. Development of the nervous system. In *Neurobiology of the Leech,* ed. K. J. Muller, J. G. Nicholls, G. S. Stent, pp. 173–95. New York: Cold Spring Harbor Lab.

Weisblat, D. A., Blair, S. S. 1984. Developmental indeterminacy in embryos of the leech *Helobdella triserialis*. *Dev. Biol.* 101: 326–35

Weisblat, D. A., Harper, G., Stent, G. S., Sawyer, R. T. 1980a. Embryonic cell lineage in the nervous system of the glossiphoniid leech *Helobdella triserialis*. *Dev. Biol.* 76: 58–78

Weisblat, D. A., Kim, S. Y., Stent, G. S. 1984. Embryonic origins of cells in the leech *Helobdella triserialis*. *Dev. Biol.* 104:65–85

Weisblat, D. A., Sawyer, R. T., Stent, G. S. 1978. Cell lineage analysis by intracellular injection of a tracer enzyme. *Science* 202:1295–98

Weisblat, D. A., Shankland, M. 1984. In preparation

Weisblat, D. A., Zackson, S. L., Blair, S. S., Young, J. D. 1980b. Cell lineage analysis by intracellular injection of fluorescent tracers. *Science* 209:1538–41

White, J. G., Southgate, E., Thompson, J. N., Brenner, S. 1976. The structure of the ventral nerve cord of *Caenorhabditis elegans*. *Philos. Trans. Roy. Soc. London Ser. B* 275:327–48

White, J. G., Southgate, E., Thompson, J. N., Brenner, S. 1983. Factors that determine connectivity in the nervous system of *C. elegans*. *Cold Spring Harbor Symp. Quant. Biol.* 48:633–40

Whitman, C. O. 1878. The embryology of Clepsine. *Q. J. Microscop. Sci.* (N.S.) 18:215–315

Whitman, C. O. 1892. *The metamerism of Clepsine. Festschrift zum 70. Geburtstage R. Leuckarts,* pp. 385–95. Leipzig: Engelmann

Whitman, C. O. 1887. A contribution to the history of germ layers in Clepsine. *J. Morphol.* 1:105–82

Whittaker, J. R. 1973. Segregation during ascidian embryogenesis of egg cytoplasmic information for tissue specific enzyme development. *Proc. Natl. Acad. Sci. USA* 70:2096–2100

Whittaker, J. R. 1979. Cytoplasmic determinants of tissue differentiation in the ascidian egg. In *Determinants of Spatial Organization,* ed. S. Subtetny, I. R. Konigsberg, pp. 29–51. New York: Academic

Whittaker, J. R. 1980. Acetylcholinesterase development in extra cells by changing the distribution of myoplasm in ascidian embryos. *J. Embryol. Exp. Morphol.* 55:343–54

Whittaker, J. R. 1982. Muscle lineage cytoplasm can change the developmental expression in epidermal lineage cells of ascidian embryos. *Dev. Biol.* 93:463–70

Whittaker, J. R., Ortolani, G., Farinella-Feruzza, N. 1977. Autonomy of acetylcholinesterase differentiation in muscle lineage cells of ascidian embryos. *Dev. Biol.* 55:196–200

Zackson, S. L. 1982. Cell clones and segmentation in leech development. *Cell* 31:761–70

Zackson, S. L. 1984. Cell lineage, cell-cell interactions, and segment formation in the ectoderm of a glossiphoniid leech embryo. *Dev. Biol.* 104:143–60 in press

zur Strassen., O. 1896. Embryonalentwicklung der Ascaris megalocephela. *Archiv Entwicklungsmechanik* 3:27–105

Ann. Rev. Neurosci. 1985. 8:71–102

CLONAL ANALYSIS AND CELL LINEAGES OF THE VERTEBRATE CENTRAL NERVOUS SYSTEM

Marcus Jacobson

Anatomy Department, University of Utah School of Medicine, Salt Lake City, Utah 84132

INTRODUCTION

Very little progress in tracing cell lineages in the CNS of vertebrates had been made before the introduction of horseradish peroxidase (HRP) as an intracellular lineage tracer in the frog embryo (Jacobson & Hirose 1978, 1981, Hirose & Jacobson 1979). Conceptual weaknesses and lack of adequate techniques reinforced one another to ensure slow progress (Jacobson 1978, p. 98). Thus, a distinction has traditionally been made between embryos that have "mosaic" development, in which lineage determines cell fate, and those with "regulative" development, in which cell interactions determine cell fate, but, in fact, most if not all embryos show a combination of both (Harrison 1933, Weiss 1939, Davidson 1976). I shall propose that there are several lineage modes with different degrees of autonomy. In addition to the *determinate lineage mode,* which is completely autonomous and in which the relationship between a given progenitor and the phenotype(s) of its descendants is all-or-none, there are others called *indeterminant lineage modes* in which the relationship between ancestral cells and the phenotypes of their descendants is contingent on external conditions to some degree. It is now clear that cell interactions can result in changes in of cellular fates in Uro- and Cephalochordates (Tung et al 1958, 1960, Nakauchi & Takashita 1983), in which there is rigorous proof that materials localized in the egg and differentially distributed into different lineages determine the ultimate fates of the differentiated cells (Whittaker 1973, 1979, Jeffery 1983, Jeffery et al 1983). Such unimpeachable evidence of extensive regulation in mosaic embryos should help to open the minds of those who have been unwilling to admit that determination

0147-006X/85/0301-0071$02.00

of cell fate by inductive cell interactions as well as determination of cell fate by cell lineages may both occur in the same embryo (cf Gimlich & Cooke 1983).

Another conceptual impediment to acceptance of a role for cell lineages in vertebrate development arises from the well known variability of cleavage patterns in vertebrate embryos. The classical experiments (reviewed by Gerhart 1980) which demonstrated that altering the cleavage patterns of amphibian embryos did not effect normal development has also been interpreted as evidence against any role for lineage in determining cell fates in amphibian embryos. But this interpretation ignored the possibility that the segregation of cytoplasmic determinants into separate lineages could be important in eggs that cleaved irregularly. This could occur if the actions of those determinants were delayed until the cells were small enough to make the prior pattern of cleavages irrelevant, provided that the cells remained coherent during those early cleavages. One of the necessary conditions for cell lineages to determine developmental programs of groups of cells is that dispersal and mingling of the cell groups or their descendants is delayed until after the time of initiation of the lineage-dependent programs (Jacobson 1982). If lineage-dependent programs in cell groups depend on cellular inheritance of materials regionally distributed in the egg, they could be initiated at any time before onset of cell mingling, which starts at the beginning of gastrulation in amphibians. There is considerable evidence that cytoplasmic determinants, regionally localized in the egg and segregated into different blastomeres, play a role in amphibian development (Fankhauser 1932, Gurdon 1966, Gurdon & Woodland 1969, Gallien 1979, reviewed by Davidson 1976). Those findings provided a basis for the concept that one of the roles of cleavages in amphibian embryos is to segregate cytoplasmic determinants into groups of blastomeres rather than into individual blastomeres, and that the determinants would not be expected to begin to act until the embryo contained hundreds of blastomeres (Jacobson 1982.)

Cell fates could be determined by lineage, induction, position, or any combination of these. In principle, disaggregation and recombination of blastomeres in different arrangements could show whether cell fates are expressed autonomously or are regulated by cell interactions, or both. In practice such experiments are difficult to interpret because the recombinations have been done at the two- to 16-cell stage, too early to test the hypothesis that lineage-dependent programs originate in groups of blastomeres with a common ancestry from a localized region of the egg. Even if recombinations were to be made when the embryo contained hundreds of blastomeres, many more cell generations are produced before patterns of cell differentiation develop, and patterning could be the result of selective cell death and selective cell reassociation as well as inductions, position-dependent cell interactions and lineage-dependent programs. The complexity of the embryo makes it difficult to reduce the problem to what seems to be a nature-nurture problem at the cellular level. Our working hypothesis is that segregation of cytoplasmic determinants during early cleavage stages results in regional specification of groups of ancestral cells that form a primary pattern and that cell interactions lead to secondary patterns at later stages (Jacobson 1983). Cellular inheritance of cytoplasmic determinants is an initial condition required for inductions (Gimlich & Gerhart 1984). This may make it difficult to separate lineage-dependant from interaction-dependent regional specification.

CLASSIFICATION OF LINEAGE MODES

Lineages of nerve cells can be classified according to whether the relationship between ancestral cell and a specified type of nerve cell is one-to-one, one-to-many, or many-to-many. A classification can also be based on the degree to which the lineage is cell-autonomous (determinate lineage, as defined by Conklin, 1897) or can be modified by external conditions (indeterminate lineage).

In order to break away from the concept of a dichotomy between mosaic and regulative development, it is necessary to conceive of modes of lineage that permit regulative cell interactions and to propose that there are modes of lineage-dependent development with different degrees of plasticity. At one extreme is an immutable one-to-one relationship between cell ancestry and cell phenotype that is called *monoclonal determinate* lineage (Figure 1). The lineages are likely to be autonomous, that is, removal of some cells or alteration of cells by a mutation would produce effects restricted to that lineage and would not alter other lineages. At the opposite extreme there is no fixed relationship between any single ancestral cell and any single cell phenotype, but the lineage relationship is between a group of ancestral cells and a range of cell phenotypes. This is called the *polyclonal indeterminate* lineage mode if the relationship between ancestral cells and phenotypes of their descendants is probablistic, many-to-many, and is to some degree contingent on external conditions. It is called the *polyclonal determinate* mode if there are fixed relationships between many ancestral cells and a number of cell phenotypes. Polyclonal determinate lineages are autonomous but removal of cells, mutations, or other changes in the lineage could produce graded effects in the descendants, depending on the number of cells in the polyclone that were altered. By contrast, in the polyclonal indeterminate mode, complete regulation can occur after removal of some ancestral cells because cell interactions enable cells within the same ancestral cell groups to substitute for one another.

Considerable evidence suggests that neither neurons nor glial cells originate by the monoclonal determinate mode in the vertebrates. Many types of nerve cells in the vertebrate CNS have to be generated in large numbers. If they

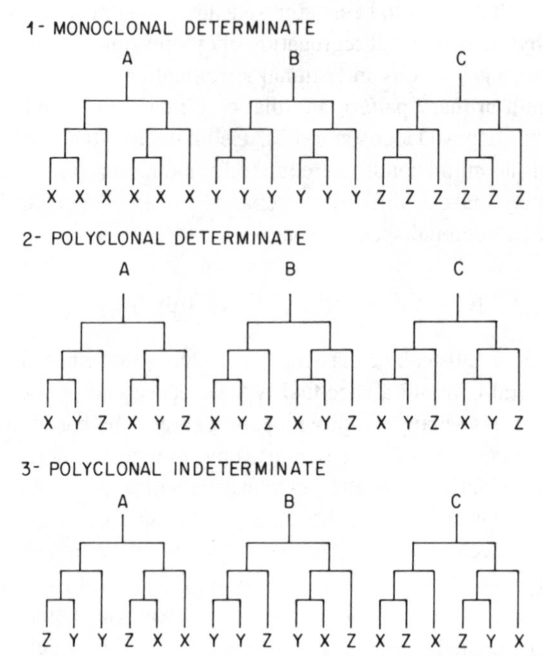

Figure 1 Diagram of three modes of lineage described in the text. The branching patterns are merely illustrative.

originated in the monoclonal determinate mode, lineages of different types of neurons would have to diverge early in development, followed by many cell generations. Evidence obtained from mouse fusion chimeras indicate the polyclonal origins of Purkinje cells (Mullen 1977, Oster-Granite & Gearhart 1981, Wetts & Herrup 1982) oligodendrocytes (Mikoshiba et al 1982), and Schwann cells (Peterson & Marler 1983). In *shiverer*-normal mouse chimeras, single axons in the CNS were myelinated by oligodendrocytes derived from both parents (Mikoshiba et al 1982), and single nerve fibers in peripheral nerves were myelinated by Schwann cells of both genotypes (Peterson & Marler 1983). These results show that at least two ancestral cells must be allocated to found the population of Schwann cells that finally associate with a single axon.

Neurons that are produced in large numbers, mainly small local circuit neurons in the vertebrate CNS, require correspondingly large numbers of progenitors and long periods of production, often extending into the postnatal period in mammals. Neurons that are produced in relatively small numbers, especially the large, principal neurons, are produced relatively early and quickly in each region, and probably originate from small groups of progenitors. There are few observations, though, regarding the numbers of progenitors and the times at which they become committed to produce each type of neuron (Jacobson 1978, Wetts & Herrup 1982, Jacobson & Moody 1984). It is not known when the divergencies occur between progenitors of (*a*) neurons and non-neuronal cells, (*b*) neurons and glial cells, and (*c*) different types of neurons in the vertebrate CNS.

It is difficult to envisage how large numbers of cells in the vertebrate CNS could originate from relatively small numbers of ancestral cells by means of a determinate lineage mode in which there are fixed relationships between ancestors and specified neuronal phenotypes. This problem has recently been raised by evidence that can be interpreted to mean that all the Purkinje cells in each half of the cerebellum in the mouse originate from 8 to 11 ancestral cells that are set aside in the neural plate, and that each of those ancestral cells gives rise to a fixed proportion of the final population (Wetts & Herrup 1982).

The polyclonal indeterminate mode of lineage is the only one that has been demonstrated in the vertebrate CNS (Jacobson & Hirose 1981, Jacobson 1983, Jacobson & Moody 1984). Even the Mauthner's neuron, which is a class of one neuron on each side of the rhombencephalon, appears to originate probabilistically from one member of a group of ancestral cells (Jacobson & Hirose 1981). The determinate lineage modes have been found in many invertebrates, whereas the only mode found in the vertebrates is the polyclonal indeterminate lineage mode. This suggests the possibility that the evolution of the invertebrate nervous system has been limited by the determinate mode of cell lineage. Evolution of the invertebrate nervous system could occur as a result of mutations that directly alter lineages (Goodman 1977, Sternberg & Horvitz 1981, Sulston & Horvitz 1981). Deletion, fusion, or extension of lineages may result from mutations, but there are a limited number of such mutations that are compatible with survival or which are advantageous. By contrast, the indeterminate mode of cell lineage may have released vertebrates from those constraints that have limited the evolution of invertebrate nervous systems. The probabilistic mode gives rise to variability upon which the forces of evolutionary selection can operate. It is likely that determinate cell divisions occur in the vertebrates but are delayed until the terminal cell generation(s) of the lineage. This delay of the invertebrate lineage mode into late stages of vertebrate development can be regarded as a form of neoteny, which is an important cause of evolutionary change (De Beer 1951, Gould 1977). In the vertebrates, the polyclonal indeterminate lineage mode results in variability based on the probabilistic relationship between nerve cell ancestry and nerve cell phenotype (Jacobson & Hirose 1981, Jacobson 1982, Jacobson & Moody 1983). This variability, which occurs as part of normal neuronal ontogeny, may have been a prerequisite for the evolution of complex behavior and intelligence.

In each of these lineage modes, cell fate is restricted by cell lineage, but those restrictions can affect cell differentiation or morphogenesis. There is evidence that clonal restrictions of cell mingling (Jacobson 1983, Jacobson & Klein 1985) and of cell proliferation (Klein & Jacobson 1985) occur during development of the frog embryo and that those clonal restrictions are related only to the overall pattern of morphology, not to local patterns of cellular phenotypes. This bears on the problem of how head and trunk or limbs become different although they contain the same types of cells.

THE CONCEPT OF CLONAL RESTRICTION

Clonal restriction may be defined as any restriction of developmental activities of cells that is determined by the cells' origins from specified progenitors and is transferred to the lineal descendants of those progenitors. Lineages have significance only if they limit or restrict development in a significant way. Restriction of a lineage to produce a unique cellular phenotype or a specified set of cellular phenotypes affects histogenesis primarily. This is the role that is usually ascribed to cell lineage, but in principle any developmental function may be restricted by lineage. Lineage restrictions of cell proliferation, survival, migration, and interaction may occur. Lineage restrictions above the level of the cellular phenotype may occur during separation of the primary embryonic germ layers. Restriction of mingling between two lineages resulting in segregation of their descendants, regardless of cellular phenotypes, will affect morphogenesis primarily. A classification of modes of lineage must at the least distinguish between those lineage modes that primarily restrict cell phenotypes and those that primarily affect morphogenesis.

In each case the causal relationship between cell lineage and the developmental restriction must be demonstrated experimentally. In the vertebrate embryo, such restrictions are often the result of cell interactions or inductions. An adequate lineage analysis should show whether there is a direct causal relationship between the lineage and the developmental restriction and show whether the restriction can occur in the absence of inductive cell interactions, that is whether it is cell autonomous. This means that the developmental restriction in question can occur in isolated cells and that no regulation occurs if the lineage is prematurely terminated by removal of the cells. However, autonomy of the lineage occurs only in the determinate mode, and even then some regulation may occur in the remaining cells. Regulation occurs in ascidians (Nakauchi & Takashita 1983), which were once regarded as typically mosaic embryos in which differentiation of cellular phenotypes is known to be determined by factors localized in the egg and distributed differentially into separate lineages (Whittaker 1979, Jeffery et al 1983). In amphibians, and probably in other vertebrates, the polyclonal indeterminate lineage mode permits considerable regulation, and removal of individual cells or even a significant fraction of polyclone does not alter the fates of the remaining cells.

Autonomy has traditionally been assayed by excising and transplanting small pieces of the embryo and observing the behavior of the transplant and of the residual cells, especially at the excision site. These studies are difficult or impossible to interpret correctly without knowing how the pieces were related to polyclonal lineages. Complete regulation occurs after removal of a piece that forms part of a polyclone provided that the lineages have not terminated but the transplanted piece may continue to develop autonomously in the ectopic position, especially if some lineages have terminated before the time of transplantation. Thus, in the amphibian embryo, small pieces of neural plate continue to undergo self-differentiation after transplantation (Lewis 1910, Spemann 1912, Boterenbrood 1970), but complete restitution of normal structures can follow removal of pieces of neural plate and neural tube (Detwiler 1944, Holtzer 1951, Corner 1963). Removal of an entire polyclonal compartment in the late blastula of the frog embryo always results in a permanent morphological deficit, whereas regulation often follows removal of pieces of the same size that do not include any complete compartment (Jacobson 1982 and unpublished observations). Regulation seems to be possible only within polyclonal lineages. It is necessary to investigate the role of inductive cell interactions in such regulation as well as the role of primary induction in determining the pattern of cellular phenotypes that differentiate within polyclones.

PROSPECTIVE CLONAL ANALYSIS BY MEANS OF HERITABLE INTRACELLULAR TRACERS

A prospective clonal analysis has been undertaken in the embryo of *Xenopus;* the clones have been traced in tailbud embryos that originated from every ancestral cell of the two-cell to 64-cell embryo and a large sample of cells from the 128- to 512-cell embryo in a series of several hundred embryos (Jacobson & Hirose 1978, 1981, Hirose & Jacobson 1979, Jacobson 1983). Horseradish peroxidase (HRP) or biotinylated HRP (Jacobson & Huang 1984) injected into a single blastomere could be traced in all the descendants at later stages at which the main parts of the CNS and many types of neurons were recognizable and the animal had complex behavior. This has proved to be the most effective method available for clonal analysis and lineage tracing in a vertebrate embryo. Fluorescent tracers are serviceable when little dilution of the tracer occurs, in short lineages and in embryos in which cells divide without growing larger (Weisblat et al 1980b, Gimlich & Cooke 1983). Unlike tracer enzymes, fluorescent tracers do not give permanent preparations, and they are applicable to tissue sections rather than to whole-mount preparations.

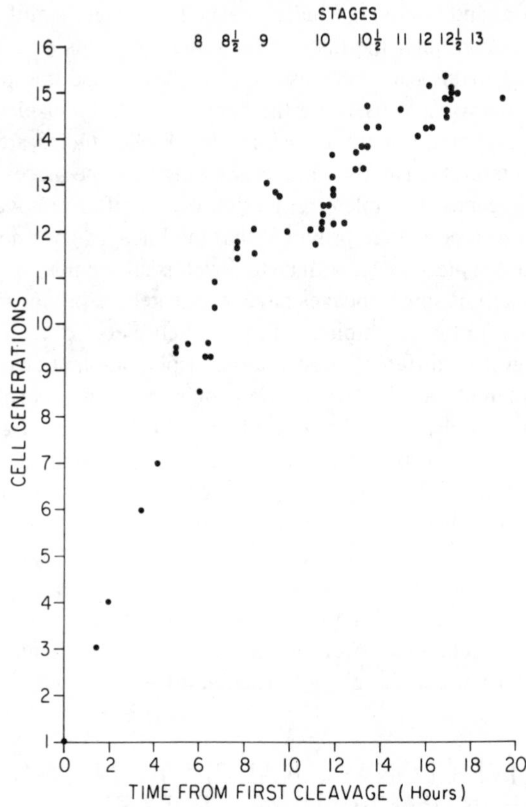

Figure 2 Clonal growth of *Xenopus* embryos from the first cleavage of the egg to the fifteenth cell generation. Each value was obtained by counting the number of labeled descendants after injecting HRP or biotinylated HRP into a single ancestral cell. The stages are from Nieuwkoop & Faber (1967). Data from Jacobson (1984b).

Their advantage is that different fluorescent tracers can be used in the same embryo.

The results that have been obtained from the preparations in which labeled descendants were traced progressively after injecting HRP into single ancestral cells in a series of *Xenopus* embryos can be summarized briefly.

1. Labeled descendants have been counted at progressive times after labeling a single cell (Figure 2). The rate of growth of clones originating from different ancestral cells has been measured (Jacobson 1984a,b, Jacobson & Moody 1984). Blastomeres divided almost synchronously, every 40 minutes at 20°C, from the first to the twelfth cell generation, which corresponded with the beginning of gastrulation. Thereafter, the mitotic cycle lengthened to an average of three hours for each of the following three cycles, and asynchrony

increased. The fifteenth generation coincided with the end of gastrulation or the early neural plate stage.

2. Spatial relationships between the positions of individual ancestral cells in the early embryo and the intermediate and final positions occupied by their clones (i.e. their *clonal domains*) have been mapped, especially in the CNS. This prospective clonal analysis has been completed for every individual blastomere up to and including the 64-cell embryo and for a large sample of blastomeres up to and including the 512-cell stage (Jacobson & Hirose 1978, 1981, Hirose & Jacobson 1979, Jacobson 1983). A fate map could be deduced from the continuous and orderly mapping relationships between positions of blastomeres and positions of their clonal domains in the late embryo (Figure 3).

3. Observations have been made of the dispersal and mingling of clones. The time of onset, time-course, and spatial pattern of dispersal of cells belonging to single clones have been determined (Jacobson 1984b). Mingling between cells belonging to different clones has been studied in vivo (Jacobson 1983) and in vitro (Jacobson & Klein 1985, Klein & Jacobson 1985). Each blastomere distributed its descendants into a region called its *clonal domain*. This clonal domain was always shared with descendants of other blastomeres. Even at the two-cell stage each blastomere contributed some descendants to both sides of the ventral parts of the head, including the ventral parts of telencephalon, diencephalon, mesencephalon, and both retinae (Figure 4). Clones descended from neighboring blastomeres of the four- and 64-cell stages occupied partially overlapping domains. From the 64-cell stage on there was a progressive increase in the sharing of the same clonal domain by neighboring blastomeres, so that at the 512-cell stage all blastomeres contributed descendants to only one of seven clonal domains, each of which consisted of the mingled descendants of a group of ancestral cells (Table 1). Little or no mingling occurred between descendants of different ancestral cell groups. Each polyclonal domain is called a *compartment* (Jacobson 1983).

Restriction of mingling between descendants of different ancestral cell groups was studied by combining two ancestral cell groups in vitro, one excised from an embryo completely labeled with HRP and another from an unlabeled embryo. Restriction of mingling was observed between descendants of two different ancestral cell groups, but cells mingled between descendants of combination of two identical ancestral cell groups (Jacobson & Klein 1984).

4. The relationships between clonal origins and cellular phenotypes have been investigated by observing the range of cellular phenotypes that differentiated in each clone and the number of cells of each type in the terminally differentiated clone (Jacobson 1983, Jacobson & Moody 1984).

5. The sequential partitioning of clones (spatial partitioning and phenotypic fractionation) has been traced from progressively later generations of progen-

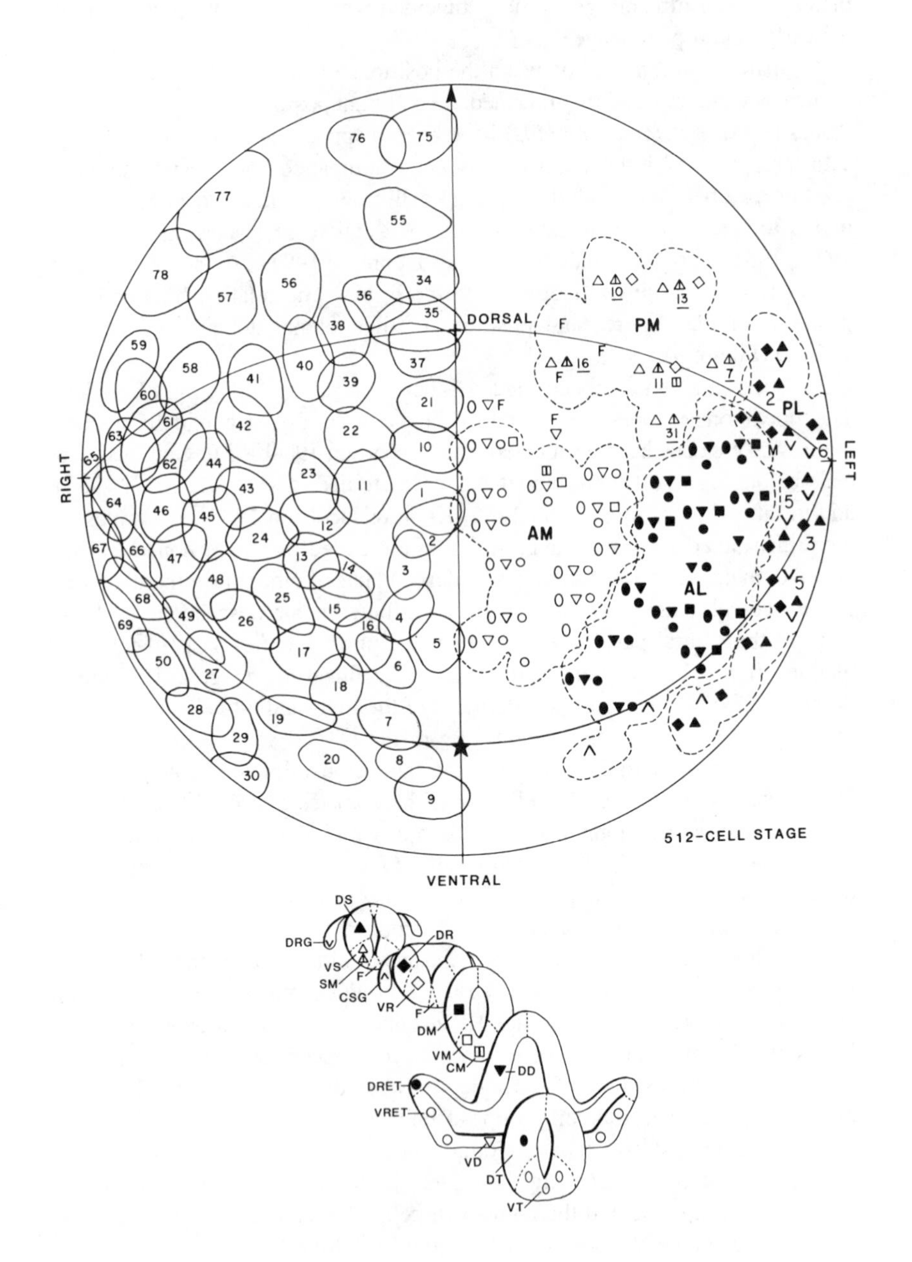
DORSAL
VENTRAL
RIGHT
LEFT
PM
PL
AM
AL
512-CELL STAGE
DS
DRG
VS
SM
CSG
DR
VR
F
DM
VM
CM
DD
DRET
VRET
VD
DT
VT

itors in the same lineage, i.e. the clonal partitioning and fractionation that occurred after each successive cleavage has been measured (Jacobson 1981a, Jacobson & Hirose 1981, Jacobson 1983, Jacobson & Moody 1984). Each type of neuron originated from a group of progenitors that also gave rise to many other cellular phenotypes in the nervous system and other systems. There were many-to-many relationships between the ancestral cell group and the different types of descendants. Nevertheless, production was biased toward the types of neurons that originated earliest. For example, counts made of the numbers of Rohon-Beard neurons and primary motoneurons that originated from single progenitors at every generation from the first to the ninth could be plotted (Figure 5). On a log-log plot, the geometric means of those counts fitted a straight line that intercepted the abscissa at the thirteenth generation for Rohon-Beard progenitors and at the sixteenth generation for the primary motoneuron progenitors. These intercepts show the mean number of generations before those types of neurons are finally produced.

6. Axons that were labeled with HRP inherited from ancestral cells have been traced from the time of initial outgrowth to the time of contacts with axonal targets in the periphery as well as within the CNS (Moody & Jacobson 1983, Jacobson & Huang 1984). Primary motoneuron axons showed a highly significant preferred association with clonally related muscle cells (Moody & Jacobson 1983). A similar preference was shown by Rohon-Beard neuron axons for associating with clonally related cells in the epidermis (Jacobson & Huang 1985). Clonal preferences were not seen in growth of axons to their

---

Figure 3 Fate map *(anterior oblique case)* of blastomeres of the 512-cell stage that contributed descendants to the CNS. Blastomere positions are shown on the right side of the embryo. CNS regions to which each blastomere contributed are shown on the left side of the embryo and in the *lower diagram* of a series of transverse sections through the larval CNS. Groups of blastomeres that all contributed descendants to the same CNS regions are enclosed by *dashed lines*. Blastomere groups: AM, anterior-median; AL, anterior-lateral; PM, posterior-medial; PL, posterior-lateral. *Numbers* on the right half of the embryo identify each blastomere. *Numbers* on the left half of the embryo are numbers of Rohon-Beard neurons that originated from each blastomere in the PL group and the numbers of primary spinal motoneurons that originated from each blastomere in the PM group *(underlined numbers)*. CM, cranial motoneurons; CSG, cranial sensory ganglia; DD, dorsal diencephalon; DM, dorsal mesencephalon; DRG, dorsal root ganglia; DR, dorsal rhombencephalon; DRET, dorsal retina; DS, dorsal spinal cord; DT, dorsal telencephalon; F, floor plate; M, Mauthner neuron; SM, spinal motoneurons; VD, ventral diencephalon; VM, ventral mesencephalon; VR, ventral rhombencephalon; VRET, ventral retina; VS, ventral spinal cord; VT, ventral telecephalon. The *star* is at the animal pole, and the *arrow* points to the vegetal pole. Diameter of the blastula is approximately 1.2 mm. *Open ellipse,* ventral telencephalon; *solid ellipse,* dorsal telencephalon; *open circle,* ventral retina; *solid circle,* dorsal retina; *open inverted triangle,* ventral diencephalon; *solid inverted triangle,* dorsal diencephalon; *open square,* ventral mesencephalon; *solid square,* dorsal mesencephalon; *open diamond,* ventral rhombencephalon; *solid diamond,* dorsal rhombencephalon; *open triangle,* ventral spinal cord; *solid triangle,* dorsal spinal cord; V, trunk neural crest; Λ, cranial neural crest.

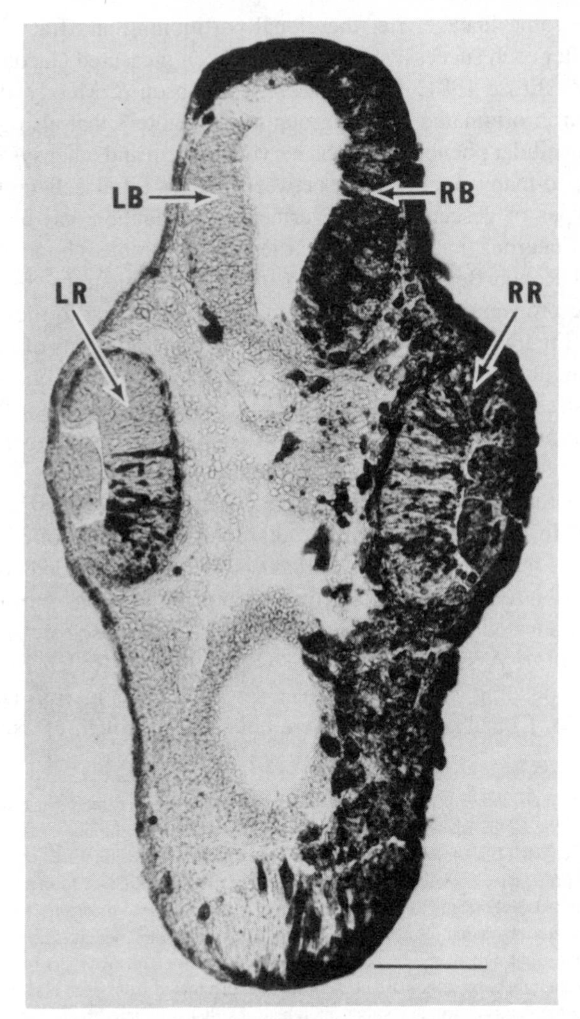

Figure 4 Section through the head of stage 30 embryo that had been given an intracellular injection of HRP into the right blastomere at the two-cell stage to show the distribution of labeled cells *(black)* and mingling across the midline between ventral halves of both retinae. Bar is 100 μm. LB, RB, left and right brain; LR, RR, left and right retina. From Hirose & Jacobson (1979).

targets in the CNS, where connections were usually between neurons in different compartments (Jacobson & Huang 1985).

The main conclusion drawn from those results were as follows:

1. Cell lineages do not determine cell phenotypic fate in the blastula stages of the frog embryo. A single blastomere gives rise to many types of cells in the nervous system and in other systems (Hirose & Jacobson 1979, Jacobson

Table 1 Compartments and their anatomical contents[a]

	Nervous System	Mesoderm	Endoderm	Ectoderm
Anterior-Median	Ventral telencephalon, ventral diencephalon, ventral mesencephalon, ventral retina.	Ventral head mesenchyme, hypobranchial muscles, intrinsic muscles of pharynx and larynx.	Ventral pharynx, ventral foregut, liver, pancreas.	Ventral head epidermis, stomodeum.
Anterior-lateral	Dorsal telencephalon, dorsal diencephalon, dorsal mesencephalon, dorsal retina, cephalic neural crest and its derivatives.	Lateral and dorsal head mesenchyme, extraocular muscles, branchiomeric muscles, facial muscles, heart and pericardium.	Lateral and dorsal foregut and pharynx.	Lateral and dorsal head epidermis, ectodermal placodes.
Posterior-medial	Ventral rhombencephalon, ventral spinal cord.	Notochord, sclerotome, ventral somitic mesoderm.	Ventral midgut and hindgut.	Ventral trunk and tail epidermis, proctodeum.
Posterior-lateral	Dorsal rhombencephalon, dorsal spinal cord, trunk neural crest and its derivatives.	Dorsal somitic mesoderm, dermatome, nephrotome, lateral plate mesoderm.	Dorsal midgut and hindgut.	Dorsal trunk and tail epidermis.

[a]Anterior–median compartment extends bilaterally across the ventral midlines.
All others are unilateral.

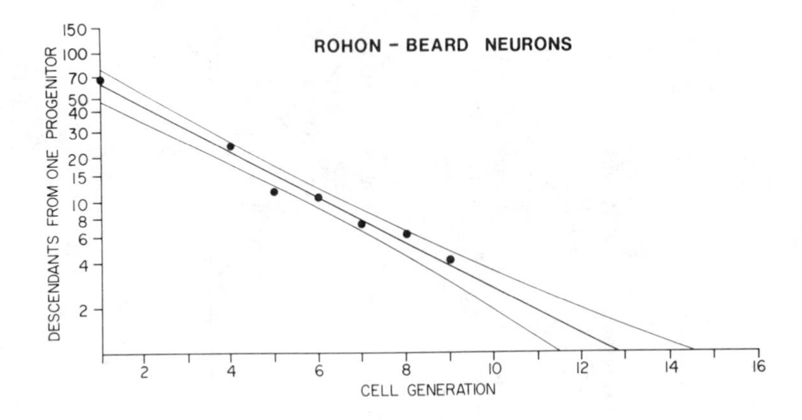

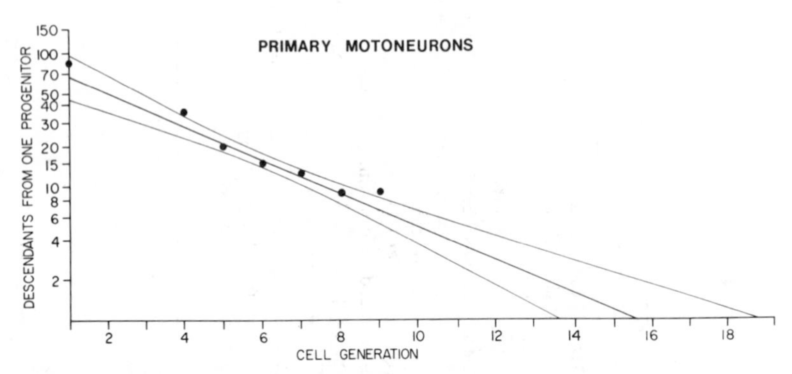

Figure 5 The numbers of Rohon-Beard neurons and primary motoneurons expected to descend from a single progenitor labeled during the first 18 cell generations are represented by a regression line *(thick central line)* plotted on log-log scales. The *points* show geometric means of the numbers of neurons derived from the single labeled progenitor at various cell generations. The *thin lines* show 95% confidence intervals. The intercept of the ordinate shows the total number of neurons present on one side of the spinal cord and the intercept on the abscissa shows the cell generation at which a single progenitor gives rise to one Rohon-Beard neuron. From Jacobson & Moody (1984).

& Hirose 1981, Jacobson 1983). The first stage at which neuronal and non-neuronal lineages separate cannot be later than the thirteenth to sixteenth cell generation following fertilization of the egg, because Rohon-Beard neurons, primary motoneurons, and other large neurons of the rhombencephalon and spinal cord are generated at that stage (Vargas-Lizardi & Lyser 1974, Lamborghini 1980, Jacobson & Moody 1984).

2. The relationship between individual ancestral cells and identified types of neurons is probabilistic. The probability could be calculated from counts

of the labeled descendants of a specified cellular phenotype (Rohon-Beard neurons or primary motorneurons) that originated from a single ancestral cell into which HRP was injected, in a series of embryos at progressively later cell generations in the lineage leading to the specified phenotype (Figure 5). These data could be expressed as a *continuation probability,* namely the probability that the daughter cells would continue in the lineage leading to that cellular phenotype. The mean continuation probability was 0.7 for the entire lineages of Rohon-Beard neurons and primary motoneurons; this indicates a continuing bias toward production of those types of neurons in lineages that also give rise to many other cell types. Those findings led to the concept that commitment to neuronal phenotypes is gradual and not an all-or-none event (Jacobson & Moody 1984).

No lineage restrictions related to differentiation of cellular phenotypes occur during blastula stages, so blastomeres can substitute completely for one another in producing Rohon-Beard neurons (Jacobson 1981a,b). Cell lineages in the early gastrula that normally do not contribute cells to the CNS can give rise to nerve cells as a result of inductive cellular interactions (Gimlich & Cooke 1983, Jacobson 1984a). Lineages of nerve cells are not cell-autonomous but can be altered by cell interactions and by changing local conditions in the embryo and even in the adult, within a certain reaction range.

3. Clonal restriction of cell activities (motility, dispersal, mingling, mitosis) occurs before phenotypic restrictions. Clonal restrictions of cell mingling that originate during midblastula stages result in development of morphological domains called *compartments,* whose boundaries cut across conventional anatomical structures in the CNS and other parts of the body. Each compartment is a polyclone populated exclusively by descendants of a group of blastomeres, starting at the 512-cell stage (Jacobson 1983). Blastomeres of a single ancestral cell group, labeled at the 512-cell stage, always contributed all or more than 95% of their descendants to one of seven compartments (Figure 3, Table 1). Blastomeres labeled at the 256-cell stage frequently contributed almost equal numbers of descendants to two compartments; this indicated that clonal restriction of mingling between descendants of different ancestral cell groups had been completed at the 512-cell stage in most cases (Jacobson 1983). When two ancestral cell groups were excised at the 512 to 1024-cell stage and combined in vitro, cell mingling occurred more freely between descendants of the same ancestral cell group than between descendants of different ancestral cell groups (Jacobson & Klein 1985).

Many types of cells, including nerve cells, differentiated in each compartment and there were no fixed relationships between ancestral cells at the time of foundation of ancestral cell groups and cellular phenotypes that later differentiated in the compartment. The compartmental fate and the phenotypic fate of any ancestral cell were independent variables. It remains to be seen whether these results can be generalized from the particular case of the frog embryo to other vertebrates.

Discovery of a *Xenopus* gene that is identical in base sequence to a region of the *Drosophila* genome located within homeotic genes and that codes for a peptide expressed in the *Xenopus* gastrula offers the possibility of studying the gene control of development of compartments in *Xenopus* and other vertebrates (Carrasco et al 1984, McGinnis et al 1984). If homeosis occurs in *Xenopus* and other vertebrates, it would have to be recognized and studied at the time at which genes were expressed that controlled the groups of ancestral cells whose descendants are compartmentally restricted. Homeotic mutations in a vertebrate embryo would be expected to transform the developmental program of one compartment into that of another at late blastula or gastrula stages when the compartmental properties are expressed as restrictions of cellular activities. These embryos would not be expected to survive to be seen as changes in patterns of cellular phenotypes. However, the changes could be studied when they occurred during early development by means of *in situ* hybridization or by means of antibodies to the homeotic gene products. Another possible way of studying such genes is to block their actions, either at the transcriptional level by means of intracellularly injected "anti-messenger" RNA or at the post-translational level by intracellular injection of antibodies to the gene products.

CELL LINEAGE ANALYSIS OF PRIMARY EMBRYONIC INDUCTION

What are the relationships betwen primary neural induction and cell phenotypes? Do cells change or lose their compartmental affiliations as a result of primary induction? These questions cannot be answered conclusively until it is known whether compartment-specific cellular properties persist, or whether, and when, they change or are lost. The time at which cells start to change in response to neural induction is also unknown. One must also bear in mind that cell commitment to a program leading to differentiation of a specific cellular phenotype can be gradual, and that the cell's phenotypic fate is not dependent on its compartmental affiliation. Therefore, primary induction may result in changes of cell phenotypes without changes in compartments and vice versa, and compartmental restrictions may persist unchanged or may have been lost before inductions have any affects.

Spemann & Mangold (1924) discovered that a second dorsal blastoporal lip grafted into the ventral marginal zone of the amphibian embryo resulted in development of duplicated embryonic structures. By using two species of *Triturus* with different cellular pigment markers as donor and host, they showed that the duplicate structures, including those of the CNS, were mosaics composed of patches of host and donor cells. Spemann & Mangold could not trace the clonal origins of the host's cells that contributed to the duplicated structures, but this can now be done by injecting HRP or another suitable tracer into single blastomeres of the host and following the fates of the labeled clones that entered the duplicated structures. These may be compared with clones that originated from the same blastomeres in normal embryos. The behavior of clones in the primary and duplicated structures should be studied with respect to their morphological restrictions into clonal domains or compartments, as well as with respect to the range of cellular phenotypes that differentiate in the clone. This distinction is vital to make because there are no fixed relationships between compartments and cellular phenotypes in the frog's CNS: until the terminal separation of different cellular phenotypes, it is expected that a range of cellular phenotypes can originate from any one of the progenitors in a compartment (Jacobson 1982, 1983).

Two separate sets of questions arise in connection with a cell lineage analysis of neural induction: one regarding the morphological compartments in which the labeled cells reside, the other regarding the cellular phenotypes of the labeled cells. These questions have at least four parts: well-designed experiments should be able to resolve them but so far they have remained largely unanswered.

1. What are the compartmental origins or affiliations of the cells before and after primary neural induction in the normal embryo? Does neural induction result in any change of cellular compartmental affiliation during normal development?
2. What happens to the compartmental affiliations of cells during a classical neural induction experiment in which the dorsal blastoporal lip (the organizer) is grafted in the ventral marginal zone or in some other ectopic site that normally does not give rise to CNS? Do cells of the donor and/or host alter their compartmental affiliations when they enter the induced CNS?
3. In a classical induction experiment, as described above, what cellular phenotypes would have developed normally from progenitors that contributed to the induced, duplicated CNS? Do cellular phenotypes alter (i.e. does prospective cellular phenotypic fate change) under influence of the organizer?
4. If answers to these questions are in the affirmative, how do changes occur in compartments or in cellular phenotypes? Are these changes the result of selection from an heterogeneous cellular population, alteration of the commitment of cells that were previously committed to different programs of differentiation, or commitment of previously uncommitted (multipotential) cells?

The answer to the first question is that the cells do not alter their compartmental affiliations during gastrulation when primary embryonic induction normally occurs: ancestral cells labeled at the 512-cell stage gave rise to clones that continued to respect lines of clonal restriction of cell mingling (i.e. stayed inside their compartment boundaries) during and after primary induction in normal embryos (Jacobson 1983). Descendants of different ancestral cell groups did not mingle in vitro, and the cells continued to behave according to their compartmental affiliation over the period in tissue culture during which primary induction would normally have occurred (Jacobson & Klein 1985). Primary neural induction is, in some unknown way, related to patterning of cellular phenotypes, particularly to neuralizing the embryonic ectoderm, but the neuralized cells do not change their compartmental affiliations during normal development. This does not rule out the possibility that the compartmental affiliations of individual cells remain labile for some time after the foundation of compartments and that those affiliations may be changed experimentally. Nor is the possibility ruled out that compartment-specific cellular properties are normally lost after large-scale cell mingling has been completed. This possibility has to be kept in mind when designing and interpreting experiments that test the effects of primary induction on compartments, especially experiments in which the inductive action of a transplanted organizer is likely to be retarded and delayed beyond the normal time of primary induction.

Recently, attempts have been made to answer the questions of whether the cells that enter the induced, duplicate CNS have altered their compartmental and cellular phenotypic programs under influence of a second organizer implanted in the ventral marginal zone (Gimlich & Cooke 1983, Smith & Slack 1984, Jacobson 1984a). When a dorsal lip from an embryo labeled totally with HRP was implanted in an unlabeled host (Figure 6, experiment 1), the resulting duplicate embryo was a mosaic composed of patches of labeled and unlabeled cells (Smith & Slack 1984, Jacobson 1984a). The structures that originated from the dorsal blastoporal lip were the notochord, ventral parts of somites, ventral parts of rhombencephalon and spinal cord, and some parts of the hindgut. These were all structures that normally develop in the posterior medial compartment (Jacobson 1982, 1983), and they retained the same compartmental affiliation in the duplicate embryos (Jacobson 1984a). The graft gave rise to the same range of cellular phenotypes in the duplicate embryo as it did during normal development (Jacobson 1984a) or when a labeled dorsal blastoporal lip was grafted to its normal position in an unlabeled host embryo (Jacobson & Xu 1985). However, such experiments cannot show whether any individual cell in the multicellular graft continued to develop according to its original fate or whether its fate was changed. Very small homotopic grafts of dorsal blastoporal lip may be confined to the posterior medial compartment (Jacobson & Xu 1985). When such small heterotopic grafts were made to the ventral marginal zone (Smith & Slack 1984, Jacobson 1984a), the compartmental affiliation of the descendants of the grafted cells had not changed in

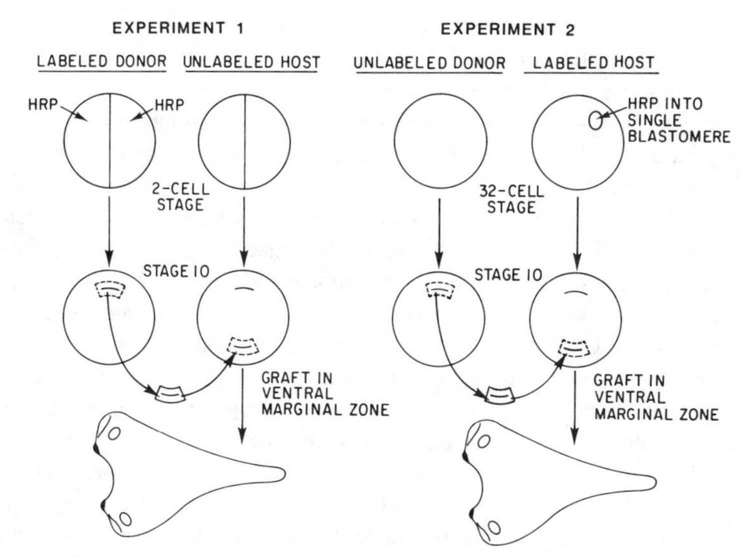

Figure 6 Cell lineage analysis of neural induction. *Experiment 1* was designed to show the cells originating from the totally labeled dorsal blastoporal lip which entered into formation of the induced structures. *Experiment 2* was designed to show cells originating from the host (descendants of blastomeres labeled at 32-cell stage) that contributed to structures induced by an unlabeled dorsal blastoporal lip. From Jacobson (1984a).

the duplicated embryo. This evidence is entirely consistent with the compartment theory (Jacobson 1980, 1982, 1983).

The compartmental origins of the host's cells that contributed to the duplicated embryo have not yet been studied in an experiment capable of resolving separate compartments. In an ideal experiment, cells belonging to a single ancestral cell group would be labeled with intracellular HRP at the 512-cell stage (or cells belonging to a single compartment would be labeled at later blastula stages), and an unlabeled dorsal blastoporal lip would be grafted into the ventral marginal zone of these labeled embryos when they reached the beginning of gastrulation. Because all the labeled cells in these host embryos would belong to a single compartment, it should be possible to see whether they had remained in the same compartment or whether they had moved into different compartments in the induced, duplicated embryo. Such experiments, now in progress in my laboratory, are designed to show whether clones originating after the foundation of ancestral cell groups retain or change their original compartmental affiliations under influence of the organizer.

Experiments designed to study the effects of the organizer on clones originating before the time of foundation of compartments cannot deal with the question of lability or stability of compartments. For example, experiments have been done (Figure 6, experiment 2) in which the embryo was labeled by intracellular injection at the 32-cell stage, followed by grafting an unlabeled dorsal blastoporal lip into the ventral marginal zone of the labeled embryos (Gimlich & Cook 1983, Jacobson 1984a). Individual blastomeres of the 32-cell *Xenopus* embryo always contributed descendants to more than one CNS compartment (Jacobson & Hirose 1981). Therefore, experiments in which the hosts were labeled at the 32-cell stage cannot resolve the problem of whether the labeled cells changed their compartmental affiliation when they entered the duplicate CNS. The claim by Gimlich & Cooke (1983) that the results of such an experiment disproved the compartment theory is based on failure to make the necessary distinction between cellular phenotypic fates and compartmental fates. Cells can alter their phenotypes without necessarily altering their compartmental affiliations: during normal development the ancestral cell groups giving rise to compartments originate before commitment of any lineages to specific cellular phenotypes. The relationships between individual ancestral cells and specific cellular phenotypes is probabilistic and, therefore, likely to change under different conditions (Jacobson 1982, 1983, Jacobson & Moody 1984). Therefore, the evidence that ancestral cells that would normally have given rise to epidermis had given rise to neurons under influence of the organizer is not in conflict with the compartment theory. Compartmentalization and neural induction are sequential steps in normal development of spatial patterns in the embryo (Jacobson 1984a). This may be another example of what Spemann (1983, pp. 92–97) called "double assurance."

TECHNICAL LIMITATIONS OF METHODS OF LINEAGE ANALYSIS

Lack of adequate techniques resulted in the failure of earlier efforts to trace cell lineages in embryos of vertebrates. Attempts were made to trace cell lineages by external application of fugitive labels to large groups of cells or by the use of stable intrinsic cellular markers to identify grafted cells. The large numbers of cells of uncertain developmental potentials in each instance precluded a rigorous analysis of cell lineages.

Vital Staining

The limitations of vital staining as a means of tracing cells from early embryonic stages through many cell generations were underestimated by proponents of that technique (Vogt 1929, Daniel & Olsen 1966, Keller 1975a,b). As a result of an excessive confidence in the vital staining technique, the fate maps obtained by using it were regarded as "fundamental and final" (Spemann 1938). Doubts about the accuracy of the method were not voiced until very recently (Løvtrup 1975, Landström & Løvtrup 1979, Jacobson 1982). The main limitations of vital staining work have recently been reviewed (Jacobson

1982) and are very briefly summarized here. The stain was applied initially to a multicellular population of uncertain extent and unknown developmental heterogeneity, so such work could not qualify as a lineage analysis. The stain faded and could not be detected in single, isolated cells, so that the amount of cell dispersal and cell mingling was seriously underestimated. Those using vital staining for tracing cell fates in *Xenopus* completely missed the migration of cells across the midline (Figure 4), which could be seen when intracellular HRP was used as a tracer (Jacobson & Hirose 1978, 1981).

Tritiated Thymidine Labeling

A recent review of the literature on use of tritiated thymidine to label neurons and to ascertain their times of origin has shown that the technique has very limited value for analysis of cell lineages (Jacobson 1978). Lineage studies with ^{3}H-thymidine injected into the whole embryo can be done, but to do so requires removal of a single cell or group of labeled cells in order to trace their fates after implantation in an unlabeled embryo (Weston 1963, Kelly 1977, Beddington 1981). Because the label is halved at mitosis and reaches undetectable levels after a few cell divisions, the method can be used only for (*a*) tracing the fates of postmitotic cells or of dividing cells that are within a few cell cycles of their final division or (*b*) tracing cells during the few divisions before dilution reduces the label to background level (Garner & McLaren 1974). For convenience it is often stated that a neuron is "born" at a certain time when its immediate progenitor incorporated ^{3}H-thymidine during the cell cycle just before production of that neuron. The technique does not show whether the progenitor completed a symmetrical and terminal division to produce two postmitotic neurons, or whether only one neuron was produced by an asymmetrical division, while the lineage continued to produce more cell generations. If only one type of nerve cell is labeled after a pulse of tritiated thymidine, or if one cell type alone remains unlabeled after cumulative labeling, it can be concluded that the progenitor of that cell type did not also give rise to other cell types at the same time. But this has little practical value because of the overlap in the times of origin of different types of nerve cells in the same region of CNS.

Cell-Specific Markers

Reliable indicators of cell phenotypes are needed to assay the range of phenotypes that differentiate within the same lineage. Cell-specific markers are also useful for showing whether cellular phenotypes change under different experimental conditions or as a result of induction. Monoclonal antibodies, antibodies raised against putative neurotransmitters or the enzymes that synthesize neurotransmitters, and antibodies raised against structural, especially cytoskeletal, components have been used as a means of identifying cell types.

One of the premises on which the use of such "cell-specific" markers is based is that there exist discrete types of nerve cells without transitional forms. This is an antiquated view of neuronal typology that is not tenable in the face of the increasing evidence that (*a*) "cell-specific" markers appear transiently during development in many types of cells that have no immediate lineage relationship and (*b*) cellular phenotypes defined by such markers can change under different local conditions during development, and even in adults (Black 1982, Jonakait et al 1984). As even more sensitive immunocytochemical techniques are used, it is likely that any one ''cell-specific" marker will be detected in an increasingly large number of cells that do not share the same lineage. This is to be expected if the control of cell differentiation is labile, subject to local conditions, and not all-or-none.

Some problems of cell lineage in the CNS may be investigated by looking at the times of appearance of neuron-specific and glial cell–specific markers during development of the nervous system. These markers could be invaluable aids for defining the phenotypes of cells in the duplicated structures that develop under the influence of a grafted organizer (Jacobson 1982). Cell-specific markers could also be used to define when neuron and glial cell lineages diverge from a common ancestral cell and when the lineages of different types of neurons and glial cells diverge from one another (reviewed in Jacobson 1978). The cellular mechanisms of divergence of lineages in the vertebrate CNS are now known. The sequence of programming events, including cell interactions and positional signalling, that ultimately result in separation of lineages of different nerve cell phenotypes remains to be investigated. One of the technical problems of studying lineages has been to identify different cellular phenotypes unambiguously from their time of origin. The use of monoclonal antibodies to identify neuron-specific antigens expressed in young neurons as well as in neuron progenitor cells has promised to be a powerful method of studying neuronal lineages. So far it has not lived up to that promise.

Ideally, cell specific markers could be used to show when different lineages diverge from a common ancestral cell. Such divergence of cell lineages could, in principle, be inferred from the observation that daughter cells each carried a different cell-specific marker or that one daughter cell did not, or that the daughter cells shared markers with their progenitor cells. Such observations are no more than correlations, and in practice such correlations are weakened by the fact that the same marker may be expressed in cells that are certainly not related by lineage. For example, tetanus toxin has frequently been used as a neuron-specific marker (Dimpfel et al 1975, 1977, Mirsky et al 1978, Abney et al 1983), but it also binds to oligodendrocyte precursor cells and to immature oligodendrocytes in culture (Abney et al 1983) and to fibrous but not to protoplasmic astrocytes (Raff et al 1983). Abney et al (1983) reported that monoclonal antibodies A4 and A2B5 bind to the majority of CNS neurons as well as to immature oligodendrocytes and their precursor cells, but not to mature oligodendrocytes. The monoclonal antibody A2B5 and tetanus toxin also bind to pancreatic islet cells (Eisenbarth et al 1982); monoclonal antibodies to human T-lymphocytes also label Purkinje neurons of many vertebrate species (Garson et al 1983). Another monoclonal antibody that labels mammalian neurons also labels *Trypanosoma cruzi* (Wood et al 1982). S-100 protein is present in very small amounts in young astrocytes but is specifically expressed in mature astrocytes of the CNS (Cicero et al 1972), enteric glial cells, and satellite cells of autonomic and sensory ganglia (Ferri et al 1982). There are no close lineage relationships between the cell types expressing S-100 protein, although they may well have shared functional relationships. These examples show that "cell-type-specific markers" (Raff et al 1979) have limited specificity, which reduces their value for establishing whether cells bearing the same marker have close lineage relationships.

Other cell specific markers are expressed too late in differentiation to be of use as lineage markers. Neurofilament protein immunoreactivity, which is a specific neuronal phenotype, appears during the terminal cell cycle of neuroepithelial cells in the chick neural tube (Tapscott et al 1981). In the mouse neural tube, tetanus toxin binding is not found on dividing cells but appears on neurons within 7 $\pm$ 1 hr following their terminal S-phase (Koulakoff et al 1983). Other neuron specific molecules also appear in the central nervous system only after cessation of DNA synthesis and withdrawal from the mitotic cycle: monoamines and their biosynthetic enzymes (Olson & Sieger 1972, Lauder & Bloom 1974); neuron specific enolase (Schmechel et al 1980, Maxwell et al 1982), and cholera-toxin binding (Willinger & Schachner 1980, Rathjen & Gierer 1981). This relationship between terminal cell cycle and acquisition of phenotypic markers is not found in neural crest derivatives in the peripheral nervous system: tetanus toxin binding sites appear in dividing cells in the spinal sensory ganglia (Koulakoff et al 1983), and monoamines appear in the central nervous system cells that are actively dividing (Black 1982, review). Catecholamine biosynthetic enzymes are present in proliferating neurons of the peripheral nervous system but not in the central nervous system (Rothman et al 1980).

Expression by several cell types of the same cell surface molecules or enzymes may indicate that they share functional properties rather than share the same ancestral cells. Nevertheless, conclusions about lineage have been drawn from such correlations. Several studies have indicated that different types of glial cells may originate directly from the same progenitor. Raff et al (1983) concluded that fibrous astrocytes and oligodendrocytes develop from the same ancestral cell because both cell types reacted with tetanus toxin and monoclonal antibody A2B5. The sequential appearance of glial fibrillary acidic protein (GFAP) immunoreactivity first in radial glial cells, then in immature oligodendrocytes, and finally in astrocytes (after its disappearance from oligodendrocytes) led to the conclusion that oligodendrocytes and astrocytes both originate directly from radial glial cells in the human fetal spinal cord (Choi & Kim 1984).

By contrast with such evidence of a common origin of different types of glial cells, other evidence indicates that neurons and glial cells originate from different progenitors. The telencephalic ventricular zone in the 80–123 day monkey fetus is composed of two classes of proliferating cells—radial neuroglial cells, which showed GFAP immunoreactivity, and GFAP-negative cells—from which it was inferred that the former gave rise to astrocytes and the latter to neurons (Levitt et al 1981, 1983). This is in agreement with the hypothesis first proposed by His (1889) that the ventricular germinal zone is composed of two different populations of progenitors for neurons and glial cells, respectively. The evidence that the proliferative cells in the ventricular zone are heterogeneous as regards GFAP immunoreactivity suggested that bifurcation of glial and neuronal cell lineages had occurred at a stage earlier than 80–123 days of gestation in the Rhesus monkey telencephalon. The study did not show when those lineages first separated. The possibility cannot be ruled out that neuronal lineages diverge first from a homogeneous proliferative population ("indifferent cells" of Schaper 1897a, b), after which bifurcation occurs into separate glial and neuronal lineages. The transitional period, during which neuronal and glial cell lineages coexist in the same region, may be related to the mean number of glial cells per neuron (glia-to-neuron ratio). The transitional period may be prolonged in species with a high glia-to-neuron ratio and in the thick-walled regions of the neural tube, where radial glial cells are required to assist migration of neurons out of the ventricular germinal zone. A short transitional period may be required in species in which the glia-to-neuron ratio is low and in the thin-walled regions of the neural tube. In the spinal cord, in which neurons are born early, there appears to be little overlap between the period of neuron production and the later period of glial cell production (Fujita 1963). In amphibians, in which a considerable fraction of spinal cord neurons originate during gastrulation before development of a neural plate or neural tube, it would be of great interest to see when GFAP-positive cells can first be detected and when those lineages separate from neuronal lineages.

Grafting Labeled Cells

Experiments in which cells labeled with a stable marker are grafted into an unlabeled host have a number of technical limitations. If the cell marker is stable and can be identified with certainty in individual cells, this technique offers the advantage of permitting identification of the descendants of host and donor cells in the chimera (Harrison 1903, Spemann & Mangold 1924, Le Douarin 1982), but the relationships between individual progenitors and their descendants cannot be traced in most cases.

It is obvious that an analysis of cell lineage cannot be made by grafting many cells of unknown heterogeneity that give rise to several different types of descendants. Even in cases in which only a single cellular phenotype originates from the graft, the uncertainty remains of whether that phenotype was the only survivor from an initially larger number of lineages. In the usual case, in which many cellular phenotypes descend from the grafted cells, the only rigorous conclusion that can be made is that some grafted cells have survived and differentiated. Any conclusions about their modes of lineage must remain speculative.

Some of the limitations can be avoided by transplanting a single labeled cell to an unlabeled host. However, interspecific or intergeneric cell transplants are limited by the tolerance of the host for the grafted cells and by other functional incompatibilities such as differences in rates of development, e.g. in chick-quail transplants. The poor viability of single cell transplants between different species suggests that considerable death of cells, and possibly clonal selection, may occur when a multicellular interspecific or intergeneric graft is made (Boucaut 1973, Boucaut & Gallien 1973). The surviving differentiated descendants of the graft may not include all the phenotypes in the chimera that would have descended from the same progenitors in the donor.

Experiments that start by grafting a multicellular population of labeled cells of unknown developmental heterogeneity and end with a mixture of several types of labeled descendants can, at best, provide material for a retrospective clonal analysis. The minimum data required for such a retrospective clonal analysis are the numbers of descendants of each type and the sizes of coherent patches of each type of descendant (West 1978). All reports of the results of grafting quail cells to chick embryos have been qualitative and do not provide the data required to do a retrospective clonal analysis. The modes of lineages of the different cellular phenotypes cannot be deduced from such qualitative studies. At best they may show the amount of cell mingling and whether the expression of the phenotypes of the grafted cells is cell-autonomous or subject to developmental control by local conditions.

Numerous difficulties arise when attempting to do a retrospective clonal analysis by using the clone sizes and distribution of the two types of cells in chimeras to estimate the numbers of ancestral cells from which the observed clones originated. These problems arise when chimeras are formed by grafting cells of one genotype into an embryo of a different genotype at a relatively late stage of development, as in making chick-quail chimeras, or when making chimeras by fusing two embryos of different genotype at an early stage of development:

1. the initial conditions of mixing of the two types of ancestral cells cannot be precisely controlled;
2. the two types of ancestral cells may not be equivalent so that selective aggregation or selective restriction of cell mixing may occur;
3. the amount of cell mixing may change during development ;
4. cell death may occur;
5. different proportions of the two types of ancestral cells may be allocated to different tissues of the same chimera.

Retrospective Clonal Analysis from Chimeras and Mosaics

In retrospective methods the position and identity of the labeled progenitors are uncertain, but they are deduced from the positions, numbers, and types of labeled descendants. For example, a small number of genetically labeled cells may be injected into the blastocele of a host embryo (Gardner 1978). These cells colonize a wide range of tissues in the host embryo, where they may be identified at later stages of development. Such injection mosaics have been made in an amphibian, *Pleurodeles* (Boucaut 1973), and in the mouse (Rossant et al 1978, Gardner & Rossant 1979, Gardner 1982). Intergeneric injection mosaics have been made between the amphibians *Pleurodeles* and *Ambystoma* (Boucaut 1973, Boucaut & Gallien 1973). Chimeras have been made between rat and mouse (Zeilmaker 1973, Gardner & Johnson 1973, 1975, Tachi & Tachi 1980). Such experiments can show the range of cellular phenotypes that differentiate in the clone that originates from the implanted cells, but they do not prove that the same cellular phenotypes originate from those ancestral cells during normal development. In principle a single injected cell may be placed at any desired position in the recipient embryo, but this is difficult to do. In practice the position of the implanted cell cannot be controlled precisely. Even greater uncertainty persists about the positions of genetically labeled cells when chimeras are made by aggregation of two embryos, usually at the eight-cell stage in the mouse (McLaren 1976). Two eight-cell mouse embryos, with different cell markers, can be fused after removal of the zona pellucida, cultured for 12 to 24 hours and then implanted in the uterus of a pseudopregnant mouse. Cellular markers, usually isozymes characteristic of each of the embryos that were originally combined, can be used to identify their descendants in the chimeric mouse.

Rather limited conclusions can be drawn from counts of various cellular phenotypes in such aggregation chimeras (McLaren 1972, 1976). Nevertheless, aggregation chimeras offer the possibilities of seeing whether the expression of a mutant phenotype is cell-autonomous, measuring the amount of cell mixing during development, and estimating the mean number of clones at the time of foundation of a tissue. These numbers are surprisingly small, ranging from 2 to 20 for many different tissues (Mintz 1970, Mintz & Sanyal 1970,

Sanyal & Zeilmaker 1974, 1976, 1977, Wetts & Herrup 1982). West (1978) has reviewed the ways in which analysis of chimeras can be done. He has shown that if extensive mixing of cells occurs, it is "impossible to determine the number of descendant clones from the distribution of two types in the adult tissue." Such extensive cell mixing has been found in the cerebellar Purkinje cells of the mouse (Dewey et al 1976, Mullen 1977). Mullen estimated that the coherent clones in the Purkinje cell layer average only 1.03 Purkinje cells. This number would preclude any calculation of the size of the Purkinje cell progenitor cell pool or of the number of clones from which the total population of Purkinje cells originates at the time at which they become "committed." An additional weakness of the methods that have been used to estimate the number of ancestral cells or of progenitor cell groups at the time of foundation of a tissue is that they depend on the unlikely assumptions that the two types of ancestral cells are randomly arranged at the time of tissue foundation and that there is no differential cell selection or cell death (McLaren 1976, West 1978).

REQUIREMENTS OF AN IDEAL LINEAGE ANALYSIS

The inherent difficulties in an analysis of cell lineages in the vertebrate embryo are variability of cleavage patterns, regulation after removal of ancestral cells, production of many cell generations before cell phenotypes are expressed, large numbers of cells, and mingling of cells. None of the methods for tracing cell lineages has been able to overcome these difficulties completely. To see how far these have fallen short of the ideal, it may be useful to define the requirements of an ideal analysis of cell lineages in the vertebrate embryo:

1. The lineage should be traced from a single identified ancestral cell, and the clonal analysis should proceed prospectively. All descendants should be identified at successive cell generations until the terminal differentiation of all descendants has occurred. This has become feasible in the vertebrate embryo only recently (Jacobson & Hirose 1978, 1981, Hirose & Jacobson 1979, Jacobson 1983).
2. A tracer should be used that does not alter normal development, is not lost from the initially labeled ancestral cell, is transferred exclusively to all its descendants, and can be identified in them after they have differentiated. Horseradish peroxidase (HRP) injected intracellularly has been found to satisfy all requirements of an ideal lineage tracer in embryos of mice (Balakier & Pedersen 1982), frogs (Jacobson & Hirose 1978, 1981, Hirose & Jacobson 1979, Jacobson 1982, 1984a,b), ascidian (Nishida & Satoh 1983), starfish (Kominami 1983), and Leech (Weisblat et al 1978, 1980a).
3. The number of cell generations from the beginning to the end of the lineage

should be counted and all cells entering or leaving the lineage should be accounted for (Jacobson & Moody 1984, Jacobson 1984b).

4. All cellular phenotypes of the descendants should be identified by means of sensitive, cell-specific markers. For an analysis of the CNS, the descendants must be identified unambiguously as neurons or glial cells.
5. Experiments should be done to show the extent to which the lineage is cell autonomous or is modified by external conditions, cell interactions, and inductions.

To this date no lineage analysis of the vertebrate embryo has satisfied all of these requirements. Hitherto, the criteria for an ideal lineage analysis have not been clearly defined, and there may not yet be a consensus on the requirements of an ideal lineage analysis in the vertebrate embryo.

Literature Cited

Abney, E. R., Williams, B. P., Raff, M. C. 1983. Tracing the development of oligodendrocytes from precursor cells using monoclonal antibodies, fluorescence-activated cell sorting, and cell culture. *Dev. Biol.* 100:166–71

Balakier, H., Pedersen, R. A. 1982. Allocation of cells to inner cell mass and trophectoderm lineages in preimplantation mouse embryos. *Dev. Biol.* 90:352–62

Beddington, R. S. P. 1981. An autoradiographic analysis of the potency of embryonic ectoderm in the 8th day postimplantation mouse embryo. *J. Embryol. Exp. Morphol.* 64:87–104

Black, I. B. 1982. Stages of neurotransmitter development in autonomic neurons. *Science* 215:1198–1204

Boterenbrood, E. C. 1970. Differentiation in small grafts of the median region of the presumptive prosencephalon. *J. Embryol. Exp. Morphol.* 23:751–59

Boucaut, J.-C. 1973. Autoradiographic analysis of *Pleurodeles waltlii* embryos injected with labeled embryo donor cells. *Differentiation* 1:413–18

Boucaut, J.-C., Gallien, L. 1973. Chimeres allopheniques intergeneriques entre Pleurodeles waltlii Michah et *Ambystoma mexicanum* Shaw (Amphibiens Urodeles). Mise en evidence du chimerisme tegmentaire. *C R Acad. Sci. Paris Ser. D* 276:1757–61

Carrasco, A. E., McGinnis, W., Gehring, W. J., De Robertis, E. M. 1984. Cloning of an *X. laevis* gene expressed during early embryogenesis coding for a peptide region homologous to *Drosophila* homeotic genes. *Cell* 37:409–14

Choi, B., Kim, R. C. 1984. Expression of glial acidic protein in immature oligodendroglia. *Science* 223:407–9

Cicero, T. J., Ferrendelli, J. A., Suntzeff, V., Moore, B. W. 1972. Regional changes in CNS levels of the S-100 and 14-3-2 proteins during development and aging of the mouse. *J. Neurochem.* 19:2119–25

Conklin, E. G. 1897. The embryology of *Crepidula*. *J. Morphol.* 13:1–226

Corner, M. A. 1963. Development of the brain of *Xenopus laevis* after removal of parts of the neural plate. *J. Exp. Zool.* 153:301–11

Daniel, J. C., Olsen, J. D. 1966. Cell movement, proliferation and death in the formation of the embryonic axis of the rabbit. *Anat. Rec.* 156:123–28

Davidson, E. 1976. *Gene Activity in Early Development*. *New York:* Academic 452 pp.

De Beer, G. R. 1951. *Embryos and Ancestors*. Oxford: Clarendon. Revised ed. 159 pp.

Detwiler, S. R. 1944. Restitution of the medulla following unilateral excision in the embryo.*J. Exp. Zool.* 96:129–42

Dewey, M. J., Gervais, A. G., Mintz, B. 1976. Brain and ganglion development from two genotypic classes of cells in allophenic mice. *Dev. Biol.* 50:68–81

Dimpfel, W., Neale, J. H., Habermann, E. 1975. ^{125}I-labeled tetanus toxin as a neuronal marker on tissue cultures derived from embryonic CNS. *Naunyn-Schmiedeberg's Arch. Pathol. Pharmak.* 290:329–33

Dimpfel, W., Huang, R. T. C., Habermann, E. J. 1977. Gangliosides in nervous tissue cultures and binding of ^{125}I-labeled tetanus toxin: A neuronal marker. *J. Neurochem.* 29:329–34

Eisenbarth, G. S., Shimizu, K., Bowring, M. A., Wells, S. 1982. Expression of receptors for tetanus toxin and monoclonal antibody A2B5 by pancreatic islet cells. *Proc. Nat. Acad. Sci. USA* 79:5066–70

Fankhauser, G. 1932. Cytoplasmic localization

in the unsegmented egg of the newt *Triturus viridescens* as shown by the development of egg fragments. *Anat. Rec.* 54:73

Ferri, G.-L., Probert, L., Cocchia, D., Michetti, F., Marangos, P. J., Polak, J. M. 1982. Evidence for the presence of S-100 protein in the glial component of the human enteric nervous system. *Nature* 297:409–10

Fujita, S. 1963. The matrix cell and cytogenesis in the developing central nervous system. *J. Comp. Neurol.* 120:37–42

Gallien, C.-L. 1979. Expression of nuclear and cytoplasmic factors in ontogenesis of amphibian nucleocytoplasmic hybrids. *Int. Rev. Cytol. Suppl.* 9:189–219

Gardner, R. L. 1978. Production of chimaeras by injecting cells or tissue into the blastocyst. In *Methods In Mammalian Reproduction,* ed. C. J. Daniel, pp. 137–65. New York: Academic

Gardner, R. L. 1982. Investigation of cell lineage and differentiation in the extraembryonic endoderm of the mouse embryo. *J. Embryol. Exp. Morph.* 68:175–98

Gardner, R. L., Johnson, M. H. 1973. Investigation of early mammalian development using interspecific chimaeras between rat and mouse. *Nature New Biol.* 246:86–89

Gardner, R. L., Johnson, M. H. 1975. Investigation of cellular interaction and development in the early mammalian embryo using interspecific chimaeras between the rat and mouse. In *Cell Patterning. Ciba Found. Symp.* 29:183–200. Amsterdam: Elsevier

Gardner, R. L., Rossant, J. 1979. Investigation of the fate of 4.5 day post coitum mouse inner cell mass cells by blastocyst injection. *J. Embryol. Exp. Morphol.* 52:141–52

Garner, W., McLaren, A. 1974. Cell distribution in chimaeric mouse embryos before implantation. *J. Embryol. Exp. Morphol.* 22:495–503

Garson, J. A., Beverly, P. C. L., Coakham. H. B., Harper, E. I. 1983. Monoclonal antibodies against human T lymphocytes label Purkinje neurones of many species. *Nature* 298:375-77

Gerhart, J. C. 1980. Mechanisms regulating pattern formation in the amphibian egg and early embryo. In *Biological Regulation and Development,* ed. R. E. Goldberger, 2:133–316. New York: Plenum

Gimlich, R. L., Cooke, J. 1983. Cell lineage and the induction of second nervous systems in amphibian development. *Nature* 306: 471–73

Gimlich, R. L., Gerhart, J. C. 1984. Early cellular interactions promote embryonic axis formation in *Xenopus laevis. Dev. Biol.* 104:117–30

Goodman, C. 1977. Neuron duplications and deletions in locust clones and clutches. *Science,* 197:1384–86

Gould, S. J. 1977. *Ontogeny and Phylogeny.* Cambridge, Mass: Belknap-Harvard Univ. Press. 501 pp.

Gurdon, J. B. 1966. The cytoplasmic control of gene activity. *Endeavour* 25:95–99

Gurdon, J. B., Woodland, H. R. 1969. The influence of the cytoplasm on the nucleus during cell differentiation, with special reference to RNA synthesis during amphibian cleavage. *Proc. R. Soc. London. Ser. B* 173:99–111

Harrison, R. G. 1903. Experimentelle Untersuchungen über die Entwicklung der Sinnesorgane der Seitenlinie bei den Amphibien. *Arch. Mikr. Anat.* 63:35–145

Harrison, R. G. 1933. Some difficulties of the determination problem. *Am. Nat.* 67:306–21

Hirose, G., Jacobson, M. 1979. Clonal organization of the central nervous system of the frog. I. Clones stemming from individual blastomeres of the 16-cell and earlier stages. *Dev. Biol.* 71:191–202

His, W. 1889. Die Neuroblasten und deren Entstehung im embryonalen Mark. *Abh. Kgl. Sach. Ges. Wiss. Math. Phys. Kl.* 15: 313–72

Holtzer, H. 1951. Reconstitution of the urodele spinal cord following unilateral ablation. Part I. Chronology of neuron regulation. *J. Exp. Zool.* 117:523–58

Jacobson, M. 1978. *Developmental Neurobiology.* New York: Plenum 562 pp.

Jacobson, M. 1980. Clones and compartments in the vertebrate central nervous system. *Trends Neurosci.* 1:3–5

Jacobson, M. 1981a. Rohon-Beard neuron origin from blastomeres of the 16-cell frog embryo. *J. Neurosci.* 1:918–22

Jacobson, M. 1981b. Rohon-Beard neurons arise from a substitute ancestral cell after removal of the cell from which they normally arise in the 16-cell frog embryo. *J. Neurosci.* 1:923–27

Jacobson, M. 1982. Origins of the nervous system in amphibians, pp. 45–99. In *Neuronal Development,* ed. N. C. Spitzer, pp. 45–99. New York: Plenum

Jacobson, M. 1983. Clonal organization of the central nervous system of the frog. III. Clones stemming from individual blastomeres of the 128-, 256-, and 512-cell stages. *J. Neurosci.* 3:1019–38

Jacobson, M. 1984a. Cell lineage analysis of neural induction: Origins of cells forming the induced nervous system. *Dev. Biol.* 102:122–29

Jacobson, M. 1984b. Quantitative lineage analysis of the frog's nervous system. II. Numbers of cells in clones contributing to the nervous system, counted at blastula and gastrula stages. In preparation

Jacobson, M., Hirose, G. 1978. Origin of the

retina from both sides of the embryonic brain: A contribution to the problem of crossing at the optic chiasma. *Science* 202: 637–739

Jacobson, M., Hirose, G. 1981. Clonal organization of the central nervous system of the frog. II. Clones stemming from individual blastomeres of the 32- and 64-cell stages. *J. Neurosci.* 1:271–84

Jacobson, M., Huang, S. 1985. Neurite outgrowth traced by means of horseradish peroxidase inherited from neuronal ancestral cells in frog embryos. *Dev. Biol.*Submitted

Jacobson, M., Klein, S. 1985. Cell mingling depends on cell ancestry in *Xenopus* embryo cell explants. In preparation.

Jacobson, M., Moody, S. A. 1984. Quantitative lineage analysis of the frog's nervous system. I. Lineages of Rohon-Beard neurons and primary motoneurons. *J. Neurosci.* 4:1361–69

Jacobson, M., Xu, W. L. 1985. In preparation

Jeffery, W. R. 1983. Messenger RNA localization and cytoskeletal domains in ascidian embryos. In *Time, Space, and Pattern in Embryonic Development,* pp. 241–59. New York: Alan Liss

Jeffery, W. R., Tomlinson, C. R., Brodeur, R. D. 1983. Localization of actin messenger RNA during early ascidian development. *Dev. Biol.* 99:408–17

Jonakait, G. M., Markey, K. A., Goldstein, M., Black, I. B. 1984. Transient expression of selected catecholaminergic traits in cranial sensory and dorsal root ganglia of embryonic rat. *Dev. Biol.* 101:51–60

Keller, R. E. 1975a. Vital dye mapping of the gastrula and neurula of *Xenopus laevis*. I. Prospective areas and morphogenetic movements of the superficial layer. *Dev. Biol.* 42:222–41

Keller, R. E. 1975b. Vital dye mapping of the gastrula and neurula of *Xenopus laevis*. II. Prospective areas and morphogenetic movements of the deep layer. *Dev. Biol.* 51: 118–37

Kelly, S. J. 1977. Studies of the developmental potential of 4- and 8-cell mouse blastomeres. *J. Exp. Zool.* 200:365–76

Klein, S., Jacobson, M. 1985. In preparation

Kominami, T. 1983. Establishment of embryonic axes in larvae of the starfish, *Asterina pectinifera*. *J. Embryol. Exp. Morphol.* 75:87–100

Koulakoff, A., Bizzini, B., Berwald-Netter, Y. 1983. Neuronal acquisition of tetanus toxin binding sites: Relationship with the last mitotic cycle. *Dev. Biol.* 100:350–57

Lamborghini, J. E. 1980. Rohon-Beard cells and other large neurons in *Xenopus* embryos originate during gastrulation. *J. Comp. Neurol.* 189:323–33

Landström, U., Løvtrup, S. 1979. Fate maps and cell differentiation in the amphibian embryo—An experimental study. *J. Embryol. Exp. Morphol.* 54:113–30

Lauder, J. M., Bloom, F. E. 1974. Ontogeny of monoamine neurons in the locus coeruleus, raphe nuclei and substantia nigra of the cat. I. Cell differentiation. *J. Comp. Neurol.* 155: 469–82

LeDouarin, N. 1982. *The Neural Crest*. London/New York: Cambridge Univ. Press. 259 pp.

Levitt, P., Cooper, M. L., Rakic, P. 1981. Coexistence of neuronal and glial precursor cells in the fetal monkey: An ultrastructural immunoperoxidase analysis. *J. Neurosci.* 1:27–39

Levitt, P., Cooper, M. L., Rakic, P. 1983. Early divergence and changing proportions of neuronal and glial precursor cells in the primate cerebral ventricular zone. *Dev. Biol.* 96:472–84

Lewis, W. H. 1910. Localization and regeneration in the neural plate of amphibian embryos. *Anat. Rec.* 4:191

Løvtrup, S. 1975. Fate maps and gastrulation in amphibia—A critique of current views. *Can. J. Zool.* 53:473–79

Maxwell, G. D., Whitehead, M. C., Connolly, S. M., Marangos, P. J. 1982. Development of neuron-specific enolase immunoreactivity in avian nervous tissue and *in vitro*. *Dev. Brain Res.* 3:401–18

McGinnis, W., Garber, R. L., Wirz, J., Kuroiwa, A., Gehring, W. J. 1984. A homologous protein-encoding sequence in *Drosophila* homeotic genes and its conservation in other metazoans. *Cell* 37:403–8

McLaren, A. 1972. Numerology of development. *Nature* 239:274–76

McLaren, A. 1976. *Mammalian Chimaeras*. London: Cambridge Univ. Press

Mikoshiba, K., Yokoyama, M., Inoue, Y., Takamatsu, K., Tsukada, Y., Nomura, T. 1982. Oligodendrocyte abnormalities in *shiverer* mouse mutant are determined in primary chimaeras. *Nature* 299:357–59

Mintz, B. 1970. Gene expression in allophenic mice. In *Control Mechanisms in the Expression of Cellular Phenotypes,* ed. H. A. Padykula, pp. 15–42. New York: Academic

Mintz, B., Sanyal, S. 1970. Clonal origin of the mouse visual retina mapped from genetically mosaic eyes. *Genetics* 64 (Suppl.): 43–44

Mirsky, R., Wendon, L. M. B., Black, P., Stolkin, C., Bray, D. 1978. Tetanus toxin: A cell surface marker for neurons in culture. *Brain Res.* 148:251–59

Moody, S. A., Jacobson, M. 1983. Compartmental relationships between anuran primary spinal motoneurons and somitic muscle fibers that they first innervate. *J. Neurosci.* 3:1670–82

Mullen, R. J. 1977. Site of pcd gene action and Purkinje cell mosaicism in the cerebella of chimaeric mice. *Nature* 270:245–47

Nakauchi, M., Takashita, T. 1983. Ascidian one-half embryos can develop into functional adult ascidians. *J. Exp. Zool.* 227: 155–58

Nieuwkoop, P. D., Faber, J. 1967. *Normal Table of Xenopus laevis (Daudin)*. Amsterdam: North-Holland

Nishida, H., Satoh, N. 1983. Cell lineage analysis in ascidian embryos by intracellular injection of a tracer enzyme. I. Up to the eight-cell stage. *Dev. Biol.* 99:382–94

Olson, L., Sieger, A. 1972. Early prenatal ontogeny of central monoamine neurons in the rat: Fluorescence histochemical observation. *Z. Anat. Entw. Gesch.* 137:301–16

Oster-Granite, M. L., Gearhart, J. 1981. Cell lineage analysis of cerebellar Purkinje cells in mouse chimeras. *Dev. Biol.* 85:199–208

Peterson, A., Marler, J. 1983. Pl Deficiency in shiverer myelin is expressed by Schwann cells in shiverer-normal mouse chimaera nerves. *Neurosci. Lett.* 38:163–68

Raff, M. C., Fields, K. L., Hakomori, S., Mirsky, R., Pruss, R. M., Winter, J. 1979. Cell-type-specific markers for distinguishing and studying neurons and the major classes of glial cells in culture. *Brian Res.* 174: 283–308

Raff, M. C., Miller, R. H., Noble, M. 1983. A glial progenitor cell that develops *in vitro* into an astrocyte or an oligodendrocyte depending on cultural medium. *Nature* 303:390–96

Rathjen, F. G., Gierer, R. 1981. Cholera-toxin binding to cells of developing chick retina analyzed by fluorescence activated cell sorting. *Dev. Brain Res.* 1:539–49

Rossant, J., Gardner, R. L., Alexandre, H. L. 1978. Investigation of the potency of cells from the postimplantation of mouse embryo by blastocyst injection: A preliminary report. *J. Embryol. Exp. Morphol.* 48:239–47

Rothman, T. P., Specht, L. A., Gershon, M. D., Joh, T. H., Teitelman, G., Pickel, V. M., Reis, D. J. 1980. Catecholamine biosynthetic enzymes are expressed in replicating cells of the peripheral but not the central nervous system. *Proc. Nat. Acad. Sci. USA* 77:6221–25

Sanyal, S., Zeilmaker, G. H. 1974. Gene action and cell lineage in retinal development in experimental chimaeric mice. *Teratology* 10:322

Sanyal, S., Zeilmaker, G. H. 1976. Comparative analysis of cell distribution in pigment epithelium and the visual cell layer of chimaeric mice. *J. Embryol. Exp. Morphol.* 36:425–30

Sanyal, S., Zeilmaker, G. H. 1977. Cell lineage in retinal development of mice studied in experimental chimaeras. *Nature* 265:731–33

Schaper, A. 1897a. Die frühesten Differenzierungsvorgänge im Centralnervensystem. *Arch. Entw. Mech. Organ.* 5:81–132

Schaper, A. 1897b. The earliest differentiation in the central nervous system of vertebrates. *Science* 5:430–31

Schmechel, D. E., Brightman, M. W., Marangos, P. J. 1980. Neurons switch from non-neuronal enolase to neuron-specific enolase during differentiation. *Brain Res.* 190:195–214

Smith, J. C., Slack, J. M. W. 1984. Dorsalization and neural induction: Properties of the organizer in *Xenopus laevis*. *J. Embryol. Exp. Morphol* 78:299–317.

Spemann, H. 1912. Über die Entwicklung umgedrehter Hirnteile bei Amphibienembroynen. *Zool. Jahrb. Supp.* 15:1–48.

Spemann, H. 1938. *Embryonic Development and Induction*. New Haven: Yale Univ. Press. 401 pp.

Spemann, H., Mangold, H. 1924. Über Induktion von Embryonanlage durch Implantation artfremden Organisatoren. *Wilhelm Roux Arch Entwicklungsmech. Org.* 123:389–517

Sternberg, P. W., Horvitz, H. R. 1981. Gonadal cell lineages of the nematode *Panagrellys redivivus* and implications for evolution by modification of cell lineage. *Dev. Biol.* 88:147–66

Sulston, J. E., Horvitz, H. 1981. Abnormal cell lineages in mutants of the nematode *Caenorhabditis elegans*. *Dev. Biol.* 82: 41–55

Tachi, S., Tachi, C. 1980. Electron microscopic studies of chimeric blastocysts experimentally produced by aggregating blastomeres of rat and mouse embryos. *Dev. Biol.* 80:18–27

Tapscott, S. J., Bennett G. S. Holtzer, H. 1981. Neuronal precursor cells in the chick neural tube express neurofilament proteins. *Nature* 292:836–38

Tung, T. C., Wu, S. C., Tung, Y. F. Y. 1958. The development of isolated blastomeres of Amphioxus. *Sci. Sinica* 7: 1280–1320

Tung, T. C., Wu, S. C., Tung, Y. Y. F. 1960. The development potencies of the blastomere layers in *Amphioxus* egg at the 32-cell stage. *Sci. Sinica* 9:119–41

Vargas-Lizardi, P., Lyser, K. M. 1979. Time of origin of Mauthner's neuron in *Xenopus laevis* embryos. *Dev. Biol.* 38: 220–28

Vogt, W. 1929. Gestaltungsanalyse am Amphibienkeim mit ortlicher Vitalfärbung. II Teil. Gastrulation und Mesodermbildung bei Urodelen und Anuren. *Wilhelm Roux Arch. Entwicklungsmech. Org.* 120:384–706

Weisblat, D. A., Sawyer, R. T., Stent, G. S. 1978. Cell lineage analysis by intracellular injection of a tracer. *Science* 202:1295–98

Weisblat, D. A., Harper, G., Stent, G. S., Sawyer, R. T. 1980a. Embryonic cell lineage in the nervous system of the glossiphoniid leech *Helobdella triserialis*. *Dev. Biol.* 76: 58–78

Weisblat, D. A., Zackson, S. L., Blair, S. S., Young, J. D. 1980b. Cell lineage analysis by intracellular injection of fluorescent tracers. *Science* 209:1538–41

Weiss, P. 1939. *Principles of Development*. New York: Henry Holt. 601 pp.

West, J. D. 1978. Analysis of clonal growth using chimeras and mosaics, In *Development in Mammals,* ed. M. H. Johnson, 3:413–460. Amsterdam: North Holland

Weston, J. A. 1963. A radioautographic analysis of the migration and localization of trunk neural crest cells in the chick. *Dev. Biol.* 6:279–310.

Wetts, R., Herrup, K. 1982. Cerebellar Purkinje cells are descended from a small number of progenitors committed during early development: Quantitative analysis of lurcher chimeric mice. *J. Neurosci.* 2:1494–98

Whittaker, J. R. 1973. Segregation during ascidian embryogenesis of egg cytoplasm information for tissue-specific enzyme development. *Proc. Nat. Acad. Sci. USA* 70:2096–2100

Whittaker, J. R. 1979. Cytoplasmic determinants of tissue differentiation in the ascidian egg. In *Determinants of Spatial Organization,* ed. S. Subtelny, I. R. Konigsberg, pp. 29–51. New York: Academic

Willinger, M., Schachner, M. 1980. GM_1 ganglioside as a marker for neuronal diffrentiation in mouse cerebellum. *Dev. Biol.* 74: 101–17

Wood, J. N., Hudson, C., Jessell, T. M., Yamamoto, M. 1982. A monoclonal antibody defining antigenic determinants on subpopulations of mammalian neurones and Trypanosoma cruzi parasites. *Nature* 296:34–38

Zeilmaker, G. 1973. Fusion of rat and mouse morulae and formation of chimaeric blastocysts. *Nature* 242:115–16

Ann. Rev. Neurosci. 1985. 8:103–24

ADENOSINE AS A NEUROMODULATOR

Solomon H. Snyder

Departments of Neuroscience, Pharmacology and Experimental Therapeutics, Psychiatry, and Behavioral Sciences, Johns Hopkins University School of Medicine, Baltimore, Maryland 21205

The criteria that neuroscientists have generally employed in evaluating a substance as a potential neurotransmitter have been based largely on the properties of acetylcholine at the neuromuscular junction. The substance should be highly localized at a specific synapse, where it should depolarize or hyperpolarize postsynaptic cells by alterations in ion permeability. Specific and selective mechanisms for the biosynthesis and inactivation of the transmitter should be present. Finally, effects of nerve stimulation should be mimicked faithfully by the candidate transmitter. The recent appreciation of amino acids and peptides as neurotransmitters in the central nervous system has altered considerably the utilization of these criteria. Amino acid transmitters often serve many general metabolic functions so that selective localization to particular synapses and specific mechanisms of biosynthesis and inactivation are not readily demonstrated. Efforts to identify neurotransmitter "pools" of amino acids by high affinity uptake systems are only partially effective, since glia as well as neurons accumulate some of the amino acid transmitters by high affinity systems. The best evidence that many neuropeptides are transmitters derives from their localization by immunohistochemical techniques to well-defined neuronal systems. Demonstrating that a particular peptide mimics effects of nerve stimulation in the brain is notoriously difficult, since many of the peptides seem to act by altering the release of other conventional transmitter molecules or by changing the sensitivity of receptors for conventional transmitters.

As a putative neurotransmitter, adenosine exemplifies all of these difficulties. It functions in multiple metabolic pools and so occurs throughout the brain. Histochemical studies have not yet been available to identify putative adenosine-containing neurons. The neurophysiologic actions of adenosine are largely inhibitory and seem primarily to involve inhibition of excitatory trans-

0147-006X/85/0301-0103$02.00

mitter release (Phillis & Wu 1981). Accordingly, it is difficult to equate effects of applied adenosine with those of activation of particular synapses. In ascertaining the role of endogenous adenosine in brain function and behavior, the researcher does possess one tool that is lacking for many amino acid and peptide transmitters, namely the availability of receptor antagonists. The xanthine drugs caffeine and theophylline are reasonably potent and fairly selective adenosine antagonists and have been widely employed in animal and human investigations.

Presence and Distribution

Chemically adenosine consists of the purine base adenine linked to ribose, hence it is designated a nucleoside. It is one of the most ubiquitous metabolic intermediates in the body. Adenosine functions in nucleic acid biosynthesis. It is also prominent in the formation of ATP and thence cyclic AMP. Less appreciated but perhaps equally important is its role in the disposition of the major methyl donor, S-adenosylmethionine. S-adenosylhomocysteine, the product of transmethylation reactions, is converted to adenosine by S-adenosylhomocysteine hydrolase.

Adenosine is not the only purine implicated as a neurotransmitter. Extensive and elegant studies by Burnstock (1981) over many years have implicated ATP as a neurotransmitter, expecially of nonadrenergic-noncholinergic neurons in the intestine. ATP is stored in specific vesicles in neurons, is released upon nerve stimulation, and exerts specific synaptic effects on smooth muscle and on neurons in the brain. Since ATP is stored together with catecholamines, acetylcholine, and possibly other transmitters in nerve terminals throughout the body, it is often unclear whether ATP provides a storage mechanism enabling the concentration of large numbers of transmitter molecules within a small vesicle, whether it is a co-transmitter, modulating the effects of other transmitters, or whether it is a neurotransmitter in its own right. When ATP is labeled by radioactive percursors, much of the radioactivity released by depolarization exists as adenosine. When researchers speak of "purinergic transmission" they refer either to ATP or adenosine. The synaptic effects of ATP and adenosine differ in numerous ways, especially in the antagonism by xanthines of adenosine but not ATP. This review is restricted to adenosine.

There certainly is no difficulty in establishing that adenosine exists in the central nervous system as well as the rest of the body. However, pinpointing exact tissue levels is difficult. Levels of ATP and adenosine vary depending on the physiological state of the organism. During hypoxia ATP levels decline greatly, and adenosine levels increase markedly. If precautions are taken to prevent such changes, such as immersing the brain rapidly in liquid nitrogen, total brain concentrations of adenosine in several species are about 2–4 μmoles/

kg. Under comparable conditions, ATP levels are about a thousand times greater (Rehncrona et al 1978). During hypoxia adenosine levels can increase a hundred to a thousandfold, with millimolar concentrations detected in the spinal fluid (Kleihues et al 1974, Berne et al 1974). Moreover, brain dialysis studies measuring free extracellular adenosine reveal basal levels of 1 μmole/kg, which increase tenfold with hypoxia (Zetterstrom et al 1982). Such alterations may have physiological consequences. The marked augmentation of adenosine levels in the heart during coronary vasoconstriction can regulate myocardial perfusion (Berne 1980). Adenosine might have similar functions in the brain following intracerebral clots or hemorrhage. The therapeutic actions of dipyridamole in angine may involve blockade of the uptake inactivation of adenosine. Similar drugs with a predilection for the central nervous system might find use in the therapy of stroke victims.

Limited studies of adenosine levels in different brain regions have not revealed major regional variations, even when care has been taken to prevent hypoxic alterations (Winn et al 1980). Of course, crude regional evaluations of ubiquitous neurotransmitters such as GABA also do not indicate marked regional differences. There is evidence that adenosine is contained selectively within neurons in some brain areas. Thus, kainic acid injections into the rat corpus striatum deplete 70% of endogenous adenosine content (Wojcik & Neff 1983a). This suggests that a major portion of adenosine in the corpus striatum is generated by neurons with cell bodies in the striatum. Destruction of the nigrostriatal pathway with 6-hydroxydopamine elicits a 25% decline in striatal adenosine. Thus, almost all the endogenous adenosine in this brain region can be accounted for by the nigrostriatal dopamine neurons and intrinsic striatal neurons. Clearly, histochemical visualization of adenosine would clarily whether a pool of adenosine is contained in particular neurons, which would favor a role for adenosine as a neurotransmitter. Our recent immunohistochemical studies with antibodies highly selective for adenosine provide some discrete localizations (Braas et al 1984). Immunoreactive adenosine is highly localized to cell groups such as pyramidal cells of the hippocampus, deep layers of the medial cerebral cortex, the caudate, and basolateral amygdala with fibers most apparent in the spinal tract of the trigeminal.

Biosynthesis

Purines are formed *de novo* in the body through a series of reactions in which components of the purine structure are donated by the amino acids glycine, glutamine, and aspartic acid, as well as formate and carbon dioxide. This sequence of reactions results in the formation of inosine monophosphate, which is then transformed to 5′-adenosine monophosphate.

Adenosine itself can be formed by a number of mechanisms. One is the condensation of adenine and ribose-1-phosphate through the agency of one of the purine nucleoside phosphorylases. This is not a likely source for most adenosine formation in the brain, since substrate availability is limited and enzyme activity is low. Moreover, this reaction tends to function largely in the direction of nucleoside degradation rather than synthesis. Adenine may also serve as a source by the action of adenine phosphoribosyltransferase generating adenosine 5′-monophosphate (5′-AMP), which in turn may be transformed to adenosine. The direct generation of adenosine from RNA is feasible in theory, but not likely. When RNA breaks down to adenosine-3′-monophosphate, dephosphorylation could contribute to adenosine formation, though there is little evidence that this reaction is a major source of adenosine in the brain.

The two likeliest sources of adenosine are the dephosphorylation of 5′-AMP by 5′-nucleotidase and the action of S-adenosylhomocysteine hydrolase upon S-adenosylhomocysteine.

5′-Nucleotidase is a predominantly membrane-bound enzyme, which is often used as a marker for membrane fractions. Its candidacy for a unique neuronal function, perhaps as a generator of endogenous adenosine, is suggested by histochemical studies showing discrete localizations. Thus, highest densities of 5′-nucleotidase occur in the molecular layer of the cerebellum (Scott 1967, Schubert et al 1979), an area that also possesses the highest densities of adenosine A_1 receptors in rat brain (Goodman & Snyder 1982). Electron-microscopic evaluation shows that cerebellar 5′-nucleotidase is associated with Purkinje cell dendrites, parallel fibers, and the synapses between parallel fibers and Purkinje cell spines (Marani 1977). The enzyme is also selectively localized in molecular and polymorphic layers of the dentate gyrus and hippocampus. A synaptic role for this enzyme is suggested by subcellular and electron microscopic studies showing the enzyme enriched in synaptosomes and axoplasm at axodendritic synapses (Marani 1977). However, 5′-nucleotidase probably has numerous functions besides possibly serving as a biosynthetic enzyme for a putative synaptic pool of adenosine. Thus, the enzyme occurs in glial cell fractions and in highly purified myelin (Cammer et al 1980).

Another possible source of synaptically active adenosine is S-adenosylhomocysteine. In most systems S-adenosylhomocysteine hydrolase functions to synthesize S-adenosylhomocysteine. However, the reverse reaction could play a major role in the brain. Such a route of adenosine formation might provide a link to synaptic methylation reactions that give rise to S-adenosylhomocysteine. Axelrod's group has suggested that methylation of phospholipids in neuronal membranes provides a general mechanism for regulating synaptic function (Hirata & Axelrod 1980). Tissue concentrations of S-adenosylhomocysteine are as low or lower than those of adenosine. Accordingly, it is conceivable that the rate of formation of adenosine might be regulated by the availability of S-adenosylhomocysteine and thus by the extent of neuronal methylation reactions.

How might one ascertain the role of various mechanisms in adenosine formation? Endogenous adenosine tends to accumulate even in extensively washed membrane preparations. Thus, the binding of [^{3}H]cyclohexyladenosine ([^{3}H]CHA) to adenosine receptors can only be detected if the membranes are thoroughly washed and residual endogenous adenosine is destroyed by adding the enzyme adenosine deaminase. In such preparations the subsequent addition of deoxycoformycin, a potent inhibitor of adenosine deaminase, produces a fairly rapid but time dependent decline in [^{3}H]CHA binding, due to the accumulation of endogenous adenosine (Bruns et al 1980). If adenosine accumulation were attributable to the action of a particular enzyme, inhibitors of such enzymes should block the deoxycoformycin-induced decline in [^{3}H]CHA binding. Thus far efforts to evaluate the source of endogenous adenosine by examining inhibitors of S-adenosylhomocysteine hydrolase of 5′-nucleotidase have not been successful (L. Toll and S. H. Snyder, in preparation).

Release

One criterion for a neurotransmitter is its release from nerve terminals by appropriate depolarization stimuli. The release of ATP, adenosine, and other purines from both peripheral and central nervous tissue is well established. In most studies endogenous purines are labeled with radioactive adenine. Even under conditions in which the bulk of radioactivity within the tissue is labeled ATP, radioactivity released by depolarization consists largely of adenosine (Fredholm & Hedqvist 1980). ATP may have been initially released and then converted to adenosine, or the conversion may have taken place just prior to release.

Adenosine release from depolarized brain slices has been well characterized (Pull & McIlwain 1977). When brain slices are labeled with [^{3}H]adenine, adenosine comprises the bulk of [^{3}H]purine released. Release can be obtained with electrical stimuli, potassium, ouabain, or veratridine. Several findings raise concerns as to whether the evoked purine release fits what one would expect from a neurotransmitter. Potassium depolarization is substantially weaker in eliciting release than ouabain or veratridine. Moreover, there is a substantial delay in the efflux of purines compared with other neurotransmitters. Hollins & Stone (1980a) found that relatively little purine was released during the period of high potassium application, while substantially greater release occurred when potassium levels were restored to normal. Of equal concern is the observation that in some experiments purine release is not dependent on external calcium concentration, though in other studies a calcium dependence has been established (Stone 1981).

Adenosine Inactivation—Uptake and Catabolism

Although the selective hydrolysis of acetylcholine by acetylcholinesterase served for many years as the paradigm for neurotransmitter inactivation, it is now clear that for most biogenic amines and amino acid neurotransmitters, high affinity, sodium-dependent uptake systems remove the transmitter from the synaptic cleft. In the case of biogenic amines, uptake occurs into the nerve terminals that had released the transmitter, while for amino acids, glial as well as neuronal uptake may play a regulatory role.

The major enzyme that degrades adenosine is adenosine deaminase, which converts adenosine to inosine. Extraordinarily potent inhibitors of this enzyme have been developed, such as deoxycoformycin and erythro-9-(2-hydroxy-3-nonyl)-adenine (EHNA), whose K_i values are in the nanomolar range. Administration of deoxycoformycin or EHNA to intact animals as well as treatment of brain tissue in vitro markedly increase endogenous levels of adenosine (Zetterstrom et al 1982). If adenosine deaminase contributes to the physiological inactivation of adenosine, then deoxycoformycin should mimic the effects of adenosine. Adenosine and derivatives of adenosine that pass the blood brain barrier, such as phenylisopropyladenosine (PIA), produce sedation and sleep in rodents. Deoxycoformycin elicits similar sedative and hypnotic effects after intravenous administrtion to humans (Major et al 1981) or intraperitoneal injections in rats (Radulovacki et al 1983). These findings clearly support a role of adenosine deaminase in the physiology of adenosine. However, they do not establish that adenosine deaminase is responsible for its synaptic inactivation. By analogy, monoamine oxidase inhibitors produce a considerable augmentation in brain levels of catecholamines and serotonin with attendant behavioral effects. Nonetheless, synaptic inactivation of the amines is related primarily to nerve terminal uptake. Synaptic inactivation of any neurotransmitter is most efficiently mediated by a process localized to external membranes of the synapse, such as transmitter uptake systems or enzymes like acetylcholinesterase. By contrast, adenosine deaminase is an intracellular cytoplasmic anzyme.

Efficient uptake systems for adenosine with K_m values in the micromolar range occur in synaptosomes and glial preparations, and several peripheral tissues. The synaptosomal uptake of adenosine has been differentiated by Bender et al (1981) into discrete rapid and slow uptake systems, and the slow system has been separated into high affinity and low affinity components with respective K_m values of 1 and 5 μM. Like many neurotransmitter uptake systems, adenosine uptake is dependent, at least in part, upon sodium ion and energy sources.

The therapeutic vasodilatation elicited by drugs such as dipyridamole, papaverine, and hexobendine is thought to be mediated by inhibition of adenosine uptake. Recently Phillis & Wu (1982) have found that many psychotropic agents inhibit synaptosomal adenosine uptake. The most potent of these are certain neuroleptic drugs, including both phenothiazines and butyrophenones. There is some relationship between potency in inhibiting adenosine uptake and antischizophrenic actions. Thus spiperone, the most potent neuroleptic, is also the most potent inhibitor of adenosine uptake. However, several exceptions to the correlation and the much greater potency of neuroleptics in blocking dopamine receptors indicate that the antischizophrenic actions of neuroleptics are more likely associated with dopamine receptor blockade (Peroutka & Snyder 1980, Seeman 1980).

It has been suggested that the pharmacologic actions of benzodiazepines derive from inhibition of adenosine uptake (Wu et al 1981, Phillis & Wu 1983). There is a partial correlation between potencies in inhibiting uptake and clinical effects. However, exceptions to this correlation do exist, and potencies in blocking adenosine uptake or influencing adenosine uptake "receptors" (Patel et al 1982) are much less than drug affinities for benzodiazepine receptor binding sites. If adenosine uptake inhibition represents the pharmacological mechanism of benzodiazepine action, then benzodiazepine antagonists should block these actions. However, Morgan et al (1983) showed that benzodiazepine antagonists fail to reverse the inhibition of adenosine uptake produced by benzodiazepines.

Adenosine Actions

Adenosine influences numerous physiological systems. Its vasodilatory effects on numerous vascular beds may mediate postischemic hyperemia of the coronary blood vessels (Berne 1980). Adenosine causes bronchial constriction. Blockade of endogenous adenosine by xanthine drugs such as theophylline may account at least in part for their antiasthmatic effects. Adenosine inhibits platelet aggregation. It is possible that the build up of adenosine following cardiac ischemia could thus influence blood clotting. Adenosine inhibits lipolysis, an action that may be exerted by endogenous adenosine, since xanthines stimulate lipolysis. Adenosine inhibits norepinephrine release from sympathetic neurons as well as acetylcholine release at the neuromuscular junction and in ganglia (Fredholm & Hedqvist 1980).

In the brain, adenosine is almost uniformly inhibitory in its actions on neuronal firing (Phillis & Wu 1981). Some direct postsynaptic inhibition can be detected. However, the predominant effects reflect presynaptic blockade of excitatory transmitter release. These actions are highly selective in that other purines and pyrimidines such as inosine, guanosine, xanthosine, cytidine, uridine, thymidine, and their nucleotides are either much weaker or inactive (Phillis & Wu 1983). Moreover, the effects of adenosine are potently blocked by xanthines.

Inhibition of neurotransmitter release by adenosine has been well characterized in brain slices (Harms et al 1979, Hollins & Stone 1980b). Adenosine inhibits the release of acetylcholine, dopamine, norepinephrine, serotonin, and GABA from brain slices, effects that are antagonized by xanthines. A role for endogenous adenosine in the modulation of transmitter release is suggested by the ability of exogenous adenosine deaminase to enhance dopamine release while blockade of adenosine deaminase with deoxycoformycin reduces dopamine release (Michaelis et al 1979). The adenosine uptake inhibitor dipyridamole diminishes GABA release (Hollins & Stone 1980b). Also, xanthines such as theophylline augment the release of acetylcholine from the intestine and cerebral cortex slices (Sawynok & Jhamandas 1976).

Inhibition of transmitter release appears attributable to a blockade of calcium influx into nerve terminals. Calcium uptake by synaptosome preparation is inhibited by adenosine and related agents quite potently (Ribeiro et al 1979, Wu et al 1982), and this action is prevented by xanthines. Moreover, presynaptic inhibitory electrophysiologic effects of adenosine in the hippocampus via one adenosine receptor subtype (A_1) involve calcium (Dunwiddie et al 1981, Dunwiddie & Fredhold 1984).

Adenosine Influences on Adenylate Cyclase

Burnstock (1981) showed that effects of adenosine and ATP involve different receptor sites. Among numerous differences between the effects of these agents, the most useful discriminator is the ability of xanthines to block adenosine but not ATP effects. Burnstock referred to two subtypes of purinergic effects as P_1 for adenosine- and P_2 for ATP-mediated actions. According to this nomenclature all adenosine actions exerted upon extracellular sites involve P_1 receptors.

The first biochemical characterization of adenosine receptors examined influences upon adenylate cyclase (Daly 1977). Sattin & Rall (1970) showed that adenosine causes an accumulation of cyclic AMP in brain slices that is blocked by xanthines. Under somewhat different conditions, adenosine can inhibit adeylate cyclase in peripheral tissues (Londos et al 1980) and brain (Van Calker et al 1979, Cooper et al 1980, Ebersolt et al 1983). The stimulatory and inhibitory effects reflect actions at different receptors, since they are readily differentiated by a variety of adenosine agonists. Receptors associated with inhibition of the cyclase have been designated A_1, while A_2 receptors mediate stimulation of the cyclase. A_1 receptors are affected by nanomolar concentrations of adenosine, whereas A_2 receptors require micromolar levels. A_1 sites demonstrate marked stereospecificity toward PIA, with the L-isomer being much more active, whereas much lesser stereoselectivity is apparent for A_2 receptors. 5′-N-ethylcarboxamide adenosine (NECA) is more potent than adenosine and PIA at A_2 sites but weaker at A_1 receptors. A_1 and A_2 receptors are antagonized with similar potencies by xanthines.

Londos et al (1979) showed that adenosine-related agents can reduce cyclic AMP accumulation via intracellular sites that differ quite markedly in their properties from A_1 and A_2 receptors. These internal sites require an intact purine ring, and so they were designated P sites. 2′,5′-Dideoxyadenosine is very potent at the P sites, while adenosine is weak and L-PIA is virtually inactive. Also unlike A_1 and A_2 adenosine receptors, the P sites are not affected by xanthines. Londos and co-workers (1979, 1980) differentiated the P sites from R receptors, which require an intact ribose ring and which correspond to adenosine A_1 and A_2 receptors. Ra refers to sites involved in adenylate cyclase activation, thus representing A_2 receptors, whereas Ri sites are equivalent to A_1 receptors. For the purposes of simplicity we restrict our classification of adenosine receptor subtypes to A_1 and A_2. Since there is little evidence for a physiological role of the P site, it is not discussed further.

In early studies, A_2 receptor activity was detected only in brain slices, and A_1 reduction of adenylate cyclase could not be detected at all (Daly 1977). To demonstrate A_1-mediated inhibition of adenylate cyclase in brain homogenates requires removal of endogenous adenosine, addition of adenosine deaminase, and inclusion of sodium ions (Cooper et al 1980, Ebersolt et al 1983). A_1 inhibition of adenylate cyclase can be demonstrated in brain slices in which cyclase activity has been enhanced with forskolin (Fredholm et al 1983a). Inclusion of adenosine deaminase also facilitates the detection of A_2 adenosine-mediated enhancement of adenylate cyclase in homogenates of the corpus striatum but of no other brain region (Premont et al 1977). Besides the regional differences between A_2 receptors observed via adenylate cyclase measurement in homogenates and slices, the two receptor preparations show different affinities both for adenosine agonists and xanthine antagonists (Daly et al 1983). Thus, on the basis of adenylate cyclase effects at least three adenosine receptors with apparent physiological relevance can be distinguished: two A_2 receptors and an A_1 site. The unique striatal A_2 receptors may have functional and clinical relevance. The A_2-selective agonist NECA elicits rotation behavior in rats when injected unilaterally into the caudate nucleus in combination with parenteral administration of the agonist apomorphine (Green et al 1982). With higher doses of apomorphine, NECA injections in the caudate produce self-mutilating behavior. The Lesch-Nyhan Syndrome involves a defect in the activity of enzyme hypoxanthine-guanine phosphoribosyl transferase (HGPRT), which forms guanosine-5′-phosphate or inosine-5′-phosphate, the latter being a precursor of adenosine. Lowered levels of adenosine exist in patients with Lesch-Nyhan Syndrome. The possibility that these effects are related to self-mutilating behavior in Lesch-Nyhan patients is tantalizing.

Adenosine Receptors Labeled by Ligand-Binding Techniques

The ability to measure neurotransmitter receptors by ligand-binding techniques has greatly facilitated studies of synaptic mechanisms, permitting refined analysis of receptor subtypes and mechanisms of action of drugs (Snyder 1984). Measurement of receptor binding sites is greatly dependent upon the selection of a ligand that binds with a relatively high affinity, sometimes the neurotransmitter itself or a related drug. In the case of adenosine, rapid metabolism by adenosine deaminase even in washed brain membranes precludes the use of [^{3}H]adenosine as a ligand. The major methylxanthines, caffeine and theophylline, have only micromolar potency at adenosine receptors, and are insufficient to serve as suitable ligands. We prepared [^{3}H]cyclohexyladenosine ([^{3}H]CHA), an adenosine derivative with a cyclohexyl group substituted on the amine moiety that enhances receptor affinity and blocks the effects of adenosine deaminase (Bruns et al 1980) (Figure 1). Systematic exploration of xanthine analogs indicated that an eight phenyl substitution as well as replacement of the methyl substituents at carbons 1 and 3 by ethyl groups would

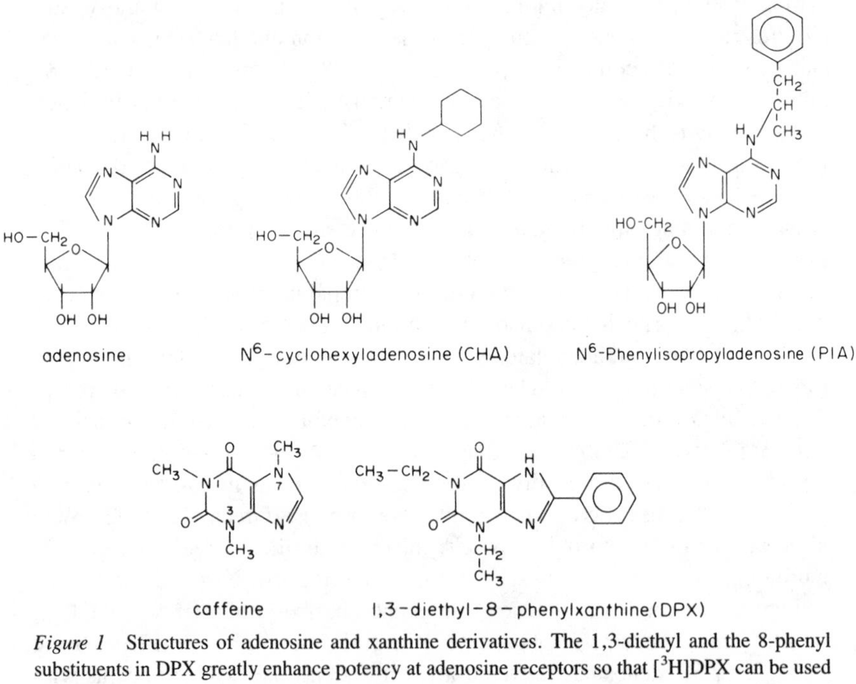

adenosine — N^6-cyclohexyladenosine (CHA) — N^6-Phenylisopropyladenosine (PIA)

caffeine — 1,3-diethyl-8-phenylxanthine (DPX)

Figure 1 Structures of adenosine and xanthine derivatives. The 1,3-diethyl and the 8-phenyl substituents in DPX greatly enhance potency at adenosine receptors so that [^{3}H]DPX can be used to label adenosine A_1 receptors. The cyclohexyl and phenylisopropyl substituents in CHA and PIA, respectively, both protect adenosine from degradation by adenosine deaminase and increase affinity for adenosine receptors. PIA and CHA have substantially greater affinity for A_1 than A_2 receptors, while NECA shows selectivity for A_2 receptors.

augment potency considerably. This derivative, [^{3}H]1,3-diethyl-8-phenylxanthine ([^{3}H]DPX), provided a useful antagonist ligand (Bruns et al 1980). Other groups developed [^{3}H]PIA (Schwabe & Trost 1980) and [^{3}H]2-chloroadenosine (Williams & Risley 1980) as ligands for adenosine receptors.

The drug competition profile for [^{3}H]CHA, [^{3}H]PIA, and [^{3}H]2-chloroadenosine binding sites reflects an interaction with A_1 receptors because of the nanomolar potency of adenosine and its derivatives as well as the pronounced stereospecificity of PIA isomers. In bovine and rat brain, [^{3}H]DPX binding also appears to involve A_1 adenosine receptors (Murphy & Snyder 1982). However, in guinea pig brain, adenosine derivatives are much less potent and L- and D-PIA have equal potencies (Bruns et al 1980). It is conceivable that a portion of [^{3}H]DPX binding to guinea pig brain membranes involves A_2 receptors, though this cannot be established definitively. Binding of [^{3}H]DPX to A_1 receptors cannot be demonstrated in guinea pig or human cerebral cortex (Murphy & Snyder 1982). These species differences reflect heterogeneity among A_1 receptors. DPX is 300 times more potent in competing for [^{3}H]CHA binding to A_1 receptors in calf cerebral cortex than in guinea pig or human receptors. Rat cerebral cortex is intermediate, with DPX displaying one-twentieth the affinity for A_1 sites in calf cerebral cortex.

Other examples of heterogeneity among adenosine A_1 receptors have been demonstrated. [^{3}H]CHA binding can be resolved into high and low affinity components in rat brain with respective K_D values of 0.4 and 4 nM (Marangos et al 1982). In the same studies [^{3}H]DPX and [^{3}H]CHA binding have been differentiated. Copper is ten times more potent in inhibiting [^{3}H]CHA than [^{3}H]DPX binding. [^{3}H]CHA binding is also more sensitive to inactivation by heat and proteolytic anzymes. Thus [^{3}H]DPX and [^{3}H]CHA appear to bind to distinct A_1 receptor subtypes or to two different conformations of the same receptor.

Quite recently it has become possible to label A_2 receptors. Yeung & Green (1984) in brain and Hütteman et al (1984) in platelets showed that [^{3}H]NECA binds both to A_1 and A_2 receptors. A_2 receptor binding can be studied selectively by treating membranes with *N*-ethylmaleimide, which degrades A_1 but not A_2 receptors.

Effects of guanine nucleotides and divalent cations establish that receptor binding sites involve adenosine receptors linked to adenylate cyclase (Goodman et al 1982). Receptor binding sites for neurotransmitters that stimulate or inhibit adenylate cyclase are regulated by guanine nucleotides with GDP and GTP decreasing agonist but not antagonist affinity. By contrast, divalent cations like manganese and magnesium selectively increase agonist affinity at such receptors. A_1 receptors labeled with [^{3}H]CHA and [^{3}H]DPX are modulated by guanine nucleotides and divalent cations (Goodman et al 1982). At micromolar concentrations GTP decreases binding of the agonist [^{3}H]CHA but not the antagonist [^{3}H]DPX. However, GTP decreases the potency of agonists in competing for [^{3}H]DPX binding, an effect that can be used to grade experimental substances along an adenosine agonist-antagonist continuum. Removal of endogenous divalent cations with chelating agents inhibits [^{3}H]CHA but not [^{3}H]DPX binding, thus suggesting that endogenous divalent cations regulate agonist affinity at adenosine receptors. [^{3}H]CHA binding is selectively augmented by manganese, magnesium, and calcium. As has been demonstrated with numerous other receptors, sodium also selectively decreases agonist binding to adenosine receptors, with lithium and potassium being inactive (Goodman et al 1982).

Actions of guanine nucleotides and divalent cations on receptor binding are thought to reflect interactions with the GTP binding of "N" proteins that couple the receptor to adenylate cyclase. If a receptor remains coupled to such proteins after solubilization from membranes, then one might anticipate the receptor regulation by quanine nucleotides or divalent cations to be retained in the soluble state. [^{3}H]CHA binding can be identified to adenosine receptors solubilized from brain membranes (Gavish et al 1982). GTP regulation of receptor binding is retained in the soluble state, whereas the effects of divalent cations are lost (Gavish et al 1982). By contrast, with solubilized histamine H_1 receptors, effects of guanine nucleotides are lost but influences of sodium and divalent cations are retained (Toll & Snyder 1982). These findings suggest that divalent cations and guanine nucleotides bind to separate protein subunits that can be differentially separated from receptors when they are solubilized.

Autoradiographic techniques have permitted a microscopic localization of numerous neurotransmitter receptors, thus clarifying considerably the specific neuronal groupings that may mediate the actions of drugs such as opiates (Pert et al 1976) and benzodiazepines (Young & Kuhar 1980). If adenosine were to serve as a neurotransmitter, one could certainly expect discrete localizations of adenosine receptors. Utilizing [^{3}H]CHA, we localized adenosine A_1 receptors by in vitro autoradiographic techniques (Goodman & Snyder 1981, 1982). Marked differences occur in various brain regions. Adenosine receptors are most concentrated in the molecular layer of the cerebellum. The molecular and polymorphic layers of the hippocampus and dentate gyrus also posses high densities. By contrast, very low receptor levels are apparent over the pyramidal cell layer. The thalamus also possesses high levels of receptors but with some variations among subdivisions. Thus, the medial, gelatinosus, and lateral nuclei have higher receptor densities than the ventral nuclei. Within the diencephalon, highest densities occur in the medial geniculate body. In contrast to the very high receptor density in the hippocampus and the thalamus, the hypothalamus appears virtually devoid of adenosine receptors. Whereas receptor density is relatively homogeneous in the thalamus, there are marked variations in the amygdala. Moderate receptor densities occur in the central nucleus, while negligible if any levels occur in other portions of the amygdala. In the cerebral cortex, adenosine receptors occur with considerable density in layers I, IV, and VI; much lower levels are apparent in layers II, III, and V. A preliminary study by others independently confirms these findings (Lewis et al 1981). Interestingly, adenosine uptake sites, visualized by autoradiography with ^{3}H-nitrobenzylthioinosine, show both similarities and differences with receptor localizations (Bisserbe et al 1984). The caudate, substantia nigra, superior colliculus, and substantia gelatinosa of the trigeminal have high densities of both sites. However, the central nucleus of the amygdala is selectively enriched in receptors, whereas uptake sites occur throughout the amygdala. In contrast to the very dense receptor distribution in the cerebellum and hippocampus, few uptake sites are visualized in these structures.

We explored the neuronal localization of the receptors in greater detail by utilizing neurologic mutant mice as well as selective brain lesions (Goodman et al 1983). We focused upon the cerebellum, which is useful for exploring the cellular localization of receptors because it contains five well-characterized cell types. The Purkinje, Golgi II, stellate, and basket cells are inhibitory. The only excitatory neuron in the cerebellum is the granule cell, whose axons give rise to parallel fibers in the molecular layer of the cerebellum that synapse upon Purkinje cells.

The extremely high density of adenosine receptors in the molecular layer could indicate an association with Purkinje cells or the parallel fibers of granule cell neurons. Neurological mutant mice that manifest selective losses of particular cerebellar cell types permit a discrimination between these alternatives. Thus, *Nervous* mice have a 90% decline of Purkinje cells, whereas other cerebellar cell types are essentially normal. The pattern and density of receptors in cerebella of *Nervous* mice are the same as litter mate controls, thus indicating that adenosine receptors in the cerebellum are not associated with Purkinje cells. By contrast, *Weaver* mice have an 80% deficiency of granule cells, with other cerebellar cell types being normal. Adenosine receptors are reduced 70–80% in the cerebella of *Weaver* mice, indicating that adenosine receptors are probably associated with parallel fibers of granule cells. Such a conclusion is supported by experiments with *Reeler* mice, in which the neural migration of granule cells does not occur so that granule cells and their axons remain in the external granular layer. In *Reeler* mice, adenosine receptors are also restricted to the external granular layer, confirming an association of receptors with axons and terminals of granule cells. Independent studies of A_1 receptors that measured inhibition of adenylate cyclase also showed receptor loss in cerebella of *Weaver* mice but not other neurologic mutants (Wojcik & Neff 1983b). That adenosine blocks neurophysiologic synaptic effects of parallel but not climbing fibers in the cerebellum indicates a functional role for granule cell adenosine receptors (Kocsis et al 1984).

Ganglion cells of the retina are the major excitatory neurons that project both to the lateral geniculate body and the superior colliculus. Their transmitter is excitatory and may correspond to one of the excitatory amino acids. Unilateral removal of the eye abolishes adenosine receptor labeling in the superior colliculus opposite to the eye removed (Goodman et al 1983). Thus, adenosine receptors seem to be associated with excitatory projections from retinal ganglion cells to the superior colliculus.

Interestingly, both granule cells of the cerebellum and ganglion cells of the retina are excitatory. However, adenosine receptors are not universally associated with terminals of excitatory pathways. Thus, destruction of the excitatory corticostriate neuronal pathway fails to alter adenosine receptor density in the corpus striatum. Similarly, destruction of the corticothalamic excitatory pathway does not influence numbers of adenosine receptors in the thalamus (Goodman et al 1983).

The association of adenosine receptors with axons and nerve terminals of at least certain excitatory neuronal pathways in the brain fits with the abundant neurophysiological and biochemical evidence that adenosine acts by reducing excitatory neurotransmitter release.

Though adenosine influences numerous organs throughout the body, thus far receptor binding has been identified only in the brain, fat cells (Trost & Schwabe 1981), and the testes (Williams & Risley 1980, Murphy & Snyder 1981). A functional role for testicular adenosine receptors is suggested by the well-known ability of caffeine to enhance sperm motility, metabolism, respiration, and the ability to penetrate the ovum (Schoenfeld et al 1975). Autoradiographic techniques similar to those utilized for [^{3}H]CHA in the brain reveal strikingly selective localization of adenosine receptors in the testes. Receptors are highly concentrated within the seminiferous tubules in close association with sperm cells. No receptors are detected in the interstitial tissue, where testosterone is produced. The localization of adenosine receptors to sperm cells is supported by the reduction in numbers of adenosine receptors by treatments that selectively decrease sperm cells, whereas adenosine receptors are not altered by manipulations that primarily influence the interstitial cells (Murphy & Snyder 1981, Murphy et al 1983).

Psychoactive Drugs and Adenosine Receptors

The ability of xanthines to block adenosine receptors suggested that these effects might be related to the central stimulant actions of caffeine and the antiasthmatic effects of theophylline. Early studies by Sutherland's group indicated that caffeine inhibits the cyclic-AMP-degrading enzyme, phosphodiesterase (Rall 1980). For many years it was assumed that caffeine acts as a stimulant by inhibiting phosphodiesterase, resulting in an accumulation of cyclic AMP in the brain. However, several inhibitors of the enzyme are substantially more potent than caffeine but lack central stimulant actions. Moreover, the concentrations of caffeine required to inhibit most forms of phosphodiesterase are 100 times greater than circulating levels of caffeine after ingestion of coffee.

To ascertain whether adenosine blockade might account for the effects of caffeine, we compared the potencies of a series of xanthines as adenosine antagonists, and in parallel studies we examined their influences on the locomotor behavior of mice (Snyder et al 1981). In general, potencies in blocking adenosine receptors correlate with the stimulation of locomotor activity in mice. Moreover, the K_i of caffeine for adenosine A_1 receptors corresponds to brain levels after a few cups of coffee. However, some xanthines that are effective adenosine antagonists, such as isobutylmethylxanthine (IBMX), do not stimulate locomotor activities. Moreover, other xanthines, such as caffeine, manifest biphasic actions, depressing locomotor activity at low doses and enhancing it at higher doses. To focus selectively upon xanthine interactions with adenosine mediated behavioral systems, we treated mice with *L*-PIA. In low doses *L*-PIA depresses locomotor activity. Not only do xanthines block this effect, but they transform it into a marked stimulation of locomotor activity, greater than that observed with the xanthines alone. IBMX, which fails to enhance locomotor activity when administered alone, is highly active in reversing the PIA-induced depression of activity. Low doses of caffeine, which by themselves elicit locomotor depression, produce locomotor stimulation when combined with a depressant dose of PIA.

The mechanism for the paradoxical reversal of xanthines of PIA-induced locomotor depression has been clarified in detailed studies of dose-response with *L*-PIA (Katims et al 1983). When still lower doses of *L*-PIA are administered to mice, locomotor enhancement is observed. Thus, *L*-PIA seems to act at a very high affinity site whereby it stimulates locomotor activity, whereas at a site of somewhat lesser affinity *L*-PIA depresses activity. According to this model, caffeine blocks effects of *L*-PIA at sites that produce locomotor depression, thus unmasking the locomotor stimulatory potential of *L*-PIA.

Another valuable animal model for behavioral effects of xanthines involves their enhancement of apomorphine-induced rotation in rats with nigrostriatal lesions (Fredholm et al 1983b). These effects appear to derive from adenosine antagonism since *L*-PIA and the adenosine deaminase inhibitor EHNA have opposite effects.

If behavioral actions in rodents can be extrapolated to man, this body of work indicates that the stimulant effects of caffeine on the central nervous system are mediated by adenosine receptor blockade. Whether the bronchodilatory therapeutic effects of theophylline also involve adenosine blockade is not clear. A comparison of the various xanthines suggests that bronchodilatory actions correlate with inhibition of A_2 receptors (Fredholm & Persson 1982). The active bronchodilator enprofylline (3-prophyxanthine) does not affect A_1 receptors, but blocks A_2 receptors in the hippocampus but not those on lymphocytes, thus suggesting A_2 receptor heterogeneity (Fredholm & Sandberg 1983, Fredholm et al 1984). A role for A_2 receptors in antiasthmatic drug action accords with observations that micromolar concentrations of adenosine are required to elicit broncho-constriction with little difference in potencies between the isomers of PIA (K. M. M. Murphy, M. J. Cooper, and S. H. Snyder, unpublished).

The apparent clinical relevance of xanthine actions via adenosine receptors suggests that novel therapeutic agents might be developed by modifying the xanthine molecule. Theophylline is the most widely used drug in the treatment of asthma. However, its use is limited by side-effects such as stimulation of the central nervous system, which can result in convulsion at blood levels as little as two to three times the therapeutic values. The known heterogeneity of adenosine receptors suggests that more potent, selective and safer xanthines effects. 1,3-Dipropyl substitution increases potency at [^{3}H]CHA-binding sites compared to the 1,3-dimethyl substitution of theophylline. Combining all these receptors (Bruns et al 1983). An 8-phenyl substituent augments potency, which is further enhanced by certain *para* and *ortho* additions to the 8-phenyl ring. Combining an *ortho* amino with a *para* chloro substituent produces maximal effects. 1,3-Dipropyl substitution increases potency at [^{3}H]CHA-binding sites compared to the 1,3-dimethyl substitution of theophylline. Combining all these substituents results in 1,3-dipropyl-8-(2-amino-4-chlorophenyl)-xanthine, a compound of extraordinary receptor affinity. Its K_i value for adenosine A_1 receptors is 22 picomolar. This compound is 70,000 times more potent than theophylline and 500,000 times more potent than caffeine in blocking A_1 reactors (Bruns et al 1983).

Though only the effects at A_1 receptors in the brain have yet been evaluated, it is quite possible that extremely potent xanthine derivatives may differentiate between receptor subtypes in various tissues. A xanthine that blocks adenosine receptors in the lung but not the brain or heart would be uniquely useful in the treatment of asthma. Historically, theophylline had been used in the treatment of congestive heart failure because of its inotropic effects. However, it declined from clinical use because of its stimulation of the central nervous system. Xanthines with selectivity for the heart might be effective cardiotonics. The behavioral stimulant effects of xanthines might be useful in certain forms of depression, but side-effects such as cardiac palpitations and diuresis interfere. Derivatives selective for brain-type receptors and that partition well into the brain might be more effective. Behavioral actions of xanthines that selectively block A_1 and A_2 receptors will be of interest. Daly et al (1984) recently have shown differential potencies of various xanthines at A_1 receptor binding sites and at the two subtypes of A_2 receptors monitored by enhancement of adenylate cyclase activity in brain slices and homogenates, respectively.

Tolerance to the behavioral stimulant effects of caffeine and withdrawal headaches are well known. These may relate to increased numbers of adenosine receptors following chronic caffeine treatment (Fredholm 1982, Murray 1982, Boulenger et al 1983, Marangos et al 1984). However, the increased receptor number is not associated with changes in the sensitivity of hippocampal adenylate cyclase to stimulation by adenosine (Fredholm 1982).

The unique behavioral effects of *L*-PIA suggest that adenosine agonists may have therapeutic value. A substantial increase of *L*-PIA dosage is required before reduction of locomotor activity is followed by sleep (Snyder et al 1981). Also, *L*-PIA is quite nontoxic, with no lethality observed at doses 10,000 times the threshold for reducing locomotor activity. Recently, Radulovacki et al (1984) have evaluated the hypnotic actions of adenosine analogs. CHA, NEC, and *L*-PIA augment rapid eye movement (REM) sleep, quite unlike barbiturates and benzodiazepines, which reduce REM sleep. Reduction of REM sleep has been associated with some of the deleterious effects of conventional sleeping medications. Thus, adenosine agonists might be safer hypnotics. Besides hypnotic effects, *L*-PIA has anticonvulsive actions, which, however, require substantially higher doses than the locomotor depressant effects (Dunwiddie & Worth 1982, Snyder et al 1981).

In evaluating the behavioral actions of adenosine derivatives such as *L*-PIA, one should be cautious about interference by hypotensive actions. NECA and *L*-PIA, respectively, produce some hypotension at doses as little as 5 and 500 nanograms/kg IV (Phillis & Wu 1983). In these studies effects of these agents on cerebral cortical neuronal firing correlates with hypotensive actions. However, with the intraperitoneal route of administration and the doses employed in behavioral work by ourselves (Katims et al 1983) and others (Crawley et al 1981), no changes have occurred in blood pressure or heart rate. Also, doses of *L*-PIA that produce threshold locomotor changes are associated with brain concentrations sufficient to occupy 50% of adenosine receptors as monitored by [^{3}H]CHA binding (Katims et al 1983). Thus, the behavioral effects of *L*-PIA appear to be centrally mediated. Such a conclusion is supported by the findings of Barraco et al (1983) showing essentially the same effects on locomotor activity of mice, whether adenosine analogs are administered directly into the cerebral ventricles or peripherally.

Several lines of evidence suggest that opiate effects may be mediated in part by adenosine. Xanthines antagonize the depressant actions of morphine on acetylcholine release from the intestine and brain (Sawynok & Jhamandas

1976). Moreover, opiate administration enhances the release of purines from cerebral cortex preparations (Fredholm & Vernet 1978). Also, xanthines produce behavioral changes in rodents that mimic in many respects the behavioral patterns observed with opiate withdrawal (Collier et al 1974).

Conclusions

In summary, evidence from many sources implicates adenosine in brain function. Whether adenosine is a neurotransmitter or neuromodulator is unclear. Indeed, the distinction between neurotransmitters and neuromodulators is becoming increasingly fuzzy. The one item of information most needed to clarify the role of adenosine in the brain is histochemical evidence of its localization. Even though metabolic pools of adenosine occur in all cells in the brain, "adenosine-ergic" neurons might possess especially high levels compared to surrounding cells. If adenosine is concentrated in specific neuronal populations, it may function as a "classical" neurotransmitter or, like the neuropeptides, it may regulate the release of other transmitters.

Alternatively, adenosine may be synthesized on neuronal membranes where it exerts its effects, functioning very much like the prostaglandins. Some evidence favors this latter view. The localization of adenosine A_1 receptors resembles that of 5′-nucleotidase, a membrane-associated enzyme. Adenosine A_1 receptors are associated with parallel fibers of cerebellar granule cells and retinal ganglion cell afferent nerve terminals to the superior colliculus. Axo-axonic synapses for putative adenosinergic neurons have not been described in either place. If 5′-nucleotidase synthesizes adenosine on membranes of cerebellar parallel fibers, then the adenosine-synthesizing system and the adenosine receptors would be contained on the same axon terminals. Endogenous adenosine accumulates rapidly even in extensively washed synaptic membranes separated from RNA- or DNA-containing particles that might generate adenosine (L. Toll and S. H. Snyder, unpublished). This further supports the notion that adenosine may be generated on membranes at the site of its actions.

On the other hand, a strong case for adenosinergic neurons comes from the lesion studies showing that in the corpus striatum 70% of endogenous adenosine occurs on intrinsic neurons and another 20% occurs on terminals of the nigrostriatal dopamine pathway (Wojcik & Neff 1983a). Resolution of this important question awaits newer experimental approaches.

ACKNOWLEDGMENTS

Supported by USPHS grants MH-18501, DA-00266, NS-16375, RSA Award DA-00074 and a grant of the International Life Sciences Institute. The manuscript was prepared by Dawn C. Dodson. The author thanks John Daly, Geoffrey Burnstock, John Phillis, Trevor Stone, Paul Marangos, Bertil Fredholm, and Richard Green for access to unpublished manuscripts.

Literature Cited

Barraco, R. A., Coffin V. L., Altman, H. J., Phillis, J. W. 1983. Central effects of adenosine analogs and locomotor activity in mice and antagonism of caffeine. *Brain Res.* 227:392–95.

Bender, A. S., Wu, P. H., Phillis, J. W. 1981. The rapid uptake and release of [^{3}H]adenosine by rat cerebral cortical synaptosomes. *J. Neurochem.* 36:651–66

Berne, R. M. 1980. The role of adenosine in the regulation of coronary blood flow. *Circ. Res.* 46:807–13

Berne, R. M., Rubio, R., Curnish, R. R. 1974. Release of adenosine from ischemic brain. Effect of cerebral vascular resistance and incorporation into cerebral adenine nucleotides. *Circ. Res.* 35:262–71

Bisserbe, J.-C., Pattel, J., Marangos, P. J. 1984. Autoradiographic localization of adenosine uptake sites in rat brain using [^{3}H] nitrobenzylthionosine. *J. Neurosci.* In press

Baoulenger, J. P., Patel, J., Post, R. M., Parma, A. M., Marangos, P. J. 1983. Chronic caffeine consumption increases the number of brain adenosine receptors. *Life Sci.* 32: 1135–42

Bruns, R. F., Daly, J. W., Snyder, S. H. 1980. Adenosine receptors in brain membranes: Binding of N^6-cyclohexyl[^{3}H]adenosine and 1,3-diethyl-8-[^{3}H]phenylxanthine. *Proc. Natl. Acad. Sci. USA* 77:5547–51

Bruns, R. F., Daly, J. W., Synder, S. H. 1983. Adenosine receptor binding: Structure-activity analysis generates extremely potent xanthine antagonists. *Proc. Natl. Acad. Sci. USA* 80: 2077–80

Burnstock, G. 1981. Neurotransmitter and trophic factors in the autonomic nervous system. *J. Physiol* 313:1–35

Cammer, W., Sirota, S. R., Zimmerman, T. R. Jr., Norton, W. T. 1980. 5′-Nucleotidase in rat brain myelin. *J. Neurochem.* 35:367–73

Collier, H. O. J., Francis, D. L., Henderson, G., Schneider, C. 1974. Quasi-morphine abstinence syndrome. *Nature* 249:471

Cooper, D. M. F., Londos, C., Rodbell, R. 1980. Adenosine receptor-mediated inhibition of rat cerebral cortical adenylate cyclase by a GTP-dependent process. *Mol. Pharmacol.* 18:598–601

Crawley, J. N., Patel, J., Marangos, P. J. 1981. Behavioral characterization of two long lasting adenosine analogs: Sedative properties and interaction with diazepam. *Life Sci.* 29:2623–30

Daly, J. W. 1977. *Cyclic Nucleotidase in th Nervous System.* New York: Plenum

Daly, J. W., Butts-Lamb, P., Padgett, W. 1984. Subclasses of adenosine receptors in the cortical nervous system: Interaction with caffeine and related methylxanthines. *Cell Mol. Neurobiol.* 3:69–80

Dunwiddie, T. V., Fredholm, B. B. 1984. Adenosine receptors mediating inhibitory electrophysiological responses in rat hippocampus are different from receptors mediating cyclic AMP accumulation. *Naunyn-Schmiedeberg's Arch. Pharmacol.* In press

Dunwiddie, T. V., Hoffer, B. J., Fredholm, B. B. 1981. Alkylxanthines elevate hippocampal excitability: Evidence for a role of endogenous adenosine. *Naunyn-Schmiedeberg's Arch. Phrmacol.* 316:326–30

Dunwiddie, T. V., Worth, T. 1982. Sedative and anticonvulsant effects of adenosine analogs in mouse and rat. *J. Pharmacol. Exp. Ther.* 220:70–76

Ebersolt, C., Premont, J., Prochiantz, A., Perez, M., Bockaert, J. 1983. Inhibition of brain adenylate cyclase by A_1 adenosine receptors: Pharmacological characteristics and locations. *Brain Res.* 267:123–29

Fredholm, B. B. 1982. Adenosine actions and adenosine receptors after 1 week treatment with caffeine. *Acta Physiol. Scand.* 115:283–86

Fredholm, B. B., Hedqvist, P. 1980. Modulation of neurotransmission by purine necleotides and nucleosides. *Biochem. Pharmacol.* 29:1635–43

Fredholm, B. B., Persson, C. G. A. 1982. Xanthine derivatives as adenosine receptor antagonists. *Eur J. Pharmacol.* 81:673–76

Fredholm, B. B., Sandberg, G. 1983. Inhibition by xanthine derivatives of adenosine receptor-stimulated cyclic adenosine 3′,5′-monophosphate accumulation in rat and guinea-pig thymocytes. *Br. J. Pharmacol.* 80:639–44

Fredholm, B. B., Vernet, L. 1978. Morphine increases depolarization induced purine release from rat cortical slices. *Acta Physiol. Scand.* 104:502–4

Fredholm, B. B., Jonzon, B., Lindstrom, K. 1983a. Adenosine receptor mediated increases and decreases in cyclic AMP in hippocampal slices treated with Forskolin. *Acta Physiol. Scand.* 117:461–63

Fredholm, B. B., Herrera-Marschitz, M., Jonzon, B., Lindstrom, K., Ungerstedt, U. 1983b. On the mechanism by which methylxanthines enhance apomorphine-induced rotation behaviour in the rat. *Pharmacol. Biochem. Behav.* 19:535–41

Fredholm, B. B., Bergman, B., Lindstrom, K. 1984. Actions of enprofylline in the rat hippocampus. *Acta Physiol. Scand.* In press

Gavish, M., Goodman, R. R., Snyder, S. H.

1982. Solubilized adenosine receptors in the brain: Regulation by guanine nucleotides. *Science* 215:1633–35

Goodman, R. R., Snyder, S. H. 1981. The light microscopic *in vitro* autoradiographic localization of adenosine (A_1) receptors. *Soc. Neurosci. Abstr.* 7:613

Goodman, R. R., Snyder, S. H. 1982. Autoradiographic localization of adenosine receptors in rat brain using [^{3}H]cyclohexyladenosine. *J. Neurosci.* 2:1230–41

Goodman, R. R., Cooper, M. J., Gavish, M., Snyder, S. H. 1982. Guanine nucleotide and cation regulation of binding of [^{3}H]cyclo hexyl-adenosine and [^{3}H]diethylphenyl xantine to adenosine A_1 receptors in brain membranes. *Mol. Pharmacol.* 21:329–35

Goodman, R. R., Kuhar, M. J., Hester, L., Snyder, S. H. 1983. Adenosine receptors: Autoradiographic evidence for their location on axon terminals of excitatory neurons. *Science* 220:967–695

Green, R. D., Proudfit, H. K., Yeung, S. -M. H. 1982. Modulation of striatal dopaminergic function by local injection of 5′-*N*-ethylcarboxamide adenosine. *Science* 218: 58–61

Harms, H. H., Wardeh, G., Mulder, A. H. 1979. Effects of adenosine on depolarization-induced release of various radiolabeled neurotransmitters from slices of rat corpus striatum. *Neuropharmacology* 18:577–80

Hirata, F., Axelrod, J. 1980. Phospholipid methylation and biological signal transmission. *Science* 209:1082–90

Hollins, C., Stone, T. W. 1980a. Characteristics of the release of adenosine from slices of rat cerebral cortex. *J. Physiol.* 303:73–82

Hollins, C., Stone, T. W. 1980b. Adenosine inhibition of γ-aminobutyric acid release from slices of rat cerebral cortex. *Br. J. Pharmacol.* 69:107–12

Hütteman, E., Ukena, D., Lenschow, V., Schwabe, U. 1984. Ra adenosine receptors in human platelets. Characterization by 5′-N-ethylcarboxamido[^{3}H]adenosine binding in relation to adenylate cyclase activity. *Nauny-Schmiedebergs Arch. Pharmacol.* 325:226–33

Katims, J. J., Annau, Z., Snyder, S. H. 1983. Interactions in the behavioral effects of methylxanthines and adenosine derivatives. *J. Pharmacol. Exp. Ther.* 227:167–73

Kleihues, P., Kobayashi, K., Hossmann, K. A. 1974. Purine nucleotide metabolism in the cat brain after one hour of complete ischemia. *J. Neurochem.* 23:417–25

Kocsis, J. D., Eng, D. L., Bhisitkul, R. B. 1984. Adenosine selectively blocks parallel fiber-mediated synaptic potentials in the rat cerebral cortex. *Proc. Natl. Acad. Sci. USA.* In press

Lewis, M. E., Patel, J., Moon Edley, S., Marangos, P. J. 1981. Autoradiographic visualization of rat brain adenosine receptors using N^6-cyclohexyl[^{3}H] adenosine. *Eur. J. Pharmacol.* 73:109–10Londos, C., Wolff, J., Cooper, D. M. F. 1979. Action of adenosine on adenylate cyclase. In *Physiological and Regulatory Functions of Adenosine and Adenine Nucleotides,* ed. H. P. Baer, G. I. Drummond, pp. 271–81. New York: Raver

Londos, C., Cooper, D. M. F., Wolff, J. 1980. Subclasses of external adenosine receptors. *Proc. Natl. Acad. Sci. USA* 77:2551–54

Major, P. P., Agarwal, R. P., Kufe, D. W. 1981. Deoxycoformycin: Neurological toxicity. *Cancer Chemother. Pharmacol.* 5:193–96

Marangos, P. J., Patel, J., Martino, A. M., Dilli, M., Boulenger, J. P. 1983. Differential binding properties of adenosine receptor agonists and antagonists in brain. *J. Neurochem.* 42:367–74

Marangos, P. J., Boulenger, J. -P., Patel, J. 1984. Effects of chronic caffeine on brain adenosine receptors: Regional ontogenetic studies. *Life Sci.* 34:899–907

Marani, E. 1977. The subcellular distribution of 5′-nucleotidase activity in mouse cerebellum. *Exp. Neurol.* 57:1042–48

Michaelis, M. L., Michaelis, E. K., Myers, S. L. 1979. Adenosine modulation of synaptosomal dopamine release. *Life Sci.* 24:2083–92Morgan, P. F., Lloyd, H. G. E., Stone, T. W. 1983. Benzodiazepine inhibition of adenosine uptake is not prevented by benzodiazepine antagonists. *Eur. J. Pharmacol.* 87:121–26

Murphy, K. M. M., Snyder, S. H. 1981. Adenosine receptors in rat testes: Labeling with [^{3}H]cyclohexyladenosine. *Life Sci.* 28:917–20

Murphy, K. M. M., Snyder, S. H. 1982. Heterogeneity of adenosine A_1 receptor binding in brain tissue. *Mol. Pharmacol.* 22:250–57

Murphy, K. M. M., Goodman, R. R., Snyder, S. H. 1983. Adenosine receptor localization in rat testes: Biochemical and autoradiographic evidence of association with spermatocytes. *Endocrinology* 113:1299–1305

Murray, T. F. 1982. Up-regulation of rat cortical adenosine receptors following chronic administration of theophylline. *Eur. J. Pharmacol.* 83:113–14

Patel, J., Marangos, P. J., Skolnick, P., Paul, S. M., Martino, A. M. 1982. Benzodiazepines are weak inhibitors of [^{3}H]nitrobenzylthionosine binding to adenosine uptake sites in brain. *Neurosci. Lett.* 29:79–82

Peroutka, S. J., Snyder, S. H. 1980. Relationship of neuroleptic drug effects at brain dopamine, serotonin, alpha-adrenergic and histamine receptors to clinical potency. *Am. J. Psychiatr.* 137:1518–22

Pert, C. B., Kuhar, M. J., Snyder, S. H. 1976. Opiate receptors: Autoradiographic localization in rat brain. *Proc. Natl. Acad. Sci. USA* 73:3729–33

Phillis, J. W., Wu, P. H. 1981. The role of adenosine and its nucleotides in central synaptic transmission. *Prog. Neurobiol.* 16:187–239

Phillis, J. W., Wu, P. H. 1983. Roles of adenosine and adenine nucleotides in the central nervous system. In *Physiological and Pharmacology of Adenosine Derivatives,* ed. J. W. Daly, Y. Kuroda, J. W. Phillis, H. Shimizu, M. Vi, pp. 219–36. New York: Raven

Premont, J., Perez, M., Bockaert, J. 1977. Adenosine-sensitive adenylate cyclase in rat striatal homogenates and its relationship to dopamine and Ca^{2+}-sensitive adenylate cyclases. *Mol. Pharmacol.* 13:662–70Pull, I., McIlwain, H. 1977. Adenine mononucleotides and their metabolites liberated from and applied to isolated tissues of the mammalian brain. *Neurochem. Res.* 2:203–16

Radulovacki, M., Virus, R. M., Djuricic-Nedelson, M., Green, R. D. 1983. Hypnotic effects of deoxycoformycin in rats. *Brain Res.* 271:392–95

Radulovacki, M., Virus, R. M., Djuricic-Nedelson, M., Green, R. D. 1984. Adenosine analogs and sleep in rats. *J. Pharmacol. Ther.* 228:268–74

Rall, T. W. 1980. Central nervous system stimulants. The xanthines. Chapter 25 In *Pharmacological Basis of Therapeutics (Sixth Ed),* ed. A. G. Gilman, L. S. Goodman, A. Gilman, Ch. 25, pp. 592–607. New York: Macmillan

Rehncrona, S., Siesjo, B. K., Westerberg, E. 1978. Adenosine and cyclic AMP in cerebral cortex of rats in hypoxia, status epilepticus and hypercapnia. *Acta Physiol. Scand.* 104:453–63

Ribeiro, J. A., Sa-Almeida, A. M., Namorado, J. M. 1979. Adenosine and adenosine triphosphate decrease ^{45}Ca uptake by synaptosomes stimulated by potassium. *Biochem. Pharmacol.* 28:1297–1300

Satin, A., Rall, T. W. 1970. The effect of adenosine and ademine nucleotides on the cyclic adenosine 3′, 5′-phospate content of guinea pig cerebral cortex slices. *Mol. Pharmacol.* 6:13-23.

Sawynok, J., Jhamandas, K. H. 1976. Inhibition of acetylcholine release from cholinergic nerves by adenosine, adenine nucleotides and morphine: Antagonism by theophylline. *J. Pharmacol. Exp. Ther.* 197:379–90

Schoenfeld, C., Amelar, R. D., Dubin, L. 1975. Stimulation of ejaculated human spermatozoa by caffeine. *Fertil. Steril.* 26:158–61

Schubert, P., Komp, W., Kreutzberg, G. W. 1979. Correlation of 5′-nucleotidase activity and selective transneuronal transfer of adenosine in the hippocampus. *Brain Res.* 168:419–24

Schwabe, U., Trost, T. 1980. Characterization of adenosine receptors in rat brain by (−)[^{3}H]N^6-phenylisopropyladenosine. *Naunyn-Schmiedeberg's Arch. Pharmacol.* 313:179–87

Scott, T. G. 1967. The distribution of 5′-nucleotidase in the brain of the mouse. *J. Comp. Neurol.* 129:97–113

Seeman, P. 1980. Brain dopamine receptors. *Pharmacol. Rev.* 32:229–313

Snyder, S. H. 1984. Drug and neurotransmitter receptors in the brain. *Science.* 224:22–31Snyder, S. H., Katims, J. J., Annau, Z., Bruns, R. F., Daly, J. W. 1981. Adenosine receptors and behavioral actions of methylxanthines. *Proc. Natl. Acad. Sci. USA* 78:3260–64

Stone, T. W. 1981. Physiological roles for adenosine and adenosine 5′-triphosphate in the nervous system. *Neuroscience* 6:523–55

Toll, L., Snyder, S. H. 1982. Solubilization and characterization of histamine H_1 receptor in brain. *J. Biol. Chem.* 257:13593–13601

Trost, T., Schwabe, U. 1981. Adenosine receptors in fat cells: Identification by (−)N^6[^{3}H]phenylisopropyladenosine binding. *Mol. Pharmacol.* 19:228–35

Van Calker, D., Muller, M., Hamprecht, B. 1979. Adenosine regulates via two different types of receptors: The accumulation of cyclic AMP in cultured brain cells. *J. Neurochem.* 33:99–1005

Williams, M., Risley, E. A. 1980. Biochemical characterization of putative purinergic receptors by using 2-chloro[^{3}H]adenosine, a stable analog of adenosine. *Proc. Natl. Acad. Sci. USA* 77:6892–96

Winn, H. R., Welsh, J. E., Rubio, R., Berne, R. M. 1980. Changes in brain adenosine during bicuculline-induced seizures in rats. Effects of hypoxia and altered systemic blood pressure. *Circ. Res.* 47:568–77Wojcik, W. J., Neff, N. H. 1983a. Location of adenosine release and adenosine A_2 receptors to rat striatal neurons. *Life Sci.* 33:755–63

Wojcik, W. J., Neff, N. H. 1983b. Adenosine A_1 receptors are associated with cerebellar granule cells. *J. Neurochem.* 41:759–63

Wu, P. H., Phillis, J. W., Bender, A. S. 1981. Do benzodiazepines bind at adenosine uptake sites in CNS? *Life Sci.* 28:1023–31

Wu, P. H., Phillis, J. W., Thierry, D. L. 1982. Adenosine receptor agonists inhibit K^+-evoked Ca^{2+} uptake by rat brain cortical synaptosomes. *J. Neurochem.* 39:700–8

Yeung, S. -M. H., Green, R. D. 1984. [^{3}H]5′-

N-ethylcarboxamide adenosine binds to both Ra and Ri adenosine receptors in rat striatum. *Naunyn-Schmiedeberg's Arch. Pharmacol.* 325:218–25

Zetterstrom, T., Ungerstedt, V. U., Tossman, U., Jonzon, B., Fredholm, B. B. 1982. Purine levels in the intact rat brain. Studies with an implanted perfused hollow fibre. *Neurosci. Lett.* 29:111–15

Ann. Rev. Neurosci. 1985. 8:125–70

BIRDSONG: FROM BEHAVIOR TO NEURON

Masakazu Konishi

Division of Biology, California Institute of Technology, Pasadena, California 91125

INTRODUCTION

The study of birdsong has made significant contributions to the development of modern ethology. Concepts such as species-specificity in animal signals, innate predisposition in learning, and sensory templates for motor development were put forth first in birdsong research (Marler 1957, 1964, Konishi 1965b, Hinde 1982). Also, it was the study of song development that elevated the much debated issue of instinct versus learning from the realm of semantic discourse and confusion to an experimentally tractable subject (Marler 1983).

The recent discovery of neural substrates for song has introduced a new dimension to the study of birdsong, making integration of behavioral and neurobiological studies feasible (Nottebohm et al 1976). Neurobiological concepts and methods are now directly applicable to this field. This integrated approach can address not only some of the outstanding issues that arose in behavioral studies and that are refractory to further behavioral analysis, but it also makes birdsong an attractive subject for the study of such basic issues as neural coding, learning, memory, developmental plasticity, and sensorimotor coordination. In this review I shall examine critically the major current issues and ideas in this field, placing special emphasis on the topics related to the development, learning, and neural control of song. Because extensive listings and reviews of recent literature on birdsongs are available (Kroodsma & Miller 1982a,b), the references cited are limited to those essential for the discussion of facts and theories on selected topics.

THE ETHOLOGY OF BIRDSONG

What is a Song?

A song is different from a call. Although the common usage "birdcall" includes song in nontechnical and old descriptions, the two types of vocalization are

0147–006X/85/0301–0125$02.00

clearly distinguished in ornithological and ethological literature. The song is usually the longest and most elaborate of the vocalizations produced by a bird. In many species only sexually mature males sing during breeding season, but in some species females sing also. Song duration varies from about 2 sec, as in many species, to tens of seconds, as in those which deliver a series of brief sounds in bouts of seconds. The delivery of songs is periodic, especially in species with short songs. Many birds can sing as frequently as one song every 7 sec. Singing can occur spontaneously; an isolated bird starts and stops singing without any external stimulus. Songbirds usually choose fixed locations such as high perches in their territories for singing. A particular body posture accompanies singing.

All of the above attributes are lacking in calls. These are usually brief, simple sounds and occur neither spontaneously nor periodically but in response to a particular stimulus, such as the presence of predators and filial members. Calls involve neither the use of specific perches nor typical body posture.

Although crowing of roosters and cooing of doves are perhaps functionally equivalent to the song of songbirds, the term, "birdsong," is reserved for the song of passerine (perching) birds. This review deals mainly with the study of passerine songs, for most of the interesting issues and ideas in this field concern them.

Acoustic Structure and Terminology of Song

Song consists of a series of sounds with silent intervals between them (Figure 1). The most elementary unit of song appears as a continuous marking on a soundspectrogram. This is called a "note" or an "element." The acoustic structure of a note can vary from a steady narrow-band sound to a complex frequency- and amplitude-modulated sound. Two or more notes may group together to form a "syllable." In some cases notes and syllables are identical, i.e. one-note syllables, which are classified as syllables because they are spaced

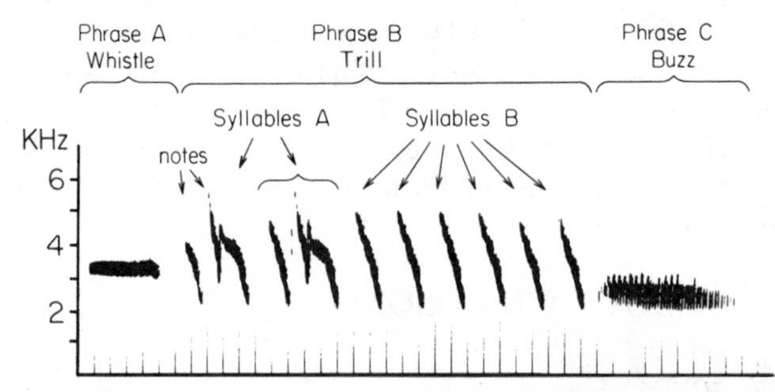

Figure 1 The terminology of birdsong. The time-frequency soundspectrogram of a white-crowned sparrow's song shows how its component sounds and their groupings are named. The ordinate and abscissa are frequency in kilohertz and time in 50 msec per division, respectively.

like multinote syllables. The component notes and their spacing characterize the structure of a syllable. Syllables are often grouped together to form a "phrase," which may contain a series of identical or different syllables. Phrases consisting of one syllable or note also occur and are identified as such by their positions in the temporal structure of song. Many birds deliver several different phrases in a fixed sequence as a unit, which is the song. Birds like mockingbirds and thrushes produce syllables in groups and in fixed or variable sequences. Completely random sequencing of phrases and syllables seldom occurs. The rules of timing and sequencing of phrases and syllables are called the syntactical rules and are usually unique to a species.

Song Repertoire

The whole set of different song types sung by a bird is called its song repertoire. Its size ranges from one, as in the white-crowned sparrow (*Zonotrichia leucophrys*), to several hundreds, as in the winter wren (*Troglodytes troglodytes*) (Kroodsma 1980). The term, "syllable repertoire," applies to those birds which deliver syllables in a continuous series instead of in a short song. Good singers usually have large syllable repertoires; for example, a brown thrasher (*Toxostomia rufum*) may have as many as 2000 syllables (Kroodsma & Parker 1977).

Species-Specificity, Dialects, and Individual Differences

Each species of bird has a unique song or songs. This rule applies both to birds with a single song type and to those with a large song repertoire (Figure 2). In many species, songs are so specific that naturalists can use them to identify birds of a single species from widely separate areas such as the East and West coasts of North America. Song sparrows (*Zonotrichia melodia*), mockingbirds (*Mimus polyglottos*), American robins (*Turdus migratorius*), and chipping sparrows (*Spizella passerina*), to name a few, can be identified by their songs anywhere in North America. However, in addition to universal features, many songs show distinct geographical variations, known as "song dialects" (Figure 3) (Marler & Tamura 1962, Baptista 1975, Mundinger 1982). These dialects are transmitted culturally: Young birds or even adult birds of some species copy the songs of their birthplace or neighbors. Song dialects are usually stable: In one of the most carefully studied species, dialects unique to small restricted areas persisted for at least 12–15 years, well beyond the life span of marked individuals (Payne et al 1981).

Birdsongs show individual differences as well as universal and local features. For example, in Oregon juncos (*Junco oreganus*) and swamp sparrows (*Zonotrichia georgiana*), all songs consist of a trill, hence a universal feature, but the syllables making up the trill vary among individuals (cf Figure 7) (Konishi 1964, Marler & Peters 1982c).

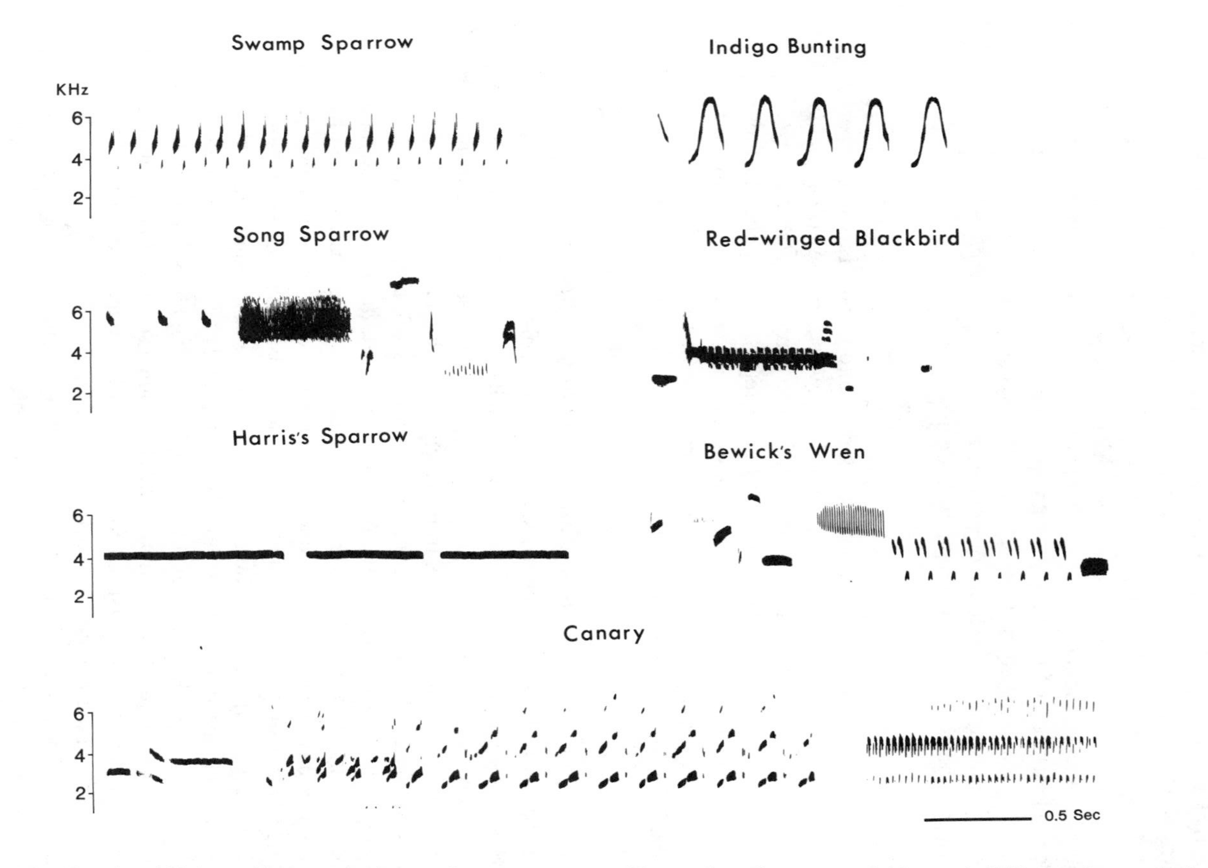

Figure 2 The diversity of birdsong. Each species has a unique song or songs. The number of songs or syllables per individual varies greatly from species to species. Most of the species in this figure sing more than one song type. Some birds sing short songs and others long series of syllables. The canary belongs to the latter group and a short segment of its song is shown here.

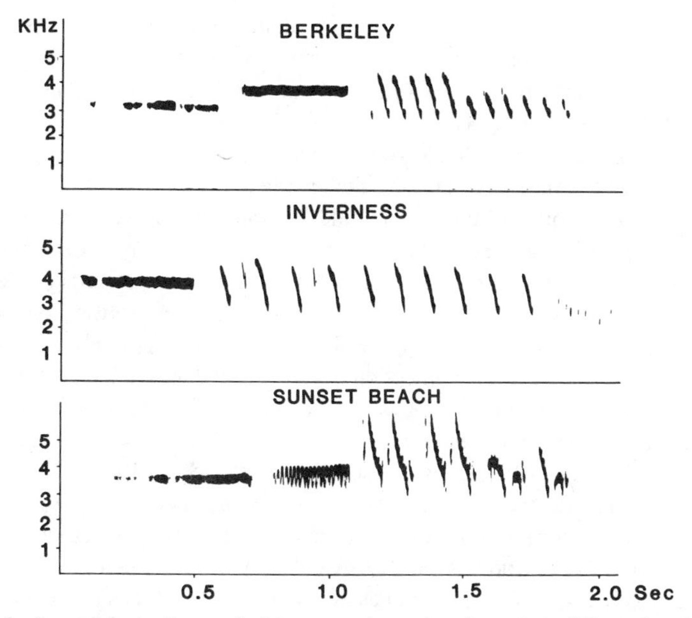

Figure 3 Song dialects. Songs of white-crowned sparrows from three different locations in coastal California. Song dialects are transmitted culturally from generation to generation and they persist for many years (from Konishi 1965b).

Songs as Social Signals

Birds sing to communicate. To a female the song may signal the presence of a potential mate, his species and individual identity, his location, his ownership of territory, his readiness to breed, and perhaps his potential as a provider for his offspring. To a male the song may convey different information, such as "neighbor" versus "stranger," in addition to species and other messages (Marler 1961).

Experimental and comparative observations are used for the study of the message contents of songs. In the experimental approach the receiver's response is the only clue to the messages conveyed by a song. If a song can carry more than one message, whether or not the receiver's response allows discrimination between different messages is a crucial question. Species identity is the message that can be most easily recognized in a song. Breeding male songbirds stake out and defend an area against male intruders of their own species. Song is one of the clues they use to recognize other males and it is thought to advertise the ownership of territories. Territorial males react to playback of recorded songs by counter-singing, aggressive display, and approach or avoidance of the speaker. Although many experiments demonstrate the bird's ability to discriminate between conspecific and alien songs, they do not prove the advertising role of songs. However, recent studies seem to demonstrate this role: Broadcasting recorded song prevents new males from trespassing and occupying vacant territories, which would otherwise be rapidly claimed (Krebs et al 1978, Yasukawa 1981).

Although the effects of male vocal signals on the behavior and reproductive physiology of female birds are well documented in some birds (e.g. Brockway 1969), what a song means to a female songbird is not well understood. In contrast to territorial males, female birds seldom show any overt response to song playback. However, as in female frogs, female indigobirds (*Vidua chalybeata*) in breeding condition approach a loudspeaker emitting the song of their own species (Payne 1973). Breeding condition brought about by manipulation of photoperiod or by injection of estrogen induces female birds to show precopulatory behavior in response to the conspecific song in white-crowned sparrows (*Zonotrichia leucophrys*) (Baker et al 1981), cowbirds (*Molothrus ater*) (West et al 1979), and swamp sparrows (Searcy et al 1981). Thus, female songbirds are selective for the songs of their own species.

The capacity for multiple song types is one of the most intriguing attributes of birds (Krebs & Kroodsma 1980). Are the songs redundant or do they have different functions? When a bird has 300 song types, as does the winter wren, their functional separation would seem unrealistic. For birds with smaller repertoires there is some indication that different song types serve different purposes, such as territorial advertisement and mate attraction (Catchpole 1983). However, such functional separation seems to be rare, although the lack of methods sensitive enough to detect it may be the main reason for this rarity. If several song types of a bird serve the same function, say, territorial advertisement, the advantage of such a scheme is an interesting issue for sociobiologists. One imaginative explanation is the Beau Geste hypothesis, which assumes, first, that territorial settlers avoid an area occupied by many individuals and, second, that many songs from an individual deceive intruders in their assessment of population density (Krebs 1977). Experiments in which small or large song repertoires were broadcast in vacant territories showed that large repertoires were more effective than small repertoires in preventing potential intruders from trespassing the territories (Krebs et al 1978, Yasukawa 1981). Despite these observations, there is no reason to believe that birds avoid a high density area or remain deceived by multiple song types for any length of time, although any deterrent effects of song that increases a bird's inclusive fitness can make a difference in evolutionary time.

Another explanation for a large song repertoire uses the concept of sexual selection: If males with larger song repertoires leave more offspring than those with smaller repertoires, the frequency of genes for large repertoires should increase. The effects of sexual selection should be much more pronounced in polygynous (many females for one male) species than in monogamous species.

This prediction seems to be true of some cases but not of others; among North American wrens, polygynous species tend to have larger song repertoires than monogamous species (Kroodsma 1977b). On the other hand, many monogamous species sing more than one song type. One of the problems in the comparative approach is its inability to distinguish cause and effect from indirect correlations. For example, in indigobirds the amount of singing, and not song repertoire, is positively correlated with mating success (Payne & Payne 1977). In the great tit (*Parus major*), although song repertoires do not influence the female's choice of mates, males with intermediate or large repertoires "are more likely to father offspring which survive to breed than males with fewer song types." Also, "males with larger repertoires produce heavier fledglings, and this may be related to territory quality" (McGregor et al 1981). The significance of song repertoires was also studied in a laboratory experiment, which avoided the difficulty of establishing a direct correlation in the field. Two groups of virgin female canaries were exposed either to a tape-recorded song made of five syllable types chosen from normal canary songs or to one made of 35 syllable types. The group that heard the more elaborate song built nests faster and laid more eggs than the other group. One interpretation of these results is that the elaborate song stimulated the females reproductive behavior more than the less elaborate song. Another interpretation is that the females responded better to the more familiar song consisting of 35 syllable types, as in the normal song, than to an unusual song containing only 5 syllable types (Kroodsma 1976). A control for the familiarity problem would be to use females that have never heard normal canary songs.

How Songs Encode Species-Specificity

A song contains many potential acoustic cues for encoding messages. Because the response of males to their species song is easy to observe and measure, most studies have addressed the encoding of species-specificity in the territorial context. The methods consisted mainly of comparing birds' responses to songs in which various properties were deleted, added, or altered. A few obvious generalizations follow from these studies:

1. Song properties common to all individuals serve in species recognition, but not all common properties are used by the birds for species recognition.
2. A small number (two to three) of song properties are usually sufficient for species recognition.
3. Different cues may be additive or redundant.
4. Different species may use different aspects of song for species recognition.

The physical properties of songs should allow certain predictions to be made about the potential cues available to the birds. Many birdsongs are composed of tonal sounds containing energy in discrete frequency bands, although noises occur in the songs of some species (cf Figure 2). In tonal songs, changes in frequency or amplitude as a function of time are potential cues. Frequency and amplitude modulations occur together or separately both within and between phrases, syllables, or notes. Syllable or note structure is important for species recognition in birds with simple trill type songs in which frequency modulation occurs only within the syllable or note; examples are swamp sparrows (Peters et al 1980), indigo buntings (*Passerina cyanea*) (Shiovitz 1975), and firecrests (*Regulus ignicapillus*) (Becker 1976). Many species sing songs in which the frequency changes between phrases or syllables. This feature is important for species recognition in European robins (*Erithacus rubecula*) (Bremond 1968), white-throated sparrows (*Zonotrichia albicollis*) (Falls 1963), and goldcrests (*Regulus regulus*) (Becker 1976), for example. The alternating pattern of sounds and silent intervals, a type of amplitude modulation, is used by some birds for species recognition, for example, ovenbirds (*Seiurus aurocapillus*) (Falls 1963), indigo buntings (Emlen 1972), song sparrows (Peters et al 1980), and red-winged blackbirds (*Agelaius phoeniceus*) (Brenowitz 1983). However, when a song contains all of the above properties, it is difficult to predict which of the properties encodes species-specificity.

Despite studies such as those cited above, the acoustical properties of song that encode species-specificity are much better defined in frogs and orthopterans than in songbirds (e.g. Capranica 1966, Gerhardt 1978, von Helversen 1971, Walker 1957). The use of purely synthetic songs in playback experiments is largely responsible for the success in the frog and insect studies. The development of methods to analyze, synthesize, and modify birdsongs rapidly will promise to elevate the level of analysis from simple rearrangement of natural song components to the testing of synthetic songs with precisely defined acoustical properties (cf Margoliash 1983). This approach, in turn, may lead to the discovery of general principles of coding.

Another problem that confounds the study of song recognition is learning; unlike frogs and insects, birds learn not only to sing but also to recognize their species song. In the species studied so far, the abnormal songs of birds raised in isolation elicited no response in territorial males (Shiovitz 1975). Also, birds can recognize different dialects and the songs of different individuals (Emlen 1971, Brooks & Falls 1975). These facts point to the importance of learning in song recognition by adult birds. If different individuals learn to use different sets of cues, finding common cues becomes difficult. Thus, the apparent lack of general rules of coding may well be due to learning.

THE ONTOGENY OF SONG

An animal's ability to select a complex stimulus or to produce a specific movement does not appear suddenly in adulthood but develops gradually. How it develops has been one of the central issues in ethology and psychology. The plasticity of song development has been known since ancient times in the Orient. The Japanese zoologist Tamiji Kawamura (1947) wrote, in his book *Science of Birdsong,* how well bird fanciers in old Japan knew about the characteristics of song development. Such properties as the impressionable phase of song learning, the effects of tutoring, subsong, imitation, song dialects, and species differences in these attributes were well known to them. Kawamura pointed out that birdsongs would offer unique opportunities to address psychological issues, particularly behavioral development. Otto Koehler (1951) in Freiburg, Germany and William Thorpe (1954, 1961) in Cambridge, England were the first to study song development under controlled laboratory conditions. Thorpe was the first to use the tape recorder and soundspectrograph in the study of song development. The questions he asked and the methods he used were adopted by many subsequent workers in this field.

Stages of Song Development

Few behaviors show developmental changes more graphically than birdsongs. The stereotyped song of an adult bird, which is called "full song," develops in several stages. The methods of study involve tracking distinct acoustic patterns in the full song from earlier stages of song development. Song development may begin with the calls of the nestling, but an objective study of this stage has not been made. There may be a long period of reduced vocal activity between the fledgling and juvenile stages. Juvenile birds start vocalizing more frequently and persistently in their first autumn when they are three to four months of age. They do not yet assume the typical singing posture of the adult, and their vocalization, termed "subsong," consists of a rambling series of sounds. Subsong tends to be longer and more variable, and its frequency range is wider and more variable, than those of full song. Some of the sounds of subsong gradually become discrete and recurrent as the birds begin to vocalize more loudly and assume the typical singing posture more frequently. The basic organization of the species song is clearly recognizable at this stage, although the syllables and song duration may still be somewhat variable. The transition from the plastic song to the full song stage usually occurs rapidly. The structure of syllables, the sequence of phrases, and the duration of song all become fixed or "crystallize" to the level of full song. Once a song is crystallized, it recurs unchanged in subsequent years in most birds.

Overproduction and Attrition of Syllables

Song development, at least in some species, is not a steady progression from precursor sounds to crystallized syllables. For example, young swamp sparrows, during their plastic song stage, produce syllable types that are never used in their final songs. These syllables simply disappear before the final stage of song crystallization (Marler & Peters 1982a,b). Although the phenomenon of overproduction and attrition of syllables has been studied closely only in one species, it may be universal (e.g. red-winged blackbirds, Marler et al. 1972, white-crowned sparrows, Marler 1970; nightingales, Todt & Hultsch 1985). Kawamura (1947) described "attrition" and "unused songs" during song development in white-eyes (*Zosteropus palpebosa*) and Japanese meadow buntings (*Emberiza cioides*). Birds tutored with alien songs sometimes produce good copies of the alien syllables during the subsong or plastic song stage, but abandon or modify the sounds by the time of song crystallization (Thorpe 1961, M. Konishi and E. Akutagawa, unpublished).

Why do birds overproduce syllables? It may be a method by which birds review and calibrate their vocal repertoires, because they cannot know what sounds they can produce before they vocalize. Birds must first try to produce various sound patterns and select those which satisfy the criteria set by genetical and experiential influences. Birds seem to develop song by modifying and selecting sounds as they produce them. Instructional and selectional processes appear to occur in series as well as in parallel at several levels of song organization. This strategy is largely responsible for the plasticity of song development.

Theory of Song Learning: An Overview

The synonymous use of "song learning" and "song development" has caused some confusion in birdsong research. A bird's use of auditory feedback in song development resembles learning by trial and error; the bird corrects errors in vocal output until it matches the intended pattern. Because feedback control of voice is essential for song development with or without a tutor model, song development inevitably contains some element of learning. However, it is useful to distinguish development from learning, because some normal features of song develop without auditory feedback in some species (Guettinger 1981, Konishi 1965a, Marler & Sherman 1982). For this reason I shall refer to copying of song as "song imitation" instead of "song learning." The term "song learning" refers explicitly to song development by auditory feedback control of voice. "Song development" will be used only in a descriptive sense.

Song imitation consists of sensory and sensorimotor stages. Birds are predisposed to learn the song of their own species. They must hear an acceptable song during a period in their youth, and they commit the song to memory without vocally reproducing it. This period will be referred to as the "impressionable phase." During the sensorimotor stage birds must hear themselves vocalize in order to match vocal output with the memory trace of the tutor song. Thus, the song memory is used like a template (cf Figure 12). When birds develop song without any song tutor, they use an innate template or an internal reference for the control of voice by auditory feedback. The innate template explains the differences between the songs of birds raised in acoustic isolation and those of birds deafened in youth (Konishi 1965b).

What is an Innate Song?

All songbirds studied so far, when raised in acoustic isolation, develop songs lacking some or most of the characteristics of their wild-type song (Figure 4). These birds are called "isolates" and their songs "isolate songs." Isolate songs are sometimes referred to as "innate songs." Interspecific differences in isolate songs are likely to be due to genetic differences between species. The source of instruction for isolate songs must be internal, because the bird hears no song to copy. These reasons justify the definition of an innate song as above. However, this practice of naming innate songs creates an interesting logical dilemma, when one considers how song learning is defined. As pointed out above, auditory feedback control of voice is regarded as learning. Because the development of innate songs requires auditory feedback, the innate songs are learned! By this logic only the sound patterns that develop without auditory feedback can be called innate.

Normal songs seldom develop in acoustic isolation, but some aspects of the song develop more normally than others. The frequency range, tonal quality, and duration of song tend to be least affected by isolation. The first two attributes are perhaps due to the properties of the vocal organ, the syrinx, itself. Also, syntactical rules of song seem less affected by isolation, because some of them are evident in isolate songs of many species. Moreover, in some species the syntactical rules resist tutoring and even interspecific hybridiza-

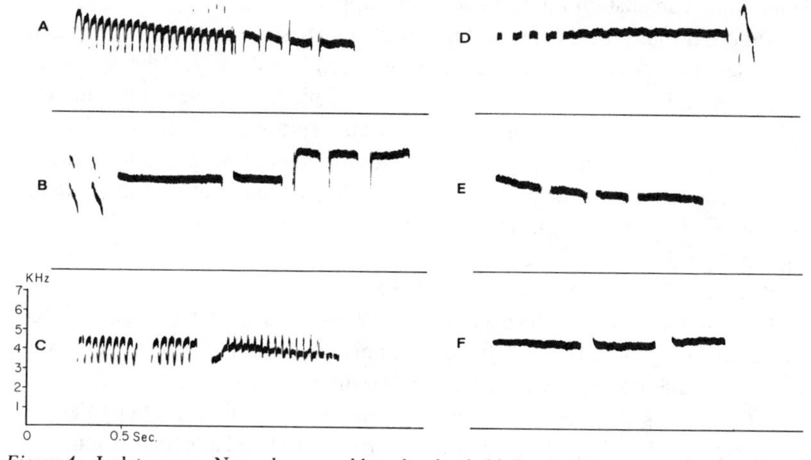

Figure 4 Isolate songs. Normal songs seldom develop in birds raised in acoustic isolation. Their songs lack some or most of the features typical of the wild-type song, as exemplified here by the songs of two white-crowned sparrows that grew up in acoustic isolation from the egg (*A*,*B*). Also, birds that failed to copy a tutor song develop isolate-like songs (*C*–*F*) (from Konishi 1978).

tion. For example, in cross-fostering experiments, a greenfinch (*Chloris chloris*) raised by canaries sang canary syllables, not by the canary rules, but by the greenfinch rules (Guettinger 1979). Hybrids between canaries and greenfinches sang syllables by the rules of the maternal species, which they could not copy from their nonsinging mothers (Guettinger et al 1978). On the other hand, isolate syllables generally contain simpler patterns of frequency modulation than those found in the wild-type song.

Song Imitation

Bird fanciers in old Japan paid fees to have young birds tutored by prize-winning singers. Both captive and wild birds served as tutors. A person would erect a small hut near a skylark singing "good" song, in order to charge admission for tutoring. If older pupils started to chirp while younger pupils were listening quietly, the chirpers were immediately removed for fear that younger pupils would learn bad sounds (Kawamura 1947). Although vocal mimicry by some wild birds such as mockingbirds and starlings (*Sturnus vulgaris*) was well known, imitation of the species' own song as a method for normal song development was not widely appreciated by professional ornithologists. Sharing of songs or syllables among birds living in close proximity, i.e. a song dialect, was considered a sign of imitation. Isolation experiments suggested the need for imitation in normal song development, but more rigorous evidence came from tests with song playback (Thorpe 1958, 1961). When young birds heard normal songs broadcast to their isolation chambers, they developed normal songs as adults, indicating that exposure to appropriate song alone can sustain normal song development.

Although many birds can imitate all characteristics of a tutor song, some rules seem to govern song imitation. As mentioned above, the syntactical rules tend to be refractory to modification by imitation. Notes in a syllable or syllables in a phrase are theoretically separable entities, yet birds usually copy a syllable or phrase as a unit. During the plastic song stage, song sparrows and swamp sparrows rearrange copied syllables without breaking them up into their component notes (Marler & Peters 1982a). Nightingales combine syllables and phrases from different tutor songs to produce new songs in addition to straight copying (Todt & Hultsch 1985).

The amount of playback necessary for copying a tutor song has not been systematically studied. It may be surprisingly small, as in the following examples. Nightingales copied a sequence of 60 different songs broadcast once per day for 20 days from a loudspeaker placed next to a silent nightingale (Todt & Hultsch 1985). Two white-crowned sparrows imitated a recorded song after exposure to it for 20 and 21 days, respectively (Marler 1970). Similarly, young swamp sparrows tutored with a recorded song for ten days reproduced the song as adults (cited in Kroodsma 1982).

The methods and conditions of tutoring greatly affect the outcome of an experiment. Training birds singly or in a group, the choice of live or tape-recorded tutors, social relationship between tutors and pupils, different photoperiods, and daily rhythm affect the success of tutoring. Birds that seldom copy even conspecific songs from playback are said to imitate live tutors (Baptista & Petrinovich 1984, Kroodsma 1982, Rice & Thompson 1968, Payne 1981). The nightingale experiment cited above shows clearly that visual cues provided by a live bird are an important factor in song imitation. The young birds that could see the caged bird during tape playback copied the tutor songs more accurately than did those prevented from seeing the live bird (Todt & Hultsch 1985). Thus, live tutors can have profound effects on various aspects of song imitation, including song preference, the accuracy of copying, the impressionable phase, and the rate of song development. The effects of live tutors on these aspects are reviewed in appropriate sections below.

Innate Preference for the Species' Own Song

Bird fanciers in old Japan looked down upon birds that incorporated the songs of other species during a singing competition (Kawamura 1947). Most species studied so far copy the song of their own species in a choice situation. Tape playback allows song preference to be tested under standardized conditions. The rate, duration, and intensity of playback can be made equal or similar for different songs to be presented. Thorpe (1958, 1961), using the chaffinch (*Fringilla coelebs*), was the first to show song preference by the tape playback method. Most experiments in this field used birds collected in the wild as nestlings that had presumably heard their fathers and other adults sing. Although five- to ten-day-old nestlings do not copy songs, hearing them might have general effects on the birds such that it biases the future choice of songs. This uncertainty was overcome by the use of canaries as foster parents to raise wild birds from the egg. Swamp sparrows raised in this manner copied the songs of their own species when tutored with both swamp and song sparrow songs (Marler & Peters 1977). White-crowned sparrows also show a preference for their own species' song. However, contrary to the earlier conclusion that white-crowned sparrows reject alien songs whether these are presented alone or with a conspecific song (Marler 1970), birds raised from the egg in isolation imitated alien songs or produced modified versions of them, either in addition to copying the conspecific song in a choice situation or in a no-choice situation involving only alien songs (Figure 5B). Nevertheless, in a choice situation, birds always copied the conspecific song, and when they imitated alien songs in addition, the imitations were seldom as complete and accurate as those of the conspecific model (Figure 5A) (M. Konishi and E. Akutagawa, unpublished).

These and other experiments indicate that young birds must be able to recognize the song of their own species solely by the particular acoustical

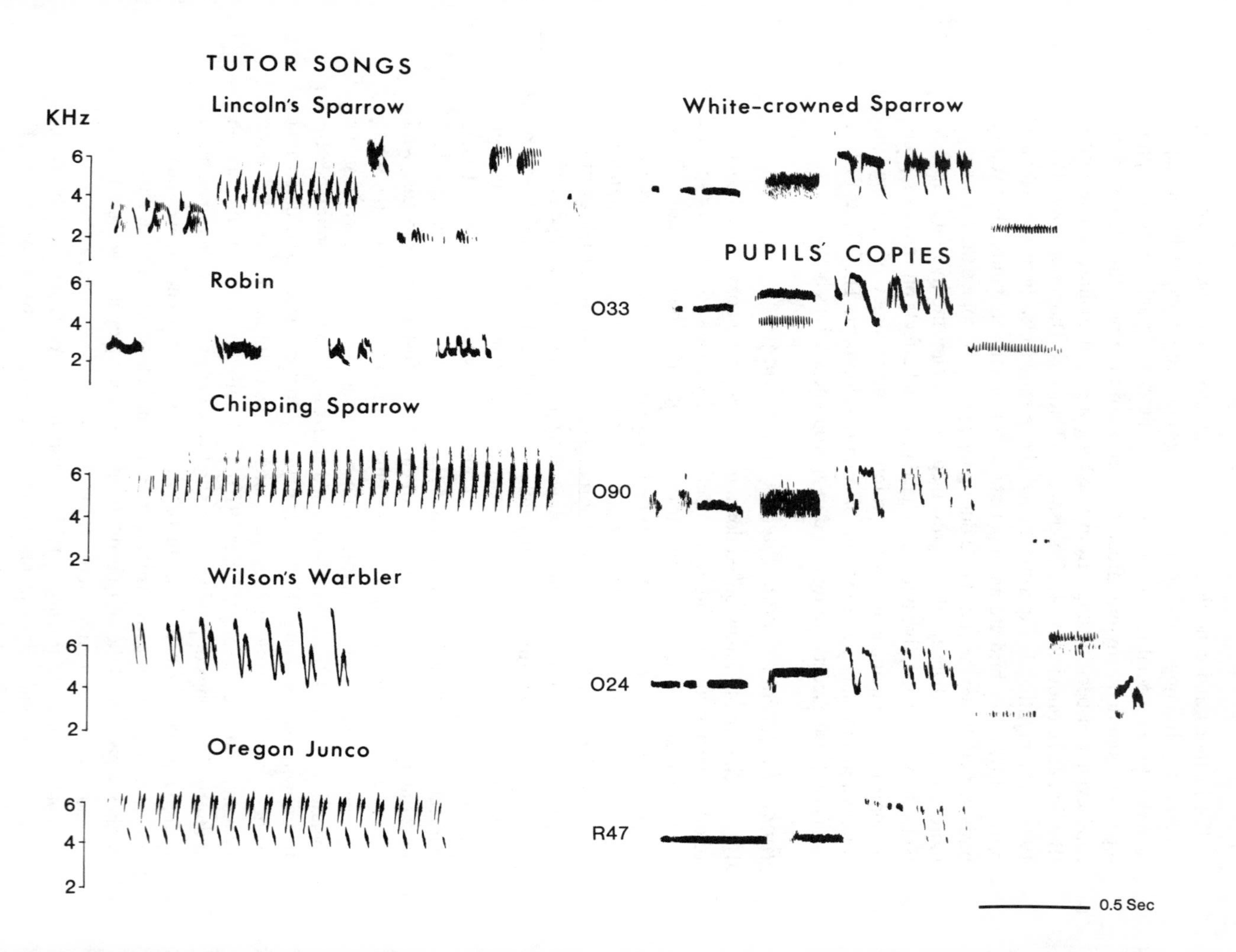
TUTOR SONGS
KHz
6
4
2
Lincoln's Sparrow
Robin
Chipping Sparrow
Wilson's Warbler
Oregon Junco
White-crowned Sparrow
PUPILS' COPIES
O33
O90
O24
R47
0.5 Sec

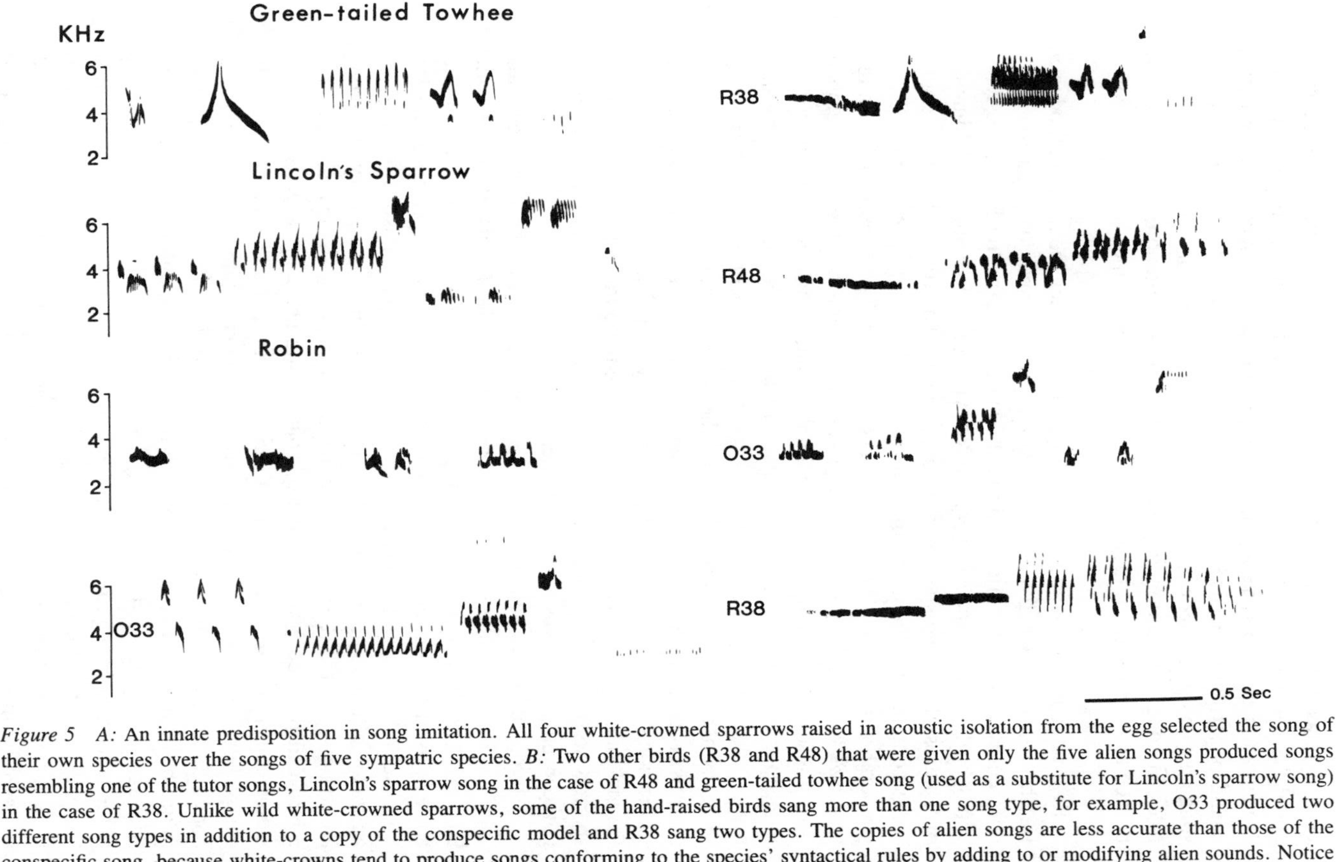

Figure 5 *A:* An innate predisposition in song imitation. All four white-crowned sparrows raised in acoustic isolation from the egg selected the song of their own species over the songs of five sympatric species. *B:* Two other birds (R38 and R48) that were given only the five alien songs produced songs resembling one of the tutor songs, Lincoln's sparrow song in the case of R48 and green-tailed towhee song (used as a substitute for Lincoln's sparrow song) in the case of R38. Unlike wild white-crowned sparrows, some of the hand-raised birds sang more than one song type, for example, O33 produced two different song types in addition to a copy of the conspecific model and R38 sang two types. The copies of alien songs are less accurate than those of the conspecific song, because white-crowns tend to produce songs conforming to the species' syntactical rules by adding to or modifying alien sounds. Notice that the songs of R38 and R48 contain an introductory whistle, which the birds could not have copied from any of the tutor songs. Unlike isolates, tutored white-crowns compose new song types by copying, improvising, and inventing, especially when they are exposed to more than one tutor song (e.g. O33 *right* and *left bottom* and R38 *right bottom*).

cues contained in it. The existing data allow few generalizations to be made about cues in song imitation; they seem to be present at all levels of song organization. In theory, a song may be acceptable because it contains particular sound patterns or because it lacks unacceptable sounds. Some parts of the song may be more important than others and different cues may play different roles in song recognition in the context of song imitation. For instance, a particular sound may promote the copying of other sounds that accompany it (Marler 1984).

Rearrangement of natural song components has been the main method used for the study of selective song imitation. Swamp and song sparrows sing different types of song often within earshot of each other (cf Figure 2). When young birds of these species were tutored with songs containing combinations of features of the two species, the young swamp sparrows selectively copied songs containing syllables from their own species' songs irrespective of their temporal arrangement (Marler & Peters 1977, 1982a). Interestingly, swamp sparrows would copy song sparrow notes and syllables, if these were embedded in a series of swamp sparrow syllables, indicating that a certain cue can induce the birds to copy normally unacceptable sounds. For young song sparrows, either the syllabic structure or the temporal pattern of song seems to serve as a cue for the recognition of the species song. Thus, young song sparrows copied a song composed of swamp sparrow syllables that were arranged in the song sparrow's temporal pattern. They would also copy a song containing song sparrow syllables arranged in the temporal pattern of the swamp sparrow song, but not a song in which swamp sparrow syllables were arranged in the swamp sparrow's temporal pattern.

A bird's ability to select its own species' song led to the hypothesis that each species has a "blueprint" of its song (Thorpe 1961). This concept became equated with that of an innate template, which was developed in another context, as discussed below. The blueprint concept implies a perceptual device. Song selectivity may be due to one such central mechanism or it may be attributable to a combination of different factors, such as auditory and motor constraints, that may be present both in the periphery and in the brain. Because a bird can produce a wide range of sound patterns, song selectivity due to motor constraints is considered unlikely. However, these factors cannot be completely excluded, because, for instance, when white-crowned sparrows copy synthetic songs that sound mechanical to the human ear, they do not sing in mechanical voice but in the voice of their normal song (M. Konishi and E. Akutagawa, unpublished). This is one aspect of motor constraints contributing to the outcome of song imitation. Other constraints might be operative at other levels of song organization. Another problem is the question of which pattern is easier or harder to sing; this cannot be easily answered. Song imitation is assessed by motor performance; i.e. imitation is said to have occurred if the bird produces a copy of the tutor song. If the bird does not produce a copy, it is difficult to assess at which stage of song imitation and why the failure occurred. Therefore, tests that do not involve singing would be useful for assessing song selectivity. One such method measures changes in heart rate: Young swamp sparrows show greater changes in heart rate when a swamp sparrow song is broadcast than when a song sparrow song is played back (Dooling & Searcy 1980).

Tutoring with recorded songs tests a bird's ability to select conspecific songs by sound alone. Live tutors can affect song selectivity. For example, young zebra finches (*Poephila guttata*) that were raised by Bengalese finches (*Lonchura striata*) copied the song of the Bengalese foster father even when they could hear adult zebra finches sing nearby. However, this result does not negate the existence of an innate preference for the species song in zebra finches. Young zebra finches raised by two female Bengalese finches, which do not sing, developed songs composed of syllables from different adult zebra finches in the room (Immelmann 1969). The white-crowned sparrow has served as a model for selective learning. Even in this species live tutors can override the innate predisposition: Young white-crowned sparrows copied the song of a strawberry finch (*Amandava amandava*) both in choice and in no-choice situations, the choice being between white-crown and finch songs (Baptista & Petrinovich 1984). Because the strawberry finch song is somewhat like an isolate song of the white-crown, its acceptance by the white-crowns is not surprising. However, in earlier tape-playback experiments young white-crowns selected their species song over the song of the Harris' sparrow (*Zonotrichia querula*), which also resembles white-crown isolate songs (Figure 2). Similarly, white-crowned sparrows did not copy the song of a song sparrow whether it was presented alone or with a white-crown song (Marler 1970). However, in another experimental paradigm in which live tutors were used, young white-crowns selected the song of a song sparrow serving as a visible tutor over conspecific songs sung by hidden white-crown tutors (L. F. Baptista and L. Petrinovich, in preparation). These findings show that a live tutor can override the white-crown's innate predisposition for the conspecific song. In nature live tutors always sing the species-specific song to which the pupils are predisposed. The innate selectivity may be part of a multifaceted system that ensures normal song development in nature.

The Impressionable Phase of Song Imitation

Another process that facilitates the selective copying of a conspecific song by young birds involves the timing of the impressionable phase of song imitation, which coincides with the period when young birds are most likely to hear their father or other adults sing. Although anecdotal accounts abound, the number of species that have been carefully studied in regard to this phenomenon is small. The white-crowned sparrow provides a classical example. Two white-crowned sparrows were tutored, one with the dialect of its birthplace, the other with an alien dialect, from post-hatching day 8 to day 28 and from day 35 to day 56, respectively. Both birds developed an approximate copy of the tutor song as adults. Two other birds that were tutored from 50 to 71 days of age did not copy the tutor song at all. Five other birds collected as fledglings of 30–100 days of age did not copy the tutor song but developed songs resembling their home dialect. Playback tutoring of a nestling and a first-year adult did not have any specific effect on their song development. These results led to the conclusion that the impressionable phase of song imitation in the white-crowned sparrow is between 10 and 50 days of age (Marler 1970). However, a recent study with live tutors disputes the above conclusion (Baptista & Petrinovich 1984). Ten white-crowned sparrows aged 50 to 54 days were exposed to visible live tutors for 50 days. All of them copied the tutor song, including a strawberry finch song sung either by a strawberry finch or by a white-crowned sparrow that had copied it earlier from the finch. Although these results indicate that a bird's age is not the sole determinant of the impressionable phase, they do not disprove the presence of such a phase. A better experiment would be to test whether the young birds that have been exposed to one live tutor before 50 days of age would copy a second live tutor after 50 days of age.

What terminates the impressionable phase? The duration of this phase does not seem to be rigidly fixed by age alone. It is becoming increasingly evident that the nature of sensory experience is an important determinant in ending the impressionable phase of song imitation.

An excellent study of the impressionable phase using live tutors was carried out with the zebra finch. Because the song of the Bengalese finch is distinctly different from that of the zebra finch, the time and extent of song copying can be conveniently studied by removing the young from their foster parents at different stages of song imitation. Young zebra finches memorize the patterns of component sounds in the song of their foster father by 40 days of age. As they start singing juvenile song at about 40 days of age and continue to develop song up to about 80 days of age, they copy the length and sequence of the component sounds as well as the overall temporal pattern of song from the same tutor. These zebra finches do not imitate conspecific songs after about the eightieth day of life, even when they are brought into close social contact with singing conspecifics for as long as four months (Immelmann 1969, Boehner 1983).

Furthermore, young zebra finches raised by their own parents for the first 35–40 days of life did not imitate the songs of other adults, but reproduced the song of their father as accurately as birds that were allowed to hear their father until the hundredth day of life. In contrast, young zebra finches reproduced the song of their Bengalese foster father less accurately when removed from the father before the fortieth day of life than when removed between the fortieth and sixtieth day (J. Boehner, in preparation). These findings suggest that the impressionable phase of song imitation ends much earlier when young birds copy a conspecific song than when they copy an alien song. It is also well known among students of song development that isolates and birds tutored with alien songs take a much longer time to crystallize song than birds tutored with their species' songs.

Some birds can copy new songs even after song crystallization in the first singing season. Red-winged blackbirds (Yasukawa et al 1981), indigo buntings (Payne 1982), saddlebacks (*Creadion carunculatus*) (Jenkins 1977), indigobirds (Payne & Payne 1977), and Bewick's wrens (*Thrymanes bewickii*) (Kroodsma 1974) copy territorial neighbors when they move to a new area.

Finally, a word of caution is in order. The results of tutoring reported by different authors differ from one another even for the same species. As mentioned above, many factors affect song imitation, and few attempts have been made to standardize the methods and conditions of tutoring. If such seemingly unimportant conditions as early social relationships, tutoring before or after feeding, and the time of day affect song imitation (Todt & Hultsch 1985), different results obtained by different methods and authors may be caused by variations in any number of experimental conditions. There are also species and individual differences in responses to tutoring. Also, the number of birds used in these experiments tends to be small, for obvious reasons. These considerations call for careful scrutiny of claims in this field.

Improvisation and Invention

Most birds do not indiscriminantly copy every feature of the tutor song, especially in playback tutoring, rather they incorporate individual styles in their final songs. Two processes contribute to the development of individual styles: A bird "improvises" a variation on the theme provided by the tutor song (cf Figure 5B). A bird also "invents" a sound pattern that is neither a copy of the tutor model nor an example of isolate sound patterns (Marler et al 1962, Marler & Peters 1982c). Song properties susceptible to improvisation and invention seem to vary among species. In white-crowned sparrows (Marler 1970, Konishi 1978), some isolate Oregon juncos (Marler et al 1962), and red-winged blackbirds (Marler et al 1972) the overall song temporal pattern as well as syllables are subject to improvisation and invention, whereas in hand-reared swamp sparrows (Marler & Peters 1982a,b) only syllable structure is varied. Improvisation and invention are perhaps another manifestation of the instructional and selectional processes operating in an integrated manner.

Memory in Song Imitation

Young birds do not imitate songs immediately like school children learning melodies from their music teacher. Birds listen to and memorize a song first and vocally reproduce it later. The interval between these two events can be as long as several months, during which the bird need neither hear nor rehearse the song. Thus, birds can remember complex acoustical patterns for a long time. In some species the two periods overlap with each other such that the bird continues to copy new sounds after the sensorimotor stage has started. The absence of rehearsal is technically difficult to demonstrate because of the number of recordings that must be examined. Swamp sparrows that had been tutored during the first 60 days of life did not sing at all or sang too infrequently to be recorded during weekly sample monitoring until about 240 days after the last tutoring session (Marler & Peters 1981).

Song Templates

When a bird learns to sing from memory, he must hear himself vocalize. Although this is expected, we must exclude the possibility that the song memory is internally translated into the corresponding vocal motor pattern. Surgical removal of the cochleae makes a bird totally deaf. When a bird becomes deaf before the onset of subsong, he can no longer vocally reproduce the tutor song (Konishi 1965b). Thus, the vocal control system can use the song memory only via the route linking vocal organ, voice, ear, auditory system, and vocal motor system: i.e. the auditory feedback loop (cf Figure 12). All songbirds studied so far develop highly abnormal songs when deafened before the onset of subsong (Konishi & Nottebohm 1969) (Figure 6). The most extreme effect of deafening is the disappearance of all the recognizable structural entities of song; notes, syllables, and phrases may be lacking in the songs of deaf birds. If deaf birds produce distinct notes and syllables at all, these tend to contain abnormal patterns of frequency modulation. Notes and syllables produced by deaf birds usually appear irregular and fuzzy on a soundspectrogram (cf Figure 7). In sharp contrast with the songs of intact birds, the notes and syllables of deaf birds are not repeated in exactly the same form either from song to song or within a song, even though their general patterns are maintained. Despite this short-term instability, deaf birds can maintain to a considerable extent the individual characteristics of their songs in successive years.

If deafness renders the song memory unusable for song development, how does it affect song development when there is no song memory, i.e. without a tutor song? If a bird raised in isolation is deafened, he develops a song different from the song of an intact isolate. The deaf bird's song typically contains significantly fewer normal characteristics than the isolate song. Thus if a bird can hear himself sing, he knows how to produce some of the normal song properties. Furthermore, the effects of deafening on song development are the same between tutored and untutored birds. Birds use hearing to prevent the developing song from deviating from the intended pattern. In other words, birds use auditory feedback to match their voice with the intended pattern serving as a template. The template can be acquired either by learning, as in tutoring, or by inheritance, as in rearing in isolation. This was how the concept of templates was formulated (Konishi 1965b, Marler 1964).

If isolates use an innate template, why should they not develop uniform song patterns? In some species, isolates show large individual differences in song (cf Figure 4) (Konishi 1978). Perhaps, the innate template only crudely defines the species song, leaving considerable freedom for imitation, improvisation, and invention. Another question about the innate template is whether it is good enough to guide the development of a normal song in any species. As mentioned before, isolates do develop some properties of the normal song in some species, but no bird produces completely normal songs in isolation. Although song sparrow isolates had been thought to develop normal songs (Mulligan 1966), a subsequent study did not support this claim (Kroodsma 1977a). The effects of deafening show large species differences; again, deaf white-crowned sparrows fail to develop most of the normal song properties,

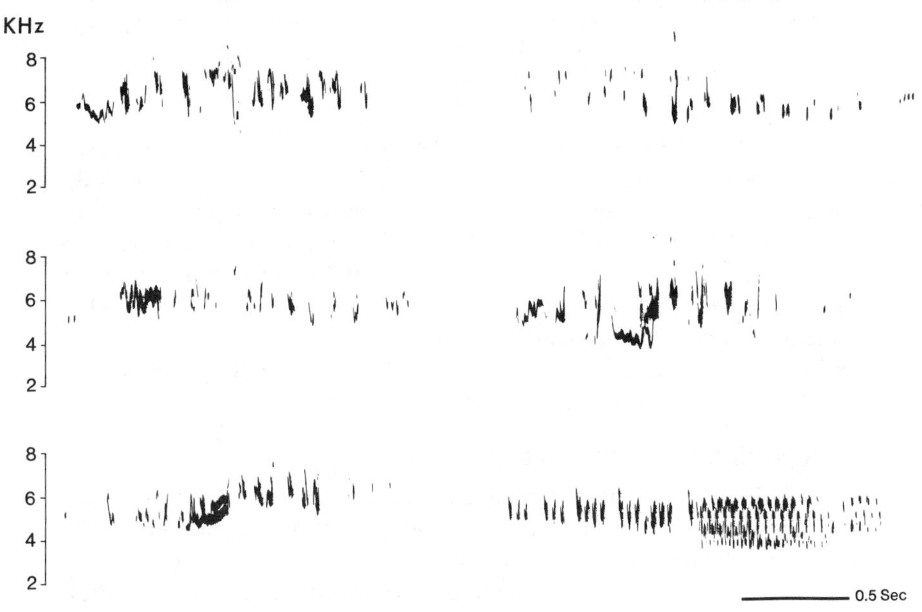

Figure 6 Songs of deaf birds. Hearing a bird's own voice is essential for normal song development in all songbirds studied so far. Deaf birds sing, but their songs are abnormal. In extreme cases none of the structural units of song, i.e. note, syllable, and phrase, develops, as exemplified here by the songs of six white-crowned sparrows. Notice that they differ from both the wild-type songs of Figure 3 and the isolate songs of Figure 4.

whereas deaf canaries (Marler & Waser 1977, Guettinger 1981) and Oregon juncos develop some of the normal syntactical rules of the species song (cf Figure 7). An Oregon junco normally has a repertoire of about four songs. Deaf Oregon juncos produced repertoires and delivered different song types in bouts like intact birds (Konishi 1964, 1965a).

Misunderstood Templates

The same innate template might underlie both the recognition of the species' song and the control of vocal development in isolation. Although there is no evidence for this dual role, the template concept came to be used in both contexts (Marler 1976). This dual usage has confused the authors of some recent publications. For example, one of them contains the following statement (Baptista & Petrinovich 1984):

> These templates are considered to guide the development of motor behavior and to explain some of the complexities of vocal learning. Studies of song development in the white-crowned sparrow have provided some of the key elements of support in attempts to understand how the template operates. Exposure to song between 10 and 50 days of age is considered to create an "engram" if the song has the characteristics of the species. If the song is not appropriate, the stimulus is rendered ineffective by a sensory gating mechanism. Thus, alien song is rejected, the mechanism is closed after 50 days of age, and the crystallized song is the result of matching vocal output to the dictates of the auditory template. Although the above explanation seems adequate with tape tutoring, our evidence suggests that it is inadequate with social tutoring unless the idea of the template is broadened considerably.

The work of these authors concerns the innate predisposition and impressionable phase of song imitation, but it has nothing directly to do with the role of templates in motor development. The concept of templates is largely a short-hand description of observed facts. All it says is that a bird memorizes a song and reproduces it from memory. The only hypothetical aspect of this theory is about the manner in which auditory feedback is used. Control of voice by auditory feedback requires a criterion by which errors in vocal output are corrected. Abnormal songs result when the errors are not corrected. The template simply refers to that criterion.

Song Maintenance Without Auditory Feedback

In several species, auditory feedback is essential for song development but not for its maintenance once crystallization has occurred (Konishi 1965b, Nottebohm 1968). The maintenance must be due either to nonauditory feedback or to the fixation of the output pattern within the central generator. Birds monitor auditory and nonauditory feedback simultaneously so as to learn the pattern of nonauditory feedback for a given sound. After this stage, the hypothetical nonauditory feedback alone becomes sufficient for the birds to evaluate the performance of their vocal organ. This possibility is difficult to exclude, because nonauditory feedback can come from many parts of the body that move during singing. Nonetheless, it is hard to imagine that fine control of membrane vibration in the syrinx can be precisely monitored by sense organs other than those in the syrinx or syringeal musculature itself. Although the tracheosyringealis nerve bundle is said to contain an afferent nerve, neither the target of its innervation nor its physiological role in the control of song is known (Bottjer & Arnold 1982). If nonauditory feedback can be excluded, we are left with the intriguing alternative that the central song generator uses auditory feedback to establish its output pattern, which later becomes independent of hearing. Does this method of establishing a central motor program exist elsewhere? Central rhythm generators are known in many animals; they were demonstrated mostly in adult animals. In only few of these cases have attempts been made to investigate whether they develop without rhythmic sensory input. For example, the basic pattern of muscular contraction underlying the song and flight of crickets develops before the wings grow, indicating that sensory feedback from wing movement is not necessary for the development of the pattern (Bentley & Hoy 1970). Similarly, the basic sequence of forelimb movement during grooming develops in mice whose forelimbs were amputated shortly after birth (Fentress 1973).

Central Song Generators

The acoustical patterns of isolate songs may be due both to the innate template and to central vocal motor generators. In most species studied so far, some deaf individuals produce noise-like songs lacking apparently all the species-specific properties. This observation led to the conclusion that the song motor control system cannot by itself generate any song-like patterns of output, meaning that the motor system needs instruction from the sensory template (Nottebohm 1968, Nottebohm et al 1976, Marler 1976, 1981, Konishi 1978, Bentley & Konishi 1978). There are species and individuals that produce some of the sound patterns of the normal song without auditory feedback, although the sounds are always abnormal. Singing different syllables in separate phrases is one of the properties of the canary song, and it is present in the songs of deaf canaries (Guettinger 1981). Similarly, the species-typical phrase structure is evident in the songs of some deaf song sparrows (Marler & Sherman 1982). Some deaf Oregon juncos (Konishi 1965a) also produced songs consisting of a simple trill typical of their normal songs (Figure 7). The most intriguing aspect of these observations is the individual differences. Different individuals mature at different rates, which may result in different rates of vocal development. Vocal patterns that developed before deafening tend to survive the operation. Therefore, if birds are deafened at different stages of song development, individual differences in songs can occur. It is important to monitor vocal development in each bird closely before it is deafened. Nevertheless,

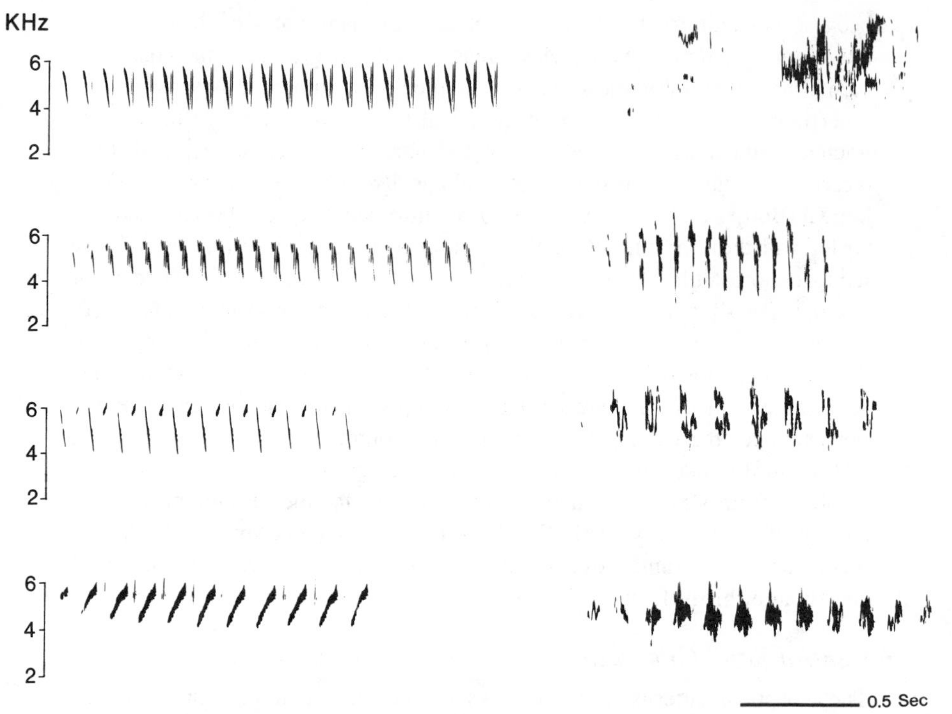

Figure 7 Normal features in deaf birds' songs. Some of the species' typical properties of song occur in deaf individuals of several species studied. The Oregon junco is an example; its song is a simple trill consisting of identical syllables (*left column*). Deaf Oregon juncos developed trill-type songs (except one song of one bird as shown at *top* in *right column*), although their syllables were abnormal.

there appear to be distinct species differences in the number of normal features that appear in the song of a deaf bird. Canaries born and raised in an ear-damaging noise and subsequently deafened before the onset of singing produced songs containing some of the normal aspects of song organization. These birds perhaps never heard themselves verbalize either before or during song development (Marler & Waser 1977).

The observation that songs of deaf birds contain normal features does not require changes in the template concept (cf Marler & Sherman 1982). The central generators for song indicate only that the postulated innate song template is not the only means to govern song development in isolation. The innate template was invoked for the explanation of the differences between the isolate and deaf songs. The theory of central generators can only account for the presence of some normal sound patterns in the songs of deaf birds, but it cannot explain why the songs of intact isolates differ from those of deaf ones.

NEURAL SUBSTRATES FOR SONG

The deafening experiment introduced a new way of thinking about song development, i.e. in terms of central generators and auditory feedback. An obvious question is where the central generator resides and how auditory feedback modulates its output. A discrete brain pathway controls song and it occupies a relatively large volume of the brain, particularly the forebrain. It consists of five nuclei in the forebrain, one in the thalamus, one in the midbrain, and one in the hindbrain (Nottebohm et al 1976, Nottebohm et al 1982, Gurney 1981). The last nucleus, the tracheosyringealis part of the hypoglossal nucleus, innervates the musculature of the vocal organ, the syrinx. The results of anatomical tracing of the vocal control system are summarized in Figure 8. The tracheosyringealis nucleus (X11ts) receives input from the robust nucleus of the archistriatum (RA) both directly and indirectly via the dorsomedial nucleus (DM) of the midbrain. RA receives input from the nucleus hyperstriatum ventrale, pars caudale (HVc) and the magnocellular nucleus of the anterior neostriatum (MAN). HVc is located just under the ependymal layer of the lateral ventricle opposite the hippocampus, which is on the dorsomedial surface of the forebrain. There is some uncertainty about the brain layer to which HVc belongs, both because it is unclear whether the hyperstriatum in songbirds extends posteriorly to encompass HVc and because this nucleus, despite its name, occurs in the neostriatum in the parakeet brain. HVc projects to RA

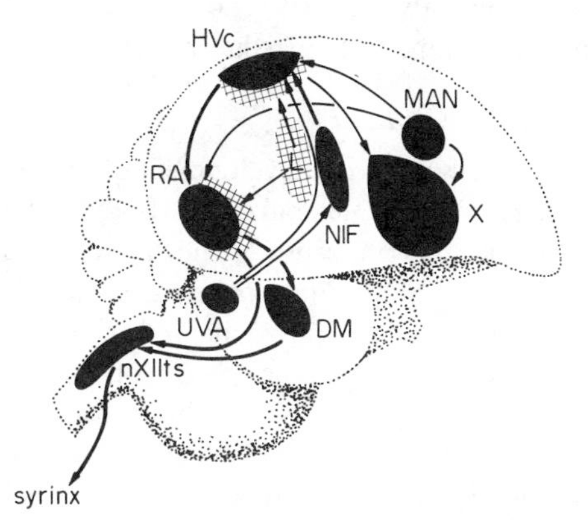

Figure 8 The vocal control system. This diagram shows most of the known nuclei and connections of the vocal control system. All *arrows* indicate anterograde connections. Neural signals for song originate in NIF (Nucleus interface) and descend the pathway to the syrinx via HVc (nucleus hyperstriatum ventrale, pars caudale), RA (nucleus robustus archistriatalis), DM (dorsomedial nucleus of nucleus intercollicularis), and X11ts (nucleus hypoglossus, pars tracheosyringealis) in that order as shown by the *thick arrows*. Other nuclei are inactive during song in adult birds. *Hatched areas* indicate the known projection zones of the telencephalic auditory area, Field L.

and area X of the parolfactory lobe and is innervated by fibers from MAN, nucleus interface (NIF) and the thalamic nucleus, nucleus uva (UVA). NIF also receives input from UVA. The forebrain auditory area, Field L, projects to the vicinity, the so-called shelf area, of both HVc and RA (Kelley & Nottebohm 1979). Intracellular recording shows that a class of neurons in HVc and RA receives auditory input, although its exact source is unknown. Furthermore, many of these auditory neurons of HVc send their axons to area X (Katz & Gurney 1981, Katz 1982).

The conclusion that the pathway controls song is supported by two types of evidence. Lesions of HVc, RA, or the hypoglossal nerve cause deterioration in the adult song. Control lesions placed elsewhere do not affect the song (Nottebohm et al 1976). Recording of neural activity from the singing bird indicates that NIF, HVc, and RA show periods of activity and inactivity corresponding, respectively, to the sounds and silent intervals of song (McCasland 1983). Furthermore, the latency between the onset of neural discharge and sound is longest in NIF, shorter in HVc, and shortest in RA, indicating the sequence of signal transmission within the forebrain nuclei. So far, a neural signal for song has not been detected before NIF in the descending pathway, and furthermore, bilateral lesions of UVA, the sole source of inputs to NIF, do not affect song. Also, UVA, MAN, and area X show no song-related neural activity at all. These nuclei are considered to be parts of the vocal control system because of their anatomical connections with it. The vocal control system regulates predominantly the ipsilateral syringeal musculature; no anatomical connection between the two sides has been found so far.

Lateralization of Song Control

Birdsongs share some attributes with development of human speech. One of them is thought to be hemispheric lateralization. In the canary, lesions of the left hypoglossal nerve or left HVc cause greater losses or more severe deterioration of syllables than similar lesions on the right side. These observations led to the hypothesis that each hemisphere independently controls the ipsilateral half of the syrinx and that the left hemisphere controls more syllables than the right one (Nottebohm et al 1976, Nottebohm & Nottebohm 1976). This hypothesis also derives support from the so-called two-voice phenomenon, which refers to the simultaneous occurrence of two harmonically unrelated frequencies in some birdsongs (Greenewalt 1968, Stein 1968). Although the two-voice theory may be correct, there are two important issues to consider. One concerns the mechanism of generating harmonically unrelated frequencies. A circular membrane, for example a drumhead, produces a family of frequencies formally described by Bessell functions. Furthermore, sounds may be produced by vortex effects as air flows through a narrow slot. Vortex-

and membrane-generated sounds can interact, resulting in a complex distribution of frequencies (Gaunt 1983). The other problem is how to produce one voice with two sound sources. Many birdsongs contain one-voice syllables. There are two mutually exclusive alternatives: One voice comes from only one side and the other side is silent, or both sides contribute to the same voice. When a canary sings with its left bronchus plugged, its song contains many of the original syllables, including the two-voice ones. This finding indicates that the right and left syringeal halves produce the same syllables (McCasland 1983).

Another important aspect of the hypothesis is the site of lateralization. The sites of effective lesions are not necessarily the loci of lateralization; direct and indirect effects of lesioning must be distinguished. Because the vocal system controls predominantly the ipsilateral syringeal musculature, a lesion of any part of the descending pathway may disable the musculature of that side. If the left half of the syrinx plays a more dominant role in song control than does the right half, then cutting the hypoglossal nerve or lesioning the HVc simply exposes the syringeal asymmetry. It does not discriminate between hemispheric and peripheral lateralization. Such a peripheral asymmetry does exist in the canary's syrinx. The syringeal musculature on the left is clearly more voluminous than that on the right (Nottebohm 1980a). Consistent with this asymmetry, the left bronchus working alone can produce a larger number of normal syllables than the right one working alone, although the right one contributes to all syllables (McCasland 1983). Some HVc-lesioned canaries showed fewer changes in their songs than hypoglossus-lesioned birds (Nottebohm 1980a, Nottebohm et al 1976, Nottebohm & Nottebohm 1976). This difference may be due to the fact that a peripheral nerve can be lesioned more cleanly than a brain nucleus. Also, the hypoglossal nerve contains neurons that cause the syringeal muscles to contract in synchrony with the respiratory rhythm (Manogue & Paton 1982). Cutting these fibers may disrupt the coordination between respiration and vocalization, whereas a lesion of HVc does not affect this aspect of vocal control. Furthermore, if the size of HVc can serve as the index of song learning, one would expect left-right anatomical differences in the vocal control nuclei. No such differences are present in any of the vocal control nuclei except the hypoglossal nucleus, where the left side is said to be larger than the right by 6% (Nottebohm & Arnold 1976). Even a detailed analysis of neuronal morphology of RA, which demonstrated clear male-female differences, did not find any left-right bias (DeVoogd & Nottebohm 1981). Recordings from the HVc of the singing canary show that both sides are active during all syllables. This observation is consistent with the result of bronchus-plugging mentioned above, but it is difficult to explain, if HVc is the site of lateralization (McCasland 1983).

How the muscular asymmetry develops in the ontogeny of an individual is an unanswered question. Some unknown central asymmetry may drive the two sides differently, or it may develop in response to differential loading by the periphery. In the adult canary, lesions of the left hypoglossus result in a shift of song control to the right side (Nottebohm et al 1979). However, this shift parallels a shift in the muscular asymmetry to the right side. Why lateralized control of song evolved is another unanswered question. There does not seem to be any correlation between the ability to imitate song and the degree of lateralization. Deaf canaries, which are not supposed to "learn" song, are left-dominant (Nottebohm et al 1976). An asymmetry is barely noticeable in the song of the zebra finch, which imitates song well (Price 1977). Thus, the analogy of song lateralization with lateralization of human speech should be viewed with caution.

Neural Theory of Song Learning

Comparative brain anatomy and neuroendocrinological studies have recently uncovered several new facts that are thought to link the vocal control system with song learning.

CROSS-TAXA COMPARISONS The forebrain nuclei HVc, RA, X, and perhaps also MAN and NIF occur only in birds capable of vocal imitation, such as songbirds, and are apparently absent in birds that lack this capability. Species in which isolates develop abnormal songs have the nuclei and those in which isolates develop normal songs, such as tyrannid flycatchers, do not (Nottebohm 1980a, Kroodsma 1984). If auditory inputs of HVc, RA, and X are necessary for the control of vocal development by auditory feedback, deafening should affect vocal development only in birds with the nuclei. This prediction is borne out within the range of species studied so far. The nuclei are present in birds that develop abnormal vocalizations when deafened and are absent in species that suffer no ill effects from deafening, such as chickens (*Gallus domesticus*) (Konishi 1963) and ring doves (*Streptopelia risoria*) (Nottebohm & Nottebohm 1971).

Interestingly, the dichotomies mentioned above seem to reflect partly taxonomic divisions. Passerines (Order Passeriformes) include oscine songbirds, such as sparrows, canaries, and starlings, and suboscine songbirds, such as tyrannid flycatchers. The forebrain nuclei occur in oscines, but not in suboscines (Nottebohm 1980a). Birds of nonpasserine orders, such as Galliformes (chickens, turkeys, etc) and Columbiformes (pigeons and doves), generally lack the forebrain nuclei. The parakeet of Order Psittaciformes (parrots) and the hermit hummingbird (*Phaethornis longuemareus*) of Order Apodiformes (swifts and hummingbirds) may be exceptions to the above rule, first, because they are the only well-documented nonpasserines that imitate sounds (Snow

1968, Wiley 1971), and, second, because structures analogous or homologous to HVc and RA are present in the parakeet's brain (Paton et al 1981). Whether or not the forebrain song nuclei evolved independently in oscine songbirds and parrots to accommodate vocal imitation is an unanswered question. The brain of the hummingbird has not been examined for the presence of the forebrain nuclei.

These taxonomic correlations are partly due to the fact that the structure of the syrinx, which is an important key in avian classification, is one of the determinants of vocal plasticity. Obviously, the vocal organ and its neural substrates must coevolve. The syrinx of nonpasserine birds, like chickens, turkeys, and ring doves, have two pairs of extrinsic muscles that cause the displacement and rotation of the whole syrinx for the production of sounds. The complex syrinx of oscine songbirds is equipped not only with extrinsic musculature but also with four to nine pairs of intrinsic muscles for the independent control of different parts of the organ (Figure 9). Intrinsic muscles also occur in the syringes of parrots and hummingbirds. Thus intrinsic muscles seem to be a necessary condition for plastic vocal development (Gaunt 1983, and personal communication). However, the converse is not always true; for instance, acoustic isolation does not affect song development in suboscine

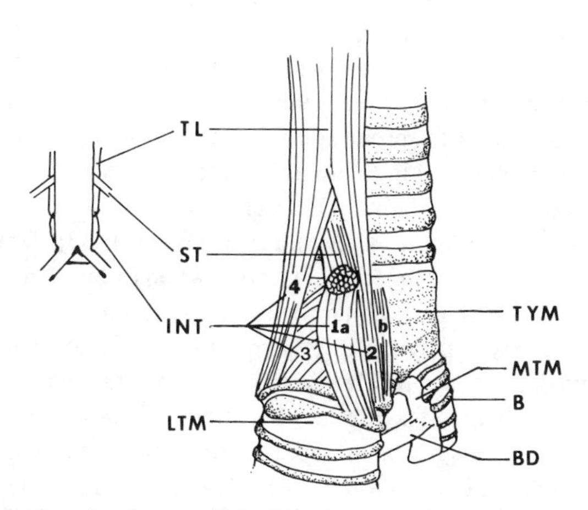

Figure 9 The syrinx of oscine songbirds. This sketch shows a frontal section and right lateral view of an idealized, oscine syrinx. The numbers indicate the presumed intrinsic muscles (INT). Muscles 1a and 1b are considered separate despite their appearance as a single mass. Muscles 2 and 4, which seem to be extensions of the tracheolateralis muscle (TL), are regarded as intrinsic muscles. B, bronchial bar; BD, bronchidesmus; TYM, tympanum; MTM, medial or internal tympanic membrane; LTM, lateral tympanic membrane; and ST, sternotrachealis muscle. TL and ST are the extrinsic muscles (from Gaunt 1983).

flycatchers that have a simple syrinx with several intrinsic muscles (Gaunt 1983).

Although the complexity of the syrinx is thought to be correlated with the complexity of vocalizations, the number of intrinsic muscles is not directly correlated with either repertoire size or imitative ability (Gaunt 1983). For example, the well known vocal mimics such as lyre birds (*Menura novaehollandiae*), Indian Hill mynahs (*Gracula religiosa*), and mockingbirds have three or four intrinsic muscles, whereas some oscines with more complex syringes can sing only a single song. Also, song repertoire size varies greatly among closely related species with similar syringeal anatomy. For example, the song sparrow with 20 song types and the white-crowned sparrow with one song type are now considered to belong to the same genus *Zonotrichia*. Thus, these differences are due perhaps to differences in the brain vocal control system (Gaunt 1983).

BRAIN ANATOMY AND SONG LEARNING Comparisons of different individuals or populations of the same species have uncovered a correlation between the size of song system nuclei and song learning. Western populations of long-billed marsh wrens (*Cistohorus palustris*) with large song repertoires have larger HVcs than Eastern populations with small repertoires (cited in Kroodsma 1982). Canaries produce a long, continuous series of syllables in their songs and the number of different syllables used by an individual can vary from about 20 to 50. Canaries with larger numbers of syllables tend to have larger HVcs and RAs than those with smaller numbers of syllables, although it is a statistically weak correlation (Nottebohm et al 1981). Also, female canaries that sing in response to testosterone administration have five to seven syllables in comparison with about 30 in the male. These females have HVcs and RAs that are less than 50% of their male counterparts' in volume (Nottebohm 1980b). However, according to a recent report (Pesch 1983), female canaries sing spontaneously after the breeding season or when isolated from other birds, and the number of syllables is about the same as that of the male.

Young canaries develop syllables by imitating their fathers and other adults and by improvising (Waser & Marler 1977). In contrast with many species that do not change their songs after the first singing season, canaries "change their song repertoires each year by adding, dropping, or modifying components" (Nottebohm & Nottebohm 1978). Canaries deafened after their first singing season fail to maintain or increase their syllable repertoires in subsequent singing seasons (Nottebohm et al 1976). Therefore, the process by which new and modified syllables are created is regarded as learning, and the number of syllables in the repertoire can serve as the index of song learning (Nottebohm 1980a). Why does the nuclear size correlation exist? The size of a brain area can vary for various reasons: In addition to neuronal components, non-neuronal components like glia, capillaries, and intercellular matrix can contribute to a size difference. It will be important to determine the ratio of neuronal and nonneuronal contributions in any such comparison. A more serious question here is how to derive cause and effect relationships from correlations. If the number of different syllables is correlated with any other variables, then it will be necessary to determine which variable is causally related with the size of HVc and RA. For example, a reasonable hypothesis is that the direct relationship is between the total amount of sound energy in the songs and the nuclear size. If this is true, then the size of a nucleus would have nothing directly to do with learning. Although canaries deafened after their first singing season produce abnormal songs, the total amount of sound energy in their songs is measurable. It would be of interest to compare the HVcs and RAs of deaf canaries that use different amounts of sound energy in singing. If there is a size correlation, then it may be due to differential usage instead of learning.

The plasticity of canary song predicts the existence of modifiable neural substrates. Testosterone, which stimulates singing in young birds, females, and castrated males, causes growth in the somata and dendrites of RA and HVc neurons (DeVoogd & Nottebohm 1981, Nottebohm 1980b). In the male canary the size of HVc and RA fluctuates in parallel with the seasonal variation in circulating testosterone. They are larger in the spring singing season than in the fall and winter (Nottebohm 1981). The growth and shrinkage of dendrites presumably contribute to this fluctuation. An increase in synaptic sites probably accompanies growth in dendrites. The view that learning necessarily involves the formation of new synapses is not generally accepted (Kandel 1979), although there is physiological evidence that in some cases it does, such as the new synapses formed in the cat's red nucleus during readjustment of forelimb coordination following cross-innervation of flexor and extensor muscles (Tsukahara 1981).

The hypothesis that the seasonal fluctuation in HVc and RA volume indicates learning and forgetting of song requires careful scrutiny (Nottebohm 1981). There is a real possibility that HVc and RA undergo seasonal volume changes in deaf canaries, because their singing fluctuates seasonally, as in intact birds. If this is true, then the significance of this phenomenon for song learning is doubtful. The normal temporal pattern of song that develops in deaf canaries is likely to be due to central pattern generator circuits, as mentioned above. Because it develops by the first singing season without auditory feedback, it would not be surprising if some of its generator circuits are annually "reformed" even in deaf birds. Therefore, the distinction between the anatomical and physiological changes associated with feedback controlled modification of song on one hand and those related to central reorganization and use-disuse phenomena on the other will be important.

A SINGLE-NEURON APPROACH TO BIRDSONG

Although some of the correlations mentioned above may link brain anatomy to song learning, the ultimate proof for it requires an understanding of the neural codes involved. How memorized songs, auditory feedback, and song motor programs are encoded is an important question that calls for physiological analysis at the cellular level.

Many studies show that special neuronal circuits generate rhythmic patterns in invertebrates (Bentley & Konishi 1978). Evidence also suggests that the spinal cord and brainstem of vertebrates contain generator circuits for locomotor coordination (Grillner 1975, Stein 1978), but little is known about the control of complex movements. The discrete nuclei and pathway of the vocal control system are convenient for the analysis of neuronal mechanisms underlying complex movements.

Multi-unit activity in the HVc of singing birds shows a unique pattern of discharge occurring before a syllable, and a quiescent period before a silent interval in the bird's song. Thus, a neural equivalent of the song manifests itself in the HVc. This nucleus precedes the hypoglossal nucleus by at least three synapses. HVc receives input from NIF and UVA, and NIF shows a similar song-related discharge pattern. Because the areas around NIF and UVA show no song-related discharge, NIF may be the source of the song pattern. An intriguing possibility is that the motor program for a copied song might reside in NIF.

Single units recorded from the HVc of mockingbirds show neuronal specializations for the generation of the song pattern (McCasland 1983). Many neurons fire for all syllables, some neurons fire long (500 msec) before the onset of song, and a few neurons produce highly stereotyped bursts of spikes for only a few syllable types. Some of these neurons (motor-specific neurons) fire selectively for one syllable type and do not fire for other very similar syllables (Figure 10).

Specializations related to the song are also evident among HVc auditory neurons. In the canary and white-crowned sparrow, the bird's own song is the most effective stimulus for eliciting multi-unit responses in HVc. Furthermore, the bird's own song broadcast backward is much less effective than when it is delivered in the normal direction, suggesting that the sequence of the component sounds is an important cue for these neurons (McCasland & Konishi 1981).

The stimulus requirements of single HVc neurons substantiates the above suggestion about stimulus cues. In the white-crowned sparrow, some HVc neurons respond selectively to a particular set of acoustic cues in the bird's own song (song-specific neurons) (Margoliash 1983). All song-specific neurons that were thoroughly studied require two consecutive sounds (phrases or parts

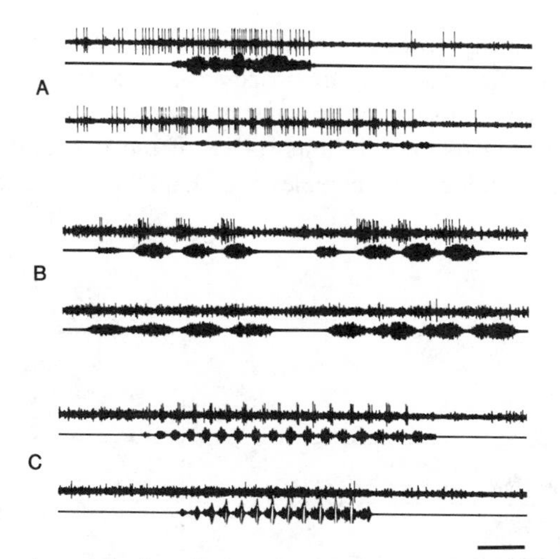

Figure 10 Neuronal specializations for song motor control. *A, B,* and *C* show segments of simultaneous recordings of neuronal HVc discharge and song syllables from a mockingbird. *A* fired before and during all syllables. *B* and *C* are motor-specific neurons, which fired only before and during a specific syllable (*upper pair of traces* in each group); they remained inactive before and during syllables (*lower pair of traces*) similar to the specific one. Scale bar indicates 100 msec.

of a phrase) of the song, each of which alone elicits little or no response (Figure 11). They detect particular acoustic cues such as changes in frequency and a silent gap between the two sounds. A bird's song-specific neurons may respond to the songs of some other individuals from the same dialect area and typically fail to respond to the songs of birds from other dialect areas. This selectivity is predictable when the relevant acoustic features in the bird's own song are compared with the corresponding features in the other songs. Song-specific neurons in birds singing abnormal songs, due to early isolation, respond selectively to these abnormal songs and fail to respond to wild-type songs of the species. When a bird's own song differs from its early tutor song, its song-specific neurons are selective not for the tutor song but for the song sung by the bird.

There have been reports of auditory neurons selectively responding to bird vocalizations. In the starling a few Field L neurons responded to a single call out of many calls presented as stimuli (Leppelsack & Vogt 1976). This finding certainly suggests the presence of selective neuronal responses. However, it does not show neuronal specificity for the call. When a neuron requires a set of physically definable acoustic properties that are unique to a call, the neuron can be named a "call-specific neuron." Song-specific neurons are "tuned" to specific acoustical parameters that are unique to the bird's own song. "Selective" neurons are those which prefer one call to other calls. Call-selective neurons are obviously more common than call-specific neurons. A match between a neuron's frequency sensitivity and a call's spectrum alone can account for the responses of call-selective neurons in some cases (Scheich et al 1979). Song-specific neurons show the importance of the temporal pattern of the stimulus in neural coding of the complex sound, an aspect that has been studied

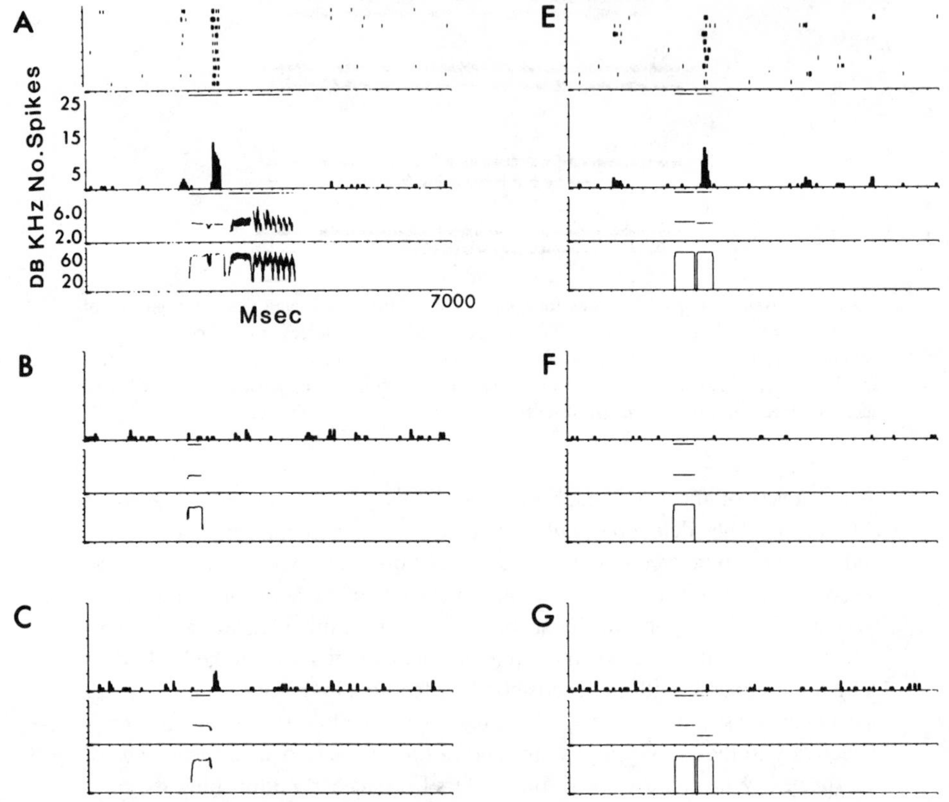

Figure 11 A song-specific neuron from the HVc of a white-crowned sparrow. Some HVc neurons respond selectively to a particular acoustic cue in the bird's own song. The pattern of transition in frequency between two consecutive phrases constitutes the cue for all song-specific neurons found in this species. In each section (*A, B*, etc), the top graph is a raster representation of unit response. Each *vertical* tick represents one spike; each row is marked on the ordinate by a horizontal tick and represents one stimulus presentation. The *middle graph* is a histogram of spike numbers. The *bottom graph* shows the pattern of frequency (*upper trace*) and amplitude (*lower trace*) modulation in the stimulus. This unit responded selectively to the bird's own song (*A*). It required the first and second parts of the introductory whistle, the first (*B*) or second (*C*) part alone eliciting little or no response. Two consecutive tones could mimic the natural trigger feature (*E*), again one tone alone having no stimulatory effect (*F*). The neuron was insensitive to variation in the interval between the two tones, but it was sensitive to differences in frequency between them (*G*) (from Margoliash 1983).

very little in auditory physiology (Suga 1982). This attribute of song-specific neurons cannot be explained by simple peripheral mechanisms such as frequency tuning and two-tone inhibition.

Whether song-specific neurons are involved in some aspect of the template matching processes is not known, although their specificities are appropriate for that which is required of the template. Because the template is established by early auditory exposure, its neural representation should be present before the onset of singing. So far, song-specific neurons have not been found in presinging birds (D. Margoliash, unpublished results). Our current hypothesis about the origin of song-specific neurons is as follows. Song-specific neurons acquire their specificity during song development; their properties become matched to the final form of song as it is shaped. A similar argument would apply to the motor-specific neurons mentioned above; their specificity may also develop during the sensorimotor phase of song learning.

Why do motor and auditory neurons occur, apparently completely mixed, within the same nucleus? The need for auditory feedback for song development indicates that the vocal control system is directly or indirectly linked to the auditory system. HVc is one of the sites of this linkage. Although the auditory and motor neurons have clearly different cell forms, the nature of anatomical connections between them is not known (Katz & Gurney 1981). However, multiunit recordings from HVc show that auditory neurons are inhibited while motor neurons are firing during song (McCasland & Konishi 1981). This inhibition perhaps occurs in HVc itself, because Field L neurons are generally not inhibited during song (McCasland 1983). Although Field L projects to the vicinity of HVc, the immediate source of auditory input to HVc is not known. Also, the properties of auditory neurons in area X and RA have not yet been studied.

A NEURAL MODEL OF SONG LEARNING: A SUMMARY

The template theory offered a simple model of song learning without any reference to brain anatomy and physiology. A slightly more realistic model that incorporates the known anatomical and physiological attributes into its design is now feasible (Figure 12). This model assigns song generator function to the forebrain nuclei, NIF, HVc, and RA, of which the last two contain sensorimotor circuits for auditory-vocal interaction. These circuits use error signals derived from matching auditory feedback with the template to regulate the output of the song generator. There must be a reciprocal relationship between the song generator and sensorimotor circuits, for these are developmentally plastic and become "fixed" as the output of the song generator crystallizes.

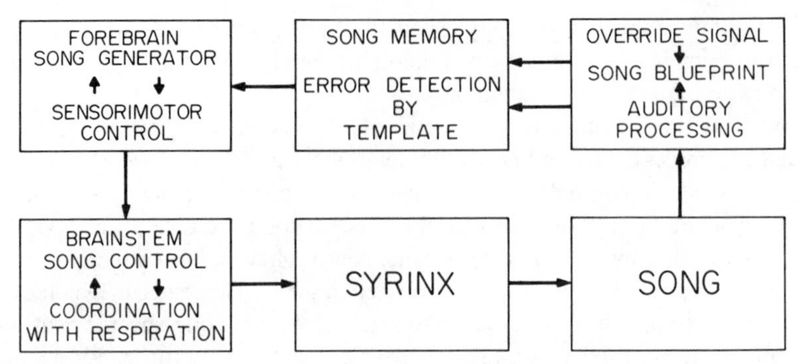

Figure 12 A neural model of song learning. This model is an attempt to integrate anatomical, physiological, and behavioral facts for the explanation of song learning (see text for discussion).

The motor-specific and song-specific neurons may be part of the circuit and result from the fixation process.

The output of the song generator exits the forebrain from RA. The brainstem song control pathway perhaps coordinates vocalization and respiration. The relationship between the two must be reciprocal: The extent to which vocal patterns can modify the respiratory rhythm must be limited; conversely, respiration certainly does not dictate patterned vocalizations. Whether respiratory control is exerted directly on the forebrain song generator is not known, but it can be accomplished automatically, for the respiratory system rejects, mechanically and metabolically, unrealizable vocal patterns, leaving those realizable for control by auditory feedback.

This model identifies unresolved issues of importance: the site and nature of the template, what constitutes the blueprint, and how visual cues override the blueprint. Finally, the coordination between the two brain hemispheres remains one of the most intriguing problems.

SEX HORMONES AND THE VOCAL CONTROL SYSTEM

Affinity for Androgens

Steroid autoradiography shows that most of the adult vocal control nuclei contain neurons that selectively absorb androgens such as testosterone and dihydrotestosterone (DHT) (Arnold et al 1976). It is not known whether NIF and UVA neurons accumulate the hormones. Although in area X, steroid autoradiography does not show a higher grain density, systemically implanted testosterone or DHT induces or increases protein synthesis, suggesting that this nucleus also contains androgen accumulating neurons (Konishi & Akutagawa 1981). The affinity for testosterone is not restricted to higher-order brain cells, but it occurs as well in the motor neurons of the tracheosyringealis part of the hypoglossal nucleus and in the syringeal musculature. Castration causes a decrease in choline acetyltransferase and acetylcholinesterase activity in this musculature, and administration of testosterone restores their activity (Luine et al 1980).

So far as tests show, the vocal control nuclei do not seem to contain neurons that accumulate estrogen (Konishi & Akutagawa 1981). On the other hand, biochemical assays show that the brain of an adult zebra finch contains both androgen and estrogen receptors. There are sex differences in the number of androgen and estrogen receptor sites; the male brain contains more androgen than estrogen receptors and the converse is true of the female brain (Siegel et al 1983). Also, neurons of the male zebra finch HVc and MAN accumulate more radioactive androgen than those of the female counterparts (Arnold & Saltiel 1979, Arnold 1980).

Anatomical Sex Differences in the Vocal Control System

In many songbirds, the male sings and the female does not. This behavioral dichotomy is due to sexual dimorphism in the vocal control system. HVc and RA in the female canary are half as large as their counterparts in the male. These differences are most pronounced in species such as the zebra finch and the Bengalese finch, in which the female does not sing even under the influence of exogenous testosterone, which induces singing in the female of many species (Nottebohm & Arnold 1976; a recent review in Arnold & Gorski 1984). In the female zebra finch, HVc, RA, MAN, NIF, and area X are either rudimentary or unrecognizable. Comparison of Golgi-stained neurons of RA shows that their somata are much smaller and their dendrites are shorter in the female than in the male. Also HVc and RA contain fewer neurons in the female than in the male. Interestingly, DM and the hypoglossal nucleus show much smaller sex differences than any of the forebrain nuclei (Gurney 1981).

Hormonal Control of Sexual Differentiation

Behavioral effects of early hormone treatment led to the theory that sex hormones exert organizing and activating effects on the nervous system; they control in youth anatomical differentiation of the brain neural circuits for sexual behavior, and they induce in adulthood sexual behavior by activating the existing neural circuits (Phoenix et al 1959). The sexual dimorphism of the brain has been thought to support the theory under the assumption that sex hormones are the cause of the dimorphism. The distinction between the two types of effects depends on the level of analysis, if morphological criteria are the only means of discriminating between the two. Sex hormones can "activate" by inducing morphological changes no matter how small they may be. Testosterone induces dendritic growth in the adult canary's RA. This effect is presumably reversible. The grown dendrites are assumed to shrink, because the dendrites of RA neurons are shorter when testosterone titer is low (DeVoogd & Nottebohm 1981). Therefore, reversibility or irreversibility should be the criterion for distinguishing one type of effect from the other. Irreversible morphological and physiological changes in the nervous system are organizing effects and reversible changes are activating effects. Because the soma size of neurons in the forebrain nuclei clearly separate the male and female zebra finch, the effects of sex hormones on the sexual differentiation of the vocal control system can be studied at the cellular level.

A subcutaneous implant of estrogen or testosterone in a newly hatched female chick induces a masculine-like differentiation of neurons in her forebrain nuclei, particularly in HVc, RA, X, and MAN (Gurney 1981, 1982, Gurney & Konishi 1980, Konishi & Gurney 1982). The somata of RA neurons in an estrogen-treated female are intermediate in size between male and normal

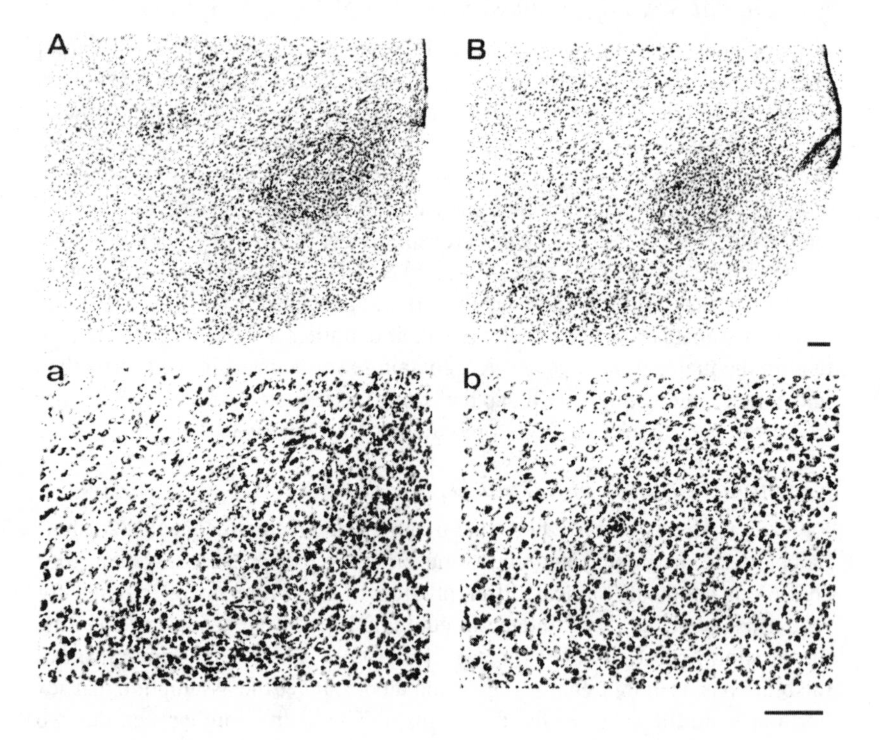

Figure 13 Brain sex differences are absent in young zebra finches. At 15 days of age, male (*A*) and female (*B*) finches show little or no sex differences in the volume and cell size of RA. *A* and *B* are low-power photomicrographs of crysl violet stained RA, and *a* and *b* are their high-power versions, respectively. Both scale bars show 100 μm.

female RA neurons. If such a bird receives an implant of testosterone or dihydrotestosterone in adulthood, RA neurons grow more and the bird sings spontaneously. Because the same treatment, estrogen followed by testosterone, induces neither cell growth nor singing in adult females, the masculinizing effect of estrogen is restricted to a specific period in development.

The above findings led to the conclusion that estrogen induces neuronal growth in both HVc and RA. However, a subsequent study suggests that the small HVc, RA, and MAN of the female are not due just to failure to grow, but to cellular atrophy. At post-hatching day 5, HVc and RA show little sex difference. By day 12, HVc has already undergone atrophy in the female, whereas both RA and MAN remain large and show little sex difference (Figure 13). Then, RA and MAN undergo atrophy in the female between days 25 and

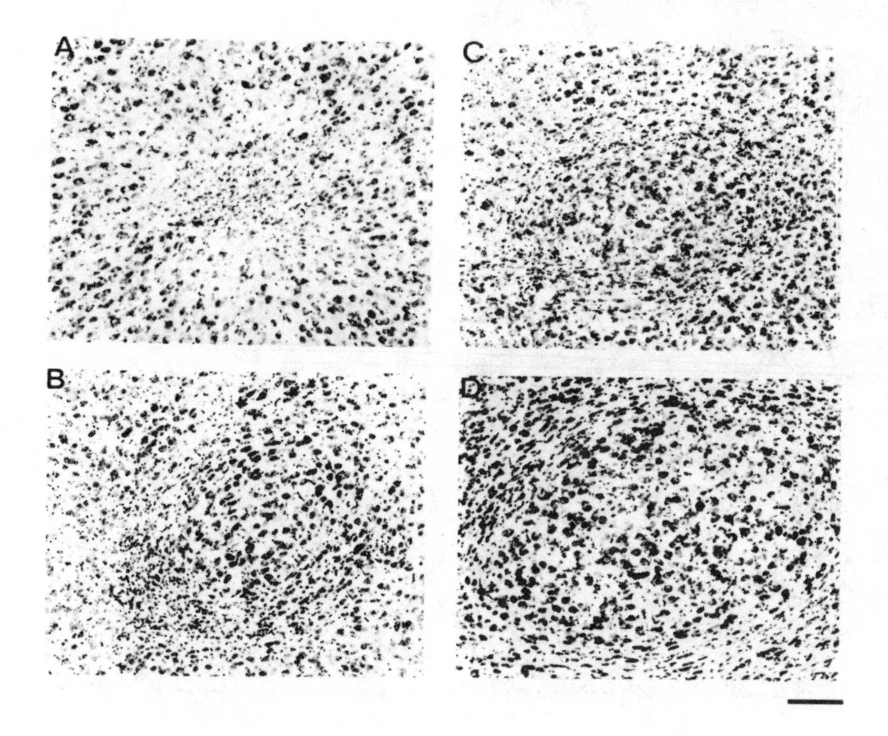

Figure 14 Hormone-induced differentiation of RA. In normal female zebra finches (*A*), RA contains only small neurons, which are not due to retarded growth but to atrophy. A central core of large neurons surrounded by a rim of small neurons characterizes a partially masculinized RA as shown in *B* and *C*, which resulted from estrogen implantation for the first 7 and 12 posthatching days, respectively. A fully masculinized RA contains only large neurons as in *D*. Administration of testosterone in adulthood to birds with *B*, *C*, or *D* condition induced singing. Scale bar 100 μm.

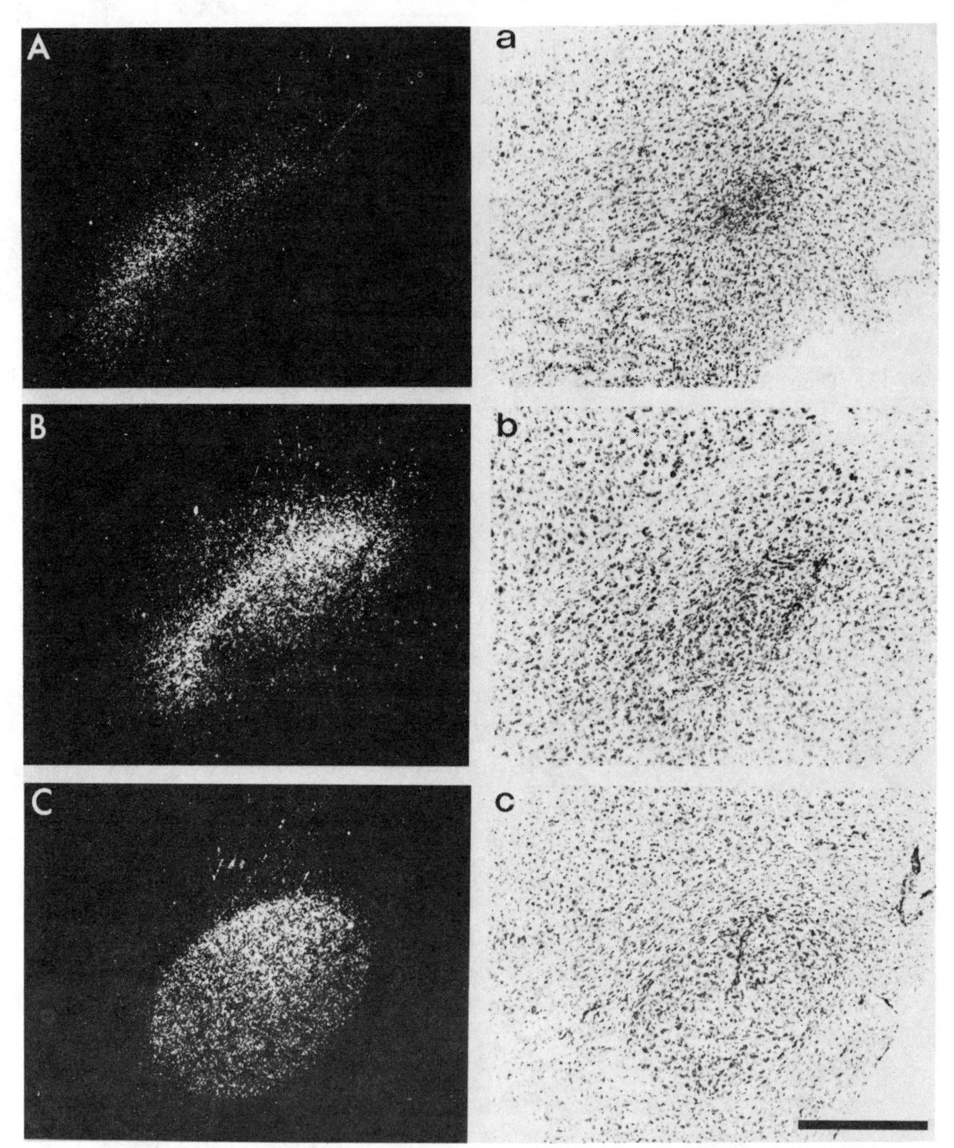

Figure 15-1 Innervation of RA in normal and hormone-treated female zebra finches. Projection of HVc axons to RA shown by amino acid autoradiography compared with the distribution of small and large cells in RA. In normal females (*A* = dark field, *a* = crysl), the axons do not terminate in RA but remain outside it. In partially masculinized females (*B*,) some of the axons appear to enter RA, but a large part of the nucleus still remains uninnervated. Notice that the crysl-violet-stained section (*b*) shows RA populated mostly by small neurons. In fully masculinized females (*C*, *c*), RA contains only large neurons and is completely innervated by HVc axons. Scale 500 μm.

35; their volume and cells become progressively smaller. Thus, sexual differentiation of the forebrain nuclei in the zebra finch vocal control system seems to involve programmed cell death with different time courses for different nuclei. How estrogen prevents cell death is an intriguing question for study.

A fully masculinized RA contains uniformly large cell bodies, whereas a partially masculinized one shows a central core of large cell bodies surrounded by a rim of small cells (Gurney 1982, M. Konishi and E. Akutagawa, unpublished) (Figure 14). The size of the central core is a function of the duration of estrogen release. Anterograde transport of tritiated amino acid and HRP shows that the axon terminals of HVc neurons innervate only these large cells, although the small uninnervated cells in the rim clearly belong to RA. The RA of an adult female does not seem to be innervated by HVc neurons; their axon terminals remain outside the well-demarcated boundaries of RA (Figure 15) (M. Konishi and E. Akutagawa, unpublished).

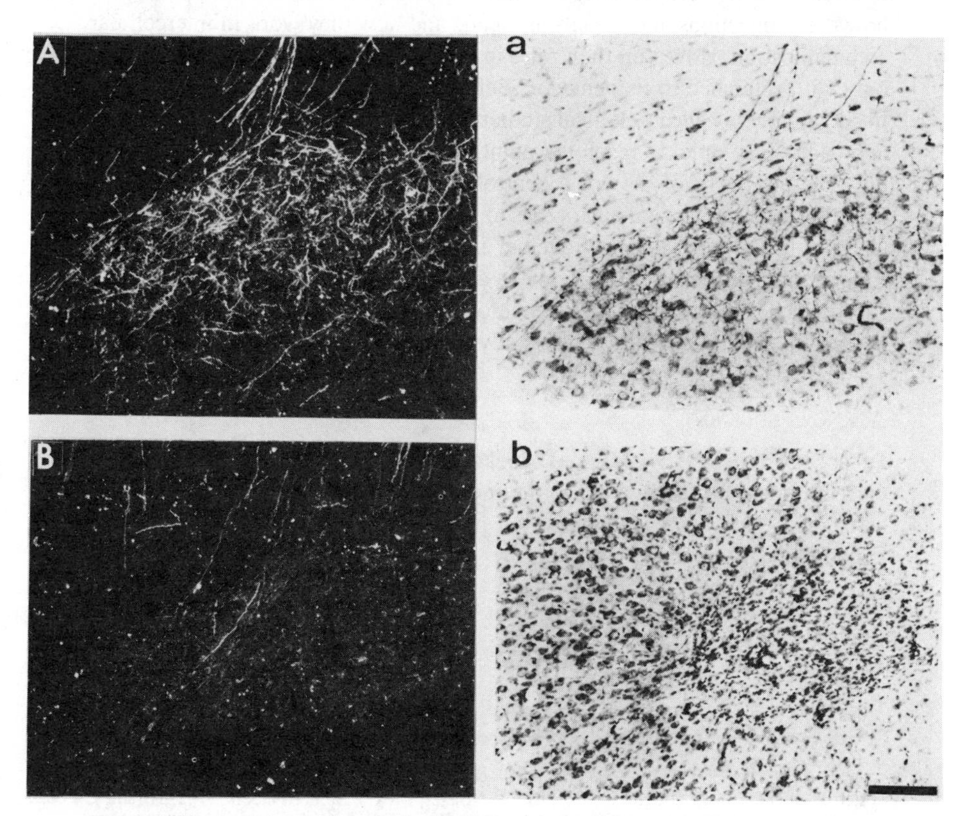

Figure 15-II Axon terminals stained with HRP + cobalt n the RA of a fully masculinized (*A*) and a normal (*B*) female. *A* and *B* are dark field pictures and the corresponding crysl violet stained sections are shown in *a* and *b*. Scale 100 μm.

What hormone controls normal differentiation in the male zebra finch is unknown. Both estrogen and testosterone, but not dihydrotestosterone, masculinize the female vocal system. This result suggests that testosterone acts on the vocal system by being converted to estrogen. However, a recent study shows that a large amount of estrogen is present in the blood of male zebra finch chicks. Therefore, there is no need to postulate the conversion of testosterone to estrogen in the male zebra finch (R. E. Hutchison, J. Wingfield, and J. B. Hutchison, personal communication).

CONCLUDING REMARKS

The ultimate goal of neurobiology is to explain behavior in terms of connections and signals between neurons. Cellular and molecular studies investigate how the connections and signals are made and how they work in intercellular communication. However, their roles in the whole organism can be understood only with reference to its behavior. Behavioral relevance is usually an undefined variable in neurobiological studies. On the other hand, purely behavioral approaches can quickly reach the point where the problems become intractable. The discovery of the vocal control system established a bridge between the cellular and behavioral approaches to birdsong. The results of recent attempts to bring the two approaches together in various studies reviewed above are encouraging. Many of the phenomena associated with birdsong and its neural substrates bear on major issues in neurobiology. Thus, we can address important problems in neurobiology by studying the beautiful songs of birds!

ACKNOWLEDGMENT

I thank Drs. Eric I. Knudsen, Daniel Margoliash, and Terry Takahashi for correcting and critically reading the manuscript, and Eugene Akutagawa for assistance. This work was supported in part by a grant from the Whitehall Foundation.

Literature Cited

Arnold, A. P. 1980. Quantitative analysis of sex differences in hormone accumulation in the zebra finch brain: Methodological and theoretical issues. *J. Comp. Neurol.* 189:421–36

Arnold, A. P., Gorski, R. A. 1984. Gonadal steroid induction of structural sex differences in the central nervous system. *Ann. Rev. Neurosci.* 7:413–42

Arnold, A. P., Nottebohm, F., Pfaff, D. W. 1976. Hormone concentrating cells in vocal control and other areas of the brain of the Zebra finch (*Poephila guttata*). *J. Comp. Neurol.* 165:487–512

Arnold, A. P., Saltiel, A. 1979. Sexual difference in pattern of hormone accumulation in the brain of a songbird. *Science* 205:702–5

Baker, M. C., Spitler-Nabors, K. J., Bradley, D. C. 1981. Early experience determines song dialect responsiveness of female sparrows. *Science* 214:819–21

Baptista, L. F. 1975. Song dialects and demes

in sedentary populations of the white-crowned sparrow (*Zonotrichia leucophrys nuttalli*). *Univ. Calif. Publ. Zool.* 105:1–52

Baptista, L. F., Petrinovich, L. 1984. Social interaction, sensitive phases and the song template hypothesis in the white-crowned sparrow. *Anim. Behav.* 32:172–81

Becker, P. H. 1976. Artkennzeichende Gesangsmerkmale bei Winter- und Sommergoldhaenchen (*Regulus regulus, R. ignicapillus*). *Z. Tierpsychol.* 42:337–42

Bentley, D., Hoy, R. R. 1970. Post-embryonic development of adult motor patterns in crickets: A neural analysis. *Science* 170:1409–11

Bentley, D., Konishi, M. 1978. Neural control of behavior. *Ann. Rev. Neurosci.* 1:35–59

Boehner, J. 1983. Song learning in the zebra finch (*Taeniopygia guttata*): Selectivity in the choice of a tutor and accuracy of song copies. *Anim. Behav.* 31:231–37

Bottjer, S. W., Arnold, A. P. 1982. Afferent neurons in the hypoglossal nerve of the zebra finch (*Poephila guttata*): Localization with horseradish peroxidase. *J. Comp. Neurol.* 210:190–97

Bremond, J. C. 1968. Reserches sur le semantique et les elements vecteurs d'information dans les signaux acoustiques du rouge-gorge (*Erithacus rubecula L.*). *Terre Vie* 2:109–20

Brenowitz, E. A. 1983. The contribution of temporal song cues to species recognition in the red-winged blackbird. *Anim. Behav.* 31:116–27

Brockway, B. F. 1969. Roles of budgerigar vocalization in the integration of breeding behaviour. In *Bird Vocalizations,* ed. R. H. Hinde, pp. 131–58. London/New York: Cambridge Univ. Press

Brooks, R. J., Falls, B. J. 1975. Individual recognition by song in white-throated sparrows. 1. Discrimination of songs of neighbors and strangers. *Can. J. Zool.* 53:879–88

Capranica, R. R. 1966. Vocal response of the bullfrog to natural and synthetic mating calls. *J. Acoust. Soc. Am.* 40:1131–39

Catchpole, C. K. 1983. Variation in the song of the great reed warbler *Acrocephalus arundinaceus* in relation to mate attraction and territorial defence. *Anim. Behav.* 31:1217–25

DeVoogd, T. J., Nottebohm, F. 1981. Sex differences in dendritic morphology of a song control nucleus in the canary: A quantitative Golgi study. *J. Comp. Neurol.* 196:309–16

Dooling, R. J., Searcy, M. H. 1980. Early perceptual selectivity in the swamp sparrow. *Dev. Psychobiol.* 13:499–506

Emlen, S. T. 1971. The role of song in individual recognition in the Indigo Bunting. *Z. Tierpsychol.* 28:241–46

Emlen, S. T. 1972. An experimental analysis of the parameters of bird song eliciting species recognition. *Behaviour* 41:120–71

Falls, J. B. 1963. Properties of bird song eliciting responses from territorial males. *Proc. 13the Int. Ornithol. Congr.,* pp. 259–71

Fentress, J. 1973. Development of grooming in mice with amputated forelimbs. *Science* 179:704–5

Gaunt, A. S. 1983. An hypothesis concerning the relationship of syringeal structure to vocal abilities. *Auk* 100:853–62

Gerhardt, H. C. 1978. Discrimination of intermediate sounds in a synthetic call continuum by female green tree frogs. *Science* 199:1089–91

Greenewalt, C. H. 1968. *Bird Song: Acoustics and Physiology.* Washington, DC: Smithsonian Inst. Press

Grillner, S., 1975. Locomotion in vertebrates: Central mechanisms and reflex interaction. *Physiol. Rev.* 55:247–304

Guettinger, H. R. 1979. The integration of learnt and genetically programmed behaviour: A study of hierarchical organization in songs of canaries, greenfinches and their hybrids. *Z. Tierpsychol.* 49:285–303

Guettinger, H. R. 1981. Self-differentiation of song organization rules by deaf canaries. *Z. Tierpsychol.* 56:323–40

Guettinger, H. R., Wolffgramm, J., Thimm, F. 1978. The relationship between species specific song programs and individual learning in song birds. *Behaviour* 65:241–62

Gurney, M. E. 1981. Hormonal control of cell form and number in the zebra finch song system. *J. Neurosci.* 1:658–73

Gurney, M. E. 1982. Behavioral correlates of sexual differentiation in the zebra finch song system. *Brain Res.* 231:153–72

Gurney, M. E., Konishi, M. 1980. Hormone induced sexual differentiation of brain and behavior in zebra finches. *Science* 208:1380–82

Hinde, R. A. 1982. Foreword. In *Acoustic Communication in Birds,* Vol. 2, *Song Learning and Its Consequences,* ed. D. E. Kroodsma, E. H. Miller. New York/London: Academic

Immelmann, K. 1969. Song development in the Zebra Finch and other estrildid finches. In *Bird Vocalizations,* ed. R. A. Hinde. London/New York: Cambridge Univ. Press

Jenkins, P. F. 1977. Cultural transmission of song patterns and dialect development in a free-living bird population. *Anim. Behav.* 25:50–78

Kandel, E. R. 1979. Cellular insight into behavior and learning. *Harvey Lecture Ser.* 73:19–92

Katz, L. C. 1982. The avian motor system for song has multiple sites and types of auditory input. *Soc. Neurosci. Abstr.* 8:1021

Katz, L. C., Gurney, M. E. 1981. Auditory responses in the zebra finch's motor system for song. *Brain Res.* 211:192–97

Kawamura, T. 1947. *Science of Birdsong*. Reprinted in 1974 by Chuokoronsha, Tokyo. (In Japanese)

Kelley, D. B., Nottebohm, F. 1979. Projections of a telencephalic auditory nucleus—field L—in the canary. *J. Comp. Neurol.* 183:455–70

Koehler, O. 1954. Der Vogelsang als Vorstufe von Musik und Sprache. *J. Ornithol.* 93:3–20

Konishi, M. 1963. The role of auditory feedback in the vocal behavior of the domestic fowl. *Z. Tierpsychol.* 20:349–67

Konishi, M. 1964. Song variation in a population of Oregon juncos. *Condor* 66:423–36

Konishi, M. 1965a. Effects of deafening on song development in two species of juncos. *Condor* 66:85–102

Konishi, M. 1965b. The role of auditory feedback in the control of vocalization in the white-crowned sparrow. *Z. Tierpsychol.* 22:770–83

Konishi, M. 1978. Auditory environment and vocal development in birds. In *Perception and Experience*, ed. R. D. Walk, H. L. Pick, pp. 105–18. New York: Plenum

Konishi, M., Akutagawa, E. 1981. Androgen increases protein synthesis within the avian brain vocal control system. *Brain Res.* 222:442–46

Konishi, M., Gurney, M. E. 1982. Sexual differentiation of brain and behaviour. *Trends Neurosci.* 5:20–23

Konishi, M., Nottebohm, F. 1969. Experimental studies in the ontogeny of avian vocalization. In *Bird Vocalizations*, ed. R. A. Hinde. London/New York: Cambridge Univ. Press

Krebs, J. R. 1977. The significance of song repertoires: The Beau Geste hypothesis. *Anim. Behav.* 25:475–78

Krebs, J. R., Ashcroft, R., Webber, M. I. 1978. Song repertoires and territory defense in the great tit. *Nature* 271:539–42

Krebs, J. R., Kroodsma, D. E. 1980. Repertoires and geographical variation in bird song. In *Advances in the Study of Behaviour*, ed. J. R. Rosenblat, R. A. Hinde, C. Beer, M.-C. Busnel, 11:143–77. New York: Academic

Kroodsma, D. E. 1974. Song learning, dialects, and dispersal in the Bewick's wren. *Z. Tierpsychol.* 35:352–80

Kroodsma, D. E. 1976. Reproductive development in a female songbird: Differential stimulation by quality of male song. *Science* 192:574–75

Kroodsma, D. E. 1977a. A re-evaluation of song development in the song sparrow. *Anim. Behav.* 25:390–99

Kroodsma, D. E. 1977b. Correlates of song organisation among North American wrens. *Am. Nat.* 111:995–1008

Kroodsma, D. E. 1980. Winter Wren singing behavior: A pinnacle of song complexity. *Condor* 82:357–65

Kroodsma, D. E. 1982. Learning and the ontogeny of sound signals in birds. See Kroodsma & Miller 1982, pp. 1–19

Kroodsma, D. E. 1984. Songs of the alder flycatcher (*Empidonax alnorum*) and willow flycatcher (*Empidonax traillii*) are innate. *Auk* 101:13–24

Kroodsma, D. E., Miller, E. H., eds. 1982a. *Acoustic Communication in Birds*, Vol. 2, *Song Learning and Its Consequence*. New York/London: Academic. 389 pp.

Kroodsma, D. E., Miller, E. H. eds. 1982b. *Acoustic Communication in Birds*, Vol. 1, *Production, Perception, and Design Features of Sounds*. New York/London: Academic. 360 pp.

Kroodsma, D. E., Parker, L. D. 1977. Vocal virtuosity in the Brown Thrasher. *Auk* 94:783–84

Leppelsack, H. J., Vogt M. 1976. Responses of auditory neurons in the forebrain of a songbird to stimulation with species-specific sounds. *J. Comp. Physiol.* 107:263–74

Luine, F., Nottebohm, F., Harding, C. McEwen, B. S. 1980. Androgen affects cholinergic enzymes in syringeal motor neurons and muscles. *Brain Res.* 192:89–107

Manogue, K., Paton, J. 1982. Respiratory gating of activity in the avian vocal control system. *Brain Res.* 257:383–87

Margoliash, D. 1983. Acoustic parameters underlying the responses of song-specific neurons in the white-crowned sparrow. *J. Neurosci.* 3:1039–57

Marler, P. 1957. Specific distinctiveness in the communication signals of birds. *Behaviour* 11:13–39

Marler, P. 1961. The logical analysis of animal communication. *J. Theoret. Biol.* 1:295–317

Marler, P. 1964. Inheritance and learning in the development of animal vocalizations. In *Acoustic Behavior of Animals*, ed. R. G. Busnel. Amsterdam: Elsevier

Marler, P. 1970. A comparative approach to vocal learning: Song development in white-crowned sparrows. *J. Comp. Physiol. Psychol.* 71:(2) 1–25

Marler, P. 1976. Sensory templates in species-specific behavior. In *Simpler Networks and Behavior*, ed. J. Fentress, pp. 314–29. Sunderland, Mass.: Sinauer Assoc.

Marler, P. 1981. Birdsong: The acquisition of a learned motor skill. *Trends Neurosci.* 4:88–94

Marler, P. 1983. Some ethological implications for neuroethology: The ontogeny of birdsong. In *Advances in Vertebrate Neuro-*

ethology, ed. J.-P. Ewert, R. R. Capranica, D. Ingle. New York/London: Plenum

Marler, P. 1984. Song learning in birds: The role of neuroselection. In *Biology of Learning*, ed. P. Marler, H. S. Terrace. Dahlem Konferenzen, Berlin

Marler, P., Kreith, M., Tamura, M. 1962. Song development in hand-raised Oregon juncos. *Auk* 79:12–30

Marler, P., Mundinger, P., Waser, M. S., Lutjen, A. 1972. Effects of acoustical stimulation and deprivation on song development in red-winged blackbirds (*Agelaius phoeniceus*). *Anim. Behav.* 20:586–606

Marler, P., Peters, S. 1977. Selective vocal learning in a sparrow. *Science* 198: 519–21

Marler, P., Peters, S. 1981. Sparrows learn adult song and more from memory. *Science* 213:780–82

Marler, P., Peters S. 1982a. Subsong and plastic song: Their role in the vocal learning process. See Kroodsma & Miller 1982, Chap. 2

Marler, P., Peters, S. 1982b. Developmental overproduction and selective attrition: New processes in the epigenesis of birdsong. *Dev. Psychobiol.* 15:369–78

Marler, P., Peters, S. 1982c. Structural changes in song ontogeny in the swamp sparrow, *Melospiza georgiana*. *Auk* 99:446–58

Marler, P., Sherman, V. 1982. Structure in sparrow song without auditory feedback: An emendation of the auditory template hypothesis. *J. Neurosci.* 3:517–31

Marler, P., Tamura, M. 1962. Song variation in three populations of white-crowned sparrow. *Condor* 64:368–77

Marler, P., Waser, M. S. 1977. The role of auditory feedback in canary song development. *J. Comp. Physiol. Psychol.* 91:1–7

McCasland, J. S. 1983. *Neuronal control of bird song production*, PhD thesis, Calif. Inst. Technol., Pasadena, Calif.

McCasland, J. S., Konishi, M. 1981. Interaction between auditory and motor activities in an avian song control nucleus. *Proc. Natl. Acad. Sci. USA* 78:7815–7819

McGregor, P. K., Krebs, J. R., Perrins, C. M. 1981. Song repertoires and lifetime reproductive success in the great tit (*Parus major*). *Am. Nat.* 118:149–59

Mulligan, J. A. 1966. Singing behavior and its development in the song sparrow, *Melospiza melodia*. *Univ. Calif. Publ. Zool.* 81:1–76

Mundinger, P. C. 1982. Microgeographic and macrogeographic variation in the acquired vocalizations of birds. See Kroodsma & Miller 1982, pp. 147–208

Nottebohm, F. 1968. Auditory experience and song development in the chaffinch, *Fringilla coelebs*. *Ibis* 110:549–68

Nottebohm, F. 1980a. Brain pathways for vocal learning in birds: A review of the first 10 years. *Prog. Psychbiol. Physiol. Psychol.* 9:85–124

Nottebohm, F. 1980b. Testosterone triggers growth of brain vocal control nuclei in adult female canaries. *Brain Res.* 289:429–36

Nottebohm, F. 1981. A brain for all seasons: Cyclical anatomical changes in song control nuclei in the canary brain. *Science* 214:1368–70

Nottebohm, F., Arnold, A. P. 1976. Sexual dimorphism in vocal control areas of the song bird brain. *Science* 194:211–13

Nottebohm, F., Nottebohm, M. E. 1971. Vocalizations and breeding behavior of surgically deafened ring doves (*Streptopelia risoria*). *Anim. Behav.* 19:313–27

Nottebohm, F., Nottebohm, M. E. 1976. Left hypoglossal dominance in the control of canary and white-crowned sparrow song. *J. Comp. Physiol.* 108:171–92

Nottebohm, F., Nottebohm, M. E. 1978. Relationship between song repertoire and age in the canary, *Serinus canarius*. *Z. Tierpsychol.* 46:298–305

Nottebohm, F., Kasparian, S., Pandazis, C. 1981. Brain space for a learned task. *Brain Res.* 213:99–109

Nottebohm, F., Kelley, D. B., Paton, J. A. 1982. Connections of vocal control nuclei in the canary telencephalon. *J. Comp. Neurol.* 207:344–57

Nottebohm, F., Manning, E., Nottebohm, M. E. 1979. Reversal of hypoglossal dominance in canaries following unilateral syringeal denervation. *J. Comp. Physiol.* 134:227–40

Nottebohm, F., Stokes, T. M., Leonard, C. M. 1976. Central control of song in the canary, *Serinus canarius*. *J. Comp. Neurol.* 165:457–86

Paton, J. A., Manogue, K. R., Nottebohm, F. 1981. Bilateral organization of the vocal control pathway in the budgerigar, *Melopsittacus undulatus*. *J. Neurosci.* 1:1276–88

Payne, R. B. 1973. Vocal mimicry of the paradise whydahs (*Viua*) and response of female whydahs to songs of their host (*Pytilia*) and their mimics. *Anim. Behav.* 21:762–71

Payne, R. B. 1981. Song learning and social interaction in indigo buntings. *Anim. Behav.* 29:688–97

Payne, R. B., Payne, K. 1977. Social organization and mating success in local song populations of village indigobirds, *Vidua chalybeata*. *Z. Tierpsychol.* 45:113–73

Payne, R. B,, Thompson, W. L., Fiala, K. L., Sweany, L. L. 1981. Local song traditions in Indigo Buntings: Cultural transmission of behavior patterns across generations. *Behaviour* 77:199–221

Pesch, A. 1983. *Gesangsorganisation bei weiblichen Kanarienvoegeln* (*Serinus canaria*)

und ihre moeglichen endokrinologischen Korrelate. Diplomarbeit, Fachbereich Biologie, Universitaet Kaiserslautern

Peters, S. S., Searcy, W. A., Marler, P. 1980. Species song discrimination in choice experiments with territorial male swamp and song sparrows. *Anim. Behav.* 28:393–404

Phoenix, C. W., Goy, R. W., Gerall, A. A., Young, W. C. 1959. Organizing action of prenatally administered testosterone prorionate on the tissues mediating mating behavior in the female guinea pig. *Endocrinology* 65:369–82

Price, P. H. 1977. *Determinants of acoustical structure in zebra finch song,* PhD thesis, Univ. Penna.

Rice, J. O., Thompson, W. L. 1968. Song development in the indigo bunting. *Anim. Behav.* 16:462–69

Scheich, H., Langner, G., Bonke, D. 1979. Responsiveness of units in the auditory neostriatum of the guinea fowl (*Numida meleagris*) to species-specific calls and synthetic stimuli. II: Discrimination of Iambus-like calls. *J. Comp. Physiol.* 132:257–76

Searcy, W. A., Peters, S., Marler, P. 1981. A test for responsiveness to song structure and programming in female sparrows. *Science* 213:926–28

Shiovitz, K. A. 1975. The process of species-specific song recognition by the indigo bunting, *Passerina cyanea,* and its relationship to the organization of avian acoustical behavior. *Behaviour* 44:128–79

Siegel, L. I., Fox, T. O., Konishi, M. 1983. Androgen and estrogen receptors in zebra finch brain. *Soc. Neurosci. Abstr.* 9:314.8

Snow, D. W. 1968. The singing assemblies of little hermits. *Living Bird* 7:47–55

Stein, P. S. G. 1978. Motor systems, with specific reference to the control of locomotion. *Ann. Rev. Neurosci.* 1:61–81

Stein, R. C. 1968. Modulation in bird sounds. *Auk* 85:229–43

Suga, N. 1982. Functional organization of the auditory cortex: Representation beyond tonotopy in the bat. In *Cortical Sensory Organization,* Vol. 3, *Multiple Auditory Areas,* ed. C. N. Woolsey, pp. 157–218. Clifton, NJ: Humana Press

Thorpe, W. H. 1954. The process of song-learning in the chaffinch as studied by means of the sound spectrograph. *Nature* 173:465

Thorpe, W. H. 1958. The learning of song patterns by birds, with especial reference to the song of the chaffinch, *Fringilla coelebs. Ibis* 100:535–70

Thorpe, W. H. 1961. *Bird-Song*. Cambridge Monogr. Exp. Biol. 12:143

Todt, D., Hultsch, H. 1985. Zum Einfluss des vokalen Lernen auf die Ausbildung gesanglicher Repertoires bei Drosselvoegeln. *Proc. Int. Symp. Verhaltensbiol.*, ed. G. Tembrock, R. Siegmund, W. Nichelmann

Tsukahara, N. 1981. Synaptic plasticity in the mammalian central nervous system. *Ann. Rev. Neurosci.* 4:351–79

von Helversen, D. 1972. Gesang des Maennchens und Lautschema des Weibchens bei der Feldheuschrecke *Chorthippus biguttulus (Orthoptera, Acrididae). J. Comp. Physiol.* 81:381–422

Walker, T. J. Jr. 1957. Specificity in the response of female tree crickets *(Orthoptera, Grillidae, Oecanthinae)* to calling songs of the males. *Ann. End. Soc. Am.* 50:626–36

Waser, M. S., Marler, P. 1977. Song learning in canaries. *J. Comp. Physiol. Psychol.* 91:1–7

West, M. J., King, A. P., Eastzer, D. H., Staddon, J. E. R. 1979. A bioassay of isolate cowbird song. *J. Comp. Physiol. Psychol.* 93:124–33

Wiley, R. H. 1971. Song groups in a single assembly of little hermits. *Condor* 73:28–35

Yasukawa, K. 1981. Song repertoires in red-winged blackbird (*Agelaius phoeniceus*): A test of the Beau Geste hypothesis. *Anim. Behav.* 24:114–25

Ann. Rev. Neurosci. 1985. 8:171–98

NEUROPEPTIDE FUNCTION: The Invertebrate Contribution

Michael O'Shea

Université de Genève, Laboratoire de Neurobiologie, CH–1211 Genève 4, Switzerland

Martin Schaffer

Department of Psychiatry, The University of Chicago, Chicago, Illinois 60637

INTRODUCTION

With the discovery and dramatic proliferation of recognized neuropeptides in vertebrate nervous systems, it has become clear that the peptides represent the largest single class of neurotransmitter substances (Iversen 1983). The important but unanswered question posted by this development: Why does the nervous system need so many different transmitters? Clearly advances in peptide identification have not been accompanied by a parallel advance in our understanding of the functions played by these new and numerous transmitters.

Understandably most of the effort in the vertebrate field has been directed at the chemical characterization and localization of neuropeptides. Immunohistochemical techniques have made the latter a relatively routine process. Consequently there now exists a large and growing literature on the cellular localization of peptides in vertebrate nervous systems. Unfortunately, these distribution studies, while they may provide general hints about peptide function, have done little to clarify our understanding of the specific roles played by peptide-containing neurons.

Although it is often an important first step, neurochemical anatomy is an indirect approach to peptide function. It cannot substitute for direct physiological experiments. The technical problems of directly studying the functions of peptide-containing neurons buried deep in the vertebrate brain are, however, enormous. By exploiting the advantage peculiar to certain invertebrates i.e. the presence of large (occasionally giant) and uniquely identifiable neurons, fundamental questions related to the functions of peptidergic neurons can be

0147–006X/85/0301–0171$02.00

studied more conveniently and more precisely. There is a rich historical precedent for predicting the success of such a strategy. Classical examples include the squid with its giant axon and giant synapse and the crayfish with its uniquely identifiable stretch receptor neuron and giant motor synapse. Together these model systems have helped establish the fundamental mechanisms of signal generation and propagation, chemical synaptic transmission in the CNS, inhibitory synaptic action, inhibition by GABA, and electrical synaptic transmission. No less impressive contributions to our understanding of neural mechanisms involved at higher levels of organization have been achieved more recently by the analysis of the role of identified neurons in behavior and learning in other invertebrates such as the leech, *Aplysia, Tritonia,* lobster, and locust.

Invertebrate neuropeptide studies are now rapidly expanding with a proliferation of newly characterized peptides and peptidergic systems. This development, however, lags behind the vertebrate peptide field. The difficulty of peptide characterization in invertebrates is primarily responsible. This is not only due to the small size of the source organism, but also to an important difference in the organization of vertebrate and invertebrate nervous systems. Each neuron in an invertebrate nervous system may be an identifiable individual that expresses a unique phenotype. At worst therefore there may be only one specific cell containing a particular peptide in the entire animal. Paradoxically, the very fact that makes invertebrates attractive for functional studies—the presence of uniquely identifiable neurons—has made advances in peptide identification difficult. The relatively slow development of functional or physiological model systems in the invertebrates can therefore be attributed to the difficulty of initial identification of peptides and chemical characterization of these systems.

This situation is rapidly changing. Technical problems have been overcome by more widespread availability of high performance liquid chromatography (hplc) purification methods, the application of powerful new mass spectroscopy technologies, and the recent application of molecular biological techniques, particularly recombinant DNA studies. In this chapter we cite examples of how this impressive array of new techniques has overcome impediments to peptide identification in invertebrates, and indicate why we anticipate the characterization of many more invertebrate neuropeptides. If the invertebrates are to make a unique contribution, however, there must be progress beyond mere peptide identification and localization. The key to progress will be the study of functional systems in which the actions of uniquely identified peptide-containing neurons can be conveniently analyzed at their identified postsynaptic target cells. The primary aims of this essay are to review current progress in the field and to assess the contribution and the potential usefulness of invertebrate systems for our understanding of how neuropeptides act and why they are employed.

PEPTIDE IDENTIFICATION AND SEQUENCING

Since the development of powerful model systems requires the complete chemical characterization of the peptides involved, it is worth reviewing how this has been accomplished in a variety of representative examples. The first neuropeptide to be sequenced from any animal, vertebrate or invertebrate, was isolated from the salivary glands of the octopus *Eledone muschata* (Erspamer et al 1962). This peptide, called eledoisin, was found in a search for substance P-like activities, and was isolated by conventional means, employing its capacity to produce atropine-resistant hypotension in mammals as a bioassay. At the time of sequencing (1962) the significance of neuropeptides was largely unrecognized, and although it was the first example, eledoisin has not been as extensively studied as some later examples.

Proctolin is arguably the best studied invertebrate neuropeptide. The first step in its discovery was an investigation by Brown (1967) of the cockroach proctodeum (hindgut) and its innervation. This led to the development of a hindgut muscle bioassay and a search for factors that caused muscle contraction. Pursuing one of several such factors, Brown & Starratt (1975) devised an extraction procedure beginning with whole cockroach, which they carried out on a kilogram scale. They used the bioassay to devise a classical peptide purification scheme employing protein precipitation, ion exchange and gel filtration chromatography, and paper electrophoresis: 125,000 cockroaches yielded 180 μg of pure proctolin. This was sequenced by the micro dansyl Edman technique and the sequence was confirmed by synthesis (Starratt & Brown 1975). Thus, proctolin was identified, purified, and sequenced by traditional methods, using very large numbers of animals.

The locust peptide adipokinetic hormone was also identified by a bioactivity, lipid release into the hemolymph (Mayer & Candy 1969). Again the assay was used to devise a purification scheme, employing column and thin layer chromatograpy (Stone et al 1976). However, in this case a highly enriched source of the peptide was identified, a neuroendocrine gland called the corpora cardiaca. This permitted a much simpler purification scheme, but sequencing held more of a challenge, since adipokinetic hormone is blocked at both its amino and carboxy termini. The sequencing was begun with a combination of amino acid analysis, a variety of enzymic digestions, and dansyl Edman degredation. The sequence was completed by applying low resolution mass spectroscopy to the parent peptide, endopeptidase-derived fragments, and various deuterated derivatives of these peptides. Three thousand animals were required for these studies.

Neuropeptides related to adipokinetic hormone, MI and MII, were discovered in cockroach by simply applying an extraction procedure similar to the one used for adipokinetic hormone to the same organ, corpora cardiaca, of a different, but closely related animal (O'Shea et al 1984). A muscle bioassay was used to identify compounds of interest, but in this case the combination of a rich source and the power of reverse phase hplc monitored by high sensitivity ultraviolet absorbance and fluorescence flow detectors permitted the development of a two-step purification scheme by directly visualizing the eluted peaks. The complete sequence of MI and MII is not yet available, but it seems likely from work in progress by the authors that the combination of amino acid analysis and high resolution, fast atom bombardment mass spectroscopy will lead to a complete structure of these blocked octapeptides. This series of three examples illustrates how the rapid improvement in technologies for peptide purification and sequencing have simplified the task of identifying and characterizing new neuropeptides based on interesting bioactivities. The utility of identifying enriched sources is also evident. Because of the as yet unexplained occurrence of peptides in families, it is likely that there will be many more examples of related but distinctly different peptides being isolated from sources such as the corpora cardiaca in different animals.

A different approach has been employed to investigate the peptide products of a set of identified neurons, the so-called R3–14 neurons, in the abdominal ganglion of *Aplysia californica*. Electrophysiological experiments on these sea hare neurons indicated their role in regulation of the circulatory system (Sawada et al 1981a,b), and morphological (Coggeshall et al 1966, Price & McAdoo 1979) and biochemical studies (Wilson 1976, Loh & Gainer 1975, Aswad 1978) suggested that these neurons employed peptides as transmitters. Despite considerable study of the neurons, the peptides were not isolated for some time. Fortunately Nambu et al (1983) made a major contribution to the understanding of this system through a molecular biological approach. These workers searched a cDNA library of abdominal ganglion poly A^+ RNA for sequences unique to R3–R14. That is, they looked for clones that hybridized with radioactive cDNA derived from R3–R14 poly A^+ RNA, but not with cDNA derived from other ganglion cells. In this way, they identified and sequenced an mRNA that appears to code for the major peptide products whose existence was suggested by earlier work. The predicted sequences should greatly facilitate the isolation of the R3–R14 peptides and the study of their physiological activities. One peptide has already been identified (E. Mayeri, personal communication). A similar approach is being employed by Buck (1983), who is studying another *Aplysia* neuron (R–15), which also appears to employ peptide transmitters. Despite the peptides' appearing to be unique to the R–15 cell, it was nonetheless possible to identify a likely peptide-related clone. Thus the power of molecular biological techniques has made it possible to predict structural information about peptides contained in just a few or even one cell. It is clear that such approaches will be increasingly important in the development of interesting invertebrate neuropeptide systems.

The cloning of the egg-laying hormone (ELH) gene from *Aplysia* illustrates another way in which new neuropeptides have been identified. ELH is itself an interesting neuropeptide and plays an important role in evoking egg-laying behavior (Strumwasser et al 1969, Arch 1976). This peptide was found in an enriched source, the bag cells, and conventionally purified with the help of a bioassay. Its sequence was determined by chemical cleavage and automated microsequencing (Chiu et al 1979). When Scheller et al (1983) determined the ELH gene sequence they came upon a strong suggestion for the existence of other important bag cell peptides. This gene apparently codes for an mRNA, which in turn gives rise to an ELH protein precursor. The inferred protein sequence contains a number of multiple basic amino acid sequences that may serve as processing sites, and this suggests the production of several peptides. One of these peptides, bag cell peptide (α-BCP), has already been identified and sequenced (Rothman et al 1983). Another, β-BCP, has been characterized as far as amino acid composition. Isolation of a third, acidic peptide was stimulated by predicted sequence, and apparently its existence was confirmed (Scheller et al 1983). Still other peptides can be predicted based on the gene sequence, and it remains to be seen whether these peptides will be found and what activities they may have. Based on this example and comparable ones from the vertebrate neuropeptide literature, it seems likely that as other neuropeptide messages and genes are sequenced, we will discover additional, previously unknown neuropeptides.

Another strategy for peptide identification deserves discussion. A number of investigators have successfully used immunocytochemistry to demonstrate that invertebrate nervous systems contain peptides that are structurally related to a variety of mammalian neuropeptides. The list of apparent homologues is long and has recently been discussed (Greenberg & Price 1983, Truman & Taghert 1983). Despite the large number of promising leads, however, there is currently only one example of peptides isolated from an invertebrate based on their homology to mammalian relatives. The example is the enkephalins, which, based on immunocytochemical and physiological studies, appear to occur widely in invertebrates. Leung & Stefano (1984) identified a particularly rich source of enkephalin-like activity, the pedal ganglion, of a mussel *Mytillus edulis* and used this as a tissue source for hplc purification of the peptides. The purification scheme was designed to isolate leu- and met-enkephalin, and the authentic biological activity was located by an opiate receptor binding assay. These activities were found to coincide with synthetic leu- and met-enkephalin. The met-enkephalin-like activity was purified by a second round of hplc to near homogeneity, and this preparation apparently gave the met-enkephalin sequence as the predominant one when it was subjected to Edman sequencing. The other opiate activity was brought to apparent homogeneity by a second round of hplc and was reported to have the leu-enkephalin sequence. Thus, procedures developed to identify mammalian opiate binding and immunoactivity led to the identification of a rich invertebrate source. Modern peptide techniques, notably hplc, provided rapid identification of these activities as met- and leu-enkephalins. The mussel pedal ganglion is a very interesting system for the study of the interaction of enkephalinergic and dopaminergic neurons (Stefano 1982) and will doubtlessly shed light on similar interactions in mammalian systems.

Immunologically identified invertebrate peptide homologues of vertebrate peptides are being pursued in other studies. For example, Duve et al (1982) made a highly enriched preparation of pancreatic polypeptide-like material from the blow fly, *Calliphora vomitoria*. Although sequence data is not yet available, amino acid analysis suggests close structural homology with pancreatic polypeptide found in mammals. In other cases homology may be less striking, as for example in the case of vasopressin-like immunoactivity in locust (Cupo & Proux 1983). Many other immunoactivities remain to be studied, and it seems inevitable that the pursuit of these activities will contribute significantly to the list of well-characterized invertebrate peptides.

The examples cited above illustrate a variety of strategies that have led to the identification of previously unrecognized neuropeptides. A list of peptides found in neural or neuroendocrine tissue that have been sequenced and that have activities on neural or muscular targets are listed in Table 1. The criteria have been liberally applied in several instances, since the extent of distribution or range of activity have not yet been investigated. Although this list is quite short in comparison with a similar list of vertebrate neuropeptides (Iversen 1983), or any reasonable estimate of the number of peptides likely to exist in invertebrates, it is sure to grow rather rapidly. Indeed, we assume that the list will be incomplete by the time it is published. We assume this based on the large number of leads that have been reported in the literature. For example, a wide variety of factors are reported to have neural activities that await isolation and characterization. Some of these, such as bursicon (Cottrell 1962, Fraenkel & Hsiao 1962), are under active study by several groups (notably Truman & Taghert 1983, Taghert & Truman 1982a,b), and full characterization is likely soon. Other activities, particularly in insects, are simply reported in the literature as such, and await investigation by chemical methods. This area has been extensively reviewed recently (Raabe 1982, Frontali & Gainer 1977). Each of the other approaches listed is also likely to make its contribution. Despite the problems of unique cells, the task of neuropeptide identification and characterization in invertebrates does not appear to be greatly different from that in vertebrates. The same opportunities and problems tend to present themselves, and the same techniques tend to lead to successful characterizations. Each case will require imaginative approaches and sufficient expenditure of resources and effort if interesting activities are to be transformed into sequenced peptides. Clearly this task is well worth the neurobiologist's time. Given good chemical and anatomical characterization, invertebrate systems routinely provide powerful model systems that are amenable to detailed analysis and that have clear implications for neuropeptide function throughout the evolutionary hierarchy. We survey some of these promising systems below.

DEVELOPMENT OF MODEL PEPTIDERGIC SYSTEMS

Model systems in invertebrate neurophysiology have been confined primarily to a small number of species representing three major invertebrate phyla, the Annelids, Arthropods, and Molluscs. Sequenced bioactive peptides are few and unevenly distributed among these phyla. If we apply our criteria (see above), there are as yet no sequenced neuropeptides among the Annelids, there are four in the Arthropods, and the remainder belong to the Molluscs. In spite of this paucity, we can expect that a useful range of model systems will be developed in each of the three phyla. Ideally the constituents of model preparations include a sequenced bioactive neuropeptide or neuropeptides, uniquely identified neurons containing and releasing the neuropeptide, and specifically identified and accessible targets.

Neuropeptides show a wide variety of activities. They may act like local transmitters or modulators near the release site within the CNS or periphery, or like hormones with local and distant sites of action. Specific actions may be transient or relatively long-lasting. Many functional descriptions are complicated by co-localization of a peptide with other neuropeptides or nonpeptide neuroeffectors. Clearly, no one model system can be expected to address all of the problems and questions related to neuropeptide function. The diversity of the invertebrate preparations currently being developed and the multidisciplinary nature of the investigations, however, offer considerable hope that significant progress will be made on a broad front.

Progress in understanding peptide function will depend on finding both centrally and peripherally located specific targets of peptidergic neurons. While postsynaptic targets within the CNS will improve understanding of the role of peptides in neural and behavioral integration, they are likely to present technical problems in understanding the subcellular or molecular modes of peptide action. Accessible and relatively large peripheral targets, like muscles for example, will, however, facilitate the analysis of both the electrophysiological and biochemical consequences of peptidergic neuronal activity. Indeed a pat-

Table 1 Sequences of invertebrate neuropeptides

Peptide	Sequence	Isolated from		Identified action[a]
		Animal	Organ	
Arthropods				
Proctolin	RYLPT[b]	*Periplaneta americana* (cockroach)	Whole animal	Contract hindgut
Adipokinetic hormone (AKH)	pQLNFTPNWGT-NH_2[c]	*Locusta migratoria* *Schistocerca gregaria* (locust)	Corpora cardiaca	Release lipid into blood
Red pigment concentrating hormone (RPCH)	pQLNFSPGW–NH_2[d]	*Pandalus borealus* (prawn)	Eyestalk	Concentration of pigment in erythrophore
Distal retinal pigment hormone (DRPH)	NSGMINSILGIPRVMTEA–NH_2[e]	*Pandalus borealus*	Eyestalk	Eye distal pigment movement
Paragonial peptide PS–1	D_L^V,SANANANNQRTAAAKPQ-ANADASS[f]	*Drosophila funebris* (fruit fly)	Whole animal	Inhibition of mating
Molluscs				
FMRFamide	FMRF–NH_2[g]	*Macrocallista nimbosa* (clam)	Ganglia	Whelk muscle contraction, clam heart stimulation
Helix FMRFamide-like peptide	pQDPFLRF–NH_2[h]	*Helix aspersa* (snail)	Ganglia	FMRFamide RIA activities like FMRF–NH_2
Met enkephalin Leu enkephalin Met enkephalin Arg^6Phe^7	YGGFM[i] YGGFL[i] YGGFMRF[j]	*Mytilus edulis* (mussel)	Pedal ganglion	Binding to opiate receptor
Snail cardioexcitatory peptide B (SCP_B)	MNYLAFPRM–NH_2[k]	*Aplysia brasiliana* (sea hare)	Central nervous system	Excitation snail heart

Egg-laying hormone (ELH)	ISINQDLKAITDMLLTDQ-IRGRQRYLADLRQRLLDK–NH_2[l,m]	*Aplysia californica*	Bag cells	Induce egg laying
α-Bag cell peptide (α-BCP)	APRLRFTSL[n]	*Aplysia californica*	Bag cells	Inhibition of LUQ abdominal ganglion cells
Peptide A	AVKLSSDGNYPFDLSKED-GAQPYFMTPRLRFYPI–NH_2[l,o]	*Aplysia californica*	Atrial gland	Induce egg laying
Peptide B	AVKSSSYGKYPFDLSKEDGA-QPYFMTPRLRFYPI–NH_2[l,o]	*Aplysia californica*	Atrial gland	Induce egg laying
R3–R14 peptide	EAEEPSAFMTRL[p]	*Aplysia californica*	R3–R14 cells	(Co-chromatography with synthetic peptide) contracts vascular smooth muscle[q]
Eledoisin	*p*QPSKDAFIGLM–NH_2[r]	*Eledone moschata* (octopus)	Salivary gland	Hypotension in mammals
Coelenterate				
Hydra head activato.	*p*QPPGGSKVILF[s]	*Anthopleura elegantissima*	Whole animal	Stimulate hydra head regeneration

[a]Action used for purification assay.
[b]Starratt & Brown 1975.
[c]Stone et al 1976.
[d]Fernlund & Josefsson 1972.
[e]Fernlund 1976.
[f]Bauman et al 1975.
[g]Price & Greenberg 1977.
[h]Price et al 1984.
[i]Leung & Stefano 1984.
[j]Stefano & Leung 1984.
[k]Morris et al 1982.
[l]The carboxyamide is inferred from carboxypeptidase resistance and gene nucleotide sequence.
[m]Chiu et al 1979.
[n]Rothman et al 1983.
[o]Heller et al 1980.
[p]E. Mayeri, personal communication.
[q]Isolated based on properties of synthetic peptide predicted from mRNA sequence.
[r]Erspamer & Anastasi 1962.
[s]Schaller & Bodenmüller 1981.

tern is emerging that indicates peptide use in nerve-muscle communication in invertebrates.

In the following we review some of the promising functional systems being developed in this growing and exciting field. Our aim is not to be comprehensive but to illustrate by selected examples the range of possibilities offered in invertebrate preparations.

Annelids

The absence of any sequenced Annelid peptides is surprising and unfortunate in view of the opportunities provided by the nervous systems of various species of leech. Pioneering work in the laboratories of J. G. Nicholls and G Stent has established the leech as a most important Annelid for the study of the cellular basis of behavior (Stent & Kristan 1981) and as a model for neuronal development (Stent et al 1982). Many of the neurons and synaptic connections of the leech central nervous system have already been uniquely identified. It seems to us that some of these identified neurons are likely to be peptidergic. The leech may therefore be one of the most attractive organisms for analyzing the behavioral role of peptides.

Why has the chemistry of leech-specific peptides not been more actively and successfully pursued? The nervous system of the leech is small, consisting of some 400 uniquely identifiable neurons per ganglion (Macagno 1980). We would expect therefore that the number of neurons in the organism containing any one specific peptide may be very few. Determining the chemical structure of a rare phenotype in such an organism presents a considerable problem. To some extent, however, the problem may simply be a lack of tradition in endocrinological studies, which in other areas have led to the identification of neuropeptides. In any case a promising approach to this problem is to exploit possible structural homologies with peptides discovered in other organisms. Structural homology, with the likely consequences of antigenic cross-reactivity and similarity of biological action, can be useful in two ways. First, the unknown leech peptide may be readily localized immunohistochemically using antibody raised against the known peptide. This is particularly attractive in leech since the localized immunoreactive cells may correspond to already identified and physiologically characterized neurons. Second, both the cross-immunoreactivity and bioactivity can provide a sensitive means of detecting the unknown peptide and are therefore useful tools in peptide isolation and purification. This general strategy is currently being adopted, with some encouraging recent developments. Identifiable neurons in the leech have been located using antibody against the mammalian peptide enkephalin (Zipser 1980), two Arthropod neuropeptides, proctolin (Li & Calabrese 1983) and AKH (Witten et al 1984), and the Molluscan peptide FMRFamide (R. L. Calabrese, personal communication). Figure 1 illustrates the presence of large, accessible and identifiable AKH-immunoreactive neurons in the leech *Hirudo medicinalis*.

Enkephalin immunoreactivity is found in one neuron in each of the posterior midbody segments of the leech *Haemopis marmorata* (Zipser 1980). To our knowledge this neuron is not one of the well characterized and identified leech neurons and nothing has been done to establish the cell's function or to characterize the enkephalin-like peptide it contains. The neurons localized with anti-proctolin and anti-AKH are more numerous and larger (Figure 1) and seem more likely to be developed into functional systems (Witten et al 1984). The specific identity of both the AKH and proctolin-immunoreactive neurons is as yet uncertain. One of the larger proctolin neurons, however, corresponds in cell body position to an identified inhibitory motoneuron and, if confirmed, this could produce a very convenient peripheral physiological preparation (Li & Calabrese 1983). Neurons immunoreactive to FMRFamide include identified efferent neurons, the HE and HA cells, that are known to be involved in

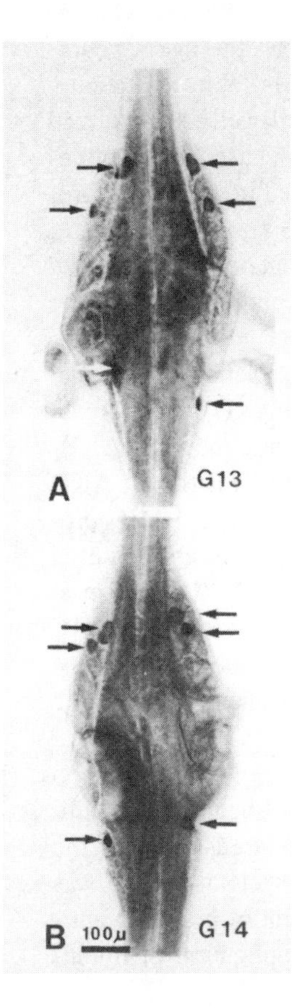

Figure 1 Adipokinetic hormone immunoreactive neurons in ganglia 13 (A) and 14 (B) of the leech *Hirudo medicinalis*. *Arrows* indicate the positions of the most prominent immunoreactive cell bodies. Note the similarity of staining in these ganglia, suggesting segmental homology between identified AKH-immunoreactive neurons. (Preparation by Jane Witten and Mary Kate Worden; see Witten et al 1984.)

the control of cardiac muscle (R. L. Calabrese, personal communication). When applied to the heart, FMRFamide produces an increase in the frequency of spontaneous myogenic contractions, an effect similar to that produced by tonic activation of the HA cell. The presence of authentic FMRFamide or proctolin in the leech or the existence of altered leech forms of these peptides has not yet been established.

These examples serve to illustrate the usefulness of initial screening for peptidergic systems by use of immunological probes. We should be aware, however, of the danger of peptide misidentification when immunochemical techniques alone are used. Further work on the identification of leech peptides and peptidergic neurons is currently in progress. We anticipate important developments in this somewhat neglected and wide open field.

Arthropods

Insects and crustaceans form the two major classes of Arthropoda. Identified peptides include proctolin, which is present in insects, crustaceans, and is probably in the Merostomata *Limulus* (Benson et al 1981), the insect neuropeptide hormone AKH, and crustacean RPCH and DRPH (Table 1). To date, proctolin is the only sequenced peptide that has provided well-characterized peptidergic systems in the Arthropods. Neural systems based on peptides related to the AKH/RPCH family are likely to be developed in the near future.

INSECTS Originally described as a metabolically important hormone released from the corpus cardiacum (Stone et al 1976), a neurosecretory organ in the locus (*Schistocerca gregaria*) head, AKH has now been detected immunochemically in neurons of the locust and cockroach CNS (Schooneveld et al 1983, Witten et al 1984). The identity and function of the AKH-immunoreactive neurons is unknown. Indirect evidence suggests neurons containing AKH-related peptides may cause skeletal muscle contraction, because authentic AKH and two related peptides newly isolated from cockroach, MI and MII, produced slow and sustained contraction of locust skeletal muscle (O'Shea et al 1984). The MI and MII peptides are also present in the CNS (O'Shea et al 1984). The pronounced myoactivity of these peptides suggests that AKH-related peptides may be found in insect skeletal motoneurons. If this is the case, the insect nerve-muscle junction will provide a convenient locus for the analysis of electrophysiological and biochemical consequences of peptide release.

Precedent for peptidergic skeletal motoneurons in insects now exists (Bishop & O'Shea 1982, O'Shea & Bishop 1982, Adams & O'Shea 1983). Until recently proctolin was considered primarily as a peptide transmitter in the insect hindgut. Its activity on other visceral targets and skeletal muscle was assumed to be mediated hormonally. With the development of proctolin antibodies (Bishop et al 1981, Eckert et al 1981) and the application of immunocytochemistry to insects, it is now clear that proctolin is widely distributed in neurons of the CNS and is probably primarily associated with skeletal motoneurons (Witten & O'Shea 1984). Two model functional systems have recently been developed based on uniquely identified proctolin-containing motoneurons. Proctolin has now been found in two large identified slow skeletal motoneurons in insects, the slow coxal depressor or *Ds* motoneuron of the cockroach *Periplaneta americana* (O'Shea & Bishop 1982), and the slow extensor of the tibia or *SETi* motoneuron of the locust *Schistocerca nitens* (Witten & O'Shea 1984). Although *SETi* is just now being studied in detail it is clear that both the *Ds* and *SETi* cell preparations have all constituents of ideal model systems: a sequenced active peptide, a uniquely identified peptidergic neuron, and an accessible postsynaptic target. Figure 2A-D illustrates some of the features of the *Ds* system.

The large, accessible proctolin-immunoreactive cell body of the *Ds* motoneuron lies laterally and dorsally in the metathoracic ganglion (Figure 2A and B). The proctolin content of this cell has been demonstrated by immunohistochemistry and by more precise methods. This cell has been dissected from the ganglion and demonstrated by reverse-phase hplc fractionation to contain the pentapeptide proctolin (O'Shea & Bishop 1982). The *Ds* axon projects to a group of postural muscles that extend the proximal segments of the hindleg—the coxal depressor muscles. Proctolin-immunoreactive nerve terminals from the *Ds* neuron can be seen ending on all the muscle cells innervated by the *Ds* neuron (Figure 2C) (Witten & O'Shea 1984). Proctolin is released onto the muscle when the *Ds* cell is stimulated, and release is calcium-dependent (Adams & O'Shea 1983). Specific stimulation of the *Ds* neuron can be achieved conveniently while monitoring both the intracellular electrophysiological response and the mechanical response of the muscle (Figure 2B and D).

The muscle response to *Ds* stimulation is biphasic (Adams & O'Shea 1983). Rapid transient contractions coupled 1:1 with motor action potentials and with excitatory junctional potentials (EJP) are followed by a delayed slow and sustained contracture (Figure 2D). Recently, the biphasic response has been attributed to the action of two transmitters, L-glutamic acid and proctolin, both of which are released by the *Ds* neuron. Proctolin produces a slow and sustained "catch" contracture of the coxal depressor muscles without depolarizing the muscle cells, whereas the transient response is produced by the glutamate-mediated transient EJP. The muscles innervated by the *Ds* neuron are also innervated by at least four other excitatory motoneurons, none of which contain proctolin. Muscle responses to the nonproctolinergic motoneurons are not biphasic but rapid, brief, and coupled 1:1 to presynaptic action potentials.

The *Ds* and *SETi* neurons are currently the only uniquely identified peptidergic skeletal motoneurons in insects. We know, however, from immunohistochemical mapping studies that perhaps 5 to 10% of the total motoneuronal

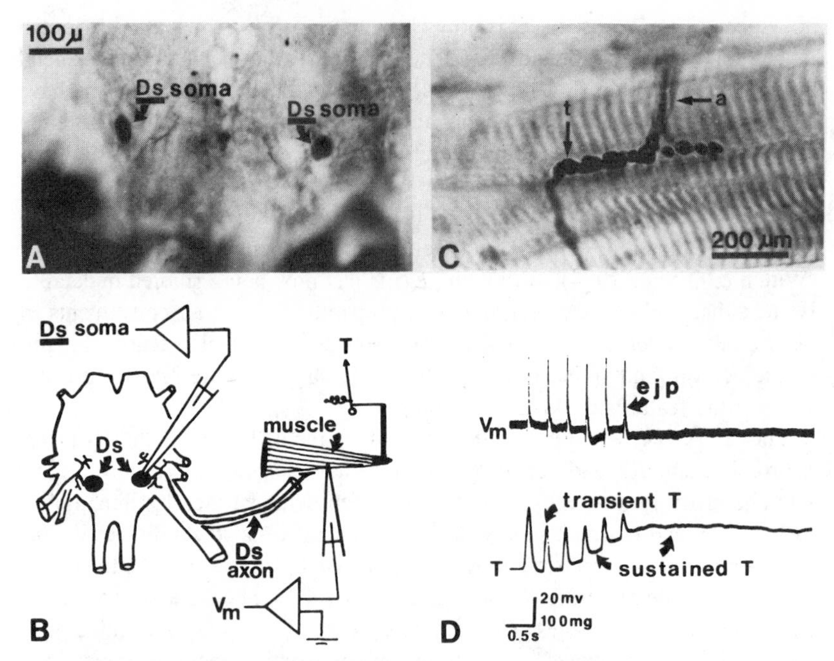

Figure 2 Identified peptidergic motoneurons in the cockroach. The proctolin-immunoreactive somata of the *Ds* motoneurons are indicated in A, which is a dorsal view of the metathoracic ganglion. In B the electrophysiological preparation is illustrated, showing microelectrodes placed in the presynaptic peptidergic neuron *Ds* and its postsynaptic muscle target. Immunoreactive *Ds* axon (a) and terminals (t) on the postsynaptic target are shown in C. The postsynaptic result of stimulating *Ds* in a brief burst of six action potentials is shown in D. The *upper trace* is an intracellular muscle recording (Vm) and the *lower trace* is a measure of muscle tension (T). Note the biphasic response in the lower trace. The sustained tension is due to peptide release and is not associated with muscle depolarization. (Staining in C provided by Jane Witten.)

pool in cockroach and locust contain proctolin (Bishop & O'Shea 1982, Witten & O'Shea 1984). These cells have yet to be identified but it is apparent from analysis of proctolin-immunoreactive terminals on identified muscles that proctolin is associated with a subpopulation of slow motoneurons (Witten & O'Shea 1984).

These findings have caused us to think about insect skeletal motoneurons in a new way. We believe the excitatory motoneurons use an excitatory transmitter such as L-glutamic acid or something structurally related to mediate the usual transient responses (Usherwood 1980). It is easy to imagine, however, that several peptidergic subpopulations of motoneurons exist. In each hypothesized subpopulation, a rapid-acting transmitter like glutamate would be associated with a different peptide co-transmitter. The existence of the other peptide neuromuscular co-transmitters has not yet been established. We are now, however, using skeletal muscle bioassays to detect and purify these putative neuromuscular peptides. Success in this approach was recently achieved when the two new myoactive peptides, MI and MII, were found (O'Shea et al 1984). It remains to be seen whether these peptides, like proctolin, actually participate in neuromuscular transmission.

In summary, the insect neuromuscular junction is partly peptidergic and is likely to provide a variety of excellent opportunities to study transmission by neuropeptides and the functional consequence of co-transmitter action.

CRUSTACEANS The presence of proctolin in crustaceans is particularly well established in lobster, crayfish, and crab (Schwarz et al 1984, Bishop et al 1984, Kingan & Titmus 1983, Sullivan 1979). Consistent with its actions in insects, proctolin has a variety of effects on several types of crustacean muscles. It increases cardiac muscle contraction in the crab *Portunus sanguinolentus* (Sullivan 1979) and the lobster *Homarus americanus* (Miller & Sullivan 1981). It enhances contractions of stomatogastric muscle in *Panulirus interruptus* (Lingle 1979) and causes long-lasting contracture of the opener muscle of the finger or dactyl of the lobster walking leg (Schwarz et al 1980, Kravitz et al 1980).

Although a variety of crustacean muscles are clearly sensitive to applied proctolin, it remains undetermined in most examples how or indeed whether the muscle is normally exposed to the peptide. Is the action mediated by proctolinergic motoneurons, as has been shown in insect muscle, or can proctolin act hormonally?

The action of proctolin on lobster skeletal muscle (the datyl opener) has some parallels with the insect *Ds* system. Proctolin causes a slow, prolonged contracture of the dactyl muscle without significant depolarization (Schwarz et al 1980). No presynaptic action of proctolin on motoneurons is measured, and from this it is concluded that the peptide acts on the muscle itself. The effects of proctolin, in particular the long time course and the absence of muscle depolarization, are very similar to the peptide-mediated effects of the proctolinergic *Ds* motoneuron in the insect nerve-muscle preparation (Adams & O'Shea 1983). This case illustrates the importance of identifying a physiological source of the peptide, since one has not been found. Proctolinergic motoneurons innervating the lobster muscle may of course exist, but an alternative view is that the dactyl muscle is the target of circulating proctolin. In this case the active agent is not likely to be the insect pentapeptide, since authentic proctolin is rapidly degraded in the hemolymph and it has not been detected in the circulation (Schwarz et al 1984, Quistad et al 1984). One possibility is that lobster employs a peptide structurally similar to proctolin that contracts the muscle but is not rapidly inactivated by proteolysis in the hemolymph. No such peptide is yet known to exist, and it may not be necessary to invoke it, or any other mechanism, if proctolinergic motoneurons can be demonstrated in the lobster. Indeed, recent work in E. A. Kravitz's laboratory on the cellular localization of proctolin in the lobster by use of immunocytochemical methods has indicated the presence of proctolinergic skeletal motoneurons (Schwarz et al 1984). It remains to be seen whether proctolin will be found in dactyl-muscle motoneurons.

Specific identification of individual proctolinergic neurons in crustaceans has now been achieved in two organisms, the lobster and the crayfish. In the lobster, proctolin has recently been found with serotonin in a large, identified neuron with targets both within the CNS and peripherally (Siwicki & Kravitz 1983). In this example the presence of proctolin in a serotonin neuron was first indicated by immunohistochemistry and then shown by hplc fractionation to be present in extracts prepared from the neuron's isolated cell body (Siwicki et al 1984). Thus, as in the insect examples, the exact molecular form of the immunoactivity of a single neuron is demonstrated quite directly. Another promising crustacean system is in the crayfish in which proctolin has been associated with three of the five identified excitatory skeletal motoneurons innervating abdominal flexor muscles (Bishop et al 1984). This system will provide an exceptionally accessible neuromuscular preparation for studying peptidergic transmission. Neuromuscular preparations in the lobster and crayfish have already contributed greatly to our understanding of classical neuromuscular transmission and neuromodulatory transmitter action mediated by octopamine, serotonin and GABA. Growing evidence indicates that they may also provide excellent model systems for studying peptidergic transmission and functional interactions between peptides and other transmitters.

These few examples represent the first specific identifications of individual peptidergic neurons in crustaceans. We anticipate that in the near future the lobster and crayfish will provide many ideal preparations for analyzing neuropeptide functions. Recently, indeed, neuronal terminals containing proctolin and FMRFamide-like peptides have been found in the stomatogastric ganglion of the lobster (Hooper & Marder 1984). This ganglion, which contains only 30 neurons, is probably the most completely understood and analyzed small neural network that exists (Selverston & Miller 1982). The network generates two rhythmic behaviors in motoneurons concerned with mastication; these are the so-called pyloric and gastric rhythms. The specific roles played by individually identified neurons and their synaptic connections in generating the pyloric motor output of the ganglion are now well understood. Questions remain, however, concerning how the neural network is activated, modulated, and turned off by other neurons projecting to this ganglion from the CNS. Recently, the peptides proctolin and FMRFamide have been implicated in this regulation (E. Marder, personal communication). Proctolin and FMRFamide immunoreactive terminals of CNS neurons have been found to ramify extensively in the neuropile of the stomatogastric ganglion. Both proctolin and FMRFamide activate the quiescent pyloric and gastric rhythms and both increase the bursting frequency of the pyloric rhythm (Marder & Hooper 1984). If the peptidergic neurons that control this pattern-generating system can be located and uniquely identified, this preparation offers exciting possibilities for understanding in precise cellular detail how a central peptidergic control system interacts with a central pattern-generating circuit.

Molluscs

FMRFamide is one of the best studied peptides in molluscs (Greenberg et al 1983). It shows promise of leading to valuable model systems in a variety of invertebrates and also appears to be one of the first examples of an invertebrate peptide having led to the identification of a mammalian homologue (Boer et al 1980, Greenberg et al 1983, Dockray et al 1981, Weber et al 1981). In molluscs FMRFamide appears to be involved widely in the control of muscle contraction. In fact, there may be molluscan systems somewhat analogous to the insect "catch" producing *Ds* proctolinergic motor neuron. For example, in the mussel *Mytilus edulis,* FMRFamide produces catch contractures in muscle that may be innervated by FMRFamide-containing neurons (Painter 1982).

Another FMRFamide-containing molluscan neuromuscular preparation more nearly meets the criteria of an ideal "model system." It is in the snail *Helix aspersa*. A specific neuron, called C3, containing FMRFamide-immunoreactivity, innervates an identified postsynaptic target, the muscle of the ipsilateral tentacle (Cottrell et al 1983). Stimulation of the C3 neuron results in contraction of the ipsilateral tentacle muscle. Similar contractions can be produced by FMRFamide and also by acetylcholine, a transmitter of some molluscan motoneurons. Although the effects of acetylcholine can be blocked by benzoquinonium chloride, the response of the muscle to stimulating C3 cannot. The peptide transmitter of the C3 neuron has not been identified definitely, because evidence is based only on immunochemistry and the similarity of the action of FMRFamide and neuronal stimulation. In fact, the snail *Helix aspersa* does not contain FMRFamide, a peptide originally isolated from a rather distantly related mollusc, the clam *Macrocallista*. Recently the sequence of a *Helix* FMRFamide-like peptide has been determined to be *p*Glu-Asp-Pro-Phe-Leu-Arg-Phe-NH_2 (Price et al 1984). It remains to be seen whether this peptide is present in the C3 neuron. The characterization of this peptide and the discovery in the future of specific neurons that use it as a transmitter will help us to understand the functional significance of the effects of FMRFamide in *Helix*. Curiously, although FMRFamide does not exist in *Helix*, it is apparently present in *Aplysia* (Lehman et al 1984). This is surprising because *Helix* and *Aplysia* are far more closely related to one another than either is to the clam *Macrocallista*. This observation emphasizes how much there is to learn about the functional and evolutionary significance of structural groupings among neuropeptides. In *Aplysia,* specific FMRFamide-containing neurons have not yet been identified, but the peptide is known to have multiple actions on neurons, including inhibition and biphasic excitatory-inhibitory responses (Stone & Mayeri 1981).

Aplysia is proving to be a remarkably favorable animal for the study of neuropeptides. Some of the systems involve actions on muscles. For example, two preparations involving newly discovered peptides not related to FMRFamide show considerable promise. The peptides are SCP_B and SCP_A, which are associated with the innervation of the accessory radular closer (ARC) muscle, and an unnamed 12-amino-acid peptide present in the R3 to R14 neurons (Table 1) that innervate vascular smooth muscle.

The sequenced SCP_B peptide (Morris et al 1982) and the related but unsequenced SCP_A peptide have been associated by immunohistochemistry with the innervation of the ARC muscle in *Aplysia* (Lloyd et al 1984). The central neurons responsible for the SCP-immunoreactive nerve terminals on the muscle have not yet been specifically identified, but this is likely to occur soon. This is an interesting example because SCP_B probably functions as a modulator of muscle contractions rather than as a direct neurotransmitter. At nanomolar concentration SCP_B enhances the force of contraction produced by the cholinergic motoneurons of the ARC muscle. The effect of SCP_B appears to be postsynaptic and directly on the muscle, and does not involve an increased release of acetylcholine from the cholinergic motoneurons. Thus SCP_B is able to potentiate the amplitude of contractions produced by constant amounts of acetylcholine applied to the muscle. Although the SCP_B-containing neurons in this system have not yet been uniquely identified, some biochemical work has already been directed at understanding the subcellular mechanisms of their action. The SCP_B-produced potentiation is correlated with a parallel stimulation of cAMP in the muscle. The threshold for the effect is about 10^{-9}M and at 10^{-7}M there is a tenfold elevation in muscle cAMP. This observation is interesting for a number of reasons. First, it contrasts with the effect of proctolin on crustacean dactyl muscle that is not associated with a stimulation of cAMP. Second, the ARC muscle is innervated by an identified serotonergic neuron, which when stimulated also potentiates neurally evoked acetylcholine contractions and also elevates muscle cAMP levels (Weiss et al 1979). The SCP_B effect on muscle contraction cannot be distinguished from serotonin. Muscle contraction in this system appears therefore to be regulated or modulated by two separate transmitter systems, serotonergic and peptidergic, apparently with parallel actions. It seems unlikely to us that no functional differences exist between these two neural systems. When the SCP_B neurons are identified, the effects of the stimulating individual neurons may reveal the distinctive functional significance of the peptidergic innervation. For example, the SCP_B-containing neurons may employ an additional neurotransmitter not found in the serotonergic innervation. This example may serve to emphasize that we cannot always expect the effects of neuron stimulation to be accounted for by the actions of a single exogenously applied neurotransmitter, even at the nerve-muscle junction (Adams & O'Shea 1983).

The 12 giant identified neurons designated R3 to R14 in the abdominal ganglion of *Aplysia* represent an intriguing peptidergic system. Progress in developing this as a model functional preparation has, however, been hampered until recently by the lack of sequence information on the peptide transmitter or transmitters. Now the sequences of peptides likely used by the R3–14 neurons have been determined by inspection of the inferred protein sequence encoded by a prevalent mRNA isolated from these neurons (see above). A 12-amino-acid peptide, predicted to be liberated by processing enzymes from a 13.4 kd protein precursor, is the first of these to be isolated from *Aplysia* (E. Mayeri, personal communication). This sequence, glu-ala-glu-glu-pro-ser-ala-phe-met-thr-arg-leu, is now known to correspond exactly to a native *Aplysia* peptide. This peptide is bioactive on one of the targets of the R3–14 neurons; it contracts the smooth muscle of the efferent-vessel of the gill. The R3–14 neurons form a rather uniform population of well-characterized efferent neurons that control or modulate the contractile activity of the circulatory system (Sawada 1981a, Price & McAdoo 1979). Neurons R3–13 innervate the heart and the distinctive R14 neuron has a more extensive innervation pattern that includes the anterior aorta, the heart, and other peripheral structures (Sawada et al 1981a,b).

The R3–14 neurons are of particular interest because they will probably provide an example of a multiple neuroeffector system. Considerable evidence has accumulated showing that glycine is released by R3–14 and is used as a modulator (Price & McAdoo 1981, Sawada et al 1984). Glycine and stimulation of R14 enhances the force of contractions of the anterior aorta induced by excitatory motoneurons. Alone, glycine produces little or no contraction, rather it modulates the actions of other transmitters. The roles of R3–14 peptides are not fully understood. The 12-amino-acid peptide acts directly on vascular smooth muscle, causing sustained contractions. Functions for this peptide and others encoded by the R3–14 message are currently being investigated (E. Mayeri, personal communication). Clearly, the R3–14 neurons form a complex neural system offering great opportunity for analysis of peptide action in a multitransmitter setting.

The molluscan examples cited above will help us to understand the specific roles of peptides during transmission at the synaptic level. These are ideal, simple neuromuscular systems for detailing the subcellular mechanisms of peptide action. In a broader functional context, however, peptides are involved in the regulation of complex behavioral arrays, and there exist excellent examples of this in invertebrates. Perhaps the most multifaceted peptide system currently being investigated in the Molluscs involves peptides that initiate and orchestrate a fixed action pattern of the egg-laying behavior in *Aplysia*. Egg-laying behavior involves cessation of locomotion and eating, increase in respiratory and circulatory function, headwaving, extrusion of eggs, and a variety of electrical effects in the CNS whose functional consequences are not entirely clear (see Strumwasser 1983 for review). This complex array can be elicited by stimulation of the so-called bag cells. These are two clusters of about 400 apparently homogeneous neurons associated with the abdominal ganglion. In each cluster, the bag cells are electrically coupled in a nonrectifying fashion (Kupfermann & Kandel 1970, Blankenship & Haskins 1979). When stimulated, the bag cells produce a synchronous discharge of action potentials that persists for ½ hr, and which results in a synchronous release of peptides from the bag cells (Dudek et al 1979, Stuart et al 1980). The physiological activator of the bag cell discharge is not entirely clear. The most likely candidates are peptides produced and release by the atrial gland (Arch et al 1978), a secretory organ associated with the reproductive tract. The atrial gland peptides are called peptides A and B (Heller et al 1980). They are structurally homologous, 34-residue peptides (Table 1). Peptides A and B produce a long (20–30 min) discharge of action potentials in the bag cells and can induce egg laying behavior in intact animals, but not when the abdominal ganglion, including bag cells, is removed (Strumwasser 1983, Heller 1980).

The bag cell peptides mediate and control the egg-laying behavior and associated physiological changes (Kupfermann 1967, Toevs 1969). The most studied among the bag-cell peptides is ELH (Chiu et al 1979). Several others, however, are encoded by the ELH gene and are also released. These include α-BCP, β-BCP, γ-BCP, and acidic peptide (Rothman et al 1983, Scheller et al 1983, E. Mayeri, personal communication). ELH acts both as a circulating hormone (Rothman et al 1980) and as a "nonsynaptic" transmitter within the CNS (Mayeri & Rothman 1982). As a circulating hormone it can directly affect the release of eggs from the ovotestis (Rothman et al 1980). In the CNS it probably diffuses widely, but its actions are localized to a specific set of identified neurons. Neuronal responses include augmentation of intrinsic bursting activity in neuron R15 (Mayeri et al 1979a,b), and prolonged excitation, prolonged inhibition (Brownell & Mayeri 1979), and transient excitation of other neurons (Stuart & Strumwasser 1980). Together these actions of ELH on central neurons probably serve to coordinate the various autonomic and locomotory components of the egg-laying behavior. Bag-cell peptides other than ELH, however, are required in order to account fully for the effects of bag-cell discharge (Rothman et al 1983, Sigvardt et al 1983). One candidate for producing the effects of bag-cell discharge in the CNS not produced by ELH is α-BCP.

Bag-cell discharge results in inhibition of the left upper quadrant (LUQ) neurons (Brownell & Mayeri 1979), an effect that is partially mimicked by application of α-BCP but not by ELH (Rothman et al 1983). In contrast to ELH, α-BCP probably does not act as a circulating hormone. This peptide is susceptible to digestion by carboxy peptidase enzymes because, unlike ELH, it does not possess a carboxy terminal amide. Amidation, by increasing its active life in the hemolymph, makes a peptide suitable for functions that require long-lasting action and action at sites remote from the site of release. The nonamidated α-BCP may provide an interesting example of a peptide's action being shaped by post-release proteolysis. The physiologically active form of α-BCP is not known. The major form isolated from the bag cells is the octapeptide α-BCP(1–8) (Rothman et al 1983), even though the ELH gene sequence predicts that dibasic cleavage and carboxy-peptidase-b-like processing would generate a nonapeptide with an additional leu at the carboxy terminus. Although small amounts of α-BCP(1–9) can be detected, this sequence holds only about 1⁄30 the activity of α-BCP(1–8) in inhibiting the LUQ neurons. α-BCP(1–7) is also more prominent in extracts and is ten times more potent than α-BCP(1–9). It is not certain that abundances in the extracts accurately reflect abundances in intact cells, and no direct characterization of the release product has yet been published. Both organ culture and cell culture systems might be helpful in clarifying this point. The interesting possibility exists that α-BCP(1–9) is the released form that is then processed into the more active forms α-BCP(1–8) and (1–7) before being functionally inactivated by further proteolysis. This may be a specific instance of an important and widespread phenomenon that functions to control the spatial and temporal parameters of neuropeptide action.

The egg laying fixed action pattern illustrates vividly how neuropeptides may be employed to orchestrate a complex behavior. Precisely how bag-cell peptides control the neural and physiological events associated with egg-laying is not yet completely understood. Clearly, however, this system provides an almost unique opportunity to understand the involvement of neuropeptides at all levels from molecular to behavioral in this process. It should be emphasized that many more peptides are likely to be involved in this activity, and that the question of modulation of the behavior by other neural inputs has just been broached. None of these open questions, however, seems particularly daunting in this elegantly accessible system. It should also be pointed out that fixed action patterns are not unique to exotic sea creatures. Indeed they seem to be quite commonly involved in reproduction, perhaps even in mammals (Moss & McCann 1973, Pfaff 1973).

Other Invertebrate Phyla

Among the remaining multicellular invertebrate phyla, the nematodes offer considerable promise as model neurobiological systems. Unfortunately, to our knowledge there has been no published work on nematode-specific neuropeptides.

SUMMARY

The following is a list of generalizations that arise from considering the present state of knowledge concerning the functions of invertebrate peptides. Some of these clearly also apply to vertebrates.

1. Invertebrate peptides can be classified into structurally related groups. Structural similarity of peptides may represent true evolutionary homology by selection acting on an original gene. Alternatively, independent evolution of similar genes may have occurred because certain amino acid sequences represent optimal solutions to complex functional problems.
2. Invertebrate neuropeptides have multiple functions. Thus, proctolin is a cardioactive peptide, a skeletal neuromuscular transmitter, a hindgut neuropeptide, a peptide of CNS interneurons (Keshishian & O'Shea 1984) and may have humoral roles.
3. Invertebrate peptides act through a variety of molecular mechanisms. Generalizations about the mechanism cannot yet be made. Thus, proctolin's action on crustacean skeletal muscle is not associated with stimulation of cyclic-AMP or protein phosphorylation, but the action of SCP_B on molluscan skeletal muscle involves elevation of cAMP.
4. Invertebrate peptide inactivation can be caused by proteolysis that can also function to enhance peptide bioactivity. Proctolin is made virtually biologically inactive by any proteolysis, but α-BCP bioactivity is enhanced by two steps of carboxy peptidase digestion before being functionally inactivated. Protease action on released peptides is not necessarily a "simple" form of transmitter inactivation. Protease action also involves functional processing whereby the temporal and spatial parameters of a peptide's action may be shaped.
5. Invertebrate neuropeptides are frequently co-localized with other neuroeffectors. Peptides may be co-localized and released with other active peptides as in the bag-cell example, or may be coactive with more conventional transmitters, as in the *Ds* motoneuron example. In such circumstances there is no reason to view either transmitter as primary or secondary.
6. Invertebrate neuropeptides are widely involved in the control of muscle contraction. These effects may be locally and directly mediated as in the *Ds* motoneuron example or may be humoral. The peptide may act directly on the muscle contractile system or function to modulate the muscles' response to other motor input. Muscle contraction may be induced by a neuropeptide without depolarization of the muscle cells, for example see proctolin.
7. Invertebrate neuropeptides are frequently involved in the control of oscillatory functions. In several examples peptides activate rhythmic myogenic

contractions of cardiac and skeletal muscle. In the stomatogastric ganglion, peptides are shown to activate a neural oscillator. The significance of this is unclear, but it may reflect a common role for peptides in regulating the subcellular mechanisms that produce rhythmic oscillatory behavior.

8. Invertebrate neuropeptides can evoke coordinated complex arrays of behavior and physiology. Events initiated by peptides A and B and the bag-cell peptides are examples.

CONCLUSIONS

Our concept of chemical neural transmission has fundamentally changed primarily as a result of advances in peptide neurobiology. Previously held criteria for transmitters and our simplifying assumptions about the number, nature, and the anatomical localization of chemical messengers have all been dashed. It seems an appropriate time to reflect on how our view had been narrowed, so that we felt quite comfortable with the previous notions of chemical transmission for so long, and what problems confront us as a result of our expanded and more complex view of chemical transmission.

We believe our previous, simplistic view of neural transmission resulted from not only the limitations of the tools available for the study of neural systems, but also the complex nature of the vertebrate systems that are most studied. These systems generally permit the examination of average effects over large fields of heterogeneous neuronal populations, and there can be no expectation that one can fully account for the phenomena observed in great detail. The experimental situation makes it very likely that minimal hypotheses will miss significant details. There are relatively simple and well characterized vertebrate preparations, the bullfrog sympathetic ganglion (Kuffler 1980, Jan & Jan 1983) for example, in which this problem can be largely overcome. But most vertebrate systems under study do not permit this kind of detailed analysis. As a consequence, the probability that significant variances from the minimal hypothesis will be observed and force us to refine and expand our theories is low. Thus it is not surprising that neurobiologists felt comfortable with a greatly oversimplified world of chemical transmission for so long.

Physiological investigations did little to expand our horizons. Pharmacological, biochemical, and anatomical studies, however, have forced us into a complex new world. We now must explain a great deal, including why we need so many transmitters. Why do transmitters employ such anatomically diverse structures? Why do neurons require multiple transmitters? What is the significance of peptide families? It seems likely that detailed physiological studies applied to systems that are well-characterized anatomically and biochemically can answer such questions. It seems unlikely that they will be answered in any other way.

What are the properties we should seek in a system used for such studies? Ideally we should aim to explain anatomically and chemically a neural event in detail. We should avoid averaging techniques that might obscure diversity, and ideally we should focus on discrete synapses of discrete neurons. This will allow us to construct a comprehensive catalog of specific examples of neural communication. The more complete these studies become, the less we will overlook in constructing our hypotheses, and the closer we will come to understanding the nature of neurochemical communication. We hope we have illustrated that invertebrate neural systems are suitable for such studies.

ACKNOWLEDGMENTS

We are indebted to many colleagues for providing us with manuscripts prior to publication and especially for sharing their thoughts and insights regarding the systems discussed. We thank Jane Witten, Mary-Kate Worden, and Haig Keshishian for valuable discussions and encouragement, and for sharing with us their unpublished observations. M. O. is supported by the NSF (BNS-8202515) and M. S. is an NIMH Research Career Development Awardee (K100325).

Literature Cited

Adams, M. E., O'Shea, M. 1983. Peptide cotransmitter at a neuromuscular junction. *Science* 221:286–89

Arch, S. 1976. Neuroendocrine regulation of egg laying in *Aplysia californica*. *Am. Zool.* 16:167–75

Arch, S., Smock, T., Gurvis, R., McCarthy, C. 1978. Atrial gland induction of egg-laying response in *Aplysia californica*. *J. Comp. Physiol.* 128:67–70

Aswad, D. W. 1978. Biosynthesis and processing of presumed neurosecretory proteins in single identified neurons of Aplysia californica. *J. Neurobiol.* 9:267–84

Bauman, H., Wilson, K. J., Chen, P. S., Humbel, R. E. 1975. The amino acid sequence of a peptide (PS–1) from *Drosophila funebris:* A paragonial peptide from males which reduces the receptivity of the female. *Eur. J. Biochem.* 52:521–29

Benson, J. A., Sullivan, R. E., Watson, W. H., Augustine, G. J. 1981. The neuropeptide proctolin acts directly on *Limulus* cardiac muscle to increase amplitude of contraction. *Brain Res.* 213:449–54

Bishop, C. A., O'Shea, M. 1982. Neuropeptide proctolin (H-Arg-Tyr-Leu-Pro-Thr-OH): Immunocytochemical mapping of neurons in the central nervous system of the cockroach. *J. Comp. Neurol.* 207:223–38

Bishop, C. A., O'Shea, M., Miller, R. J. 1981. Neuropeptide proctolin (H-Arg-Tyr-Leu-Pro-Thr-OH): Immunological detection and neuronal localization in the insect central nervous system. *Proc. Natl. Acad. Sci. USA* 78:5899–6002

Bishop, C. A., Wine, J. J., O'Shea, M. 1984. Neuropeptide proctolin in postural motoneurons of the crayfish. *J. Neurosci.* 4:2001–9

Blackenship, J. E., Haskins, J. T. 1979. Electronic coupling among neuroendocrine cells in *Aplysia*. *J. Neurophysiol.* 42:347–55

Boer, H. H., Schot, L. P. C., Veenstra, J. A., Reichelt, D. 1980. Immunocytochemical identification of neural elements in the central nervous system of the snail, some insects, a fish, and a mammal with an antiserum to the molluscan cardio-excitatory tetrapeptide FMRF-amide. *Cell Tissue Res.* 213:21–27

Brown, B. E. 1967. Neuromuscular transmitter substance in insect visceral muscle. *Science* 155:595–97

Brown, B. E., Starratt, A. N. 1975. Isolation of proctolin, a myotropic peptide from *Periplaneta americana*. *J. Insect Physiol.* 21:1879–81

Brownell, P., Mayeri, E. 1979. Prolonged inhibition of neurons by neuroendocrine cells in *Aplysia*. *Science* 204:417–20

Buck, L. 1983. *Cell specific gene expression of identified cells*. Presented at Ann. Meet. Soc. Neurosci. 13th, Boston

Chiu, A. Y., Hunkapillar, M. W., Heller, E., Stuart, D. K., Hood, L. E., Strumwasser, F. 1979. Neuropeptide egg-laying hormone of *Aplysia:* Purification and primary structure. *Proc. Natl. Acad. Sci. USA* 76:6656–60

Coggeshall, R. E., Kandel, E. R., Kupfermann, I., Waziri, R. 1966. A morphological and functional study on a cluster of identifiable neurosecretory cells in the abdominal ganglion of Aplysia californica. *J. Cell Biol.* 31:363–68

Cottrell, C. B. 1962. The imaginal ecdysis of blowflies. Detection of the blood-borne darkening factor and determination of some of its properties. *J. Exp. Biol.* 39:413–30

Cottrell, G. A., Schot, L. P. C., Dockray, G. J. 1983. Identification and probable role of a single neuron containing the neuropeptide *Helix* FMRFamide. *Nature* 304:638–40

Cupo, A., Proux, J. 1983. Biochemical characterization of a vasopressin-like neuropeptide in *Locusta migratoria*. Evidence of a high molecular weight protein encoding vasopressin sequence. *Neuropeptides* 3:309–18

Dockray, G. J., Vaillant, C., Williams, R. G. 1981. New vertebrate brain-gut peptide related to a molluscan neuropeptide and an opioid peptide. *Nature* 293:656–57

Dudek, F. E., Cobbs, J. S., Pinsker, H. M. 1979. Bag cell electrical activity underlying spontaneous egg laying in freely behaving *Aplysia brasiliana*. *J. Neurophysiol.* 42:804–17

Duve, H., Thorpe, A., Lazarus, N. R., Lowry, P. J. 1982. A neuropeptide of the blowfly *Calliphora vomitoria* with an amino acid composition homologus with vertebrate pancreatic polypeptide. *Biochem. J.* 201:429–32

Eckert, M., Agricola, H., Penzlin, H. 1981. Immunocytochemical identification of proctolin-like immunoreactivity in the terminal ganglion and hindgut of the cockroach *Periplaneta americana* (L.). *Cell Tissue Res.* 217:633–45

Erspamer, V., Anastasi, A. 1962. Structure and pharmacological actions of eledoisin, the active endecapeptide of the posterior salivary gland of Eledone. *Experientia* 18:58–61

Fernlund, P. 1976. Structure of a light-adapting hormone from the shrimp, *Pandalus borealis*. *Biochim. Biophys. Acta* 439:17–25

Fernlund, P., Josefsson, L. 1972. Crustacean color-change hormone: Amino acid sequence and chemical synthesis. *Science* 177:173–75

Fraenkel, G., Hsiao, C. 1962. Hormonal and nervous control of tanning in the fly. *Science* 138:27–29

Frontali, N., Gainer, H. 1977. Peptides in invertebrate nervous systems. In *Peptides in Neurobiology*, ed. H. Gainer, New York/London: Plenum. 464 pp.

Greenberg, M. J., Painter, S. D., Doble, K. E., Nagle, G. T., Price, D. A., Lehman, H. K. 1983. The molluscan neurosecretory peptide FMRFamide: Comparative pharmacology and relationship to the enkephalins. *Fed. Proc.* 42:82–86

Greenberg, M. J., Price, D. A. 1983. Invertebrate neuropeptides: Native and naturalized. *Ann. Rev. Physiol.* 45:271–88

Heller, E., Kaczmarck, L. K., Hunkapiller, M. W., Hood, L. E., Strumwasser, F. 1980. Purification and primary structure of two neuroactive peptides that cause bag cell after discharge and egg-laying in *Aplysia*. *Proc. Natl. Acad. Sci. USA* 77:2328–32

Hooper, S. L., Marder, E. 1984. Modulation of a central pattern generator by two neuropeptides, proctolin and FMRFamide. *Brain Res.* 305:186–91

Iversen, L. L. 1983. Neuropeptides—what next? *Trends Neurosci.* 6:293–94

Jan, Y. N., Jan, L. Y. 1983. A LHRH-like peptidergic neurotransmitter capable of 'action at a distance' in autonomic ganglia. *Trends Neurosci.* 6:320–25

Keshishian, H., O'Shea, M. 1984. The distribution of a peptide neuropeptide in the post embryonic grasshopper CNS. *J. Neurosci.* Submitted

Kingan, T. G., Titmus, M. 1983. Radioimmunologic detection of proctolin in Arthropods. *Comp. Biochem. Physiol. C* 74:75–78

Kravitz, E. A., Glusman, S., Harris-Warrick, R. M., Livingstone, M. S., Schwarz, T. L., Goy, M. F. 1980. Amines and a peptide as neurohormones in lobster: Action on neuromuscular preparations and preliminary behavioral studies. *J. Exp. Biol.* 89:159–75

Kuffler, S. W. 1980. Slow synaptic responses in autonomic ganglia and the pursuit of a peptidergic transmitter. *J. Exp. Biol.* 89:257–86

Kupfermann, I. 1967. Stimulation of egg laying: Possible neuroendocrine function of bag cells of abdominal ganglion of *Aplysia californica*. *Nature* 216:814–15

Kupfermann, I., Kandel, E. R. 1970. Electrophysiological properties and functional interconnections of two symmetrical neurosecretory clusters (bag cells) in abdominal ganglion of *Aplysia*. *J. Neurosci.* 33:865–76

Lehman, H. K., Price, D. A., Greenberg, M. J. 1984. The FMRFamide-like peptide of *Aplysia* is FMRFamide. *Nature*. In press

Leung, M. K., Stefano, G. B. 1984. Isolation and identification of enkephalins in pedal ganglia of *Mytilus edulis* (mollusca). *Proc. Natl. Acad. Sci. USA* 81:955–58

Li, C., Calabrese, R. L. 1983. Evidence for proctolin-like substances in the central nervous system of the leech. *Soc. Neurosci.* 9(1):76 (Abstr.)

Lingle, C. J. 1979. *The effects of acetylcholine, glutamate, and biogenic amines on muscle and neuromuscular transmission in the stomatogastric system of the spiny lobster,* Panulirus interruptus. PhD thesis, Univ. Oregon, Eugene

Lloyd, P. E., Kupfermann, I., Weiss, K. R. 1984. Evidence for parallel actions of a molluscan neuropeptide (SCP_B) and serotonin in mediating arousal in *Aplysia*. *Proc. Natl. Acad. Sci. USA* 81:2934–37

Loh, Y. P., Gainer, H. 1975. Low molecular weight specific proteins in identified molluscan neurons. I. Synthesis and storage. *Brain Res.* 92:181–92

Macagno, E. R. 1980. Number and distribution of neurons in the leech segmental ganglion. *J. Comp. Neurol.* 190:283–302

Marder, E., Hooper, S. L. 1984. Neurotransmitter modulation of the stomatogastric ganglion of decapod crustaceans. In *Model Neural Networks and Behavior,* ed. A. A. Selverston. New York/London: Plenum. In press

Mayer, R. J., Candy, D. J. 1969. Control of hemolymph lipid concentration during locust flight: An adipokinetic hormone from the corpora cardiaca. *J. Insect Physiol.* 15:611–20

Mayeri, E., Brownell, P., Branton, W. D. 1979a. Multiple, prolonged actions of neuroendocrine bag cells on neurons in *Aplysia*. II. Effects on beating pacemaker and silent neurons. *J. Neurophysiol.* 42:1185–97

Mayeri, E., Brownell, P., Branton, W. D., Simon, S. B. 1979b. Multiple, prolonged actions of neuroendocrine bag cells on neurons in *Aplysia*. I. Effects on bursting pacemaker neurons. *J. Neurophysiol.* 42:1165–84

Mayeri, E., Rothman, B. S. 1982. Nonsynaptic peptidergic neurotransmission in the abdominal ganglion of *Aplysia*. In *Neurosecretion: Molecules, Cells, Systems,* ed. D. S. Farner, K. Lederis, pp. 305–415. New York/London: Plenum

Miller, M. W., Sullivan, R. E. 1981. Some effects of proctolin on the cardiac ganglion of the Maine lobster, *Homarus americanus* (Milne Edwards). *J. Neurobiol.* 12:629–39

Morris, H. R., Panico, M., Karplus, A., Lloyd, P. E., Riniker, B. 1982. Elucidation by FAB-MS of the structure of a new cardioactive peptide from *Aplysia*. *Nature* 300:643–45

Moss, R. C., McCann, S. M. 1973. Induction of mating behavior in rats by luteinizing hormone-releasing factor. *Science* 181:177–79

Nambu, J. R., Taussig, R., Mahon, A. C., Scheller, R. H. 1983. Gene isolation with cDNA probes from identified *Aplysia* neurons: Neuropeptide modulators of cardiovascular physiology. *Cell* 35:47–56

O'Shea, M., Bishop, C. A. 1982. Neuropeptide proctolin associated with an identified skeletal motoneuron. *J. Neurosci.* 2:1242–51

O'Shea, M., Witten, J., Schaffer, M. 1984. Isolation and characterization of two myoactive neuropeptides: Further evidence of an invertebrate peptide family. *J. Neurosci.* 4: 521–29

Painter, S. D. 1982. FMRFamide catch contractures of a molluscan smooth muscle: Pharmacology, ionic dependence and cyclic nucleotides. *J. Comp. Physiol.* 148:491–501

Pfaff, D. W. 1973. Luteinizing hormone-releasing factor potentiates lourdosis behavior in hypophysectomized ovariectomized female rats. *Science* 182:1148

Price, C. H., McAdoo, D. J. 1979. Anatomy and ultrastructure of the axons and terminals of neurons R3–14 in *Aplysia*. *J. Comp. Neurol.* 188:647–77

Price, C. H., McAdoo, D. J. 1981. Localization of axonally transported ^{3}H glycine in vesicles of identified neurons. *Brain Res.* 219:307–15

Price, D. A., Cottrell, G. A., Doble, K. E., Greenberg, M. J., Lehman, H. K., Riehm, J. P. 1984. The sequence of the FMRFamide-like peptide in the pulmonate snail *Helix aspersa*. *Nature*. In press

Price, D. A., Greenberg, M. J. 1977. Structure of a molluscan cardioexcitatory neuropeptide. *Science* 197:670–71

Quistad, G. B., Adams, M. E., Scarborough, R. M., Carney, R. L., Schooley, D. A. 1984. Metabolism of proctolin, a pentapeptide neurotransmitter in insects. *Life Sci.* 34:569–76

Raabe, M. 1982. *Insect Neurohormones*. New York: Plenum. 352 pp.

Rothman, B. S., Mayeri, E., Brown, R. O., Yuan, P.-M., Shively, J. E. 1983. Primary structure and neuronal effects of α-bag cell peptide, a second candidate neurotransmitter encoded by a single gene in bag cell neurons of *Aplysia*. *Proc. Natl. Acad. Sci. USA* 80:5753–57

Rothman, B. S., Weir, G., Dudek, F. E. 1980. Neurally active peptide, ELH, releases eggs from isolated ovotestis of *Aplysia*. *Soc. Neurosci.* 6:623 (Abstr.)

Sawada, M., Blankenship, J. E., McAdoo, D. J. 1981a. Neural control of a molluscan blood vessel, anterior aorta of *Aplysia*. *J. Neurophysiol.* 46:967–86

Sawada, M., McAdoo, D. J., Blankenship, J.

E., Price, C. H. 1981b. Modulation of arterial muscle contraction in *Aplysia* by glycine and neuron R14. *Brain Res.* 207:486–90

Sawada, M., McAdoo, D. J., Ichinose, M., Price, C. H. 1984. Influences of glycine and neuron R14 on contraction of the anterior aorta of *Aplasia. J. Neurosci.* In press

Schaller, H. C., Bodenmüller, H. 1981. Isolation and amino acid sequence of a morphogenetic peptide from hydra. *Proc. Natl. Acad. Sci. USA* 78:7000–4

Scheller, R. H., Jackson, J. F., McAllister, L. B., Rothman, B. S., Mayeri, E., Axel, R. 1983. A single gene encodes multiple neuropeptides mediating a stereotyped behavior. *Cell* 32:7–22

Schooneveld, H., Tesser, G. I., Veenstra, J. A., Romberg-Privee, H. 1983. Adipokinetic hormone and AKH-like peptide demonstrated in the corpora cardiaca and nervous system of *Locusta migratoria* by immunocytochemistry. *Cell Tissue Res.* 230:67–76

Schwarz, T. L., Harris-Warrick, R. M., Glusman, S., Kravitz, E. A. 1980. A peptide action in a lobster neuromuscular preparation. *J. Neurobiol.* 11:623–28

Schwarz, T. L., Lee, G. M. H., Siwicki, K. K., Standaert, D. G., Kravitz, E. A. 1984. Proctolin in the lobster: The distribution, release and characterization of a likely neurohormone. *J. Neurosci.* 4:1300–11

Selverston, A., Miller, J. P. 1982. Application of cell inactivation technique to the study of a small neural network. *Trends Neurosci.* 5:120–23

Sigvardt, K., Rothman, B. S., Mayeri, E. 1983. Analysis of inhibition produced by the candidate neurotransmitter, α-bag cell peptide, in identified neurons of *Aplysia. Soc. Neurosci.* 9:311 (Abstr.)

Siwicki, K. K., Beltz, B., Kravitz, E. A. 1984. Co-localization of proctolin and serotonin in an identified neuron in the lobster. In preparation.

Siwicki, K. K., Kravitz, E. A. 1983. Proctolin in lobsters: General distribution and colocalization with serotonin. *Soc. Neurosci.* 9(1):313 (Abstr.)

Starratt, A. N., Brown, B. E. 1975. Structure of the pentapeptide proctolin, a proposed neurotransmitter in insects. *Life Sci.* 17:1253–56

Stefano, G. B. 1982. Comparative aspects of opoiod-dopamine interaction. *Cell. Mol. Neurobiol.* 2:167–78

Stefano, G. B., Leung, M. 1984. Presence of met-enkephalin Arg^6-Phe^7 in molluscan neural tissue. *Brain Res.* 298:362–65

Stent, G. S., Kristan, W. B. 1981. Neural circuits generating rhythmic movements. In *Neurobiology of the Leech,* eds. K. J. Muller, J. G. Nicholls, G. S. Stent, pp. 113–46. New York: Cold Spring Harbor Lab. 320 pp.

Stent, G. S., Weisblat, D. A., Blair, S. S., Zackson, S. L. 1982. Cell lineage in the development of the leech nervous system. In *Neuronal Development,* ed. N. Spitzer, p. 1–44. New York/London: Plenum. 558 pp.

Stone, L. S., Mayeri, E. 1981. Multiple actions of FMRFamide on identified neurons in the abdominal ganglion of *Aplysia. Soc. Neurosci.* 7:636 (Abstr.)

Stone, J. V., Mordue, W., Batley, K. E., Morris, H.R. 1976. Structure of locust adipokinetic hormone that regulates lipid utilisation during flight. *Nature* 263:207–11

Strumwasser, F. 1983. Peptidergic neurons and neuroactive peptides in molluscs: From behavior to genes. In *Brain Peptides,* ed. D. T. Krieger, M. J. Brownstein, J. B. Martin, pp. 183–215. New York: Wiley. 1032 pp.

Strumwasser, F., Jacklet, J., Alverez, R. B. 1969. A seasonal rhythm in the neural extract induction of behavioral egg-laying in *Aplysia. Comp. Biochem. Physiol.* 29:197–206

Stuart, D. K., Chiu, A. Y., Strumwasser, F. 1980. Neurosecretion of egg-laying hormone and other peptides from electrically active bag cell neurons of *Aplysia. J. Neurophysiol.* 43:488–98

Stuart, D., Strumwasser, F. 1980. Neuronal sites of action of a neurosecretory peptide, egg-laying hormone, in *Aplysia californica. J. Neurophysiol.* 43:499–519

Sullivan, R. E. 1979. A proctolin-like peptide in crab pericardial organs. *J. Exp. Zool.* 210:543–52

Taghert, P. H., Truman, J. W. 1982a. The distribution and molecular characteristics of the tanning hormone bursican, in the tobacco hornworm *Manduca sexta. J. Exp. Biol.* 98:373–84

Taghert, P. H., Truman, J. W. 1982b. Identification of the bursican-containing neurons in abdominal ganglia of the tobacco hornworm *Manduca sexta. J. Exp. Biol. 98:385–402*

Toevs, L. 1969. *Identification and characterization of the egg-laying hormone from neurosecretory bag cells of Aplysia*. PhD thesis, Calif. Inst. Technol., Pasadena, Calif.

Truman, J. W., Taghert, P. H. 1983. Neuropeptides in insects. In *Brain Peptides,* ed. D. T. Krieger, M. J. Brownstein, J. B. Martin, pp. 165–82. New York: Wiley. 1032 pp.

Usherwood, P. N. R. 1980. Neuromuscular transmitter receptors of insect muscle. In *Receptors for Neurotransmitters, Hormones and Pheromones in Insects,* ed. D. B. Sattelle, L. M. Hall, J. G. Hildebrand, pp. 141–52. Amsterdam/New York/Oxford: Elsevier/North-Holland. 310 pp.

Weber, E., Evans, C. J., Samuelsson, S. J., Barchas, J. D. 1981. Novel peptide neuronal system in the rat brain and pituitary. *Science* 214:1248–51

Weiss, K. R., Mandelbaum, D. E., Schonberg, M., Kupfermann, I. 1979. Modulation of buccal muscle contractility by serotonergic metacerebral cells in *Aplysia:* Evidence for a role of cyclic adenosine monophosphate. *J. Neurophysiol.* 42:791–803

Wilson, D. L. 1976. Alteration of protein metabolism in individual identified neurons from *Aplysia*. *J. Neurobiol.* 7:407–16

Witten, J., O'Shea, M. 1984. Proctolin associated with slow insect skeletal motoneurons. In preparation

Witten, J., Worden, M. K., Schooneveld, H., O'Shea, M. 1984. Adipokinetic hormone-like immunoreactivity in the leech, locust and cockroach central nervous systems. In preparation

Zipser, B. 1980. Identification of specific leech neurones immunoreactive to enkephalin. *Nature* 283:857–58

Ann. Rev. Neurosci. 1985. 8:199–232

APPLICATIONS OF MONOCLONAL ANTIBODIES TO NEUROSCIENCE RESEARCH

Karen L. Valentino, Janet Winter, and Louis F. Reichardt

Department of Physiology, University of California, School of Medicine, San Francisco, California 94143

INTRODUCTION

The introduction of immunological methods has revolutionized neurobiology. Anatomical studies have profited from immunocytochemical localization of neurotransmitters and peptides for a better understanding of the organization of the nervous system. Antibodies have also been useful in identifying different cell classes in the nervous system, and antibodies to subcellular organelles have contributed to our knowledge of neuron function. Monoclonal antibodies are increasingly used by neurobiologists to build on this knowledge.

This review consists of four major sections. In the first three, applications of monoclonal antibodies are reviewed with regard to the cell biology of the neuron and the anatomy and development of the nervous system. The final section is an overview of the strategies involved in isolating and applying monoclonal antibodies to studies of the nervous system. This is not intended to be a comprehensive review of the literature of these subjects, and, where possible, reviews of individual subjects are indicated in the text.

In some areas, such as peptide and transmitter research, the main contributions of monoclonal antibodies have been to make scarce antibodies more widely available and to make antibodies to antigens that have been difficult to purify. In developmental neurobiology, monoclonal antibodies have made a promising beginning in identifying molecules that might be involved in cell-cell recognition and cell adhesion. Some of the most interesting contributions of monoclonal antibodies have been in determining protein structure (e.g. acetylcholine receptor and sodium channel), and we discuss these areas in greater detail.

0147-006X/85/0301-0199$02.00

NEURONAL CELL STRUCTURE AND FUNCTION

Channels, Pumps, and Receptors

SODIUM CHANNEL Voltage sensitive Na^+ channels play a key role in conducting the action potential and are the one channel type that has been significantly purified and studied with traditional biochemical methods (reviewed in Reichardt & Kelly 1983). The Na^+ channel has been purified substantially from *Electrophorus* electroplax, rat brain, and rat muscle. Purified channels from the latter two preparations have been successfully reconstituted in lipid bilayers. All channel preparations contain a very large (ca 270 kDa) glycoprotein. Smaller subunits have been seen in the purified channel preparations from rat brain and rat muscle. During the last two years, conventional antisera and monoclonal antibodies have been raised that bind the large glycoprotein subunit of the channel, both from *Electrophorus* electroplax and rat muscle (Moore et al 1982, Ellisman & Levinson 1982, Casadei et al 1983). A monoclonal antibody that binds the *Electrophorus* channel subunit has been used to complete the purification of that subunit by affinity chromatography (Nakayama et al 1982). Another monoclonal antibody has shown that Na^+ channels are concentrated on the electrically excitable caudal face of the electrocytes and not on the inexcitable rostral face (Fritz & Brockes 1983). There was no detectable binding with this antibody to sections of *Electrophorus* spinal cord, muscle, or nerve, or to electrically excitable tissues of other species. In contrast, a polyclonal antiserum to a highly but not completely purified *Electrophorus* Na^+ channel preparation binds to both the innervated surfaces of electrocytes and to nodal regions in myelinated axons (Ellisman & Levinson 1982). The monoclonal antibody appears to recognize an antigenic determinant found only on the Na^+ channel in the electroplax. The polyclonal serum appears to recognize a more widely shared determinant. The specificity of the polyclonal antiserum for the Na^+ channel, however, has not been assessed using total membrane proteins. The distribution of Na^+ channels in dendrites, axons, and nerve terminals is functionally important in modulating integration of different synaptic inputs, initiation and conduction of the action potential, and neurotransmitter release. Detailed studies using these antibodies and antibodies to other channels will be very important in understanding the electrical properties of excitable cells. Monoclonal antibodies to the small Na^+ channel subunits should be useful in understanding the role of these subunits in Na^+ channel function.

SODIUM PUMPS The major mechanism for removing internal Na^+ and restoring internal K^+ after action potentials is the ATP driven Na^+–K^+ exchange pump. It is detected in biochemical assays as a Na^+, K^+-ATPase, which contains two polypeptide subunits of approximate M_r 120,000 and 50,000.

Distribution of the Na^+, K^+-ATPase on neurons in fish brain has been examined with immunocytochemical methods using a conventional antiserum (Wood et al 1977). On neurons, it is restricted to the plasmalemma, where it is distributed over the surface of cell somata and dendrites. It appears to be concentrated in nodes of Ranvier in myelinated axons. Two classes of synaptic terminals were observed, one with high and one with lower concentrations of the ATPase. Differences in concentration between different nerve terminals are likely to be important in regulating the response properties of these terminals to stimulation.

There is biochemical evidence for multiple forms of the Na^+, K^+-ATPase in neurons (Sweadner 1979). One form is the standard form (α) that is seen in kidney, muscle, and astrocytes in addition to cultured sympathetic neurons. A specific neuronal form (α^+) is found in both vertebrate and invertebrate brain and is the only form found in myelinated axons. Recently, a monoclonal antibody to the Na^+, K^+-ATPase has been isolated that suggests the existence of an even wider diversity of Na^+-pumps (Fambrough & Bayne 1983). This antibody binds the Na^+, K^+-ATPase in chicken neurons, muscle fibers, and kidney tubules, but not in Schwann cells, fibroblasts, or erythrocytes. This distribution is not the same as that of α^+ in rodent tissues. The antibody binds to nonglycosylated antigen, so the heterogeneity revealed by the antibody is unlikely to reflect posttranslational modification. The most likely explanation is molecular heterogeneity, arising either from differential expression of members of a closely related gene family or differences in subunit composition in different cell types. The monoclonal antibody has been used to study regulation of Na^+ pump levels in response to demand for pumping and to study distribution of the pump in chick neurons (Fambrough 1983). In contrast to fish brain, substantial concentrations of the Na^+ pump appear to exist in the internodal regions of chick myelinated axons.

Ca^{2+}-PUMPS Cytoplasmic Ca^{2+} levels regulate many facets of neuronal metabolism, most notably the release of neurotransmitters by exocytosis. Cytoplasmic Ca^{2+} levels are regulated by the activity of voltage-sensitive Ca^{2+} channels and several Ca^{2+} removal systems, including a Na^+:Ca^{2+} antiporter in the plasma membrane, a Ca^{2+} porter in the mitochondrial inner membrane, and several distinct ATP-dependent Ca^{2+} uptake systems in the plasma membrane, smooth endoplasmic reticulum, and synaptic vesicles (reviewed in Reichardt & Kelly 1983). The major neuronal ATP-dependent Ca^{2+} translocators have been purified (Goldin et al 1983, Chan et al 1984). An M_r^+ 140,000 translocator, regulated by Ca^{2+}-calmodulin, has the same molecular weight as the Ca^{2+}-calmodulin-regulated (Ca^{2+} + Mg^{2+}) ATPases in red blood cells and many other tissues. A second, M_r^+ 94,000 Ca^{2+} translocator is not regulated by Ca^{2+}p-calmodulin and is similar in size to the Ca^{2+}

translocator in muscle sarcoplasmic reticulum. A monoclonal antibody to this neuronal translocator, however, does not cross-react with either the sarcoplasmic reticulum or red blood cell translocators (Chan et al 1984), thus suggesting that this is a nerve-cell-specific protein. The antibodies to this translocator should facilitate studies on the regulation of its activity. Its ultrastructural localization will also be very interesting.

ACETYLCHOLINE RECEPTOR The acetylcholine (ACh) receptor has been intensely studied with monoclonal antibodies, yielding much information about structure and function. The aCh receptor is a pentameric protein complex that forms a transmembrane channel that is opened by binding to ACh (reviewed in Conti-Troncini & Raftery 1982). The receptor consists of four different, but homologous, subunits that form a pentameric rosette around a central channel with stoichiometry $\alpha_2\beta\gamma\delta$. It is the primary target in myasthenia gravis and in an animal disease model, experimental autoimmune myasthenia gravis (EAMG).

Several groups have isolated monoclonal antibodies to the ACh receptor, purified from *Torpedo* electroplax, *Electrophorus* electroplax, or mammalian skeletal muscle (Mochley-Rosen et al 1979, Tzartos & Lindstrom 1980, Conti-Troncini et al 1981). Monoclonal antibodies that bind determinants unique for each subunit have been identified (Tzartos & Lindstrom 1980). Among these are a few examples of antibodies that block ion flux and snake α-toxin binding (Mochley-Rosen et al 1979, Gomez et al 1979). Antibodies that recognize homologous determinants shared by the α and β, or γ and δ subunits, respectively, have also been isolated and provide evidence for homologies between these subunits (Tzartos & Lindstrom 1980). Previous studies using conventional antisera and comparative peptide analysis had not revealed these subunit homologies, which have since been confirmed by sequencing completely the four genes that encode these subunits (e.g. Noda et al 1983). Nine antigenic determinants, some sequence dependent and some conformation dependent, have been recognized on the *Electrophorus* receptor (Tzartos et al 1981). Twenty-eight determinants have been recognized on denatured *Torpedo* receptor subunits (Gullick & Lindstrom 1983). F_{ab} fragments of an antibody that binds the α-subunit have been used in high resolution electron-microscopic studies to map the position of the α-subunits in *Torpedo* receptor dimers in which the δ-subunits link the two monomers (Kristofferson et al 1984). The results show clearly that the two α-subunits are not adjacent to each other, but are separated by an intercalated β or γ-subunit. Examination of the binding of F_{ab} fragments specific for other subunits should reveal the order in which the subunits surround the central ion channel. Antibodies that bind to different antigenic determinants are being used to probe the structure of each receptor subunit. In many cases these have been shown to bind different peptides (Gullick et al 1981, Gullick & Lindstrom 1983). Some, but not all, of the epitopes in the δ- and β-subunits have been ordered along the peptide sequence by examining the sensitivity to limited proteolysis of in vitro synthesized subunits that have been co-translationally inserted into microsomes (Anderson et al 1983). Extracellular and cytoplasmic domains on the β subunit have been identified. In general, these results are consistent with models of subunit structure derived from hydrophobicity of the amino acid sequence, but are not sufficiently refined to determine whether a fifth amphipathic α-helix crosses the lipid bilayer (Finer-Moore & Stroud 1984). Four and five transmembrane domain models of the acetylcholine receptor could in theory be distinguished by determining whether antibodies specific for the C-terminus of a receptor subunit bind to the cytoplasmic or extracellular surface. Recently, an antibody to a synthetic peptide corresponding to the C-terminus of the δ-subunit has been shown to bind to the cytoplasmic side of the membrane, arguing persuasively for a five trans-membrane domain model (Young et al 1984).

The acetylcholine receptor contains a major immunogenic determinant that is located on the extracellular surface of the α-subunit and is conserved between species (Tzartos & Lindstrom 1980; Tzartos et al 1981). Antibodies to this determinant do not prevent ion flux through the channel, but do induce passive EAMG when injected into rats (Tzartos et al 1981). The ability of different monoclonal antibodies to induce EAMG has shown to correlate with their ability to aggregate solubilized receptor in vitro (Conti-Troncini et al 1981). As expected, only antibodies to the α subunit, two copies of which are included in each receptor, are able to aggregate receptor into complexes larger than dimers. Only those anti-α antibodies that induce the formation of large aggregates efficiently induce EAMG. These antibodies are also the only ones to increase the turnover of ACh receptor on the surface of cultured myotubes. Monoclonal antibodies have been used in competitive binding assays to determine the specificity of anti-receptor antibodies in the serum of patients with myasthenia gravis. Patients with myasthenia gravis produce antibodies to the same regions on receptor as produced by animals immunized with receptor purified from fish electric organ (Tzartos et al 1982). Most of these bind to the main immunogenic region. These results suggest that antibody production in myasthenia gravis is stimulated by receptor, not by a cross-reacting antigen. No correlation was found between the titer of antibody to a specific determinant and the severity of the disease state. This suggests that other factors are important in determining the severity of illness. Monoclonal antibodies will make it possible to investigate these factors in detail.

RECEPTORS TO OTHER LIGANDS Monoclonal antibodies to the β-adrenergic receptor have been isolated and used to map its distribution in the brain, where it is concentrated in postsynaptic densities (Strader et al 1983). Monoclonal antibodies to purified rat brain muscarinic acetylcholine receptor have been isolated and have been used to show that all identified antigenic epitopes on these receptors are conserved in all tissues and species examined, including both vertebrates and invertebrates (Ventner et al 1984). These data suggest that there is only one form of the receptor and that it has been extraordinarily conserved throughout evolution. Monoclonal antibodies to the α-adrenergic receptor have been isolated by immunizing with rat liver plasma membranes (Ventner et al 1984). Some of these antibodies cross-react with the muscarinic acetylcholine receptor. The receptors may share antigenic sites because they modulate common effectors. These results show that plasma membrane preparations can be used to obtain monoclonal antibodies specific for receptors, provided that the antibodies can be screened effectively. Plasma membranes have been used to obtain monoclonal antibodies specific for both the epidermal and nerve growth factor receptors (Richert et al 1983, Chandler et al 1984).

Internal Organelles and Cytoskeletal Elements

The complex subcellular anatomy of neurons has been examined with conventional and monoclonal antibodies. The studies have revealed that specific molecules are concentrated in anatomically distinct regions of the cell.

CYTOSKELETAL ELEMENTS Neurons contain the same cytoskeletal elements as other cells—actin, tubulin, and intermediate filament networks. The role of these filaments in neural function is poorly understood, although fast transport of membrane vesicles has been closely associated with microtubules in vivo and in vitro and requires the integrity of the microtubule network (reviewed in Reichardt & Kelly 1983). Studies initiated with conventional antibodies and extended with monoclonal antibodies have examined the distribution of actin, tubulin, and intermediate filament-associated proteins in cultured neurons and in sections of the brain. Antibodies specific for each subunit of the neurofilament triplet have shown that all three subunits are found only in neurons (e.g. Shaw et al 1981a, Lee et al 1982). Some classes of neurons, for example the granule cells in the cerebellum, lack neurofilaments and do not bind antibodies to any of the neurofilament subunits (e.g. Shaw et al 1981b). In electron-microscopic examination, each antibody can be shown to stain the same filament (Sharp et al 1982). The M_r 200,000 subunit appears to have a periodic distribution in the filaments, while the other two smaller subunits are distributed continuously (e.g. Sharp et al 1982). While the three neurofilament subunits always appear to coexist in the same neurons, the M_r 200,000 subunit does not appear to be found in neurofilaments in all parts of some classes of neurons (Hirokawa et al 1984). For example, a monoclonal antibody specific for the M_r 200,000 subunit does not bind to dendrites in pyramidal cells, even though these dendrites have high concentrations of the M_r 68,000 subunit (Debus et al 1982). More recent work has raised the possibility, though, that some of the observed restrictions in distribution of the M_r 200,000 subunit may reflect differences in its state of phosphorylation. Many monoclonal antibodies to the subunit have been shown to be specific for either phosphorylated or nonphosphorylated forms of the subunit (Goldstein et al 1983, Sternberger & Sternberger 1984). The two classes of antibodies stain different neurons and parts of neurons. Their staining pattern suggests, for example, that the M_r 200,000 subunit is more heavily phosphorylated in axons than in dendrites of Purkinje cells.

A differential distribution of other cytoskeletal proteins has also been shown in pyramidal and Purkinje cells. A conventional antiserum to the high molecular weight microtubule-associated-proteins (MAP1 and MAP2) stains dendrites but not axons, whereas antibodies to tubulin bind to both regions (Matus et al 1981). More recent studies with a MAP2 monoclonal antibody suggest that MAP2 is restricted to dendrites, while MAP1 is probably not restricted in its distribution (Caceres et al 1983, 1984, Vallee & Davis 1983, Huber & Matus 1984). MAP2 has been shown to be a substrate for both cAMP-dependent and Ca^{2+}-calmodulin dependent protein kinases, and it has a binding site for the type II regulatory subunit (RII) of the cAMP-dependent kinase. Binding sites for the RII subunit are concentrated in dendrites and are almost indetectable in axons. This is essentially the same distribution reported for MAP2 (Miller et al 1982). A differential distribution of actin and β-tubulin has also been shown in the Purkinje cell dendritic shaft and dendritic spine. A monoclonal antibody to β-tubulin stained only the dendritic shaft, whereas antibodies to MAP2 and actin stained both the shaft and the spine (Caceres et al 1983). One monoclonal antibody to actin, however, did not stain actin in the spines. These results suggest that there may be different functional states of actin and that the interactions between cytoskeletal components may vary in different compartments of the cell.

Conventional antibodies to brain-specific and erythrocyte-specific spectrins, proteins that bind to both the membrane and cytoskeleton, have been used to show that both forms are present in neurons, but have different distributions (Lazarides & Nelson 1983a,b, Lazarides et al 1984). The erythrocyte form appears only at a terminal state in differentiation and is restricted to the cell soma and dendrites, while the brain-specific form, also known as fodrin, appears early in neuronal development and is distributed throughout the neuron. It seems likely that the spectrins function to coordinate domains in the neuronal membrane and cytoskeleton.

As illustrated above, monoclonal antibodies have the potential for identifying and probing the function of modified forms of proteins that are associated with particular regions of the neuron. As one example, injection of a monoclonal antibody specific for the tyrosylated form of α-tubulin has been shown to have dramatic effects on organelle movement and cytoplasmic organization in fibroblasts. This suggests that addition of tyrosine to tubulin may modulate its activity within this cell (Wehland & Willingham 1983). Intriguingly, monoclonal antibodies specific for the tyrosylated and nontyrosylated forms of α-tubulin stain the axons of cerebellar granule cells differently, depending on their age, in patterns that suggest that tyrosine addition to tubulin is reduced as axons mature (Cummings et al 1984). Monoclonal antibodies to different epitopes on actin and calmodulin stain different subcellular structures in neurons and fibroblasts, respectively, suggesting that these proteins interact with different binding proteins in different compartments (Caceres et al 1983, Pardue et al 1983). Thus, monoclonal antibodies make it possible to study the specific interactions of these proteins with other cellular macromolecules. An individual monoclonal antibody, however, cannot be assumed to reveal all the sites to which its antigen is localized.

ORGANELLES Synaptic vesicles and mitochondria are prominent organelles in nerve terminals. A monoclonal antibody directed against a neuron-specific mitochondrial protein has been isolated using rat brain as an immunogen (Hawkes et al 1982a). This antibody binds a M_r 23,000 protein in neuronal mitochondria that is absent from other cell types in the brain. The antigen appears in neurons only after the termination of cell division. This suggests that changes in mitochondrial proteins may be important in neuronal differentiation.

Monoclonal procedures have been very useful for identifying antigens shared by many classes of synaptic vesicles. Two monoclonal antibodies to a M_r 65,000 vesicle membrane protein bind vesicles in many and possibly all types of nerve terminals (Matthew et al 1981). The antibodies have been used successfully to purify synaptic vesicles from crude brain homogenates by immunoprecipitation and to confirm the presence of norepinephrine and several peptide transmitters in vesicle fractions. The vesicle antigen is highly conserved throughout the vertebrate phylogeny. A second vesicle-specific protein, the phosphoprotein synapsin I, was initially studied with polyclonal antibodies and was shown also to exist in virtually all types of nerve terminals and throughout the vertebrate phylogeny (e.g. Goelz et al 1981). More recently, monoclonal antibodies have been isolated to this phosphoprotein and these should make possible more detailed studies on the structure and function of synapsin I in the future (Nestler & Greengard 1983). The distributions of these two antigens have been used to monitor differentiation of the nerve axon and nerve terminal during development in vivo or in vitro. Dramatic changes with development are seen both in the brain and culture dish (Chun & Schatz 1983; Matthew & Reichardt 1982).

Cholinergic synaptic vesicles from fish electric organs contain a specific proteoglycan that has been defined by both conventional and monoclonal antisera (Jones et al 1982, Buckley et al 1983). Antibodies have been used to show that this proteoglycan is released into the cleft by exocytosis during stimulation and is recycled during rest (Jones et al 1982). Studies with a monoclonal antibody show that the antigen is found on the surface of the cholinergic nerve terminals in the vicinity of active zones. Evidence suggests that a fraction remains in the cleft for extended periods during rest, possibly immobilized in the extracellular matrix (Buckley et al 1983).

SYNAPSE CONSTITUENTS Chemical synapses exist in a diversity of morphological conformations, usually containing one or more of the following specializations: presynaptic vesicle clusters, presynaptic membrane densities, and postsynaptic dense material. Conventional and monoclonal antibodies have provided a promising approach to identifying the molecules in the specializations associated with the synapse. Few antibodies to antigens concentrated in the presynaptic nerve terminal have been found so far. The only well-defined antigens are components of synaptic vesicles (discussed under Organelles), which include neurotransmitters and neuropeptides (see next section).

Antibodies to antigens in the postsynaptic specializations are more common. The ACh receptor has been localized at the neuromuscular junction with monoclonal antibodies to the receptor (Z. W. Hall, personal communication), and has also been shown by use of α-bungarotoxin to be in the membrane at the top and sides of the postsynaptic folds (Fertuck & Salpeter 1976). Antibodies to a M_r 43,000 protein that copurifies with ACh receptor have been used to localize the protein at the neuromuscular junction (Froehner et al 1981). In electron-microscopic examination, the M_r 43,000 protein appears to form a dense bar of contrasted material that is coextensive with ACh receptor (Sealock 1982, Sealock et al 1984). It can be cross-linked to the acetylcholine receptor with a heterobifunctional reagent. Monoclonal antibodies to the M_r 43,000 protein and four acetylcholine receptor subunits have been used to show that the cross-linked product is a β-receptor subunit-43 kDa dimer (Burden et al 1983). The same anti-43 kDa protein monoclonal antibody immunoprecipitates a M_r 43,000 protein that binds ATP and has protein kinase activity (Gordon et al 1983, A. S. Gordon, personal communication), so this monoclonal antibody has made it possible to show rigorously that the M_r 43,000 polypeptide associated with the acetylcholine receptor has protein kinase activity. Concentrated between the folds at the neuromuscular junction are 10 nM filaments, almost certainly intermediate filaments. An amorphous, electron-dense material is between these filaments and the membrane. Recently, a monoclonal antibody has been described that binds an antigen, related antigenically to intermediate filaments, that is concentrated in this amorphous material (Burden 1982). Other antibodies have been used to localize to the neuromuscular junction antigens related to actin and three actin-binding proteins—α-actinin, vinculin, and filamin (Bloch & Hall 1983).

Classical CNS synapses have somewhat similar postsynaptic specializations. Antibodies are not available to most of the receptors in the CNS. Development of monoclonal reagents to these receptors is likely to be one of the major contributions of modern immunology in the next few years. Where receptor antibodies are available, such as to the β-adrenergic receptor, they have localized the receptor to the postsynaptic density (Strader et al 1983). Microfilaments and microtubules have been visualized in the lattice of the postsynaptic density (e.g. Matus 1981). Recently, monoclonal antibodies to β-tubulin, actin, and the high molecular weight microtubule-associated-protein MAP2 have localized these molecules to the postsynaptic density (Caceres et al 1983). The major postsynaptic density protein, a M_r 50 kDa polypeptide of previously unknown function, has recently been shown, by using monoclonal antibodies specific for this kinase (Kennedy et al 1983, Kelly et al 1984), to be the 50 kDa subunit of a Ca^{2+}-calmodulin-dependent protein kinase. Immunocytochemistry with these antibodies shows that this subunit is not localized exclusively to the postsynaptic density, but is found also in neuronal cell bodies, axons, and terminals in all or nearly all areas of the brain (McGuinness et al 1983). Conventional antisera have been used to demonstrate the presence in the postsynaptic density of a M_r 95,000 protein (Nieto-Sampedro et al 1982), fodrin (Carlin et al 1983), calmodulin (Wood et al 1980), and calcineurin, which is the Ca^{2+}-calmodulin regulated protein phosphatase 2B (Wood et al 1980). Monoclonal antibodies to these and other antigens will be crucial for delineating the structure and composition of this organelle.

Extracellular Matrix Components

The extracellular matrix between the pre- and postsynaptic elements of the neuromuscular junction has become a subject of intense interest. Acetylcholinesterase is inserted into this matrix and is important in removing acetylcholine from the synapse. Acetylcholinesterase occurs in several molecular forms: soluble or membrane bound, intra- or extracellular, with or without a collagen-like tail (reviewed in Massoulie & Bon 1982). The occurrence of forms with a collagen-like tail is correlated with innervation. With the objective of understanding the relationship between the different forms of esterase, Fambrough et al (1982b) isolated five monoclonal antibodies to the plasma-membrane-associated acetylcholinesterase of purified erythrocytes. All of these antibodies recognized an antigen in the neuromuscular junction; two of these were shown to bind to the form of esterase with a collagen-like tail. The results argue that there is a high degree of homology between different forms of the esterase.

It is now clear that components in the synaptic extracellular matrix induce synaptic specializations in both nerve and muscle during regeneration (reviewed in Burden 1982). Conventional antibodies to anterior lens capsule, two collagen extracts, and a substantially purified factor that induces aggregation of ACh receptors in cultured myotubes have been shown to distinguish synaptic and nonsynaptic regions of basal lamina (Sanes & Hall 1979, Nitkin et al 1983). More recently, antibodies to the major components of basal lamina have been screened for binding specificity (Sanes 1982). Although collagen IV, laminin, and fibronectin are found in both synaptic and nonsynaptic regions, collagen V appears to be excluded from synaptic regions. Monoclonal procedures have been used to generate several antibodies directed to synapse-specific components of the basal lamina (Chiu & Sanes 1982, Fambrough et al 1982a). One antigen has been identified as a heparan sulfate proteoglycan synthesized by aneural muscle cells that is concentrated within the synaptic cleft at the neuromuscular junction (Anderson & Fambrough 1983). This proteoglycan is unlikely to be related to the proteoglycan defined by monoclonal antibodies that is found in synaptic vesicles and on the surface of cholinergic nerve terminals in fish electric organ (Buckley et al 1983). These and other synapse-specific antigens could be important in synapse function, structure, or development.

NERVOUS SYSTEM ANATOMY

Major Cell Types

One important application that hybridoma technology has had in neurobiology is the generation of probes that recognize different subsets of cells in the nervous system. These range from antibodies that distinguish neurons from glial cells, and peripheral neurons from central neurons, to those which recognize very small subsets of neurons. Antibodies to cell surface antigens are particularly important for isolation of a cell type by positive or negative selection (reviewed by Basch et al 1983). Positive selection procedures include adhesion to antibody columns or magnetic beads (e.g. Meier et al 1982) and cell sorting (Dangl & Herzenberg 1982). Selenium- and iron-conjugated monoclonal antibodies can be used to select antigen-positive cells by growing cell mixtures in defined media that are missing these essential nutrients (Block & Bothwell 1983). Antigen-positive cells can also be selected in hydrogen-peroxide-containing media by using peroxidase-conjugated antibodies to protect these cells from the deleterious effects of peroxide (Basch et al 1983). Negative selection procedures include antibody-directed complement- or toxin-mediated killing of unwanted cell types (e.g. Brockes et al 1979, Vitetta et al 1983).

The major cell types in the peripheral and central nervous systems can now be distinguished using conventional antisera. Polyclonal antisera have been produced that distinguish astrocytes, Schwann cells, oligodendrocytes, fibroblasts and neurons (see Table 1 for references). Monoclonal antibodies have

Table 1 Major cell class markers

Cell and antigen	References[a]	Cell and antigen	References[a]
Neurons		Astrocytes	
Tetanus toxin	1	Glial fibrillary	
Neurofilament proteins	2, 3, 4	acidic protein (GFAP)	19, 20, 13
Neuron-specific enolase	5	S-100	19, 21, 22
N-CAM	6	α_2-Glycoprotein	23
A4	7	Non-neuronal enolase	24
38/D7	8	M1, C1	25, 26
NILE glycoprotein	9	Ran-2	27
BSP-2	10	Vimentin	2, 3, 28
GQ ganglioside	11, 12	Stage specific embryonic	
Thy-1	13, 14	antigen (SSEA-1)	28
Oligodendrocytes		Glutamine synthetase	29
Galactocerebroside	13, 15	(Thy-1)[b]	30
01-04	16	(Tetanus toxin)[b]	13, 31
Schwann cells		(GQ ganglioside)[b]	31, 11, 14
Ran-1	17, 13	Epithelial cells, fibro-	
(GFAP)[b]	3	blasts, ependymal cells	
(GAL-C)[b]	18	Vimentin	11, 12
		Fibronectin	13
		(Thy-1)[b]	32
		(GFAP)[b]	2

[a]1. Mirsky et al (1978). 2. Shaw et al (1981b). 3. Yen & Fields (1981). 4. Shaw et al (1981a). 5. Marangos et al (1982). 6. Rutishauser et al (1982). 7. Cohen & Selvendran (1981). 8. Vulliamy et al (1981). 9. Salton et al (1983). 10. Hirn et al (1981). 11. Berg & Schachner (1982). 12. Eisenbarth et al (1979). 13. Raff et al (1979). 14. Mirsky (1982). 15. Ranscht et al (1982). 16. Sommer & Schachner (1981). 17. Brockes et al (1977). 18. Mirsky et al (1980). 19. Ludwin et al (1976). 20. Bignami & Dahl (1977). 21. Ghandour et al (1981). 22. Haan et al (1982). 23. Langley et al (1982). 24. Langley & Ghandour (1981). 25. Lagenaur et al (1980). 26. Sommer et al (1981). 27. Bartlett et al (1981). 28. Solter & Knowles (1978). 29. Norenberg & Martinez-Hernandez (1979). 30. Pruss (1979). 31. Raff et al (1983). 32. Brockes et al (1979).

[b]-( )- Shared markers.

been made to some of these cell specific markers, and some new antigens have also been discovered such as M1 and C1 for astrocytes (Lagenaur et al 1980). The astrocyte markers M1 and C1 are expressed at different times during development, with C1 appearing at embryonic day 10 in the mouse. Cell type specific antibodies have been used to study a wide range of other questions, including the requirements for synthesis of myelin components by Schwann cells and oligodendrocytes (Mirsky et al 1980) and the effects of neurons on glial cell differentiation (Holton & Weston 1982).

For the majority of cells, the markers shown in Table 1 are restricted in their distribution. However, some markers are found on more than one cell type, although in some cases distribution of antigen may be altered by artifacts of fixation (Ludwin et al 1976, Ghandour et al 1981). It has been shown that some Schwann cells may express glial fibrillary acidic protein (GFAP) (Yen & Fields 1981). Also, there is a population of astrocytes in white matter cultures with a neuron-like morphology that expresses two neuronal epitopes: the A_2B_5 antigen (GQ ganglioside) and tetanus toxin receptor (Raff et al 1983). These oligodendrocytes and neuron-like astrocytes are derived from a common precursor that is an A_2B_5-positive, GFAP-negative cell. Culture conditions determine which phenotype is eventually expressed by this precursor (Raff & Miller 1983). These precursor cells have been purified from embryonic optic nerve by using a fluorescence-activated cell sorter to positively select A_2B_5+ cells (Abney et al 1983). The purified cells develop into oligodendrocytes in vitro in the absence of living neurons.

Neuronal Subpopulations

Conventional and monoclonal antibodies have been used in a wide variety of experiments to detect, categorize, and monitor the development of antigens that distinguish one type of neuron from another. Antibodies that distinguish neuronal subpopulations fall into two classes: (*a*) antibodies to neurotransmitters, neuropeptides, or enzymes involved in their metabolism; (*b*) monoclonal antibodies to random antigens identified by screening with immunocytochemical procedures supernatants from large numbers of hybridoma clones.

Neurotransmitters, Neuropeptides, and Transmitter Enzymes

The reagents most widely used to distinguish different neurons have been antibodies to neurotransmitters. During the past decade, immunocytochemical examination has shown that more than 30 peptides are localized within specific neurons, coexisting in many cells with other transmitters (e.g. Hökfelt et al 1980, Iversen 1983). Even larger numbers of peptides have been revealed within the genes for these, and tissue-specific RNA splicing and prohormone processing have further expanded the numbers of possible peptide transmitters. Antibodies capable of distinguishing between related peptides should be particularly valuable for future work. Recently, it has become possible to visualize several classical transmitters directly with antibodies raised to serotonin, glutamic acid, and γ-amino butyric acid coupled to protein carriers (Cuello et al 1982, Storm-Mathisen et al 1983, Seguela et al 1984). These classical transmitters have also been found to be localized within specific subpopulations of neurons.

The number of available monoclonal antibodies to neurotransmitters, neuropeptides, and transmitter-related enzymes is expanding rapidly. Monoclonal antibodies are proving particularly useful in making reagents of defined specificity available in large quantities to many laboratories. We discuss only a few examples of special interest.

Monoclonal antibodies specific for substance P and serotonin have been used in a number of creative applications of monoclonal technology. They have been internally labelled by growth of the hybridomas in the presence of [^{3}H]-lysine. These internally labelled antibodies have been used to localize each transmitter in nerve terminals by high resolution autoradiography (Cuello et al 1982). Internally labelled antibodies offer a number of advantages in immunocytochemistry. First, the antibodies are not chemically coupled and are fully active. Second, molecules as small as F_{ab} fragments can be used to penetrate sections, so it is possible to use these antibody fragments with tissues that are more thoroughly fixed for good preservation of ultrastructure than is possible with antibody conjugates. It is also possible to combine autoradiography and immunoperoxidase to detect two antigens simultaneously using the electron microscope. These procedures have been used to investigate the coexistence of two putative transmitters in individual neurons, e.g. substance P and enkephalin-containing terminals in the *substantia gelatinosa* (Cuello et al 1982), and substance P and serotonin in the nucleus raphe magnus of the rat (Cuello et al 1982).

Another new method utilizes bispecific antibodies in which one antigen-binding site is directed against a neurotransmitter and the other is directed against horseradish peroxidase (Milstein & Cuello 1983). These antibodies are synthesized in cell lines derived from the fusion of transmitter-specific and peroxidase-specific hybridomas. Their use minimizes the size of reagents that must diffuse into tissue sections.

Monoclonal antibodies specific for several neurotransmitter enzymes have been isolated, including tyrosine hydroxylase, dopamine-β-hydroxylase, and choline acetyltransferase (CAT) (Ross et al 1981a,b, Crawford et al 1982, Levey et al 1983, Eckenstein & Thoenen 1982). The generation of monoclonal antibodies to the latter enzyme has been particularly useful, since CAT is difficult to purify and is not very immunogenic. Recently, monoclonal antibodies specific for CAT have been obtained using partially purified preparations of bovine and rat CAT (Levey et al 1983, Crawford et al 1982, Eckenstein & Thoenen 1982). These antibodies immunoprecipitate CAT, bind to at least three non-overlapping sites on the enzyme, and bind cholinergic neurons in sections of brain. Although some of the antibodies are species-specific, others bind CAT in a wide variety of species. These antibodies provide the first unambiguous markers for cholinergic neurons.

Other Antibodies that Define Neuronal Subclasses

The most successful strategy for isolating monoclonal antibodies that bind subclasses of neurons has been to immunize mice with material from one part of the nervous system and screen the antibodies secreted by the hybridomas on sections or cultures. Antibodies specific for peripheral and central neurons have been identified (Vulliamy et al 1981, Cohen & Selvendran 1981). Antibodies to retinal membranes have been identified that distinguish photoreceptors, other neurons, and Muller cells from each other (Barnstable 1980, Barnstable et al 1983, Trisler et al 1983). Antibodies to other regions of the mammalian CNS distinguish many of the different neurons within these regions. Striking among these are a series of antibodies, originally isolated using *Drosophila* nervous tissue as an immunogen, that cross-react with the human nervous system (Miller & Benzer 1983). Some of these antibodies stain subpopulations of neurons and others stain particular areas within individual neurons. When analyzed in immunoblots, the *Drosophila* and human antigens often are closely similar, suggesting an extraordinary conservation of antigenic epitopes (see also Ventner et al 1984). Another striking antibody, CAT 301, isolated using cat spinal cord as an immunogen, binds in the cerebellum only to a surface or extracellular matrix protein associated with the cells of Lugaro, a very rare cell type (McKay & Hockfield 1982, Hockfield & McKay 1983a). This antibody also binds elsewhere in the CNS, including pyramidal cells in the hippocampus and a variety of cells in area 17 of the visual cortex. This antibody exhibits particularly intriguing binding specificity to cells in the visual system of monkeys and cats (Hendry et al 1984). In the monkey lateral geniculate nucleus, binding is primarily restricted to magnocellular, not parvocellular, layers. In area 17, patches of stained cells are seen in layers III, IVB, and VI. These patches line up radially with each other and with the centers of ocular dominance columns. The distribution of the CAT 301 antibody suggests that it distinguishes cell groups with particular functions in the lateral geniculate nucleus and visual cortex. Conceivably, injection of this antibody coupled to ricin or another cytotoxic agent could be used to kill these cell populations in order to examine in more detail their roles in processing visual information.

Specific glycolipids, defined by antibodies, have been found to be associated with subpopulations of neurons (Richardson et al 1982), most notably subsets of sensory neurons (Dodd et al 1984). As these antigens are resistant to proteases, they should be particularly useful for separating neurons.

Neuron-specific Antigens in Invertebrates

Monoclonal antibodies to neural antigens in invertebrates, especially the leech and insects, have revealed a large degree of antigenic diversity among neurons (e.g. Zipser & McKay 1981). Many of the antibodies directed to antigens in the leech nervous system define small groups of cells, for example subsets of sensory neurons and motoneurons. The antibodies distinguish a very large number of different sets of neurons, some of which are nested within others (Hogg et al 1983). The antibodies have made it possible to identify all the neurons of a particular type in the leech nervous system and have revealed relationships between neurons that were not previously recognized (Zipser 1982, Zipser et al 1983). Examination of the position of axons stained by individual antibodies has shown that they have consistent positions in connectives (Hockfield & McKay 1983b). In some cases, antigens mark axons clustered into fascicles that are derived from scattered cell somata; in other cases, antigens mark clusters of cell somata whose axons are dispersed into different fascicles. Thus, the factors important for aggregating cell somata and nerve bundles must be different.

Monoclonal antibodies directed against *Drosophila* and grasshopper neural tissues also reveal a large diversity in the antigenic properties of different neurons (e.g. Fujita et al 1982, Aceves-Pina et al 1982, Kotrla & Goodman 1982). Antigens specific for the neurophil, cortex, and nerve fibers are readily distinguished (Fujita et al 1982). Focusing on the compound eye, Benzer & colleagues (Fujita et al 1982) isolated antibodies that labelled the lens, the underlying crystalline cone, the secretory cone cells, and the photoreceptors. Several of these were used to study the development of the compound eye (Zipursky et al 1984). The M_r's of many of these antigens have been successfully estimated by immunoblot analysis.

MONOCLONAL ANTIBODIES THAT PROBE DEVELOPMENT OF THE NERVOUS SYSTEM

Determination of Neural Crest Cells

Multiple cell types, including neurons and Schwann cells in the peripheral nervous system, are derived from the neural crest (reviewed in Le Douarin et al 1981). Transplantation experiments have shown that populations of crest cells can differentiate differently depending on the environmental stimuli that they encounter. An important issue, now being addressed using monoclonal antibodies, is whether cells in the crest are truly pluripotent or whether subpopulations with restricted fates are generated before crest cell migration. A monoclonal antibody, termed NC-1, that binds to a cell surface antigen on avian neural crest cells has recently been isolated and used to examine neural crest cell migration (Vincent et al 1983). With this antibody, it is possible to examine neural crest cell migration in normal animals. Previously, this required experimental manipulation, usually transplantation. Non-neural derivatives of the neural crest, such as melanocytes and mesenchymal cells in the branchial arches, do not bind the NC-1 antibody, but do appear to be derived from NC-1–positive cells (Vincent & Thiery 1984), a result consistent with a pluripotent precursor population of crest cells. A complementary monoclonal antibody, E/C8, that binds mesenchymal cells in the branchial arches, but not neural crest cells, has been isolated by immunizing with avian sensory neurons (Ciment & Weston 1982). It has been possible to show that E/C8-positive mesenchymal cells from the third and fourth branchial arches can develop into neurons, but not melanocytes in vitro (Ciment & Weston 1983). They can invade aneural gut to form enteric ganglionic neurons in organ culture, but are not competent to form melanocytes in developing embryos. It should be possible to determine whether loss of the NC-1 antigen or acquisition of the E/C8 antigen is related to the loss by cells in the branchial arches of competence to differentiate into melanocytes. The E/C8 antibody also binds neuronal cells in both the central and peripheral nervous systems. It binds to neuronal precursors at an earlier stage than any other known neuronal marker. For example, it appears in the future sites of sensory ganglia a day before condensation of these ganglia. It should consequently be very useful in studying early steps in neuronal commitment and differentiation.

Although the previous results suggest that neural crest cells are homogeneous before migration, monoclonal antibodies specific for ciliary neurons have been isolated that reveal heterogeneity in early neural crest cell cultures (Barald 1982). Two antibodies bind a small percentage of cultured neural crest cells provided the cells come from the region of crest that gives rise to the ciliary ganglion. Two different antigenic determinants have been defined by these antibodies, increasing the probability that they are related to cell lineage. This suggests that heterogeneity arises among crest cells at an early stage and results in the appearance of cell types that are partially restricted in their developmental potential. Positive and negative selections using antibodies to surface antigens that define neural crest subpopulations should be helpful for exploring crest cell lineages.

Neuronal Cell Adhesion Molecules (N-CAM and Ng-CAM)

Both Ca^{2+}-dependent and Ca^{2+}-independent adhesion systems have been detected in embryonic chick neural tissues (Brackenbury et al 1981). N-CAM, a glycoprotein that mediates Ca^{2+}-independent adhesion between chick neural cells, has been purified as a molecule that neutralizes antibodies, prepared against neuronal membranes, that inhibit neural cell adhesion (see Edelman 1983). Conventional antibodies to this glycoprotein were used to demonstrate the presence of N-CAM on the surface of the somata and processes of neurons, but not on other cell types in the chick brain. The antibodies prevented the formation of axonal bundles in sensory neuron cultures and disrupted histogenesis of developing retina in organ culture (Rutishauser et al 1978, Buskirk et al 1980). Very recently, the injection of N-CAM antibodies into the developing retina has been shown to disrupt the topological organization of retinal ganglion cell axons in the optic nerve and to prevent some of these axons from contacting their normal target in the optic tectum (Thanos et al 1984). These results argue that N-CAM is important in mediating interactions among neural cells in development. Monoclonal antibodies to N-CAM have recently been isolated (Rutishauser et al 1982). Using these antibodies it has become possible to purify milligram quantities of N-CAM for biochemical investigations, including binding studies (Rutishauser et al 1982), and chemical characterization (Hoffman et al 1982). Using affinity-purified N-CAM, it has been possible to detect differences in the carbohydrate attached to embryonic and adult brain N-CAM molecules and show that these differences alter the adhesive properties of N-CAM (Rothbard et al 1982, Hoffman & Edelman 1983). Monoclonal antibodies with cross-species reactivity have also been used to detect and purify a very similar N-CAM from rodent brain (Chuong et al 1982), and to show that mice homozygous for the neurological mutation *staggerer* do not synthesize the adult form of N-CAM in the cerebellum, the site at which this mutation affects histogenesis (Edelman & Chuong 1982).

More recently, similar strategies using monoclonal antibodies have been used to purify and characterize other molecules that mediate cell adhesion. A partially purified tryptic fragment of a molecule that mediates neuronal adhesion to glia has been used to raise monoclonal antibodies that inhibit the binding of neuronal membranes to glia (Grumet et al 1984). The monoclonal antibodies have been used to purify the intact, unproteolyzed form of this glycoprotein, termed Ng-CAM, which differs from N-CAM in having a M_r of 135,000 and a different pattern of peptide fragments generated by proteolysis. Ng-CAM can be detected by immunofluorescence on neurons, but not on glia. It is found on the same neurons as N-CAM. Despite the chemical differences between N-CAM and Ng-CAM, some monoclonal antibodies bind both glycoproteins, and the conserved determinants do not appear to be on carbohydrate: This suggests that there is a shared protein domain on these two different cell adhesion molecules. A trypsin fragment of a cell adhesion molecule, liver-CAM (L-CAM), has been purified by the same strategy with monoclonal antibodies (Gallin et al 1983). This Ca^{2+}-dependent adhesion molecule has a wide but specific distribution in both the embryo and adult (Edelman et al 1983, Thiery et al 1984). It appears to be lost from cells when they become committed to neuronal differentiation.

Neuronal Cell Migration

Studies with conventional antibodies have demonstrated strong spatial and temporal correlations between the appearance of fibronectin and the migration of granule cells to form the external granule cell layer in the cerebellum (Hatten et al 1982) and migration of neural crest cells to form the variety of tissues derived from the crest (Thiery et al 1982a,b). When cerebellar granule cells or neural crest cells are cultured in vitro, their abilities to adhere to and migrate on fibronectin substrata correlate temporally with their migratory behaviors in vivo (Hatten et al 1982, Rovasio et al 1983). In addition, N-CAM appears at an appropriate time in vivo to be important in condensation of the sympathetic and sensory ganglia from migrating crest cells (Thiery et al 1982b). Migration of cerebellar granule cells from the outer to inner granule cell layer can occur in vitro in brain slices. In a promising start in identifying molecules important in this migration, Schachner and colleagues have shown that mono- and polyclonal antibodies to L1, a neuronal cell surface glycoprotein, reduce the movement of granule cells in cultured slices of developing cerebellum (Lindner et al 1983). The polyclonal, but not monoclonal, antibodies to L1 also block a Ca^{2+}-independent aggregation of granule cells in vitro (Rathjen & Schachner 1984). L1 antisera bind to antigens of three distinct M_r, each of which corresponds in size to an antigen detected with antisera to Ng-CAM. If these two molecules are in fact the same, Ng-CAM is likely to guide granule cell migration.

Axon Growth and Pathfinding

A remarkable monoclonal antibody, I5, has been used in a series of experiments on the grasshopper nervous system to study the mechanisms by which axons grow along stereotypic pathways to reach their targets. This antibody stains a specific subset of neurons, including peripheral pioneer neurons (Ho & Goodman 1982). Since it stains all parts of these neurons, even before the initiation of axon growth, it has been possible to use this antibody to study the pathways and contacts made by pioneer fiber growth cones. The I5 antibody stains an array of previously unseen pioneer neurons that lay down different segments of the early fiber tracts. Identification of these cells has made it clear that pioneer neurons and other landmark cells are quite close to each other during initial fiber growth—within the reach of filopodia in neighboring growth cones. Consequently, pathways may be constructed by sequential movement of growth cones from one of these landmark cells to the next. Evidence for preferential filopodial adhesion to landmark cells has been obtained by comparing at the electron microscope level the number of filopodia adhering to landmark vs other cells, using Lucifer Yellow to fill the filopodia of individual neurons and an antibody to Lucifer Yellow to visualize the filled processes (Taghert et al 1982). The I5 monoclonal antibody also stains a previously unrecognized class of mesodermal cells that arise early in development (Ho et al 1983). Their pattern of growth and development, revealed by staining with the I5 antibody, indicates that they form a scaffold that guides the later development of tendons, muscles, and nerves. The nature of the antigen defined by the I5 antibody is not yet known, but it is clearly associated with a set of cells with important roles in development.

Studies in both vertebrates and invertebrates have shown that the growth cones of different neurons will make divergent choices at branch points (e.g. Raper et al 1983a,b). A monoclonal antibody has recently been isolated that inhibits axon growth by avian retinal ganglion cells, but not sensory neurons (Henke-Fahle & Bonhoeffer 1983). The antigen defined by this antiserum could be one member of the family of molecules that guide these disparate decisions.

Trophic Factors

Many factors have been described that regulate the expression of differentiated functions in neurons. These include substances essential for survival, proliferation, transmitter choice, neurite growth, and steps in synapse formation. Immunological studies will clearly be critical in establishing their biological relevance.

Nerve growth factor (NGF) is a protein that is essential for normal development of both sensory and sympathetic neurons. It is the one neurotrophic factor, of many believed to exist, that has been purified, characterized, and demonstrated to have a role in normal development in vivo. The presence of NGF antibodies during development results in massive neuronal death in the sensory ganglia (Gorin & Johnson 1979). Injection of NGF antibodies at any time destroys the sympathetic nervous system (Levi-Montalcini & Booker 1960). Although mouse NGF acts on fish and avian neurons, antisera to NGF show strong species specificity (e.g. Harper et al 1983). Recently, several groups have isolated monoclonal antibodies to NGF and most of these prevent NGF binding to its receptor and NGF-dependent neurite outgrowth in vitro (e.g. Warren et al 1980, Zimmermann et al 1981, H. Thoenen, personal communication). The monoclonal antibodies also appear to show strong species specificity (H. Thoenen, personal communication). Recently, these antibodies have been used in two-step, sandwich-type antigen assays to measure for the first time the endogenous levels of NGF in tissues innervated by sympathetic neurons (Korsching & Thoenen 1983a). The same assay has been used to show that endogenous NGF is transported retrogradely from targets and is accumulated in sympathetic ganglia and on the distal side of nerve ligatures (Korsching & Thoenen 1983a,b). Since NGF is present at only a few nanograms per grams of tissue, it is very difficult to demonstrate that the antigen present in the tissue is NGF and not a cross-reacting antigen. Different monoclonal antibodies to NGF give similar results. If these antibodies can be shown to bind different epitopes on NGF, it will strengthen the argument that NGF and not a cross-reacting antigen is being detected.

Glial growth factor (GGF), a mitogen for Schwann cells and astrocytes, was found in pituitary and brain extracts (Brockes et al 1980). Although it was initially purified over 4000-fold by biochemical fractionation, the yield was very low. Monoclonal antibodies to GGF, prepared using a partially purified factor as an immunogen, have been invaluable for isolating larger quantities of the factor (Brockes et al 1981) and should facilitate future studies on its distribution and function. The monoclonal antibodies do not directly block the mitogenic activity of GGF, either because they do not bind a site essential for activity or because their affinity is lower than that of GGF receptors. To remove activity, GGF must be precipitated with a second antibody.

Markers of Cell Position

Topological arrangements are preserved in the pattern of synapses formed between groups of neurons in many parts of the nervous system and are important for conveying spatial information. When axons from ganglion cells in the retina form synapses in the tectum, a point-to-point representation of the visual field is conserved. Sperry (1963) postulated that selective chemoaffinity generated by two orthogonal gradients of molecules provided the molecular basis for preserving this spatial order. In an effort to test this theory, Trisler et al (1981) have searched for antibodies that define cell position in the retina. A cell surface antigen, termed TOP, was identified that is distributed in a bilaterally symmetric dorsal-ventral gradient in the retina. At least 30-fold more antigen was found in dorsal than in ventral retina. The antigen was found on all cell types in fresh retinal dissociates. It appears in the optic cups of 2 day-old chick embryos, and the gradient is established before day 4. Thus a gradient is established early in retinal development, during the period of neuroblast proliferation, and is maintained after neurogenesis is completed. TOP has now been purified by affinity chromatography and shown to be a M_r 47,000 cell surface glycoprotein (Trisler et al 1983). It remains to be determined whether the molecule has an important role in generating spatial order in the connections made by retinal ganglion cells with other parts of the brain.

A monoclonal antibody that recognizes a surface antigen whose expression is position-dependent has also been identified by using a grasshopper membrane fraction as an immunogen (Kotrla & Goodman 1982). The antibody, termed epi-1, stains an antigen on a broad strip of ectodermal epithelial cells near the midline in the blastula stage and a narrower strip of the same cells in the grasshopper gastrula. In later development, the antigen is restricted to two of the seven rows of neuroepithelial cells in each segment. The antigen is also positionally expressed on strips of cells near the distal end of each limb bud. In summary, screens for antibodies defining antigens that mark cell position have produced dramatic results in both vertebrate and invertebrate species and suggest that monoclonal antibodies to such molecules will be very useful in developmental studies.

STRATEGIES FOR MAKING MONOCLONAL ANTIBODIES

In this section, we review aspects of particular interest to neuroscientists regarding the methods for producing, screening, and productively using hybridomas and monoclonal antibodies. Numerous books and reviews on monoclonal antibodies have been published in the last few years (e.g. Kennett et al 1980, Hurrell 1982).

Isolation of Specific Hybridoma Cell Lines

Detailed methods for producing monoclonal antibodies have been published elsewhere (e.g. Fazekas de St. Groth & Scheidegger 1980, Stahli et al 1980). Differences in procedures can result in selections of antibodies with different allotypes and specificities. It has therefore proven well worth the effort to prepare the immunogen, choose the immunization protocol, and select a screening procedure to optimize the probability of obtaining antibodies of the greatest interest and utility for future studies.

Recent work has shown that the probability of obtaining hybridoma cell lines is proportional to the strength of the immune response to a particular antigen (Lake et al 1979). Therefore, the first requirement for obtaining specific monoclonal antibodies is to induce a strong response to the desired antigen. Since inbred mouse strains differ significantly in their responsiveness to defined antigens, screening a variety of strains has proven useful when testing for an immune response. The original immunization protocols were optimized for injections of cells and membranes, and typically it was found that an optimal time for fusion of spleen and myeloma cells was three to four days after a final i.v. injection (Lake et al 1979). It now appears that hybridomas specific for soluble antigens are produced more efficiently by daily i.v. injections for three to four days prior to fusion, probably because soluble antigens are rapidly removed from the circulation (Stahli et al 1980).

Different allotypes are preferable for different purposes. IgM antibodies are excellent cytotoxic and agglutinating reagents, but are not as useful as IgG antibodies for many other purposes, including immunocytochemistry, which is crucial for many applications in neuroscience. Specific protocols for producing a high frequency of hybridoma antibodies of a specific allotype have been published (e.g. Colwell et al 1982). Immunization protocols of sufficient length to generate a high frequency of IgG-secreting hybridomas are usually preferable to short immunization regimens. Allotype-specific reagents can also be used in an initial screening to select for antibodies likely to be useful in future applications.

A major goal of neuroscientists has been to obtain cell-type-specific antibodies. The most successful fusions seeking these reagents have used fixed neural tissues as immunogens and have screened fixed tissue sections (Zipser & McKay 1981, McKay & Hockfield 1982). This strategy (*a*) minimizes the diffusion of antigens that can occur in unfixed tissue sections, (*b*) optimizes the morphology of cells, which may be unrecognizable in unfixed tissues, and (*c*) increases the probability that antibodies can be used in ultrastructural investigations. Since fixation can generate epitopes not present on unfixed antigens, not all antibodies generated in this way will be useful for binding unfixed antigens, either on living cells or in immunoblots.

Interpretation of antibody binding patterns is almost always strengthened by identifying the molecules bound in each tissue by a particular monoclonal antibody. The most common procedures for doing this assay involve antibody binding to denatured antigens separated by size on polyacrylamide gels (e.g. Towbin et al 1979). To increase the probability of antibodies recognizing such antigens, denatured antigens have been injected in one step of some immunization protocols (Zipser & McKay 1981, McKay & Hockfield 1982, Flaster et al 1983).

Antibodies that cross species efficiently are naturally more widely applicable than those that bind only the species from which a particular antigen was derived. To increase the probability of obtaining such antibodies, neural material from more than one species has been used during the sequence of immunizations (e.g. Aceves-Pina et al 1982).

Another problem has been to identify monoclonal antibodies to minor epitopes in an antigen mixture. Monoclonal antibodies to the dominant antigens have been coupled to a resin and used to deplete an antigen mixture of previously identified antigens (Springer 1981). In another approach, cyclophosphamide, an alkylating agent known to kill antigen-stimulated lymphocytes preferentially, has been injected into previously immunized mice in an effort to kill lymphocytes specific for dominant antigenic determinants. Minor antigens have been recognized subsequently in surviving splenic lymphocytes that were challenged in vitro and fused to produce hybridomas (Matthew & Patterson 1983).

An often challenging problem has been to obtain a strong immune response to a minute amount of a neuronal antigen or to a functionally important and conserved part of such a molecule to which the mouse may be effectively tolerant. Microsequencing of natural proteins and chemical synthesis of milligram quantities of peptide sequences in these proteins has been one successful approach to overcoming such problems (Hunkapiller & Hood 1983, Sutcliffe et al 1983a,b). Immunization protocols in which lymphocytes are challenged with antigens in vitro provide another promising approach because they often require dramatically less antigen (picogram to nanogram) and can yield hybridoma lines derived from B lymphocytes that would be suppressed in vivo (e.g. Luben & Mohler 1980, Reading 1982, Pardue et al 1983). By including γ-interferon, it may prove possible in the future to obtain hybridoma clones in vivo and in vitro using even lower amounts of antigens (Nakamura et al 1984). In vitro protocols also have potential problems. The in vitro responses are often primitive compared to those in animals and a high proportion of the antibodies are IgM's. Too much antigen can easily result in isolation of hybridomas secreting low affinity antibodies. Prolonged in vitro culture of lymphocytes with repeated challenges with low doses of antigen has been reported to produce a higher frequency (80%) of IgG-allotype-secreting hybridoma lines (Pardue et al 1983).

Identification of Specific Hybridomas

The most common screening procedures for monoclonal antibodies use enzyme-linked immunosorbent assays (ELISA) in which antibody binding to antigens adsorbed to solid supports is measured. These procedures can be modified to detect fewer than 100,000 molecules of bound IgG, a very high level of sensitivity (Shalev et al 1980). Many ELISA assays, though, require as much as a microgram of bound antigen, because much of the antigen is not accessible to antibody (Kennel 1982). As a result, whereas modern immunization procedures have made it possible to obtain hybridoma-secreting antibodies to subnanogram quantities of an antigen, it may be impractical to assay for antibodies binding such antigens in standard ELISA assays. NGF, receptors, and ion-selective channels, present typically in nanomolar concentrations, are examples of molecules that are too rare in most crude preparations to be detected in routine ELISA assays.

One way of overcoming this problem has been to precede hybridoma selection with a partial or complete purification of the molecule of interest. Antibodies to Nerve Growth Factor, the acetylcholine receptor, beta-adrenergic receptor, sodium channel, and choline acetyltransferase have been isolated in this way (Zimmermann et al 1981, Tzartos & Lindstrom 1980, Strader et al 1983, Moore et al 1982, Crawford et al 1982). A second approach has been to use substrates that bind antigens more efficiently and in more accessible forms than PVC plastic. Application of neuronal antigen mixtures to nitrocellulose provides an assay that requires about tenfold less antigen than standard solid phase assays (Hawkes et al 1982). Assays using a beta-galactosidase-immunoglobin conjugate and a fluorogenic substrate have increased the sensitivity of assays for NGF and several other proteins approximately 100-fold (e.g. Korsching & Thoenen 1983a).

In many cases, solid phase immune assays have been replaced with a more specific assay. The most specific assays have involved tests of function or enzymatic activity. Antibodies to choline acetyltransferase and glial growth factor were identified by their abilities to precipitate each molecule (Crawford et al 1982, Brockes et al 1980). Monoclonal antibodies to the Na^+ channel, muscarinic acetylcholine receptor, and α_1-adrenergic receptor were identified by their abilities to precipitate [^{3}H]-saxitoxin, [^{3}H]-propylbenzilcholine mustard, and [^{3}H]-phenoxybenzamine, respectively, from partially purified preparations (Moore et al 1982, Ventner et al 1984). These screens are most successful when dealing with a purified antigen.

The most sensitive assays for cell-type specific monoclonal antibodies have utilized histological screens of sections of nervous tissues (e.g. Zipser & McKay 1981, Hockfield & McKay 1981). In adequately preserved tissues, antibodies that bind specific subsets of cells and specific organelles can be recognized in sections in the light microscope. The most successful screens in sections of mammalian nervous systems have used frozen or Vibratome sections of fixed tissues (McKay & Hockfield 1982). Although time consuming, these screens are the only means of recognizing antibodies directed against specific subsets of neurons. Discussions of immunocytochemical procedures can be found in several reviews and are beyond the scope of this article (Sternberger 1979, Jones & Hartman 1978, Vaughn et al 1981).

SUMMARY

The preceding discussion documents the diverse ways in which monoclonal antibodies have contributed to neuroscience research. They provide highly specific reagents to membrane-associated proteins, such as pumps, channels, receptors, and cell-adhesion molecules, that are useful for purifying these proteins, studying their structures at high resolution, and mapping their distributions. In many cases, the specific reagents were obtained using only partially purified antigens. Monoclonal antibodies to cytoskeletal proteins, organelles, and protein kinases have revealed that specific molecules are concentrated in anatomically distinct regions of the cell. A protein kinase has been shown to be a major postsynaptic constituent in many synapses. Individual proteins, such as actin, tubulin, and calmodulin appear to have different antigenic epitopes shielded in different parts of the cell.

Monoclonal antibodies have provided a diversity of cell-type-specific reagents in both vertebrate and invertebrate nervous systems. They seem likely to be useful in identifying functionally related subpopulations of neurons and describing neural cell lineages. They will also serve to identify molecules that are important in regulating cell migration in the cerebellum, in marking cell position in the retina, and directing axon growth.

This review also documents many purposes for which monoclonal antibodies are poorly suited or must be used with caution:

1. A monoclonal antibody to a protein does not always reveal every place where that molecule is located. Pre- or post-translational microheterogeneity can expose different epitopes on the protein, such as may occur on the Na^+-channel. Other proteins within the cell may shield antigenic sites on proteins such as calmodulin.
2. Monoclonal antibodies can bind to epitopes on unrelated molecules (Nigg et al 1982, Lane & Koprowski 1982). This is revealed in some cases as multiple bands on immunoblots. Some cross-reactivity, however, may have

a functional basis. For example, structural homology is clearly the basis for the antigenic epitopes that are shared among the five classes of intermediate filaments (Pruss et al 1981). The epitope that appears to be shared between the muscarinic and α_1-adrenergic receptors may be conserved because the two receptors modulate common effectors. The cross-reactivity between these receptors was only recognized because very specific and sensitive assays exist for each. It is quite possible that these same antibodies also bind sites on many other types of receptors. Mapping the distribution of this epitope may therefore have little relationship to the actual distribution of the muscarinic receptor. (Antibody binding to proteins in polyacrylamide gels is probably not sensitive enough to detect these additional receptors.) If two antibodies to different epitopes on the same protein show the same pattern of binding, it greatly strengthens the interpretation of histochemical data. Antibodies to different peptides in the same protein have been used for this type of analysis (Sutcliffe et al 1983a,b). With this same objective, antigens identified by monoclonal antibodies on polyacrylamide gels have been excised and used to generate additional monoclonal antibodies with the same binding specificity (Flaster et al 1983).

3. Many molecules in the nervous system can be anticipated to be detected by immunocytochemistry, but not by other currently available procedures, so it may not be possible to identify the antigens defined by many monoclonal antibodies. Many other molecules of importance, such as the Ca^{2+}-dependent K^+ channel and NGF, exist at such low concentrations that the best current immunocytochemical procedures do not have the specificity and sensitivity needed to discern their distributions in the nervous system.

Literature Cited

Abney, E. R., Williams, B. P., Raff, M. C. 1983. Tracing the development of oligodendrocytes from precursor cells using monoclonal antibodies, fluorescence-activated cell sorting and cell culture. *Dev. Biol.* 100:166–71

Aceves-Pina, E., Barbel, S., Evans, L., Jan, Y. N., Jan, L. Y. 1982. Immunocytochemical studies of neural development in *Drosophila. Soc. Neurosci. Abstr.* 8:15

Anderson, D. J., Blobel, G., Tzartos, S., Gullick, W., Lindstrom, J. 1983. Transmembrane orientation of an early biosynthetic form of acetylcholine receptor *delta* subunit determined by proteolytic dissection in conjunction with monoclonal antibodies. *J. Neurosci.* 3:1773–84

Anderson, M. J., Fambrough, D. M. 1983. Aggregates of acetylcholine receptors are associated with plaques of a basal lamina heparan sulfate proteoglycan on the surface of skeletal muscle fibers. *J. Cell Biol.* 97:1396–1411

Barald, K. F. 1982. Monoclonal antibodies to embryonic neurons, cell-specific markers for chick ciliary ganglion. In *Neural Development*, ed. N. C. Spitzer, pp. 101–9. New York: Plenum

Barnstable, C. J. 1980. Monoclonal antibodies which recognize different cell types in the rat retina. *Nature* 286:231–35

Barnstable, C. J., Akagawa, K., Hofstein, R., Horn, J. P. 1983. Monoclonal antibodies that label discrete cell types in the mammalian nervous system. *Cold Spring Harbor Symp. Quant. Biol.* 48:863–76

Bartlett, P. F., Nobel, M. D., Pruss, R. M., Raff, M. C., Rattray, S., Williams, C. A. 1981. Rat neural antigen-2 (RAN-2): A cell surface antigen on astrocytes, ependymal cells, Muller cells and leptomeninges defined by a monoclonal antibody. *Brain Res.* 204:339–51

Basch, R. S., Berman, J. W., Lakow, E. 1983. Cell separation using positive immunoselection. *J. Immunol. Methods* 56:269–80

Berg, G. J., Schachner, M. 1982. Electron microscopic localization of A2B5 cell surface antigen in monolayer cultures of murine cerebellum and retina. *Cell Tissue Res.* 224:637–45

Bignami, A., Dahl, D. 1977. Specificity of the glial fibrillary acidic protein for astroglia. *J. Histochem. Cytochem.* 25:466–99

Bloch, R. J., Hall, Z. W. 1983. Cytoskeletal components of the vertebrate neuromuscular junction: Vinculin, *alpha*-actinin and filamin. *J. Cell Biol.* 97:217–23

Block, T., Bothwell, M. 1983. Use of iron- or selenium-coupled monoclonal antibodies to cell-surface antigens as a positive selection system for cells. *Nature* 301:342–44

Bloom, G. S., Schoenfeld, T. A., Vallee, R. B. 1984. Widespread distribution of the major polypeptide component of MAP1 (Microtubule-associated protein) in the nervous system. *J. Cell Biol.* 98:320–30

Brackenbury, R., Rutishauser, U., Edelman, G. 1981. Distinct calcium independent and calcium-dependent adhesion systems of chicken embryo cells. *Proc. Natl. Acad. Sci. USA* 78:387–91

Brockes, J. P., Lemke, G., Balzer, D. R. Jr. 1980. Purification and preliminary characterization of a glial growth factor from the bovine pituitary. *J. Biol. Chem.* 255:8374–77

Brockes, J. P., Fields, K. L., Raff, M. C. 1977. A surface antigenic marker for rat Schwann cells. *Nature* 266:364–66

Brockes, J. P., Fields, K. L., Raff, M. C. 1979. Studies on cultured rat Schwann cells. I. Establishment of purified populations from cultures of peripheral nerve. *Brain Res.* 165:105–18

Brockes, J. P., Fryxell, K., Lemke, G. E. 1981. Studies on cultured Schwann cells—The induction of myelin synthesis and the control of their proliferation by a new growth factor. In *Glial-Neurone Interactions,* ed. J. E. Treherne. London/New York: Cambridge Univ. Press

Buckley, K. M., Schweitzer, E. S., Miljanich, G. P., Clift-O'Grady, L., Kushner, P. D. et al. 1983. A synaptic vesicle antigen is restricted to the junctional region of the presynaptic plasma membrane. *Proc. Natl. Acad. Sci. USA* 80:7342–46

Burden, S. J. 1982. Identification of an intracellular postsynaptic antigen at the frog neuromuscular junction. *J. Cell Biol.* 94:521–30

Burden, S. J., DePalma, R. L., Gottesman, G. S. 1983. Crosslinking of proteins in acetylcholine receptor-rich membranes: Association between the beta-subunit and the 43 kd subsynaptic protein. *Cell* 35:687–92

Buskirk, D. R., Thiery, J.-P., Rutishauser, U., Edelman, G. M. 1980. Antibodies to a neural cell adhesion molecule disrupt histogenesis in cultured chick retinae. *Nature* 285:488–89

Caceres, A., Binder, L. I., Payne, M. R., Bender, P., Rebhun, L., Steward, O. 1984. Differential subcellular localization of tubulin and the microtubule associated protein MAP_2 in brain tissue as revealed by immunocytochemistry. *J. Neurosci.* 4:394–410

Caceres, A., Payne, M. R., Binder, L. I., Steward, O. 1983. Immunocytochemical localization of actin and microtubule-associated protein MAP2 in dendritic spines. *Proc. Natl. Acad. Sci. USA* 80:1738–42

Carlin, R. C., Bartelt, D. C., Siekevitz, P. 1983. Identification of fodrin as a major calmodulin binding protein in post synaptic density preparations. *J. Cell Biol.* 96:443–81

Chan, S. Y., Hess, E. J., Rahaminoff, H., Goldin, S. M. 1984. Purification and immunological characterization of a calcium pump from bovine brain synaptosomal vesicles. *J. Neurosci.* 4:1468–78

Chandler, C. E., Parsons, L. M., Hosang, M., Shooter, E. M. 1984. A monoclonal antibody modulates the interaction of nerve growth factor with PC12 cells. *J. Biol. Chem.* 259:6882–89

Chiu, A. Y., Sanes, J. R. 1982. Differentiation of basal lamina is an early event in the development of the neuromuscular junction *in vivo*. *Soc. Neurosci. Abstr.* 8:128

Chun, J. J. M., Schatz, C. J. 1983. Immunohistochemical localization of synaptic vesicle antigens in developing cat cortex. *Soc. Neurosci. Abstr.* 9:692

Chuong, C.-M., McClain, D. A., Streit, P., Edelman, G. M. 1982. Neural cell adhesion molecules in rodent brain isolated by monoclonal antibodies with cross-species reactivity. *Proc. Natl. Acad. Sci. USA* 79:4234–38

Ciment, G., Weston, J. A. 1982. Early appearance in neural crest and crest-derived cells of an antigenic determinant present in avian neurons. *Dev. Biol.* 93:355–67

Ciment, G., Weston, J. A. 1983. Enteric neurogenesis by neural crest-derived branchial arch mesenchymal cells. *Nature* 305:424–27

Cohen, J., Selvendran, S. Y. 1981. A neuronal surface antigen is found in the CNS but not in peripheral neurons. *Nature* 291:421–23

Colwell, D. E., Gollahon, K. A., McGhee, J. R., Michalek, S. M. 1982. IgA hybridomas: A method for generation in high numbers. *J. Immunol. Methods* 54:259–60

Conti-Troncini, B., Raftery, M. 1982. The nicotinic cholinergic receptor. *Ann. Rev. Biochem.* 51:491–50

Conti-Troncini, B., Tzartos, S., Lindstrom, J. 1981. Monoclonal antibodies as probes of acetylcholine receptor structure. 2. Binding to native receptor. *Biochemistry* 20:2181–91

Crawford, G. D., Correa, L., Salvaterra, P. M. 1982. Interaction of monoclonal antibodies with mammalian choline acetyltransferase. *Proc. Natl. Acad. Sci. USA* 79:7031–35

Cuello, A. C., Priestley, J. V., Milstein, C. 1982. Immunocytochemistry with internally labeled monoclonal antibodies. *Proc. Natl. Acad. Sci. USA* 79:665–69

Cummings, R., Burgoyne, R. D., Lytton, N. A. 1984. Immunocytochemical demonstration of *alpha*-tubulin modification during axonal maturation in the cerebellar cortex. *J. Cell Biol.* 98:347–51

Dangl, J. L., Herzenberg, L. A. 1982. Selection of hybridomas and hybridoma variants using the fluorescence activated cell sorter. *J. Immunol. Methods* 52:1–14

Debus, E., Flugge, G., Weber, K., Osborn, M. 1982. A monoclonal antibody specific for the 200K polypeptide of the neurofilament triplet. *EMBO J.* 1:41–45

Dodd, J., Solter, D., Jessel, T. M. 1984. Monoclonal antibodies against carbohydrate differentiation antigens identify subsets of primary sensory neurones. *Nature* 311: 469–72

Eckenstein, F., Thoenen, H. T. 1982. Production of specific antisera and monoclonal antibodies to choline acetyltransferase: Characterization and use for identification of cholinergic neurons. *EMBO J.* 1:363–68

Edelman, G. M. 1983. Cell adhesion molecules. *Science* 219:450–57

Edelman, G. M., Chuong, C.-M. 1982. Embryonic to adult conversion of neural cell adhesion molecules in normal and staggerer mice. *Proc. Natl. Acad. Sci. USA* 79:7036–40

Edelman, G. M., Gallin, W. J., Delouvee, A., Cunningham, B. A., Thiery, J.-P. 1983. Early epochal maps of two different cell adhesion molecules. *Proc. Natl. Acad. Sci. USA* 80: 4384–88

Eisenbarth, G. S., Walsh, F. S., Nirenberg, M. 1979. Monoclonal antibody to an plasma membrane antigen of neurons. *Proc. Natl. Acad. Sci. USA* 76:4913–17

Ellisman, M. H., Levinson, S. R. 1982. Immunocytochemical localization of sodium channel distribution in the excitable membranes of *Electrophorus electricus*. *Proc. Natl. Acad. Sci. USA* 79:6707–11

Fambrough, D. M. 1983. Studies of the Na^+-K^+ ATPase of skeletal muscle and nerve. *Cold Spring Harbor Symp. Quant. Biol.* 48:297–304

Fambrough, D. M., Bayne, E. K., Gardner, J. M., Anderson, M. J., Wakshull, E., Rotundo, R. 1982a. Monoclonal antibodies to skeletal muscle cell surface. *Neuroimmunology*, ed. J. Brockes, pp. 49–89. New York: Plenum

Fambrough, D. M., Engel, A. G., Rosenberg, T. L. 1982b. Acetylcholinesterase of human erythrocytes and neuromuscular junctions: Homologies revealed by monoclonal antibodies. *Proc. Natl. Acad. Sci. USA* 79:1078–82

Fambrough, D. M., Bayne, E. K. 1983. Multiple forms of $(Na^+ + K^+)$-ATPase in chicken. Selective detection of the major nerve, skeletal muscle, and kidney form by a monoclonal antibody. *J. Biol. Chem.* 258:1926–35

Fazekas de St. Groth, S., Sheidegger, D. 1980. Production of monoclonal antibodies: Strategy and tactics. *J. Immunol. Methods* 35:1–21

Fertuck, H. C., Salpeter, M. M. 1976. Quantitation of junctional and extrajunctional acetylcholine receptors by electron microscope autoradiography after ^{125}-*alpha*-bungarotoxin binding at mouse neuromuscular junctions. *J. Cell Biol.* 69:114–58

Finer-Moore, J., Stroud, R. M. 1984. Amphipathic analysis and possible formation of the ion channel in acetylcholine receptor. *Proc. Natl. Acad. Sci. USA* 81:155–59

Flaster, M. S., Schley, C., Zipser, B. 1983. Generating monoclonal antibodies against excised gel bands to correlate immunocytochemical and biochemical data. *Brain Res.* 277:196–99

Fritz, L. C., Brockes, J. P. 1983. Immunocytochemical properties and cytochemical localization of the voltage-sensitive sodium channel from the electroplax of the eel (*Electrophorus electricus*). *J. Neurosci.* 3:2300–9

Froehner, S. C., Gulbrandsen, V., Hyman, C., Jeng, A. Y., Neubig, R. R., Cohen, J. B. 1981. Immunofluorescence localization at the mammalian neuromuscular junction of the M_r 43,000 protein of *Torpedo* postsynaptic membranes. *Proc. Natl. Acad. Sci. USA* 78:5230–34

Fujita, S. C., Zipursky, S. L., Benzer, S., Ferrus, A., Shotwell, S. L. 1982. Monoclonal antibodies against the *Drosophila* nervous system. *Proc. Natl. Acad. Sci. USA* 79:7929–33

Gallin, W. J., Edelman, G. M., Cunningham, B. A. 1983. Characterization of *L*-CAM, a major cell adhesion molecule from embryonic liver cells. *Proc. Natl. Acad. Sci. USA* 80:1038–42

Ghandour, M. S., Langley, O. K., Labourdette, G., Vincedon, B., Gombos, G. 1981. Specific and artifactual cellular localizations of S100 protein: An astrocyte marker in rat cerebellum. *Dev. Neurosci.* 4:68–78

Goelz, S. E., Nestler, E. J., Chehrazi, B., Greengard, P. 1981. Distribution of Protein I in mammalian brain as determined by a detergent-based radioimmune assay. *Proc.*

Natl. Acad. Sci. USA 78:2130–34

Goldin, S. M., Chan, S. Y., Papazian, D. M., Hess, E. J., Rahamimoff, H. 1983. Purification and characterization of ATP-dependent calcium pumps from synaptosomes. *Cold Spring Harbor Symp. Quant. Biol.* 48:287–96

Goldstein, M. E., Sternberger, L. A., Sternberger, N. H. 1983. Microheterogeneity ("Neurotypy") of neurofilament proteins. *Proc. Natl. Acad. Sci. USA* 80:3101–5

Gomez, C. M., Richman, D. P., Berman, P. W., Burres, S. A., Arnason, B. G. A., Fitch, F. W. 1979. Monoclonal antibodies against purified nicotinic acetylcholine receptor. *Biochem. Biophys. Res. Commun.* 88:575–82

Gordon, A. S., Milfay, D., Diamond, I. 1983. Identification of a molecular weight 43,000 protein kinase in acetylcholine receptor-enriched membranes. *Proc. Natl. Acad. Sci. USA* 80:5862–65

Gorin, P. D., Johnson, E M. Jr. 1979. Experimental autoimmune model of nerve growth factor deprivation: Effect on developing peripheral sympathetic and sensory neurons. *Proc. Natl. Acad. Sci. USA* 76:5382–86

Grumet, M., Hoffman, S., Edelman, G. M. 1984. Two antigenically related neuronal cell adhesion molecules of different specificities mediate neuron-neuron and neuron-glia adhesion. *Proc. Natl. Acad. Sci. USA* 81:267–71

Gullick, W. J., Lindstrom, J. M. 1983. Mapping the binding of monoclonal antibodies to the acetylcholine receptor from *Torpedo californica*. *Biochemistry* 22:3312–20

Gullick, W. J., Tzartos, S., Lindstrom, J. 1981. Monoclonal antibodies as probes of acetylcholine receptor structure. I. Peptide mapping. *Biochemistry* 20:2173–80

Haan, E. A., Boss, B. D., Cowan, W. M. 1982. Production and characterization of monoclonal antibodies against the "brain-specific" proteins 14-3-2 and S-100. *Proc. Natl. Acad. Sci. USA* 79:7585–89

Harper, G. P., Barde, Y.-A., Edgar, D., Ganten, D., Hefti, F., et al. 1983. Biological and immunological properties of the nerve growth factor from bovine seminal plasma: Comparison with the properties of mouse nerve growth factor. *Neuroscience* 8:375–87

Hatten, M. E., Furie, M. B., Rifkind, D. B. 1982. Binding of developing mouse cerebellar cells to fibronectin: a possible mechanism for the formation of the external granular layer. *J. Neurosci.* 2:1195–1206

Hawkes, R., Niday, E., Gordon, J. 1982. A dot-immunobinding assay for monoclonal and other antibodies. *Analyt. Biochem.* 119:142–47

Hawkes, R., Niday, E., Matus, A. 1982a. MIT-23: A mitochondrial marker for terminal neuronal differentiation defined by a monoclonal antibody. *Cell* 28:253–58

Hendry, S. H. C., Hockfield, S., Jones, E. G., McKay, R. 1984. Monoclonal antibody that identifies subsets of neurones in the central visual system of monkey and cat. *Nature* 307:267–69

Henke-Fahle, S., Bonhoeffer, F. 1983. Inhibition of axonal growth by a monoclonal antibody. *Nature* 303:65–67

Hirn, M., Pierres, M., Deagostini-Bazin, H., Hirsch, M., Goridis, C. 1981. Monoclonal antibody against cell surface glycoprotein of neurons. *Brain Res.* 214:433–39

Hirokawa, N., Glicksman, M. A., Willard, M. B. 1984. Organization of mammalian neurofilament polypeptides within the neuronal cytoskeleton. *J. Cell Biol.* 98:1523–36

Ho, R. K., Goodman, C. S. 1982. Peripheral pathways are pioneered by an array of central and peripheral neurons in grasshopper embryos. *Nature* 297:404–6

Ho, R. K., Ball, E. E., Goodman, C. S. 1983. Muscle pioneers: Large mesodermal cells that erect a scaffold for developing muscles and motoneurones in grasshopper embryos. *Nature* 301:66–69

Hockfield, S., McKay, R. 1983a. A surface antigen expressed by a subset of neurons in the vertebrate central nervous system. *Proc. Natl. Acad. Sci. USA* 80:5758–61

Hockfield, S., McKay, R. 1983b. Monoclonal antibodies demonstrate the organization of axons in the leech. *J. Neurosci.* 3:369–75

Hoffman, S., Sorkin, B. C., White, P. C., Brackenbury, R., Mailhammer, R., Rutishauser, U., Cunningham, B. A., Edelman, G. M. 1982. Chemical characterization of a neural cell adhesion molecule purified from embryonic brain membranes. *J. Biol. Chem.* 257:7720–29

Hoffman, S., Edelman, G. M. 1983. Kinetics of homophilic binding by embryonic and adult forms of the neural cell adhesion molecule. *Proc. Natl. Acad. Sci. USA* 80:5762–66

Hogg, N., Flaster, M., Zipser, B. 1983. Cross reactivities of monoclonal antibodies between select leech neuronal and epithelial tissues. *J. Neurosci. Res.* 9:445–57

Hökfelt, T., Johansson, O., Ljungdahl, A., Lundberg, J. M., Schultzberg, M. 1980. Peptidergic neurons. *Nature* 284:515–21

Holton, B., Weston, J. A. 1982. Analysis of glial cell differentiation in peripheral nervous tissue. II. Neurons promote S100 synthesis by purified glial precursor cell populations, *Dev. Biol.* 89:72–81

Huber, G., Matus, A. 1984. Differences in the cellular distributions of two microtubule associated proteins, MAP_1 and MAP_2, in rat brain. *J. Neurosci.* 4:151–60

Hunkapiller, M. W., Hood, L. E. 1983. Protein sequence analysis: Automated microsequencing. *Science* 219:650–59

Hurrell, J. G. R. 1982. *Monoclonal Hybridoma Antibodies: Techniques and Applications*. Boca Raton, Florida: CRC Press

Iversen, L. L., 1983. Neuropeptides—What next? *Trends Neurosci.* 6:293–94

Jones, E. G., Hartman, B. K. 1978. Recent advances in neuroanatomical methodology. *Ann. Rev. Neurosci.* 1:215–96

Jones, R. T., Walker, J. H., Stadler, H., Whittaker, V. P. 1982. Further evidence that glycosaminoglycan specific to cholinergic vesicles recycles during electrical stimulation of the electric organ of *Torpedo marmarata*. *Cell Tissue Res.* 224:685–88

Kelly, P. T., McGuinness, T. L., Greengard, P. 1984. Evidence that the major postsynaptic density protein is a component of a Ca^{+} calmodulin-dependent protein kinase. *Proc. Natl. Acad. Sci. USA* 81:945–49

Kennedy, M. B., Bennett, M. K., Erondu, N. E. 1983. Biochemical and immunochemical evidence that the "major postsynaptic density protein" is a subunit of a calmodulin-dependent protein kinase. *Proc. Natl. Acad. Sci. USA* 80:7357–61

Kennel, S. J. 1982. Binding of monoclonal antibody to protein antigen in fluid phase or bound to solid supports. *J. Immunol. Methods* 55:1–12

Kennett, R. H., McKearn, T. J., Bechtol, K. 1980. *Monoclonal Antibodies*. New York: Plenum

Korsching, S., Thoenen, H. 1983a. NGF in sympathetic ganglia and corresponding target organs of the rat: Correlation with density of sympathetic innervation. *Proc. Natl. Acad. Sci. USA* 80:3513–16

Korsching, S., Thoenen, H. 1983b. Quantitative demonstration of the retrograde axonal transport of endogenous NGF. *Neurosci. Lett.* 39:1–4

Kotrla, K., Goodman, C. S. 1982. Positional expression of a surface antigen on cells in the neuroepithelium and appendages of grasshopper embryos. *Soc. Neurosci. Abstr.* 8:256

Kristofferson, D., Desmeules, P., Lindstrom, J., Stroud, R. M. 1984. Acetylcholine receptor *alpha* subunits located using monoclonal antibody F_{ab} fragments. (In preparation)

Lagenaur, C., Sommer, I., Schachner, M. 1980. Subclass of astroglia in mouse cerebella recognized by monoclonal antibody. *Dev. Biol.* 79:367–78

Lake, P., Clark, E. A., Khorshidi, M., Sunshine, G. H. 1979. Production and characterization of cytotoxic Thy-1 antibody-secreting hybrid cell lines. Detection of T cell subsets. *Eur. J. Immunol.* 9:875–86

Lane, D., Koprowski, H. 1982. Molecular recognition and the future of monoclonal antibodies. *Nature* 296:200–2

Langley, O. K., Ghandour, M. S.. 1981. An immunocytochemical investigation of non-neuronal enolase in cerebellum: A new astrocyte marker. *Histochem. J.* 13:137–48

Langley, O. K., Ghandour, M. S., Vincendon, G., Gombos, G., Warecka, K. 1982. Immunoelectron microscopy of *alpha*$_2$-glycoprotein: An astrocyte-specific antigen. *J. Neuroimmunol.* 2:131–43

Lazarides, E., Nelson, W. J. 1983a. Erythrocyte and brain forms of spectrin in cerebellum: Distinct membrane cytoskeletal domains. *Science* 200:1295–96

Lazarides, E., Nelson, W. J. 1983b. Erythrocyte form of spectrin in cerebellum: Appearance at a specific stage in the terminal differentiation of neurons. *Science* 222:931–33

Lazarides, E., Nelson, W. J., Kasamatsu, T. 1984. Segregation of two spectrin forms in the chicken optic system: A mechanism for establishing restricted membrane-cytoskeletal domains in neurons. *Cell* 36:269–78

LeDouarin, N. 1982. *The Neural Crest*. Cambridge: Cambridge Univ. Press

Le Douarin, N. M., Smith, J., LeLievre, C. S. 1981. From the neural crest to the ganglia of the peripheral nervous system. *Ann. Rev. Physiol.* 43:653–71

Lee, V., Wu, H. L., Schlaepfer, W. W. 1982. Monoclonal antibodies recognize individual neurofilament triplet proteins. *Proc. Natl. Acad. Sci. USA* 79:6089–92

Levey, A. I., Armstrong, D. M., Atweh, S. F., Terry, R. D., Wainer, B. H. 1983. Monoclonal antibodies to choline acetyltransferase: Production, specificity, and immunohistochemistry. *J. Neurosci.* 3:1–9

Levi-Montalcini, R. M., Booker, B. 1960. Destruction of the sympathetic ganglia in mammals by an antiserum to nerve growth factor. *Proc. Natl. Acad. Sci. USA* 46:384–91

Lindner, J., Rathjen, F., Schachner, M. 1983. L1 mono- and polyclonal antibodies modify cell migration in early postnatal mouse cerebellum. *Nature* 305:427–30

Luben, R. A., Mohler, M. A. 1980. *In vitro* immunization as an adjunct to the production of hybridomas producing antibodies against the lymyphokine osteoclast activating factor. *Mol. Immunol.* 17:635–39

Ludwin, S. K., K sek, J. C., Eng, L. F. 1976. The topographical distribution of S-100 and GFA protein in the adult rat brain: An immunohistochemical study using horseradish peroxidase-labelled antibodies. *J. Comp. Neurol.* 165:197–208

McGuinness, T. L., Kelly, P. T., Ouimet, C. C., Greengard, P. 1983. Studies on the subcellular and regional distribution of calmodulin-dependent protein kinase II in rat brain. *Soc. Neurosci. Abstr.* 9:1029

McKay, R. D. G., Hockfield, S. J. 1982. Monoclonal antibodies distinguish antigenically discrete neuronal types in the vertebrate central nervous system. *Proc. Natl. Acad. Sci. USA* 79:6747–51

Marangos, P. J., Polak, J. M., Pearse, A. G. E. 1982. Neuron-specific enolase. A probe for neurons and neuroendocrine cells. *Trends Neurosci.* 5:193–96

Massoulie, J., Bon, S. 1982. The molecular forms of cholinesterase and acetylcholinesterase in vertebrates. *Ann. Rev. Neurosci.* 5:57–106

Matthew, W. D., Patterson, P. H. 1983. The production of a monoclonal antibody which blocks the action of a neurite outgrowth promoting factor. *Cold Spring Harbor Symp. Quant. Biol.* 48:625–31

Matthew, W. D., Reichardt, L. F. 1982. Monoclonal antibodies used to study the interaction of nerve and muscle cell lines. *Prog. Brain Res.* 58:375–81

Matthew, W. D., Tsavaler, L., Reichardt, L. F. 1981. Identification of a synaptic vesicle-specific membrane protein with a wide distribution in neuronal and neurosecretory tissue. *J. Cell Biol.* 91:257–69

Matus, A., 1981. The postsynaptic density. *Trends Neurosci.* 4:51–53

Matus, A., Bernhardt, R., Hugh-Jones, T. 1981. High molecular weight microtubule-associated proteins are preferentially associated with dendritic microtubules in brain. *Proc. Natl. Acad. Sci. USA* 78:3010–14

Meier, D. H., Lagenauer, C., Schachner, M. 1982. Immunoselection of oligodendrocytes by magnetic beads. I. Determination of antibody coupling parameters and cell binding conditions. *J. Neurosci. Res.* 7:135–45

Miller, C. A., Benzer, S. 1983. Monoclonal antibody cross-reactions between *Drosophila* and human brain. *Proc. Natl. Acad. Sci. USA* 80:7641–45

Miller, P., Walter, U., Theurkauf, W. E., Vallee, R. B., DeCamilli, P. 1982. Frozen tissue sections as an experimental system to reveal specific binding sites for regulatory subunit of type II cAMP-dependent protein kinase in neurons. *Proc. Natl. Acad. Sci. USA* 79:5562–66

Milstein, C., Cuello, A. C. 1983. Hybrid hybridomas and their use in immunohistochemistry. *Nature* 305:537–40

Mirsky, R. 1982. The use of antibodies to define and study major cell types in the central and peripheral nervous system. In *Neuroimmunology*, ed. J. Brockes, pp. 141–82. New York: Plenum

Mirsky, R., Wendon, L. M. B., Black, P., Stolkin, C., Bray, D. 1978. Tetanus toxin: A cell surface marker for neurones in culture. *Brain Res.* 148:251–59

Mirsky, R., Winter, J., Abney, E. R., Pruss, R. M., Gavrilovic, J., Raff, M. C. 1980. Myelin-specific proteins and glycolipids in rat Schwann cells and oligodendrocytes in culture. *J. Cell Biol.* 84:483–94

Mochley-Rosen, D., Fuchs, S., Eshhar, Z. 1979. Monoclonal antibodies against defined determinants of acetylcholine receptor. *FEBS Lett.* 106:389–92

Moore, H.-P. H., Fritz, L. C., Raftery, M. A., Brockes, J. P. 1982. Isolation and characterization of a monoclonal antibody against the saxitoxin-binding component from the electric organ of the eel *Electrophorus electricus*. *Proc. Natl. Acad. Sci. USA* 79:1673–77

Nakamura, M., Manser, T., Pearson, G. D. N., Paley, M. J., Gefter, M. L. 1984. Effect of IFN-gamma on the immune response *in vivo* and on gene expression *in vitro*. *Nature* 306:381–82

Nakayama, H., Withy, R. M., Raftery, M. A. 1982. Use of a monoclonal antibody to purify the tetrodotoxin binding component from the electroplax of *Electrophorus electricus*. *Proc. Natl. Acad. Sci. USA* 79:7575–79

Nestler, E. J., Greengard, P. 1983. Unpublished results

Nieto-Sampedro, M., Bussineau, C. M., Cotman, C. W. 1981. Postsynaptic density antigens: Preparation and characterization of an antiserum against postsynaptic densities. *J. Cell Biol.* 90:675–86

Nigg, E. A., Walter, G., Singer, S. J. 1982. On the nature of crossreactions observed with antibodies directed to defined epitopes. *Proc. Natl. Acad. Sci. USA* 79:5939–43

Nitkin, R. M., Wallace, B. G., Spira, M. E. Godfrey, E. W., McMahan, U. J. 1983. Molecular components of the synaptic basal lamina that direct differentiation of regenerating neuromuscular junctions. *Cold Spring Harbor Symp. Quant. Biol.* 48:653–65

Noda, M., Takahashi, H., Tanabe, T., Toyosato, M., Kikyotani, S., et al. 1983. Structural homology of *Torpedo californica* acetylcholine receptor subunits. *Nature* 302:528–32

Norenberg, M. D., Martinez-Hernandez, A. 1979. Fine structural localization of glutamine synthetase in astrocytes of rat brain. *Brain Res.* 161:303–10

Pardue, R. L., Brady, R. C., Perry, G. W., Dedman, J. R. 1983. Production of monoclonal antibodies against calmodulin by *in vitro* immunization of spleen cells. *J. Cell Biol.* 96:1149–54

Pruss, R. 1979. Thy-1 antigen on astrocytes in long term cultures of rat central nervous system. *Nature* 280:688–90

Pruss, R. M., Mirsky, R., Raff, M. C., Thorpe, R., Dowding, A. J., Anderton, B. H. 1981.

All classes of intermediate filaments share a common antigenic determinant defined by a monoclonal antibody. *Cell* 27:419–28

Raff, M. C., Abney, E. R., Cohen, J., Lindsay, R., Noble, M. 1983. Two types of astrocytes in cultures of developing rat white matter: Differences in morphology, surface gangliosides and growth characteristics. *J. Neurosci.* 3:1289–1300

Raff, M. C., Fields, K. L., Hakamori, S., Mirsky, R., Pruss, R. M., Winter, J. 1979. Cell-type-specific markers for distinguishing and studying neurons and the major classes of glial cells in culture. *Brain Res.* 174:283–308

Raff, M. C., Miller, R. H. 1983. A glial progenitor cell that develops *in vitro* into an astrocyte or an oligodendrocyte depending on the culture medium. *Nature* 303:390–96

Ranscht, B., Clapshaw, P. A., Price, J., Noble, M., Seifert, W. 1982. Development of oligodendrocytes and Schwann cells studied with a monoclonal antibody against galactocerebroside. *Proc. Natl. Acad. Sci. USA* 79:2709–13

Raper, J. A., Bastiani, M., Goodman, C. S. 1983a. Pathfinding by neuronal growth cones in grasshopper embryos. I. Divergent choices made by the growth cones of sibling neurons. *J. Neurosci.* 3:20–30

Raper, J. A., Bastiani, M., Goodman, C. S. 1983b. Pathfinding by neuronal growth cones in grasshopper embryos. II. Selective fasciculation onto specific axonal pathways. *J. Neurosci.* 3:31–41

Rathjen, F. G., Schachner, M. 1984. Monoclonal antibody L1 recognizes neuronal cell surface glycoproteins mediating cellular adhesion. In *Neuroimmunology*, ed. P. O. Behan, F. Spreafico, pp. 79–88. New York: Raven

Reading, C. L. 1982. Theory and methods for immunization in culture and monoclonal antibody production. *J. Immunol. Methods* 53:261–91

Reichardt, L. F., Kelly, R. B. 1983. A molecular description of nerve terminal function. *Ann. Rev. Biochem.* 52:871–926

Richardson, P. J., Walker, J. H., Jones, R. T., Whittaker, V. P. 1982. Identification of a cholinergic-specific antigen Chol-1 as a ganglioside. *J. Neurochem.* 38:1605–14

Richert, N. D., Willingham, M. C., Pastan, I. 1983. Epidermal growth factor characterization of a monoclonal antibody specific for the receptor of A431 cells. *J. Biol. Chem.* 258:8902–7

Ross, M. E., Reis, D. J., Joh, T. H. 1981a. Monoclonal antibodies to tyrosine hydroxylase: Production and characterization. *Brain Res.* 208:493–98

Ross, M. E., Baetze, E. E., Rees, D. J., Joh, T. J. 1981b. Monoclonal antibodies to catecholamine-neurotransmitter-synthesizing enzymes can be used for immunocytochemistry and immunohistochemistry. In *Monoclonal Antibodies to Neural Antigens*, ed. R. McKay, M. C. Raff, L. F. Reichardt, pp. 101–108. Cold Spring Harbor, NY: Cold Spring Harbor Lab.

Rothbard, J. B., Brackenbury, R., Cunningham, B. A., Edelman, G. M. 1982. Differences in the carbohydrate structures of neural cell-adhesion molecules from adult and embryonic chicken brains. *J. Biol. Chem.* 257:11064–69

Rovasio, R. A., Delouvee, A., Yamada, K. M., Timpl, R., Thiery, J.-P. 1983. Neural crest migration: Requirements for exogenous fibronectin and high cell density. *J. Cell Biol.* 96:462–73

Rutishauser, U., Gall, W. E., Edelman, G. M. 1978. Adhesion among neural cells of the chick embryo. IV. Role of the cell surface molecule CAM in the formation of neurite bundles in cultures of spinal ganglia. *J. Cell Biol.* 79:382–93

Rutishauser, U., Hoffman, S., Edelman, G. M. 1982. Binding properties of a cell adhesion molecule from neural tissue. *Proc. Natl. Acad. Sci. USA* 79:685–89

Salton, S. R. J., Richter-Landsberg, C., Greene, L. A., Shelanski, M. J. 1983. Nerve growth factor-inducible large external (NILE) glycoprotein: Studies of a central and peripheral neuronal marker. *J. Neurosci.* 3:441–54

Sanes, J. R. 1982. Laminin, fibronectin and collagen in synaptic and extrasynaptic portions of muscle fiber basement membrane. *J. Cell Biol.* 93:442–51

Sanes, J. R., Hall, Z. W. 1979. Antibodies that bind specifically to synaptic sites on muscle fiber basal lamina. *J. Cell Biol.* 83:357–70

Sealock, R. 1982. Visualization at the mouse neuromuscular junction of a submembrane structure in common with *Torpedo* postsynaptic membranes. *J. Neurosci.* 2:418–23

Sealock, R., Wray, B. E., Froehner, S. S. 1984. Ultrastructural localization of the M_r 43,000 protein and the acetylcholine receptor in *Torpedo* postsynaptic membranes using monoclonal antibodies. *J. Cell Biol.* 98:2239–44

Seguela, P., Gefford, M., Buijs, R. M., LeMoal, M. G. 1984. Antibodies against *gamma*-aminobutyric acid. Specificity studies and immunocytochemical results. *Proc. Natl. Acad. Sci. USA* 81:3888–92

Shalev, A., Greenberg, A. H., McAlpine, P. J. 1980. Detection of attograms of antigen by a high sensitivity enzyme-linked immunoadsorbent assay (HS-ELISA) using a fluorogenic substrate. *J. Immunol. Methods* 38:125–39

Sharp, G. A., Shaw, G., Weber, K. 1982.

Immunoelectronmicroscopical localization of the three neurofilament triplet proteins along neurofilaments of cultured dorsal root ganglion neurons. *Exp. Cell Res.* 137:403–13

Shaw, G., Osborne, M., Weber, K. 1981a. Arrangement of neurofilaments, microtubules and microfilament-associated proteins in cultured dorsal root ganglia cells. *Eur. J. Cell Biol.* 24:20–27

Shaw, G., Osborne, M., Weber, K. 1981b. An immunofluorescence microscopical study of the neurofilament triplet proteins, vimentin and glial fibrillary acidic protein within the adult rat brain. *Eur. J. Cell Biol.* 26:68–82

Solter, D., Knowles, B. B. 1978. Monoclonal antibody defines a stage-specific mouse embryonic antigen (SSEA-1). *Proc. Natl. Acad. Sci. USA* 75:5565–69

Sommer, I., Lagenaur, C., Schachner, M. 1981. Recognition of Bergmann glial and ependymal cells in the mouse nervous system by monoclonal antibody. *J. Cell Biol.* 90:448–58

Sommer, I., Schachner, M. 1981. Monoclonal antibodies (01 to 04) to oligodendrocyte cell surfaces: An immunocytological study in the central nervous system. *Dev. Biol.* 83:311–27

Sperry, R. W. 1963. Chemoaffinity in the orderly growth of nerve fiber patterns and connections. *Proc. Natl. Acad. Sci. USA* 50:703–10

Springer, T. A. 1981. Monoclonal antibody analysis of complex biological systems: Combination of cell hybridization and immunoadsorbents in a novel cascade procedure and its application to the macrophage cell surface. *J. Biol. Chem.* 256:3833–39

Stahli, C., Staehelin, T., Miggiano, V., Schmidt, J., Haring, P. 1980. High frequencies of antigen specific hybridomas: Dependence on immunization parameters and prediction by spleen cell analysis. *J. Immunol. Methods* 32:297–304

Sternberger, L. A. 1979. *Immunocytochemistry.* New York: Wiley. 2nd ed.

Sternberger, L. A., Sternberger, N. H. 1983. Monoclonal antibodies distinguish phosphorylated and non-phosphorylated forms of neurofilaments *in situ. Proc. Natl. Acad. Sci USA* 80:6126–30

Storm-Mathisen, J., Leknes, A. K., Bore, A. T., Vaaland, J. L., Edminson, P., Huag, F. M. S., Ottersen, O. P. 1983. First visualization of glutamate and GABA in neurons by immunocytochemistry. *Nature* 301:917–20

Strader, C. D., Pickel, V. M., Joh, T. H., Strohsacker, M. W., Shorr, R. G. L., et al. 1983. Antibodies to the *beta*-adrenergic receptor: Attenuation of catecholamine-sensitive adenylate cyclase and demonstration of postsynaptic receptor localization in brain. *Proc. Natl. Acad. Sci. USA* 80:1840–44

Sutcliffe, J. G., Milner, R. J., Shinnick, T. M., Bloom, F. E. 1983a. Identifying the protein products of brain-specific genes using antibodies to chemically synthesized peptides. *Cell* 33:671–82

Sutcliffe, J. G., Shinnick, T. M., Green, N., Lerner, R. A. 1983b. Antibodies that react with predetermined sites on proteins. *Science* 219:660–65

Sweadner, K. J. 1979. Two molecular forms of ($Na^+ + K^+$)-stimulated ATPase in brain. *J. Biol. Chem.* 254:6060–87

Taghert, P. H., Bastiani, M. J., Ho, R. K., Goodman, C. S. 1982. Guidance of pioneer growth cones: Filopodial contacts and coupling revealed with an antibody to Lucifer yellow. *Dev. Biol.* 94:391–99

Thanos, S., Bonhoeffer, F., Rutishauser, U. 1984. Fiber-fiber interaction and tectal clues influence the development of the chicken retinotectal projection. *Proc. Natl. Acad. Sci. USA* 81:1906–10

Thiery, J.-P., DeLouvee, A., Gallin, W. J., Cunningham, B. A., Edelman, G. M. 1984. Ontogenetic expression of cell adhesion molecules: L-CAM is found in epithelia derived from the three primary germ layers. *Dev. Biol.* 102:61–78

Thiery, J.-P., Duband, J. L., DeLouvee, A. 1982a. Pathways and mechanisms of avian trunk neural crest cell migration and localization. *Dev. Biol.* 93:324–43

Thiery, J.-P., Duband, J.-L., Rutishauser, U., Edelman, G. M. 1982b. Cell adhesion molecules in early chick embryogenesis. *Proc. Natl. Acad. Sci. USA* 79:6737–41

Towbin, H., Staehelin, T., Gordon, J. 1979. Electrophoretic transfer of proteins from polyacrylamide gels to nitrocellulose sheets: Procedure and some applications. *Proc. Natl. Acad. Sci. USA* 76:4350–54

Trisler, D., Grunwald, G. B., Moskal, J., Darveniza, P., Nirenberg, M. 1984. Molecules that identify cell type or position in the retina. In *Neuroimmunology and Neural Disease*, ed. P. Behan, F. Spreafico, pp. 89–97. New York: Raven

Trisler, G. D., Schneider, M. D., Nirenberg, M. 1981. A topographic gradient of molecules in retina can be used to identify neuron position. *Proc. Natl. Acad. Sci. USA* 78:2145–49

Tzartos, S. J., Lindstrom, J. M. 1980. Monoclonal antibodies used to probe acetylcholine receptor structure. Localization of the main immunogenic region and detection of similarities between subunits. *Proc. Natl. Acad. Sci. USA* 77:755–59

Tzartos, S. J., Rand, D. E., Einarson, B. L.,

Lindstrom, J. M. 1981. Mapping of surface structures of *Electrophorus* acetylcholine receptor using monoclonal antibodies. *J. Biol. Chem.* 256:8635–45

Tzartos, S. J., Seybold, M. E., Lindstrom, J. M. 1982. Specificities of antibodies to acetylcholine receptors in sera from myasthenia gravis patients measured by monoclonal antibodies. *Proc. Natl. Acad. Sci. USA* 79:188–92

Vallee, R. B., Davis, S. B. 1983. Low molecular weight microtubule-associated proteins are light chains of microtubule-associated protein (MAP1). *Proc. Natl. Acad. Sci. USA* 80:1342–46

Vaughn, J. E., Barber, R. P., Ribak, C. E., Houser, C. R. 1981. Methods for the immunocytochemical localization of proteins and peptides involved in neurotransmission. *Curr. Trends Morphol. Tech.* 3:33–70

Ventner, J. C., Eddy, B., Hall, L. M., Fraser, C. M. 1984. Monoclonal antibodies detect the conservation of muscarinic cholinergic receptor structure from *Drosophila* to human brain and detect possible structural homology with $alpha_1$-adrenergic receptors. *Proc. Natl. Acad. Sci. USA* 81:272–76

Vitetta, E. S., Cushley, W., Uhr, J. W. 1983. Synergy of ricin A chain-containing immunotoxins and ricin B chain-containing immunotoxins in *in vitro* killing of neoplastic human B cells. *Proc. Natl. Acad. Sci. USA* 80:6332–35

Vulliamy, T., Rattray, S., Mirsky, R. 1981. Cell-surface antigen distinguishes sensory and autonomic peripheral neurones from central neurones. *Nature* 291:418–20

Warren, S. L., Fanger, M., Neet, K. E. 1980. Inhibition of biological activity of mouse-*beta*-nerve growth factor by monoclonal antibody. *Science* 210:910–12

Wehland, J., Willingham, M. C. 1983. A rat monoclonal antibody reacting specifically with the tyrosylated form of *alpha*-tubulin. II. Effects on cell movement, organization of microtubules, and arrangement of Golgi elements. *J. Cell Biol.* 97:1476–90

Wood, J. G., Jean, D. H., Whitaker, J. N., McLaughlin, B. J., Albers, R. W. 1977. Immunocytochemical localization of the sodium, potassium activated ATPase in knifefish brain. *J. Neurocytol.* 6:571–81

Wood, J. G., Wallace, R. W., Whitaker, J. N., Cheung, W. Y. 1980. Immunocytochemical localization of calmodulin and a heat-labile calmodulin-binding protein (CaM-BP 80) in basal ganglia of mouse brain. *J. Cell Biol.* 84:66–76

Yen, S.-H., Fields, K. L. 1981. Antibodies to neurofilament, glial filament and fibroblast intermediate filament proteins bind to different cell types in the nervous system. *J. Cell Biol.* 88:115–26

Young, E. F., Blake, J., Ramachandran, J., Ralston, E., Hall, Z. W., Stroud, R. M. 1984. Localization of the C-terminus of the acetylcholine *delta* subunit on the membrane. Implications for the ion channel. *Proc. Natl. Acad. Sci. USA.* In press

Zimmermann, A., Sutter, A., Shooter, E. 1981. Monoclonal antibodies against nerve growth factors and their effects on receptor binding and biological activity. *Proc. Natl. Acad. Sci. USA* 78:4611–15

Zipser, B. 1982. Complete distribution patterns of neurons with characteristic antigens in the leech central nervous system. *J. Neurosci.* 2:1453–64

Zipser, B., McKay, R. 1981. Monoclonal antibodies distinguish identifiable neurons in the leech. *Nature* 289:549–54

Zipser, B., Stewart, R., Flanagan, T., Flaster, M., Macagno, E. R. 1983. Do monoclonal antibodies stain sets of functionally related leech neurons. *Cold Spring Harbor Symp. Quant. Biol.* 48:551–56

Zipursky, S. L., Venkatesh, T. R., Teplow, D. B., Benzer, S. 1984. Neuronal development in the *Drosophila* retina: Monoclonal antibodies as molecular probes. *Cell* 36:15–26

Ann. Rev. Neurosci. 1985. 8:233–61

CENTRAL PATTERN GENERATORS FOR LOCOMOTION, WITH SPECIAL REFERENCE TO VERTEBRATES

Sten Grillner and Peter Wallén

Department of Physiology III, Karolinska Institutet, Lidingövägen 1, S-114 33 Stockholm, Sweden

INTRODUCTION

There is a gap to be bridged in vertebrate neurobiology, a gap between our knowledge of cellular mechanisms and behavior. We have achieved a widely expanded knowledge of how individual neurons function, and about synaptic transmission and membrane channels. We also increasingly understand the overall organization of different patterns of behavior, which principles apply, and which parts of CNS are important. We know much about how cells in the visual cortex respond, but little about how we perceive visual images; we know much about neuronal activity in motor cortex in relation to hand movements, but little about how they are actually generated. One great challenge is to bridge this gap to understand how individual CNS neurons interact to generate different patterns of behavior, or make us perceive the environment, or memorize what we experience.

Invertebrate neurobiologists have had considerable success in describing some very important aspects of the cellular bases of behavior, such as associative learning in molluscs (Kandel & Schwartz 1982) and basic patterns of behavior (Getting 1983a, Bullock 1982). In this chapter we deal with one limited aspect: how the central pattern generators that control locomotion function. Although the review is general, it emphasizes the vertebrates. We try to outline not only what we know, but also what directions seem most promising for gaining insight into how the CNS controls behavior.

0147-006X/85/0301-0233$02.00

To discuss central pattern generation of locomotion in isolation is to discuss only one aspect of a complex control system, one which includes as an integral part a variety of sensory mechanisms. These sensory mechanisms can influence not only motoneurons via different pathways, but also the operation of the central networks themselves. For example, the feedback mechanisms probably correct not only for unexpected events but also optimize the motor output during phases of acceleration, deceleration, or turning (Grillner 1975, 1981). Despite all clear-cut demonstrations of powerful feedback mechanisms interacting with a complex central network, voices are still heard that claim, often loudly, that one or the other is the most important. In our opinion, this is as fruitful as trying to decide whether the hip or the foot is of the greatest significance in walking. In this review, however, we deal with only the central part of this control system. Although striking similarities exist between pattern generators controlling different types of rhythmic behavior, such as respiration and mastication (Grillner 1977, Feldman & Grillner 1983), we limit this chapter to an analysis of the neural organization underlying locomotion in some of the few vertebrates and invertebrates that have been studied.

OVERALL ORGANIZATION OF LOCOMOTOR BEHAVIOR

First, we briefly outline the general features of the control system responsible for generating locomotion in order to provide a background for the discussion of the pattern generator network. Behaviorally meaningful locomotor movements require that the central nervous system generates

1. the movements underlying the stereotypic synergy that provides the propulsion;
2. an adequate equilibrium control during the movements;
3. adaptation of the locomotor synergy to (*a*) the goals of the animal, (*b*) the environment, i.e. to avoid obstacles, (*i*) anticipatory control (visuomotor, etc), (*ii*) compensation for actual perturbations.

Decorticate cats have a surprisingly complex behavioral repertoire, including goal-directed locomotion (i.e. number 3 above). After a transection through the caudal end of diencephalon, cats can still be made to execute well-coordinated locomotor movements, but they are now "robot-like" in character and not goal directed (see Grillner 1981). The control system generating the basic movement synergy (i.e. number 1 above) is our prime concern here (see Figure 1).

In decerebrate primates, cat, turtle, fish and cyclostomes, locomotion can be elicited by stimulation of circumscribed brainstem areas (see Figure 1). In

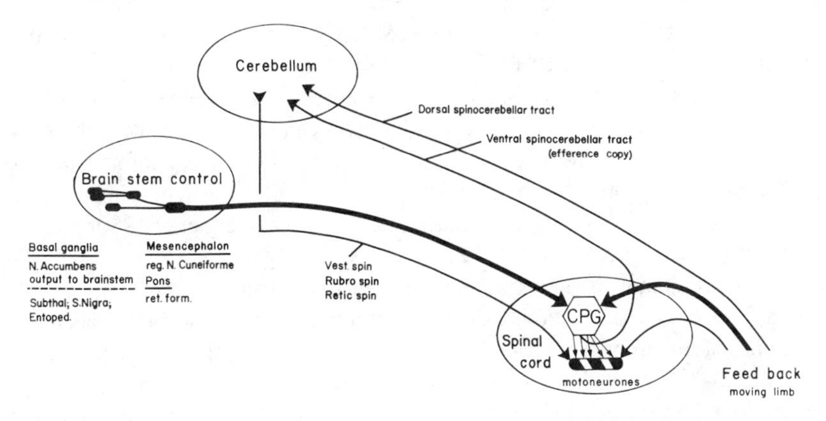

Figure 1 Schematic diagram of the control system for locomotion in vertebrates, based mainly on data from mammals. See text for details.

a cat, a low strength electrical stimulation (e.g. 10 μA, 20 Hz) may elicit walking, but a somewhat higher strength induces trotting and galloping. That is, the movement synergy can be regulated with a very simple control signal (Shik et al 1966, Shik & Orlovsky 1976, Shik 1980). In tetrapods, at least, these brain regions for motor control are homologous from reptiles to primates, a finding that suggests phylogenetically old and unchanged systems. Regions in pons and mesencephalon, as well as the output portions of the basal ganglia [N. subthalamicus, N. entopenduncularis, Substantia nigra (magnocellular part)], are involved (Garcia-Rill 1983, Garcia-Rill et al 1983a,b, Eidelberg et al 1981, Orlovsky 1969, see Grillner 1981).

Brainstem loci control the spinal motorcircuits via pathways in the ventral funiculus (Jordan et al 1980). The spinal cord itself contains neuronal circuits that can coordinate the muscles to produce locomotor movements either if activated pharmacologically, or only after an acute or chronic spinal cord transection, depending on the species (Grillner 1974, Forssberg et al 1980a,b, Wallén & Williams 1984). Mammals with spinal transections at the lower thoracic level can walk without external weight support on a treadmill, or even over ground, if the loss of equilibrium control is compensated for by artificially maintaining the balance of the animal. The isolated spinal cord deprived of descending control by the spinal cord transection and of phasic sensory input (see below) can still produce a complex motor output. Thus, central rhythm-generating networks located within the spinal cord exist, i.e. spinal central pattern generators (see below) (Brown 1911, 1914, Jankowska et al 1967a,b, Grillner & Zangger 1974, 1979). That animals with spinal transection can adapt to the treadmill speed clearly demonstrates the existence of sensory feedback control (cf Figure 1). In this context, sensory signals elicited in the proximal joints, as well as load receptor information, seem to be of crucial importance (Grillner & Rossignol 1978, Andersson & Grillner 1981, 1983, Duysens & Pearson 1980). The central network in conjunction with the sensory mechanisms constitutes one control system. The sensory information is probably an integrated part of the control system, and of particular importance during rapid changes of speed or during turning (Forssberg et al 1980b, Grillner 1981, Grillner & Wallén 1982).

Although spinal feedback systems can provide detailed compensation for perturbations, the perfection of the locomotor movements, such as the accurate placing of the foot on ground, depends on the cerebellum's being intact (Arshavsky et al 1983). The cerebellum receives, via spino-cerebellar pathways, detailed information concerning the movements at each joint and also detailed copies of commands issued from the central pattern generators (CPGs), i.e. efference copy information (Arshavsky et al 1983). In each step cycle, the cerebellum provides phasic corrections via vestibulo-, reticulo-, and rubrospinal pathways (Orlovsky 1972, Orlovsky & Pavlova 1972). Although the signals from the cerebellum clearly contribute to the final output of all motor nuclei (Figure 1), we limit the discussion here to central pattern generators.

COMMENTS ON THE TERM "CENTRAL PATTERN GENERATOR"

The motor pattern produced by a nervous system deprived of all sensory feedback is clearly generated within this nervous system. The neurons responsible for creating the particular motor pattern are referred to herein as constituting the central pattern generator (CPG), regardless of whether all aspects of the motor pattern of the intact animal are produced or some part is missing. Neurons responsible only for an output function, as often is the case with motoneurons, are not considered part of the pattern generator: nor are neurons that provide input to the central network, such as those belonging to feedback systems or systems for initiation of locomotion, part of the pattern generator. A neuron may often have several functions. In such cases, trying to force the cell conceptually into one category or the other seems unimportant; rather, it is sufficient to detail its genuine role. Needless to say, neurons within the CPG differ in function. Some may be important in the rhythm-generating circuitry, others only significant for some aspect of the output pattern, such as a phase shift of a motor burst. The term "central pattern generator" refers to function, not a circumscribed anatomical entity: The individual neurons that constitute the CPG may in principle be located in widely separate parts of the central nervous system.

Another linguistic convention to consider is whether one should say that we have a single CPG for locomotion called upon to activate all parts of the body, or a group of functionally distinct CPGs. Which alternative to choose is arbitrary, but here we use the latter. The cat locomotor network, for example, can be subdivided into several parts, one for each limb and separate parts for the trunk. These are combined in different ways to produce the motor pattern used in walk, trot and gallop, each requiring different types of interlimb coordination. It is convenient to refer to the parts that control each limb, for instance, as the CPG for that limb.

HOW DETAILED IS CENTRAL PATTERN GENERATION?

The first level of approach to this question is to establish the effect of suppressing sensory inflow during movement. This can be produced (*a*) by transecting all afferent fibers while recording the motor pattern as movement, muscle activity, or neurograms, or (*b*) by immobilizing the animal with curarizing agents and recording the motor pattern as neuronal activity. In mammals like a decerebrate cat, the immediate effect of a dorsal root transection on the walking movements is relatively minor. All phases of the walking movements can be unchanged and the pattern of muscle activity can remain complex (Figure 2), but at the same time it should be noted that the pattern becomes more fragile and in fact can break down altogether (Grillner & Zangger 1975, 1984). Similar conclusions have been reached in experiments with curarizing agents (Perret & Cabelguen 1980). Previous hypotheses, suggesting that the central pattern generator controlling the cat's hindlimb is restricted to generating a pure, alternating flexor and extensor activity (Lundberg 1969, 1980),

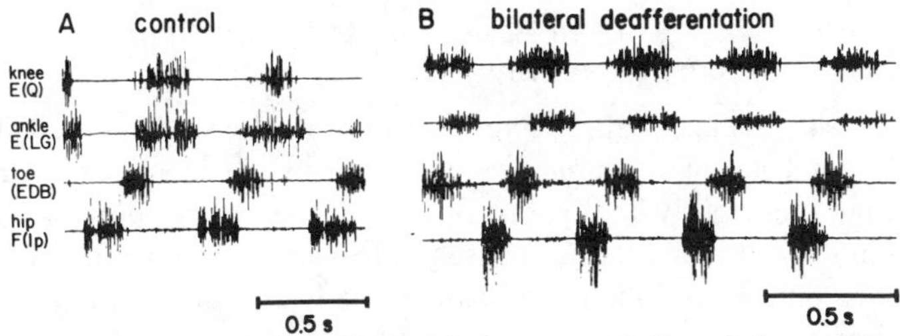

Figure 2 Locomotor pattern of the hindlimb in the mesencephalic cat before and after deafferentation. *A*. Control recording with afferents intact of EMG activity in four hindlimb muscles during walking on a treadmill belt at 1.2 m/s. Muscles recorded from were the knee extensor (E) quadriceps (Q), the ankle extensor lateral gastrocnemius (LG), the short toe dorsiflexor extensor digitorum brevis (EDB), and the hip flexor (F) iliopsoas (Ip). *B*. Same type of recording from the same muscles as in *A*, but after bilateral transection of the dorsal roots supplying the hindlimbs. Same preparation as in *A*. Note that the general pattern of activity is maintained after deafferentation. Adapted from Grillner & Zangger (1984).

must thus be rejected. Other rhythmic motor acts coordinated within the spinal cord such as the scratch reflex in cat and turtle (see Stein 1983) are organized in the same way, with a more complex pattern being generated centrally.

In more primitive vertebrates such as the lamprey, a fish-like cyclostome, the motor pattern is maintained even in an isolated spinal cord (or brainstem-spinal cord) preparation kept under in vitro conditions (Wallén & Williams 1984). Compared to tetrapod locomotion, this motor pattern is of a quite different type. Under normal conditions it results in a mechanical wave traveling down the body, which pushes the animal forward in the water. The motoneurons on the two sides of a segment exhibit alternating activity, while the onset of motoneuron activity in the different segments is progressively delayed from head to tailfin, with about 1% of the cycle duration for each segment. In relation to the most rostral segment, segment number 75 would thus be activated after a delay of 75% of the cycle duration. In the in vitro preparation, the motor pattern is coordinated in the same way (Figure 3). The frequency range overlaps to a significant extent with that of the intact animal, although it may also be lower (Wallén & Williams 1984). Presumably this discrepancy is related to the absence of afferent barrage in the isolated nervous system.

These two cases show that the CNS of vertebrates at different ends of the phylogenetic ladder have complex pattern generators that contribute to the production of locomotor movements. Other well-studied vertebrates are the tadpoles, in which one species exhibits a good central coordination (Kahn & Roberts 1982a,b), but another, the bullfrog tadpole, has been reported to lose its phase lag along the body when afferent activity is suppressed (Stehouwer & Farel 1980, 1981).

Interpreting Findings from Experiments Involving Suppression of Sensory Inflow

If a motor pattern remains after suppression of phasic sensory inflow, it clearly demonstrates that central neuronal networks are able to coordinate the motor pattern, but it obviously does not allow any conclusion regarding the role of sensory inflow when present. Conversely, when abolition of a sensory inflow causes a modification of the motor pattern, the interpretation is not straightforward either.

Let us scrutinize two critical experimental situations:

1. If all sensory fibers are transected, the CPG must operate without any interaction from sensory elements. If the motor pattern is modified—let us say if one burst is omitted—it might mean either that (*a*) this burst is generated purely by afferent activity elicited by the ongoing movement, or (*b*) that a central network generates the burst but only if a certain type of sensory information is available. In the latter case the sensory signals may play a specific,

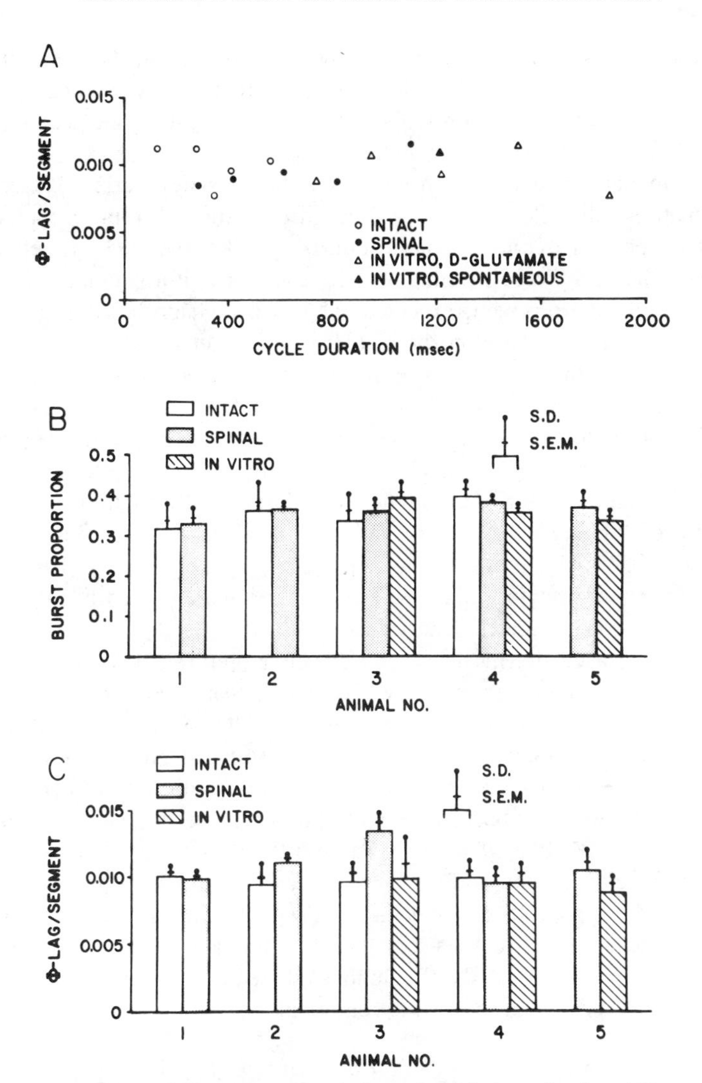

Figure 3 Comparison of the motor patterns during real and fictive swimming. *A* illustrates that the phase lag per segment remains a constant fraction of the cycle duration (about 1%, or 0.01 if the normalized cycle duration is set to 1.0) for different rates of locomotion. This holds true for intact and spinal animals swimming in a swim-mill, as well as for the in vitro preparation during fictive swimming induced by amino acid application (D-glutamate in this case). A value from a spontaneously active in vitro preparation has also been included. In *B* is shown the mean values of the proportion of the cycle occupied by the locomotor burst for each type of preparation in five different animals. Also for this parameter there is no significant difference between preparations. *C* is a similar representation of the mean phase lag per segment (cf *A*) for the different preparations. In two cases (animals 3 and 4), data from all three types of preparation were obtained from the same animal. Adapted from Wallén & Williams (1984).

perhaps permissive, role, triggering the network to generate the burst, or they may just provide additional background excitation. Many researchers have chosen the first interpretation without even acknowledging the second, thus unduly simplifying the interpretation.

2. If movement-related sensory inflow is abolished by curarization, sensory information will still reach the CNS both from skin and from receptors that signal the position of different parts of the body. When the CPGs are activated, there will be a conflict of information: Efference copy information will indicate that a motor pattern is being executed, while sensory signals will suggest that the body is kept in a fixed position. If the motor pattern is modified, it may mean either (*a*) that the burst pattern somehow is critically dependent on the sensory inflow arising during movement (see 1. above), or (*b*) that the abnormal sensory inflow, caused by the body's being kept in one position, produces a change in the motor pattern. The very powerful feedback systems acting on the CPGs can thus operate during fictive locomotion if the curarization is not combined with an extensive denervation. A somewhat similar situation with a biased abnormal sensory inflow can be present after a partial denervation if, for example, receptors activated by limb flexion are intact while those that signal extension have been denervated.

Clearly, great caution must be exercised in interpreting experiments designed to eliminate effects of sensory inflow. The finding that the intersegmental phase lag disappears in one type of tadpole but not in the other does not therefore necessarily justify the conclusion that the phase lag is produced by sensory inflow exclusively (see above). Moreover, the finding that marked changes in the motor pattern may occur after partial denervation in a variety of insects (Bässler 1983, Wendler 1974, 1978, Pearson 1976) cannot be interpreted without a subtle combination of different types of experiments and considerations. Although the hypothesis of a general organization with a control system composed of both central and sensory elements probably will hold true in all species, the particular family of solutions utilized may differ between groups of animals and different modes of progression.

INTRINSIC MECHANISMS OF CENTRAL PATTERN GENERATORS

The existence of CPGs controlling locomotion is well established in a variety of systems (Delcomyn 1980), but their intrinsic function is reasonably well understood only in a few systems.

Invertebrate Systems

Getting and colleagues (1977, 1980, 1981, 1983a,b, Hume & Getting 1982, Hume et al 1982) have unraveled most aspects of the CPG controlling escape swimming in *Tritonia diomedea,* a sea slug. A skin stimulus elicits an initial reflex withdrawal followed by a burst of 2–20 alternating dorsi- and ventroflexions of the slug's slender body. This comparatively simple movement pattern helps the animal escape by swimming away from the stimulus. The motor pattern is elicited by an interneuronal CPG that provides alternating activation of dorsal and ventral flexion motoneurons. The CPG interneurons are of four types: the C2 ($n = 1$), dorsal swim interneurons (DSI, $n = 3$), and ventral swim interneurons (VSI) of A and B type (one of each). The dorsal swim interneurons (DSI) activate dorsiflexion motoneurons and ventral swimming interneurons (VSI-B) activate ventroflexion motoneurons. In Figure 4, the pattern of interneuron activity during a bout of swimming is shown to the left. When swimming is initiated, all interneurons receive a ramp depolarization, which slowly decreases as the bout reaches its termination. At the onset, all interneurons are excited, and subsequently the pattern is generated. A characteristic feature of this network is that chemical synapses dominate (only electrical synapses between DSI-interneurons), but the time courses of synaptic actions differ markedly between synapses (over 30 times). Furthermore, there are dual and triple function synapses with, for instance, excitation followed by inhibition between C2 and DSI, or inhibition followed by excitation between C2 and VSI-A (Figure 4). C2 and DSI have a somewhat similar pattern, which is reciprocal to that of VSI-A and B. The dual function synapses have a time course that agrees with the general pattern. In the early phase of the C2 activity it excites its "co-worker" DSI and inhibits the antagonist VSI-A, but somewhat later the same C2 cell will instead inhibit DSI and excite VSI-A. This effect will act to terminate DSI and initiate VSI-A activity. When VSI-A starts, it will excite VSI-B, and both will inhibit DSI. This intricate pattern of synaptic connectivity is the basis of the swimming motor pattern, but some other factors are very significant: (*a*) the pattern of adaptation of the spike frequency in different cells, (*b*) the high threshold for activation of the C2-cell, and (*c*) an intrinsic mechanism for delayed excitation in VSI-B. In this case the synaptic depolarization activates an intrinsic potassium current (A-current) that slowly inactivates. This transient K-current will delay the synaptically driven action potentials and lead to a delayed excitation.

With these data at hand, Getting (1983a,b) has modeled the network and can account for the entire pattern. To understand the network, it proved important to consider not only the pattern of connectivity but also the detailed membrane properties of the individual cells and the time course of the synaptic potentials. It is noteworthy that information about the firing pattern of the individual neurons was not sufficient to provide insight into the mechanism.

The second best understood locomotor network is that controlling leech swimming (Kristan & Weeks 1983). The CNS of the leech is composed of a

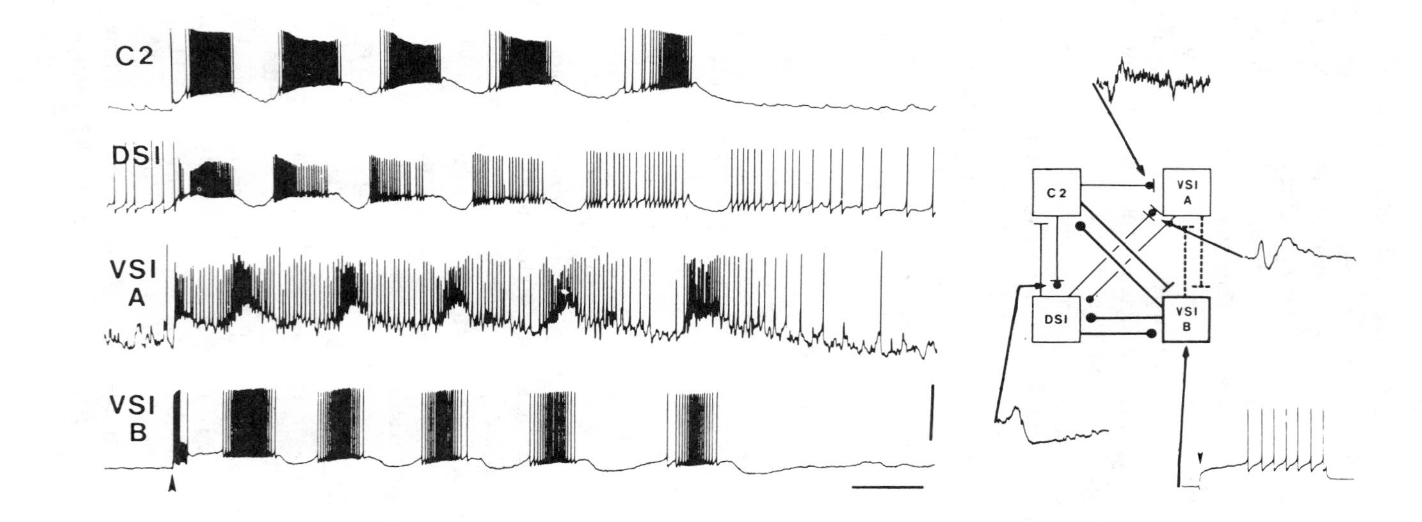

Figure 4 Generation of the motor pattern underlying escape swimming in *Tritonia*. The *left-hand panel* shows simultaneous intracellular records from four premotor interneurons that have been suggested as members of the CPG-network (abbreviations, see text). The swim sequence was elicited by electrical stimulation (10 Hz for 1 s) of a peripheral nerve (*arrow*) in an isolated brian preparation. Voltage calibration: 50 mV for all records except VSI-A, for which it corresponds to 25 mV. Time scale, 5 s. To the *right* is shown a circuit diagram with the synaptic connections between the different classes of interneurons. Recordings of multicomponent, monosynaptic PSPs have been included, as indicated by the *arrows*. The *lower right* record illustrates the delayed excitation occurring in VSI-B (*arrow* indicates onset of a depolarizing pulse). The identified interneurons comprise two bilaterally symmetrical networks, with each neuron type electrically coupled to its homolog neuron on the contralateral side. Modified from Getting (1983a,b).

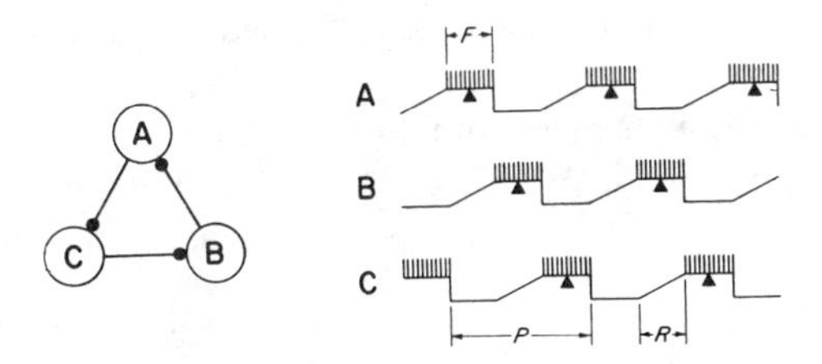

Figure 5 *Left:* Ring model of 3 neurons. *Right:* inhibitory trough of one neuron (e.g. *C*) coincides with spike train (*F*) of another neuron (e.g. *A*), which will disinhibit neuron *B*. Duration of recovery time (*R*) is influenced by the excitability of the neurons. *P*, period length. From Friesen & Stent (1978).

series of ganglia, each of which can generate alternating activity. Most interneurons responsible for pattern generation appear to have been identified. The principle of oscillation is thought to depend on inhibitory interaction between interneurons connected in a ring (see Szekely 1976, Friesen & Stent 1978). Basically, each interneuron can assume three states: (*a*) spiking, (*b*) actively inhibited, (*c*) in recovery phase after cessation of inhibition but before resuming spiking. The activity starts in the network when the excitability level is raised, so the neurons will resume spiking when not inhibited (Figure 5).

The only pattern generator for locomotion in which most aspects of the burst pattern can be accounted for by a realistic model is that in *Tritonia,* but CPGs for some other patterns of behavior have been well studied, such as the lobster stomatogastric system and the leech heart CPG (e.g. Selverston et al 1976, Eisen & Marder 1982, Miller & Selverston 1982, Calabrese 1979). Classical preparations such as the crayfish swimmeret system and that of locust flight, in which much of the basic general organization has been worked out (e.g. Stein 1978, Wilson 1964, 1966, Wiersma & Ikeda 1964, Pearson et al 1983, see also Burrows 1980), have not yet provided sufficient information to make us understand the intrinsic function of their CPGs, although much new information exists on circuitry and feedback mechanisms.

Vertebrate Systems

The most studied vertebrate CPG systems in this context are those of cat, turtle, tadpole, dogfish, and lamprey. In none as yet is there a satisfactory description of the intrinsic function of the CPGs, but in most systems a rapid increase in knowledge has been achieved about a variety of aspects of presumed importance for the network function. In view of the detailed knowledge necessary to understand the simple CPG network of *Tritonia,* it would be surprising if the complex CPGs of higher vertebrates could be understood simply by recording the discharge pattern of different types of neurons during fictive or real locomotion. A much more detailed mesh of knowledge must certainly be acquired.

Below we detail different pieces of information available about the intrinsic function of vertebrate CPGs. We discuss new general findings that will probably be of relevance, as well as what type of evidence is absolutely required.

DETAILED KNOWLEDGE OF THE MOTOR OUTPUT Such knowledge must be available within the entire working range of the intact animal and under the experimental conditions used. This type of information has been acquired in several preparations (Engberg & Lundberg 1969, Grillner & Zangger 1979, 1984, Halbertsma 1983, Kahn et al 1982, Kahn & Roberts 1982a,b, Stein & Grossman 1980, Stein 1983, Deliagina et al 1975, Wallén & Williams 1984).

KNOWLEDGE OF THE OVERALL ORGANIZATION Comprehensive knowledge of the control systems, location of CPGs, organization of systems of CPGs, e.g. limb CPGs, mechanisms for initiation and feedback (see above) is required.

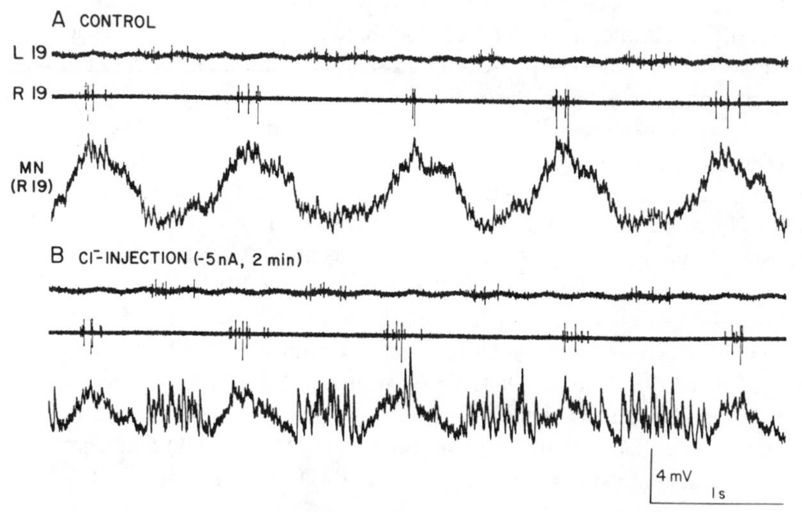

Figure 6 *A*. Oscillations of membrane potential of a motoneuron (MN) in an in vitro preparation of the lamprey spinal cord, coupled with rhythmic activity in ventral roots (D-glutamate 0.3 mM). Ventral root recordings (suction electrodes) are from the right (*R*) and left (*L*) sides of segment 19 in a preparation containing 38 segments. The motoneuron was located on the right side of segment 19. *B*. Reversal of hyperpolarizing phase by intracellular injection of chloride ions. Continuation of traces in *A*, after injection of chloride ions by passing −5 nA hyperpolarizing current for 2 min. Microelectrodes were filled with 3 M potassium chloride. Voltage calibration applies to intracellular traces. From Russell & Wallén (1983).

DEFINITION OF THE CIRCUITRY Details regarding circuitry, with exact input and output of all relevant CPG neurons, is necessary. The input signals to α-motoneurons in general consist of a phasic excitatory drive from interneurons alternating with active postsynaptic inhibition, as demonstrated in lamprey (Figure 6; Russell & Wallén 1983), turtle (Stein 1983), cat (Edgerton et al 1976, Shefchyk et al 1981, Perret & Cabelguen 1980, Perret 1983), and tadpole (Soffe & Roberts 1982).

Direct knowledge of identified premotor interneurons that participate in CPG function is scarce. Information about such cells is technically difficult to obtain, as it requires simultaneous intracellular recordings from interneurons and motoneurons. In the cat the established Ia inhibitory interneuron contributes (Edgerton et al 1976, Feldman & Orlovsky 1975), but this interneuron is used both by the CPG and Ia afferents and also by a number of other systems. In the cat, the Renshaw cells are driven as well (McCrea et al 1980) and two identified inhibitory premotor interneurons in the lamprey have been shown to take part (Buchanan & Cohen 1982, cf. Grillner & Wallén 1984). It is striking that the excitatory premotor interneurons remain to be identified. Knowledge about premotor interneurons in other species is still lacking.

In the spinal cord, a large number of segmental interneurons that are part of or driven by the CPG are active during fictive locomotion (Edgerton et al 1976, Arshavsky et al 1983, Orlovsky & Feldman 1972, Sigvardt & Grillner 1981). Neither the specific input nor the projections of these interneurons have been identified. They may be active or have their peak activity in any phase of the locomotor cycle. Such experimental data are useful in a general sense, when trying to relate interneuronal activity patterns to anatomical loci and motor bursts, for example, but presently they do not contribute to our understanding of the intrinsic function of the CPG. To this category also belong the interneurons in the cat spinal cord active during the late reflex discharges that occur after intravenous L-DOPA (Jankowska et al 1967a,b). One group of interneurons fire during flexor discharges and another during extensor bursts; when one group is active the other is silenced and vice versa. Since noradrenergic drugs may release locomotion (see Grillner 1981), these interneurons may be part of the CPG. Their pattern of activity suggests that reciprocal inhibition may play a role within the network.

A combination of neuropharmacological and histochemical tools can help us identify neurons that are part of the CPG circuitry (Grillner & Wallén 1980) and thereby aid the analysis of connectivity. The possibility of combining immunohistochemistry with intracellularly identified Lucifer yellow injected neurons will also prove important (P. A. M. van Dongen et al, unpublished).

PROPERTIES OF INDIVIDUAL NEURONS The spike frequency and its adaptation during a constant stimulus is obviously of central importance for CPG function. Discharge characteristics are affected by the afterhyperpolarization following the action potential (e.g. Gustafsson 1974), which is regulated mainly by calcium-activated potassium channels (cf Meech 1978). The magnitude and duration of the hyperpolarization after each action potential will influence the interval to the subsequent spike. Moreover, the degree of summation of the afterhyperpolarization in a train of action potentials is important, since it will affect the adaptation. In an extreme case a constant depolarizing signal in a neuron may elicit a series of action potentials. The summed afterhyperpolarization can terminate the burst and silence the neuron for a while, until it recovers and a new burst is initiated, which in turn may be terminated in the same way. In the hippocampus, the duration of the afterhyperpolarization can in fact be controlled synaptically by an indirect action on the calcium-dependent potassium channel by noradrenaline (Madison & Nicoll 1982). Another factor that will contribute to adaptation is the changing threshold at which an action potential is initiated. The first action potential may be elicited closer to the resting membrane potential than subsequent ones (Schwindt & Crill 1982).

The level of the resting potential and the "mix" of membrane channels in soma and dendrites and their degree of activation will decide how a neuron reacts to a given input. For instance, the VSI-B neuron in Getting's circuitry (Figure 4) responds with an activation of a potassium-current (A-current) on depolarizing current, which is subsequently inactivated. During this initial period the excitation of the neuron is delayed. This feature proved to be one important aspect of the pattern-generating circuitry.

Most neurons respond under resting conditions to a positive pulse with a depolarization that is approximately identical in duration to that of the pulse. Transmitters may change this condition; for instance, *N*-methyl-aspartate activates NMDA-receptors (aspartate) to induce a region of negative slope conductance in the current-voltage relationship of a cell. This negative conductance can lead to a bistable membrane potential. Under these conditions the membrane potential at a given synaptic current can take on two stable values, e.g. -65 and -50 mV. A short depolarizing pulse (EPSP) can then move the membrane potential from -65 to -50 mV or vice versa for a hyperpolarizing pulse of short duration (IPSP). The membrane potential may also oscillate between the two states (Flatman et al 1983, MacDonald et al 1982).

It is noteworthy that bath-applied *N*-methyl-aspartate induces activity in the pattern generators used for swimming in the in vitro preparation of the lamprey spinal cord (Figure 7; Grillner et al 1981). This membrane property thus seems likely to be important in the pattern-generating circuitry. Moreover, NMDA induces rhythmic oscillating activity in lamprey neurons functionally isolated from each other by TTX-blockade of the action potentials (Sigvardt & Grillner

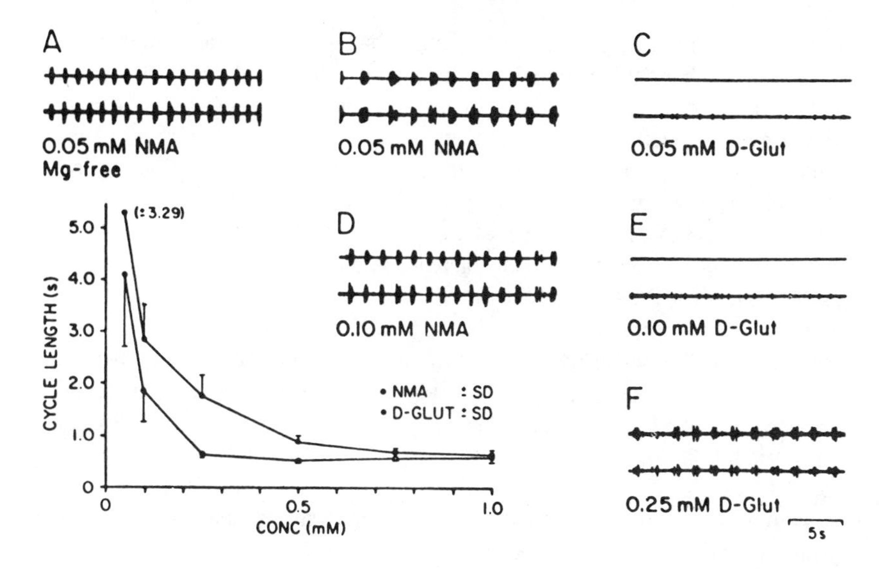

Figure 7 Induction of fictive locomotion by bath application of amino acids. The effects of different concentrations of *N*-methyl-aspartate (NMA) and D-glutamate were compared by recording the efferent activity in two adjacent ventral roots (*B-F*). At a concentration of 0.05 and 0.10 mM, NMA is effective (*B,D*) but only at 0.25 mM D-glutamate has an effect (*C,E,F*). Time calibration in *F* applies to all records. The graph illustrates that the cycle duration is always longer with D-glutamate (*upper curve*) than with NMA (*lower curve*) at the same concentration. In this case the shortest cycles were about 0.5 s, i.e. a burst rate of 2 Hz. NMA itself is more potent in Mg^{2+}-free solution (*A*). From Grillner et al (1981).

1981, Grillner et al 1982, 1983; unpublished). That transmitters such as dopamine and noradrenaline can convert "ordinary" neurons in invertebrates to pacemaker cells or vice versa has been known for some time (Ducreux 1976, Boisson & Gola 1976). The role of plateau potentials (Russell & Hartline 1978, 1982) in invertebrate pattern generation is well established. For a reader with a background in vertebrate neurobiology, it is worth noting that groups of neurons with an inherent tendency to oscillate (cf Gola & Selverston 1981) may well be coordinated in larger groups by mutual excitation combined with reciprocal inhibition or dual function synapses. Recall also that 5-HT can induce swimming in the leech (Willard 1981) and that 5-HT and noradrenaline precursors elicit the motor activity correlated to locomotion in cat and rat (see Grillner 1981).

THE POSTSYNAPTIC EFFECT A variety of postjunctional events can occur. Inhibitory and excitatory synaptic potentials may be elicited through mechanisms that cause either an increase or a decrease in conductance for relevant ions. A transmitter may give rise to a pure change in membrane conductance or activate a channel that is voltage dependent (MacDonald et al 1982). The PSP-duration can vary over 30-fold in the same pattern generator (Getting 1983a,b) and the PSP may have dual or triple functions (Figure 4). These multiple functions may result from the same transmitter activating separate channels or, when two transmitters coexist in the same terminals, from simultaneous release of distinct transmitters (Hökfelt et al 1983). Synaptic effects can modify a neuron to generate plateau potentials or membrane potential oscillations (cf NMDA above), or modify the adaptation process by interfering with the afterhyperpolarization (Madison & Nicoll 1982). In view of these complexities, the modeling that used to be carried out with neuromimes—in which networks were represented by standard "neurons" connected through standard inhibitory and excitatory synapses—is now interesting mainly from a historical point of view.

POSSIBLE PRESYNAPTIC FACTORS It cannot be assumed that the membrane properties of the presynaptic terminals are similar to those of the cell body or the dendrites. In invertebrate nonspiking interneurons, for example, the amount of transmitter release is graded and continuously dependent on the level of membrane potential (Siegler & Burrows 1980). To what extent such graded release occurs in vertebrate motor systems is unknown. One reason for our lack of knowledge in this area may be that a reasonably rigorous demonstration of a graded release requires paired intracellular recordings of the pre- and postsynaptic neuron. It is noteworthy, however, that the lamprey spinal cord possesses neurons without apparent axon that do generate spikes (Sigvardt & Grillner 1981, and in Grillner et al 1983). On the other hand, some locust local interneurons are spiking and others are nonspiking (Siegler & Burrows 1980, Burrows 1980).

The possibility of presynaptic mechanisms regulating directly or indirectly the efficiency of the Ca-channels must also be born in mind.

NONSYNAPTIC FACTORS During all neuronal activity the level of extracellular potassium increases to some extent. This causes a depolarization and can in turn affect a variety of neuronal functions, including synaptic transmission. Potassium accumulation has long been considered to be of possible importance in rhythmogenesis (Brown 1914, Jankowska et al 1967b). With the finding that a nerve volley in hindlimb nerves can increase the extracellular potassium levels from 3 to 9 mM in the dorsal horn, this potential factor became a serious candidate mechanism for rhythm generation (Kriz et al 1974). This mechanism was tested in the lamprey during fictive locomotion by recording the extracellular potassium levels with potassium electrodes within the grey matter.

Indeed, phasic oscillations in potassium levels were found to occur with each locomotor burst, but the concentration changed by only about 0.2 mM (Figure 8), an amount that would cause only a 1 mV fluctuation in membrane potential (Wallén et al 1984a). Thus, the phasic potassium concentration changes presumably cannot account for any major effect. The low-level oscillations represent local overall changes and show how a silent neuron may be affected by the activity in nearby neurons. In a certain critical location, functionally important effects of phasic changes in the extracellular potassium level are still conceivable (cf Yarom et al 1982).

In certain experimental models of bursting (Jefferys & Haas 1982), large burst-related variations in extracellular fields occur; these may give rise to large changes in the transmembrane potential, which may in turn recruit new neurons into action in synchrony with already active neurons. This mechanism is not important in locomotion, since in neither the lamprey (Grillner & Wallén 1984) nor the cat spinal cord (V. R. Edgerton, S. Grillner, A. Sjöström, unpublished) are there significant extracellular burst-related fields. Presumably, the extracellular fields of individual neurons cancel each other.

To Summarize

In invertebrates the gap between knowledge of the function of individual neurons and behavior has been bridged in several instances, and we can rightly

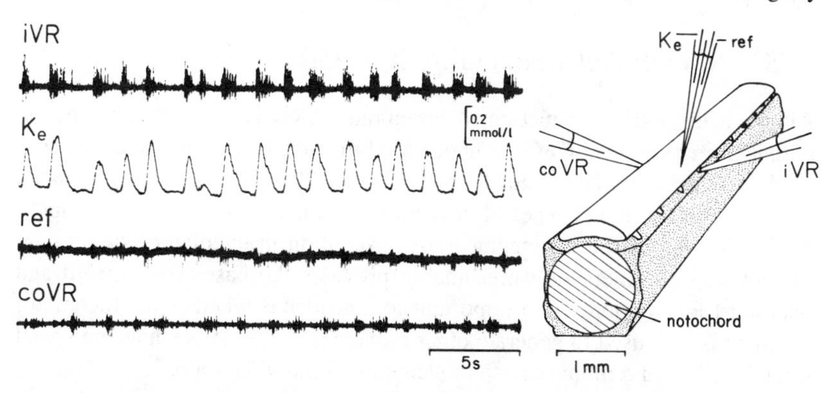

Figure 8 Phasic variations in the extracellular potassium concentration $(K^+)_e$ during fictive locomotion. The *traces* show from above the efferent activity in an ipsilateral ventral root (iVR), the associated phasic changes in $(K^+)_e$, for comparison the recording from the reference electrode (ref), and below the contralateral ventral root (coVR). The bath contained 0.2 mM NMA. Note time and concentration calibrations. The schematic drawing to the *right* depicts the spinal cord—notochord preparation with the recording arrangement used. Ventral root activity was recorded *en passant* with suction electrodes and $(K^+)_e$ was monitored with a double-barreled microelectrode, filled with a Corning ion exchanger resin (477317) in the ion-sensitive barrel and with 1 M magnesium acetate in the reference barrel. From Wallén et al (1984a).

claim that the cellular basis of behavior is known in some cases. However, we clearly are not yet in such a position with vertebrates. One major stumbling block is the lack of knowledge of the connectivity between individual components and the properties of synapses. This difficulty can probably be minimized by using lower vertebrate preparations such as the tadpole or the lamprey. That sufficient understanding can be reached directly with present techniques in higher mammals like the cat is improbable, in view of the detailed knowledge needed about a simpler system like *Tritonia*. The lamprey spinal cord provides a favorable medium in which to learn how a vertebrate CPG operates, since the rhythmic activity underlying locomotion can be elicited in small pieces in vitro. But can learning the organization of the pattern generation for locomotion in one primitive vertebrate provide any relevant knowledge for an understanding of mammalian locomotion? We are optimistic that it can, because the overall neuronal organization of vertebrate locomotion, as well as the CNS in general, seems to be quite similar throughout the vertebrate phylum: This implies that the same neural organization has probably been modified to suit the changing conditions during evolution. The circuitry controlling paired fins probably evolved from the circuitry generating alternating contractions in a segment of the body wall in a cyclostome. The circuitry controlling paired fins presumably gave rise to that controlling tetrapod limbs (cf tadpole) and ultimately primate limbs.

SYSTEMS OF PATTERN GENERATORS

Although the intrinsic function of vertebrate CPGs is not fully known, the output from individual CPGs is understood in some detail, as is their general mode of interaction (see Stein 1978, Grillner 1981). Different CPGs interact to produce the different types of interlimb coordination used in tetrapod locomotion as well as the intersegmental coordination in undulatory swimming; moreover, such interaction presumably provides the bases for forward and backward locomotion. Also important to consider is whether the locomotor circuitry is also used to generate other movement patterns, such as the spinal scratch reflex, and more precise movements of individual joints.

Undulatory Locomotion

All parts of the spinal cord in both dogfish and lamprey and presumably tadpole have the inherent ability to generate alternating activity; This ability is thus distributed throughout the spinal cord (Grillner 1974, Cohen & Wallén 1980, Grillner et al 1982). The minimal circuitry that can generate alternation is close to one segment (see Grillner et al 1983) and is possibly even smaller. Independent circuits probably occur within each hemisegment, which normally are reciprocally coupled to provide the alternation (Cohen & Wallén 1980, Cohen & Harris-Warrick 1984, Roberts & Kahn 1982). Moreover, the output signal to motoneurons within one segment that supply different parts of the body wall may differ considerably, thus suggesting that the circuitry does not function as a single unit (Grillner & Wallén 1984, Wallén et al 1984b). One tempting hypothesis is that the organization is modular with unit pattern generators controlling a hemisegment or a part thereof. The intersegmental phase lag then presumably results from the coordination between the unit pattern generators along the spinal cord.

During forward swimming this phase lag is from rostral to caudal, but in backward swimming (or struggling) it is reversed, with the caudal segments leading (Grillner 1974, Wallén & Williams 1984, Roberts & Kahn 1982). During resting, it may happen that, say, segment number 16 is leading and both rostral and caudal segments have a phase lag in relation to this segment (Grillner 1974). Any small part of the spinal cord has the ability to generate a phase lag in the two directions, and also this ability is thus distributed. It is tempting to believe that each unit pattern generator can be coordinated via local interaction to adjacent unit pattern generators, and presumably also interact with more distant pattern generators via propriospinal neurons. The exact neuronal mechanisms that will produce an intersegmental constant phase lag of 1% of the cycle duration, regardless of whether the cycle duration is 0.1 s or 2 s, can probably only be identified when we know the intrinsic mechanisms of the unit CPG. An alternative, but remote, possibility is that two separate networks exist in the spinal cord, one for forward and one for backwards swimming. For a variety of reasons of versatility and economy of components (see e.g. Grillner 1974, 1981), however, the former possibility is more attractive. If so, we have to postulate that each unit generator is connected by a set of rostral to caudal and caudal to rostral coordinating neurons (cf Stein 1978) with the adjacent rostral and caudal unit pattern generator. In relation to the leading segment, all other segments in either the rostral or caudal direction will have a progressively increasing phase lag. It is noteworthy that the lamprey may in addition perform C-shaped bending movements along the body when all segments are in phase, i.e. the phase lag is zero.

The propulsive machinery would thus appear very simple and versatile—a series of unit pattern generators normally reciprocally coupled over the midline and connected via different coordinating neuronal elements.

Tetrapod Coordination

The different limbs are controlled by separate CPGs, each of which can generate the appropriate motor pattern for its limb. The different CPGs are in principle coordinated in two ways, with either the fore and hindlimbs strictly

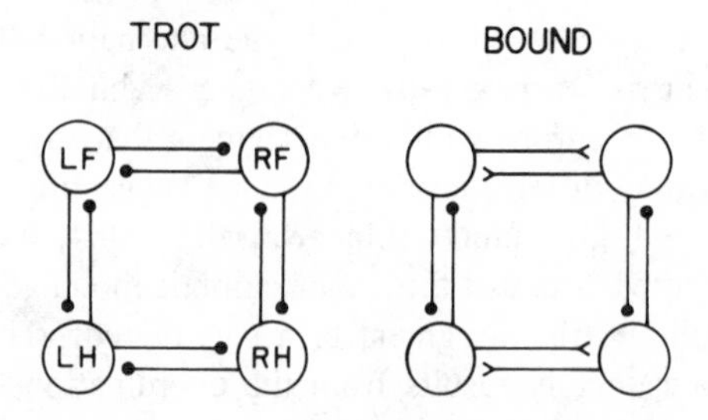

Figure 9 Presumed mechanisms of interlimb coordination in two different gaits. The central networks (CPGs) of left (LF) and right (RF) forelimbs and left (LH) and right (RH) hindlimbs are represented. Inhibitory connections between centers are represented by *lines with filled circles* in the end and excitatory connections with a *"fork."* In the "trot pattern" to the left, reciprocal inhibition between the forelimbs and the hindlimbs, respectively, will assure that each pair will alternate. Correspondingly, the reciprocal connections between fore and hindlimbs at the same side will make these limbs alternate. By changing just one set of connections between the two forelimbs and the two hindlimbs from inhibitory to excitatory (*right diagram*), each pair of limbs will instead become active in synchrony as in a bound, or approximately as in a gallop. Modified from Grillner (1975).

alternating as in walk and trot, or more in synchrony as in half bound, bound, or the different types of gallop (see Grillner 1981). It can be assumed that the hind- or forelimb CPGs during alternating gaits are reciprocally coupled, whereas the same CPGs mutually excite each other during nonalternating gaits such as gallop (see Grillner 1981). The coordination between fore and hindlimbs can be achieved in a similar way (see Figure 9). The trunk is presumably controlled by a separate network coordinated with the limb CPGs (Zomlefer et al 1984). The stiffness of the trunk appears to be the main factor controlled during walk and trot, since the erector spinae muscles on both sides of the trunk are coactivated with two bursts per movement cycle (Carlson et al 1979, English 1980, Zomlefer et al 1984), a pattern also found in man (Thorstensson et al 1982). In gallop, an important part of the movement synergy is the flexion and extension movements of the spine. The extension of the spine contributes to the overall thrust of the hindlimbs; under these conditions, only one erector spinae burst occurs per cycle, coordinated with the hindlimb extensor activity.

With the presumed design of the control system, the transition from trot to gallop, for instance, can be achieved in a very simple way: just by switching from one set of coordinating neurons to another, from one set utilizing reciprocal inhibition to another, utilizing mutual excitation. This would be a very simple and efficient control strategy from the point of view of higher centers, requiring little extra information.

Modifiability of the Motor Pattern Within a Limb

The motor pattern of a limb during locomotion must be modified continuously to suit the demands of the environment. Small changes occur during steady locomotion forward, but become gradually larger if we (man, cat, or turtle) choose to step sideways or backwards, which of course is part of our normal behavioral repertoire (see Grillner 1981). In addition to changes of the relative amplitudes of motor bursts in different muscles, the phase relation between muscles must change. Consider for instance backward locomotion, during which the hip flexes during the support phase, whereas in forward locomotion hip, knee, and ankle are synergists during both swing and support phase. Here the phase relationships change 180° between forward and backward locomotion (see Grillner 1981). The switch between forward and backward locomotion can also be made by a spinal cat walking on a treadmill, and the switch can consequently be achieved by the spinal cord circuitry without descending interaction.

Stein and his associates (see Stein 1983) have studied a similar type of experimental condition in another spinal motor act, the scratch reflex. If an object irritates the skin, the hindlimb will be brought toward that spot and it will try to rub the irritant away. This will happen in skin regions both rostral and caudal to the limb. The motor pattern is strikingly different in these different "scratch reflexes," with different phase relationships between different muscles.

Both in locomotion and in the scratch reflex, a high degree of modifiability is required, which presumably results from modifications within the central

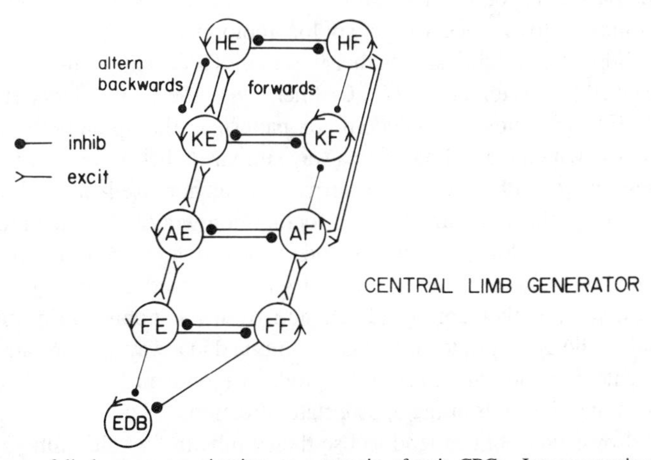

Figure 10 Scheme of limb generator circuitry as a mosaic of unit CPGs. Interconnections between unit generators decide the relative phase of different muscle groups and how they are used during locomotion. Connections (inhibitory or excitatory) that could produce forward and backward locomotion have been indicated. Abbreviations: E, extensor; F, flexor; H, hip; K, knee; A, ankle; FE, foot extensor; FF, foot flexor; EDB, short toe flexor, extensor digitorum brevis. Modified from Grillner (1981).

pattern generating network. The alternative possibility is that the CNS is equipped with a family of different pattern generators, each recruited during slightly different conditions. The former possibility appears more likely and is compatible with present evidence (see Grillner 1981). It has been suggested that each muscle group is controlled by a unit CPG and that the coordination between different muscles and joints results from the interaction between these unit CPGs (Figure 10). CPGs controlling synergistic muscles would be mutually excitatory, those that alternate reciprocally inhibitory, and specific phase relations could be achieved by other forms of interaction. In analogy with the reasoning above regarding interlimb coordination, a switch from alternation to synchrony can be achieved by switching from one set of interconnections to another. If hip flexors and extensors are coactive with those of the lower limb, the animal will walk forward, but if the connection between the unit CPGs are switched, the animal will change direction instantaneously.

In conclusion, it is conceivable that the locomotor system in each animal can be reduced to a number of unit pattern generators grouped together in somewhat larger entities controlling individual limbs or the trunk. By changing their interconnections we would obtain a very versatile system that could easily produce backward or forward walking, trotting, and galloping.

Are the Presumed Unit CPGs for Locomotion Used in Other Patterns of Behavior?

If the CPGs for locomotion can be subdivided into unit CPGs that can be combined to produce a variety of locomotor patterns (backward-forward; walk-gallop), one might ask whether these unit CPGs can be used in other contexts as well (Edgerton et al 1976, Grillner 1981, 1982, Arshavsky et al 1983, Stein 1983). The closest nonlocomotor pattern is the spinally controlled scratch reflex, which is modifiable in a way similar to that of locomotion (Stein 1983, see above). If these two patterns of behavior used part of the same CPG circuitry, the command to elicit locomotion would activate certain unit CPGs to a specific level of activity and, in addition, select the appropriate set of interconnections between the unit CPGs to produce the type of locomotion desired. Another command, elicited from the itching skin stimulus, would select the appropriate unit CPGs, combined in a way appropriate for scratching the itching spot, and in addition produce a tonic bias in joints like the hip joint to direct the limb in the appropriate direction.

Lower vertebrates tend to use their limbs in "whole-limb" synergies (e.g. locomotion, scratching) involving mainly flexion and extension synergies. The transition from movement to a maintained flexion and extension is smooth, and an animal can freeze the joint in any position whenever necessary. The more evolved, in the phylogenetic sense, the more versatile the motor system

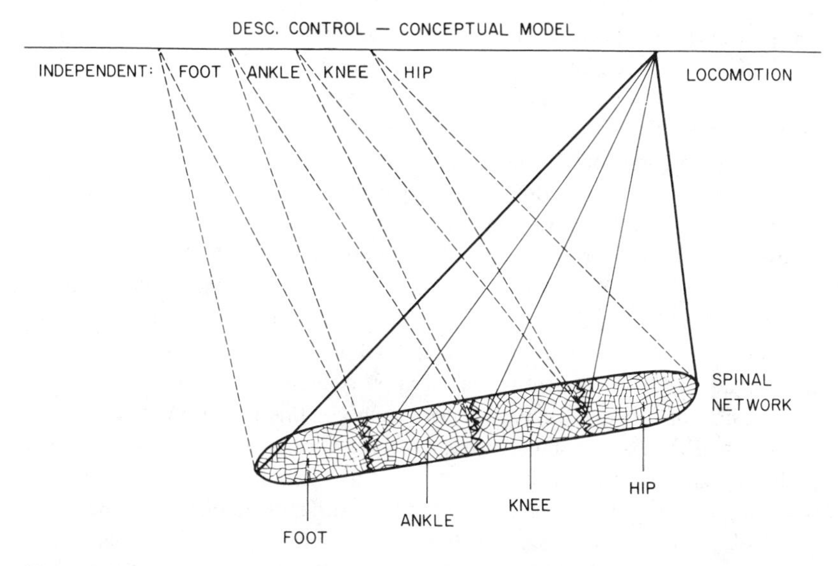

Figure 11 Fractionation of the central pattern generator network for locomotion. Schematic representation showing that an activation of the entire network will give rise to locomotion, but that an independent control of parts of this network could be utilized (in for instance the volitional control of independent ankle, knee, or foot movements and even in the activation of individual muscle groups). From Grillner (1982).

tends to be. Birds and mammals often have a complex prehensile ability with their feet. A falcon or an eagle may grasp a victim with as high a precision as a tit may use its feet to grasp the thin branch of a bush. Moreover, some birds use their feet in learned motor tasks to grasp or pull on certain objects (Thorpe 1963). A mammal like a rat, a cat, or a monkey will grasp objects with its forelimbs and has the ability to perform precise movements of individual joints. This ability to fractionate the motor pattern to smaller components of the whole limb synergy develops gradually in phylogeny from an independent control of the large joints to a precise control of the individual finger joints, as in primates. In the baboon, a transection of the direct corticospinal fibers will cause a permanent loss of independent finger movements (Lawrence & Kuypers 1968a) with such small effects on the general behavioral repertoire that they are barely detectable. If a lesion including the rubrospinal tract is added, the baboon will in addition lose independent wrist movements, and it will have to use the limb with entire flexion or extension synergies when grasping objects such as food (Lawrence & Kuypers 1968a,b). During ontogeny the reverse development occurs in the human infant (cf Forssberg 1983). These fractionated movement patterns could be elicited either (*a*) by motor systems that have evolved separately from the motor systems controlling the coarse movement synergies, not utilizing the segmental motor apparatus used in whole limb movements, or (*b*) by descending control systems that acquire a separate control over part of the spinal network (Grillner 1982), for instance over separate unit CPGs or their components. Under such conditions the unit CPG for the big toe could be used in wiggling the big toe and that of the elbow in other tasks. The descending control over the segmental apparatus in lower tetrapods is thus relatively stereotyped, but a more refined control develops gradually during evolution. Descending systems could gradually acquire under these conditions a more specific control over parts of the spinal circuitry (Figure 11). The refined precision movements referred to as "voluntary" would thus use part of the old segmental motor apparatus, which would provide ready-made modules for activation of appropriate motor nuclei and possible suppression of others. The explanatory value of this possibility would seem greater than the former alternative, which requires that new systems evolve each time a new type of voluntary movement is acquired. As far as we know it is compatible with all available experimental data (see Grillner 1982).

All movements commanded volitionally, regardless of the type, should be designated as "voluntary," irrespective of whether they represent locomotion or learned precision movements with the fingers. Precision movements might be elicited by a specifically timed activation of small portions of a fragmented or mosaic segmental motor apparatus (Grillner 1981, 1982). This line of reasoning can be extended to the relation between the pattern generators for respiration and chewing, or for vocalization and speech (Grillner 1982).

CONCLUDING REMARKS

We have limited this review to locomotor CPGs, and have discussed in only the most general sense how activity in CPGs is initiated and maintained, and what effects the feedback control may have on the CPGs. Much knowledge has been acquired about both these topics over the last few years, and they could in themselves be subjects for separate reviews. Our understanding of locomotor CPGs in general and their overall activity has markedly expanded over the last decade. Although our knowledge of intrinsic CPG mechanisms is still limited, we have begun unravelling the different mechanisms. Is it really important to understand how a CPG operates? Would it not be sufficient to accept that localized groups of nerve cells can produce a patterned output and to describe the detailed behavior of the network itself? It is clearly not sufficient, since in order to make progress in understanding how motor acts are coordinated, how higher centers control movements, and how feedback signals interact with the CPGs, we must know what is inside the black box that we call a vertebrate CPG.

ACKNOWLEDGMENTS

The dedicated and never-failing help of Mrs. I. Klingebrant is gratefully acknowledged, as well as the support from the Swedish Medical Research Council (Project number 3026).

Literature Cited

Andersson, O., Grillner, S. 1981. Peripheral control of the cat's step cycle. I. Phase dependent effect of ramp-movements of the hip during "fictive locomotion." *Acta Physiol. Scand.* 113:89–101

Andersson, O., Grillner, S. 1983. Peripheral control of the cat's step cycle. II. Entrainment of the central pattern generators for locomotion by sinusoidal hip movements during "fictive locomotion." *Acta Physiol. Scand.* 118:229–39

Arshavsky, Yu., Gelfand, I. M., Orlovsky, G. N. 1983. The cerebellum and control of rhythmical movements. *Trends Neurosci.* 6:417–22

Bässler, U. 1983. Neural basis of elementary behavior in stick insects. In *Studies of Brain Function*, ed. H. B. Barlow, T. H. Bullock, E. Florey, O. J. Grüsser, A. Peters, 10:1–167. Berlin/Heidelberg: Springer

Boisson, M., Gola, M. 1976. Dual effects of catecholamines on the burst production by Aplysia neurons. In *Abnormal Neural Discharges*, ed. C. Chalazonitis, M. Boisson, pp. 263–70. New York: Raven

Brown, T. G. 1911. The intrinsic factors in the act of progression in the mammal. *Proc. R. Soc. B London Ser.* 84:308–19

Brown, T. G. 1914. On the nature of the fundamental activity of the nervous centres; together with an analysis of the conditioning of rhythmic activity in progression, and a theory of the evolution of function in the nervous system. *J. Physiol. London* 48:18–46

Buchanan, J. T., Cohen, A. H. 1982. Activities of identified interneurons, motoneurons, and muscle fibers during fictive swimming in the lamprey and effects of reticulospinal and dorsal cell stimulation. *J. Neurophysiol.* 47:948–60

Bullock, T. H. 1982. Electroreception. *Ann. Rev. Neurosci.* 5:121–70

Burrows, M. 1980. The control of sets of motoneurones by local interneurones in the locust. *J. Physiol.* 298:213–34

Calabrese, R. L. 1979. The roles of endogenous membrane properties and synaptic interaction in generating the heartbeat rhythm of the leech, *Hirudo medicinalis*. *J. Exp. Biol.* 82:163–76

Carlson, H., Halbertsma, J., Zomlefer, M. 1979. Control of the trunk during walking in the cat. *Acta Physiol Scand.* 105:251–53

Cohen, A. H., Harris-Warrick, R. M. 1984. Strychnine eliminates alternating motor output during fictive locomotion in the lamprey. *Brain Res.* 293:164–67

Cohen, A. H., Wallén, P. 1980. The neuronal correlate of locomotion in fish. "Fictive swimming" induced in an in vitro preparation of the lamprey spinal cord. *Exp. Brain Res.* 41:11–18

Delcomyn, F. 1980. Neural basis of rhythmic behavior in animals. *Science* 210:492–98

Deliagina, T. G., Feldman, A. G., Gelfand, I. M., Orlovsky, G. N. 1975. On the role of central program and afferent inflow in the control of scratching movements in the cat. *Brain Res.* 100:297–313

Ducreux, C. 1976. Pharmacological alteration of the ionic currents underlying the slow potential wave generation in molluscan neurons. In *Abnormal Neural Discharges*, ed. C. Chalazonitis, M. Boisson, pp. 271–86. New York: Raven

Duysens, J., Pearson, K. G. 1980. Inhibition of flexor burst generation by loading ankle extensor muscles in walking cats. *Brain Res.* 187:321–32

Edgerton, V. R., Grillner, S., Sjöström, A., Zangger, P. 1976. Central generation of locomotion in vertebrates. In *Neural Control of Locomotion*, ed. R. Herman, S. Grillner, P. Stein, D. Stuart, 18:439–64. New York: Plenum

Eidelberg, E., Walden, J. G., Nguyen, L. H. 1981. Locomotor control in macaque monkeys. *Brain* 104:647–64

Eisen, J. S., Marder, E. 1982. Mechanisms underlying pattern generation in lobster stomatogastric ganglion as determined by selective inactivation of identified neurons. III. Synaptic connections of electrically coupled pyloric neurons. *J. Neurophysiol.* 48:1392–1415

Engberg, I., Lundberg, A. 1969. An electromyographic analysis of muscular activity in the hindlimb of the cat during unrestrained locomotion. *Acta Physiol. Scand.* 75:614–30

English, A. W. M. 1980. The functions of the

lumbar spine during stepping in the cat. *J. Morphol.* 165:55–66

Feldman, J. L., Grillner, S. 1983. Control of vertebrate respiration and locomotion: A brief account. *Physiologist* 26:310–16

Feldman, A. G., Orlovsky, G. N. 1975. Activity of interneurons mediating reciprocal Ia inhibition during locomotion. *Brain Res.* 84:181–94

Flatman, J. A., Schwindt, P. C., Crill, W. E., Stafstrom, C. E. 1983. Multiple actions of N-*Methyl*-D-*asparate* on cat neocortical neurones in vitro. *Brain Res.* 266:169–73

Forssberg, H. 1983. Centralnervösa reglerIngsmekanismer för handens motorik. In *Utveckling av centralnervösa regleringsmekanismer för människans motorik,* pp. 181–207. Stockholm: RBU-Folksam

Forssberg, H., Grillner, S., Halbertsma, J. 1980a. The locomotion of the low spinal cat. I. Coordination within a hindlimb. *Acta Physiol. Scand.* 108:269–81

Forssberg, H., Grillner, S., Halbertsma, J., Rossignol, S. 1980b. The locomotion of the low spinal cat. II. Interlimb coordination. *Acta Physiol. Scand.* 108:283–95

Friesen, W. O., Stent, G. S. 1978. Neural circuits for generating rhythmic movements. *Ann. Rev. Biophys. Bioeng.* 7:37–61

Garcia-Rill, E. 1983. Connections of the mesencephalic locomotor region (MLR). III. Intracellular recordings. *Brain Res. Bull.* 10:73–81

Garcia-Rill, E., Skinner, R. D., Gilmore, S. A., Owings, R. 1983a. Connections of the mesencephalic locomotor region (MLR). II. Afferents and efferents. *Brain Res. Bull.* 10:63–71

Garcia-Rill, E., Skinner, R. D., Jackson, M. B., Smith, M. M. 1983b. Connections of the mesencephalic locomotor region (MLR). I. Substantia nigra afferents. *Brain Res. Bull.* 10:57–62

Getting, P. A. 1977. Neuronal organization of escape swimming in *Tritonia. J. Comp. Physiol.* 121:325–42

Getting, P. A. 1981. Mechanisms of pattern generation underlying swimming in *Tritonia.* I. Neuronal network formed by monosynaptic connections. *J. Neurophysiol.* 46:65–79

Getting, P. A. 1983a. Mechanisms of pattern generation underlying swimming in *Tritonia.* II. Network reconstruction. *J. Neurophysiol.* 49:1017–35

Getting, P. A. 1983b. Mechanisms of pattern generation underlying swimming in *Tritonia.* III. Intrinsic and synaptic mechanisms for delayed excitation. *J. Neurophysiol.* 49:1036–50

Getting, P. A., Lennard, P. R., Hume, R. I., 1980. Central pattern generator mediating swimming in *Tritonia.* I. Identification and synaptic interactions. *J. Neurophysiol.* 44:151–64

Gola, M., Selverston, A. 1981. Ionic requirements for bursting activity in lobster stomatogastric neurons. *J. Comp. Physiol.* 145:191–207

Grillner, S. 1974. On the generation of locomotion in the spinal dogfish. *Exp. Brain. Res.* 20:459–70

Grillner, S. 1975. Locomotion in vertebrates: Central mechanisms and reflex interaction. *Physiol. Rev.* 55:247–304

Grillner, S. 1977. On the neural control of movement—a comparison of different basic rhythmic behaviors. In *Dahlem Workshop on Function and Formation of Neural Systems,* ed. G. S. Stent, pp. 197–224. Berlin: Dahlem Konferenzen

Grillner, S. 1981. Control of locomotion in bipeds, tetrapods and fish. *Handb. Physiol. Sect. 1, Vol. 2, The Nervous System, Motor Control,* ed. V. B. Brooks, pp. 1179–1236. Am. Physiol. Soc. Maryland: Waverly Press

Grillner, S. 1982. Possible analogies in the control of innate motor acts and the production of sound in speech. In *Speech Motor Control,* ed. S. Grillner, B. Lindblom J. Lubker, A. Persson. *WennerGren Cent. Int. Symp. Ser.* 36:217–29

Grillner, S., McClellan, A., Sigvardt, K., Wallén, P., Wilén, M. 1981. Activation of NMDA-receptors elicits "fictive locomotion" in lamprey spinal cord in vitro. *Acta Physiol. Scand.* 113:549–51

Grillner, S., McClellan, A., Sigvardt, K., Wallén, P., Williams, T. 1982. On the neural generation of "fictive locomotion" in a lower vertebrate nervous system, in vitro. In *Brain Stem Control of Spinal Mechanisms,* ed. B. Sjölund, A. Björklund, pp. 273–95, Fernström Found. Ser. No. 1. New York/Oxford: Elsevier

Grillner, S., Rossignol, S. 1978. On the initiation of the swing phase of locomotion in chronic spinal cats. *Brain Res.* 146:269–77

Grillner, S., Wallén, P. 1980. Does the central pattern generation for locomotion in lamprey depend on glycine inhibition? *Acta Physiol. Scand.* 110:103–5

Grillner, S., Wallén, P. 1982. On peripheral control mechanisms acting on the central pattern generators for swimming in the dogfish. *J. Exp. Biol.* 98:1–22

Grillner, S., Wallén, P. 1984. How does the lamprey CNS make the lamprey swim? *J. Exp. Biol.* In press

Grillner, S., Wallén, P., McClellan, A., Sigvardt, K., Williams, T., Feldman, J. 1983. The neural generation of locomotion in the lamprey—an incomplete account. In *Neural Origin of Rhythmic Movements,* ed. A. Roberts, B. Roberts, pp. 285–303. Brighton: Symp. Soc. Exp. Biol.

Grillner, S., Zangger, P. 1974. Locomotor movements generated by the deafferented spinal cord. *Acta Physiol. Scand*. 91:38–39A

Grillner, S., Zangger, P. 1975. How detailed is the central pattern generator for locomotion? *Brain Res*. 88:367–71

Grillner, S., Zangger, P. 1979. On the central generation of locomotion in the low spinal cat. *Exp. Brain Res*. 34:241–61

Grillner, S., Zangger, P. 1984. The effect of dorsal root transection on the efferent motor pattern in the cat's hindlimb during locomotion. *Acta Physiol. Scand*. 120:393–405

Gustafsson, B. 1974. Afterhyperpolarization and the control of repetitive firing in spinal neurones of the cat. *Acta Physiol. Scand. Suppl*. 416

Halbertsma, J. 1983. The stride cycle of the cat: The modelling of locomotion by computerized analysis of automatic recordings. *Acta Physiol. Scand. Suppl*. 521

Hökfelt, T., Skagerberg, G., Skirboll. L., Björklund, A. 1983. Combination of retrograde tracing and neurotransmitter histochemistry. In *Handbook of Chemical Neuroanatomy, Vol. 1. Methods in Chemical Neuroanatomy*, ed. A. Björklund, T. Hökfelt, pp. 228–85. Amsterdam: Elsevier

Hume, R. I., Getting, P. A. 1982. Motor organization of *Tritonia* swimming. II. Synaptic drive to flexion neurons from premotor interneurons. *J. Neurophysiol*. 47:75–90

Hume, R. I., Getting, P. A., Del Beccaro, M. A. 1982. Motor organization of *Tritonia* swimming. I. Quantitative analysis of swim behavior and flexion neuron firing patterns. *J. Neurophysiol*. 47:60–74

Jankowska, E., Jukes, M. G. M., Lund, S., Lundberg, A. 1967a. The effect of DOPA on the spinal cord. 5. Reciprocal organization of pathways transmitting excitatory action to alpha motoneurones of flexors and extensors. *Acta Physiol. Scand*. 70:369–88

Jankowska, E., Jukes, M. G. M., Lund, S., Lundberg, A. 1967b. The effect of DOPA on the spinal cord. 6. Half-centre organization of interneurones transmitting effects from the flexor reflex afferents. *Acta Physiol. Scand*. 70:389–402

Jefferys, J. G. R., Haas, H. L. 1982. Synchronized bursting of CAI hippocampal pyramidal cells in the absence of synaptic transmission. *Nature* 300:448–50

Jordan, L. M., Pratt, C. A., Menzies, J. E. 1980. Intraspinal mechanisms for the control of locomotion. In *Regulatory Functions of the CNS Motion and Organization Principles*, ed. J. Szentagothai, M. Palkovits, J. Hamori. *Adv. Physiol. Sci*. 1:183–85

Kahn, J. A., Roberts, A. 1982a. The central nervous origin of the swimming motor pattern in embryos of Xenopus laevis. *J. Exp. Biol*. 99:185–96

Kahn, J. A., Roberts, A. 1982b. The neuromuscular basis of rhythmic struggling movements in embryos of Xenopus laevis. *J. Exp. Biol*. 99:197–205

Kahn, J. A., Roberts, A., Kashin, S. 1982. The neuromuscular basis of swimming movements in embryos of the amphibian Xenopus laevis. *J. Exp. Biol*. 99:175–84

Kandel, E. R., Schwartz, J. H. 1982. Molecular biology of learning: Modulation of transmitter release *Science*. 218:433–43

Kristan, W. B., Weeks, J. C. 1983. Neurons controlling the initiation, generation, and modulation of leech swimming. See Perret 1983, pp. 243–60

Kriz, N., Sykova, E., Ujec, E., Vyklicky, L. 1974. Changes of extracellular potassium concentration induced by neuronal activity in the spinal cord of the cat. *J. Physiol. London* 238:1–15

Lawrence, D. G., Kuypers, H. G. J. M. 1968a. The functional organization of the motor system in the monkey. I. The effects of bilateral pyramidal lesions. *Brain* 91:1–14

Lawrence, D. G., Kuypers, H. G. J. M. 1968b. The functional organization of the motor system in the monkey. II. The effects of lesions of the descending brainstem pathways. *Brain* 91:15–36

Lundberg, A. 1969. The excitatory control of the Ia inhibitory pathway. In *Excitatory Synaptic Mechanisms*, ed. P. Andersen, J. K. S. Jansen, pp. 333–40. Oslo: Universitetsforlaget

Lundberg, A. 1980. Half-centres revisited. In Regulatory Functions of the CNS. Motion and Organization Principles, ed. J. Szentagothai, M. Palkovits, J. Hamori, *Adv. Physiol. Sci*. 1:155–67. Budapest: Akademiai Kiado

Madison, D. V., Nicoll, R. A. 1982. Noradrenalin blocks accommodation of pyramidal cell discharge in the hippocampus. *Nature* 299:636–38

McCrea, D. A., Pratt, C. A., Jordan, L. M. 1980. Renshaw cell activity and recurrent effects on motoneurons during fictive locomotion. *J. Neurophysiol*. 44:475–88

MacDonald, J. F., Porietis, A. V., Wojtowicz, J. M. 1982. L-aspartic acid induces a region of negative slope conductance in the current-voltage relationship of cultured spinal cord neurons. *Brain Res*. 237:248–53

Meech, R. W. 1978. Calcium-dependent potassium activation in nervous tissues. *Ann. Rev. Biophys. Bioeng*. 7:1–18

Miller, J. P., Selverston, A. I. 1982. Mechanisms underlying pattern generation in lobster stomatogastric ganglion as determined by selective inactivation of identified neurons. II. Oscillatory properties of pyloric neurons. *J. Neurophysiol*. 48:1378–91

Orlovsky, G. N. 1969. Spontaneous and induced locomotion of the thalamic cat. *Biofizika* 14:1095–1102. (Eng. transl. 1154–62)

Orlovsky, G. N. 1972. Activity of rubrospinal neurons during locomotion. *Brain Res.* 46:99–112

Orlovsky, G. N., Feldman, A. G. 1972. Classification of lumbosacral neurons according to their discharge patterns during evoked locomotion. *Nejrofisiologia* 4:410–17 (in Russian), (Engl. transl. 311–17)

Orlovsky, G. N., Pavlova, G. A. 1972. Response of Deiters' neurones to tilt during locomotion. *Brain Res.* 42:212–14

Pearson, K. G. 1976. The control of walking. The nerve mechanisms that generate the leg movements are similar in the cat and the cockroach. *Sci. Am.* 235:72–87

Pearson, K. G., Reye, D. N., Robertson, R. M. 1983. Phase-dependent influences of wing stretch receptors on flight rhythm in the locust. *J. Neurophysiol.* 49:1168–81

Perret, C., Cabelguen, J.-M. 1980. Main characteristics of the hindlimb locomotor cycle in the decorticate cat with special reference to bifunctional muscles. *Brain Res.* 187:333–52

Perret, C. 1983. Centrally generated pattern of motoneuron activity during locomotion in the cat. In *Neural Origin of Rhythmic Movements,* ed. A. Roberts, B. Roberts, pp. 405–22. Brighton: Symp. Soc. Exp. Biol.

Roberts, A., Kahn, J. A. 1982. Intracellular recordings from spinal neurones during "swimming" in paralysed amphibian embryos. *Philos. Trans. R. Soc. London Ser. B* 296:213–28

Russell, D. F., Hartline, D. K. 1978. Bursting neural networks: A reexamination. *Science* 200:453–56

Russell, D. F., Hartline, D. K. 1982. Slow active potentials and bursting motor patterns in pyloric network of the lobster, *Panulirus interruptus. J. Neurophysiol.* 48:914–37

Russell, D. F., Wallén, P. 1983. On the control of myotomal motoneurones during "fictive swimming" in the lamprey spinal cord in vitro. *Acta Physiol. Scand.* 117:161–70

Schwindt, P. C., Crill, W. E. 1982. Factors influencing motoneuron rhythmic firing: Results from a voltage-clamp study. *J. Neurophysiol.*48:875–90

Selverston, A. I., Russell, D. F., Miller, J. P., King, D. G. 1976. The stomatogastric nervous system: Structure and function of a small neural network. *Progr. Neurobiol.* 7:215–90

Shefchyk, S. J., Menzies, J. E., Jordan, L. M. 1981. Evidence for rhythmic excitation and inhibition of flexor and extensor motoneurons during fictive locomotion. *Soc. Neurosci. Abstr.* 7:687

Shik, M. L. 1980. Control of locomotion. In *Regulatory Functions of the CNS. Motion and Organization Principles,* ed. J. Szentagothai, M. Palkovits, J. Hamori, *Adv. Physiol. Sci.,* pp. 143–48. Budapest: Akademiai Kiado

Shik, M. L., Orlovsky, G. N. 1976. Neurophysiology of locomotor automatism. *Physiol. Rev.* 56:465–501

Shik, M. L., Orlovsky, G. N., Severin, F. V. 1966. 11:1011–19 (Eng. transl.)

Siegler, M. V., Burrows, M. 1980. Non-spiking interneurones and local circuits. *Trends Neurosci.* 3:73–77

Sigvardt, K. A., Grillner, S. 1981. Spinal neuronal activity during fictive locomotion in the lamprey. *Soc. Neurosci.* 7:362

Soffe, S. R., Roberts, A. 1982. Tonic and phasic synaptic input to spinal cord motoneurons during fictive locomotion in frog embryos. *J. Neurophysiol.* 48:1279–88

Stehouwer, D. J., Farel, P. B. 1980. Central and peripheral controls of swimming in Anuran larvae. *Brain Res.* 195:323–35

Stehouwer, D. J., Farel, P. B. 1981. Sensory interactions with a central motor program in Anuran larvae. *Brain Res.* 218:131–40

Stein, P. S. G. 1978. Motor systems with specific reference to the control of locomotion. *Ann. Rev. Neurosci.* 1:61–82

Stein, P. S. G. 1983. The vertebrate scratch reflex. See Perret 1983, pp. 383–403

Stein, P. S. G., Grossman, M. L. 1980. Central program for scratch reflex in turtle. *J. Comp. Physiol.* 140:287–94

Szekely, G. 1976. Developmental aspects of locomotion. In *Neural Control of Locomotion,* ed. R. Herman, S. Grillner, P. Stein, D. Stuart, 18:735–58. New York: Plenum

Thorpe, W. H. 1963. *Learning and Instinct in Animals.* Cambridge: Harvard Univ. Press

Thorstensson, A., Carlson, H., Zomlefer, M. R., Nilsson, J. 1982. Lumbar back muscle activity in relation to trunk movements during locomotion in man. *Acta Physiol. Scand.* 116:13–20

Wallén, P., Grafe, P., Grillner, S. 1984a. Phasic variations of extracellular potassium during fictive swimming in the lamprey spinal cord in vitro. *Acta Physiol. Scand.* 20:457–63

Wallén, P., Grillner, S., Feldman, J., Bergelt, S. 1984b. Dorsal and ventral myotome motoneurons and their input during fictive locomotion in lamprey. *J. Neurosci.* In press

Wallén, P., Williams, T. L. 1984. Fictive locomotion in the lamprey spinal cord in vitro compared with swimming in the intact and spinal animal. *J. Physiol.* 347:225–39

Wendler, G. 1974. The influence of proprioceptive feedback in locust flight co-ordination. *J. Comp. Physiol.* 88:173–200

Wendler, G. 1978. Lokomotion: das Ergebnis central-peripherer Interaktion. *Verh. Dtsch. Zool. Ges.* 80–96

Wiersma, C. A. G., Ikeda, K. 1964. Interneurons commanding swimmeret movements in the crayfish, *Procambarus clarki (Girard)*. *Comp. Biochem. Physiol.* 12:509–25

Willard, A. L. 1981. Effects of serotonin on the generation of the motor program for swimming by the medicinal leech. *J. Neurosci.* 1:936–44

Wilson, D. M. 1964. The origin of the flight-motor command in grasshoppers. In *Neuronal Theory and Modelling*, ed. R. F. Reiss, pp. 331–45. Stanford: Stanford Univ. Press

Wilson, D. M. 1966. Insect walking. *Ann. Rev. Entomol.* 11:103–22

Yarom, Y., Grossman, Y., Gutnick, M. J., Spira, M. E. 1982. Transient extracellular potassium accumulation produced prolonged depolarizations during synchronized bursts in picrotoxin-treated cockroach CNS. *J. Neurophysiol.* 48:1089–97

Zomlefer, M. R., Provencher, J., Blanchette, G., Rossignol, S. 1984. Electromyographic study of lumbar back muscles during locomotion in acute high decerebrate and in low spinal cats. *Brain Res.* 290:249–60

Ann. Rev. Neurosci. 1985. 8:263-305

REAL-TIME OPTICAL MAPPING OF NEURONAL ACTIVITY: From Single Growth Cones to the Intact Mammalian Brain

*Amiram Grinvald**

Department of Neurobiology, The Weizmann Institute of Science, Rehovot, 76100, Israel

INTRODUCTION

Electrophysiology is at present the most commonly used approach with both the spatial and the temporal resolution required to investigate the real time function of individual neurons or of neuronal networks. However, it suffers from two major limitations: for the investigation of single neurons, the use of microelectrodes usually cannot provide simultaneous recordings from more than two sites on the neuron's arborization, thus severely hampering understanding of the detailed integration function of even the basic computational elements. Similarly, studying the cellular basis for function of a neuronal network requires the simultaneous recording form many individual cells, but electrophysiological methods can give information about the behavior of only a few neurons. Other approaches do not share some of the powerful advantages of electrophysiological techniques; for example, the 2-deoxyglucose method (Sokoloff 1977), which permits high resolution localization of active neurons (Sejnowski et al 1980, Lancet et al 1982), has a time resolution of minutes or hours rather than milliseconds. Positron-emission tomography (Raichle 1979), which offers three-dimensional localization of active regions in the live brain, lacks both time and spatial resolution. A newly developed magnetic imaging technique (Kaufman & Williamson 1982) also shares some of these drawbacks.

*Current address: The Rockefeller University, Laboratory of Neurobiology, 1230 York Avenue, New York, New York 10021.

0147-006X/85/0301-0263$02.00

A powerful alternative approach is offered by monitoring neuronal activity optically with voltage-sensitive dyes. The initial efforts of Tasaki and his collaborators (1968), of Patrick et al (1971), and primarily of L. B. Cohen and his colleagues, have laid the foundation for this promising technique (Davila et al 1973, 1974, Salzberg et al 1973, 1977, Cohen et al 1974, Ross et al 1974, 1977, Grinvald et al 1977, 1981a). The availability of suitable voltage-sensitive dyes and arrays of photodetectors has recently facilitated the optical monitoring of electrical activity in the processes of single nerve cells simultaneously from more than 100 sites, both in culture and in invertebrate ganglia. This method also provides the unique ability of detecting activity in many individual neurons in an entire invertebrate ganglion, while a particular behavioral response is elicited. The in vitro activity of individual populations of neuronal elements (cell bodies, axons, dendrites, or nerve terminals) at many neighboring loci in mammalian brain slices or isolated but intact brain structures has been investigated. Dynamic patterns of electrical activity evoked in the intact vertebrate or mammalian brain, by a natural stimulus, were also recently monitored optically. By employing computerized optical recording and a display processor, video-displayed images of neuronal elements can be superimposed on the corresponding patterns of the optically detected electrical activity, thus allowing the spatio-temporal patterns of intracellular activity to be visualized in slow motion.

Developments and applications of the technique in new directions in the last five years are described below. For earlier and extensive reviews, see Conti 1975, Waggoner & Grinvald 1977, Cohen et al 1978, Cohen & Salzberg 1978, Salzberg 1983. The applications to skeletal (e.g. Vergara & Bezanilla 1976) and cardiac muscle (e.g. Salama & Morad 1976, Fujii et al 1981a–c) are beyond the scope of this review (see Baylor 1982). For reviews concerning the use of slow permeant dyes (Waggoner 1976) to study small cells and organelles in suspension, see Waggoner & Grinvald 1977, Cohen & Salzberg 1978, Waggoner 1979, and Freedman & Laris 1981.

THE PRINCIPLE OF OPTICAL MONITORING OF NEURONAL ACTIVITY

The optical technique is simple in principle. Voltage-sensitive probe molecules are used to stain vitally a preparation. The bath-applied dye molecules bind to the external surface of excitable membrane, and act as molecular transducers that transform changes in membrane potential into optical signals. The mechanisms of this transduction are discussed elsewhere (Cohen et al 1974, Tasaki & Warashina 1976, Waggoner & Grinvald 1977, Ross et al 1977, Dragsten & Webb 1978, Gupta et al 1981, Loew & Simpson 1981, Loew et al 1984). The resulting changes in the absorption or the emitted fluorescence of the stained cells correlated with their electrical activity are monitored with light-measuring devices. The amplitude of the optical signal is often linearly related to the membrane potential change, with a response time usually in the microsecond range (Davila et al 1974, Cohen et al 1974, Ross et al 1977). The similarity between optical and intracellular recording is illustrated in the inset of Figure 1, which depicts an optical recording (noisy trace) of an action potential and an electrical recording (smooth trace). Note that the time course

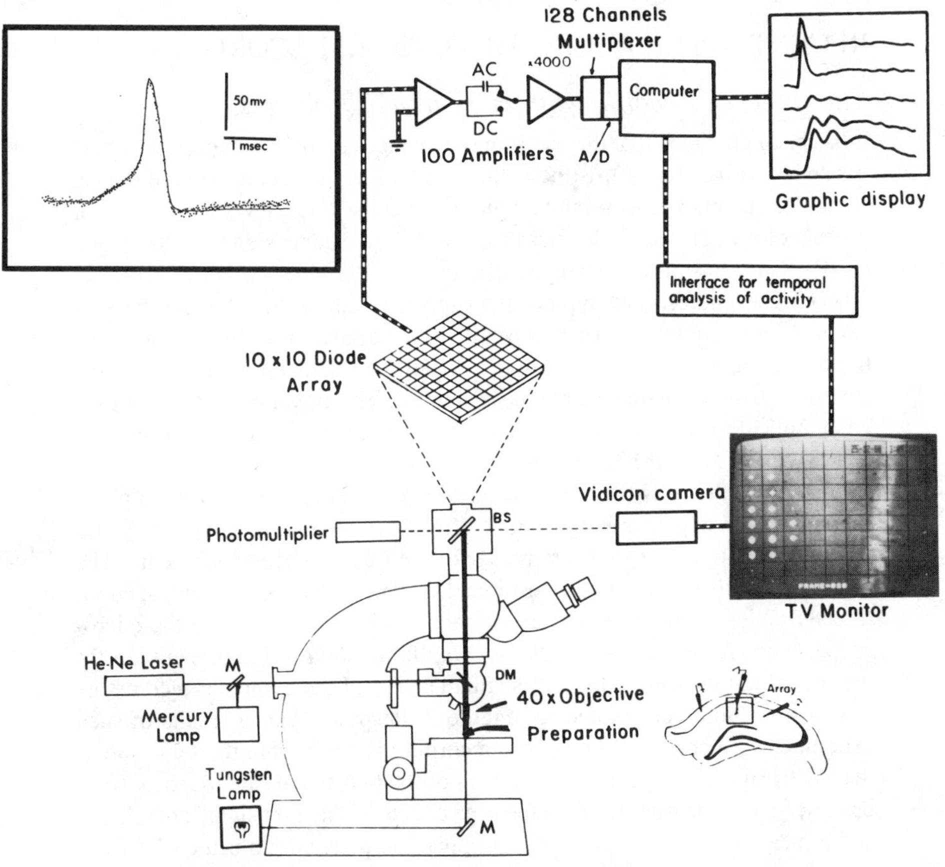

Figure 1 The computer-controlled optical apparatus used to record electrical activity from multiple locations. In transmission measurements, the preparation is illuminated by a tungsten lamp, and optically detected electrical activity is monitored with a 10 × 10 photodiode array. In fluorescence experiments, a He–Ne laser or a 100 W mercury lamp are also employed as light sources. The TV screen shows a slice preparation. The bright hexagon illustrates the spatial pattern of the electrical activity 0.8 msec after the stimulation. *Inset:* comparison between optical and intracellular recordings from squid giant axon. (Figure modified from Grinvald et al 1982c, Ross et al 1977).

of the optical recording is nearly identical to that of the electrode recording. Evidently, by using an array of photodetectors positioned in the microscope image plane, the activity of many individual targets can be detected simultaneously. The first computerized optical monitor of neuronal activity from 100 sites was constructed by Cohen and his colleagues (Grinvald et al 1981a). Figure 1 depicts a similar but improved apparatus (Grinvald et al 1981b, 1982c). The technical aspects of the apparatus and its improvements required in the future are discussed in detail below.

RECENT APPLICATIONS OF OPTICAL RECORDING

Fluorescence Recordings from Neuronal Processes

It has been predicted that the recording of changes in fluorescence, rather than transmission (see below), is the method of choice for obtaining large optical signals from small processes of single nerve cells. [Whenever the number of probe molecules is small, fluorescence changes give better signal-to-noise ratio (SNR) than transmission changes (Rigler et al 1974, Waggoner & Grinvald 1977).] The fluorescence approach is especially appropriate for the study of isolated cells maintained in monolayer culture, because under these conditions, the background fluorescence is minimal and the main source of the fluorescence is dye-bound to the membrane of interest. This prediction was confirmed when optical measurements were attempted on cultured neurons (Grinvald et al 1981b, 1982a,b, 1983, Grinvald & Farber 1981).

To prepare the cells for optical measurements, they are stained with physiological solution containing $1-100 \times 10^{-6}$ M of the fluorescent probe. After one to five minutes the cells are ready for the fluorescence experiments. The part of the cell of interest is positioned at the center of the field of view, directly under the laser or mercury monitoring beam, by moving the microscope stage. An electronic shutter allows illumination of the preparation for the period during which the cell is stimulated, and the computer records the changes in fluorescence intensity. Figure 2 illustrates typical results of such experiments. When the fluorescence recording is made from the same site as the electrical recording, the time courses of the two signals are identical (Figure 2A). Thus, measurements of the time course of the light signal are reliable and can be also used to monitor "intracellular" responses from neuronal regions that are not easily accessible to microelectrode recording, such as very small neurons, neurites (Figure 2B), or growth cones.

Figure 2C illustrates the first simultaneous recording of Ca^{2+} action potentials obtained from the cell body (by means of an electrode) and from a growth cone (using fluorescence) in the presence of tetrodotoxin (TTX) and tetraethylammonium (TEA). Even though the Ca^{2+} action potential was evoked at the soma (by passing current through the electrode), the growth cone Ca^{2+}

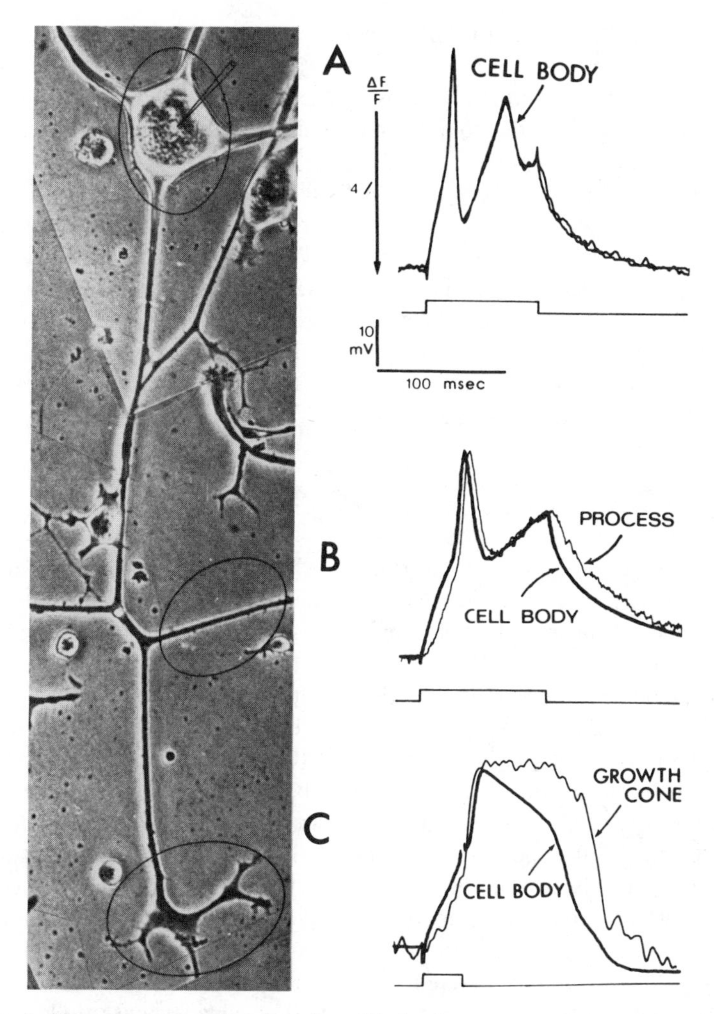

Figure 2 Fluorescence recordings from the cell body, a process, and a growth cone. (*a*) Comparison between electrical recording (*thin trace*) and fluorescence recording (*noisy trace*) from the soma. (*b*) Comparison between the time course of action potential in the soma (*thick trace*) and in a process (2 μm thick) of another cell. (*c*) Comparison of the time course of Ca^{2+} action potential in the cell body (*thick trace*) and in the growth cone recorded from a third cell. The scale for the amplitude of the optical signal, in this and the following figures, shows the fractional change in light intensity. The photograph shows a typical N1E–115 neuroblastoma cell. (Figure modified from Grinvald & Farber 1981, Grinvald et al 1982b, 1983.)

action potential slightly preceded the cell body action potential. (Compare the rising phases of the corresponding action potentials.) These results imply that low threshold voltage-sensitive Ca^{2+} channels are present at or near the growth cone (Grinvald & Farber 1981). Further experiments have shown that under conditions that permit Ca^{2+} entry, growth cones expand or elongate more rapidly than under control conditions, suggesting that Ca^{2+} ions may act as one of the triggers in the complex biochemical chain of events underlying neurite elongation (Anglister et al 1982).

Similar measurements were made on dissociated cells cultured from *Aplysia* abdominal ganglia (I. C. Farber, R. Bodmer, I. Levitan, and A. Grinvald, unpublished observations), bag cells from *Aplysia,* chick ciliary ganglion cells, and rat superior cervical ganglion neurons (J. Pine and A. Grinvald, unpublished results). Simultaneous fluorescence recordings from many individual cells using the diode array have not been made as yet. However, the technical improvements of the fluorescence apparatus to be discussed below (Grinvald et al 1983) now permit such recordings.

Simultaneous Transmission Recordings from Multiple Sites on the Arborization of Single Cells

OPTICAL RECORDINGS IN DISSOCIATED CELL CULTURES The activity of many individual targets can be detected simultaneously using transmission measurements with an array of photodetectors positioned in the microscope image plane. Ross & Reichardt (1979) have pioneered this approach. They failed to find large optical signals from their tissue culture cells, however, because they tested only five dyes. Subsequently, Ross tested about 40 dyes, and discovered the appropriate ones for such experiments (Grinvald et al 1981b). Figure 3 shows the projected image of a large neuroblastoma cell superimposed upon the array. The cell body was stimulated with a microelectrode. The optically detected electrical responses from multiple sites are displayed at their appropriate locations. Fifty trials were averaged in this experiment. The SNR could not be further improved, because with the dye (WW–401) and preparation used, photodynamic damage (dye-related phototoxic effect, discussed below) and bleaching limited the duration of the experiment.

The stimulation of the cell evoked a fast all-or-none response, followed by a second, smaller deflection, known to be due largely to the entry of Ca^{2+} ions. Inspection of the optical recordings (not shown) indicated that the second deflection was smaller in the thin processes than in the cell body and thick processes. This observation was the first hint that Ca^{2+} channels are less abundant or not functional along some of the processes of neuroblastoma cells. This finding was confirmed later by extracellular patch recordings (Anglister et al 1982). In this type of optical experiment, it was discovered that some of the processes of neuroblastoma cells are inexcitable. The conduction velocity and the space constants for the processes were estimated (Grinvald et al 1981b). The same type of experiment was also used to determine synaptic connectivity

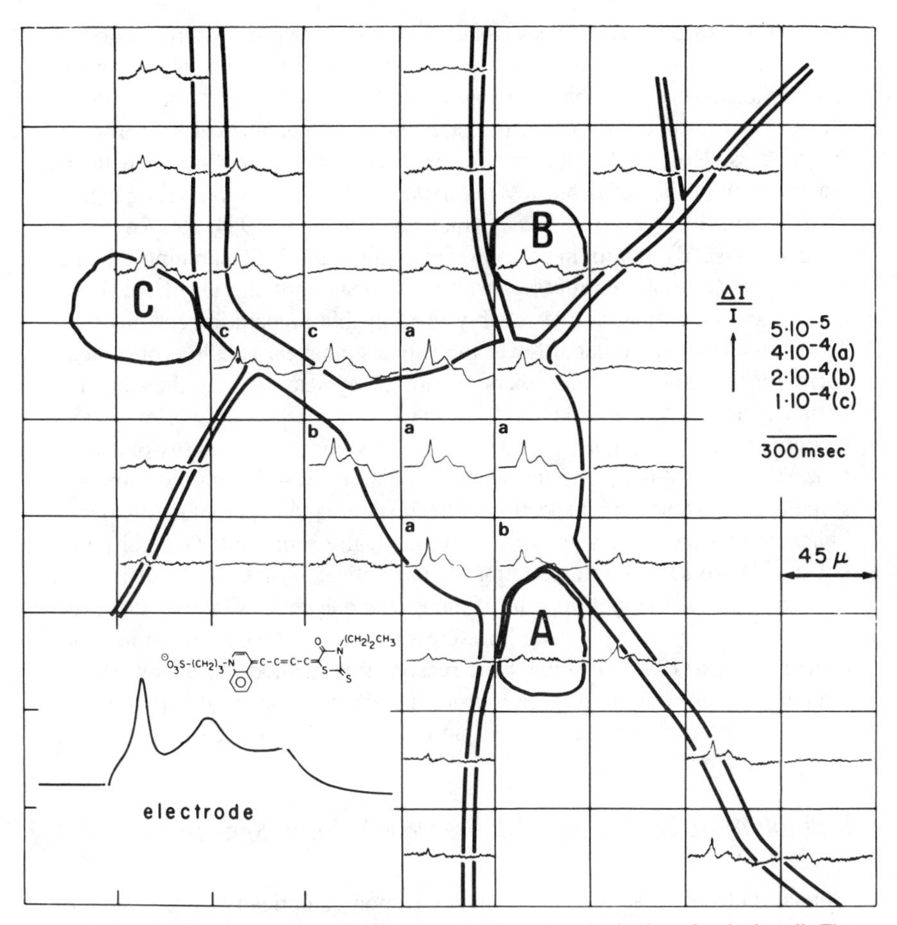

Figure 3 Simultaneous optical recordings changes from the arborization of a single cell. The array is positioned over the real magnified image of a stained N1E–115 neuroblastoma cell. Traces in the individual boxes show the optically detected activity. Some records that do not lie over the cell or its processes are presented as a control at a high gain to show the spatial resolution. The electrical recording and the dye structure are shown in the *inset*. (Figure modified from Grinvald et al 1981b.)

between an electrically stimulated cell and other neurons in the microscope's field of view. It was found that cultured *Aplysia* neurons from the abdominal ganglion were usually electrically coupled, although, interestingly, absence of electrical coupling was frequently detected as well (and then confirmed with microelectrodes). This result suggests some degree of specificity in the formation of electrical synapses (I. Levitan and A. Grinvald, unpublished results). Optical recording from cultured identified leech neurons were also successfully made (A. L. Obaid and B. M. Salzberg, personal communication).

OPTICAL RECORDING FROM PROCESSES IN INVERTEBRATE GANGLIA The feasibility of similar experiments in the invertebrate central nervous system was first demonstrated in the buccal ganglion of *Navanax* (Grinvald et al 1981a). However, the authors have not investigated any physiological properties of the process. Ross & Krauthamer investigated the regional variation in the electrical properties of dendrites and axons in the supraesophageal ganglion of the barnacle *Balanus nabilus* (Krauthamer & Ross 1984; Ross & Krauthamer 1984). By improving the time resolution and by combining the anatomical structure of the processes with the optical mapping, they have shown that optical recording can be reliably used in this preparation for this task. They found that some dendrites in the supraesophageal ganglion of the barnacle are excitable, whereas others are not. They also detected the spike initiation zone along the axons and estimated the space constants of various neuronal processes. Figure 4 shows the evidence for the excitability of a small process, presumably a dendrite, which fired first, even though the soma was stimulated. Furthermore inspection of the responses to hyperpolarizing pulses indicated that the three sites were electrotonically separated. Generally (but not in this work), the interpretation of the optical signals obtained in such experiments is not straightforward if more than one cell was active, however, because all the cells in the ganglion were stained (see the discussion in Krauthamer & Ross 1984). For the same reason, this method is not suitable for monitoring postsynaptic responses from the arborization of identified single cells. An approach I believe to be more suitable for such circumstances is described below.

Iontophoretic Injection of Fluorescent Voltage-Sensitive Probes

To record electrical activity and synaptic responses optically from the site of the synapses of single nerve cells of intact CNS preparations (rather than in isolated cells in culture), voltage sensitive dyes can be intracellularly injected, thereby staining only the cell under investigation. [Signals may be obtained when the membrane-bound probe is inside or outside, but such signals are of opposite sign (Cohen et al 1974, Gupta et al 1981).] This approach was pioneered by B. M. Salzberg; however, all the dyes that he tested were inadequate for the recording of optical signals from the processes. To this end we have designed and synthesized voltage-sensitive fluorescent probes optimized for iontophoretic injection (Agmon et al 1982, Grinvald et al 1982a). We tried only fluorescence rather than absorption probes because theoretical calculations (A. Agmon, unpublished results) predict that the fluorescence measurement would provide better SNR. When leech neurons were injected with such a probe (e.g. RH–461), the dye diffused through the processes without adversely affecting

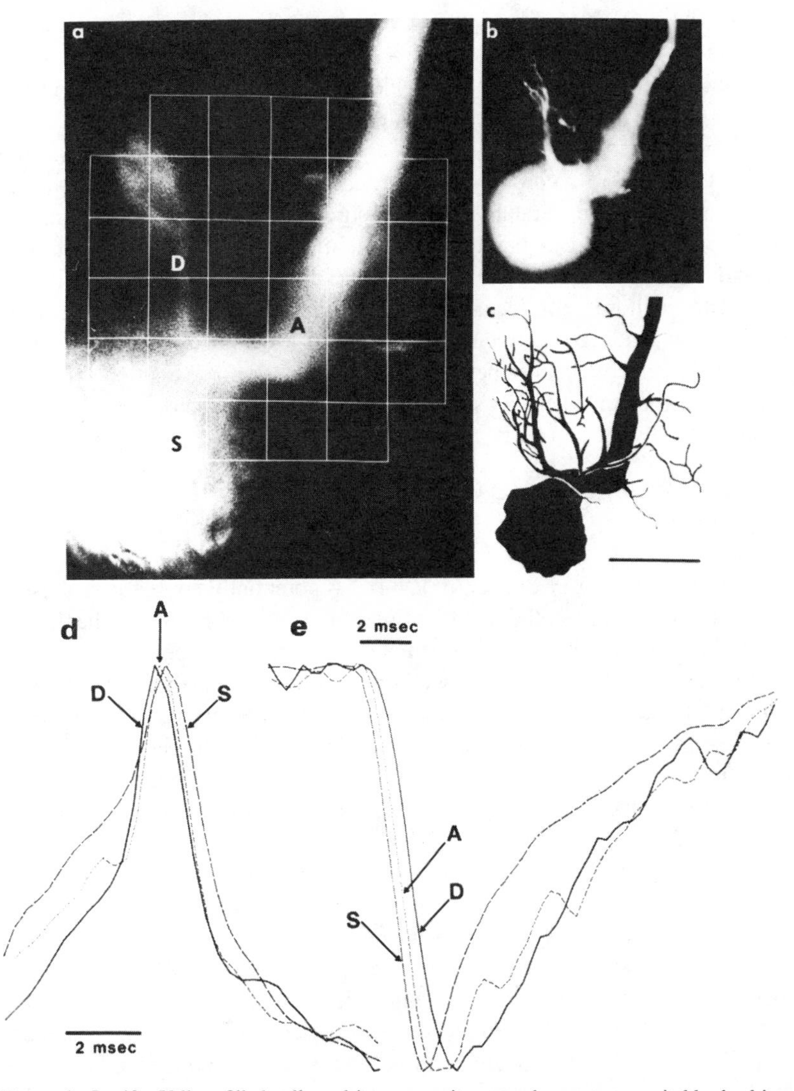

Figure 4 Lucifer Yellow filled cell used in an experiment to demonstrate excitable dendrites. (*a*) Unfixed ganglion viewed with fluorescence optics. The overlay shows the positions of the elements of the photodiode array. S, A, and D refer to the locations of the recordings shown below and indicate soma, axon, and dendrite. (*b*) Fluorescence photograph of the fixed and cleared ganglion. (*c*) Reconstruction of the same cell based on a series of fluorescence photographs. (*d*) Simultaneous absorption changes corresponding to action potentials recorded on photodiodes positioned over the soma (S), dendritic field (D), and axon (A). (*e*) Absorption changes corresponding to hyperpolarizing pulses at the same three positions. (Figure modified from Krauthamer & Ross 1984.)

the cell viability, electrical properties, or synaptic input. Changes in membrane potential in a given illuminated segment of the neuronal arborization in the neuropile were monitored optically, during cell body stimulation. Furthermore, even small (~1–4 mV) synaptic responses, such as those at the site of the synapses between the P sensory cell and the injected AE or L motor neurons, were also recorded. As expected, the neuropile synaptic potential had a faster time course relative to that recorded in the cell body (A. Grinvald, B. M. Salzberg, A. L. Obaid, R. Hildesheim, and V. Lev-Ram, unpublished results). This technique is more difficult to use than recording from the processes of isolated culture cells, because the background fluorescence from all the intracellular binding sites reduces the SNR 10–20-fold. Nevertheless, intracellular dye injection may be indispensible for the investigation of remote dendrites, which might exhibit semi-autonomous electrical behavior, not reflected at all in the distant cell body.

Localization of Neurons Controlling a Behavioral Response in Invertebrate CNS

MAPPING OF THE SINGLE CELLS INVOLVED A potentially powerful application of the method has been the investigation of the cellular basis of behavior and its modification by experience in simple invertebrates. When suitable invertebrate central nervous systems with large cells and relatively few neurons are studied, the activity of most of the neurons controlling a given behavioral response of the animal can be detected optically and their cell bodies can be identified. This application was pursued originally by Cohen and his colleagues (Salzberg et al 1973, 1977, Grinvald et al 1977, 1981a, Boyle et al 1983, London et al 1985).

The giant barnacle, *Balanus nubilus,* has a stereotypic defensive withdrawal behavior known as the shadow response. The barnacle responds to passing shadows by terminating feeding behavior, rapidly withdrawing its cirri, and closing its opercular plates. Figure 5 depicts the experiments carried out to locate large motor neurons in the ganglion that fire in response to reduced illumination of the ocellus. Analysis of the optical recordings from one hemiganglion led to the identification of five neurons that responded to turning off the light. Many experiments were attempted to locate the small interneurons that mediate the response of the motor neurons (Stuart & Ortel 1978), unfortunately without success, presumably because the cells were small relative to the low magnification and the size of the detector used and they are also nonspiking. The possibility that the dye did not stain these interneurons or did not respond to voltage change was not ruled out. These disappointing results (Grinvald et al 1981a) illustrate the difficulties in locating all of the neurons in the network, even in a relatively simple preparation.

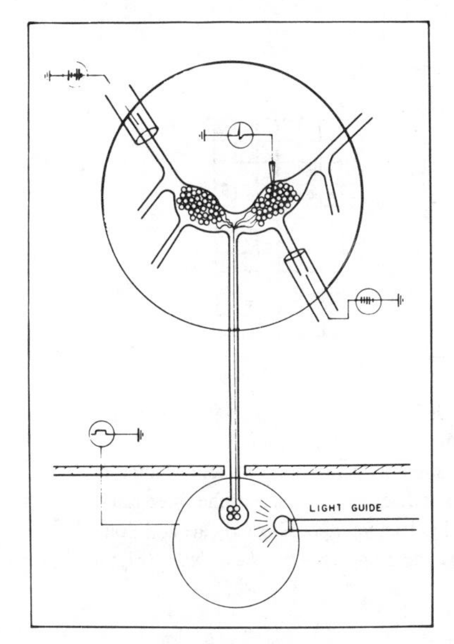

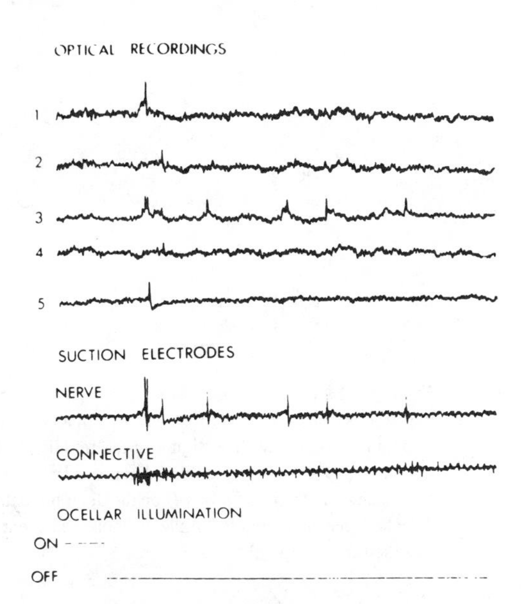

Figure 5 Detection of motor neurons controlling the shadow reflex. (*Left*) Schematic diagram of the experimental arrangement illustrating the supraesophageal ganglion, the occelar nerve, the photoreceptors, and the position of the array. (*Right*) Detection of optical signals (*top traces*) from five neurons that responded to reduced illumination (*bottom trace*) of the ocellus. Suction electrode recordings from the ipsilateral connective and antennular nerve are shown below the optical traces. (Figure modified from Grinvald et al 1981a.)

EVALUATION OF SYNAPTIC CONNECTIVITY The technique was also used to search for "all" of the cells postsynaptic to a given neuron by the detection of small (~1 mV) postsynaptic responses following the stimulation of that neuron. Extensive signal-averaging of 100–200 trials was employed to increase the sensitivity (Grinvald et al 1981a). Figure 6 illustrates the optical detection of a postsynaptic cell in the *Navanax* ganglion. Unfortunately, this powerful capability of optical recording has not yet been used for actual evaluation of neuronal networks. However, progress was achieved in (*a*) the selection of a suitable preparation (Boyle et al 1983), (*b*) computer programming for continuous data acquisition for 20 seconds from 128 channels, and (*c*) the analysis of the huge amounts of information-rich but extremely complicated data (20 million bits) obtained in such recordings (i.e., the sorting of individual neurons when multiple spikes appear on a single optical trace, and the detection of synaptically activated single cells by cross-correlations and by off-line spike-triggered averaging of the responses to each of the single cells that fired).

Figure 7 illustrates the results of an experiment performed on *Aplysia,* after data reduction. The buccal and circumesophageal ganglia were isolated in a dish together with the buccal mass. The cerebral buccal connective was stim-

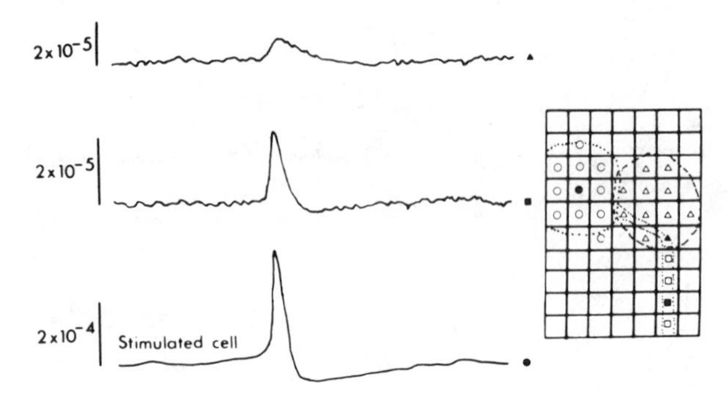

Figure 6 Mapping of postsynaptic cells. Optically recorded synaptic potential (*top trace*) resulting from action potentials (*bottom trace*) in a neuron of the *Navanax* buccal ganglion in a signal-averaging experiment. The signal in the middle trace presumably represents the action potential in a large process of the stimulated neuron. Positions of the detectors for each of the three traces are indicated in the *filled symbols* on the diagram to the right. The *open symbols* indicate additional detectors that had signals similar to those represented by the respective *filled symbols*. (Figure modified from Grinvald et al 1981a.)

ulated at the times indicated on the bottom trace. Tension that developed in the buccal mass was measured simultaneously with the optical measurement of the spike activity in the buccal ganglion. The optically detected patterns of spike activity in 48 neurons are shown at the top. Evidently, these optical data (London et al 1975) demonstrate the capability of the technique to locate many of the neurons involved in the control of a particular behavior. Such measurements may be useful for the design and testing of models that account for how the neurons interact to control behavior and its modification by experience.

Recently, a complementary optical method to identify *presynaptic* neurons (rather than postsynaptic) by laser photostimulation of individual neurons (externally stained with a photostimulation probes, e.g. RH–500) was developed. This technique should also speed up the evaluation of synaptic connectivity in invertebrate ganglia (Farber & Grinvald 1983).

Visualization of Spatio-Temporal Patterns of Population Activity In Vitro

OPTICAL RECORDING FROM MAMMALIAN BRAIN SLICES The application of voltage-sensitive dyes to the investigation of population activity in mammalian brain slices was attempted with rat and guinea pig hippocampus slices (Grinvald et al 1982c, Kuhnt & Grinvald 1982). The hippocampus was chosen for the first application of the technique in this direction because of the high degree of stratification among its various neuronal elements. Such stratification is readily observed in transverse slices (Yamamoto 1972). In such a "three-

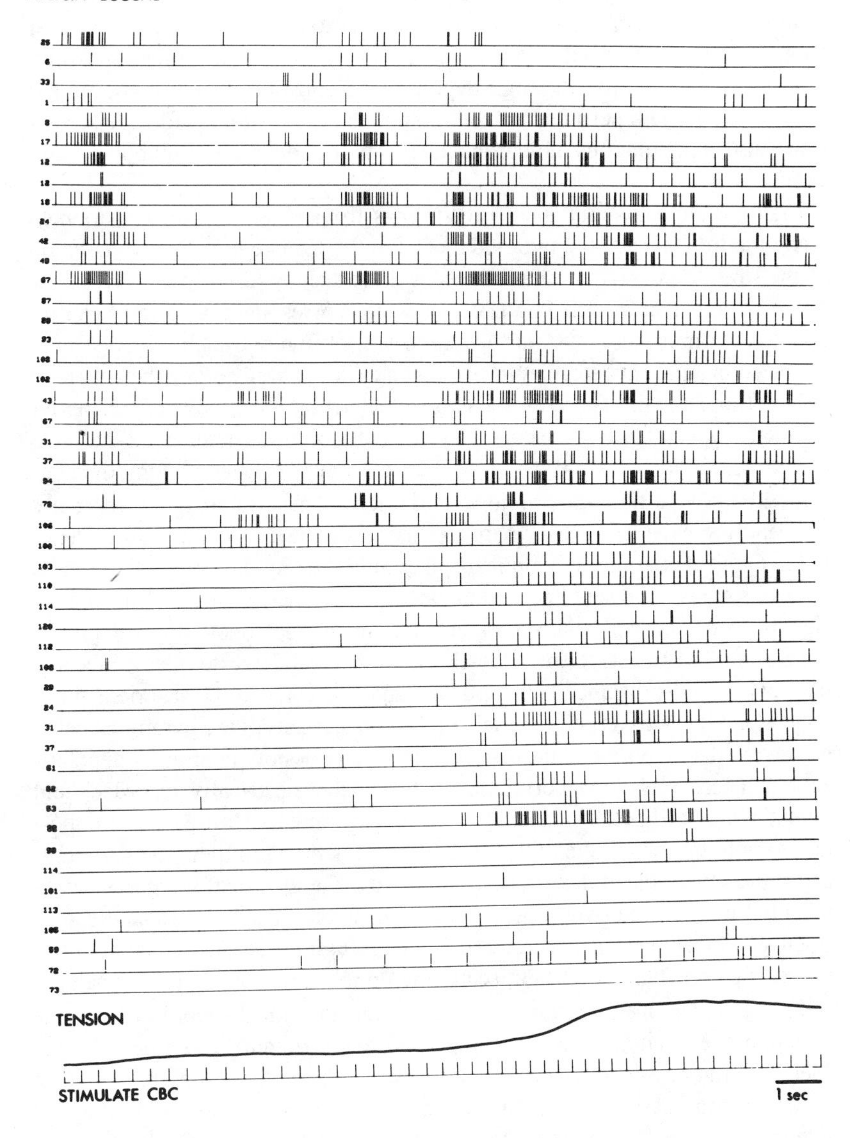

Figure 7 Firing patterns of 46 neurons in the *Aplysia* buccal ganglia. The bars represent spike in individual neuron. A computer program and manual operation was used to sort out these spikes from the original optical data. Many of these neurons are probably involved in the control of feeding behavior. The tension developed in the buccal mass is shown at the bottom. This firing was evoked by repetitive stimulation of the cerebral buccal connective. (From London et al 1985.)

dimensional" preparation, however, a given photodetector will receive light from many presynaptic and/or postsynaptic neuronal elements. Thus, optical monitoring of slice activity leads to recordings of the "intracellular population activity." Nevertheless, the high degree of organization of the hippocampal transverse slice still permits intracellular population recordings form relatively well-defined populations of neuronal elements, i.e. dendrites, cell bodies, and axons.

Figure 8 depicts simultaneous recordings from 10 photodetectors oriented along the long axis of the pyramidal cells in the CA1 region. The stimulation of the stratum (st.) radiatum produced a short latency (2–4 msec) fast (3.5–6 msec) signal in a strip 90–180 μm wide, corresponding to the Schaffer collateral-commissural system (*two top traces* of Figure 8). The signals could be traced in adjacent detectors for a long distance, but only along the collateral system (not shown). These fast signals reflect action potentials in the Schaffer collateral axonal fibers. A second wave of excitation was detected in the same region with a latency of 4–15 msec, but this wave "traveled" along the dendritic tree toward the st. pyramidale and oriens. These slow signals represent the dendritic excitatory postsynaptic potentials (EPSPs). The dendritic depolarization triggered multiple optical spikes at the pyramidal cell bodies. These optical signals represent the action potential discharges there. The optical signals propagated into the st. oriens with an average conduction velocity of 0.1 m/sec. Their passive spread back into the apical dendrites was noticeable only over a short distance. The optical signals detected at the st. oriens probably reflect the activity of the basal dendrites and axons of the pyramidal cells, as well as the activity of interneurons there and possibly neuroglial depolarization. There was good correlation between the electrically recorded field potential and the optical signals from the st. pyramidale. However, it is important to note that while the optical signals are restricted to their site of origin, and represent intracellular population activity, the electrically recorded field potential may spread over a large, and often unpredictable, area, and it represents extracellular current flow. (But see discussion, below, on the effect of microscope resolution and light scattering on the spread of optical signals.)

Figure 9 demonstrates the main advantage to optical recording, i.e. the feasibility of simultaneous recording from many (hundreds of) neighboring loci. The figure illustrates the spread of a focal excitation along the Schaffer collateral commissural axons. Subsequent postsynaptic spread down the apical dendrites of the CA1 neurons initiated the multiple discharge at the axonal hillock region of the pyramidal cells, which continued toward the st. oriens. Occasionally, the dendritic depolarization was not the largest in size at the location of the synapses in the st. radiatum (e.g. top of columns 5–7). One possible explanation is that Ca^{2+} action potentials were evoked by the EPSPs. The slow after-hyperpolarization and the shape of the dendritic responses

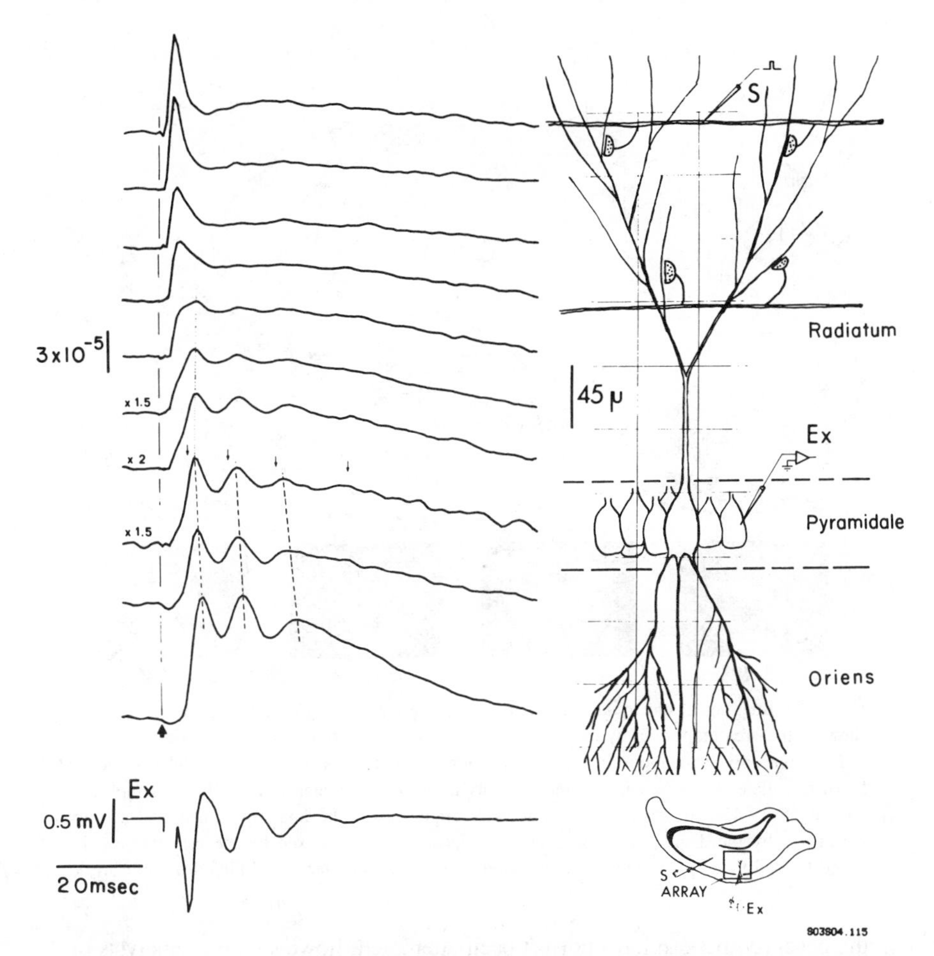

Figure 8 Optical recordings from ten loci along the long axis of CA1 pyramidal cells. The scheme shows the various areas along the cell axis, the position of the stimulating (S) and recording (Ex) electrodes, and the relative positions of the ten individual photodetectors (*light squares*). A scheme of the slice in the bottom right corner shows the optically monitored area (*square*) and the approximate position of stimulating and recording electrodes. Twenty trials were averaged. (Modified from Grinvald et al 1982c.)

support this interpretation, in line with other recent observations (Schwartzkroin & Slawsky 1977, Wong et al 1979, Llinas & Sugimori 1980). The experiment illustrated in Figure 9 indicates that for the first time optical recording has facilitated the detection of dynamic patterns of electrical activity of cell populations; moreover, these patterns are heterogeneous even in a relatively small area in the highly ordered CA1 region. (The experiments required to shed light on the synaptic networks and morphological correlates underlying

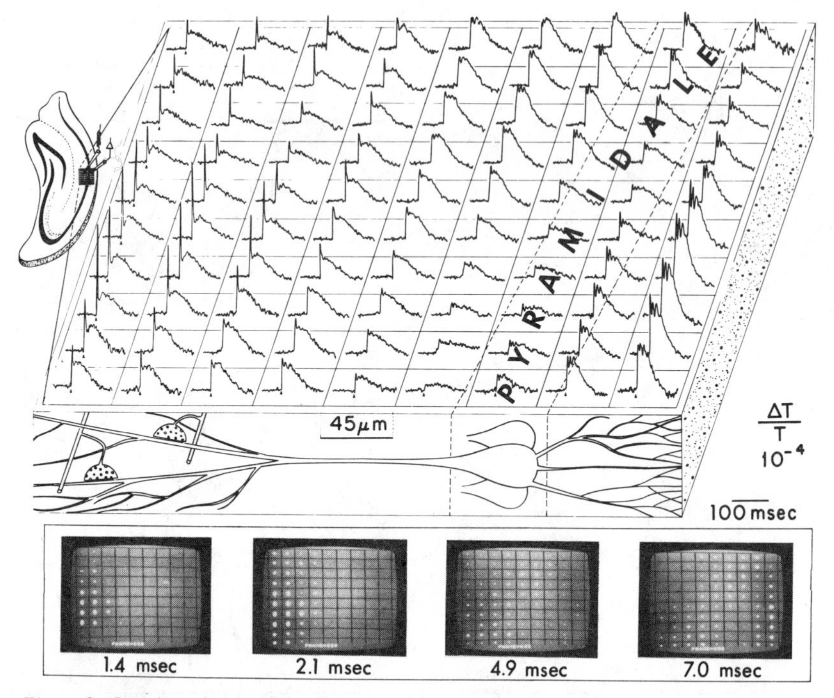

Figure 9 Imaging of electrical activity in a hippocampal slice. *Top left:* Scheme of the slice showing the area monitored by the array. *Top:* Optical traces from 9 × 10 detectors are shown at their appropriate locations in the field of view. Activity was evoked by stimulation of the Schaffer collateral-commissural system. Twenty trials were averaged. Each detector samples an area of 45 × 45 μm. *Bottom:* The four video frames show the imaging of neuronal activity performed by the display processor. The appropriate time is marked below each frame. The superimposed video picture of the preparation was omitted for clarity. (Modified from Grinvald et al 1982c.)

the observed pattern have not yet been attempted, however.) Fast analysis of the data presented in Figure 9 to obtain spatio-temporal information was facilitated by the display processor, which affords an on-line, slow motion visualization of the electrical activity. The imaging of electrical activity at four different time intervals is shown in the bottom of Figure 9.

Other findings of these studies (Grinvald et al 1982c, Kuhnt & Grinvald 1982) are as follows:

1. Externally applied TEA broadens the action potential in the nonmyelinated Schaffer collaterals in a reversible manner.
2. Twenty μM of 4-aminopyridine (4-AP) induces broadening of the presynaptic action potentials in st. radiatum. This broadening depends on the external Ca^{2+} concentration. In normal solution it led to a dramatic increase in the postsynaptic response and in its facilitation by a second stimuli.

3. 4-AP also broadens the action potentials in the myelinated axons in the alveus and oriens, suggesting that potassium channels are active at the nodal region.
4. The conduction velocity in the nonmyelinated fibers was determined.
5. Picrotoxin increases postsynaptic response at the st. pyramidale as well as in the st. radiatum, suggesting that inhibitory synapses are located there also.
6. In preliminary experiments it was found that long-term potentiation (LTP) is correlated with both pre- and postsynaptic changes as well as reduction of the inhibition (M. Segal and A. Grinvald, unpublished results).

Local circuits can probably be investigated in slices by optical recording of the responses to intracellular stimulation of single identified cells in a manner similar to the investigations performed on invertebrate ganglia (Grinvald et al 1981a, Ross & Krauthamer 1984). Such potentially useful experiments have not yet been reported, however.

OPTICAL RECORDING FROM ISOLATED AND INTACT BRAIN STRUCTURES IN-VITRO Experiments similar to those described above were also carried out on isolated, relatively transparent brain structures. In such preparations, one would expect signals to be even larger than in slices, because a larger fraction of cells and synaptic connections are intact. Recently, Orbach & Cohen (1983) demonstrated that the isolated salamander olfactory bulb can be investigated using both transmission and fluorescence measurements. Salzberg et al (1983) investigated the ionic mechanisms underlying the electrical activity of a population of nerve terminals in the isolated neurohypophysis of *Xenopus*. This preparation offers a unique opportunity for the study of both excitation-secretion coupling and the electrical properties of vertebrate nerve terminals, because the majority of its excitable membrane area is composed of the terminals themselves. Figure 10 depicts these experiments. A single sweep optical recording of an action potential from a population of nerve terminals is shown in Figure 10b. The effect of various concentrations of external Ca^{2+}, Cd^{2+}, and Ni^{2+} are shown at the right. The effects of other sodium and calcium channel blockers were also investigated. Evidently, the action potential in these nerve terminals has both Ca^{2+} and Na^{+} components and probably also a calcium-mediated potassium conductance.

Another important conclusion drawn from the above experiments is that with a successful choice of preparation, very large signals can be obtained from a population of synchronously active vertebrate nerve terminals. With such preparations the experiments are both easy and powerful. Furthermore, they can be carried out with an inexpensive optical detection system. The construction of such an optical set up has been described (Cohen et al 1974, Salzberg et al 1977).

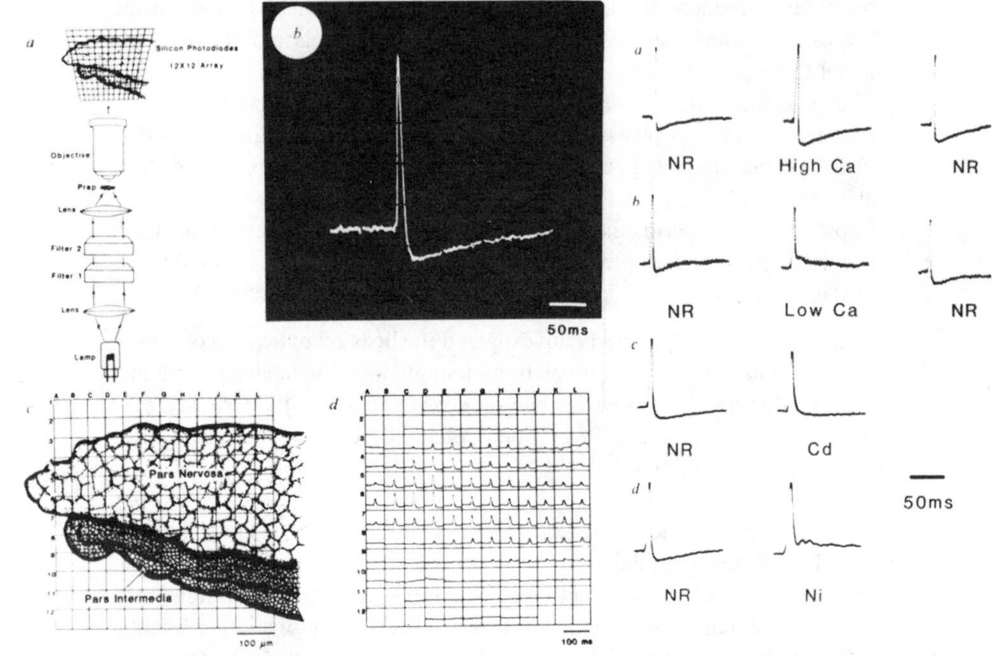

Figure 10 *a*. Schematic diagram of apparatus. *b*. Single sweep optical recording of action potentials from a population of nerve terminals. *c*. The real magnified image of the preparation superimposed on the 12 × 12 photodiode array. *d*. The optical signals that were recorded in the experimental arrangement shown in *c*. *Right:* Calcium dependence of the after-hyperpolarization of the action potential recorded from nerve terminals in the neurohypophysis. (Modified from Salzberg et al 1983.)

Real-Time Imaging of Naturally Evoked Responses in the Intact Brain In Vivo

The detection of a pattern of population activity in mammalian brain slices suggested that (*a*) the optical technique could be applied to the investigation of mammalian brain tissue and (*b*) that intracellular population recordings still provide meaningful information (Grinvald et al 1982c). Orbach & Cohen (1983) were the first to demonstrate that large fluorescence signals can be obtained in vivo after topical application of recently synthesized styryl dyes (Grinvald et al 1984). They observed responses in the salamander olfactory bulb after electrical stimulation of the whole olfactory nerve. Their results suggested that the optical method could prove a powerful approach for studying functional organization in the intact vertebrate brain as well. Orbach, Cohen & Grinvald exploited the technique to investigate the rat visual (Orbach et al 1982) and the somatosensory cortex (Orbach et al 1983). Furthermore, they have shown that meaningful optical signals can also be detected in response to a natural physiological stimulus.

MAPPING OF ELECTRICAL ACTIVITY IN THE MAMMALIAN CORTEX In vivo measurements on the mammalian cortex are considered difficult because of the large noise from movements of the brain due to heartbeat and respiration, and because of the relative opacity of the cortex. The dense packing of the neuronal and glial elements may prevent proper staining with sensitive but hydrophobic optical probes. The very large noise from the heart pulsation ($\Delta I/I = 10^{-2}$–10^{-3}) was much greater than the evoked optical signals. However, because this noise was synchronized with the heartbeat and the electrocardiogram (ECG), it was relatively easy to reduce it by subtracting the result of a trial with a stimulus from a subsequent trial without a stimulus. (Both trials were triggered by the peak of the electrocardiogram.) This procedure reduced the noise by about a factor of ten. Noise from breathing movement was reduced by holding the rat head tightly with a snout clamp and earbars. In preliminary experiments on the cat visual cortex, respiration noise was further reduced by stopping the respirator for a few seconds during data acquisition. In signal-averaging experiments, further improvement was achieved by a computer program that allowed the rejection of exceptionally noisy trials from the accumulated average (Grinvald et al 1982b, Orbach et al 1985).

In preliminary experiments on the rat visual cortex (Orbach et al 1982), fluorescence signals were detected from a large area of the cortex when a diffused flash of white light or a flashing checkerboard were used as a stimulus. The response was relatively uniform in the entire field of view, even when a ten-degree light flash was used. Because more appropriate visual stimuli were not tried, it is impossible to conclude whether the method can be useful for the investigation of the rat visual cortex. More appropriate physiological stimuli were used in preliminary experiments on the cat visual cortex (moving bars of variable orientation, black and white stripes). Reliable optical signals were not obtained in these experiments (H. S. Orbach, L. B. Cohen, A. Grinvald, C. Gilbert, and T. N. Wiesel, unpublished results). However, when the cortical inhibition was partially blocked by topical application of bicuculline, relatively large responses were recorded. These results suggest that even though the sensitivity of the dye may be similar in the rat and cat cortices, the noise level and the sensitivity of the dye tested were not sufficient to resolve the signals from the small fractions of neurons that were active in the cat experiments.

The whisker barrels (see Woolsey & van Der Loos 1970) in the rat somatosensory cortex clearly proved to be amenable to optical investigation. Relatively large signals were obtained in response to a small whisker deflection of 0.3 mm. Under the best conditions, such signals could be detected without signal averaging and about 200 trials could be reliably repeated in the same cortical area. Figure 11 illustrates the results of an experiment in which 124 adjacent regions of the cortex were simultaneously monitored with the diode array. The exposed cortex (4 × 4 mm diameter craniotomy) was stained by topical application of 1 mM RH–414 for 120 minutes. (The transparent dura was left intact.) When the tip of the whisker B1 (Simons 1978) was moved by 1 mm, optical signals were recorded in the center of the field of view. The average diameter of the response area in such experiments was about 1300 μm. This area is much larger than that of the whisker barrels in layer IV (300–600 μm). Nevertheless, these are the expected results if the optical signals indeed originate from the underlying barrel:

1. Most of the optical signals originate from layers I–III, and it is known that the processes from the somata in a single barrel extend to at least two neighboring barrels.
2. Stimulation of one whisker evoked activity of neurons in neighboring barrels (Simons 1978).
3. Light scattering and out-of-focus fluorescence tend to increase the apparent area of the evoked activity (Orbach & Cohen 1983).

Stimulation of two whiskers that are relatively far apart (e.g. A1 and D4) evoked responses in two circumscribed areas. A large overlap in the response area was obtained when neighboring whiskers were stimulated, but the responses could be resolved if they were activated at different times. Other findings of this study are as follows:

1. In most experiments, signals in the center zone were faster than those in the periphery (see Figure 11c), suggesting that the periphery signals indeed originate from processes extending out of the barrel, and from interneurons or postsynaptic cells in neighboring barrels.
2. Larger amplitudes of whisker deflection evoked larger activity, which also spread over larger cortical areas.
3. Activity evoked by one whisker often inhibited the activity evoked by a neighboring whisker when the interstimulus interval was about 20–120 msec.

The optical signals from the rat cortex were relatively slow and smooth in contradistinction to the optical signals originating from visually evoked responses in the frog optic tectum (see below) or the electrically evoked responses in the salamander olfactory bulb (Orbach & Cohen 1983). Thus, at present, optical signals recorded from mammalian cortex have far less information content than those recorded from lower vertebrates. The slow, smooth nature of the cortical optical signals could be explained either by assuming that the intracellular population activity should appear smeared and slow because several relay stations are involved or by there being a dominant contribution from

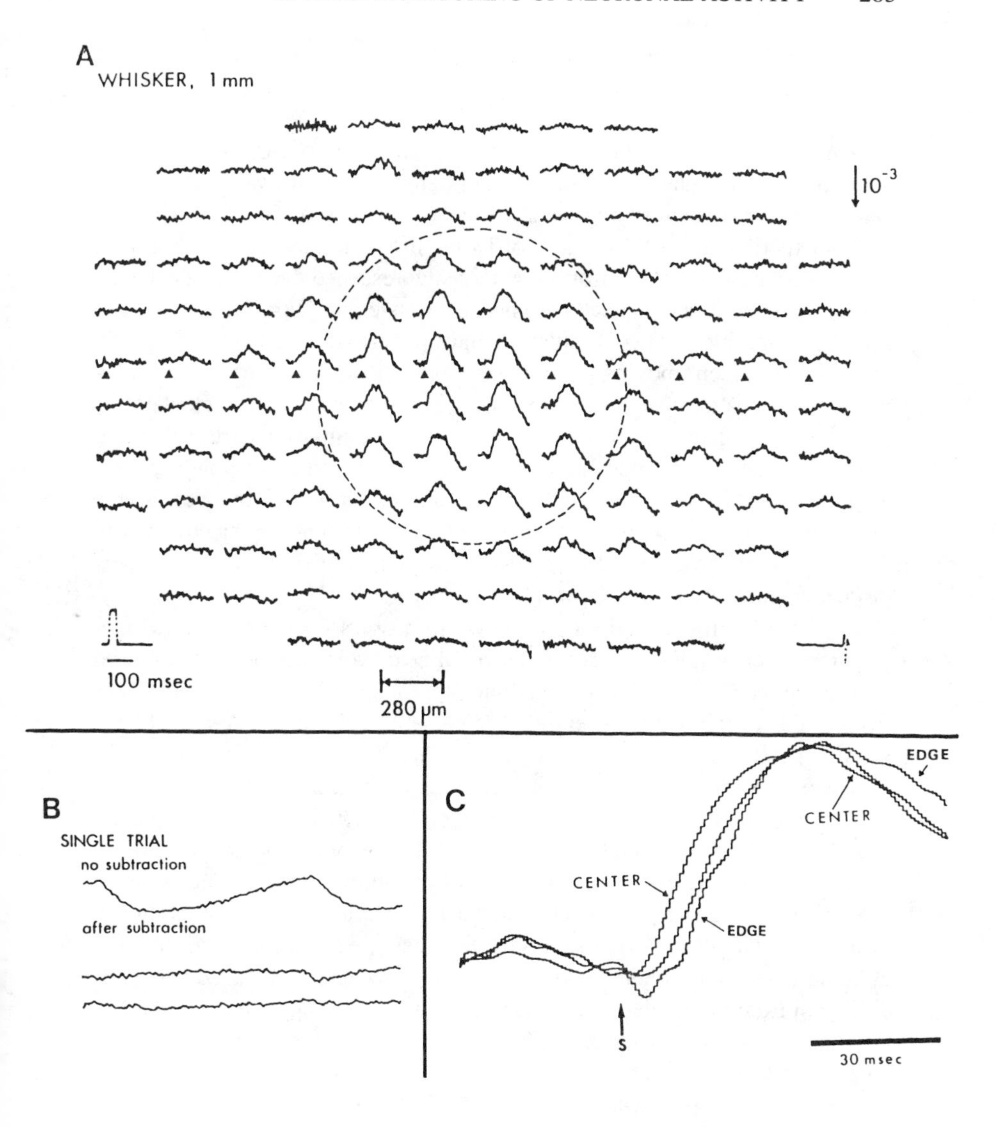

Figure 11 (*a*) Optical detection of a whisker barrel in the rat somatosensory cortex that results from a 1 mm whisker movement. The *inset at the lower left* shows the timing of the current to the galvanometer (stimulation). The *inset at the lower right* shows the results of the electrocardiogram subtraction. Thirty-two trials were averaged. The interval between stimuli was 9 seconds. (From Orbach et al 1985.) (*b*) Removal of the heart beat noise by subtraction. (*c*) Comparison of the time course of the optical signals in the center of the barrel and its periphery.

glial depolarization. A distinction between these two alternatives has not yet been made.

MAPPING OF VISUALLY EVOKED RESPONSES IN THE FROG OPTIC TECTUM Optical studies of the intact frog optic tectum (Anglister & Grinvald 1983, Grinvald et al 1984) suggest that such a preparation may be more suitable for optical experiments partly because of its relative transparency. When a flash of light was presented to the frog's eye, large fluorescence signals were detected from the exposed and topically stained contralateral tectum (without signal averaging; see Figure 12). Display of light patterns to the frog's eye on an oscilloscope screen (moving or stationary spots, bars, or annuli) evoked complex response patterns in the tectum. The optical signals from a single detector were highly structured; the number of resolved components from a single detector was similar to the number of components in the corresponding electrically recorded evoked potential (Figure 12, *bottom left*). Moreover, in this preparation, optical recording of the activity at different depths, by simply changing the microscope focus, allowed a partial laminar analysis to be performed (Figure 12, *bottom right*).

The retinotectal connections are known to possess a large degree of topographical order. A point stimulus in the visual field evokes activity in a restricted area in the tectum. We characterized the present spatial resolution and sensitivity of the technique by presenting discrete and weak stimuli: A spot of light was first positioned on a corner of a monitor screen for 100 msec and then after a delay of 210 msec in the opposite corner (26° displacement) for another 150 msec. Figure 13 illustrates that two excitation foci (*circled area*) at the tectum could be resolved. Both had a latency of ~100 msec with respect to the onset of each light stimulus. Additional experiments showed that inhibitory interactions shaped the size of the excitation vertical column; blocking of inhibitory interactions with drugs led to a considerable increase (ten-fold or more) of the excitation area. The conduction velocity of the biccuculine-induced spread of excitation from the excited foci was found to be faster along the rostral caudal axis than the medial lateral one. This behavior may reflect asymmetry in the long-range synaptic interactions in the tectum. Other explanations have not yet been ruled out (Grinvald et al 1984).

TECHNICAL ASPECTS OF OPTICAL MEASUREMENTS

The development of the optical technique has required multidisciplinary efforts involving organic chemistry, spectroscopy, optics, electronics, computer programming, and neurophysiology. In practice the use of the technique is not always easy. Technical aspects of optical measurements and further improvements are described below.

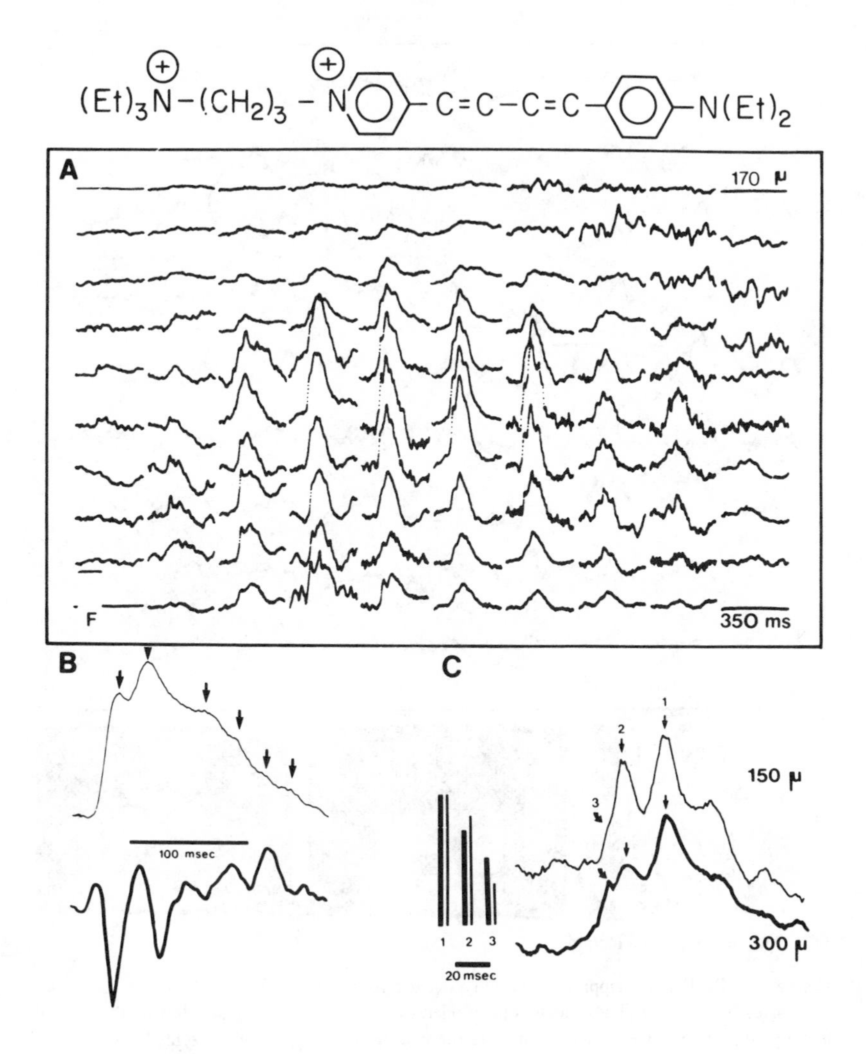

Figure 12 (*a*) Fluorescence recording of naturally evoked responses in the tectum. The output of 96 photodetectors is shown at their relative locations. Signal averaging was not used. A 150 msec light flash stimulus (shown in *F*) was presented to the contralateral eye. (*b*) Comparison of optical intracellular population recording with the electrically recorded surface-evoked potential. (*c*) Partial laminar analysis. Comparison between the output of a photodetector in two consecutive experiments at two different depths of 150 μm and 300 μm. A 40× water immersion objective was used. (From Grinvald et al 1984.)

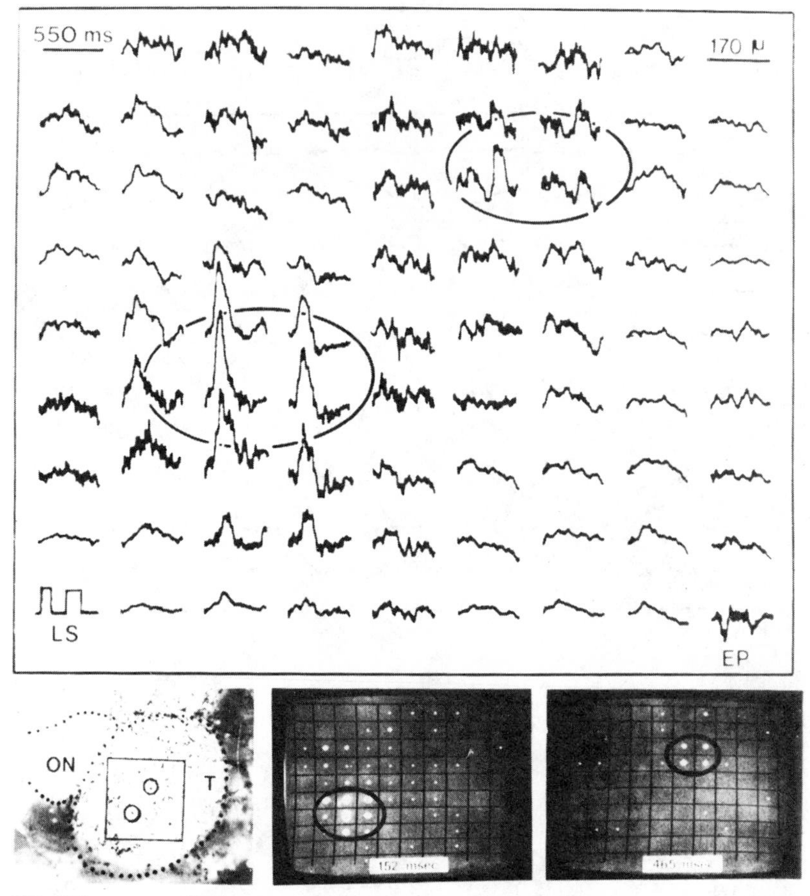

ON: Optic nerve T: Tectum

Figure 13 Real time mapping of visually evoked responses in the intact frog tectum. *Top:* Responses from 9 × 9 photodetectors positioned over the tectum image are shown. Two loci of activity could be resolved corresponding to the two different light spots presented one after the other to the contralateral eye. Forty trials were averaged. *Bottom:* Visualization of the spatial patterns of activity, at two different times, depicting the two loci of activity. The superimposed picture of the tectum was omitted for clarity and is shown at the left. (From Grinvald et al 1984.)

The Apparatus

Figure 1 depicts the computerized apparatus (Grinvald et al 1981b, Grinvald et al 1982c). A Zeiss Universal microscope is rigidly mounted on a vibration-isolation table. A 12 V/100 W tungsten/halogen lamp is used for transmission experiments. In fluorescence experiments, either the same lamp or brighter light sources, such as a He–Ne laser or a mercury lamp, are employed. The preparation is viewed through a microscope objective (4× to 80×, depending upon the required magnification), which forms a real magnified image of the preparation in the microscope image plane. Changes in transmitted light or fluorescence are detected in this plane by a 10 × 10 (or 12 × 12) square array of photodiodes, each 1.4 × 1.4 mm. Each photodiode receives light from a 10 × 10 μm^2 to 400 × 400 μm^2 area of the preparation. Each diode is coupled to a current-to-voltage converter and an amplifier. The amplifiers filter the optical data by means of two resistance-capacity (RC) filters having a time constant of 0.2 or 0.4 msec. The signals are AC coupled with a switch-selected time constant of 100 msec or 900 msec and amplified again (× 1000–4000). In order to be able to measure a DC signal (e.g. Senseman & Salzberg 1980, Nakajima & Gilai 1980), a new bank of 128 amplifiers was constructed (e.g. Grinvald et al 1982b). Each individual amplifier contains an analog sample-and-hold circuit to measure the light intensity at the onset of the measurements, after the optical shutter is opened. This value is then differentially subtracted from the light signal before it is further amplified (×1–4000). In this way a high-gain auto-DC-offset recording can be obtained rather than an AC recording. Furthermore, the shutter can be opened only for the duration of the measurements, thus minimizing bleaching and photodynamic damage (discussed below). The analog sample and hold circuit suffers from droop, which becomes significant when small signals are measured. (The droop can be corrected by a subtraction procedure that must be implemented anyway in DC recordings, in order to correct for the bleaching time course.) Another alternative is to use a digital sample-and-hold circuit utilizing back-to-back analog-to-digital (A/D) and digital-to-analog (D/A) conversions (e.g. Senseman & Salzberg 1980). Amplifiers with an auto-zero-tracking analog-to-digital conversion are now being designed (W. Singer and J. Meyer, personal communication). The output of the amplifiers is multiplexed and digitized by two multiplexer 8-bit A/D converter cards, and deposited in the PDP 11/34 computer memory. The time resolution thus obtained is 0.6 msec for each trace. Twelve-bit A/d conversion has been used also but it requires twice the computer memory (for 16 bit computers). Unless a digital multiplexer is used, it is also two times slower. An improved time resolution (0.3 msec) was recently achieved by the construction of a digital multiplexer collecting sequential data from four 32-channel A/D cards (A. Grinvald, unpublished results) or by the use of faster A/D multiplexers (e.g. Data Translation, W. N. Ross, personal communication). In some low-light-level fluorescence experiments, a single photodiode or a photomultiplier was used instead of the array (Salzberg et al 1973, Grinvald et al 1982b, 1983). A five-fold reduction in the photodiode-amplifier noise was obtained by the use of a large feedback resistor on the first stage of the current-to-voltage converter (200 MΩ–1GΩ, see Grinvald et al 1983; J. Pine and A. Grinvald, unpublished observation).

In order to combine electrophysiological recordings with the optical experiment, up to four hydraulic Narashige micromanipulators are mounted on the movable stage of the Universal microscope. Recently, we have also used an inverted microscope for transmission (Kuhnt & Grinvald 1982, U. Kuhnt, in preparation) and fluorescence (Grinvald et al 1983) experiments. The inverted microscope was mounted on an X–Y positioner (Calvet & Calvet 1981) and its stage replaced by a large flat stage rigidly attached to the vibration isolation table top. This arrangement offers four advantages:

1. Heavy manipulators can be mounted on the fixed stage. The insertion of microelectrodes into the preparation is done with the aid of a swing-in stereomicroscope that replaces the swing-out condenser of the inverted microscope. With a 0.63 numerical aperture (NA) condenser, the microelectrode angle can be as large as 45° and the working distance is 10 times larger (instead of the 27° and 1.6 mm working distance permitted by the water immersion objective used in the Universal microscope).
2. It is easy to change the objectives under the preparation to attain the optimal magnification without interfering with intracellular recordings.
3. High numerical aperture objectives with relatively short working distances can be used, because the electrode impalements are made from the side of the condenser. (Custom-made condensers having even a larger working distance can also be obtained.) The use of such objectives provides an improvement in both the SNR and the spatial resolution (see below).
4. Under these conditions, the optical measurements can be made also during fast perfusion of the preparation with various physiological solutions.

Our experience suggests that the use of the inverted microscope is preferable whenever applicable (e.g. for tissue cultures, brain slices, small and thin invertebrate ganglia, etc). However, for optical recording from the intact brain of an anesthetized animal, or from large invertebrate ganglia, an upright microscope must be used.

In the optical experiments, it is important to know which part of the preparation is viewed by each photodetector element. Therefore, the preparation is also viewed with a Vidicon camera; the video image is stored on videotape to further identify the location of the elements of the photodiode array with respect to the preparation image. Inspection of the large amount of optical data is time consuming and slows the experiment. For fast analysis of the observed patterns of activity, a computer-to-video imaging processor was designed (Manker 1982, Grinvald et al 1982c) that permits visualization of the pattern of electrical activity on a TV screen. It projects a calibrated outline of the array on the TV monitor, superimposed on the television picture of the preparation. In addition, a bright symbol is displayed in each of the 100 elements of the picture. The sizes of these symbols are proportional to the amplitude of the electrical activity at a given time. The interface displays the temporal activity from all the detectors "simultaneously" for each time interval, thus providing the imaging of neuronal activity. Detailed information about the hardware and the software used for the various optical measurements are available (L. B. Cohen, B. M. Salzberg, W. N. Ross and A. Grinvald, personal communication).

PRESENT DIFFICULTIES AND FUTURE IMPROVEMENTS OF THE TECHNIQUE

Limitations of the Present Optical Probes

There are some difficulties with the use of the optical technique that warrant detailed discussion. Intimate understanding of them is imperative for successful and optimal use of the technique.

SIGNAL SIZE The size of the optical signals is often small. Therefore, whenever intracellular recording is easy, it is preferable. For example, in the present brain slice or salamander absorption experiments, spontaneous activity of single cells could not be unequivocally identified. It has been reported that the sensitivity of an optical probe may depend on the preparation (Ross & Reichardt 1979, Gupta et al 1981). Thus, more sensitive optical probes probably already exist for mammalian brain slices, and additional search for better dyes is worthwhile. In other type of optical experiments the signals are satisfactorily large (e.g. Grinvald et al 1977, 1981b, 1982b, Salzberg et al 1977, 1983, Grinvald & Farber 1981, Boyle et al 1983, London et al 1985, Ross & Krauthamer 1984, Senseman et al 1983).

LIGHT SCATTERING If the light absorption or fluorescence signals are small, the activity-dependent (dye-unrelated) light-scattering signals from the preparation (Cohen et al 1968, 1972, Cohen & Keynes 1971, Lipton 1973) may distort the voltage-sensitive optical signals. Large and slow light-scattering signals were observed in unstained hippocampal slices (Grinvald et al 1982c) and in the frog optic tectum. Such signals were also observed in at least some loci in the salamander olfactory bulb (Orbach & Cohen 1983). Light-scattering signals reflect some physiological processes and therefore they can be useful. However, they interfere with optical recordings of the dye-related signals. A solution to this problem was provided by subtracting the light-scattering signals from the optical response (Grinvald et al 1982c). Furthermore, in the guinea pig experiment with the more sensitive dye, RH–155, the light-scattering signals were small relative to the optical signals (Kuhnt & Grinvald 1982; U. Kuhnt, in preparation). Similarly, Salzberg and co-workers (1983) did not observe any contribution of light-scattering signals to the large dye-related signals in the *Xenopus* neurohypophysis experiments.

PHARMACOLOGICAL SIDE EFFECTS Light-independent pharmacological side effects can be expected whenever extrinsic probe molecules are bound to the neuronal membrane, especially if high concentrations of dye are used. The dye binding may change the threshold, the specific ionic conductances, synaptic transmission, membrane resistance, etc. After careful choice of the proper dye, significant pharmacological side effects were not observed in many of the above studies. However, more stringent tests using intracellular recording are required to assess pharmacological effects, and careful controls are always essential. More than 200 voltage-sensitive probes are now available, with very different chemical structures and net charges. It is unlikely that all of them will cause similar pharmacological side effects. An instance has indeed been reported (Grinvald et al 1981a) in which proper choice of the voltage-sensitive probe eliminated pharmacological side effects on a behavioral reflex mediated by a polysynaptic pathway. It was also reported that pharmacological side effects may often be minimized if the staining is done with a modified saline solution containing high concentrations of divalent cations (Senseman et al 1983).

PHOTODYNAMIC DAMAGE The dye molecules, in the presence of intense illumination, sensitize the formation of reactive singlet oxygen and other free radicals. These reactive radicals attack membrane components and damage the cells (Pooler 1972, Cohen et al 1974, Ross et al 1977). Such photodynamic damage limits the duration of the experiments. Using the present transmission probes, continuous measurements can be made for one to five minutes without marked damage. If the duration of each trial is 50 msec (DC recording only), then 1200–6000 trials can be carried out before significant damage occurs. This is precisely the reason that it is important to construct DC amplifiers as described in the instrumentation section. In fluorescence experiments with brighter light sources, the duration of a reliable experiment is often limited to a total of five to ten seconds only. The use of antioxidants (radical scavengers) has been suggested (Oxford et al 1978), but in one tested case it also reduced the signal size (Salzberg 1978). Because very large numbers of radical scavengers exist, including enzymes, but only a few were tested (L. B. Cohen, W. N. Ross, and B. M. Salzberg, personal communication), this approach is still promising. Thus, a systematic evaluation of the usefulness of radical scavengers in optical experiments may be rewarding. It is important to note that the extent of photodynamic damage (or bleaching) may depend also on the preparation. (Compare for example Ross & Reichardt 1979 or Grinvald et al 1981b with Ross & Krauthamer 1984, or Salzberg et al 1973 with Salama and Morad 1976.)

DYE BLEACHING Bleaching during the optical measurement also limits the duration of reliable experiments. To minimize bleaching, the exposure time of the preparation to light should be reduced to a minimum. The dye bleaching may also affect the time course of the optical signals, especially in fluorescence experiments when bright light sources are used. A solution to this problem has been described (Grinvald et al 1982b). Subtraction procedures of the type used to remove the heartbeat noise are effective as well.

Design and Synthesis of Improved Optical Probes

Because of the above difficulties, the key to the success of optical monitoring of neuronal activity is the design of adequate voltage-sensitive probes. Of more than 1800 dyes already tested (Cohen et al 1974, Ross et al 1977 Gupta et al 1981, Grinvald et al 1982b, 1983), less than 200 have been proven to be sensitive indicators of membrane potential while causing minimal pharmacological side effects or light-induced photochemical damage to the neurons. Considerable improvement has recently been achieved in the quality of fluorescent probes, especially styryl dyes (Cohen et al 1974, Grinvald et al 1978, Loew et al 1978, 1979, 1984, Grinvald et al 1982b, 83). For example, the best fluorescence dye for neuroblastoma cells is designated RH–421; the change in fluorescence intensity with this dye is 25%/100 mV of membrane potential change (Grinvald et al 1983). This value is 120 times greater than the fractional change obtained with leech neurons in the pioneering experiments of Salzberg et al (1973). (But sensitivity depends on the preparation, see below.) Furthermore, the photodynamic damage with this dye was reduced by a factor of 200 relative to that of merocyanine–540 (Salzberg et al 1973). Another useful family of the probes are asymmetrical oxonols, analogues of WW–781 designed recently (Grinvald et al 1978, Gupta et al 1981). In the design of improved transmission voltage-sensitive probes, progress has been slower, presumably because the sensitivity of the best probes is already close to the theoretical maximum (Waggoner & Grinvald 1977). A significant limitation of the optical probes is that a dye's sensitivity can vary from one preparation to the next and even among different species of the same genus (Ross & Krauthamer 1984). Thus, for a new preparation, careful selection of the best probe is an important prerequisite. The availability of several "kits" of close analogues of the best types of dyes has greatly facilitated such a selection. Additional efforts at synthesis are worthwhile for both absorption and fluorescence. Table 1 summarizes the useful dyes discovered for a variety of preparations.

Because the quality of the voltage sensitive probes is the limiting factor in the widespread application of this technology, it is unfortunate that only in the laboratory of Dr. A. Waggoner were voltage-sensitive dyes synthesized in the past (about 500); and now only R. Hildesheim is synthesizing many such dyes (about 200). Much faster progress might be expected if the following were achieved:

1. more laboratories would synthesize probes;

Table 1 Dyes that provide large absorption or fluorescence signals in a variety of preparations

	Preparation	Dye name[a] (source)	S/N[b]	ΔI/I	N[c]	References
Cephalopods	Squid giant axon	RGA–84(A)[d]	80	1×10^{-3}	1800	Gupta et al 1981, Ross et al 1977, Cohen et al 1974, Hildesheim, Cohen, Salzberg, and Grinvald, unpublished
		RH–428(F)[e]	26	1×10^{-2}		
Invertebrate ganglia	Leech	WW–433(A)	20	2×10^{-4}	20	Grinvald et al 1981a
		RH–237(F)	8	6×10^{-3}	4	Grinvald et al 1982b, Salzberg et al 1973
	(Inside dyes)	RH–461(F)	15	1×10^{-3}	8	Grinvald et al 1982a, Agmon et al 1982, Obaid et al 1982
	Barnacle	NK–2367(A)	30	1×10^{-3}	20	Grinvald et al 1977, 1981a, Salzberg et al 1977, Ross & Krauthamer 1984
	Pluorobranchea	RGA–525(A)	20	5×10^{-4}	5	Cohen & Orbach 1983, London et al 1985
	Navanax	RH–155(A)	30	6×10^{-4}	19	Cohen & Orbach 983, London et al 1985
	Aplysia	RGA–525(A)	15	4×10^{-4}	6	Cohen & Orbach 1983, London et al 1985
	Other species (16)	WW–433(A) NK–2367 (A)	15	4×10^{-4}	2	Boyle et al 1983
	Helisoma salivary gland	NK–2367(A)	80	7×10^{-3}	1	Senseman & Salzberg 1980, Senseman et al 1983
Tissue culture cells	Neuroblastoma	WW–401 (A)	50	4×10^{-4}	35	Grinvald et al 1981b
		RH–421(F)	200	2.3×10^{-1}	40	Grinvald et al 1982b, Grinvald et al 1983

	Aplysia	WW–401(A)	25	2×10^{-4}	10	Levitan and Grinvald, unpublished
		RH–376(F)	100	1×10^{-1}	15	Bodmer, Farber, and Grinvald, unpublished
	SCG (rat)	RH–421(F)	20	6×10^{-2}	15	Pine and Grinvald, unpublished
	DRG (chick)	WW–802(F)	15	5×10^{-2}	3	Anglister and Grinvald, unpublished
	Leech	NK–2367(A)	10	2×10^{-4}	1	Salzberg and Obaid, unpublished results
Brain slices	Hippocampus	RH–155(A)	7	5×10^{-4}	40	Grinvald et al 1982c, Segal, Kuhnt, and Grinvald, unpublished
Isolated and intact brain structures	Myelinated optic nerve (rat)	RH–414(F)	5	3×10^{-4}	10	
		RH–155(A)	6	5×10^{-4}	1	Lev-Ram and Grinvald, unpublished
	Olfactory bulb (salamander)	WW–375(A)	40	5×10^{-3}	10	Orbach & Cohen 1983
		RH–160 (F)	20	2×10^{-2}	7	
	Neurohypophysis (*Xenopus*, mouse)	NK–2761(A) (RGA–84)	60	2.5×10^{-3}	3	Salzberg et al 1983
	Hypothalamus (*Xenopus*, mouse)	NK–2761(A)	40	1×10^{-3}	3	Salzberg et al 1983
Intact vertebrate brain	Olfactory bulb	WW–375(A)	30	4×10^{-3}	4	Orbach & Cohen 1983
		RH–414(F)	15	2×10^{-2}	2	
	Optic tectum (frog)	RH–414(F) RH–376(F)	10	4×10^{-3}	29	Grinvald et al 1984
	Mammalian cortex (rat, cat)	RH–414(F)	5	2×10^{-3}	21	Orbach et al 1983

[a]WW & RGA dyes are available from Dr. Alan Waggoner; RH dyes are available from A. Grinvald; NK dyes are commercially available (see Gupta et al 1981).
[b]The signal-to-noise ratio was not corrected for bandwidth or instrumental factors; it was normalized to the optical signals for action potentials (about 100 mv).
[c]Number of dyes tested.
[d]A = Absorption.
[e]F = Fluorescence.

2. synthesis of dyes having different structures were attempted (other than cyanine, merocyanine, and oxonol);
3. dye screening would be done on the relevant preparations rather than on mostly squid giant axons (e.g. the useful dye RH–414 has a very small fluorescence signal on the squid experiment, but see Table 1);
4. synthesis and testing were done simultaneously for quick feedback;
5. theoretical approaches for probe design of the type introduced by Loew et al (1978) were incorporated [However, this theory was developed for electrochromic absorption probes, but those dyes did not prove useful in absorption experiments. The very large fluorescence signals provided by some styryl dyes do not result from electrochromic effect (A. Grinvald and L. B. Cohen, unpublished results; Loew et al 1984)];
6. theoretical work on the mechanisms underlying the probe signal and their relation to signal size were to be performed (e.g. Waggoner & Grinvald 1977);
7. more investigations of the biophysical mechanisms underlying the probe sensitivity were carried out (e.g. Ross et al 1977, Waggoner & Grinvald 1977, Dragsten & Webb 1978, Loew & Simpson 1981, Loew et al 1984).

Difficulties with the Interpretation of Optical Signals

Optical signals and intracellular electrical recordings follow the same time course (ignoring the series resistance problem in electrical recordings; see Brown et al 1979, Salzberg & Bezanilla 1983, Ross & Krauthamer 1984). However, the resting potential cannot readily be measured or manipulated. The inability to evaluate reversal potentials for various EPSPs and IPSPs limits the interpretation of some observed responses. Furthermore, difficulties exist with the interpretation of intracellular population recordings.

AMPLITUDE CALIBRATION The size of optical signals is related to the membrane area, the extent of binding, and the sensitivity of the dye for a given membrane. For example, in hippocampal slices, differences exist in "concentrations" of membrane elements across the slice. In distal dendrites, many processes occur and therefore there is a larger membrane area. There is much less membrane in the somata layer, and the amplitude of the optical signals from the st. pyramidale is estimated to be roughly three to four times smaller than those from equal potential changes within the st. radiatum. Furthermore, a lower density of bound dye molecules in some parts of the tissue will result in a smaller signal size. There is no evidence that dye molecules bound to different cell types, or even different segments of the same cell, will provide equal signals. Thus, direct comparison of the amplitudes of optical signals in different regions may not be straightforward; only comparison of signals recorded from the same area under different experimental conditions are reliable. Otherwise, the interpretation must rely principally on the time course of the signals. In the investigation of single cells, however, amplitude information can often be interpreted. (For examples see Grinvald et al 1981b, and Krauthamer & Ross 1984.)

ANALYSIS OF INTRACELLULAR POPULATION ACTIVITY The analysis of the intracellular population recording may be difficult in three-dimensional preparations with heterogenous neuronal elements viewed by a single detector. For example, although the hippocampus was selected because of its clear stratification, neurons of various types coexist within strata, and signals from these neurons reaching the same detector can undoubtedly obscure signals from any particular population of neuronal elements, i.e. axons, dendrites, pyramidal somata, etc. Thus, the activity detected at the st. oriens reflects the activity of the CA1 pyramidal axons and the basal dendrites, as well as that of the interneurons there. However, proper manipulation of stimulus size, location, frequency, and pharmacological treatment, or varying the ionic composition, etc, may permit a separation of the various components. Furthermore, in such preparations, changes in membrane potential in glial cells may contribute significantly to the optical recording. Grinvald et al (1982c) discuss a possible glial contribution to the slow signals observed in slices. Salzberg et al (1983) discuss a possible glial origin for the slow signals detected from the hypothalamus. Orbach & Cohen (1983) concluded that in their salamander experiment, the contribution from glia was small. However, data to substantiate that claim for their slow signals were not presented or discussed. Lev-Ram & Grinvald (1984) have recently recorded very large slow signals (longer than 1 sec) from the rat optic nerve. These slow signals were sensitive to the extracellular potassium concentration and probably represent glia depolarization. Thus, caution should be exercised in the interpretation of slow signals. Development of optical probes that are specific to a given cell type, or the possible iontophoretic injection of a suitable fluorescent dye into single cells, or the investigation of evoked activity in single cells impaled with a microelectrode, should resolve this difficulty.

The Spatial Resolution of Optical Recording

THE MICROSCOPE RESOLUTION The spatial resolution of optical recording from flat, two-dimensional preparations is excellent ($\sim 1\ \mu m$); however, that for three-dimensional preparations is relatively poor (Salzberg et al 1977, Orbach & Cohen 1983, Grinvald & Segal 1983). For example, in experiments on the frog optic tectum (Grinvald et al 1984) spatial resolution is about 200 μm with a $10\times$ objective and about 80 μm with a $40\times$ objective. These results are not surprising because conventional microscopes do not resolve images in a thick preparation well, and therefore the three-dimensional resolution of the optical signals is hampered also. The spatial resolution may be further improved in at least three ways: (*a*) by the design of custom-made long-working distance objectives with high numerical aperture; (*b*) by mathematical deconvolution of results obtained from measurements at different focal planes and by the use of the mathematical equation for the point spread function (i.e. the defocus blurring function) of a given objective (e.g. Agard & Sedat 1983); (*c*) by the use of a confocal detection system (e.g. Egger & Petran 1967), and optical recording with a laser microbeam (Grinvald & Farber 1981), and scanning (Dillon & Morad 1981) in three dimensions instead of continuous illumination of the whole field under investigation. No one has yet implemented the existing optical approaches and mathematical analysis to improve three-dimensional resolution, but at least one laboratory is planning to perform this important task (P. Saggau and G. ten Bruggencate, personal communication).

THE EFFECT OF LIGHT SCATTERING ON SPATIAL RESOLUTION Light scattering from cellular elements, especially in preparations that seem opaque, leads to the deterioration of images resolved by conventional microscope optics. Thus, light scattering both blurs the images of individual targets and causes an expansion of the apparent area of detected activity. These effects of light scattering on optical recordings were investigated recently (Orbach & Cohen 1983). A solution was found for this problem that is quite significant for the optical investigation of cortex-like structures. Egger & Petran (1967) constructed a modified microscope for confocal imaging. If only a small spot in the preparation is illuminated at a given time, and coincident detection is employed at the image plane from a small spot where the unscattered image should appear, the effects of light scattering are considerably diminished. Using their modified microscope, Egger & Petran were able to resolve visually single neurons residing over the roof of the ventricle in the optic tectum, 500 μm below the surface. This approach can be implemented for optical recording as well by using laser scanning and coincident random access detection with a photodiode array. Attempts to pursue this rather difficult approach have already begun (P. Saggau, personal communication).

Another, simpler, solution to the problem of three dimensional resolution is to stain only a very small area in the preparation by iontophoretic application or by pressure injection of the dye. Specific staining restricted to the deep layer below the surface would also increase the depth of the loci susceptible to the optical measurements. Alternatively, if the mathematical equation for the light scattering point spread function were determined, iterative deconvolution procedures could be used to refine the data.

OPTIMALIZATION OF THE OPTICAL EXPERIMENTS

The Noise in Optical Experiments

The signal to noise ratio obtained in optical measurement is one of the most important factors determining the feasibility of certain physiological experiments. The purpose of this section is to provide readers with tools to predict the maximum of the expected SNR in a given experimental situation.

SOURCES OF NOISE IN OPTICAL MEASUREMENTS Five sources of noise exist in optical experiments:

1. Light shot noise due to random fluctuation in the photocurrent originating from the quantal nature of light and electricity: The shot noise is proportional to the square root of the detected light intensity. Therefore, increase of the illumination intensity will reduce the relative noise level (Cohen et al 1974, Ross et al 1977, Grinvald et al 1983).
2. Dark noise of the detectors: This noise is fairly constant. It is exceptionally low in photomultipliers; in the photodiode-amplifier combination it is much higher, equivalent to the shot noise originating from 10^7–10^8 photons per millisecond. Therefore, for the detection of lower level fluorescence signals (e.g. from segments of neuronal process), photomultipliers are preferable (Grinvald et al 1982b).
3. Fluctuations in the stability of the light sources: This depends both on the nature of the light source and its power supply. The stability of a tungsten-halogen lamp is in the range of 10^{-6} (10 Hz–1 khz) when optimally operated. Mercury arc lamps can be stabilized to 5.10^{-5} to 1–10^{-4} with a negative feedback circuit (J. Pine and A. Grinvald, unpublished results).
4. Noise that originates from movement of the preparation due to blood circulation (Grinvald et al 1984), heartbeat (Orbach et al 1985), cilia beat (Senseman et al 1983), etc.
5. Vibrational noise originating from relative movements of any of the components along the optical light path including the preparation (Salzberg et al 1977). Vibrational noise is especially large if the preparation is not uniform optically. For example, the frog optic tectum is covered by dark pigments overlying the blood vessels. Because of the normal blood flow these pigments move all the time; therefore, signals from detectors covering blood vessels may be ten times noisier than those covering a pigment-free area of the tectum. This noise and the heartbeat noise were totally eliminated when the frog was perfused with a laminar flow of saline through the aorta (Grinvald et al 1984, K. Kamino and A. Grinvald, unpublished results).

Methods for minimizing the noise are discussed in the manuscripts quoted above. Three important rules must be kept in mind:

1. If the signals are small, it is very important to minimize the contribution of all the noise sources, and then increase the light intensity as much as possible up to the level that light shot noise is equal to the light source nonrandom noise.
2. However, when other types of noise (not the light shot noise) are predominant, the light level should be reduced to the point at which the shot noise is equal to that noise. Lowering the light level would minimize photodynamic damage and bleaching, and would thus permit more extensive signal averaging to improve the SNR.
3. If constant periodical noise is predominant (e.g. heartbeat), it is relatively easy to correct it by proper synchronization and subtraction procedures (Orbach et al 1985, Grinvald et al 1984).

Estimation of the Expected Signal-to-Noise Ratio

I shall discuss some preparation-dependent factors that affect the SNR in absorption and fluorescence measurements (Rigler et al 1974, Waggoner & Grinvald 1977). This should help in making the decision about which mode is preferable in a given situation and should also help to provide estimates for the expected SNR in a given situation.

The SNR in transmission experiments is given by

$$(S/N)_T = (\Delta T/T) \cdot (2q \cdot \tau)^{1/2} \cdot T^{1/2} \quad (1)$$

where τ is the rise time of the detector circuit (τ = ¼Δf, Δf is the power bandwidth, T is the transmitted light intensity reaching the detector, q is the quantum efficiency of the detector, and $\Delta T/T$ is the fractional change in transmission. In fluorescence experiments the SNR is given by

$$(S/N)_F = (\Delta F/F)\,(2\tau q)^{1/2} \cdot (gF)^{1/2} \quad (2)$$

where F is the fluorescence intensity originating from the preparation (it is linearly proportional to the illumination intensity), g is a geometrical factor related to the collection efficiency of the fluorescence detector, and $\Delta F/F$ is the fractional change in fluorescence.

T, the transmitted light intensity, is usually larger than F by 3–4 orders of magnitude, for equal illumination intensity. However, from the above equations, it is evident that if $\Delta F/F$ is much larger than $\Delta T/T$, fluorescence measurements can give a better SNR. The largest observed $\Delta F/F$ from a single cell was $2.5–10^{-1}$ (Grinvald et al 1983), but the largest $\Delta T/T$ was only $5.10^{-4}–10^{-3}$ (Grinvald et al 1977, Ross & Reichardt 1979).

The SNR in an optical experiment depends on the efficiency of the illumination and detection system, but the fractional changes $\Delta T/T$ and $\Delta F/F$ are inherent properties of the dye and the preparation. In multilayer preparations having n layers of neurons, to a first approximation $(\Delta T/T) = n \cdot (\Delta T/T)$. As long as the total dye absorption is a few percent, T is relatively insensitive to n; thus the SNR will increase linearly with n (Eq. 1). On the other hand, in fluorescence experiments

$$(\Delta F/F)_n = \frac{(n \cdot \Delta F)}{(n \cdot F)} = \Delta F/F5.$$

Because $\Delta F/F$ is relatively independent of n in multilayer preparations, the SNR will increase only as $\sqrt{n}$ (because only F will increase) (see Eq. 2). Therefore, for the detection of activity of a population of neurons in multilayer preparations (large n), transmission measurements may be more suitable than fluorescence (whenever applicable), and the signals may be huge (e.g. Salzberg et al 1983).

Transmission measurements are also much less sensitive to nonspecific binding of the dye to the nonactive membranes or other binding sites; such binding does not affect ΔT, and it affects T only moderately. In fluorescence experiments, nonspecific dye fluorescence does not affect ΔF but it affects F linearly according to the number of nonspecific binding sites. Therefore, the SNR will deteriorate as the square root of the nonspecific binding. For example, in a hypothetical situation in which 25 neurons, on top of each other, are viewed by a photodetector but only one is active, the SNR in transmission will deteriorate very little (relative to the situation of only one active cell being present there), but the SNR in fluorescence will deteriorate by a factor of five.

On the other hand, in transmission measurements, it is important that the detector have the same shape and size as the target image, otherwise the SNR will deteriorate as the square root of the extra area viewed by the detector. This does not happen in fluorescence if only the target is fluorescing (e.g. processes of stained single cells).

A useful equation for estimating the expected SNR is given by

$$(S/N) = \Delta I/I^{1/2} = (\Delta I/I) \cdot (I)^{1/2} \tag{3}$$

where I_p is the photocurrent of flux reaching the detector in electrons/msec; $\Delta I/I$ is the expected fractional size of the signal that should be estimated.

Thus, from the above discussion, one should be able to predict the expected fractional changes and SNRs for absorption or fluorescence measurements. Using the reasonable assumption that proper probes for the given preparation can be found, the fractional change for a single cell is $0.5-2\cdot10^{-1}$ and $1-10\cdot10^{-4}$ (per 100 mV potential change) for fluorescence and transmission, respectively. This should be then corrected for the assumed nonspecific binding, the fraction of active cells, times the expected change in their membrane potential, and the target area relative to the detector area. An estimate of the

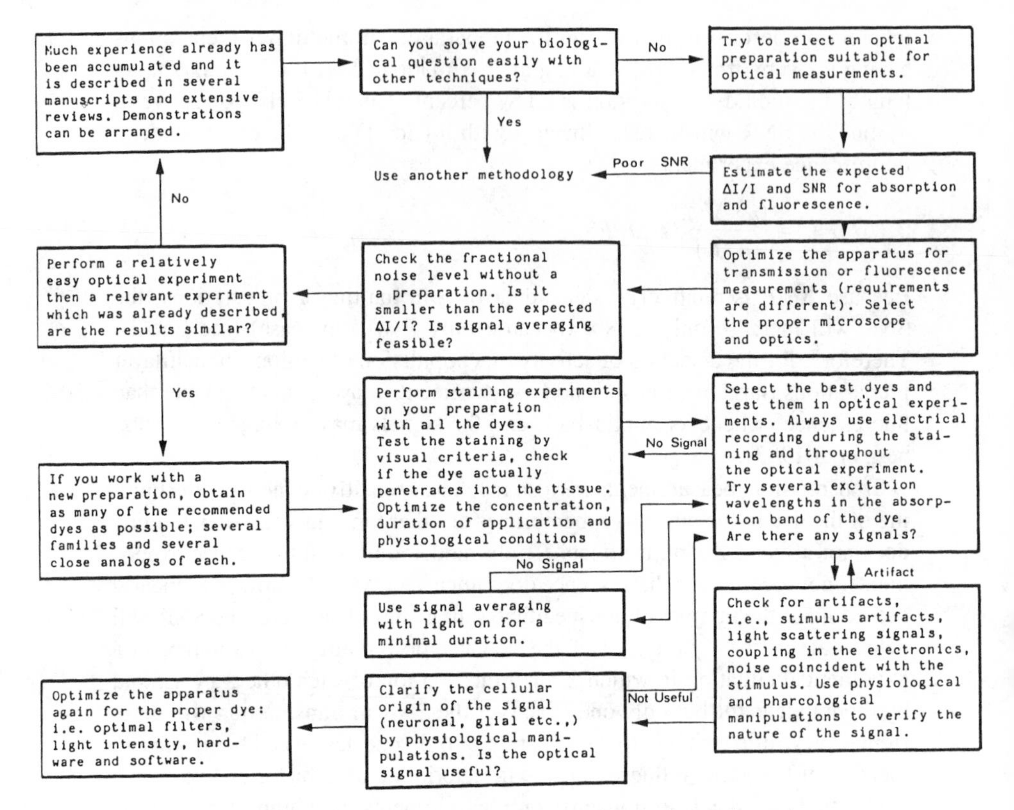

Figure 14 A flow chart suggesting how to approach a novel optical experiment.

intensity that will reach the photodetector is also required; it can be made from a preparation properly stained with a voltage sensitive dye without measuring any signals, and even if the dye is not optimal for the preparation. The simple current-to-voltage amplifier required for such measurements was described (Cohen et al 1974). The output voltage of the amplifier V can be used to estimate the photocurrent I_p. ($I_p = V/R$, where R is the negative feedback resistor of the amplifier.) A flowchart suggesting how to approach a novel optical experiment is shown below (Figure 14).

CONCLUSIONS

Compared with electrical recording, optical methods have several inherent advantages.

1. There is no lower limit to the size of the neuronal element recorded, provided a voltage-sensitive dye can attach to its membrane. (However, a

large number of such elements may have to be synchronously active or, alternatively, a large number of responses may have to be averaged, in order to detect responses from small elements.)

2. This method is a noninvasive one: there is no need to impale the membrane and risk injuring it in order to record. For this reason, a recording can be obtained form the same elements for a considerable length of time but not continuously.
3. The major advantage is that a large number of detectors can be placed side by side to monitor simultaneously the electrical patterns of activity from hundreds of loci, thus affording real time imaging of neuronal activity and facilitating the localization of functional units.

Considerable progress has also been made in designing arrays of extracellular electrodes that provide complementary information (Thomas et al 1972, Pickard & Welberry 1976, Gross 1979, Pine 1980, Freeman 1977, Kruger & Bach 1981, Bowler & Llinas 1982). The combination of optical and electrical techniques would most probably provide another missing dimension to the investigation of brain function.

Many difficulties have been associated with optical measurements. The four groups active in this field have focused mostly on improvements of the technology and its exploitation in new directions. The technique is now far more developed for widespread usage. It has been quite rewarding to notice that most of the predictions made in an earlier review (Cohen et al 1978) have been realized. The cost of the equipment is decreasing despite its improved performance, and the number of optical monitors in use has almost tripled during the last year.

Some difficulties still remain. I have outlined some possible solutions. New methods to handle and analyze the large amount of data should be developed as well as novel conceptual frameworks for the interpretation of such unique data. These advances call for additional multidisciplinary efforts. If the above improvements are successful, the optical recording technique will provide a tomography-like tool for imaging neuronal activity in three dimensions (up to a depth of 400–1000 μm). The submillisecond time resolution and 30–60 μm spatial resolution provided by this method could be especially helpful in understanding both developmental changes and information processing in various CNS structures and also for investigating the integration properties of single cells and the function of local circuits in simpler preparations.

ACKNOWLEDGMENTS

I would like to thank Drs. I. Z. Steinberg, V. I. Teichberg, M. Segal, L. B. Cohen, B. M. Salzberg, and W. N. Ross for their critical and constructive comments on the manuscript. I am indebted to our chemist Rina Hildesheim, without whose dyes we could not have pursued much of the above research.

My work was supported by grants from the NIH (NS 14716), the US-Israel Binational Science Foundation, the Muscular Dystrophy Association, the Israel Academy of Sciences, The Psychobiology Foundation, and the March of Dimes.

Literature Cited

Agard, D. A., Sedat, J. W. 1983. Three-dimensional architecture of a polytene nucleus. *Nature* 302:676–81

Agmon, A., Hildesheim, R., Anglister, L., Grinvald, A. 1982. Optical recordings from processes of individual leech CNS neurons iontophoretically injected with new fluorescent voltage-sensitive dye. *Neurosci. Lett.* 10:S35

Anglister, L., Farber, I. C., Shahar, A., Grinvald, A. 1982. Localization of voltage-sensitive Ca^{++} channels along developing neurites: Their possible role n regulating neurite growth. *Dev. Biol.* 94:351–65

Anglister, L., Grinvald, A. 1983. Real time visualization of the spatiotemporal responses in the optic tectum of vertebrates. *Israel J. Med.*(described in Grinvald & Segal 1983)

Baylor, S. M. 1982. Optical studies of excitation-contraction coupling using voltage-sensitive and calcium-sensitive probes. *Handb. Physiol., Section on Skeletal Muscle,* ed. L. D. Peachey, R. H. Adrian, pp. 355–79. Bethesda, Md.: Am. Physiol. Soc.

Bowler, J., Llinas, R. 1982. Simultaneous sampling and analysis of the activity of multiple, closely adjacent purkinje cells. *Neurosci. Abstr.* 8:830

Boyle, M. B., Cohen, L. B., Macagno, E. R., Orbach, H. S. 1983. The number and size of neurons in the CNS of gastropod molluscs and their suitability for optical recording of activity. *Brain Res.* 266:305–17

Brown, J. E., Harary, H. H., Waggoner, A. 1979. Isopotentiality and optical determination of series resistance in *Limulus* ventral photoreceptor. *J. Physiol.* 296:357–72

Calvet, J., Calvet, M. C. 1981. A simple device for making a standard inverted phase-contrast microscope moveable. *J. Neurosci. Methods* 4:105–8

Cohen, L. B., Keynes, R. D. 1971. Changes in light-scattering associated with the action potential in crab nerve. *J. Physiol.* 212:259–75

Cohen, L. B., Keynes, R. D., Hille, B. 1968. Light scattering and birefringence changes during nerve activity. *Nature* 218:438–41

Cohen, L. B., Keynes, R. D., Landowne, D. 1972. Changes in axon light-scattering that accompany the action potential: Current dependent components. *J. Physiol.* 224:727–52

Cohen, L. B., Salzberg, B. M. 1978. Optical measurement of membrane potential. *Rev. Physiol. Biochem. Pharm.* 83:35–88

Cohen, L. B., Salzberg, B. M., Davila, H. V., Ross, W. N., Landowne, D., et al. 1974. Changes in axon fluorescence during activity: Molecular probes of membrane potential. *J. Membr. Biol.* 19:1–36

Cohen, L. B., Salzberg, B. M., Grinvald, A. 1978. Optical methods of monitoring neuron activity. *Ann. Rev. Neurosci.* 1:171–82

Conti, F. 1975. Fluorescent probes in nerve membranes. *Ann. Rev. Biophys. Bioeng.* 4:287–310

Davila, H. V., Cohen, L. B., Salzberg, B. M., Shrivastav, B. B. 1974. Changes in ANS and TNS fluorescence in giant axons from *Loligo. J. Membr. Biol.* 15:29–46

Davila, H. V., Salzberg, B. M., Cohen, L. B. 1973. A large change in axon fluorescence that provides a promising method for measuring membrane potential. *Nature New Biol.* 24:159–60

Dillon, S., Morad, M. 1981. Scanning of the electrical activity of the heart using a laser beam with acousto-optics modulators. *Science* 214:453–56

Dragsten, P. R., Webb, W. W. 1978. Mechanism of membrane potential sensitivity of fluorescent membrane probe merocyanine 540. *Biochemistry* 17:5228–40

Egger, M. D., Petran, M. 1967. New reflected light microscope for viewing unstained brain and ganglion cells. *Science* 157:305–7

Farber, I. C., Grinvald, A. 1983. Identification of presynaptic neurons by laser photostimulation. *Science* 222:1025–27

Freedman, J. C., Laris, P. C. 1981. Electrophysiology of cells and organelles: Studies with optical potentiometric indicators. *Int. Rev. Cytol. Suppl.* 12:177–245

Freeman, J. A. 1977. Possible regulatory function of acetylcholine receptor in maintenance of retinotectal synapses. *Nature* 269:218–22

Fujii, S., Hirota, A., Kamino, K. 1981a. Optical recording of development of electrical activity in embryonic chick heart during early phase of cardiogenesis. *J. Physiol.* 311:147–60

Fujii, S., Hirota, H., Kamino, K. 1981b. Optical indications of pace–maker potential and rhythm generation in early embryonic chick heart. *J. Physiol.* 312:253–63

Fujii, S., Hirota, A., Kamino, K. 1981c. Action potential synchrony in embryonic precontractile chick heart: Optical monitoring with potentiometric dyes. *J. Physiol.* 319:529–41

Grinvald, A., Anglister, L., Freeman, J. A., Hildesheim, R., Manker, A. 1984. Real time optical imaging of naturally evoked electrical activity in the intact frog brain. *Nature* 308:848–50

Grinvald, A., Cohen, L. B., Lesher, S., Boyle, M. B. 1981a. Simultaneous optical monitoring of activity of many neurons in invertebrate ganglia, using a 124 element 'Photodiode' array. *J. Neurophysiol.* 45:829–40

Grinvald, A., Farber, I. 1981. Optical recording of Ca^{2+} action potentials from growth cones of cultured neurons using a laser microbeam. *Science* 212:1164–69

Grinvald, A., Fine, A., Farber, I. C., Hildesheim, R. 1983. Fluorescence monitoring of electrical responses from small neurons and their processes. *Biophys. J.* 42:195–98

Grinvald, A., Hildesheim, R., Agmon, A., Fine, A. 1982a. Optical recording from neuronal processes and their visualization by iontophoretic injection of new fluorescence voltage-sensitive dyes. *Neurosci. Abstr.* 8:491

Grinvald, A., Hildesheim, R., Farber, I. C., Anglister, L. 1982b. Improved fluorescent probes for the measurement of rapid changes in membrane potential. *Biophys. J.* 39: 301–8

Grinvald, A., Kamino, K., Lesher, S., Cohen, L. B. 1978. Larger fluorescence and birefringence signals for optical monitoring of membrane potential. *Biophys. J.* 21:82a

Grinvald, A., Manker, A., Segal, M. 1982c. Visualization of the spread of electrical activity in rat hippocampal slices by voltage sensitive optical probes. *J. Physiol.* 333:269–91

Grinvald, A., Ross, W. N., Farber, I. 1981b. Simultaneous optical measurements of electrical activity from multiple sites on processes of cultured neurons. *Proc. Natl. Acad. Sci. USA* 78:3245–49

Grinvald, A., Salzberg, B. M., Cohen, L. B. 1977. Simultaneous recording from several neurons in an invertebrate central nervous system. *Nature* 268:140–42

Grinvald, A., Segal, M. 1983. Optical monitoring of electrical activity: Detection of spatiotemporal patterns of activity in hippocampal slices by voltage-sensitive probes. In *Brain Slices*, ed. R. Dingledine, pp. 227–61. New York: Plenum

Gross, G. W. 1979. Simultaneous single unit recording *in vitro* with a photoetched laser deinsulated gold multimicroelectrode surface. *IEEE Trans. Biomed. Eng. BEE* 26:273–79

Gupta, R., Salzberg, B. M., Cohen, L. B., Grinvald, A., Kamino, K., et al. 1981. Improvements in optical methods for measuring rapid changes in membrane potential. *J. Membr. Biol.* 58:123–38

Kaufman, L., Williamson, S. J. 1982. Magnetic location of cortical activity. *Ann. NY Acad. Sci.* 388:197–213

Krauthamer, V., Ross, W. N. 1984. Regional variations in excitability of barnacle neurons. *J. Neurosci.* 4:673–82

Kruger J., Bach, M. 1981. Simultaneous recording with 30 microelectrodes in monkey visual cortex. *Exp. Brain Res.* 41:191–94

Kuhnt, U., Grinvald, A.. 1981. 4–AP induced presynaptic changes in the hippocampal slices as measured by optical recording and voltage-sensitive probes. *Pflugers Arch. Suppl.* 394:R45

Lancet, D., Greer, C. A., Kaner, J. S., Shepherd, G. M. 1982. Mapping of odor related neuronal activity in the olfactory bulb by high resolution 2-deoxyglucose autoradiography. *Proc. Natl. Acad. Sci. USA* 79:670–74

Lev-Ram, V., Grinvald, A.. 1984. Is there a potassium dependent depolarization of the paranodal region of myelin sheath? Optical studies of rat optical nerve. *Neurosci. Abstr.* 10:948

Lipton, P. 1973. Effects of membrane depolarization on light scattering by cerebral slices. *J. Physiol.* 231:365–83

Llinas, R., Sugimori, M. 1980. Electrophysiological properties of *in vitro* Purkinje cell dendrites in mammalian cerebellar slices. *J. Physiol.* 305:197–213

Loew, L. M., Bonneville, G. W., Surow, J. 1978. Charge shift probes of membrane potential theory. *Biochemistry* 17:4065–71

Loew, L. M., Scully, S., Simpson, L., Waggoner, A. S. 1979. Evidence for a charge shift electrochromic mechanism in a probe of membrane potential. *Nature* 281:497–99

Loew, L. M., Simpson, L. L. 1981. Charge shift probes of membrane potential. *Biophys. J.* 34:353–63

Loew, L. M., Cohen, L. B., Salzberg, B. M., Obaid, A. L., Benzanilla, F. 1984. Charge shift probes of membrane potential. Characterization of aminostyrylpyridinium dyes on the squid giant axon. *Biophys. J.* In press

London, J. A., Zecevic, D., Cohen, L. B. 1985. Simultaneous monitoring of activity of many neurons in buccal ganglia of *Pleurobranchaea Aplysia* and *Navanax*. In *Optical Methods in Cell Physiology*, ed. P. De Weer, B. M. Salzberg. New York: Wiley. In press

Manker, A. 1982. *Design of a display processor to visualize neuronal activity.* M.Sc. thesis, Feinberg Grad. Sch., Weizmann Inst. of Sci., Rehovot, Israel

Nakajima, S., Gilai, A. 1980. Action potentials of isolated single muscle fibers recorded by potential sensitive dyes. *J. Gen. Physiol.* 76:729–50

Obaid, A. L., Shimizu, H., Salzberg, B. M. 1982. Intracellular staining with potentiometric dyes; optical signals from identified leech neurons and their processes. *Biol. Bull.* 163:388

Orbach, S. H., Cohen, L. B. 1983. Simultaneous optical monitoring of activity from many areas of the salamander olfactory bulb. A new method for studying functional organization in the vertebrate CNS. *J. Neurosci.* 3:2251–62

Orbach, H. S., Cohen, L. B., Grinvald, A. 1982. Optical recording of evoked activity in the visual cortex of the rat. *Biol. Bull.* 163:389

Orbach, H. S., Cohen, L. B., Grinvald, A. 1985. Optical mapping of neuronal activity in the mammalian sensory cortex. Submitted for publication

Orbach, H. S., Cohen, L. B., Grinvald, A., Hildesheim, R. 1983. Optical monitoring of neuron activity in rat somatosensory and visual cortex. *Neurosci. Abstr.* 9:39

Oxford, G. S., Pooler, J. P., Narahashi, T. 1977. Internal and external application of photodynamic sensitizers on squid giant axons. *J. Membr. Biol.* 36:159–73

Patrick, J., Valeur, B., Monnerie, L., Changeux, J. P. 1971. Changes in extrinsic fluorescence intensity of the electroplax membrane during electrical excitation. *J. Membr. Biol.* 5:102–20

Pickard, R. S., Welberry, T. R. 1976. Printed circuit microelectrodes and their application to honeybee brain. *J. Exp. Biol.* 64:39–44

Pine, J. 1980. Recording action potentials from cultured neurons with extracellular microcircuit electrodes. *J. Neurosci. Methods* 2:19–31

Pooler, J. P. 1972. Photodynamic alteration of sodium currents in lobster axon. *J. Gen. Physiol.* 60:367–87

Raichle, M. E. 1979. Quantitative *in vivo* autoradiography with position emission tomography. *Brain Res. Rev.* 1:47–68

Rigler, R., Rable, C. R., Jovin, T. M. 1974. A temperature jump apparatus for fluorescence measurements. *Rev. Sci. Instrum.* 45:581–87

Ross, W. N., Krauthamer, V. 1984. Optical measurements of potential changes in axons and processes of neurons of a barnacle ganglion. *J. Neurosci.* 4:659–72

Ross, W. N., Reichardt, L. F. 1979. Species-specific effects on the optical signals of voltage sensitive dyes. *J. Membr. Biol.* 48:343–56

Ross, W. N., Salzberg, B. M., Cohen, L. B., Grinvald, A., Davila, H. V., et al. 1977. Changes in absorption, fluorescence, dichroism and birefringence in stained axons: Optical measurement of membrane potential. *J. Membr. Biol.* 33:141–83

Ross, W. N., Salzberg, B. M., Cohen, L. B., Davila, H. V. 1974. A large change in dye absorption during the action potential. *Biophys. J.* 14:983–86

Salama, G., Morad, M. 1976. Merocyanine 540 as an optical probe of transmembrane electrical activity in the heart. *Science* 191:485–87

Salzberg, B. M. 1978. Optical signals from squid giant axons following perfusion or superfusion with potentiometric probes. *Biol. Bull.* 155:463–64

Salzberg, B. M. 1983. Optical recording of electrical activity in neurons using molecular probes. In *Current Methods in Cellular Neurobiology,* ed. J. L. Barker, pp. 139–87. New York: Wiley

Salzberg, B. M., Benzanilla, F. 1983. An optical determination of the series resistance in Laligo. *J. Gen. Physiol.* 82:807–17

Salzberg, B. M., Davila, H. V., Cohen, L. B. 1973. Optical recording of impulses in individual neurons of an invertebrate central nervous system. *Nature* 246:508–9

Salzberg, B. M., Grinvald, A., Cohen, L. B., Davila, H. V., Ross, W. N. 1977. Optical recording of neuronal activity in an invertebrate central nervous system: Simultaneous monitoring of several neurons. *J. Neurophysiol.* 40:1281–91

Salzberg, B. M., Obaid, A. L., Senseman, D. M., Gainer, H. 1983. Optical recording of action potentials from vertebrate nerve terminals using potentiometric probes provides evidence for sodium and calcium components. *Nature* 306:36–39

Schwartzkroin, P. A., Slawsky, M. 1977. Probable calcium spikes in hippocampal neurons, *Brain Res.* 135:157–61

Sejnowski, T. J., Reingold, S. C., Kelley, D., Gelperin, A. 1980. Localization of [H^3]-2-deoxyglucose in single molluscan neurons. *Nature* 287:449–51

Senseman, D. M., Salzberg, B. M. 1980. Electrical activity in an exocrine gland: Optical recording with a potentiometric dye. *Science* 208:1269–71

Senseman, D. M., Shimizu, H., Horwitz, I. S., Salzberg, B. M. 1983. Multiple-site optical recording of membrane potential from a salivary gland. *J. Gen. Physiol.* 81:887–908

Simons, D. J. 1978. Response properties of vibrisa units in rat SI somatosensory cortex. *J. Neurophysiol.* 41:798–820

Sokoloff, L. 1977. Relation between physiological function and energy metabolism in the central nervous system. *J. Neurochem.* 29:13–26

Stuart, A. E., Ortel, D. 1978. Neuronal properties underlying processing of visual information in the barnacle. *Nature* 275:187–190

Tasaki, I., Warashina, A. 1976. Dye membrane interaction and its changes during nerve excitation. *Photochem. Photobiol.* 24:191–207

Tasaki, I., Watanabe, A., Sandlin, R., Carnay, L. 1968. Changes in fluorescence turbidity and birefringence associated with nerve excitation. *Proc. Natl. Acad. Sci. USA* 61:883–88

Thomas, C. A. Jr., Springer, P. A., Loeb, G. E., Berwald-Netter, Y., Okun, L. M. 1972. A miniature microelectrode array to monitor the bioelectric activity of cultured cells. *Exp. Cell Res.* 74:61–66

Vergara, J., Bezanilla, F. 1976. Fluorescence changes during electrical activity in frog muscle stained with merocyanine. *Nature* 259:684–86

Waggoner, A. S. 1976. Optical probes of membrane potential. *J. Membr. Biol.* 27:317–34

Waggoner, A. S. 1979. Dye indicators of membrane potential. *Ann. Rev. Biophys. Bioengr.* 8:47–63

Waggoner, A. S., Grinvald, A. 1977. Mechanisms of rapid optical changes of potential sensitive dyes. *Ann. NY Acad. Sci.* 303:217–42

Wong, R. K. S., Prince, D. A., Basbaum A. I. 1979. Intradendritic recordings from hippocampal neurons. *Proc. Natl. Acad. Sci. USA* 76:986–90

Woolsey, T. A., van Der Loos, H. 1970. The structural organization of layer IV in the somatosensory region (SI) of mouse cerebral cortex. *Brain Res.* 17:205–42

Yamamoto, C. 1972. Intracellular study of seizure-like after discharges elicited in thin hippocampal sections *in vitro*. *Exp. Neurol.* 35:154–64

Ann. Rev. Neurosci. 1985. 8:307–37

BRAINSTEM CONTROL OF SACCADIC EYE MOVEMENTS

A. F. Fuchs, C. R. S. Kaneko, and C. A. Scudder

Department of Physiology & Biophysics, and Regional Primate Research Center, University of Washington, Seattle, Washington 98195

INTRODUCTION

Saccades are rapid shifts in the direction of gaze. They include the fast (reset) phases of nystagmus generated by vestibular or optokinetic stimuli, the catch-up movements required in the pursuit of a small moving target, and the scanning movements used to explore a stationary visual scene. All these rapid eye movements have a similar trajectory composed of a rapid acceleration to peak velocity followed by a deceleration that brings the eye to its final position. Primate saccades usually last between 15 and 100 msec and can reach peak velocities in excess of 500 deg/sec. Because they have such similar time courses, all rapid eye movements are thought to share a common neural saccade generator.

If a saccade is made to a visual target, several steps are presumed to precede the activation of the saccade generator. After the target and its visual surroundings fall on the retina and are detected by the visual system, the target is selected from its surroundings, its location is calculated, and a decision is made as to whether the target is worthy of a saccade. The saccade is triggered and appropriate commands are passed first to brainstem centers, then to ocular motoneurons, and finally to the extraocular muscles. Although we know relatively little about the early stages of the process that leads to a saccade, experiments in the last 15 years have provided an understanding of the brainstem circuitry involved in saccade production. The relative simplicity of the oculomotor periphery and the easy access to premotor elements in the brainstem have provided a unique opportunity to probe into the more central structures that generate movement.

In this review we summarize what is known about the brainstem saccade generator by starting at the motor periphery and moving centrally. We emphasize new data, particularly from primates, and structure our discussion in terms

0147–006X/85/0301–0307$02.00

of what is known, points of disagreement, current hypotheses, and fruitful directions for future research.

THE OCULOMOTOR PERIPHERY

Mechanics of the Globe

The discharge patterns of ocular motoneurons reflect the load that must be driven. Westheimer (1954) first proposed that saccades were the responses of the orbital load (composed of mass and viscous and elastic elements) to a step change in net muscle tension. When Robinson (1964) actually applied a step of force externally to the fixating eye, the eye did not move in a saccadic trajectory, but rather its motion resembled the sum of at least two exponential curves with a short (12 msec) and a long (about 300 msec) time constant. The existence of the long time constant element suggests that the mechanical events in the orbit continue for several hundred milliseconds after the saccade is over (in the monkey, a 20 deg saccade lasts less than 50 msec). By measuring the forces that humans generate in attempts to produce saccades against a variety of loads, Robinson demonstrated that the eye requires not only a step of force to hold it in its eccentric position, but also a pulse of force in excess of the step to generate a saccade. This pulse-step pattern of net muscle force is employed for all saccades larger than 5 deg. Since loading the eye with a 100-fold increase in mass barely affects the time course of the saccade, the pulse of force seems to be required largely to overcome the viscoelastic properties of the extraocular muscles, the globe, and its suspensory ligaments. In addition, Robinson showed that as the size of the saccade increases, the pulse amplitude does not increase proportionately but instead is applied for longer periods of time. These data led him to suggest that a separate group of motor units is called into play only to generate the pulse of force that produces the saccade.

Extraocular Muscles

Investigators using a variety of approaches to study whether there are separate motor units dedicated to saccades have reported conflicting results. Morphological studies of extraocular muscles suggest that such units could exist since a rich variety of muscle fiber types have been described in most mammals (Peachy 1971). In the cat (Alvarado & Van Horn 1975), monkey (Mayr et al 1966, Miller 1967), and man (Mukuno 1968, Dietert 1965, Ringel et al 1978), at least five separate fiber types have been distinguished based on morphological and histochemical characteristics such as fiber diameter, innervation patterns, amount of sarcoplasmic reticulum, extent of the T-tubule system, number and size of mitochondria, and the level of oxidative and glycolytic enzyme activity. Despite considerable disagreement regarding the classification of fiber types both among investigators studying the same species and among those studying different species, certain similarities emerge that allow us to make the following generalizations.

At one end of the fiber spectrum are large, singly innervated fibers that possess glycolytic enzymes, extensive internal membrane systems, and relatively few mitochondria. Their small, well-defined myofibrils surrounded by abundant sarcoplasm produce the characteristic "Fibrillenstruktur" when cut in cross section. These fibers are concentrated in the global layer of the muscle (adjacent to the globe) where, in the cat, they comprise up to 60% of its cross-sectional area (Alvarado & Van Horn 1975). Such fibers resemble the well-studied fast-twitch fibers in skeletal muscles. Indeed, fibers with fast-twitch properties have been studied electrophysiologically in the extraocular muscles of the rat (Chiarandini & Stefani 1979) and in the cat (Lennerstrand 1974, Goldberg et al 1976). In the cat, such fibers have very short contraction times (5–6 msec) and high fusion frequencies (between 150 and 250/sec on average), typically produce tetanic tensions from 100 to 400 mg, and conduct action potentials.

At the other end of the spectrum are small-diameter, multiply innervated fibers that are deficient in internal membrane systems but rich in mitochondria and oxidative enzymes. Their large blocks of poorly delineated fibrils produce the characteristic "Felderstruktur" pattern. These fibers are also located in the global layer where, in the cat, they comprise 16% of its cross-sectional area (Alvarado & Van Horn 1975). Since these fibers do not have the membrane specializations necessary to conduct action potentials, they most resemble the slow, non-twitch fibers of amphibia (Kuffler & Vaughan Williams 1953). Non-twitch fibers that do not conduct action potentials have been recorded in cat extraocular muscle (Hess & Pilar 1963, Lennerstrand 1974, Goldberg et al 1981) and in the rat they have been shown to be multiply innervated (Bondi & Chiarandini 1983). They have fusion frequencies ranging from 10 to 50/sec and tetanic tensions ranging from 10 to 100 mg. Less than 6% of all the fibers recorded in these cat studies had non-twitch characteristics.

At least three additional types of fibers have been distinguished with intermediate morphologies. In general, these fibers can be singly or multiply innervated, can have a sarcoplasmic reticulum that ranges from well developed to spare, and can be found in either the global or orbital (adjacent to the bony orbit) layers of the muscle. It is unclear whether any of these types differ in their electrophysiological or mechanical properties. At least one multiply innervated fiber type in the cat exhibits a slower twitch (contraction time of about 10 msec), fuses at lower frequencies (100 to 250/sec), and produces lower tetanic tensions (30 to 350 mg) than the fast-twitch fiber (Lennerstrand 1974, Goldberg et al 1976, Goldberg et al 1981).

Despite the variety of muscle fiber types based on morphological criteria, it seems clear that the overwhelming majority of all fibers must be of the twitch variety. Probably all of the singly innervated fibers—84% of all muscle fibers in the monkey (Mayr et al 1966) and 78% in the cat (Alvarado & Van Horn 1975)—and at least some of the multiply innervated fibers produce twitches.

Ocular Motoneurons

The diversity of fiber types in the extraocular muscles is consistent with Robinson's (1964) suggestion that different fibers might participate in different types of movement. Specifically, it might be expected that fast-twitch fibers would be activated phasically (i.e. only during saccades) and that the non-twitch or slower twitch fibers would be activated tonically (i.e. only during fixations). However, electrophysiological recordings from ocular motoneurons in behaving monkeys and cats have revealed only one pattern of firing for all motoneurons.

During steady fixation, the typical motoneuron discharges at regular tonic rates that increase linearly with eye position in an optimal direction (the "on-direction") (Figure 1), which corresponds to the pulling direction of the muscle that the motoneuron innervates. The slope (K) of the relation between discharge rate and eye position varies from motoneuron to motoneuron, as does the eye-position threshold at which different motoneurons are recruited into tonic firing (Fuchs & Luschei 1970, Keller 1971, Robinson 1970). Most motoneurons have thresholds in excess of 15 deg in the "off-direction" and therefore participate in all fixations within the normal oculomotor range. Units with thresholds in the on-direction tend to have steeper slopes (Robinson 1970, Keller 1971). For all motoneurons, K values range from 1 to 12, with 3.5 spikes/sec/deg being typical. The firing rate associated with a given eye position is very regular but may differ slightly (Goldstein 1983), depending on whether that position is approached from the on- or off-direction.

About 5 to 8 msec before a saccade in the on-direction, the firing rate jumps rapidly to a higher frequency, then decreases slightly during the course of the saccade and drops rapidly about 10 msec before the saccade ends. This burst activation of the muscles provides the pulse of force that drives the eye during a saccade. The duration of the saccade is proportional to the duration of the burst (Figure 1). For small saccades, the firing rate is proportional to saccade size, but for saccades greater than 10 deg, firing rate saturates. During the saccadic burst of spikes, discharge rates often average 400/sec and can reach peak rates in excess of 800/sec. There is no burst for saccades in the orthogonal direction. When the eye makes an off-direction saccade, most motoneurons exhibit a complete cessation of activity (pause; Figure 1), which begins about 10 msec before the saccade and lasts until just before the saccade ends (Keller 1971, Fuchs & Luschei 1970, Robinson 1970). There seems to be no clear evidence that following the pause there is a burst that could actively brake the eye (Goldstein 1983).

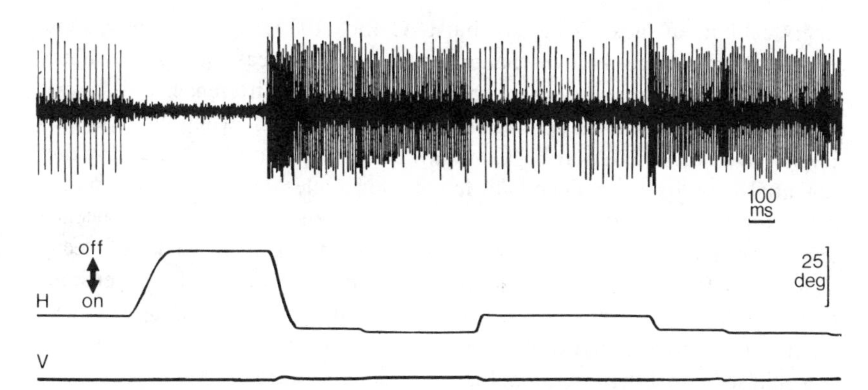

Figure 1 Discharge pattern of an identified motoneuron in the abducens nucleus of the monkey.

Accurate saccades require a precise coordination of the burst and tonic components of motoneuron discharge. If a muscle is impaired by disease or experimental intervention, this coordination can be altered. If the burst is too small relative to the tonic component, the saccade will fall short and then slide forward to reach the target. If the burst is relatively too large, the eye will pass the target and then drift back. Adaptive mechanisms, apparently involving the cerebellum, can adjust the burst and tonic components independently to restore the proper coordination (Optican & Robinson 1980).

We believe that all motoneurons have a burst-tonic firing pattern. Putative motoneurons have been reported that exhibit only the burst but never tonic firing or vice versa (Henn & Cohen 1972, 1973). In our laboratory, however, we have identified over 150 abducens motoneurons from spike-triggered averages of lateral rectus muscle potentials, and all exhibited both burst and tonic discharge components when trained animals were required to use their entire oculomotor range. Furthermore, the spectrum of motoneuron size (17 to 28 μm; Büttner-Ennever et al 1983) is not so large that we were likely to miss any because of their small size.

Even if all motoneurons exhibit a burst-tonic discharge pattern, it is still possible that the different types of muscle fibers play slightly different functional roles. In humans, for instance, where electromyographic (EMG)[1] recordings of single motor units also reveal only burst-tonic discharges, orbital fibers have lower thresholds and smaller K values than global fibers, whereas global fibers exhibit a prominent burst for large eccentric saccades, but orbital

[1]List of abbreviations: D-burster, direction burst neuron; DME, dynamic motor error; DSS, desired saccade size; EBN, excitatory burst neuron; EMG, electromyogram; EPSP, excitatory postsynaptic potential; HRP, horseradish peroxidase; IBN, inhibitory burst neuron; IPSP, inhibitory postsynaptic potential; LLBN, long-lead burst neuron; MLF, medial longitudinal fasciculus; OPN, omnipause neuron; PPRF, paramedian pontine reticular formation; riMLF, rostral interstitial nucleus of the medial longitudinal fasciculus; TN, tonic neuron; TRIG, trigger burst neuron; V-burster, vector burst neuron; VOR, vestibuloocular reflex.

fibers exhibit, at most, a modest burst (Collins 1975). Consequently, most orbital fibers change their discharge for fixation, whereas most global fibers change their discharge during saccades. Although the differences in behavior among these twitch fibers are minor, the behavior of non-twitch fibers, which might have been missed in EMG studies because they do not conduct action potentials, could be significantly different. Nonetheless, the slow development of tension in non-twitch fibers argues against their participation in saccade generation. Furthermore, their overall contribution to the maintenance of steady gaze is problematic, since they are few in number and develop lower tensions than twitch-type fibers. Because most twitch fibers do not fatigue (Lennerstrand 1974) and have an exquisite relation between firing rate and eye position, it would seem that they are more than adequate to control fixations accurately. In summary, no diversity of functions to match the rich variety of muscle fiber types has yet been demonstrated.

GENERATION OF MOTONEURON DISCHARGE

The burst-tonic discharge pattern in motoneurons could be generated by a variety of mechanisms, including specialized motoneuron membrane properties, local circuits, and dedicated premotor networks involving the reticular formation.

Barmack (1974) suggested that the membrane properties of motoneurons could effectively differentiate a step change in synaptic input to produce both burst and tonic components. As in spinal motoneurons, the injection of a steady current into cat abducens and trochlear motoneurons produces a transient increase in firing that exceeds the steady-state discharge. Usually, however, only the first 3 to 5 spikes have shorter interspike intervals and the steady state is always achieved within 50 msec (Grantyn & Grantyn 1978). Because the duration of the transient increase in frequency is too short to produce the long durations of cat saccades (Evinger & Fuchs 1978), it is unlikely that the generation of burst-tonic discharges depends primarily on the membrane properties of the motoneuron.

Local circuits involving either the motoneurons themselves or muscle afferents also apparently are not used in creating the firing pattern for saccades. Axon collaterals of abducens motoneurons are probably lacking in the cat (Highstein et al 1982) and are rare in the monkey (Strassman et al 1982). Even in motoneurons of the oculomotor nucleus that have anatomically demonstrable collaterals (Evinger et al 1981a), no recurrent activation or inhibition has been demonstrated (Sasaki 1963). Although human extraocular muscles have spindles (see Dietert 1965, for a beautiful picture), neither those of the cat nor the monkey do. The musculotendon endings that are present in the monkey (Ruskell 1978) appear to have no short-latency role in eye movements, since brief stretches applied to the muscle during fixation cause no change in motoneuronal firing (Keller & Robinson 1971) and sectioning the muscle afferents has no apparent effect on saccades (Guthrie et al 1982). However, muscle afferents appear to be important in the long term for maintaining conjugate gaze in the monkey (Guthrie et al 1982).

Since the burst in motoneurons does not arise from membrane properties or local circuits acting on a step input, it must originate in afferents to the motoneuron pools. There are two different afferent patterns:

1. For most motoneuron pools, the afferents provide separate burst and tonic inputs that are generated in the neighboring reticular formation.
2. For medial rectus motoneurons whose discharge moves the eye nasally, the major afferent pathway provides an already combined burst-tonic input that originates in "internuclear neurons" lying in the contralateral abducens nucleus.

The existence of internuclear neurons has been demonstrated in both cat and monkey by injection of the retrograde label horseradish peroxidase (HRP) into the oculomotor nucleus (Graybiel & Hartwieg 1974, Steiger & Büttner-Ennever 1979). Labeled neurons comprise about one-third of the somata within the boundary of the contralateral abducens of the cat (Steiger & Büttner-Ennever 1978). Internuclear neurons may make up a similar proportion in the monkey, as only 60% of abducens neurons are motoneurons (Porter et al 1983). Axons of these cells ascend in the contralateral medial longitudinal fasciculus (MLF) and make powerful excitatory synapses on medial rectus motoneurons (Highstein & Baker 1978). This pathway is functionally important, since bilateral transection of the monkey MLF rostral to the abducens nucleus produces an inability to move either eye beyond the midline and causes a clear reduction in the speed of nasalward saccades (Evinger et al 1977). The burst-tonic discharge of internuclear neurons appears to originate from the same inputs as that of abducens motoneurons. These inputs include disynaptic excitation and inhibition from the contralateral and ipsilateral vestibular nerves, respectively (Baker & Highstein 1975), and monosynaptic excitation from the ipsilateral pontine reticular formation (Highstein et al 1976). In the cat, at least, the firing patterns of internuclear neurons are similar but not identical to those of motoneurons, as internuclear neurons have higher K values, higher burst rates, and begin their burst earlier than do motoneurons (Delgado-Garcia et al 1983). Therefore, the discharge properties of internuclear neurons might compensate for the delay in MLF transmission, ensuring that the same neural signal reaches the horizontal rectus muscles of each eye to produce conjugate saccades.

Several lines of evidence suggest that the synaptic potentials that produce the burst-tonic discharge in motoneurons and internuclear neurons are due to neural circuits that lie in the reticular formation near the ocular motor nuclei. First, in the monkey, both electrolytic (Goebel et al 1971) and kainic acid (Lang et al 1982) lesions of the paramedian pontine reticular formation (PPRF) rostral to the abducens nucleus, ventral to the MLF, and dorsal to the nucleus reticularis tegmenti pontis produce a complete and enduring paralysis of eye movements directed toward the side of the lesion. Deficits of vertical eye movements can be produced by unilateral lesions in the mesencephalic reticular formation, rostral to the interstitial nucleus of Cajal, an area identified as the rostral interstitial nucleus of the MLF (riMLF) in the monkey (Büttner-Ennever & Büttner 1978) and the nucleus of the prerubral field in the cat (Graybiel 1977). Second, in both the monkey and the cat, anatomical studies have shown that parts of both the pontine and medullary reticular formations project directly into the abducens nucleus (Büttner-Ennever 1977, Graybiel 1977), while neurons in the riMLF project directly to the ipsilateral oculomotor nucleus (Steiger & Büttner-Ennever 1979, Büttner-Ennever & Lang 1978). Finally, three types of neurons (described below) within the reticular formation have been identified that have discharge patterns related to saccades.

Taken together, these various experimental results suggest that the burst-tonic discharge pattern of motoneurons is assembled from the outputs of several types of brainstem neurons. Two of these, the burst and omnipause neurons,are thought to interact to produce the burst input to motoneurons; a third type, the tonic neuron, may provide the tonic input to motoneurons, and may also interact with the other neurons to generate saccades. We first describe the discharge patterns and connections of these neurons and then discuss how those related to horizontal saccades could be connected in a circuit to produce the firing pattern of motoneurons.

SACCADE-RELATED BRAINSTEM NEURONS

Burst Neurons

Burst neurons discharge an intense "burst" of spikes for saccades and remain silent during fixation, pursuit, and vergence movements. Their discharge patterns depend only on the difference between initial and final eye position, and not on initial and final position per se. Some, the short-lead burst neurons, begin firing an average of 8 to 10 msec before the saccade, while others, the long-lead burst neurons, have an irregular prelude of activity that can begin more than 100 msec before the saccade. (The short-lead burst neurons originally were called "medium-lead" since their burst began before the shorter-lead burst in motoneurons. Because the medium-lead designation implies the existence of burst neurons with a shorter lead, we prefer our designation, which refers the discharge in burst neurons to that in the motoneuron.) Although originally they were thought to fall into two separate groups (Luschei & Fuchs 1972), burst neurons with intermediate lead times have been reported in both the cat (Kaneko et al 1981) and the monkey (Scudder et al 1982), suggesting a continuum rather than a dichotomy. Most burst neurons in the pons and the medulla begin discharging earlier and discharge most vigorously for ipsilateral horizontal saccades. Saccades in other directions are accompanied by less intense bursts that begin closer to saccade onset. For contralateral saccades, there is frequently no burst at all or a few spikes near the end of the saccade. Many of these "horizontal" burst neurons also exhibit some discharge for vertical saccades, so the directional "tuning curves" are rather broad. A smaller population of pontine neurons have on-directions that are within 20 deg of vertical. Although burst neurons with true oblique on-directions have been reported (Luschei & Fuchs 1972; Keller 1974), they are scarce and inadequately studied.

Two groups of brainstem burst neurons have been distinguished that, because of their anatomical projections, must have different functional roles. These two groups have been revealed by injections of HRP into the abducens nuclei of both cats and monkeys (Langer & Kaneko 1984a, Maciewicz et al 1977). One group lies in the ipsilateral dorsal PPRF. In the cat, there is strong circumstantial evidence that both short- and long-lead burst neurons in this region make monosynaptic excitatory connections with ipsilateral abducens motoneurons (Sasaki & Shimazu 1981); consequently, they have been named excitatory burst neurons (EBNs). Since they discharge for ipsilateral saccades, EBNs must be the source of, or contribute to, the burst in abducens motoneurons. In the monkey, the homologous region also contains both short- and long-lead burst neurons, with a clear tendency for long-lead burst neurons to occur more frequently in rostral regions, particularly the nucleus reticularis pontis oralis (Luschei & Fuchs 1972, Raybourn & Keller 1977, Hepp & Henn 1983).

The second group of retrogradely labeled cells lies in the dorsomedial medullary reticular formation, caudal and ventromedial to the abducens nucleus. Burst neurons in this area of the cat produce monosynaptic inhibitory post synaptic potentials (IPSPs) in an estimated 60% of contralateral abducens motoneurons and, therefore, have been called inhibitory burst neurons (IBNs; Hikosaka et al 1978). Since they discharge when the contralateral eye moves medially, they are thought to produce the pause in abducens motoneuron firing. Both short- and long-lead burst neurons have been identified in this region in the behaving cat (Kaneko & Fuchs 1981) and monkey (Scudder et al 1982), and several indirect tests suggest that medullary burst neurons also inhibit contralateral abducens neurons in the monkey (Scudder et al 1982).

Burst neurons with vertical on-directions are present in the riMLF (Büttner & Henn 1977, King & Fuchs 1979). Their firing pattern resembles that of pontine short-lead burst neurons, but they have shorter lead times and lower burst rates. Burst neurons within the riMLF have either upward or downward on-directions and could serve as either IBNs or EBNs since unilateral stimulation of the prerubral fields in the cat produces both IPSPs and excitatory postsynaptic potentials (EPSPs) in oculomotor neurons (Nakao & Shiraishi 1983). Although bilateral electrical stimulation produces conjugate downward movements from more lateral and rostral portions of the riMLF and upward movements from more medial and caudal portions (Kömpf et al 1979), upward and downward burst neurons appear to be intermixed (Büttner & Henn 1977, King & Fuchs 1979).

Most short-lead burst neurons apparently have nearly horizontal or nearly vertical preferred directions. Büttner et al (1977) believe that burst neuron discharge is encoded along the pulling planes of the extraocular muscles, since the average on-direction of mesencephalic burst neurons is tilted by 10 to 20 deg from the vertical and therefore corresponds to the pulling directions of the vertical recti. On the other hand, Robinson & Zee (1981) argued that because the saccadic system also appears to generate the quick phases of vestibular nystagmus, it would be "reasonable" for it to be organized according to the planes of the semicircular canals to ensure that each quick phase is exactly opposite to the preceding slow phase. Because of the rough correspondence between muscle pulling directions and canal planes in many species (Graf & Simpson 1981), it is currently impossible to decide between these two alternatives.

Two approaches have been used to determine how the discharge of burst neurons controls the course of the saccade in space and time. Our approach has been to analyze the correlations between the various parameters that characterize the burst (e.g. burst duration, initial firing rate, peak firing rate, number of spikes) and those that characterize the saccade (e.g. saccade duration, peak velocity, size). Three highly significant correlations ($r > 0.9$) emerge consistently and therefore are considered to reflect causal relations. One correlation is between burst duration and saccade duration. For monkey saccades with durations greater than 20 msec (approximately 5 deg) this relation is linear with a slope near 1 for both putative EBNs (Luschei & Fuchs 1972) and IBNs (Scudder et al 1982). Since saccades of less than 5 deg have a roughly constant duration and since some burst neurons discharge weakly for off-direction saccades, the correlation and slope drop significantly when all saccades are considered. The burst duration for long-lead neurons is longer than saccade duration. The second correlation is between the peak firing rates of monkey short-lead burst neurons and peak saccadic velocity; the slope of the relation is steep (3 spikes/sec/deg/sec) for saccades less than 10 deg (300 deg/sec), but shallow (0.4 spikes/sec/deg/sec) for larger saccades because the firing rate saturates (Keller 1974, van Gisbergen et al 1981, C. A. Scudder and A. F. Fuchs, unpublished observations). Finally, the number of spikes in the burst of short-lead burst neurons is highly correlated with saccade size in the monkey (Keller 1974, King & Fuchs 1979, van Gisbergen et al 1981, Scudder et al

1982) and cat (Kaneko & Fuchs 1981, Yoshida et al 1981). The slope of this relation is 0.6 spikes/deg in monkey short-lead burst neurons (van Gisbergen et al 1981), but it can be as high as 2 to 3 for the most directionally selective burst neurons (Keller 1974). Slopes of 1.3 to 1.4 spikes/deg are typical for monkey short-lead IBNs (Figure 2, EBN or IBN), whereas long-lead IBNs have slopes ranging from 0.6 to 0.8 spikes/deg.

In the second approach, Henn and colleagues (Henn & Cohen 1976) represented saccades as vectors with a magnitude and angle, and found that all the pontine neurons from which they recorded had activity related to one or more parameters of the vector. Three neuron types were identified. The most common type, "Δ-position" cells, are qualitatively and quantitatively similar to the short-lead burst neurons described above. They encode saccade size along their on-directions by the number of spikes in the burst. Their second most prevalent type encodes the vector magnitude of the saccade by the number of spikes regardless of saccade direction, while a third type, encountered infrequently, encodes saccade angle independent of magnitude. We have subjected our cat short- and long-lead burst neurons and monkey IBNs to similar vector analyses and found none that coded saccade angle or magnitude uniquely. However, the burst neurons that we have recorded are quite different from the

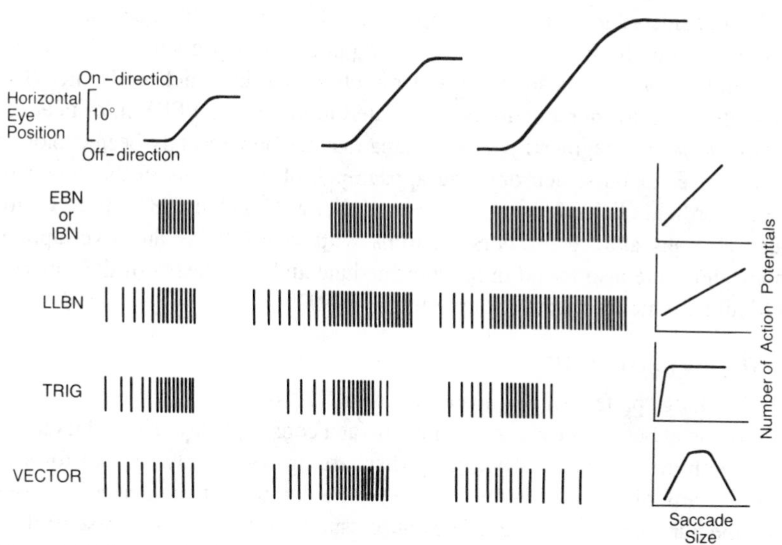

Figure 2 Schematic discharge patterns of the different types of burst neurons for three sizes of horizontal on-direction saccades. To the *right*, the number of action potentials in the burst is plotted as a function of saccade size for each unit type. The units are an excitatory (EBN) or inhibitory (IBN) burst neuron, a long-lead burst neuron (LLBN), a trigger burst neuron (TRIG), and a vector burst neuron (VECTOR). It is not known whether the EBN and IBN have identical discharges as depicted.

saccade magnitude or angle neurons of Henn & Cohen (1976), which appear to have much lower peak discharge rates and can discharge tonically during fixation.

More recently, Hepp & Henn (1983) extended their findings to include a rich variety of long-lead burst neurons in the PPRF. Their "direction" or "D-burster" neuron exhibits a monotonic increase in the number of spikes with saccade size in an on-direction (Figure 2, LLBN). Some D-bursters have a linear relation between number of spikes and saccade size for all ipsilateral saccades and therefore resemble the long-lead burst neurons described above (see also Luschei & Fuchs 1972). Others discharge a relatively constant number of spikes for all ipsilateral saccades except for the very smallest, and have been called TRIGGER burst neurons (Figure 2, TRIG) for reasons that will become clear later. A second type of long-lead burst neuron, the "vector" or "V-burster," fires only for saccades to a circumscribed region of the visual field (Figure 2, VECTOR), which is called the cell's "motor field"; the size and location of a motor field, like those of visual receptive fields, vary from unit to unit. Hence V-bursters have an optimal saccade size as well as direction (i.e. an optimal vector) and deviations from either will cause the number of spikes to decline. Since the optimal vector specifies the change in the direction of gaze, the motor field is always in the same location relative to the fovea and is considered to be in "retinotopic" coordinates. Finally, because there is nothing unique in the temporal discharge pattern of any given V-burster, the only indication of saccade size and direction lies in which cells fire. The majority of vector burst neurons are located in the rostral PPRF in or near the nucleus reticularis tegmenti pontis, an area that has only recently been explored. In some vector burst neurons, the appearance of a stimulus in the center of the motor field elicits a visual response (Keller & Crandall 1981). Pure vector burst neurons and vector burst neurons with visual fields and overlapping motor fields are also found in the intermediate and deep layers of the superior colliculus, which projects strongly to the PPRF.

Omnipause Neurons

The discharge pattern of omnipause neurons (OPNs) is the complement of that of short-lead burst neurons, since they fire at a constant rate during all fixations or smooth eye movements but cease discharging completely (pause) for saccades in any direction (Figure 3, OPN). The duration of the pause increases with the duration of a saccade. The pause usually begins about 16 msec before the saccade (Raybourn & Keller 1977) and ends with, or slightly before, the saccade. Between saccades, OPNs in both the cat and monkey fire at constant rates that frequently exceed 100/sec.

In both monkeys and cats, the vast majority of OPNs are a homogeneous population of neurons confined to a thin, dorsoventrally oriented nucleus, symmetric to the midline, and rostral to the abducens nucleus. Since electrical

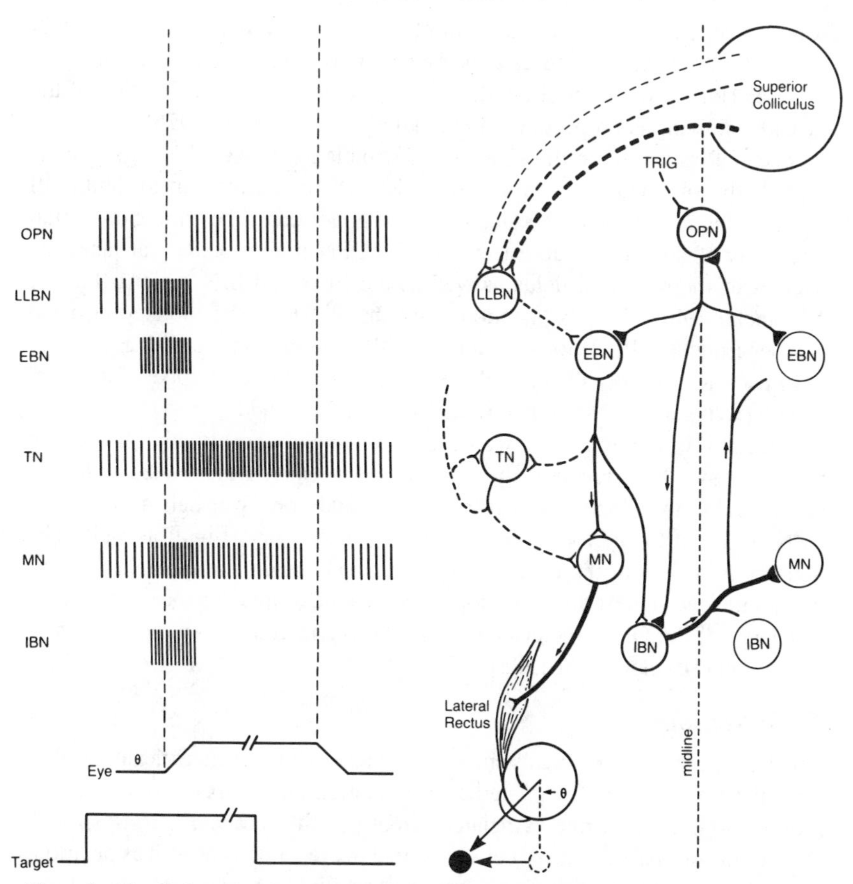

Figure 3 The discharge patterns and connections of neurons in the horizontal burst generator. *Left:* Firing patterns for an on-direction (first *vertical dashed line*) and off-direction (*second dashed line*) horizontal saccade of size θ to a target step (schematized in the eye and target traces below). *Right:* Excitatory connections are shown as *open endings*, inhibitory connections are shown as *filled triangles*, and axon collaterals of unknown destination (revealed by intracellular HRP injections or postulated in models) are shown without terminals. Connections known with certainty are represented by *thick lines*, uncertain connections by *thin lines*, and hypothesized connections by *dashed lines*. A complete description of the behavior of this neural circuit is found in the text. The abbreviations here and in Figure 4 identify excitatory (EBN) and inhibitory (IBN) burst neurons, long-lead burst neurons (LLBN), trigger input neurons (TRIG), omnipause neurons (OPN), tonic neurons (TN), and motoneurons (MN).

stimulation in this area completely eliminates saccades for the duration of the stimulus, Keller (1974) suggested that OPNs must inhibit burst neurons. This suggestion is supported by the obvious reciprocal relation between OPN and burst neuron firing patterns (Figure 3). For example, when a monkey becomes drowsy and its eyes roll around in their orbits, OPNs cease firing completely whereas burst neurons discharge continually. OPNs clearly exert an exquisite control over saccades since a train of electrical stimuli delivered to the OPN region during a saccade causes a marked decrease in its velocity within 14 to 15 msec (long trains even drive the velocity to zero), and the duration of the saccadic slowing is proportional to stimulus duration (King 1976).

Since OPNs pause for all saccades and stimulation in the OPN region interrupts both horizontal and vertical saccades, OPNs are thought to inhibit all short-lead burst neurons. This suggestion is supported by the finding that small injections of an orthograde tracer into the OPN region of the cat label the nucleus of the prerubral fields as well as the EBN and IBN regions (Langer & Kaneko 1983). Larger injections into the PPRF of the monkey also led Büttner-Ennever (1977) to conclude that OPNs project to the riMLF.

In the cat, virtually every OPN exhibits a brief burst of spikes for visual stimuli (Evinger et al 1982). Both small spots and full field striped backgrounds moving in any direction with a wide range of velocities are effective at eliciting responses over large receptive fields. Very few monkey OPNs tested under the same conditions exhibit visual responses. In addition, over half of the feline OPNs respond to auditory and some to somatosensory stimuli as well. The nonspecific nature of these various sensory inputs and their apparent absence in the monkey suggest that these inputs contribute to maintenance of the tonic discharge of the OPN, rather than having a role in controlling the metrics of the saccade (Evinger et al 1982).

Tonic Neurons

The third saccade-related brainstem unit (Figure 3, TN) characteristically exhibits a regular tonic rate that is related to eye position during fixation but exhibits little or no change in rate associated directly with saccades. During smooth pursuit, the discharge of these cells is related to eye velocity as well as position. The relation between firing rate and eye position for many of these tonic neurons is frequently very nonlinear, with units discharging at a nearly fixed rate over some part of the horizontal oculomotor range (Keller 1974).

True horizontal burst-tonic neurons also can be found in the monkey PPRF, although there is some controversy as to whether they are scarce (Keller 1974) or plentiful (Hepp & Henn 1983); in the cat PPRF, they represent the dominant eye movement neuron (C. R. S. Kaneko, unpublished observations). Burst-tonic, but apparently not pure tonic, units with vertical on-directions can be found in the mesencephalic reticular formation and the accessory oculomotor nuclei (e.g. interstitial nucleus of Cajal; King et al 1981), and those areas may project directly to oculomotor neurons (Steiger & Büttner-Ennever 1979).

An input that maintains steady fixation during the intersaccadic intervals must impinge on motoneurons. Such a signal proportional to eye position could be provided by tonic or burst-tonic neurons. Unfortunately, neurons in the PPRF that have tonic eye-position activity apparently are widely scattered, and nothing is known about their connectivity in either the monkey or the cat. Moreover, both the nucleus prepositus hypoglossi and medial vestibular nucleus project directly to the abducens nucleus and contain burst-tonic neurons that could also provide all or part of an eye-position signal.

SACCADIC BURST GENERATOR

The reticular formation neurons described above are shown schematically in Figure 3 along with their saccade-related discharge patterns and those connections that are known with certainty (*thick lines*), those that are likely on the basis of indirect electrophysiological experiments (*fine lines*), and those that have been hypothesized (*dashed lines*). IBNs must certainly project to contralateral abducens motoneurons since a monosynaptic IPSP is produced in motoneurons consequent to a spike in an IBN (Hikosaka et al 1978). OPNs very likely inhibit IBNs, since weak electrical stimulation of the OPN region produces IPSPs in IBNs at monosynaptic latencies (Nakao et al 1980) and intracellular HRP injections of OPNs produce labeled terminal boutons on neurons in the IBN area (Evinger et al 1980). Unfortunately, a recent study purporting to show monosynaptic OPN inhibition of IBNs directly by spike-triggered averaging of intracellular potentials (Furuya & Markham 1982) was unconvincing. OPNs also probably inhibit EBNs, since weak electrical stimulation of the EBN region antidromically activates OPNs and averaged field potentials in the EBN region can be obtained using OPN action potentials as a trigger (Curthoys et al 1984). Probable projections of EBNs to the abducens nucleus similarly were revealed by antidromic activation and by the recording of averaged field potentials (Sasaki & Shimazu 1981). Unfortunately, these EBN connections are uncertain, because electrical stimulation might affect fibers in passage and it is unclear whether the averaged field potentials indeed reflect synaptic potentials. With the same caveats, EBNs probably also project to IBNs, since they can be activated antidromically by low currents from the IBN region and IBNs can be activated orthodromically by low-intensity stimuli of the ipsilateral EBN area. Finally, IBNs might inhibit OPNs, since electrical stimulation of the OPN region antidromically activates IBNs (Nakao et al 1980), and neurons in the IBN region are retrogradely labeled following HRP injections in the OPN region (Langer & Kaneko 1984b).

With the neurons and connections diagrammed in Figure 3, it is now possible to describe the likely events that produce a saccade. The heart of the saccade generator is envisioned to be a high gain loop composed of burst neurons and OPNs. During fixation, the OPNs discharge at a high tonic rate and, because of their strong inhibitory connections to both EBNs and IBNs, they prevent burst neurons from firing. To make a saccade, an excitatory signal proportional to desired saccade size is fed to the appropriate EBNs; at the same time, an inhibitory trigger signal is thought to silence the OPNs, thereby disinhibiting EBNs and allowing them to respond to the excitatory drive with their characteristic burst of action potentials. The EBN discharge drives ipsilateral motoneurons to produce the burst component of their discharge. Since EBNs also excite ipsilateral IBNs, which feed back to inhibit the OPNs, OPNs are inhibited for the duration of the saccade. Therefore, even if the trigger signal is brief, the EBN burst and the saccade are allowed to go to completion.

To create the tonic portion of the motoneuron discharge, Robinson (1975) suggested that EBNs also project to the tonic neurons (Figure 3, TN), where their burst discharge is integrated (in the mathematical sense) to generate an eye-position signal. An integrator was first invoked for the oculomotor system in order to explain the operation of the vestibulo-ocular reflex (VOR), where the sensory input (from the semicircular canals) is roughly proportional to head *velocity,* and the oculomotor output is the mirror image of head *position* (Robinson 1968). Robinson (1973) also suggested that an integrator could be used in the saccade generator. This idea is reasonable since the integral of the discharge rate over the duration of a burst (spikes/sec × sec) is the number of spikes in the burst, which, as discussed above, is highly correlated with saccade size. In Figure 3 this integrator is shown schematically as a single tonic neuron with recurrent excitatory feedback, but it is most probably a group of neurons that may be distributed widely. Indeed, if an inability to maintain eccentric gaze after lesions is any indication, large parts of the brainstem and cerebellum are involved in the integration.

Although the brainstem circuit shown in Figure 3 can generate bursts, it is unclear how bursts appropriate for saccades of different sizes are generated on command. To address these issues, we take up the following four questions in the next sections. First, what is the source and nature of the trigger signal that is hypothesized to turn off the OPNs? Second, what is the source and nature of the excitatory signal that drives the burst neurons? Third, are these signals adequate to generate a saccade that gets on target? Fourth, what models have been hypothesized to generate saccades?

The OPN Trigger Signal

Anatomical studies have shown that neurons in three areas probably project to the OPN region and could perhaps provide the trigger signal. The most likely area is the superior colliculus, since tritiated amino acids injected there label the OPN region of both cat (Graham 1977) and monkey (Harting 1977), and HRP injected into the cat OPN region labels neurons in the intermediate and deep layers of the colliculus (Langer & Kaneko 1984b). The latter study also shows that two other "oculomotor" regions, the rostral PPRF and the rostral mesencephalic reticular formation, can provide OPN afferents.

The electrophysiological evidence implicating neurons in any of these three areas is inconclusive. Raybourn & Keller (1977) showed that most OPNs were excited, not inhibited, by single-pulse stimulation of the colliculus, with 44% of OPNs driven at monosynaptic latencies. However, by increasing the number of pulses, they found that the initial excitation was followed by a suppression of ongoing firing. When they applied triple pulse stimuli, the suppression began within 6 to 8 msec at low currents that elicited no visible eye movement, but the clearest pause in firing occurred with higher stimulus intensities that also evoked a saccade. We have repeated these experiments with similar results in both ketamine-anesthetized (Kaneko & Fuchs 1982b) and alert cats.

Büttner et al (1977b) and Raybourn & Keller (1977) were the first to suggest that long-lead burst neurons in the rostral PPRF might be intermediate neurons in the superior colliculus pathway to the OPNs. This suggestion was attractive for two reasons. First, as noted above, collicular stimulation produced OPN inhibition only after a delay of 6 to 8 msec, suggesting that at least one additional neuron was intercalated in the pathway. Second, single-pulse stimulation of the colliculus evoked discharges in every contralateral long-lead burst neuron, with 40% being activated at monosynaptic latencies. The type of long-lead burst neuron that exhibits a similar discharge for all ipsilateral saccades regardless of size has been proposed as the trigger burst neuron (Figure 2, TRIG). Interposing trigger burst neurons between the colliculus and OPNs could explain why saccades can still be elicited following collicular ablation, since inputs from other structures could address pontine trigger burst neurons directly to initiate saccades. In fact, the rostral pons does receive an input from the frontal eye fields (Leichnetz 1981), which must be lesioned in addition to the colliculus to produce enduring saccadic deficits (Schiller et al 1980).

Cohen et al (1982) suggested that the saccadic trigger signal may originate in the mesencephalic reticular formation, since stimulation of a restricted region lateral to the oculomotor complex evokes contralateral horizontal saccades. However, this region appears to have a larger role than just triggering saccades, because stimulation at different sites within it induces saccades of different sizes. Moreover, horizontal burst neurons with complicated discharge patterns can be recorded there (Waitzman 1982, Kaneko & Fuchs 1982a). Such cells discharge a burst of spikes before both targeting and spontaneous saccades to their motor fields and, in addition, respond if a visual target falls in or near their motor fields whether or not a saccade occurs; some exhibit a tonic rate proportional to eye position, and some maintain an increased discharge when the target and fovea are not aligned following the appearance of the target in the motor field (i.e. when a retinal error exists) (Kaneko & Fuchs 1982a). It is difficult to imagine that so complicated a discharge pattern would be used only to trigger saccades.

Although these candidate regions probably project to the OPN area, we know nothing about the firing patterns of OPN afferents or whether their discharge could act as a suitable trigger. However, the time course of the putative trigger signal has been studied indirectly in behavioral experiments in which the OPN region was stimulated with trains of electrical pulses timed to have their effect during the course of a saccade. In the monkey, a 40-msec train delivered at the onset of a 20-deg saccade interrupted the saccade in midflight, whereas a longer train truncated the saccade completely, and a separate, corrective saccade occurred after a delay of over 100 msec (Becker et al 1981). Such data suggested to Keller (1980) that the trigger signal must have ended 40 msec after saccade onset, since by that time it was impossible to restart the interrupted saccade. Since the tiny back-to-back saccadic oscillations seen in monkeys pursuing small target steps could be modeled by assuming that OPNs were inhibited for about 40 msec, van Gisbergen et al (1981) also suggested that the trigger signal lasts about 40 msec. However, delaying the OPN stimulation to occur as late as 80 msec after the onset of large saccades often produces an interrupted trajectory that restarts almost immediately after the train (Becker et al 1981), thus suggesting that the inhibitory trigger signal must still be present at that time to allow the saccade to continue. A reasonable explanation of these stimulation results is that the duration of the trigger signal is nearly as long as the duration of the saccade, as might be expected if the trigger were obtained from a long-lead burst neuron. This interpretation, however, assumes that electrical stimulation excites enough of the OPN pool that any residual EBN activity is inadequate to restart the burst without a trigger. It is also possible, of course, that interrupted saccades are restarted by some as yet unanticipated mechanism that is not related to the trigger.

Excitatory Input

Both the superior colliculus and the frontal eye fields of the cerebral cortex could provide excitatory inputs to the burst generator since stimulation of either structure evokes saccades at short latencies, and both project directly to the PPRF. Because more is known about the discharge patterns of its cells, the superior colliculus has usually been chosen as the prototypical source of input signals. Of those neurons in its deep and intermediate layers, the saccade-related cells (Mays & Sparks 1980), which discharge only when the monkey makes a saccade, appear to be the most likely candidates to drive the brainstem burst generator, since the discharge of other collicular cells with visual sensitivity (visuomotor or quasi-visual neurons) can be dissociated from the saccade. Like vector burst neurons in the PPRF, saccade-related neurons discharge only for saccades to a restricted motor field. They are topographically organized in the colliculus, with those having motor fields near the fovea lying rostrally, and those with peripheral motor fields lying caudally. Since nothing in the discharge of individual saccade-related cells reflects saccade size (Sparks & Mays 1980), an excitatory drive to the burst generator proportional to desired saccade size must be extracted from this topographic map. One hypothesis is that saccade-related neurons project to the burst generator with a synaptic efficacy that is weighted in proportion to the eccentricity of their motor fields. Therefore, units with peripheral motor fields would provide more excitation, which would presumably result in bigger bursts. This suggestion is supported by anatomical findings of a heavier projection to the PPRF from the caudal colliculus (Edwards & Henkel 1978) and the observation that PPRF neurons can be driven from many sites in the colliculus (Raybourn & Keller 1977). Wurtz & Albano (1980) observed that since caudal units have larger motor fields, more of them would be active for large than for small saccades, again resulting in more excitation to the burst generator.

Because monkey short-lead burst neurons are not activated directly by collicular stimulation but many long-lead burst neurons do receive monosynaptic excitation, Raybourn & Keller (1977) suggested that long-lead burst neurons are intercalated between the superior colliculus and EBNs. On the other hand, van Gisbergen et al (1981) suggested that long-lead burst neurons play an unknown role in caccade generation since, for the ten neurons that they examined, "information about saccade size, saccade duration, or saccade velocity is not contained in any obvious way in the discharge patterns of individual long lead burst neurons." This conclusion seems unwarranted, since several other investigators have found long-lead burst neurons in the PPRF (Luschei & Fuchs 1972, Büttner et al 1977b, Hepp & Henn 1983) and the medullary reticular formation (Scudder et al 1982) that encode saccade parameters in their firing patterns. Serial processing from long- to short-lead burst neurons is also suggested in Hepp & Henn's (1983) most recent model. Furthermore, they suggest that vector burst neurons drive long-lead burst neurons, but this seems unlikely since the nucleus reticularis tegmenti pontis, which contains most of the vector burst neurons, projects primarily to the cerebellum (Keller & Crandall 1981).

Other visuomotor structures, such as the frontal eye fields, must also have access to the burst generator since animals with lesions of the superior colliculus can initiate accurate saccades. The frontal eye field projection to the rostral PPRF may provide this access via long-lead burst neurons.

Programming a Saccade

Early investigators thought that the burst generator was controlled by a retinal error signal (i.e. position of the target on the retina relative to the fovea) and that once a saccade was initiated, its course would be immutable. As we saw in the previous section, the former assumption seemed reasonable since the most likely input to the saccade generator from the superior colliculus provides the motor equivalent of a retinal error signal. The suggestion that saccades are ballistic was supported by the observations that the saccade's short duration seemed to preclude visual feedback control and that humans were unable to slow the speed of a saccade voluntarily. However, recent experiments suggest that eye position as well as retinal error is important in programming some saccades and that saccades may be under continuous internal feedback control.

It has long been apparent that an internal representation of target position in space must be available to the central nervous system since we can accurately point to a visual target that is presented briefly. Behavioral studies in man and monkey reveal that such a signal also is used to generate saccades. In one study, a target step was presented to human subjects, and while an initial saccade was in progress, the target was flashed at a new location. After an initial saccade to the first target location, the subjects made a second saccade which accurately placed the eyes on target (Hallet & Lightstone 1976). Similarly, monkeys presented with two targets that were flashed sequentially and extinguished before any movement had occurred made successive saccades to both the first and then the second target (Mays & Sparks 1980). In both these experiments, the retinal error at the time of the second target presentation was not equivalent to the size of the saccade that finally placed the eyes on target. Therefore, eye position must be used in conjunction with retinal error to generate saccades, suggesting that target position in space rather than retinal error is the driving signal for saccades.

Other experimenters question whether saccades are ballistic, since in the same "double-step" paradigm, the first saccade can be affected by the second target step. Becker & Jürgens (1975, 1979) presented human subjects with a target that jumped to one location, and then jumped forward or backward to a second location before any eye movement had begun. On trials in which the reaction time was long, subjects made an initial saccade that often landed between the first and second target locations before making a second saccade, which landed on target. On some trials, the initial saccade appeared to turn around in midflight. When patients with spinocerebellar disease were subjected to this task, their slow gaze shifts also were adjusted in midflight, with no saccadic reaction time, so the eye landed accurately on target (Zee et al 1976). Unfortunately, those patients shifted their gaze so slowly (80 deg/sec in contrast to over 400 deg/sec in normal humans) that their eye movements might not have been saccades.

These data have been explained in two ways. Becker & Jürgens (1979) favor the possibility that the altered initial saccade is a preprogrammed response to a modified target location. In their scenario, a decision to make a saccade of a certain direction is made first and its size is computed later, but before the saccade begins. In parallel, but somewhat later, the saccadic system programs the second saccade with the knowledge of where the the first saccade will land. In this scheme, a signal representing target position in space is used only in those cases in which a second saccade is needed, e.g. for large target steps where a corrective saccade is usually required (or in experiments in which subjects expect the target to step twice). The second explanation, discussed below, assumes that the saccade begins normally but reacts to the second target step while the first saccade is in flight by a mechanism that maintains continuous control (Robinson 1975).

Models of the Saccade Generator

THE ROBINSON MODEL In 1975, Robinson proposed an attractively simple model that accounted for the experimental results mentioned above by using, for the first time, a *target position in space* signal. In his model (Figure 4A), retinal error is processed by two important feedback loops that use the neural replica of eye position from tonic neurons. First, eye position is *added* to retinal error (the left-most summing junction) to create a neural representation of *target position in space*. This signal is passed through a memory element (the 0.2-sec delay), which, among other things, ensures that the neural replica of *target position in space* seen at the second summing junction (the EBN) does not change during the course of a normal saccade. Second, eye position is *subtracted* at the EBN to produce a *motor error* signal in retinal coordinates. For single-step stimuli, the initial motor error or *desired saccade size* (Figure 4A inset) is exactly equal to retinal error. Before the second saccade in a double-step paradigm, however, the initial motor error is the difference between the neural replicas of final *target position in space* and current eye position, or, once again, *desired saccade size*. This signal is not equivalent to the retinal error that existed at the time the second target briefly appeared.

In addition to creating a motor error signal in retinal coordinates, the second (or "local") feedback loop serves another important function, namely, it terminates the saccade. As the eye approaches the target during a saccade, *dynamic motor error* (the difference between the neural replicas of *target position in space* and eye position) gradually declines and ultimately reaches zero, whereupon the EBNs have no drive and cease firing. Thus, control of burst duration is neatly accomplished since the burst will continue only as long as necessary to bring the motor error to zero. If a feedback loop controls the saccade, the saccade cannot be ballistic, because the feedback loop will correct for an unexpected perturbation and allow the eyes to reach the target.

The results of several recent experiments are consistent with the Robinson model. Since the model uses a *target position in space* signal to program saccades, it clearly can account for the behavioral data from the "double-step" paradigms. The results of other experiments are consistent with the existence

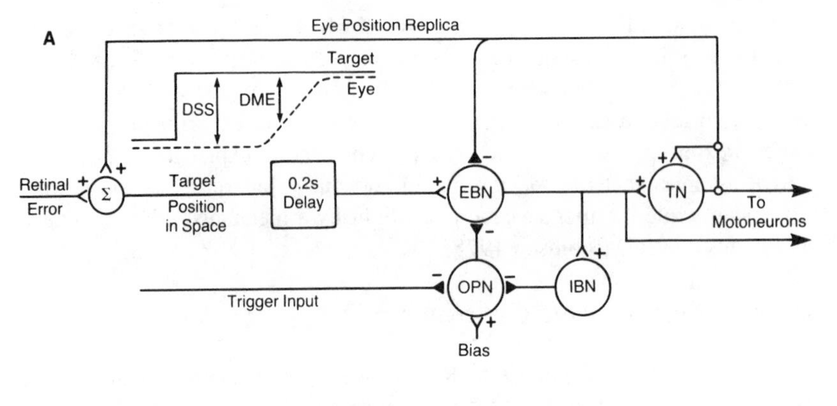

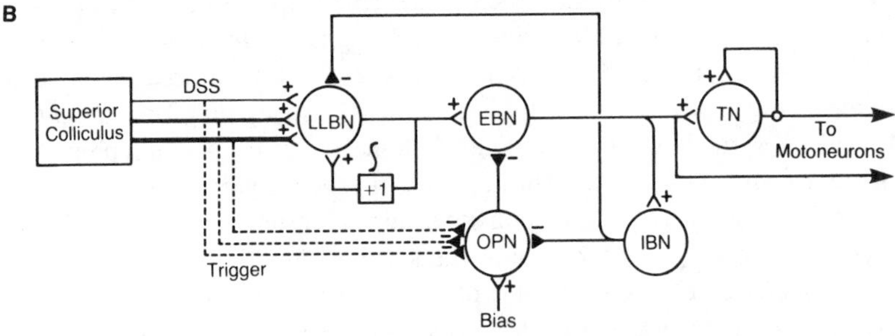

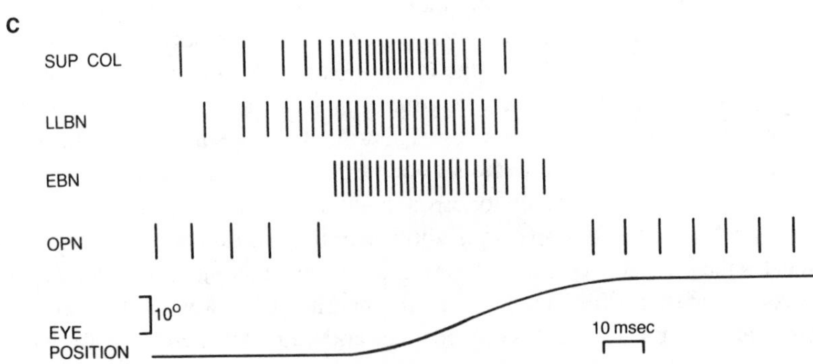

Figure 4 Models of the saccade generator. *A:* The Robinson model, in which a neural replica of eye position is added to retinal error to create target position in space, and later subtracted to generate motor error at the EBN membrane. The physical variables that exist as neural replicas in the model include (*inset*) target and eye position relative to the head, desired saccade size (DSS, the motor error that exists before the saccade begins), and dynamic motor error (DME, the motor error that exists while the saccade is in progress). The bias signal to the OPN produces its steady firing between saccades. *B:* The Scudder model, in which the difference between an excitatory input from the superior colliculus and an inhibitory feedback from the IBN is integrated (schematically realized by recurrent positive feedback) to provide the drive for the EBN. The topographical weighting of the collicular projection is indicated by lines of different thickness. The colliculus also provides the OPN with a weak trigger signal (*dashed lines*), which is not

of local feedback. For example, the OPN stimulation experiments mentioned above in which saccades landed accurately on target after being interrupted in midflight argue strongly that saccades are under feedback control (Becker et al 1981). Other supportive evidence was obtained recently by Sparks & Mays (1983), who stimulated the superior colliculus in trained monkeys and thereby deflected the eye to a new location immediately before a "targeting" saccade was to be made to an eccentric target. Rather than producing a movement with the amplitude and direction of the original "targeting" saccade, a saccade was produced that compensated for the deflection and landed on target, although with less accuracy than control saccades. However, these results do not require that current eye position be used to control the saccade in midflight, but only that current eye position be known at the time the compensating saccade is initiated. Finally, if the driving signal on the EBN is *motor error,* EBN firing rate should decrease as the eye approaches the target. Of 18 horizontal short-lead burst neurons tested by van Gisbergen et al (1981), 16 exhibited a consistent relation between instantaneous firing rate and *motor error,* and the relation was similar for saccades of all sizes. However, the best relation between firing rate and *motor error* was obtained if firing rate was considered as the difference between ipsilateral (on-direction) EBN discharge and contralateral (off-direction) IBN discharge. Since monkey IBNs had not been studied at that time, van Gisbergen et al (1981) assumed that EBN and IBN discharges were similar. This may not be valid since in the cat the discharge patterns of EBNs and IBNs are in fact quite different (Kaneko & Fuchs 1981).

If local feedback indeed exists, it is important to know how and where the motor error signal is created. Robinson (1975) suggested that *motor error* is generated at the EBNs. This seems unlikely since the superior colliculus (the probable input to the burst generator) does not project directly to short-lead burst neurons (Raybourn & Keller 1977), and the colliculus already encodes initial motor error (*desired saccade size*) in the topographic distribution of its neurons. Consequently, Keller (1980) proposed that the neural replica of eye position is sent to the superior colliculus, where it is compared with target position to generate *motor error.* This proposal was supported by the findings of Sparks & Porter (1983), who stimulated the colliculus to drive the eyes

---

topographically weighted; this latter pathway probably has one or more intercalated neurons. *C:* Simulated firing patterns of neurons in the Scudder model. The envelope of the collicular discharge (SUP COL) was assumed to be nearly Gaussian but with some added leading spikes, and the IBN discharge (not shown) was assumed to be the same as that of the EBN. The EBN was assumed to have a saturating nonlinearity in its input-output characteristics to account for the burst frequency-saccade velocity saturation described in text. The 20-deg saccade was simulated using 12-msec and 200-msec time constants for the orbital mechanics and a slope of 1.3 spikes/deg for the EBN (see Scudder 1984 for further details).

eccentrically just before a targeting saccade. Not only did the resulting saccade compensate for the perturbation, but the collicular neurons always discharged in relation to the compensating saccade and not the saccade that was originally intended. If the colliculus does provide the *motor error* signal as defined in the model, its weighted output must reflect *dynamic motor error* (DME, Figure 4A) as well as *desired saccade size* (DSS). There are two ways this could occur. *Dynamic motor error* could also be coded topographically, in which case neurons with decreasingly eccentric motor fields would discharge sequentially as the saccade approached the target. Therefore, a cell with a motor field located near the fovea would discharge vigorously for every larger saccade that passed through its motor field. No cells in the colliculus have been recorded with such behavior. Alternatively, *desired saccade size* could be coded topographically, but *dynamic motor error* could be reflected in the temporal discharge pattern of those cells that were active. Unfortunately, the firing rate of single saccade-related cells in individual trials appears to peak too late and drop off too rapidly for this purpose, but averages over many trials may reveal a more suitable discharge (cf Figure 2, Sparks & Porter 1983). To resolve this issue, the discharge of saccade-related cells needs to be examined in more detail.

Although the Robinson model appears to have strong experimental support, it cannot account for some of the details of the available data. As noted previously, in the double-step experiments of Becker & Jürgens (1975, 1979), some of the initial saccades landed between the locations of the first and second targets. Moreover, the initial saccadic velocity of the first saccade differed from that normally elicited in single-step experiments as though the modified trajectory were preprogrammed. The Robinson model allows the first saccade to be modified by the second target step, but predicts that under most conditions the saccade would land at one or the other target location but not in between. Moreover, the Robinson model alters the initial saccade in midflight, rather than producing the properly adjusted initial velocity. The Robinson model also has difficulty generating realistic oblique saccades. In a cat, monkey or human making an oblique saccade, the duration of the smaller component (horizontal or vertical) is increased and its velocity is decreased so that, in two-dimensional space, the eye moves in a straight line from its initial to its final point (Evinger et al 1981b, W. M. King, S. G. Lisberger, and A. F. Fuchs, unpublished observations). If the most recent version of the Robinson model (van Gisbergen et al 1981) is expanded to include a horizontal and a vertical burst generator coupled only through the OPNs, our simulations show that the duration of the smaller component is not increased, the saccade trajectory is curved, and the saccade does not reach the target. Although the parameters of the model can be modified (e.g. increasing h to 1.5 in their Figure 10) so that the saccade reaches the target, the trajectory remains curved.

Finally, it is now clear that any model that uses *target position in space* and local feedback to control burst neuron firing has as much experimental support as Robinson's model. For example, the Becker & Jürgens model (1979) uses *target position in space* and therefore can also account for the data of Sparks & Mays (1983) if the stimulated saccade is considered comparable to the initial saccade in double-step paradigms. Similarly, the transitory interruption of a saccade by OPN stimulation is explained by models that use local feedback of any kind.

SCUDDER MODEL We propose a modification of the Robinson model (Scudder 1984) that has many of its advantages but more accurately reflects our current knowledge of the connections and discharge patterns of the saccade-related brainstem neurons (Figure 4B). Our model also uses the familiar high-gain loop involving EBNs, IBNs, and OPNs; however, the EBN is driven by a long-lead burst neuron, which is wired to integrate mathematically its two synaptic inputs. This integration is accomplished schematically with recurrent excitation (Figure 4B, +1 feedback loop). One input, which represents the topographically weighted output of the superior colliculus, provides a net excitatory burst proportional to *desired saccade size*. The envelope of this burst, which was chosen to resemble that of collicular SR neurons, does not depend on events in the saccadic burst generator. Only its size changes, according to which location in the colliculus is active. The other input is the "local feedback" pathway that arises from the IBN, whose discharge is hypothesized to reflect that of the EBN. The collicular input burst and the IBN feedback burst can be viewed as being integrated separately and then subtracted, although the subtraction actually occurs before the integration in the model. When the number of spikes added by the colliculus equals the number of spikes subtracted by the IBN feedback, this difference (the output of the long-lead burst neuron) drops to zero and the EBN stops firing. In this model, therefore, saccade size is coded by the colliculus as the total (weighted) number of spikes and not by total (weighted) firing rate. Although we use the colliculus as the prototypical input, it need not be the only input, nor must it project directly to the integrating long-lead burst neuron.

Unlike the case in previous models, the trigger signal to the OPNs is not pivotal in initiating a saccade, since our OPNs provide a weaker inhibition of EBNs. Consequently, a burst can be initiated automatically when the growing output of the long-lead burst neuron exceeds the OPN inhibition of the EBN. The onset of EBN firing then causes the IBN to fire, and this produces a sharp inhibition of OPN firing to allow a further rapid rise in EBN discharge. We have, however, also included a trigger signal (shown as originating from the superior colliculus) to assist in the initiation of a burst by reducing OPN discharge just as the long-lead burst neuron discharge is growing. This assistance allows the model to produce a more realistic envelope for the firing pattern of burst neurons by permitting their bursts to be triggered closer to the peak of the colliculus discharge. By relying less on the trigger signal than previous models, and assuming only that the horizontal and vertical burst generators are coupled through the OPNs, our model permits the duration of the smaller component of an oblique saccade to be stretched. This occurs because the short-lead neurons for the larger component will begin to fire first, inhibiting the OPNs and allowing the short-lead neurons of the smaller component to fire earlier than if the smaller component had occurred by itself. Consequently, the integrator/long-lead burst neuron discharge for the smaller component is lower than normal and, furthermore, is kept lower by the inhibitory feedback from its own IBN. The resultant lower discharge rate of the short-lead neurons serving the smaller component guarantees that it will have a longer duration than usual.

Our model can simulate many of the discharge patterns already described for brainstem neurons (Figure 4C). The time course of the input signal (Figure 4C, SUP COL) accurately reflects that of collicular saccade-related cells, which exhibit a gradual increase in rate to reach a maximum near saccade onset (Sparks & Mays 1980). This gradual increase causes a gradual increase in the discharge of the long-lead burst neuron (Figure 4C, LLBN). For the long-lead burst neuron, the number of spikes and peak firing rate increase with saccade size, but the precise relations depend on the input waveform and its timing relative to the OPN trigger. Since these signals probably vary, even for saccades of the same size, the long-lead burst neuron in our model, like those actually recorded, has a poorer correlation between saccade and burst parameters than do short-lead neurons. The gradual rise of the collicular discharge also causes a slight decrease in OPN discharge just before the frank pause, a pattern that we also find in actual neurophysiological recordings.

By eliminating eye position as the signal used by the local feedback pathway, our model has two additional advantages. First, the superior colliculus provides a signal proportional to *desired saccade size*, which is consistent with the known firing properties of collicular saccade-related neurons. We do not have to assume that the colliculus also codes *dynamic motor error*. Second, while retaining the advantages of using local feedback, the feedback does not depend on the tonic neurons, whose location and connections are the least well understood of all oculomotor brainstem neurons.

Our model is incomplete in some respects. First, it may not be readily adaptable to the vertical system, since vertical long-lead burst neurons have yet to be recorded in the mesencephalon and have been encountered in the pons only infrequently. Second, our model would benefit greatly from a bilateral representation, like that of van Gisbergen et al (1981), where interaction between symmetric burst generators improves the details of the simulated burst. Finally, it deals only with the burst generator per se, and does not address how *desired saccade size* is generated. The scheme of Becker & Jürgens (1979) does, however, provide a signal for *desired saccade size* that, in combination with our burst generator, could constitute a complete model of saccade generation.

CONCLUSION

Although we know very little about how a movement is produced in most mammalian motor systems, even at the most peripheral level of the final common path, this review has shown that our understanding of the control of saccadic eye movements is quite advanced. A combination of behavioral, neurophysiological, and anatomical studies not only has revealed the signals carried by motoneurons but has identified premotor neurons carrying these signals and, in many cases, has shown how these premotor neurons are connected. The early application of quantitative models to describe the premotor circuitry of the saccade generator has been instrumental in generating further experiments and in forcing the oculomotor community to think in terms of testable hypotheses. In this review, we have concluded by using one such model (Robinson 1975) as a framework to evaluate the nature of the signals that must drive the brainstem generator, and we have suggested a modified version of that model (Scudder 1984).

A consideration of these models has pointed out several gaps in our knowledge. First, in order to evaluate differences between our model and Robinson's, we need to know more details about the firing patterns of collicular saccade-related cells and how they influence their target brainstem neurons. In particular, do saccade-related cells encode *motor error* as well as *desired saccade size*? Second, the recently discovered vector burst neurons of the rostral pons must be tested as another possible component of the saccade generator. Specifically, do they play a role similar to that of saccade-related cells? Third, more information is required about the various types of long-lead burst neurons in the pons. Does one type provide the link between vector burst neurons and EBNs while another triggers the OPNs? Finally, the role of the OPNs should be investigated by determining the nature of the signal that drives them. Answers to questions like these should give us some insight into the processes by which the oculomotor system transforms a sensory stimulus into a motor command.

ACKNOWLEDGMENTS

Preparation of this manuscript, and the studies performed in our laboratory, were supported by National Institutes of Health grants RR00166 and EY00745, and by National Science Foundation grant BNS 8218901.

We are very grateful to Josh Wallman, Tom Langer, Mike Mustari, Jenny McFarland, Jim Phillips, Wolfgang Becker, and Dom Finocchio for useful comments on substance and style, to Sue Schaefer for word processing, and to Kate Schmitt for editorial help.

Literature Cited

Alvarado, J. A., Van Horn, C. 1975. Muscle cell types of the cat inferior oblique. See Lennerstrand & Bach-y-Rita 1975, pp. 15–45

Baker, R., Berthoz, A., eds. 1977. *Control of Gaze by Brain Stem Neurons. Dev. Neurosci.* Vol. I. New York: Elsevier/North-Holland. 514 pp.

Baker, R., Highstein, S. M. 1975. Physiological identification of interneurons and motoneurons in the abducens nucleus. *Brain Res.* 91:292–98

Barmack, N. 1974. Saccadic discharges evoked by intracellular stimulation of extraocular motoneurons. *J. Neurophysiol.* 37:395–412

Becker, W., Jürgens, R. 1975. Saccadic reactions to double step stimuli: Evidence for model feedback and continuous information uptake. See Lennerstrand & Bach-y-Rita 1975, pp. 519-24.

Becker, W., Jürgens, R. 1979. An analysis of the saccadic system by means of double step stimuli. *Vision Res.* 19:967–83

Becker, W., King, W. M., Fuchs, A. F., Jürgens, R., Johanson, G., Kornhuber, H. H. 1981. Accuracy of goal-directed saccades and mechanisms of error correction. See Fuchs & Becker 1981, pp. 29–37

Bondi, A. Y., Chiarandini, D. J. 1983. Morphologic and electrophysiologic identification of multiply innervated fibers in rat extraocular muscles. *Invest. Ophthal. Vis. Sci.* 24:516–19

Büttner, U., Henn, V. 1977. Vertical eye movement related unit activity in the rostral mesencephalic reticular formation of the alert monkey. *Brain Res.* 130:239–52

Büttner, U., Hepp, K., Henn, V. 1977. Neurons in the rostral mesencephalic and paramedian pontine reticular formation generating fast eye movements. See Baker & Berthoz 1977, pp. 309–18

Büttner-Ennever, J. A. 1977. Pathways from the pontine reticular formation to structures controlling horizontal and vertical eye movements in the monkey. See Baker & Berthoz 1977, pp. 89–98

Büttner-Ennever, J. A., Büttner, U. 1978. A cell group associated with vertical eye movements in the rostral mesencephalic reticular formation of the monkey. *Brain Res.* 151:31–47

Büttner-Ennever, J. A., d'Ascanio, P., Sakai, H., Schnyder, H. 1983. Separation of large and small motoneurons in the oculomotor nuclei of the monkey. *Soc. Neurosci. Abstr.* 9:1087

Büttner-Ennever, J. A., Lang, W. 1978. Connections of a vertical eye movement area in the rostral mesencephalic tegmentum of the monkey. *Soc. Neurosci. Abstr.* 4:161

Chiarandini, D. J., Stefani, E. 1979. Electrophysiological identification of two types of fibres in rat extraocular muscles. *J. Physiol.* 290:453–65

Cohen, B., Matsuo, V., Raphan, T., Waitzman, D., Fradin, J. 1982. Horizontal saccades induced by stimulation of the mesencephalic reticular formation. See Roucoux & Crommelinck 1982, pp. 325–44

Collins, C. C. 1975. The human oculomotor control system. See Lennerstrand & Bach-y-Rita 1975, pp. 145–80

Curthoys, I. S., Markham, C. H., Furuya, N. 1984. Direct projection of pause neurons to nystagmus-related excitatory burst neurons in the cat pontine reticular formation. *Exp. Neurol.* 83:414–22

Delgado-Garcia, J. M., Del Pozo, F., Baker, R. 1983. The properties of antidromically identified abducens motoneurons in the alert cat. *Soc. Neurosci. Abstr.* 9:1088

Dietert, S. E. 1965. The demonstration of different types of muscle fibers in human extraocular muscle by electron microscopy and cholinesterase staining. *Invest. Ophthal.* 4:57–63

Edwards, S. B., Henkel, C. K. 1978. Superior colliculus connections with the extraocular motor nuclei in the cat. *J. Comp. Neurol.* 179:457–67

Evinger, C., Baker, R., McCrea, R. A., Spencer, R. 1981a. Axon collaterals of oculomotor nucleus motoneurons. See Fuchs & Becker 1981, pp. 263–70

Evinger, C., Fuchs, A. F. 1978. Saccadic, smooth pursuit, and optokinetic eye movements of the trained cat. *J. Physiol.* 285:209–29

Evinger, C., Fuchs, A. F., Baker, R. 1977. Bilateral lesions of the medial longitudinal fasciculus in monkeys: Effects on the horizontal and vertical components of voluntary and vestibular induced eye movements. *Exp. Brain Res.* 28:1–20

Evinger, C., Kaneko, C. R. S., Fuchs, A. F. 1981b. Oblique saccadic eye movements of the cat. *Exp. Brain Res.* 41:370–79

Evinger, C., Kaneko, C. R. S., Fuchs, A. F. 1982. Activity of omnipause neurons in alert cats during saccadic eye movements and visual stimuli. *J. Neurophysiol.* 47:827–44

Evinger, C., McCrea, R. A., Baker, R. 1980. Intracellular injection of HRP into omnipause neurons in the alert cat. *Soc. Neurosci. Abstr.* 6:16

Fuchs, A. F., Becker, W., eds. 1981. *Progress in Oculomotor Research. Dev. Neurosci.* Vol. 12. New York: Elsevier/North-Holland. 685 pp.

Fuchs, A. F., Luschei, E. S. 1970. Firing patterns of abducens neurons of alert monkeys in relationship to horizontal eye movement. *J. Neurophysiol.* 33:382–92

Furuya, N., Markham, C. H. 1982. Direct inhibitory synaptic linkage of pause neurons with burst inhibitory neurons. *Brain Res.* 245:139–43

Goebel, H. H., Komatsuzaki, A., Bender, M. B., Cohen, B. 1971. Lesions of the pontine tegmentum and conjugate gaze paralysis. *Arch. Neurol.* 24:431–40

Goldberg, S. J., Clamann, H.-P., McClung, J. R. 1981. Relation between motoneuron position and lateral rectus motor unit contraction speed: An intracellular study in the cat abducens nucleus. *Neurosci. Lett.* 23:49–54

Goldberg, S. J., Lennerstrand, G., Hall, C. D. 1976. Motor unit responses in the lateral rectus muscle of the cat: Intracellular current injection of abducens nucleus neurons. *Acta Physiol. Scand.* 96:58–63

Goldstein, H. P. 1983. *The neural encoding of saccades in the rhesus monkey.* PhD Thesis. Johns Hopkins Univ., Baltimore, Md.

Graf, W., Simpson, J. I. 1981. Relations between the semicircular canals, the optic axis, and the extraocular muscles in lateral-eyed and frontal-eyed animals. See Fuchs & Becker 1981, pp. 409–17

Graham, J. 1977. An autoradiographic study of the efferent connections of the superior colliculus in the cat. *J. Comp. Neurol.* 173:629–54

Grantyn, R., Grantyn, A. 1978. Morphological and electrophysiological properties of cat abducens motoneurons. *Exp. Brain Res.* 31:249–74

Graybiel, A. M. 1977. Organization of oculomotor pathways in the cat and rhesus monkey. See Baker & Berthoz 1977, pp. 79–88

Graybiel, A. M., Hartwieg, E. A. 1974. Some afferent connections of the oculomotor complex in the cat: An experimental study with tracer techniques. *Brain Res.* 81:543–51

Guthrie, B. L., Porter, J. D, Sparks, D. L. 1982. Role of extraocular muscle proprioception in eye movements studied by chronic deafferentation of intraorbital structures. *Soc. Neurosci. Abstr.* 8:156

Hallet, P., Lightstone, A. D. 1976. Saccadic eye movements to flashed targets. *Vision Res.* 16:107–14

Harting, J. 1977. Descending pathways from the superior colliculus: An autoradiographic analysis in the rhesus monkey (*Macaca mulatta*). *J. Comp. Neurol.* 173:583–612

Henn, V., Cohen, B. 1972. Eye muscle motor neurons with different functional characteristics. *Brain Res.* 45:561–68

Henn, V., Cohen, B. 1973. Quantitative analysis of activity in eye muscle motoneurons during saccadic eye movements and positions of fixation. *J. Neurophysiol.* 36:115–26

Henn, V., Cohen, B. 1976. Coding of information about rapid eye movements in the pontine reticular formation of alert monkeys. *Brain Res.* 108:307–25

Hepp, K., Henn, V. 1983. Spatio-temporal recording of rapid eye movement signals in the monkey paramedian pontine reticular formation (PPRF). *Exp. Brain Res.* 52:105–20

Hess, A., Pilar, G. 1963. Slow fibres in the extraocular muscles of the cat. *J. Physiol.* 169:780–98

Highstein, S. M., Baker, R. 1978. Excitatory termination of abducens internuclear neurons on medial rectus motoneurons: Relationship to syndrome of internuclear ophthalmoplegia. *J. Neurophysiol.* 41:1647–61

Highstein, S. M., Karabelas, A., Baker, R., McCrea, R. A. 1982. Comparison of the morphology of physiologically identified abducens motor and internuclear neurons in the cat: A light microscopic study employing the intracellular injection of horseradish peroxidase. *J. Comp. Neurol.* 208:369–81

Highstein, S. M., Maekawa, K., Steinaker, A., Cohen, B. 1976. Synaptic input from the pontine reticular nuclei to abducens motoneurons and internuclear neurons in the cat. *Brain Res.* 112:162–67

Hikosaka, O., Igusa, Y., Nakao, S., Shimazu, H. 1978. Direct inhibitory synaptic linkage of pontomedullary reticular burst neurons with abducens motoneurons in the cat. *Exp. Brain Res.* 33:337–52

Kaneko, C. R. S., Evinger, C., Fuchs, A. F. 1981. Role of cat pontine burst neurons in generation of saccadic eye movements. *J. Neurophysiol.* 46:387–408

Kaneko, C. R. S., Fuchs, A. F. 1981. Inhibitory burst neurons in alert trained cats: Comparison with excitatory burst neurons and functional implications. See Fuchs & Becker 1981, pp. 63–70

Kaneko, C. R. S., Fuchs, A. F. 1982a. Mesencephalic neurons that discharge for target movements, saccades and eye position. *Soc. Neurosci. Abstr.* 8:157

Kaneko, C. R. S., Fuchs, A. F. 1982b. Connections of cat omnipause neurons. *Brain Res.* 241:166–70

Keller, E. L. 1971. *Abducens unit activity in*

the alert monkey during vergence movements and extraocular muscle stretch. PhD Thesis. Johns Hopkins Univ., Baltimore, Md.

Keller, E. L. 1974. Participation of medial pontine reticular formation in eye movement generation in monkey. *J. Neurophysiol.* 37:316–32

Keller, E. L. 1980. Oculomotor specificity within subdivisions of the brain stem reticular formation. In *The Reticular Formation Revisited*, ed. J. A. Hobson, M. A. B. Brazier, pp. 227–80. New York: Raven

Keller, E. L., Crandall, W. F. 1981. Neural activity in the nucleus reticularis tegmenti pontis in the monkey related to eye movements and visual stimulation. *NY Acad. Sci.* 374:249–61

Keller, E. L., Robinson, D. A. 1971. Absence of a stretch reflex in extraocular muscles of the monkey. *J. Neurophysiol.* 34:908–19

King, W. M. 1976. *Quantitative analysis of the activity of neurons in the accessory oculomotor nuclei and the mesencephalic reticular formation of alert monkeys in relation to vertical eye movements induced by visual and vestibular stimulation*. PhD Thesis, Univ. Washington, Seattle

King, W. M., Fuchs, A. F. 1979. Reticular control of vertical saccadic eye movements by mesencephalic burst neurons. *J. Neurophysiol.* 42:861–76

King, W. M., Fuchs, A. F., Magnin, M. 1981. Vertical eye movement-related responses of neurons in midbrain near interstitial nucleus of Cajal. *J. Neurophysiol.* 46:549–62

Kömpf, D., Pasik, T., Pasik, P., Bender, M. B. 1979. Downward gaze in monkeys. Stimulation and lesion studies. *Brain* 102:527–58

Kuffler, S. W., Vaughan Williams, E. M. 1953. Properties of the "slow" skeletal muscle fibres of the frog. *J. Physiol.* 121:318–40

Lang, W., Henn, V., Hepp, K. 1982. Gaze palsies after selective pontine lesions in monkeys. See Roucoux & Crommelinck 1982, pp. 209–18

Langer, T. P., Kaneko, C. R. S. 1983. Efferent projections of the cat oculomotor reticular omnipause neuron region: An autoradiographic study. *J. Comp. Neurol.* 217:288–306

Langer, T. P., Kaneko, C. R. S. 1984a. Brainstem afferents to the abducens nucleus in the monkey. *Soc. Neurosci. Abstr.* 10(2):987

Langer, T. P., Kaneko, C. R. S. 1984b. Brainstem afferents to the omnipause region in the cat: A horseradish peroxidase study. *J. Comp. Neurol.* In press

Leichnetz, G. R. 1981. The prefrontal cortico-oculomotor trajectories in the monkey. *J. Neurol. Sci.* 49:387–96

Lennerstrand, G. 1974. Electrical activity and isometric tension in motor units of the cat's inferior oblique muscle. *Acta Physiol. Scand.* 91:458–74

Lennerstrand, G., Bach-y-Rita, P., eds. 1975. *Basic Mechanisms of Ocular Motility and Their Clinical Implications*. Oxford: Pergamon

Luschei, E. S., Fuchs, A. F. 1972. Activity of brain stem neurons during eye movements of alert monkeys. *J. Neurophysiol.* 35:445–61

Maciewicz, R. J., Eagen, K., Kaneko, C. R. S., Highstein, S. 1977. Vestibular and medullary brain stem afferents to the abducens nucleus in the cat. *Brain Res.* 123:229–40

Mayr, R., Stockinger, L., Zenker, W. 1966. Electronenmikroskopische Untersuchungen an unterschiedlich innervierten Muskelfasern der äusseren Augenmuskulatur des Rhesusaffen. *Z. Zellforschung.* 75:434–52

Mays, L. E., Sparks, D. L. 1980. Dissociation of visual and saccade-related responses in superior colliculus neurons. *J. Neurophysiol.* 43:207–32

Miller, J. E. 1967. Cellular organization of Rhesus extraocular muscle. *Invest. Ophthalmol.* 6:18–39

Mukuno, K. 1968. The fine structure of the human extraocular muscles. Report 2. The classification of muscle fibers. *Jpn. J. Ophthalmol.* 12:9–20

Nakao, S., Curthoys, I. S., Markham, C. H. 1980. Direct inhibitory projection of pause neurons to nystagmus-related pontomedullary reticular burst neurons in the cat. *Exp. Brain Res.* 40:283–93

Nakao, S., Shiraishi, Y. 1983. Excitatory and inhibitory synaptic inputs from the medial mesodiencephalic junction to vertical eye movement-related motoneurons in the cat oculomotor nucleus. *Neurosci. Lett.* 42:125–30

Optican, L. M., Robinson, D. A. 1980. Cerebellar-dependent adaptive control of primate saccadic system. *J. Neurophysiol.* 44:1058–76

Peachy, L. 1971. The structure of the extraocular muscle fibers of mammals. In *The Control of Eye Movements*, ed. P. Bach-y-Rita, C. C. Collins, J. E. Hyde, pp. 47–66. New York: Academic

Porter, J. D., Guthrie, B. L., Sparks, D. L. 1983. Innervation of monkey extraocular muscles: Localization of sensory and motor neurons by retrograde transport of horseradish peroxidase. *J. Comp. Neurol.* 218:208–19

Raybourn, M. S., Keller, E. L. 1977. Colliculoreticular organization in primate oculomotor system. *J. Neurophysiol.* 40:861–78

Ringel, S. P., Wilson, W. B., Barden, M. T., Kaiser, K. K. 1978. Histochemistry of human extraocular muscle. *Arch. Ophthalmol.* 96:1067–72

Robinson, D. A. 1964. The mechanics of human

saccadic eye movement. *J. Physiol.* 174:245–64

Robinson, D. A. 1968. The oculomotor control system: A review. *Proc. IEEE* 56:1032–49

Robinson, D. A. 1970. Oculomotor unit behavior in the monkey. *J. Neurophysiol.* 33:393–404

Robinson, D. A. 1973. Models of the saccadic eye movement control system. *Kybernetik* 14:71–83

Robinson, D. A. 1975. Oculomotor control signals. See Lennerstrand & Bach-y-Rita 1975, pp. 337–74

Robinson, D. A., Zee, D. S. 1981. Theoretical considerations of the function and circuitry of various rapid eye movements. See Fuchs & Becker 1981, pp. 3–9

Roucoux, A., Crommelinck, M., eds. 1982. *Physiological and Pathological Aspects of Eye Movements*. The Hague: Junk. 448 pp.

Ruskell, G. L. 1978. The fine structure of innervated myotendinous cylinders in extraocular muscles of rhesus monkeys. *J. Neurocytol.* 7:693–708

Sasaki, K. 1963. Electrophysiological studies on oculomotor neurons of the cat. *Jpn. J. Physiol.* 13:287–302

Sasaki, S., Shimazu, H. Reticulovestibular organization participating in generation of horizontal fast eye movement. *Ann. NY Acad. Sci.* 374:130–43

Schiller, P. H., True, S. D., Conway, J. H. 1980. Deficits in eye movements following frontal eye-field and superior colliculus ablations. *J. Neurophysiol.* 44:1175–89

Scudder, C. A. 1984. A different local feedback model of the saccadic burst generator. *Soc. Neurosci. Abstr.* 10(2):910

Scudder, C. A., Langer, T. P., Fuchs, A. F. 1982. Probable inhibitory burst neurons in the monkey. *Soc. Neurosci. Abstr.* 8:157

Sparks, D. L., Mays, L. E. 1980. Movement fields of saccade-related burst neurons in the monkey superior colliculus. *Brain Res.* 190:39–50

Sparks, D. L., Mays, L. E. 1983. Spatial localization of saccade targets. I. Compensation for stimulation-induced perturbations in eye position. *J. Neurophysiol.* 49:45–63

Sparks, D. L., Porter, J. D. 1983. Spatial localization of saccade targets. II. Activity of superior colliculus neurons preceding compensatory saccades. *J. Neurophysiol.* 49:64–74

Steiger, H.-J., Büttner-Ennever, J. A. 1978. Relationship between motoneurons and internuclear neurons in the abducens nucleus: A double retrograde tracer study in the cat. *Brain Res.* 148:181–88

Steiger, H.-J., Büttner-Ennever, J. A. 1979. Oculomotor nucleus afferents in the monkey demonstrated with horseradish peroxidase. *Brain Res.* 160:1–15

Strassman, A., McCrea, R. A., Highstein, S. M. 1982. Structure function correlation of abducens motor and internuclear neurons in relation to eye movements: An intracellular HRP study in the alert squirrel monkey. *Soc. Neurosci. Abstr.* 8:156

van Gisbergen, J. A. M., Robinson, D. A., Gielen, S. 1981. A quantitative analysis of generation of saccadic eye movements by burst neurons. *J. Neurophysiol.* 45:417–42

Waitzman, D. M. 1982. *Burst neurons in the mesencephalic reticular formation (MRF) of the rhesus monkey associated with saccadic eye movement*. Thesis, City Univ. of New York, NY

Westheimer, G. 1954. Mechanism of saccadic eye movements. *AMA Arch. Ophthalmol.* 52:719–24

Wurtz, R. H., Albano, J. E. 1980. Visual-motor function of the primate superior colliculus. *Ann. Rev. Neurosci.* 3:189–226

Yoshida, K., McCrea, R., Berthoz, A., Vidal, P. P. 1981. Properties of immediate premotor inhibitory burst neurons controlling horizontal rapid eye movements in the cat. See Fuchs & Becker 1981, pp. 71–80

Zee, D. S., Optican, L. M., Cook, J. D., Robinson, D. A., Engel, W. K. 1976. Slow saccades in spinocerebellar degeneration. *Arch. Neurol.* 33:243–51

Ann. Rev. Neurosci. 1985. 8:339–67

PHOTOTRANSDUCTION IN VERTEBRATE RODS

E. A. Schwartz

Department of Pharmacological and Physiological Sciences, The University of Chicago, Chicago, Illinois 60637

Rods elaborate a special organelle for converting light into an electrical signal. Their outer segment contains a long stack of flattened sacs, or discs, covered by the plasmalemma. Embedded in the disc membrane are rhodopsin molecules; within the plasmalemma are channels that allow the entry of Na ions during darkness. The absorption of light by rhodopsin initiates a series of events that cross the cytoplasm and stop the plasmalemmal current. Recent studies have revealed the first biochemical steps initiated by light, properties of the plasmalemmal current, and tantalizing bits concerning how the early steps are connected to the subsequent closure of membrane channels. Accordingly, I have divided this review into four parts: (*a*) properties of the plasmalemmal current, (*b*) the early events produced by light, (*c*) observations that place restrictions on how light controls the current, and (*d*) a speculative synthesis.

Both Ca and cGMP[1] have been suggested to have an important role in transduction. In one scheme, Ca is stored within discs and is released by light to diffuse across the cytoplasmic space and to block channels (Hagins 1972). In another scheme, cGMP keeps channels open and light activates a phosphodiesterase that reduces cGMP concentration (see review by Hubbell & Bownds 1979). These two ideas dominate the transduction landscape. Their shadow is seen in every part of the discussion.

[1] Abbreviations: ATP, adenosine triphosphate; BICINE, *N,N* bis-(2-hydroxyethyl) glycine; cAMP, adenosine-3′,5′-cyclic monophosphate; EGTA, ethylene glycol-bis (β-aminoethylether)-*N,N*-tetracetic acid; cGMP, guanosine-3′,5′-cyclic monophosphate; GDP, guanosine diphosphate; GMP, guanosine monophosphate; GTP, guanosine triphosphate; HEPES, N-2-hydroxyethylpiperazine-*N*′-ethane-sulfonic acid; IBMX, isobutylmethylxanthine; TEA, tetraethylammonium ion; TTX, tetrodotoxin.

0147–006X/85/0301–339$02.00

THE PLASMALEMMAL CURRENT AND ION MOVEMENTS IN THE OUTER SEGMENT

Recently there have been attempts to inject Ca or cGMP into rods of the intact retina to see whether these agents influence the voltage in a manner consistent with either hypothesis (Brown et al 1977, Nicol & Miller 1978, Miller 1982, Oakley & Pinto 1983, Waloga 1983). The results are claimed to support each hypothesis. However, two difficulties hamper interpretation of the results. First, Ca and cGMP may affect both the current entering the outer segment and voltage- and calcium-activated currents of the inner segment. The net result could be confusing. Second, it is not clear how activity will be altered when an agent is injected into one of many electrically interconnected rods. The recorded voltage will be determined by both the injected cell and its uninjected neighbors. Interpretation of results is easier for experiments in which the current controlled by light is directly measured.

Methods for Recording the Generator Current

The current entering the outer segment can be measured by three techniques.

1. The extracellular voltage produced by the flow of current along outer segments is measured with two extracellular pipettes, one at the tip of the outer segments and a second at their base (Hagins et al 1970). The procedure requires outer segments arranged in a regular array and averages the activity produced by many rods. Accordingly, it cannot record the activity produced by a single rod. Therefore, current fluctuations and single photon events (see below) cannot be observed.
2. A rod is positioned so that its outer segment fits snugly into a large pipette and the inner segment is left protruding from the opening (Baylor et al 1979a). The current flowing between the part drawn into the pipette and the part left outside flows through an amplifier and is measured. This preparation is very stable and membrane currents can be measured with great sensitivity for relatively long periods. A disadvantage is that membrane voltage is not controlled.
3. Individual rods are isolated and maintained as solitary cells that can be penetrated by micropipettes. A voltage clamp apparatus can control the membrane voltage (Bader et al 1979). Pharmacologic agents can be injected into the rod from the intracellular micropipette (MacLeish et al 1984). When they can be compared, all three techniques produce similar results.

Pore or Carrier?

It is not yet certain whether the generator current passes through a pore or a carrier. Preliminary information comes from observing either the intracellular voltage or the generator current at high gain. A continuous fluctuation in the voltage or current is produced by spontaneous variation in the opening and closing of either individual channels or groups of channels. Because of this uncertainty it is convenient to describe the fluctuation as being produced by a transduction event that is either a single channel opening or the coordinate opening of a group of channels. The amplitude of the fluctuation when compared with the mean value provides an estimate of the size of the transduction event. Three measurements indicate that the transduction event is small: the voltage recorded from a rod in the turtle retina indicates an amplitude of 20 fA (Schwartz 1977); the current recorded from a toad outer segment indicates an amplitude of 2.6 fA (Baylor et al 1980); the current recorded from a small patch of lizard outer segment indicates an amplitude of 12 fA (Detwiler et al 1982). The average is approximately 10 fA. By comparison, a single voltage-activated Ca pore in the membrane of a chromaffin cell exposed to 1 mM Ca has a conducting state of 25 fA (Fenwick et al 1982; see also Tsien 1983). A carrier with a turnover time of 10 μsec, equal to the fastest acting enzymes, would produce a current of 16 fA. Thus, the observed size of the fluctuation is compatible with current passing through either a pore or carrier.

The power density spectrum of the fluctuation indicates that the transduction events are relatively long lived. The half power frequencies are approximately 4 Hz for turtle rods (Schwartz 1977) and 2 Hz for toad rods (Baylor et al 1980). Thus, each transduction event lasts for several hundred milliseconds. The exact shape of the power density spectrum can provide information concerning the shape of the transduction event. Because the transduction events are slow, very long records must be analyzed. In the best data from toad rods the power density declines faster than would be predicted if the noise were due to square events whose duration was exponentially distributed. This might indicate that the channel exists in three or more states. Another possibility, suggested by Baylor et al (1980), is that fluctuations are due to spontaneous bursts in the production of an intracellular compound that controls the opening of channels. Each channel would be much smaller than the 10 fA amplitude and 200–500 msec duration of a transduction event. The shape of the power density spectrum would be determined by processes controlling the concentration of the compound.

P. B. Detwiler and R. D. Bodoia are now studying fluctuations in the generator current with cell-attached and whole-cell patch recording (personal communication). Their results confirm the previous findings. In addition, the low noise of these newer techniques has allowed them to observe an additional, smaller component. The shape of the power density spectrum for this component differs from that expected to be produced by a carrier but could be produced by pores that are 20 fA in amplitude and have exponentially distributed life times with a mean duration of 0.7 msec in the frog and 2.5 msec in the lizard. Consequently, this component is probably produced by the opening and closing of individual pores. Transduction events may be produced when an intracellular compound interacts with these pores to change their open-closed transition.

The Number of Channels

The number of open channels in an outer segment can be estimated from the observed noise and certain assumptions. The 50 pA current observed during darkness in a salamander rod (Bader et al 1979) can be increased by the injection of cGMP to more than 700 pA (MacLeish et al 1984). These experiments will be discussed below; for the moment we need only a number for the maximum size of the current. Fluctuation analysis of membrane voltage and current (see above) indicates that individual channels provide a current of approximately 20 fA. If the size of individual channels is the same before and after cGMP is injected, then dividing the maximum current by the estimated current of an individual channel yields 700 pA/20 fA $= 3.5 \times 10^4$ open channels in an outer segment. If cGMP increases the current through single channels without changing the number of open channels, the number of open channels would be 50 pA/20 fA $= 2.5 \times 10^3$. The total number of open and closed channels must be greater than these estimates. Consequently, a rough estimate of approximately 10–100 channels opposite each disc is reasonable. By comparison, embedded in or associated with one disc there are approximately 10^6 rhodopsin, 10^5 G-protein (Godchaux & Zimmerman 1979, Kühn 1980), and 10^4 phosphodiesterase molecules (Baehr et al 1979); within one interdisc space (aproximately 0.6 μm^3) there are approximately 40 H ions (equivalent to a pH of 7.0; see Hagins & Yoshikami 1977), 400 Ca ions (a concentration of approximately 1 μM; see Hagins & Yoshikami 1975), and 1.6×10^4 cGMP molecules (a concentration of 40 μM; see Table 2 in Woodruff & Fain 1982), and less than 10^3 cAMP molecules (20-fold less than the cGMP concentration; see Orr et al 1976).

Current-voltage Relation

The magnitude of the generator current is relatively constant as the potential is changed from -120 to -40 mV, decreases as depolarization continues toward 0 mV, reverses polarity between 0 and 10 mV, and then rapidly increases as an outward current as membrane potential is made positive (Bader et al 1979). Thus, the current should not be modeled as a light controlled resistor in series with a fixed battery. The nonlinear behavior of the current appears to be "instantaneous." When voltage is changed and current is observed on a fast time base, the current assumes its new value as soon as the capacitative transient required by the change is completed. Thereafter, a slow change takes place during several seconds. The change is slight during physiological conditions but can be made more dramatic by reducing the external Ca concentration (MacLeish et al 1984). The slow change has been attributed to a partial inactivation that is secondary to changes in internal Ca concentration (see below).

Inactivation

Injecting Ca into an outer segment decreases the generator current; injecting the Ca chelater EGTA has the opposite effect (MacLeish et al 1984). The simplest interpretation is that internal Ca blocks the generator-current channels. Other manipulations that change internal Ca may also be expected to block the current. Two types of experiments have been interpreted in this way. The size of the generator current has been measured while making a change in the concentration of external ions or changing membrane potential. After raising the external Na or lowering external Ca concentration there follows a large increase in the amplitude of the generator current, which then subsides to a lesser steady value (Yau et al 1981). Similar but smaller transient changes occur when the membrane is hyperpolarized (MacLeish et al 1984). These overshoots (or undershoots when making the opposite ion or polarization changes) are suggestive of a regulatory system controlled by a feedback system involving internal ion concentrations. The decline following the initial transient has in each case been attributed to a readjustment of internal Ca concentration and Ca's ability to block or inactivate the channel.

Blocking Action of Polyvalent Cations

The shape of the current-voltage relation is altered by adding Co to the external medium (MacLeish et al 1984). Co has little effect at potentials more depolarized than −40 mV, but increasingly inhibits the current as the potential is hyperpolarized through and beyond the physiological range. The voltage-dependent ability of Co to block can be described by its ability to bind a site on the external surface in a manner that becomes more effective as membrane hyperpolarization is increased. Lanthanum has also been reported to decrease the generator current (Cervetto & McNaughton 1983). Unfortunately, the voltage-dependence of its blocking action is not known. The binding site for external polyvalent cations is of interest since, as described below, it appears to be modified by cGMP.

A Role for cGMP

During darkness the current is normally 30–50 pA (Bader et al 1979). Injecting cGMP into a rod increases the current to more than 700 pA (MacLeish et al 1984). The increase is specific for cGMP. The hydrolyzed analogue, GMP, has no effect. Cyclic AMP has only a slight effect, as though it were a weak agonist. The current-voltage curve provides information on how cGMP modifies the current. If cGMP increases the current without changing the shape of the current-voltage curve, then the simplest interpretation would be a change in the number of identical channels. If the shape is changed, then a more complicated explanation would be necessary. Experiments to measure the shape of the current-voltage relation were performed after pharmacologic agents (Co, TEA, and Cs) were added to the external medium to block voltage-dependent currents of the inner segment. Co also produces a voltage-dependent block of the generator current that becomes appreciable as the membrane is hyperpolarized beyond -60 mV. Injecting cGMP changes the shape of the current-voltage curve by removing the blocking action of Co. Inactivation during depolarizing steps also disappears. Thus, cGMP reduces the ability of external Co and internal Ca to block. This presumably requires a structural change in the channel.

Permeability

Normally, the generator current is carried by Na ions (Sillman et al 1969, Brown & Pinto 1974, Hagins & Yoshikami 1975). The ability of monovalent cations to replace Na can be summarized by the permeability sequence $P_K/P_{Rb}/P_{Cs}/P_{Li}/P_{Na}$ = .02: .015: .01: .2: 1 (Yau et al 1981). Replacement of external Na by choline abolishes the light response in a rapid and reversible manner. A different picture is seen after exposing rods to the phophodiesterase inhibitor IBMX (Capovilla et al 1983). Replacement of external Na by choline then reduces but does not eliminate the generator current. The polarity of the current then depends upon the presence of divalent cations. If a divalent cation remains in the bathing medium, the current is inward at a physiological voltage; if all the external divalent cations are removed, then the current is outward. The ability of divalent cations to carry current after a rod is treated with IBMX and Na removed is in the sequence Mn>Ba> (Ca, Co, Mg, Sr). Rather similar effects occur when the external Ca concentration is buffered to 10^{-8}M (Yau et al 1981). After external Na is removed, a small generator current remains. Removing both Na and Mg reverses the polarity of the current. These effects produced by a very low external Ca concentration or adding IBMX might result from a common mechanism of action. Both procedures increase the internal cGMP concentration (Cohen et al 1978, Kilbride 1980, Woodruff & Fain 1982). A small, but significant, divalent cation permeability may exist at physiological concentrations. An elevated internal cGMP concentration may then increase the divalent cation permeability, making it electrically detectable.

As described above, injecting cGMP into a rod reduces the ability of external Co and internal Ca to block the current (MacLeish et al 1984). These observations and the phenomena described in this section can be summarized by saying that cGMP decreases the ability of divalent cations to block and increases their ability to permeate.

Reversal Potential and Intracellular Na Concentration

As described above, the generator current is carried predominantly by Na ions and has a reversal potential of 0 to +10 mV (Bader et al 1978). If the current is carried entirely by Na, then the internal concentration predicted from the Nernst equation is approximately 90 mM, an unusually high intracellular Na concentration. The Na content of outer segments in dark-adapted frog retinae has been measured with an electron microprobe and found to be 68 mM (Hagins & Yoshikami 1975). The intracellular Na concentration in rat rods increases from 40 to 110 mM when the amplitude of the generator current is increased by reducing Ca in the external medium from 1 mM to 1 μM (Yoshikami et al 1983). Continuous entry of Na into the outer segment may be partly responsible for the high intracellular Na concentration. However, even in the absence of Na entry, internal Na is regulated to maintain a high concentration. An inward rectifying current of the inner segment is carried equally by Na and K ions (Bader et al 1982). It can be studied in solitary inner segments, which lack an outer segment, or in solitary intact rods in which the generator current has been stopped by bright illumination. In these preparations the inward rectifying current of the inner segment reverses at −30 mV. The same reversal potential is observed when the cytoplasm is replaced with a solution containing 45 mM K and 50 mM Na (Bader & Bertrand 1974), indicating that these concentrations are near the normal values. Thus, rods regulate their internal Na concentration at a level higher than found in many other neurons. A high internal Na concentration may affect processes coupled to the Na gradient (see next section).

Ca Extrusion

The activity of a Na-Ca exchanger has been suggested to regulate the internal Ca concentration of the outer segment (Yau et al 1981). The difference between internal and external Ca concentrations would then be limited by the energy of the Na electrochemical gradient. If the system is readily reversible:

$$z_{\mathrm{Ca}}(E_{\mathrm{Ca}} - E) = r z_{\mathrm{Na}}(E_{\mathrm{Na}} - E) \tag{1}$$

where z_{Ca} and z_{Na} are the valences of Ca and Na, r is the number of Na ions transported per Ca ion, E_{Ca} and E_{Na} are the Nernst potential for Ca and Na, and E is membrane voltage. From this it follows that

$$\frac{[\mathrm{Ca}]_0}{[\mathrm{Ca}]_i} = \left[\frac{[\mathrm{Na}]_0}{[\mathrm{Na}]_i}\right]^r e^{(2-r)EF/RT} \tag{2}$$

where F, R, and T have their usual meaning. In a number of cells, r has been found to be 3 (see Blaustein & Nelson 1982). If $[\mathrm{Na}]_0 = 110$ mM, $[\mathrm{Na}]_i =$

50 mM, $E = -40$ mV, and $r = 3$, then the maximum ratio of calcium concentrations that can be produced is 53:1. Consequently, if the external, physiological Ca concentration is 1.8 mM, then the internal concentration must be greater than 34 μM. This value is perhaps 100-fold higher than usually estimated for an intracellular Ca concentration. The Ca concentration could be reduced tenfold if r were 6. However, there is no precedent in other cells for $r > 4$. Either the simple model of Na-Ca exchange must be changed or the internal Ca concentration is higher than is commonly assumed.

Two additional arguments indicate that a simple, reversible Na-Ca antiporter may not be sufficient to regulate internal Ca concentration to a submicromolar concentration. The first starts from the observation that light increases the Ca concentration around outer segments (Gold & Korenbrot 1980, Yoshikami et al 1980). In the experiments of Gold & Korenbrot the external Ca rose 10^4 ions per photon absorbed, even though the external Na concentration was either 27 or 7 mM. Inspection of Eq. 2 indicates that these concentrations are unlikely to support Na-Ca exchange. The second argument starts from the observation of a large outward generator current when the membrane potential is increased to positive values (Bader et al 1979). When the membrane is depolarized and the Na electrochemical gradient reversed, the direction of flux through a Na-Ca exchanger should reverse, Ca should enter the outer segment, and the current should rapidly be blocked. Voltage clamp records so far do not indicate this behavior (Bader et al 1979, MacLeish et al 1984). Consequently, there are reasons to doubt that the simple scheme for Na-Ca exchange outlined above could have a prominent role in regulating the internal Ca concentration of outer segments.

Ca can also be exported from cells by a Ca-ATPase pump. An ATP-dependent uptake of Ca into membrane vesicles prepared from outer segments has been observed (Puckett et al 1984; for a report concluding that outer segments lack a Ca-ATPase see Berman et al 1977). The role of a Ca pump in Ca extrusion from outer segments is not certain.

Outer Segment Membrane Resistivity

After the generator current is suppressed by light, the resistance of the outer segment is approximately 4×10^6 ohm cm^2 (Baylor & Lamb 1982), a value sufficiently high to preclude the presence of even a few open pores. Evidently, the five voltage and Ca-activated channels that operate in the inner segment (Bader et al 1982) do not spill over into the outer segment plasmalemma. When the five currents are blocked with pharmacological agents, the inner segment resistance is approximately 3×10^4 ohm cm^2 (Bader et al 1982, Corey et al 1984). Part of the residual conductance may be due to the presence of a few "leakage" channels, perhaps Ca-activated K and Cl channels held open by the internal Ca concentration. When both internal and external permeant ions are replaced with large impermeant ions, the inner segment resistivity increases to 2×10^5 ohm cm^2 (Corey et al 1984). These numbers indicate that the "passive" membrane resistivity of the outer segment is at least tenfold higher than that of the inner segment.

The difference in passive membrane resistivity in inner and outer segments has an important effect on the physiology of rods. Any current that enters through a channel must find an exit from the cell to complete a current loop. Because the outer segment contains only generator current channels (whose slope conductance at physiological voltages is small, see page 342) and a membrane with a very high passive resistivity, most of the current that enters the outer segment exits through the inner segment membrane. In contrast, current that enters through a channel in the inner segment exits through inner segment membrane. As described above, the generator current is nearly independent of voltage within the physiological range (-35 to -70 mV). Thus, the outer segment behaves as an ideal current source injecting into the inner segment a current that does not depend upon membrane voltage. In contrast, the inner segment behaves as an integrating center that receives the injected current and produces a voltage in proportion to the resistance determined by the activity of its own localized currents.

EARLY EVENTS

The first step in vision is the absorption of light by rhodopsin. Subsequent events control transduction, the slower process of adaptation, regeneration of photolyzed rhodopsin, and the circadian shedding of discs from the outer segment tip. It is not clear how each phenomenon diverges from the events initiated by the first step. The metabolism of Ca and cGMP have been intensively investigated in the hope that each is involved in the transduction process. Recently, protons have also been proposed as intermediaries.

Rhodopsin, G-protein, and Phosphodiesterase

One path can be followed from the absorption of light to the hydrolysis of cGMP. I shall briefly summarize biochemical events, which have been mapped with considerable detail.

Rhodopsin molecules diffuse freely in disc membranes (Brown 1972, Cone 1972, Poo & Cone 1974, Liebman & Entine 1974). The absorption of one photon isomerizes retinal and produces an activated opsin molecule. As the activated molecule diffuses in the disc membrane, it encounters G-protein, a peripheral membrane protein comprised of three subunits (Baehr et al 1982, Fung 1983). The encounter allows the G(α) subunit to exchange a bound GDP molecule for a GTP molecule (Fung & Stryer 1980, Fung et al 1981). The rhodopsin-G-protein-GTP complex dissociates, leaving the still activated rhodopsin to find, bind, and activate another G-protein, and the G(α)-GTP subunit to find and activate a phosphodiesterase by displacing an inhibitory subunit. The activated phosphodiesterase hydrolyzes cGMP. Thus, the first four steps in this path are as follows:

1. diffusion of an activated rhodopsin molecule in the disc membrane to find a G-protein molecule,
2. GDP-GTP exchange and disassembly of G-protein,
3. activation of phosphodiesterase by G(α)-GTP, and
4. cGMP hydrolysis.

One activated rhodopsin molecule has been estimated to activate approximately 100 phosphodiesterase molecules during the first half second following a flash (Fung & Stryer 1980, Liebman & Pugh 1981); this is in agreement with the number of GTP molecules hydrolyzed (Robinson & Hagins 1979). Purified phosphodiesterase without its inhibitory subunit hydrolyzes 45 cGMP sec^{-1} (Baehr et al 1979). Consequently, a significant fraction of the cGMP content of an inderdisc space (1.6×10^4 molecules) may be depleted. The activity of the phosphodiesterase may be modulated by changes in Ca concentration. Decreasing the Ca concentration decreases phosphodiesterase activation in a preparation of discs and disrupted outer segments (Robinson et al 1980, Kawamura & Bownds 1981).

Activated rhodopsin and phosphodiesterase (PDE) must both be inactivated. The catalytic activity of the G(α)-GTP-PDE complex is terminated by the spontaneous hydrolysis of GTP and disassembly into G(α)-GDP, inactive phosphodiesterase, and phosphate. Activated rhodopsin is a substrate for a kinase (Kühn & Dreyer 1972, Bownds et al 1972, Shichi & Somers 1978) that is inhibited by high concentrations of cGMP (Hermolin et al 1982, Shuster & Farber 1984). After an activated rhodopsin molecule becomes multiply phosphorylated (Wilden & Kühn 1982), it may cease to activate the enzyme cascade (Liebman & Pugh 1980, Sitaramayya & Liebman 1983). Dephosphorylation occurs slowly (Kühn 1974), presumably after rhodopsin has been reisomerized. This elegant scheme has been resolved with considerable biochemical detail. But its place in the overall process of transduction is not yet certain.

Attempts have been made to correlate the hydrolysis of cGMP with closure of membrane channels in isolated outer segments. Relatively gentle agitation snaps and detaches the rigid, cylindrical outer segment at its base. A large number can then be harvested. With gentle treatment, many of the amputated outer segments reseal the broken end. When brightly illuminated, their cGMP content declines. A steady light bleaching 5×10^5 Rh sec^{-1} or a 2 msec flash bleaching 10^3 Rh were calculated to hydrolyze 10^4 cGMP molecules per photon absorbed (Woodruff & Bownds 1979), similar to the figure that can be calculated from the number of phosphodiesterase molecules activated and their rate of cGMP hydrolysis. However, the extent and time course of cGMP depletion may be different in intact rods (see below). The isolated organelle lacks mitochondria and the ability to synthesize nucleotide triphosphates. After isolation, outer segments rapidly lose GTP (Bignetti et al 1978, Biernbaum & Bownds 1979, Robinson & Hagins 1979) and slowly swell (Bownds & Brodie 1975, Brodie & Bownds 1976). Swelling, produced by the accumulation of osmotic particles as ions cross the plasmalemma, has been used as a measure of membrane permeability. Comparing the "permeability" of intact rods and isolated outer segments reveals a difference in their sensitivity to light. Although the generator current of an intact rod can be half suppressed by a light that bleaches 8 Rh sec^{-1} (Baylor et al 1980), swelling of isolated outer segments was half suppressed by a light that bleached 5000 Rh sec^{-1} (Brodie & Bownds 1976). The relative insensitivity of swelling in metabolically depleted outer segments indicates that they may not reproduce the normal behavior of intact rods.

An interesting picture of cGMP metabolism in rods of the intact rabbit retina comes from measurements made using a new technique. Cyclic nucleotide phosphodiesterase catalyzes 3′–P–O bond cleavage and the insertion of a single nonexchangeable atom of ^{18}O from [^{18}O] water. Accumulation of ^{18}O into the phosphoryl of a 5′-nucleotide can be measured by gas chromatography–mass spectroscopy. Using this technique, Goldberg et al (1983) have measured the cGMP metabolic flux in outer segments of rabbit retina. A high metabolic flux during darkness is increased to even higher levels by light. The flux during darkness is calculated to turn over the entire cGMP content of an outer segment in 3 sec. This would require a rate of hydrolysis 100-fold slower than that observed with the purified enzyme (see above). The activity of phosphodiesterase within the outer segment is inhibited during darkness by its γ subunit. A moderate light intensity activating a few percent of the phosphodiesterase molecules increases the total flux fivefold, without a large decrease in cGMP concentration. The relative constancy of cGMP concentration indicates that an increase in hydrolysis is matched by an increase in synthesis. The results of Goldberg et al (1983) agree with previous studies that report that physiological light intensities produce only slight changes in cGMP concentration (Kilbride & Ebrey 1979, Govardovskii & Berman 1981) and bright lights, beyond the physiological range, reduce the cGMP content of intact rods by approximately 50% (Goridis et al 1977, Cohen et al 1978, Govardovskii & Berman 1981, DeAzeredo et al 1981).

Light probably increases cGMP hydrolysis by the enzyme cascade described above. There need not be a similar, direct light-initiated pathway for the stimulation of synthesis. A balance between synthesis and hydrolysis could be maintained if cGMP regulates its own synthesis through a feedback pathway. Cyclic GMP could be under homeostatic control with light increasing the shuttling of molecules through a synthesis-degradation cycle.

Guanylate Cyclase

Troyer et al (1978) report that mouse retinae contain three forms of guanylate cyclase. Two of these, a soluble and a particulate form, display characteristics similar to soluble and particulate forms of guanylate cyclase found in other mammalian tissues. A second particulate form is found in normal retinae but absent in retinae of mice with photoreceptor dystrophy. This form, unlike the others, is inhibited by high concentrations of Ca. A guanylate cyclase with similar characteristics has been isolated from bovine outer segments (Krishman et al 1978). Activity requires either Mn or Mg; Mn is reported to be tenfold more effective than Mg. No other divalent cation will stimulate activity. Ca inhibits activity; 100 μM Ca inhibits approximately 20% in the presence of 500 μM Mn. Unfortunately, this enzyme has received less attention than the phosphodiesterase. It would be interesting to know the effect of calcium concentrations in the range 0.1–10 μM in the presence of physiological concentrations of GTP (400 μM) and Mg (1–2 mM).

To understand the distribution of guanylate cyclase it is necessary to recall that outer segments are modified cilia. At their base is an axoneme with a characteristic 9+0 arrangement of tubules (Sjostrand 1959). Tubules leave the axoneme and run between the disc edges and the plasmalemma in grooves created by indentations or clefts in the disc rims. Discs in primates (Young 1971a,b, Hogan et al 1971) and amphibians (Brown et al 1963) have approximately two dozen shallow clefts; discs in carnivores have only a few (Steinberg & Wood 1975); and discs in rodents have a single deep cleft (Cohen 1972, Dowling & Gibbons 1961). When there is only one cleft, all nine microtubules lie opposite, and when there are multiple clefts one or two microtubules face individual clefts. Tubules extend close to the tip of outer segments in mammals but extend only about half the 60 μm length of outer segments in frog rods. Axonemes and tubules, isolated by differential solubilization and organelle fractionation, contain 90% of outer segment guanylate cyclase activity (Fleischman & Denisevich 1979, Fleischman 1981). These observations together with those of the preceding section indicate a spatial separation in cGMP metabolism. Phosphodiesterase hydrolyzes cGMP in an interdisc space. Guanylate cyclase synthesizes cGMP just beneath the plasmalemma. Nucleotides synthesized in the inner segment should enter the outer segment and diffuse along its length in the space in which tubules also run. This would be the ideal place for an enzyme requiring GTP as a substrate.

Ca and Nucleotide Movement Within the Outer Segment

At the base of the outer segment is the synthetic and metabolic machinery of the rod. A ball of mitochondria immediately beneath the outer segment synthesizes trinucleotides, which move into and along the outer segment where they supply hydrolytic enzymes. Nucleotides appear to move easily between the two parts of a rod. Cyclic GMP injected into either location produced a rapid and large increase in the generator current. In contrast, the diffusion of Ca within a rod appeared to be limited by intracellular buffering and export across the plasmalemma. Ca injected into an inner segment had no effect on the generator current, whereas Ca injected directly into the outer segment suppressed the current (MacLeish et al 1984). Evidently, Ca does not readily move between inner and outer segments. Even when injected directly into an outer segment, Ca's ability to suppress the current may be limited by its ability to diffuse from the injection site.

Ca Movement Across the Plasmalemma

The Ca concentration around outer segments transiently rises following a flash (Gold & Korenbrot 1980, Yoshikami et al 1980). The increase can be interpreted to indicate a sequence predicted by the Ca hypothesis: Light releases Ca from discs; the Ca concentration in the cytoplasm rises; Ca is extruded from the outer segment; the Ca concentration around the outer segment rises. However, another interpretation is possible. The following observations will help the explanation.

Lowering the external Ca concentration increases the size of the generator current (Yoshikami & Hagins 1973, Yau et al 1981, MacLeish et al 1984). When the external Ca concentration is rapidly changed, the current alteration lags the solution change by less than 0.5–1 sec (Yau et al 1981). The rapidity of Ca's effect indicates that either Ca has its action at an external site, or it easily passes across the membrane and has its action at an internal site.

Experiments in which EGTA is injected into an outer segment indicate that Ca continuously enters and has an internal site of action (MacLeish et al 1984). Injecting EGTA increases the current. When the injection is stopped, the current quickly returns to its original level. The quick return indicates that the buffering capacity of intracellular EGTA is rapidly consumed. This could occur if Ca is continuously appearing in the cytoplasm during darkness. Possibly, Ca is continuously entering through generator-current channels. Continuous Ca entry is also indicated by events that occur when an injection of EGTA commences. As EGTA is *continuously* injected, the current quickly increases to a *steady* level. The new level is not a maximum, for experiments with cGMP demonstrate that the current can be much larger. This behavior could occur if Ca entry increased as the generator current became larger. The effect of EGTA may then be explained as follows. As EGTA was injected, the generator current increased and more Ca entered until the added influx of Ca balanced the continuous injection of EGTA.

One way to explain the preceding observations is for Ca to enter the outer segment continuously through generator-current channels. The entry of one Ca ion with, for, example, one thousand Na ions would not be detected in the measurement of a reversal potential. Nonetheless, a small component carried by Ca would have a profound effect. An inward current of 0.05 pA (compared with a 50 pA amplitude for the total generator current) would completely replace the free Ca of an outer segment (1 pl volume and 1 μM concentration) every 4 sec. The permeation of Ca with Na through generator current channels should not be a surprising suggestion. Biological channels are not infinitely selective. For example, calcium passes through the TTX-sensitive Na pore in squid nerve, the permeability being roughly $P^{Ca}/P^{Na} = .01$ (Hodgkin & Keynes 1957, Baker et al 1971).

The transient rise in Ca concentration around an outer segment that follows a flash can now be interpreted as the end of the following sequence: Ca enters through generator-current channels and is continuously removed; light stops entry and for a moment extrusion continues; the internal concentration falls and the external concentration rises. Thus, the rise in external Ca following illumination can be interpreted in one of two ways. The difference is whether Ca enters the cytoplasm from discs during illumination or from the external space during darkness. One mechanism predicts that illumination is followed by a rise in cytoplasmic concentration; the other predicts a fall.

Is Ca Stored in Discs and Released by Light?

The Ca hypothesis proposed that Ca is stored in discs, is released by light into the cytoplasm, and blocks channels. Consequently, isolated discs should have a mechanism for accumulating Ca and initiating release when illuminated. Because, as is described below, one absorbed photon closes at least several hundred plasmalemmal channels, one photon should release, at least, several hundred Ca ions. A release of this magnitude has so far not been detected from isolated discs (Liebman 1974, Shevchenko 1976, Smith & Bauer 1979, Noll et al 1979, Szuts 1980; see review by Kaupp & Schnetkamp 1982). The largest flux observed from isolated discs is less than one Ca per rhodopsin bleached.

A very different conclusion has been drawn in a recent study of broken outer segments. George & Hagins (1983) incubated broken outer segments in media supplemented with nucleotide triphosphates and observed the Ca activity outside the broken outer segments with a miniaturized Ca electrode. After adding cGMP, the Ca activity declined. Light then induced an increase in free

Ca. The estimated release of Ca by a single photon was tenfold slower and tenfold smaller than expected for transduction in intact rods. The difference might be ascribed to the loss of important soluble factors after outer segments are broken. The results have been interpreted to indicate that discs release Ca when illuminated and require cGMP (and nucleotide triphosphates) for reaccumulation. This view would strongly support the Ca hypothesis. However, there is another interpretation suggested by the difference in ability of cyclic nucleotides and Ca to move into and along the outer segment. The methods and photographs in the report by George & Hagins (1983) indicate that most of their preparations were not individual discs, but outer segments ruptured open at the end(s). *N,N′*-didansylcystine readily moved through the fractured ends to stain the entire organelle. Presumably, nucleotides also entered and diffused throughout the organelle. But Ca may behave differently. Entry and exit of Ca across the plasmalemma may be faster than diffusion of Ca into and along the organelle. Thus, broken outer segments may, if supplied by nucleotides through broken ends, reproduce some of the physiology of an intact rod. Extracellular Ca may rise following illumination; supplying cGMP may open generator current channels and allow Ca in the external space to flow into the outer segment. In this view, experiments with broken outer segments would reproduce phenomena seen earlier with intact cells (Gold & Korenbrot 1980, Yoshikami et al 1980) and may not indicate the properties of isolated discs.

Cyclic GMP has been reported to increase the permeability of vesicles made from outer segment membrane (Caretta et al 1979, Caretta & Cavaggioni 1983). The permeability is increased for both monovalent and divalent cations but not anions. Unfortunately the magnitude of the permeability normalized to rhodopsin content was not reported and the significance of the apparent permeability change cannot be evaluated. Furthermore, the contribution of pores, carriers, and pumps derived from the plasmalemma cannot be evaluated. The phenomena might be produced by a small percentage of plasmalemmal vesicles or the fusion of plasmalemmal vesicles with discs.

Although outer segments and isolated discs contain between 2 and 5 mmol/l of Ca (Hagins & Yoshikami 1975, Szuts & Cone 1977, Schröder & Fain 1984), the free Ca concentration in the cytoplasm of the outer segment is estimated to be 1 μM (Hagins & Yoshikami 1974). Therefore, most of the Ca is either sequestered within discs or bound to sites on the cytoplasmic surface of discs. There is still no experimental method for clearly deciding whether Ca is within or on the surface of discs. The Ca content, equal to 1–2 ions per rhodopsin molecule, could be bound to phospholipids (see McLaughlin & Brown 1981) that are asymmetrically distributed in the disc bilayer (Drenthe et al 1980, Miljanich et al 1981). Phosphotidylcholine is preferentially contained in the inner membrane leaflet; phosphotidylserine, which has a net negative charge at physiological pH and is able to bind divalent ions, is almost entirely contained in the outer or cytoplasmic membrane leaflet. Ca absorbed onto the disc surface may act to buffer the cytoplasmic concentration. A decrease in the cytoplasmic Ca concentration may be followed by loss from the bound pool.

Schröder & Fain (1984) have measured Ca content in outer segments before and during continuous illumination. Their method, burning a small area with a laser and detecting the vaporized ions with a mass spectrometer, measured total Ca content. Since the free concentration is less than 1/2000 of the total, they could not detect changes in free concentration or changes in plasmalemmal permeability. Bleaching 68 Rh sec^{-1} induced the loss of 8×10^5 Ca sec^{-1}, or about 10^4 Ca per Rh. This is the same number of Ca ions observed to appear in the extracellular space by Gold & Korenbrot (1980). Taken together these studies suggest that light induces the net transfer of Ca from a bound or sequestered pool to the extracellular space. This may occur if Ca is released from discs into the cytoplasm and subsequently removed across the plasmalemma. Another possibility is that Ca stops entering the outer segment, internal Ca concentration falls, and Ca is slowly removed from buffering sites to replenish the intracellular concentration deficit. This second scheme does, however, encounter one difficulty. Ca depletion appears to increase with light intensities beyond that sufficient to block the generator current (Miller & Korenbrot 1984, Schröder & Fain 1984). The scheme could certainly be modified to patch this failure; but, for the moment, any particular patch would be selected arbitrarily.

Do Protons Liberate Ca?

The hydrolysis of cGMP produces a proton and GMP (Yee & Liebman 1978). Liebman et al (1984) have suggested that this proton is an intermediary in the transduction process. They studied light-initiated changes in the generator current after changing extracellular pH between 3.5 and 10.5. In general, increasing the proton concentration had effects roughly equal to increasing external Ca concentration. The ability of protons to block could be described by a titration cure. The $pK_{1/2}$ depended upon the external Ca concentration. Lowering Ca from 1 mM to 5 μM shifted the apparent $pK_{1/2}$ from pH 4.8 to 4.0.

The ability of protons to block the current need not be surprising. Large changes in proton concentration often affect a wide spectrum of enzymes and may affect channels. For example, protons block Na channels in frog myelinated fibers from either the internal or external solutions at a site that can be titrated with an apparent pK of 5.4 (Woodhull 1973). Yet one does not think of protons as being a specific regulator of Na channels.

concentration and pH. The extent of buffering in outer segments is not certain. The pH buffer capacity of the cytoplasm in many cells in 10—50 mEq/pH unit (Roos & Boron 1981). From this one can calculate that hydrolysis of all the cGMP in an outer segment would change the pH by less than .01. Protons might, however, displace Ca ions from buffer sites. Increasing the proton buffer capacity should dilute this effect. Incubating cells (that contain carbonic anhydrase) in a bicarbonate-buffered medium approximately doubles their buffer capacity (Roos & Boron 1981). An increase in buffer capacity should attenuate the responses recorded from rods. Responses to a dim light recorded from rods bathed in a medium buffered with bicarbonate are smaller than responses recorded from rods bathed in a medium buffered with HEPES (Lamb et al 1981, Lamb 1984). Exposure to bicarbonate decreases the amplitude by a factor of approximately 3 and shortens the time to peak by a factor of approximately 2. On the other hand, increasing the buffer capacity of outer segments by bathing a retina in imidazole is reported not to change the relation between light intensity and current amplitude (Yoshikami & Hagins 1984). Also, injecting protons or the pH buffer BICINE into individual outer segments was without effect on the amplitude of the generator current (MacLeish et al 1984). Similar experiments in which Ca, EGTA, or cGMP were injected produced a noticeable change. However, the actual quantities injected in each experiment were not known. Needed are exact experiments in which known quantities are injected and intracellular concentrations measured.

Cyclic GMP-dependent Kinase

I have avoided discussing experiments in which active agents were injected into rods in intact retinae. Yet, the results of one experiment, although not entirely convincing, may be mentioned at this point. The injection of cGMP-dependent kinase purified from bovine lung has been claimed to depolarize rods, and its depolarizing effect to be reversed by light (Shimoda et al 1983). If corroborated by appropriate experiments, the result might indicate that cGMP mediates control of the plasmalemmal current through a kinase rather than as a substrate for the production of protons.

Although a cyclic-nucleotide-dependent kinase has not yet been identified in the outer segment, its existence may be inferred from the cyclic-nucleotide-dependent phosphorylation of several low-molecular-weight, peripheral membrane proteins (Polans et al 1979). These proteins are phosphorylated during darkness and dephosphorylated during light. At present, these are the only cyclic nucleotide dependent phosphorylations that have been reported. Their role in the transduction process should be studied.

There is also preliminary and incomplete evidence that cGMP may have effects not mediated by a kinase. Caretta & Cavaggioni (1983) have reported that cGMP increases the Ca uptake of outer segment membrane vesicles (including disc and plasmalemmal membranes) that have been washed free of soluble proteins and small molecules. Since nucleotide triphosphates were believed to have been removed, the authors concluded that the action of cGMP did not involve a kinase but instead was a direct interaction of cGMP with a Ca transport molecule.

RESPONSE TO LIGHT

Electrophysiologists, like systems engineers, have delivered various inputs, flashes or steps of light, and carefully measured an output, a change in the generator current that enters the outer segment. This quantitative method has produced results that place significant restrictions on the nature of the overall transduction process.

One Photon Closes Many Channels

Analyzing responses to single photons indicates that one absorbed photon produces a chain of events that finally closes many channels. When a dim light is flashed onto the retina many times, the amplitude of the voltage fluctuates. Variation can come from two sources: (*a*) the number of photons in each dim flash is given by a Poisson distribution, and (*b*) each absorbed photon may produce a slightly different response. If each response initiated by a single photon were built by the contribution of a large number of small units, a few units more or less would make little difference to the total, and variation from this source would be small. Analysis of the variance at each moment during the response indicates that nearly all of the fluctuation is due to variation in the number of photons absorbed and that each absorbed photon produces a nearly identical response with a shape like the average (Schwartz 1975, 1976). The relatively constant shape implies that each response is produced by the coordinate activation of a large number of elementary events. Comparison of the average response to one photon with the spontaneous light-sensitive voltage fluctuation indicates that the response to one photon is composed of 100–300 transduction events (Schwartz 1977).

Responses to single photons can be directly observed while recording the current flowing into an individual outer segment (Baylor et al 1979b, 1980). During continuous exposure to a very dim light, single photons are absorbed at random intervals and produce responses with a stereotyped image that is, again, like the average. When a large number of individual responses are compared, their amplitudes have a relatively small variance; this indicates that a minimum of 100 discreet events contribute to each single photon response.

Thus, experiments performed with two very different techniques indicate that the absorption of one photon produces a shower of several hundred events at the membrane that together produce a response with a characteristic shape and amplitude. Similar responses rarely occur when a rod is left in complete darkness. Spontaneous responses occur at a rate of 1 per 50 sec (at 20°C) or 1 event per disc in 21 hr (Baylor et al 1980). Transduction produces a stereotyped, all-or-nothing response with great reliability and very few errors.

Time Course

The ability of extrinsic agents to alter response shape may provide clues to the mechanism of transduction. The ideal experiment would be to observe changes in response shape during a voltage clamp. Agents could be added to the external medium or injected into solitary cells. Unfortunately, a complete set of experiments of this type has not yet been done. Some results are available from studies of the generator current. For additional results we must consider experiments in which voltage was recorded from rods in the intact retina. In this situation the best estimate for the time course of the generator current is the voltage produced by a small spot centered on the impaled rod (Schwartz 1976). Electrical coupling (Schwartz 1973, 1975, 1976, Fain et al 1976, Copenhagen & Owen 1976, Leeper et al 1978, Detwiler et al 1980, Griff & Pinto 1981) and voltage-dependent currents of the inner segment (Schwartz 1976, Fain et al 1978, Bader et al 1979, 1982, Detwiler et al 1980, Attwell & Wilson 1980) alter this shape when the spot is enlarged to illuminate adjacent rods. It is easier, however, to use a large spot which need not be centered or focused. Detwiler et al (1980) calculated that voltage-dependent currents shorten the time-to-peak of the voltage by 40% (e.g. from 1.91 to 1.21 sec, see their Figure 17). Actually, a response of about 1–2 mV amplitude is shortened approximately 15% (Baylor et al 1984, Figures 2, 4) and this difference is reduced as the amplitude is decreased (see Schwartz 1976, Figure 3). Consequently, voltage responses produced by a large spot can be used to estimate the time course of the current if the intensity of the light is sufficiently dim.

The response to a dim flash begins with a sigmoidal onset, smoothly rises to a peak, and then gradually decays. Responses recorded from rods in the dark-adapted toad retina reach their peak in an average time of 0.65 sec (see Capovilla et al 1983). Currents recorded with a suction pipette from toad rods have so far been observed to be slower and reach their peak in 0.6 to 4 sec (Baylor et al 1979a). The slowest cells are frequently the most sensitive, producing the largest responses for an absorbed photon. Responses often lengthen as the condition of a cell deteriorates (Baylor et al 1979a, Lamb 1984). Cells that are particularly sensitive and produce responses with an abnormally long duration may be in a nonphysiological condition.

The response produced by each absorbed photon becomes smaller when rods are continuously illuminated. Although the kinetics of responses during adaptation have not yet been studied in detail, the general trend is clear. The duration of a response is shortened, the time to peak decreasing from 0.7 to 0.2 sec, and the shape changes so that recovery is relatively faster. Simultaneously, the amplitude is decreased. The attenuation can be more than 100-fold.

As an aside, the response during adaptation may become smaller than the transduction events that produce a fluctuation in the generator current during complete darkness. Consequently, during adaptation the size of a transduction event must be significantly reduced.

The primary effect of transduction and the slower effect of adaptation have often been ascribed to separate messengers. For example, Ca has been assumed to mediate one and cGMP the other (see for example Bastian & Fain 1979). Yet there is no clear evidence for this simple dichotomy. The absorption of one photon stops the generator current from entering along approximately 1% of outer segment length (see below). The effect of adaptation spreads over a similar distance (Lamb et al 1981). Activity in one region does not appear to desensitize neighboring regions. The process of adaptation is perhaps inextricably woven into the transduction process.

Manipulations believed to change the internal Ca concentration alter the shape of responses to a flash. Increasing external Ca concentration slows the recovery phase (Yau et al 1981). Lowering external Na has a similar effect. Yau et al suggest that internal Ca concentration is controlled by Na–Ca exchange and that both procedures increase internal Ca concentration (see Eq. 2). This could occur if Ca were removed by Na–Ca exchange, and inhibiting the removal of Ca slowed recovery. This sequence also explains the slowing of recovery during depolarization (MacLeish et al 1984), another procedure expected to depress Na–Ca exchange. Although Na–Ca exchange provides a unified explanation, the simplest scheme of reversible exchange is energetically unlikely and would need to be modified.

The ability of cGMP to alter the time course of responses has not been studied. Instead, the effects of a series of phosphodiesterase inhibitors have been studied (Capovilla et al 1983). These are effective when added to the external solution. All have the same effect and IBMX is the most potent. At 30 μM the effect of IBMX is stable and reversible. The normal time-to-peak of 0.65 sec is increased to 5.3 sec. Contraction of the time base demonstrates that the response is slowed but unchanged in shape. It is as if an internal metronome ticked at ⅛ the normal rate. The greatly prolonged response can be shortened to match the original time course by adapting with a dim, continuous background. These results prompted the interesting suggestion that the internal cGMP concentration determines the rate with which the transduction mechanism proceeds.

Thus, in partial summary, one photon produces a response with a characteristic shape. During adaptation this response becomes faster and changes shape. The ability to change shape and time scale are important properties of the transduction process. Changing intracellular Ca concentration appears to produce changes in shape. Changing cGMP concentration appears to produce relatively large changes in time scale.

Is Messenger Concentration Proportional to Light Intensity?

Comparing the shapes of responses produced by dim and bright lights indicates how the effects of simultaneously absorbed photons sum up. But before considering this subject, it is necessary to describe a complication. The voltage responses to dim and bright lights have different shapes. An early explanation attributed the difference in shape to an increasing change in the transduction process as more and more photons were absorbed. Cervetto et al (1977) formalized this conjecture with an autocatalytic model suggested by Baylor et al (1974). The model hypothesizes that stimulation by many photons produces a delayed change in the transduction process. It is now known that the peculiar shape of the voltage response elicited by a bright light is due in large part to voltage- and time-dependent currents in the inner segment (Schwartz 1976, Fain et al 1978, Bader et al 1979, 1982, Detwiler et al 1980, Attwell & Wilson 1980). Their effect has been demonstrated by recording voltage and current in the same solitary rod (Bader et al 1979), or recording the voltage from one rod and the current flowing into the outer segment of a nearby rod (Nunn et al 1980). Both experiments give the same result: Current and voltage responses to a bright light are different in shape. The difference is due to voltage-dependent currents activated during a normal response but not activated when voltage is maintained constant by a voltage clamp.

The amplitude of the current change produced by light can be measured at a fixed time after a flash. As a first approximation, the amplitude at each moment during a response increases with light intensity in a fixed manner. Thus, it is not necessary to assume that a large autocatalytic change takes place during a response. The response to a nonsaturating flash can be described by assuming that each absorbed photon independently activates the same biochemical system. Although an effect of adaptation is seen during the terminal phase of a response produced by bright light (see below), it is small when compared to the change in shape proposed in the autocatalytic model.

The relationship between peak amplitude and light intensity can be approximately described by the Michaelis equation (Penn & Hagins 1972, Baylor et al 1979a, Bader et al 1979),

$$\frac{i}{i_{\max}} = \frac{I}{I + k} \qquad (3)$$

where i is the change in current, i_{max} its maximum value, I the light intensities, and k is a constant. This equation has been used often in retinal physiology after being introduced by Naka & Rushton (1966) to describe responses of horizontal cells. However, careful measurements demonstrate small deviations from the Michaelis description. During the early phase of a response, amplitude increases slightly more steeply than predicted (Lamb et al 1981). Late in a response a small effect of adaptation flattens the curve so that amplitudes increase more gradually than predicted. Even though these deviations will later suggest other interpretations it is instructive to consider a model that yields a Michaelis relation and then consider objections to the model. A possible model is as follows. Light produces a messenger that closes channels. The concentration of the messenger is M. There are a total of N channels, of which C are closed and $N-C$ are opened. If the time course for generation of the messenger is slow so that the instantaneous relation between bound and free messenger is given by the Law of Mass Action, and the outer segment is a single, well-mixed intracellular space, then

$$K = \frac{C}{(M-C)(N-C)} \tag{4}$$

where K is an equilibrium constant. If the affinity of messenger for the channel is low then there is always a large excess of M and

$$K \doteqdot \frac{C}{M(N-C)} \tag{5}$$

rearranging

$$\frac{C}{N} \doteqdot \frac{M}{M + 1/K} \tag{6}$$

which is the same as Eq. 3 if the concentration of messenger rises in proportion to the light intensity.

Several comments should be made about the assumptions necessary for deriving the equation. The equation describes how channels that are all opened at zero messenger concentration are closed as messenger is produced. The results of experiments in which cGMP was injected indicate that either 95% of channels are already closed during darkness or the size of individual channels can be changed. In either case, the model fails. Furthermore, the outer segment is unlikely to be a well-stirred volume. A dense stack of discs is likely to hinder diffusion. Indeed, observations of how activation spreads along the outer segment indicate that transduction is localized.

A narrow length of outer segment can be illuminated with a bar of light to stop the current in a limited region. By carefully drawing different lengths of outer segment into an extracellular pipette and shifting the region that is illuminate, the profile of current in the illuminated and adjacent unilluminated regions can be mapped (Lamb et al 1981; see also Hemilä & Reuter 1981). The profile can be described by a space constant of less than 7 μm (resolution was limited by scattered light). One way to visualize this behavior is to believe that events within one disk surface and one interdisc space activate a messenger that diffuses from the interdisc space to impinge on the nearby plasmalemma, where it finds channels to which it immediately binds. A ring of plasmalemma opposite the rim of an intradisc space avidly absorbs messenger. If the affinity of messenger for the channel is great, then channels in one region would be totally occluded before the messenger could diffuse to adjacent regions. As the quantity of messenger increases, the width of the ring in which channels are occluded would broaden. The preceding description has been phrased for a messenger whose concentration is increased by light. Of course it can be rephrased for a messenger whose concentration is decreased, or for a modulator that regulates inactivation (see below). The entire scheme can be formalized by a model of total occlusion (described in the report of Lamb et al 1981). Let the mean number of isomerizations per unit length of outer segment be n. If each disc has an equal and independent probability of absorbing a photon, then the probability that there is no isomerization in a region Δ centered on the channel is $e^{-n\Delta}$ and the intensity-response relation has the form

$$B = 1 - e^{-n\Delta} \tag{7}$$

where B is the normalized amplitude of a response. Equation 7 is a good description of amplitudes measured during the rising phases of responses produced by a wide range of light intensities; Δ is a function of time and increases from 0.039 μm (approximately the length of one interdisc space) at 0.2 sec to nearly 1 μm (approximately the length occupied by 25 discs) at the peak of a response. Occluding all of the current entering along 1 μm of a 60 μm outer segment would allow one photon to stop $1/60 = 0.017$ of the total current.

The model of total occlusion leading to Eq. 7 may be compared with the Michaelis model leading to Eq. 6. The Michaelis model assumes a well-mixed internal space and a weak binding of messenger to the channel. The total occlusion model assumes a localized change in messenger concentration and a high affinity binding to the channel; in the absence of strong binding the messenger would diffuse along the outer segment length and closure of channels would not be localized.

Thus, in partial summary, each photon may produce a nearly complete closure of channels over a short length of outer segment. The concentration of messenger within an interdisc space may not change in proportion to light intensity. The interaction of the messenger and the channel is likely to be one of high affinity.

SYNTHESIS

Neither the Ca nor cGMP hypotheses, rivals for more than a decade, explain all the observations of the preceding sections. If neither alone is adequate, an obvious alternative is to blend the two. One attempt is the "proton" hypothesis. Unfortunately this model still leaves many observations unexplained. In this situation, I have been unable to resist the temptation to confect another model for phototransduction. The central ideas are as follows:

1. cGMP concentration is regulated by a homeostatic feedback system. Hydrolysis is controlled by phosphodiesterase and can be increased by light through an enzyme cascade. Synthesis is controlled by guanylate cyclase, which can be activated by decreases in cytoplasmic Ca concentration.
2. The entry of Ca is controlled by a homeostatic system. Ca enters along with Na ions through generator current channels. This channel is blocked when Ca binds at a specific site on its internal surface.
3. cGMP regulates the affinity of the internal site for Ca. A fall in cGMP concentration increases Ca binding.
4. Each absorbed photon transiently upsets the balance created by the feedback system in items 1–3 and closes all of the channels within a short length of outer segment.

Since light in this scheme operates on a feedback system, it would not be correct to describe transduction by a linear sequence of action. Nonetheless, an approximate description is as follows. During darkness a high rate of Ca influx is balanced by a continuous efflux. A photon bleaches a rhodopsin molecule, which then activates approximately 100 phosphodiesterase molecules. A subsequent decline in cGMP concentration increases the ability of internal Ca to block generator current channels. The influx of Na and Ca is stopped while Ca efflux continues unabated. The result is a transient fall in internal and a rise in external Ca concentrations. A decline in internal Ca concentration activates guanylate cyclase and decreases activation of phosphodiesterase. Both effects act to oppose the decline in cGMP concentration initiated by phosphodiesterase activation. Control of Ca entry though generator-current channels is part of a feedback system that operates to minimize changes in cGMP concentration. If the feedback loop has a large gain, an absorbed photon might always cause the feedback loop to saturate momentarily, that is, all of the channels in a short region to close. This homeostatic hypothesis postulates that nothing is released from discs. The enzymes of the initial cascade are located on the disc surface. A fall in cGMP concentration is initially limited to an interdisc space. Discs limit longitudinal diffusion and localize the effect of light to a short length of outer segment.

The model can explain the ability of Ca and cGMP to affect each other's concentration. It can also explain why physiological light intensities reduce the cGMP concentration of GTP-starved, isolated outer segments although similar declines are not seen in intact rods. Furthermore, an unstable feedback mechanism may explain why the current change produced by a flash terminates with a damped oscillation when rods are superfused with a medium containing a reduced Ca concentration (see records in Hagins & Yoshikami 1977). Aside from these abilities, the principal virtue of the homeostatic hypothesis may be its ability to act as a counterweight against other models. For example, homeostatic and Ca hypothesis can take the same set of observations and make very different interpretations and predictions. From the tension between their divergent views perhaps a better picture will emerge.

Literature Cited

Attwell, D., Wilson, M. 1980. Behaviour of the rod network in the tiger salamander retina mediated by membrane properties of individual rods. *J. Physiol.* 309:287–315

Bader, C. R., Bertrand, D. 1984. Effect of changes in intra- and extracellular sodium on the inward (anomalous) rectification in salamander photoreceptors. *J. Physiol.* 347:611–31

Bader, C. R., Bertrand, D., Schwartz, E. A. 1982. Voltage-activated and calcium-activated currents studied in solitary rod inner segments from the salamander retina. *J. Physiol.* 331:253–84

Bader, C. R., MacLeish, P. R., Schwartz, E. A. 1978. Responses to light of solitary rod photoreceptors isolated from tiger salamander retina. *Proc. Natl. Acad. Sci. USA* 75:3507–11

Bader, C. R., MacLeish, P. R., Schwartz, E. A. 1979. A voltage-clamp study of the light response in solitary rods of the tiger salamander. *J. Physiol.* 296:1–26

Baehr, W., Devlin, M. J., Applebury, M. L. 1979. Isolation and characterization of cGMP phosphodiesterase from bovine rod outer segments. *J. Biol. Chem.* 254:11669–77

Baehr, W., Morita, E. A., Swanson, R. J., Applebury, M. L. 1982. Characterization of bovine rod outer segment G-protein. *J. Biol. Chem.* 257:6452–60

Baker, P. F., Hodgkin A. L., Ridgway, E. B. 1971. Depolarization and calcium entry in squid giant axons. *J. Physiol.* 218:709–55

Bastian, B. L., Fain, G. L. 1979. Light adaptation in toad rods: Requirement for an internal messenger which is not calcium. *J. Physiol.* 297:493–520

Baylor, D. A., Hodgkin, A. L., Lamb, T. D. 1974. Reconstruction of the electrical responses of turtle cones to flashes and steps of light. *J. Physiol.* 242:759–91

Baylor, D. A., Lamb, T. D. 1982. Local effects of bleaching in retinal rods of the toad. *J. Physiol.* 328:49–71

Baylor, D. A., Lamb, T. D., Yau, K.-W. 1979a. The membrane current of single rod outer segments. *J. Physiol.* 288:589–611

Baylor, D. A., Lamb, T. D., Yau, K.-W. 1979b. Responses of retinal rods to single photons. *J. Physiol.* 288:613–34

Baylor, D. A., Mathews, G., Yau, K.-W. 1980. Two components of electrical dark noise in toad retinal rod outer segments. *J. Physiol.* 309:591–621

Baylor, D. A., Mathews, G., Yau, K.-W. 1983. Temperature effects on the membrane current of retinal rods of the toad. *J. Physiol.* 337:723–34

Baylor, D. A., Mathews, G., Nunn. B. J. 1984. Location and function of voltage-sensitive conductances in retinal rods of the salamander, *Ambystoma tigrinum*. *J. Physiol.* 354:203–23

Berman, A. L., Azimova, A. M., Gribakin, F. G. 1977. Localization of Na^+, K^+-ATPase and Ca^{2+}-activated Mg^{2+}-dependent ATPase in retinal rods. *Vision Res.* 17:527–36

Biernbaum, M. S., Bownds, M. D. 1979. Influence of light and calcium on guanosine 5′-triphosphate in isolated frog rod outer segments. *J. Gen. Physiol.* 74:649–69

Bignetti, E., Cavaggioni, A., Sorbi, R. T. 1978. Light-activated hydrolysis of GTP and cyclic GMP in the rod outer segments. *J. Physiol.* 279:55–69

Blaustein, M. P., Nelson, M. T. 1982. Sodium-calcium exchange: Its role in the regulation of cell calcium. In *Membrane Transport of Calcium*, ed. E. Carafoli. New York: Academic

Bownds, D., Brodie, A. E. 1975. Light-sensitive swelling of isolated frog rod outer segments as an *in vitro* assay for visual transduction and dark adaptation. *J. Gen. Physiol.* 66:407–25

Bownds, D., Dawes, J., Miller, J., Stahlman, M. 1972. Phosphorylation of frog photoreceptor membranes induced by light. *Nature New Biol.* 237:125–27

Brodie, A. E., Bownds, D. 1976. Biochemical correlates of adaptation processes in isolated frog photoreceptor membranes. *J. Gen. Physiol.* 68:1–11

Brown, J. E., Coles, J. A., Pinto, L. H. 1977. Effects of injections of calcium and EGTA into the outer segments of retinal rods of *Bufo marinus*. *J. Physiol.* 269:707–22

Brown, J. E., Pinto, L. H. 1974. Ionic mechanism for the photoreceptor potential of the retina of *Bufo marinus*. *J. Physiol.* 236:575–91

Brown, P. K. 1972. Rotational diffusion of rhodopsin in the visual receptor membrane. *Nature New Biol.* 236:35–38

Brown, P. K., Gibbons, I. R., Wald, G. 1963. The visual cells and visual pigment of the mudpuppy, *Necturus*. *J. Cell Biol.* 19:79–106

Capovilla, M., Caretta, A., Cervetto, L., Torre, V. 1983. Ionic movements through light-sensitive channels of toad rods. *J. Physiol.* 343:295–310

Capovilla, M., Cervetto, L., Torre, V. 1983. The effect of phosphodiesterase inhibitors on the electrical activity of toad rods. *J. Physiol.* 343:277–94

Caretta, A., Cavaggioni, A. 1983. Fast ionic flux activated by cyclic GMP in the membrane of cattle rod outer segments. *Eur. J. Biochem.* 132:1–8

Caretta, A., Cavaggioni, A., Sorbi, R. T. 1979. Cyclic GMP and the permeability of the disks of the frog photoreceptors. *J. Physiol.* 295:171–78

Cervetto, L., McNaughton, P. A. 1983. Inhibition of the light-sensitive current in vertebrate rods by La^{3+} ions. *J. Physiol.* 341:75–76P

Cervetto, L., Pasino, E., Torre, V. 1977. Electrical responses of rods in the retina of *Bufo marinus*. *J. Physiol.* 267:17–51

Cohen, A. I. 1972. Rods and cones. *Handb. Sensory Physiol.* 7(2):63–110

Cohen, A. I., Hall, I. A., Ferrendelli, J. A. 1978. Calcium and cyclic nucleotide regulation in incubated mouse retinas. *J. Gen. Physiol.* 71:595–612

Cone, R. A. 1972. Rotational diffusion of rhodopsin in the visual receptor membrane. *Nature New Biol.* 236:39–43

Copenhagen, D. R., Owen, W. G. 1976. Functional characteristics of lateral interactions between rods in the retina of the snapping turtle. *J. Physiol.* 259:251–82

Corey, D. P., Dubinsky, J. M., Schwartz, E. A. 1984. The calcium current in inner segments of rods from the salamander *(Ambystoma tigrinum)* retina. *J. Physiol.* 354:557–75

DeAzeredo, F. A. M., Lust, W. D., Passonneau, J. V. 1981. Light-induced changes in energy metabolites, guanine nucleotides and guanylate cyclase within frog retinal layers. *J. Biol. Chem.* 256:2731–35

Detwiler, P. B., Conner, J. D., Bodoia, R. D. 1982. Gigaseal patch clamp recordings from outer segments of intact retinal rods. *Nature* 300:59–61

Detwiler, P. B., Hodgkin, A. L., McNaughton, P. A. 1980. Temporal and spatial characteristics of the voltage response of rods in the retina of the snapping turtle. *J. Physiol.* 300:213–50

Dowling, J., Gibbons, J. R. 1961. The effect of vitamin A deficiency on the fine structure of the retina. In *The Structure of the Eye*, ed. G. K. Smeleser. New York: Academic

Drenthe, E. H. S., Klompmakers, A. A., Bonting, S. L., Daemen, F. J. M. 1980. Transbilayer distribution of phospholipids in photoreceptor membrane studied with trinitrobenzenesulfonate alone and in combination with phospholipase D. *Biochem. Biophys. Acta* 603:130–41

Fain, G. L., Gold, G. H., Dowling, J. E. 1976. Receptor coupling in toad retina. *Cold Spring Harbor Symp. Quant. Biol.* 40:547–61

Fain, G. L., Quandt, F. N., Bastian, B. L., Gerschenfeld, H. M. 1978. Contribution of a caesium-sensitive conductance increase to rod response. *Nature* 272:467–69

Fenwick, E. M., Marty, A., Neher, E. 1982. Sodium and calcium channels in bovine chromaffin cells. *J. Physiol.* 331:599–635

Fleischman, D. 1981. Rod guanylate cyclase located in axonemes. *Curr. Topics Membr. Transport.* 15:109–19

Fleischman, D., Denisevich, M. 1979. Guanylate cyclase of isolated bovine retinal rod axonemes. *Biochemistry* 18:5060–66

Fung, B. K. K. 1983. Characterization of transducin from bovine retinal rod outer segments. I. Separation and reconstitution of the subunits. *J. Biol. Chem.* 258:10495–1502

Fung, B. K. K., Hurley, J. B., Stryer, L. 1981. Flow of information in the light triggered cyclic nucleotide cascade of vision. *Proc. Natl. Acad. Sci. USA* 78:152–56

Fung, B. K. K., Stryer, L. 1980. Photolyzed rhodopsin catalyzes the exchange of GTP for bound GDP in retinal rod outer segments. *Proc. Natl. Acad. Sci. USA* 77:2500–4

George, J. S., Hagins, W. A. 1983. Control of Ca^{2+} in rod outer segment disks by light and cyclic GMP. *Nature* 303:344–48

Godchaux, W., Zimmerman, W. F. 1979. Membrane-dependent guanine nucleotide binding and GTPase activities of soluble protein from bovine rod cell outer segments. *J. Biol. Chem.* 254:7874–84

Gold, G. H., Korenbrot, J. I. 1980. Light-induced calcium release by intact retinal rods. *Proc. Natl. Acad. Sci. USA* 77:5557–61

Goldberg, N. D., Ames, A., Gander, J. E., Walseth, T. F. 1983. Magnitude of increase in retinal cGMP metabolic flux determined by ^{18}O incorporation into nucleotide α-phosphoryls corresponds with intensity of photic stimulation. *J. Biol. Chem.* 258:9213–19

Govardovskii, V. I., Berman, A. L. 1981. Light-induced changes of cyclic GMP content in frog retinal rod outer segments measured with rapid freezing and microdissection. *Biophys. Struct. Mech.* 7:125–30

Goridis, C., Urban, P. F., Mandel, P. 1977. The effect of flash illumination on the endogenous cyclic GMP content of isolated frog retinae. *Exp. Eye Res.* 24:171–77

Griff, E. R., Pinto, L. H. 1981. Interactions among rods in the isolated retina of *Bufo marinus*. J. Physiol. 314:237–54

Hagins, W. A. 1972. The visual process: Excitatory mechanisms in the primary receptor cells. *Ann. Rev. Biophys. Bioeng.* 1:131–58

Hagins, W. A., Penn, R. D., Yoshikami, S. 1970. Dark current and photocurrent in retinal rods. *Biophys. J.* 10:380–412

Hagins, W. A., Yoshikami, S. 1974. A role of Ca^{2+} in excitation of retinal rods and cones. *Exp. Eye Res.* 18:299–305

Hagins, W. A., Yoshikami, S. 1975. Ionic mechanisms in excitation of photoreceptors. *Ann. NY Acad. Sci.* 264:314–25

Hagins, W. A., Yoshikami, S. 1977. Intracellular transmission of visual excitation in photoreceptors: Electrical effects of chelating agents introduced into rods by vesicle fusion. In *Vertebrate Photoreception,* ed. H. B. Barlow, P. Fatt. London: Academic

Hemilä, S., Reuter, T. 1981. Longitudinal spread of adaptation in the rods of the frog's retina. *J. Physiol.* 310:501–28

Hermolin, J., Karell, M. A., Hamm, H. E., Bownds, M. D. 1982. Calcium and cyclic GMP regulation of light-sensitive protein phosphorylation in frog photoreceptor membranes. *J. Gen. Physiol.* 79:633–55

Hodgkin, A. L., Keynes, R. D. 1957. Movements of labelled calcium in squid giant axons. *J. Physiol.* 138:253–81

Hogan, M. J., Alvarado, J. A., Weddell, J. E. 1971. *Histology of the Human Eye*. Philadelphia: Saunders

Hubbell, W. L., Bownds, M. D. 1979. Visual transduction in vertebrate photoreceptors. *Ann. Rev. Neurosci.* 2:17–34

Kaupp, U. B., Schnetkamp, P. P. M. 1982. Calcium metabolism in vertebrate photoreceptors. *Cell Calcium* 3:83–112

Kawamura, S., Bownds, M. D. 1981. Light adaptation of the cyclic GMP phosphodiesterase of frog photoreceptor membranes mediated by ATP and calcium ions. *J. Gen. Physiol.* 77:571–91

Kilbride, P. 1980. Calcium effects on frog retinal cyclic guanosine-3′:5′-monophosphate levels and their light-initiated rate of decay. *J. Gen. Physiol.* 75:457–65

Kilbride, P., Ebrey, T. G. 1979. Light-initiated changes of cyclic guanosine monophosphate levels in the frog retina measured with quick-freezing techniques. *J. Gen. Physiol.* 74:415–26

Krishman, N., Fletcher, R. T., Chader, G. J., Krishna, G. 1978. Characterization of guanylate cyclase of rod outer segments of the bovine retina. *Biochem. Biophys. Acta.* 523:506–15

Kühn, H. 1974. Light-dependent phosphorylation of rhodopsin in living frogs. *Nature* 250:588–90

Kühn, H. 1980. Light- and GTP-regulated interaction of GTPase and other proteins with bovine photoreceptor membranes. *Nature* 283:587–89

Kühn, H., Dreyer, W. J. 1972. Light dependent phosphorylation of rhodopsin by ATP. *FEBS Lett.* 20:1–6

Lamb, T. D. 1984. Effects of temperature changes on toad rod photocurrents. *J. Physiol.* 346:557–78

Lamb, T. D., McNaughton, P. A., Yau, K.-W. 1981. Spatial spread of activation and background desensitization in toad rod outer segments. *J. Physiol.* 319:463–96

Leeper, H. F., Normann, R. A., Copenhagen, D. R. 1978. Evidence for passive electrotonic interactions in red rods of toad retina. *Nature* 275:234–36

Liebman, P. A. 1974. Light-dependent Ca^{++} content of rod outer segment disc membranes. *Invest. Ophthalmol.* 13:700–1

Liebman, P. A., Entine, G. 1974. Lateral diffusion of visual pigment in photoreceptor disk membranes. *Science* 185:457–59

Liebman, P. A., Mueller, P., Pugh, E. N. Jr. 1984. Protons suppress the dark current of frog retinal rods. *J. Physiol.* 347:85–110.

Liebman, P. A., Pugh, E. N. Jr. 1980. ATP mediates rapid reversal of cyclic GMP phosphodiesterase activation in visual receptor membranes. *Nature* 287:734–36

Liebman, P. A., Pugh, E. N. Jr. 1981. Control of rod disk membrane phosphodiesterase and a model for visual transduction. *Curr. Topics Membr. Transport.* 15:157–70

MacLeish, P. R., Schwartz, E. A., Tachibana, M. 1984. Control of the generator current in solitary rods of the *Ambystoma tigrinum* retina. *J. Physiol.* 348:645–64

McLaughlin, S., Brown, J. 1981. Diffusion of calcium ions in retinal rods: A theoretical calculation. *J. Gen. Physiol.* 77:475–87

Miljanich, G. P., Nemes, P. P., White, D. L.,

Dratz, E. A. 1981. The asymmetric transmembrane distribution of phosphatidylethanolamine, phosphatidylserine, and fatty acids of the bovine retinal rod outer segment disk membrane. *J. Membr. Biol.* 60:249–55

Miller, W. H. 1982. Physiological evidence that light-mediated decrease in cyclic GMP is an intermediary process in retinal rod transduction. *J. Gen. Physiol.* 80:103–23

Miller, D. L., Korenbrot, J. I. 1984. Simultaneous photocurrent and calcium release measurements from single rod photoreceptors. *Biophys. J.* 45:341a

Naka, K. I., Rushton, W. A. H. 1966. S-potentials from colour units in the retina of fish *Cyprinidae*. *J. Physiol.* 185:536–55

Nicol, G. D., Miller, W. H. 1978. Cyclic GMP injected into retinal rod outer segments increases latency and amplitude of response to illumination. *Proc. Natl. Acad. Sci. USA* 75:5217–20

Noll, G., Stieve, H., Winterhager, J. 1979. Interaction of bovine rhodopsin with calcium ions. II: Calcium release in bovine rod outer segments upon bleaching. *Biophys. Struct. Mech.* 5:43–53

Nunn, B. J., Mathews, G. G., Baylor, D. A. 1980. Comparison of voltage and current responses of retinal rod photoreceptors. *Fed. Proc.* 39:2066

Oakley, B., Pinto, L. H. 1983. Modulation of membrane conductance in rods of *Bufo marinus* by intracellular Ca ion. *J. Physiol.* 339:273–98

Orr, H. T., Lowry, O. H., Cohen, A. I., Ferrendelli, J. A. 1976. Distribution of 3′:5′-cyclic AMP and 3′-5′-cyclic GMP in rabbit retina *in vivo:* Selective effects of dark and light adaptation and ischemia. *Proc. Natl. Acad. Sci. USA* 73:4442–45

Penn, R. D., Hagins, W. A. 1972. Kinetics of the photocurrent of retinal rods. *Biophys. J.* 12:1073–94

Polans, A. S., Hermolin, J., Bownds, M. D. 1979. Light-induced dephosphorylation of two proteins in frog rod outer segments. *J. Gen. Physiol.* 74:595–613

Poo, M.-M., Cone, R. A. 1974. Lateral diffusion of rhodopsin in the photoreceptor membrane. *Nature* 247:438–41

Puckett, K. L., Aronson, E. T., Goldin, S. M. 1984. ATP-dependent calcium uptake into native disc membranes. *Biophys. J.* 45:338a

Robinson, P. R., Kawamura, S., Abramson, B., Bownds, M. D. 1980. Control of the cyclic GMP phosphodiesterase of frog photoreceptor membranes. *J. Gen. Physiol.* 76:631–45

Robinson, W. E., Hagins, W. A. 1979. GTP hydrolysis in intact retinal rod outer segments and the transmitter cycle in visual excitation. *Nature* 280:398–400

Roos, A., Boron, W. F. 1981. Intracellular pH. *Physiol. Rev.* 61:296–434

Schröder, W. H., Fain, G. L. 1984. Light-dependent calcium release from photoreceptors measured by laser micro mass analysis. *Nature*. 309:268–70

Schwartz, E. A. 1973. Responses of single rods in the retina of the turtle. *J. Physiol.* 232:503–14

Schwartz, E. A. 1975. Rod-rod interaction in the retina of the turtle. *J. Physiol.* 246:617–38

Schwartz, E. A. 1976. Electrical properties of the rod syncytium in the retina of the turtle. *J. Physiol.* 257:379–406

Schwartz, E. A. 1977. Voltage noise observed in rods of the turtle retina. *J. Physiol.* 272:217–46

Shichi, H., Somers, R. L. 1978. Light-dependent phosphorylation of rhodopsin. *J. Biol. Chem.* 253:7040–46

Shimoda, Y., Miller, W. H., Lewis, R. M., Nairu, A. C., Greengard, P. 1983. Injection of cyclic GMP-dependent kinase into rod outer segments (ROS) causes depolarization that is inhibited by illumination. *Soc. Neurosci. Abstr.* 9:165

Shevchenko, T. F. 1976. Change in the activity of calcium ions on illumination of a suspension of the fragments of the external segments of visual cells. *Biophysics* 21:327–30

Shuster, T. A., Farber, D. B. 1984. Phosphorylation in sealed rod outer segments: Effects of cyclic nucleotides. *Biochemistry* 23:515–21

Sillman, A. J., Ito, H., Tomita, T. 1969. Studies on the mass receptor potential of the isolated frog retina. II. On the basis of the ionic mechanism. *Vision Res.* 9:1443–51

Sitaramayya, A., Liebman, P. A. 1983. Mechanism of ATP quench of phosphodiesterase activation in rod disc membranes. *J. Biol. Chem.* 258:1205–9

Sjostrand, F. S. 1959. The ultrastructure of the retinal receptors of the vertebrate eye. *Ergeb. Biol.* 21:128–60

Smith, H. G., Bauer, P. J. 1979. Light-induced permeability changes in sonicated bovine disks: Arsenazo III and flow system measurements. *Biochemistry* 18:5067–73

Steinberg, R. H., Wood, I. 1975. Clefts and microtubules of photoreceptor outer segments in the retina of the domestic cat. *J. Ultrastruct. Res.* 51:397–403

Szuts, E. Z. 1980. Calcium flux across disk membranes: Studies with intact rod photoreceptors and purified disks. *J. Gen. Physiol.* 76:253–86

Szuts, E. Z., Cone, R A. 1977. Calcium content of frog rod outer segments and discs. *Biochem. Biophys. Acta* 468:194–208

Troyer, E. W., Hall, I. A., Ferrendelli, J. A.

1978. Guanylate cyclases in CNS: Enzymatic characteristics of soluble and particulate enzymes from mouse cerebellum and retina. *J. Neurochem.* 31:825–33
Tsien, R. W. 1983. Calcium channels in excitable cell membranes. *Ann. Rev. Physiol.* 45:341–58
Waloga, G. 1983. Effects of calcium and guanosine 3′-5′ cyclic monophosphoric acid on receptor potential of toad rods. *J. Physiol.* 341:341–57
Wilden, U., Kühn, H. 1982. Light-dependent phosphorylation of rhodopsin: Number of phosphorylation sites. *Biochemistry* 21:3014–22
Woodhull, A. M. 1973. Ionic blockage of sodium channels in nerve. *J. Gen. Physiol.* 61:687–708
Woodruff, M. L., Bownds, M. D. 1979. Amplitude, kinetics, and reversibility of a light-induced decrease in guanosine 3′-5′-cyclic monophosphate in frog photoreceptor membranes. *J. Gen. Physiol.* 73:629–53
Woodruff, M. L., Fain, G. L. 1982. Ca^{2+}-dependent changes in cyclic GMP levels are not correlated with opening and closing of the light-dependent permeability of toad photoreceptors *J. Gen. Physiol.* 80:537–55
Yau, K.-W., McNaughton, P. A., Hodgkin, A. L. 1981. Effect of ions on the light-sensitive current in retinal rods. *Nature* 292:502–5
Yee, R., Liebman, P. A. 1978. Light-activated phosphodiesterase of the rod outer segment: Kinetics and parameters of activation and deactivation. *J. Biol. Chem.* 253:8902–9
Yoshikami, S., Foster, M. C., Hagins, W. A. 1983. Ca^{++} regulation, dark current control, and Na^{+} gradient across the plasma membrane of retinal rods. *Biophys. J.* 41:342a
Yoshikami, S., George, J. S., Hagins, W. A. 1980. Light-induced calcium fluxes from outer segment layer of vertebrate retinas. *Nature* 286:395–98
Yoshikami, S., Hagins, W. A. 1973. Control of the dark current in vertebrate rods and cones. In *Biochemistry and Physiology of Visual Pigments,* ed. H. Langer. Berlin: Springer-Verlag
Yoshikami, S., Hagins, W. A. 1984. Phototransduction in rods does not require a change in cytoplasmic pH. *Biophys. J.* 45:339a
Young, R. W. 1971a. The renewal of rod and cone outer segments in the rhesus monkey. *J. Cell Biol.* 49:303–18
Young, R. W. 1971b. Shedding of discs from rod outer segments in the rhesus monkey. *J. Ultrastruct. Res.* 34:190–203

Ann. Rev. Neurosci. 1985. 8:369–406

MOBILITY AND LOCALIZATION OF PROTEINS IN EXCITABLE MEMBRANES

Mu-ming Poo

Department of Physiology and Biophysics, University of California, Irvine, California 92717

INTRODUCTION

Studies of the cell membrane in the past two decades have led to the realization of the dynamic nature of membrane organization. Under favorable conditions, membrane components are capable of rotational and translational motions as well as extensive lateral interactions within the membrane matrix. This dynamic property of the cell membrane bears important consequences regarding the cellular mechanisms responsible for development and maintenance of membrane organization. For example, it is well known that mature nerve and muscle cells usually have a specific pattern of protein distribution in their plasma membranes: Transmitter receptors and ionic channels are frequently segregated or localized in specific regions of the membrane. If the proteins are laterally mobile in the membrane, how does a nonrandom protein topography develop? How is it maintained in the face of metabolic turnover of the proteins? Is protein topography modifiable in the mature membrane? The answers to these questions are directly relevant for understanding the function and the plasticity of excitable membranes.

The main subjects to be covered in the present review are (*a*) The lateral mobility of membrane proteins in the developing and mature excitable cells; (*b*) the nonrandom distribution of proteins in excitable membranes and possible mechanisms responsible for the development and maintenance of such distribution; (*c*) the implications of protein mobility in the functions of developing and mature excitable cells. For other aspects of the structure and organization of excitable membranes, a number of recent reviews may be consulted (Strichartz 1977, Patrick et al 1978, Pfenninger 1978, Rogart 1981).

0147–006X/85/0301–0369$02.00

PROTEIN MOBILITY IN NERVE AND MUSCLE MEMBRANES

Protein mobility in cell membranes has been a subject of a number of recent reviews (Cherry 1979, Shinitzky & Henkart 1979, Elson & Schlessinger 1979, Jacobson 1980, Peters 1981, Edidin 1981, Axelrod 1983, McCloskey & Poo 1984), which should be consulted for the details of various experimental methods and results. The following provides only a brief summary of the findings that may be relevant to nerve and muscle cells.

Free and Restricted Protein Diffusion

The notion of membrane fluidity originated in the studies of physical properties of pure lipid bilayer systems conducted in the 1960s. In one of the earliest attempts to extend this notion to biological membranes, Hubbell & McConnell (1968) showed that certain excitable membrane systems—nonmyelinated nerves and muscle fibers—contain liquid-like hydrophobic regions of low viscosity, as suggested by the motions of a spin label incorporated into the membranes. The possibility that a protein embedded in such a fluid lipid matrix could actually undergo long-range lateral diffusion was soon demonstrated by Frye & Edidin (1970), who observed a rapid intermixing of cell surface histocompatibility antigens after fusion of mouse and human cells. Measurements of the rate of protein diffusion were later carried out for surface antigens on cultured muscle fibers (Edidin & Fambrough 1973) and for rhodopsin in rod photoreceptor disk membranes (Poo & Cone 1974, Liebman & Entine 1974). The lateral diffusion coefficients (D) obtained from these measurements were in the range of 1 to 5×10^{-9} cm^2/s, corresponding to a rate of translation of about 5 to 10 μm within 1 min. This rate of lateral diffusion is a few hundredfold smaller than that for a typical globular protein in aqueous solution, suggesting that membrane proteins encounter an environment with a viscosity of about a few poise, or that of olive oil. Hydrodynamic analysis of the rotational and translational diffusion of an integral membrane protein in a two-dimensional homogeneous lipid matrix (Saffman & Delbrück 1975) suggests that the rates of rotational and translational diffusion of rhodopsin in photoreceptor disk membrane (Cone 1972, Poo & Cone 1974) are consistent with the notion of free diffusion of a protein in a two-dimensional fluid with the viscosity determined by the lipid matrix and with negligible viscous constraint from the aqueous phase of the membrane.

This simple picture of free protein diffusion was soon clouded by results from diffusion measurements on other cell types, using the method of fluorescence photobleaching recovery (FPR or FRAP, Peters et al 1974, Axelrod et al 1976, Edidin et al 1976, Jacobson et al 1976). In the most popular version of this method, an isolated cell is stained with a fluorescently labeled ligand that binds specifically to the plasma membrane protein under study. The fluorescence on the cell surface is then irreversibly bleached in a local spot on the surface by brief exposure to a high intensity laser beam. The rate of recovery of fluorescence at the bleached spot is used to calculate the diffusion coefficient, assuming that the recovery is due to lateral diffusion of proteins from the nearby unbleached regions of the membrane. When measurements were carried out for a number of cell surface proteins in a variety of cell types, the D values obtained (10^{-9} to 10^{-11} cm^2/s) were in general one to two orders of magnitude smaller than that expected for the free diffusion of proteins, e.g. for rhodopsin in photoreceptor disk membranes. Moreover, the recovery of fluorescence after photobleaching is rarely complete, suggesting that a substantial fraction of the labeled proteins are immobile over the period of measurement (up to tens of minutes). Even though uncertainty still remains in the interpretation of fluorescence photobleaching results, e.g. the possible impeding effect of fluorescent ligands used to label the proteins and the meaning of incomplete recovery after photobleaching (see discussion in McCloskey & Poo 1984), it is now widely accepted that plasma membrane proteins in most cell types are restrained in their lateral diffusion, and mechanisms responsible for their lack of mobility are generally assumed to involve structures external to the lipid matrix. (See section on stabilization mechanisms.)

Protein Mobility in Embryonic and Cultured Cells

MEASUREMENTS OF LATERAL DIFFUSION In isolated embryonic and cultured cells, there is now no doubt that proteins may exhibit long-range movements in the plane of the plasma membrane. For example, recovery of acetylcholine (ACh) sensitivity following local inactivation of ACh receptors by a pulse of α-bungarotoxin showed that ACh receptors in *Xenopus* myoblast membrane undergo lateral diffusion over the entire muscle cell surface, with an average diffusion coefficient of about 3×10^{-9} cm^2/s (Poo 1982), close to that for the free diffusion of an integral membrane protein. From the measurements of the rate of recovery of uniform receptor topography after the receptors were first redistributed asymmetrically by an external electric field, *D* values within the range of 1 to 5×10^{-9} cm^2/s were obtained for cell surface concanavalin A and soybean agglutinin receptors (Poo et al 1979, Poo 1981, Chow & Poo 1982) on the myoblasts and embryonic neurons of *Xenopus laevis* (Patel & Poo 1982, Patel 1984). Fluorescence photobleaching measurements were also performed on a number of excitable cells in culture: In cultured rat myotube, Axelrod et al (1976) found that ACh receptors labeled with fluorescent α-bungarotoxin diffuse with *D* of 5×10^{-11} cm^2/s in the region of membrane, thus showing diffuse fluorescence staining, but those in the ACh receptor clusters were essentially immobile. Using a fluorescent conjugate of nerve growth factor (NGF), Levi et al (1980) have measured the rate of lateral diffusion of NGF-receptor complexes in the plasma membranes of chick sensory neurons and of pheochromocytoma (PC-12) cells, and obtained a D value of 8×10^{-10} cm^2/s. The lateral mobility of proteins may decrease with the maturation and the age of the cell: There is a reduction of both the electrokinetic mobility (Orida & Poo 1980) and diffusional mobility (Stya & Axelrod 1983b) of the ACh receptor in embryonic muscle membranes with the age of the culture.

LIGAND-INDUCED REDISTRIBUTION The finding that binding of multivalent ligands, e.g. lectins and antibodies, can induce a "patching" or "capping" of membrane receptors has long been used as evidence of lateral mobility of membrane proteins (see review by de Petris 1976), although no quantitative information on the rate of protein movement was deduced from these observations. Examples of ligand-induced patching of receptors or proteins in excitable membranes have also been reported. Levi et al (1980) have shown that a fluorescent conjugate of NGF binds initially to diffusely distributed mobile receptors, but at 37°C the NGF-receptor complexes become clustered into immobile receptor patches. Diffusely distributed ACh receptors on the myotube surface redistribute into clusters or patches after addition of certain neurally derived factors (Christian et al 1978) or multivalent ligands that bind to ACh receptors (Heinemann et al 1977, Axelrod 1980). (Na^+-K^+)-ATPase also undergoes extensive lateral redistribution in the chick myotube membrane from a smooth dispersed distribution into numerous dots and patches of protein clusters when the myotube is first labeled with fluorescent monoclonal antibody to the ATPase and then incubated with anti–mouse-IgG (Pumplin & Fambrough 1983).

OTHER MIGRATORY MOVEMENTS Many other qualitative observations indicate lateral mobility of membrane proteins in embryonic and cultured cells, although the movements observed are not necessarily diffusional in nature. For example, contact of a neurite with the *Xenopus* myoblast induces a redistribution of myoblast surface ACh receptors toward the site of contact (Anderson et al 1977). ACh receptor clusters in cultured rat myotube membrane become dispersed after treatment with metabolic inhibitors, and clusters can be reformed from the diffuse receptors after the drugs are removed (Bloch 1979). Diffusely distributed ACh receptors in the myoblast membrane undergo exchange with those in the ACh receptor clusters (Stya & Axelrod 1983a). The motion of small particles attached to the surface of the growth cone (Bray 1970) or the neurite of cultured neurons (Koda & Partlow 1976) suggests the existence of gross membrane flow in at least some local regions of the developing neuron. Similar membrane flow has been observed in fibroblasts (Abercrombie et al 1970, Dembo & Harris 1981) and is attributed to a polarized membrane (or lipid) turnover process, in which the incorporation of the new and the uptake of old membrane occur at different sites of the plasma membrane (Bretscher 1976). Such membrane flow could induce directional migration of proteins in the membrane and is likely to influence the steady-state protein topography.

DIFFUSION WITHIN INTACT TISSUE Due to limitations of the current experimental techniques, mobility of membrane proteins has been studied almost exclusively in isolated cell preparations, frequently in cultures. In fact, the notion of laterally mobile proteins in a fluid-mosaic structure of the cell membrane (Singer & Nicolson 1972) originated entirely from studies on isolated cells. Even now we know very little about the mobility of membrane proteins within intact tissue, where the presence of extensive extracellular matrix and cell– cell contacts could conceivably produce effects not detectable in isolated cells. Recently Young & Poo (1983a) used a pulse of α-bungarotoxin to inactivate the extrajunctional ACh receptors on the exposed surface of the developing *Xenopus* tadpole muscle. They observed rapid recovery of ACh sensitivity in the inactivated region, which suggested rapid lateral diffusion of functional ACh receptors from the unexposed muscle surface buried within the muscle tissue. The *D* value obtained (1 to 4 $\times$ 10^{-9} cm^2/s) for the ACh receptor was very close to that found in isolated muscle cells from the same tissue (Poo 1982), suggesting that membrane proteins within intact developing tissue can equally undergo rapid lateral diffusion. It remains to be seen whether this can be generalized to other proteins or other tissues.

Protein Mobility in Mature Membrane

There are numerous examples of nonrandom distribution of membrane proteins in mature differentiated cells; this clearly suggests at least some proteins are not free to diffuse laterally. However, localization or nonrandom distribution is not necessarily equivalent to a true immobilization in which the proteins are individually "anchored." For example, proteins may be segregated into local domains by intra- or extra-membranous barriers (e.g. cytoplasmic filament networks, extracellular matrix, or tight-junctional structures) but still be fully mobile within and outside the domain (see Koppel et al 1981, Wolf & Ziomek 1983, Evans 1980). To elucidate the mechanism underlying the nonrandom distribution, direct measurement is thus required, and a distinction must also be made between the long-range and the short-range mobility. Unfortunately, the existing experimental techniques are incapable of measuring the rate of protein diffusion within a restricted domain with a size less than about 1 μm. There is indirect evidence for the short-range (<1 μm) lateral mobility of receptors in mature membranes: Binding of multivalent lectins induces local clustering of the lectin receptors in myelin fragments and the synaptosmal membrane of rat neocortex (Matus et al 1973) and in membrane fragments of the rabbit hippocampus and cerebellum (Kelly et al 1976) at extrasynaptic regions of the membranes.

Technical difficulty in obtaining isolated cells from mature tissue without the use of extensive enzyme treatment has long prevented relevant studies in this area. A unique contribution was recently made by Almers and co-workers on the lateral mobility of voltage-dependent Na^+ channels in adult frog muscle membrane in the absence of enzyme treatment. Stuhmer & Almers (1982) found that after local irreversible inactivation of sodium channels by focal UV irradiation, no recovery of the channel activity was detectable electrophysiologically for 1 hr at the irradiated area (diameter $\sim$ 10 μm). This suggests that the Na^+ channels are essentially immobile ($D < 10^{-12}$ cm^2/s) over a distance of 5–10 μm. A subsequent study (Almers et al 1983b) showed that application of an electric field along the surface of the muscle failed to cause lateral migration of the Na^+ channels, i.e. the channels are also electrokinetically immobile, unlike many membrane proteins in embryonic *Xenopus* muscle membrane (Poo 1981). Since no mobility measurement was made on other proteins in the adult frog muscle membrane, we do not know whether the lack of lateral mobility of the Na^+ channel is a selective restraint on a specific population of the membrane proteins or reflects a general nonspecific immobilization of all proteins in the membrane. Whether a similar situation also exists in mature neuronal membranes is also unknown.

Protein Mobility and Membrane Metabolism

Despite the lack of information on protein mobility in mature plasma membranes, we may consider some functional consequences of the immobility of proteins. A general immobility of membrane proteins would be useful for the maintenance of the pattern of protein topography established during cellular differentiation and embryonic development. But at least a limited lateral mobility in the plasma membrane appears indispensable for many processes involved in membrane metabolism. For example, the incorporation of newly synthesized membrane proteins, which presumably involves the fusion of post-Golgi vesicles with the plasma membrane (see review Dautry-Versat & Lodish 1983), may require lateral displacement and clearing of the proteins at the site of membrane fusion, as suggested by freeze-fracture studies of exocytosis in secretory cells (Orci et al 1977, Lawson et al 1977) and by physicochemical analysis of the fusion process (Gingell & Ginsberg 1978). Exchange of new and old proteins also requires at least a limited lateral mobility, especially at regions where membrane proteins are highly concentrated, e.g. the nodes of Ranvier and the neuromuscular endplates. Furthermore, regardless of protein mobility in the plasma membrane, proteins in the intracellular membranes of the endoplasmic reticulum and Golgi apparatus must be free to move. Sorting of various newly synthesized proteins that are destined for other cellular organelles and the plasma membrane, the budding and fusion of vesicles at various membrane compartments, and the assembly of multi-subunit proteins whose individual subunits are translated from separate mRNAs, as in the case of ACh receptor (see review Anderson 1983), all require substantial lateral mobility of the proteins in intracellular membranes.

Local membrane flow is also likely to occur at regions of the plasma membrane undergoing extensive vesicular secretion. At the presynaptic motor nerve terminal, ultrastructural evidence suggests a membrane-recycling in which the exocytosis of synaptic vesicles and the endocytosis (membrane re-uptake) occur in separate zones of the plasma membrane (see review Heuser & Reese 1977). This recycling will lead to a local membrane flow at the nerve terminal during synaptic activity. It is interesting to note that von Wedel et al (1981) found that antibodies directed specifically against the vesicle membrane components show increased binding to the intact terminal after stimulation of massive ACh release by lanthanum, but no increased binding could be detected after electrical stimulation of the terminal. Since the amount of vesicle membrane incorporated into the plasma membrane is quite substantial even under moderate synaptic activity, this result suggests that either there is an extremely rapid lateral migration of the vesicular membrane components from the site of exocytosis to that of the endocytosis, or the turnover of synaptic vesicles at physiological conditions does not involve complete fusion and mixing of vesicular membrane with the plasma membrane.

Experimental studies relating to the problem of protein topography during the process of protein turnover have only recently been reported. When developing frog end-plates were labeled with ^{125}I-labeled α-bungarotoxin one or two days after treatment with unlabeled toxin, annuli of radioactivity were found at the labeled end-plates (Weinberg et al 1981a), suggesting that the newly synthesized ACh receptors appear first at the perimeter of the endplate ACh clusters. The turnover of the entire endplate presumably involves the exchange of proteins between the perimeter and the center of the endplate. Stya & Axelrod (1983a) found that diffuse ACh receptors can exchange with those of the ACh receptor clusters in cultured myotube, and the exchange also appears to proceed from the perimeter of the cluster inward. These results suggest that the turnover of highly clustered membrane proteins, e.g. those at muscle endplate and nodes of Ranvier, is likely to involve a lateral exchange between newly inserted, diffuse proteins with those in the clusters. It is not clear, however, whether the newly synthesized proteins are inserted at regions in close proximity to the preexisting receptor clusters or randomly over the entire plasma membrane. So far, no evidence exists that the newly synthesized proteins are inserted preferentially at or near specific sites of the plasma membrane. This would be required if the protein in the mature membrane is not capable of long-range lateral motion.

Summary

Based upon the available evidence, one may reach the following tentative conclusion: Proteins are highly mobile and capable of long-range movements in the plasma membrane of embryonic cells, with diffusion coefficients approaching that for unrestricted protein diffusion (up to 5×10^{-9} cm^2/s). However, as differentiation and maturation of the cell proceeds, many proteins become localized or immobilized at specific regions of the membrane through constraints imposed by intra- and extramembranous structures. Membrane proteins could be localized or segregated into local domains without true immobilization. Consideration of the problems of protein biosynthesis and turnover suggests that short-range lateral mobility is obligatory even in the plasma membrane of mature differentiated cells, where membrane proteins are known to exhibit nonrandom topography and are incapable of long-range lateral motions.

NONRANDOM DISTRIBUTION OF MEMBRANE PROTEINS

Nonrandom distribution of both integral and peripheral plasma membrane components is found in nearly all nerve and muscle cells that have been examined. The patterned topography in the plasma membrane is in fact a reflection of the differentiated state of the cell. Selected examples of patterned topography will be presented in the following. Many of the proteins to be discussed have not been purified, their existence usually can only be inferred from electrophysiological, pharmacological, or cytochemical evidence. Examples not discussed in the text can be found in the references listed in Table 1.

Nonrandom Distribution of Proteins in Developing Neurons

The neuritic (dendritic or axonal) processes of developing neurons in culture are electrically excitable (Hild & Tasaki 1962, Okun 1972, Dichter & Fischbach 1977, Willard 1980). Detailed mapping of the distribution of ion channels has not been performed, but from the reported data it appears that the distribution of voltage-dependent channels is relatively uniform along the neurite. This is supported by the finding of the relatively uniform distribution of scorpion toxin binding sites, presumably voltage-dependent Na^+ channels, along the neuritic processes of cultured spinal cord neurons and pheochromocytoma (PC 12) cells (Catterall 1981). Non-uniformity appears at two ends of the neurite: Catterall (1981) observed a higher number of scorpion toxin binding

Table 1 Nonrandom distribution of ionic channels and neurotransmitter receptors

Cell type	Method of mapping	Inferred distribution	Refs.
Voltage-dependent Na^+ channel			
Myelinated nerve	Radioactive saxitoxin binding, immunocytochemical staining	Highly concentrated at the nodes of Ranvier, absent at internodes	Ritchie & Rogart 1977 Ellisman & Levinson 1982
Motor nerve terminal	Drug effects on extracellular action currents	Accumulation at preterminal regions near the myelin, absent at terminal branches	Brigant & Mallart 1982
Demyelinated nerve	Extracellular recording of inward action current	"Node-like" high density channel aggregates	Smith et al 1982
Cultured neurons	Radioactive scorpion toxin labeling and autoradiography	Uniform density along the neurite, higher density at initial segment	Catterall 1981
Frog skeletal muscle	Patch voltage-clamp recording of inward current	Channel densities vary up to 3-fold over distances 10–30 μm	Almers et al 1983a
Denervated rat diaphragm muscle	Extracellular recording of inward action current	Discrete high density patches about 100 μm in size	Purves & Sackman 1974
Denervated frog skeletal muscle	Extracellular recording of inward action current	Channel densities vary up to 10-fold over distances ~500 μm	Lehouelleur & Schmidt 1980
Electrocytes of eel electric organ	Immunocytochemical staining	Preferential localization on the innervated face of the electrocyte	Ellisman & Levinson 1982 Fritz & Brockes 1983
Voltage-dependent K^+ channel			
Myelinated nerve	Voltage-clamp recording of outward K^+ current, effects of K^+ channel blockers on action potential waveform	Absence at the nodes of Ranvier, presence in the paranodal and internodal membrane	Chiu et al 1979, Brismar, Kocsis & Waxman 1980, Bostock et al 1981, Chiu & Ritchie 1980, 1981, 1982

Table 1 Nonrandom distribution of ionic channels and neurotransmitter receptors (*Continued*)

Cell type	Method of mapping	Inferred distribution	Refs.
Voltage-dependent Ca^{2+} channel			
Presynaptic nerve terminal	Optical recording of Ca^{2+} influx using Ca^{2+}-sensitive dye	Localized within the last 50–100 μm of the terminal arborization	Stockbridge & Ross 1984
Cerebellar Purkinje neuron	Intradendritic recording of Ca^{2+} spikes	Localized along the dendrite at discrete spots	Llinás & Hess 1976
Frog motoneuron	Intracellular recording of Ca^{2+} spikes	Presence in the somatal membrane	Barrett & Barrett 1976
Cultured dorsal root ganglion neuron	Intracellular recording of Ca^{2+} spikes	Presence in the somatal membrane, absent along the neurite	Dichter & Fischbach 1977
Cultured neuroblastoma	Optical recording of Ca^{2+} spike using voltage-sensitive dye	Presence at or near growth cone	Grinvald & Farber 1981
Aplysia ganglion neuron	Voltage-clamp patch recording of inward Ca^{2+} current	Higher density at somatic membrane away from the axon	Kado 1973
Acetylcholine (ACh) receptor			
Vertebrate skeletal muscle	Iontophoresis of ACh, labeling with fluorescent or radioactive α-bungarotoxin	Highly concentrated at the endplates	del Castillo & Katz 1955 Fertuck & Salpeter 1974 Anderson & Cohen 1974
Frog parasympathetic ganglion neuron	Iontophoresis of ACh	Highly concentrated at the synaptic areas	Harris et al 1971
Denervated skeletal muscle	Fluorescent α-bungarotoxin labeling	Discrete high-density clusters at extrajunctional area	Ko et al 1977
Embryonic skeletal muscle	Iontophoresis of ACh, fluorescent or radioactive α-bungarotoxin labeling	Discrete high-density clusters at extrajunctional area and on the surface of noninnervated muscle in culture	Fischbach & Cohen 1973 Sytkowski et al 1973 Anderson et al 1977 Chow & Cohen 1983

Midbrain & hypothalamic neurons	Horseradish peroxidase-α-bungarotoxin labeling	Concentrated at postsynaptic membrane of a small group of synapses	Lentz & Chester 1977
Glutamate receptor			
Crayfish muscle	Iontophoresis of glutamate	Highly concentrated in postjunctional areas of excitatory nerve terminals	Takeuchi & Takeuchi 1964
Drosophila larval muscle	Iontophoresis of glutamate	Highly concentrated in postjunctional areas	Jan & Jan 1976
Locust muscle	Iontophoresis of glutamate and analogue	Concentration of D-type receptors at junctional area, extra-junctional distribution of DH-type receptors	Cull-Candy 1976
Rabbit hippocampal pyramidal neuron	Iontophoresis of glutamate	Discrete localization at basal & apical dendrites	Schwartzkroin & Andersen 1975
Guinea-pig cerebellar Purkinje neuron	Iontophoresis of glutamate	Higher density at dendrites than at soma	Chujo et al 1975
Cat spinal cord motoneuron	Iontophoresis of glutamate	Higher density at dendrites than at soma	Zieglgansberger & Champagnant 1979
Cultured mouse spinal cord neuron	Iontophoresis of glutamate	Higher density at discrete spots on the soma and neurites	Ransom et al 1977
Gamma-aminobutyric acid (GABA) receptor			
Crayfish muscle	Iontophoresis of GABA	Highly concentrated in postjunctional area of inhibitory nerve terminals	Takeuchi & Takeuchi 1965
Goldfish Mauthner neuron	Iontophoresis of GABA	Higher density near postsynaptic area receiving large club endings	Diamond 1965

sites at the initial segment of the neurite. Optical recording showed the presence of Ca^{2+} action potentials at or near the growth cones of cultured neuroblastoma cell (Grinvald & Farber 1981), suggesting the localization of voltage-dependent Ca^{2+} channels there. The protein composition of the somatic membrane may be different from that of the neuritic membrane: There are voltage-dependent Ca^{2+} channels in the soma but not in the neurites of the chick dorsal root ganglion neuron (Dichter & Fischbach 1977). Willard (1980) found that in cultured *Xenopus* neural-plate neurons, the ions carrying most of the inward action current change from Ca^{2+} to Na^+ at an earlier time in the neurite than that at the soma (see also Spitzer 1979), thus suggesting that Na^+ channels first appear in the neurite. The preferential incorporation of new membrane at the growth cone of the neuron (Bray 1970) may lead to the first appearance of newly expressed membrane proteins in the neurite. On the other hand, Pfenninger & Bunge (1974) found that there is a decreasing gradient in the density of intramembranous particles, presumably a subpopulation of membrane proteins, from the cell body of cultured neurons to the growing tip of the neurite; this suggests that the newly-incorporated membrane near the growth cone is relatively free of particles.

Recent finding of the endogenous DC currents near the tip of growing neurites suggests interesting local segregation of proteins in the growth cone membrane: Using an oscillating microelectrode, Freeman et al (1981, 1982) found that the growth cones of goldfish retinal neurons generate a DC current that flows inward across the plasma membrane at the tip of the growth cone, and back outward at the base. The current loop can only occur when the ion pumps and leaks are segregated in the plasma membrane. It was shown that this current is carried primarily by Ca^{2+} ions, and it appears that there are leaky Ca^{2+} channels located at the filapodia and active Ca^{2+} pumps at the base of the growth cone. The function of this current remains a mystery.

Segregation of Voltage-dependent Na^+ and K^+ Channels

The distribution of voltage-dependent Na^+ and K^+ channels is relatively uniform along the squid axon, as indicated by the relatively small spatial variations in the local Na^+ and K^+ action currents in voltage-clamp experiments (Taylor et al 1960). In myelinated nerve fibers, Na^+ channels in the axon membrane are highly concentrated at the nodes of Ranvier and not detectable at the internodal regions (Ritchie & Rogart 1977). K^+ channels are scarce or completely absent at the nodes (Brismar 1979, Chiu et al 1979), but distributed at the paranodal and internodal regions of the axon membrane (Chiu & Ritchie 1980, 1981, 1982). At the nonmyelinated terminal, there is evidence again for a segregation of Na^+ and K^+ channels: In mouse motor axon, Na^+ channels are exclusively located at the preterminal part close to myelin, whereas K^+ channels are present only at the terminal part of the endings (Brigant & Mallart 1982).

Interesting observations have recently been made on the distribution of Na^+ channels after experimental demyelination. Axons demyelinated with diphtheria toxin showed regions of continuous conduction with evenly distributed inward currents (Bostock & Sears 1978), suggesting that Na^+ channel distribution becomes uniform over some demyelinated regions. Smith et al (1982) found that micro-injection of lysolecithin also induces local demyelination, but in this case the inward current within the demyelinated internode is not evenly distributed; rather it is concentrated into a series of current foci. These authors suggest that these foci of inward current represent sites of Na^+ channel aggregates that are precursors of the nodes of Ranvier formed during the remyelination. It is interesting that the position of these node-like foci bears no apparent relation to the distribution of the surrounding Schwann cells, indicating that perhaps the position of the node is determined prior to the myelination. There is evidence that both the redistribution of old channels and the incorporation of new channels are involved in the reorganization of Na^+ channel topography during demyelination and remyelination. Regenerating nerve fiber shows a marked increase in the number of nodes per unit fiber length (Cragg & Thomas 1964). Saxitoxin binding studies showed that the incorporation of new Na^+ channels occurs during remyelination (Ritchie et al 1981), and the Na^+ channel content in the regenerating axon increases so that the number of Na^+ channels per node remains roughly constant (Ritchie 1982). However, since there is no evidence for the appearance of new Na^+ channels during the early phase of demyelination (Ritchie et al 1981), when continuous conduction may develop (Bostock & Sears 1978), a spreading of the old nodal Na^+ channels may also occur. It is not clear how much and where the additional newly synthesized channels are incorporated into the membrane and what fraction of the original nodal Na^+ channels undergo lateral redistribution vs metabolic degradation. Presently, we have little information on the turnover rate of the Na^+ channels in either the clustered or diffuse state. Such information is indispensable in understanding the mechanism underlying the development and modulation of Na^+ channel topography.

Localization of Voltage-dependent Ca^{2+} channels

Studies of synaptic transmission have long implicated the existence of presynaptic voltage-dependent Ca^{2+} channels (Katz & Miledi 1969, Llinás et al 1976, Ross & Stuart 1978). Using Ca^{2+}-sensitive dye, Arsenazo III, Miledi & Parker (1981) found higher levels of Ca^{2+} signal near the distal ends of the presynaptic terminal of the squid giant synapse in response to nerve impulses. Higher spatial resolution for the location of Ca^{2+} channels was obtained recently by Stockbridge & Ross (1984), who showed that the depolarization-induced entry of Ca^{2+} at the terminal of the barnacle photoreceptor axon is confined within the last 50–100 μm of the terminal arborization. Regenerative Ca^{2+} spikes have been observed also at the growth cone region of cultured neuroblastoma cells (Grinvald & Farber 1981) and at the dendrites of cerebellar Purkinje neurons (Llinás & Nicolson 1971) and hippocampal pyramidal neurons (Wong et al 1979). These Ca^{2+} spikes appear to be initiated only at restricted "hot spots" along the dendrite, perhaps at dendritic branch points, due to discrete localization of Ca^{2+} channels (Llinás & Hess 1976).

Ca^{2+} channels found at the dendrite and at the terminals of both growing and mature nerve processes may be ontogenetically related: The appearance of Ca^{2+} conductance in postnatal development of Purkinje cells is associated invariably with the initiation of main dendrite and a prominent dendritic growth cone (Llinás & Sugimori 1979), and the Ca_{2+}-dependent secretion of transmitter and recycling of vesicular membrane at the mature nerve terminal appear to be analogous to the "isometric" growth of the growth cone, whereby the exocytotic and endocytotic activities become equal (see Llinás 1979). Spontaneous and electrically evoked secretion of acetylcholine from isolated growth cones of growing neurites have recently been detected (Young & Poo 1983c, Hume et al 1983), further supporting the idea that growth and secretion of nerve terminals are closely linked processes, perhaps both regulated by similar Ca^{2+} channels. Activity-dependent influx of Ca^{2+} at the dendrites or nerve terminals, which are the most likely controlling points for signal transmission in the nervous system, could be a basis of neuronal plasticity (see Llinás 1979, Kandel 1981).

Localization of Postsynaptic Transmitter Receptors

The localization (or clustering) of ACh receptors at the vertebrate neuromuscular junction is well characterized by a number of electrophysiological, cytochemical, and ultrastructural studies (see reviews Fambrough 1979, Barnard 1979). At the mature junctional region, the density of the muscle surface ACh receptors can be 10^3 to 10^4 times higher than that of the extrajunctional regions (Fambrough 1974, Fertuck & Salpeter 1976). In the embryonic muscle before innervation, ACh receptors are distributed at relatively high concentration over the entire surface of the muscle cell. As the synapse matures, the concentration of the ACh receptors at the endplates greatly increases, while that of extrajunctional regions decreases to a barely detectable level (see Dennis 1981). This developmental change of the ACh receptor's topography after innervation can be observed in cell culture systems (Anderson & Cohen 1977, Anderson et al 1977, Frank & Fischbach 1979). Results from *Xenopus* nerve-muscle preparation suggest that increased density of ACh receptors at the nerve contact site is due to a redistribution of preexisting muscle surface ACh receptors, and the effect is nerve-specific (Anderson et al 1977, Cohen & Weldon 1980). Clustering of ACh receptors can occur without innervation: ACh receptor clusters were observed at extra-junctional regions before and during the early phase of innervation (Chow & Cohen 1983) or after denervation (Ko et al

1977). Cultured muscle not contacted by the neurites also forms receptor clusters or "hot patches" (Sytkowski et al 1973, Fischbach & Cohen 1973, Anderson & Cohen 1977). These clusters are frequently located at sites coincident with the "adhesion plaques" where the cells attach to the culture substratum (Bloch & Geiger 1980, Moody-Corbett & Cohen 1982).

Innervation of a central neuron by heterogeneous nerve terminals does not occur at random. In the case of the cerebellar Purkinje cell and hippocampal pyramidal neuron, different classes of inputs innervate defined regions of the neuron (see Shepherd 1979). Localization of various transmitter receptors and the associated ionic channels in the immediate postsynaptic membrane of the appropriate nerve terminal would not only be advantageous for synaptic transmission, but would also be obligatory for keeping order in synaptic integration and local neuronal interactions. In the few cases that have been examined, there is evidence for at least a preferential distribution of the receptors to the regions of postsynaptic membrane receiving the appropriate inputs. For example, excitatory afferent inputs on hippocampal pyramidal neurons are located exclusively on the dendritic spines (Andersen et al 1966), and the dendritic region indeed shows much higher sensitivity to the excitatory transmitter L-glutamate (Schwartzkroin & Andersen 1975). There is no doubt that regional differentiation of the neuron must occur for mature neuronal function to take place. The question is whether it occurs before or after the regional innervation; or alternatively, whether the regional differentiation, which is likely to involve the localization of specific membrane proteins, and the selective innervation are in fact mutually reinforcing processes in the development of the nerve connections. (See later section on Selective Cell–cell Adhesion.)

Segregation of Ionic Pumps and Channels

In renal tubule cells and duct cells of the duck supraorbital gland, (Na^+-K^+)-ATPase is known to be concentrated in the basolateral membrane, while relatively few ATPases are found in the apical membrane (Kyte 1976, Ernst & Mills 1977, Mazurkiewicz et al 1978). Recently, an immunocytochemical study using a monoclonal antibody that selectively binds to (Na^+-K^+)-ATPase has shown relatively uniform distribution of this enzyme over the entire muscle and nerve membranes, including the internodal axonal membrane of the myelinated nerve (Fambrough & Bayne 1983, Pumplin & Fambrough 1983). Using tritiated ouabain and electron-microscopic autoradiographic techniques, Bok has shown that (Na^+-K^+)-ATPase molecules in the plasma membrane of the vertebrate photoreceptor cells are localized only at the distal region of the inner segment (see Fain & Lisman 1982).

(Na^+-K^+)-ATPase catalyzes a transmembrane exchange of three Na^+ ions for two K^+ ions and is thus electrogenic in nature. If passive ion flux (through open channels or "leaks") occurs at membrane sites different from that for the flux pumped by the (Na^+-K^+)-ATPase, a transcellular current will be generated. This is indeed the case for the vertebrate photoreceptor. The plasma membrane of the photoreceptor outer segment is highly permeable to Na^+ in the dark. Sodium ions flow into the outer segment, diffuse to the inner segment, and are then pumped out by the (Na^+-K^+)-ATPase located at the plasma membrane of the inner segment. This results in a steady transcellular current, namely the "dark current," that can be detected extracellularly in a dark-adapted retina (Hagins et al 1970, Baylor et al 1979). Steady currents have also been detected extracellularly near the mammalian neuromuscular junction (Betz et al 1980, Betz & Caldwell 1984), the growth cone of cultured goldfish retinal neurons (Freeman et al 1981), and a variety of developing cells (see review Jaffe 1979). Given the ubiquitous non-uniformity of channel distribution found in all types of excitable cells and that a codistribution of all pumps and channels seems unlikely, it is not surprising that the existence of steady transcellular current is a wide-spread phenomenon. Such current may play an active role in the development or the function of the cell (see Jaffe 1979). Hagins et al (1970) suggested that the dark current, besides serving as a DC signal carrier to deliver the light-induced electrical response to the synaptic sites located at the base of the photoreceptor, might also act to transport materials in the cytoplasm by electrophoresis.

Colocalization of Membrane Components

Localized distribution has been observed for many peripheral membrane components on both the cytoplasmic and the extracellular faces of the excitable membrane, a situation best exemplified by the pre- and postsynaptic densities observed ultrastructurally at various types of synapses (see review Pfenninger 1978). The identity and function of these molecules are now only beginning to be explored. Among them, acetylcholinesterase (AChE) is the best characterized. Two forms of the enzyme have been identified, and the tailed form that is thought to contain collagen-like sequences has been found, in some species, only in endplate regions of both adult and developing muscles where ACh receptors are concentrated (Hall 1973, Vigny et al 1976, Weinberg et al 1981a,b). Besides AChE, immunospecific basal lamina components or cytoplasmic proteins (e.g. actin, vinculin, α-actinin, and filamin) have also been found to colocalize with high-density ACh receptor clusters at muscle endplates or in isolated cultured myotubes (Sanes & Hall 1979, Froehner et al 1981, Hall et al 1981, Burden 1982, Anderson & Fambrough 1983, Bloch & Hall 1983; see reviews Kelly & Hall 1982, Fambrough et al 1982). These components may be involved in the development or the maintenance of ACh receptor cluster, or in the regulation of ACh receptor-channel function.

The intracellular pathway for the synthesis and transport of integral membrane proteins is thought to be in common with that for the secretory proteins destined to the extracellular space (Palade 1975). In some cases, including that for AChE and ACh receptor in chick myotubes (Devreotes et al 1977, Rotundo & Fambrough 1980a,b), secretory proteins and plasma membrane proteins are found to be transported together, probably via the same post-Golgi vesicle (see review Dautry-Versat & Lodish 1983). If the site of vesicle incorporation into the plasma membrane is localized and lateral diffusion on the cell surface is restricted, colocalization of various extracellular macromolecules (proteins or proteoglycans) and integral membrane proteins could be a direct result of cotransport from cytoplasm to plasma membrane, in the absence of any specific affinity between the colocalized components. So far, little is known about the route of vesicle incorporation, the affinity between the colocalized components, and the possible lateral mobility of the peripheral membrane components. Such information is required for understanding the mechanism of colocalization.

LOCALIZING MOVEMENTS AND STABILIZATION MECHANISM

In considering the cellular mechanism responsible for the localization of proteins in cell membranes, it is useful to distinguish two different aspects of the problem: (*a*) the localizing movement whereby the protein translocates from the site of synthesis within the cytoplasm to a specific region of the plasma membrane, and (*b*) the mechanism by which the protein is kept localized after it has arrived there. Mechanisms underlying these two processes may be different. At present, we do not know (for any protein) whether the newly synthesized protein is incorporated into the plasma membrane at random or at selected sites; the localizing movement could thus occur in the plane of the plasma membrane, or within the cytoplasm, or both. Mechanisms of localization within the cytoplasm are beyond the scope of the present review. The following discussion pertains only to the possible mechanisms that induce localizing movements of proteins in the membrane plane and that stabilize the protein topography. Localization of the ACh receptor at the neuromuscular junction will frequently be used as an example, since this is the only case for which substantial evidence is available.

Diffusion-mediated Trapping

The simplest cellular process that could account for both the localizing movement and the stabilization of localized proteins is a diffusion-mediated trapping: Freely diffusing proteins are localized or "trapped" in a region of the membrane where they bind to other localized molecules. The trap could be formed by molecules on the cytoplasmic face of the plasma membrane, in the extracellular matrix, or on the surface of a neighboring cell that is in close contact with the membrane. It could also be a small nucleus of protein clusters within the membrane itself, trapping diffusing proteins by a process analogous to the crystallization of proteins in aqueous solution. Edwards & Frisch (1976) first used the idea of diffusion-mediated trapping to account for the turnover of ACh receptors at the mature neuromuscular endplate. These authors suggested that the turnover can be accomplished by random insertion of newly synthesized ACh receptors in the extrajunctional membrane of the muscle, followed by diffusion to the junction. For the trapping to be efficient, the receptors must be relatively mobile; Edwards & Frisch (1976) used a lateral diffusion coefficient of 2×10^{-9} cm^2/s in their calculation. Recently, diffusion-mediated trapping was also shown to be a plausible mechanism for the nerve-induced clustering of ACh receptors during synaptogenesis (Poo 1982, Young & Poo 1983a): The lateral diffusion rate of functional ACh receptors ($D = 1-4 \times 10^{-9}$ cm^2/s) in the embryonic *Xenopus* muscle membrane is rapid enough to account for the time course of nerve-induced ACh receptor clustering both in culture (Anderson et al 1977) and in vivo (Chow & Cohen 1983). Accumulation of surface glycoproteins induced by muscle–muscle contact in culture (Chow & Poo 1982) was also consistent with the diffusion-mediated trapping of surface components by specific binding with ligands on the contacting surfaces. Chao et al (1981) have presented a quantitative analysis of the molecular trapping by a perfect sink on the surface of a spherical cell, and Weaver (1983) recently extended the analysis to include imperfect traps where the molecule has a non-zero chance to escape from the trap.

Interaction with Extracellular Matrix

Molecules bound in the cell surface coat or extracellular matrix are the best candidates to serve as traps for plasma membrane components. Burden et al (1979) found that ACh receptors on the regenerating muscle cell are induced to cluster at the original site of endplate by remnants of the old synaptic basal lamina in the absence of nerve (see also McMahan et al 1980). Nitkin et al (1983) have purified a protein from the extracellular matrix of the electric organ of *Torpedo,* which induces clustering of preexisting diffuse ACh receptors on the surface of cultured myotube. As mentioned in a previous section, many membrane-associated extracellular components show colocalization with ACh receptor clusters at the endplate of adult muscle fibers or in cultured muscle cells. It would be of great interest to determine whether any causal relationship exists between the localization of these peripheral membrane components and that of the ACh receptor. Weinberg et al (1981b) have shown that during ectopic endplate formation in adult muscle, the appearance of synapse-specific basal lamina components and AChE at the new synapse occurred distinctly later than the clustering of ACh receptors, suggesting that these components are probably not responsible for the localization of ACh receptors.

Anderson & Fambrough (1983) found that the heparan sulphate proteoglycan on the muscle surface exhibits a complex surface organization that is closely correlated with the dense accumulation of ACh receptors at the endplate, but its highly concentrated patches extend beyond that of the ACh receptor clusters. The localization of ACh receptors clearly cannot be accounted for simply by the interaction of ACh receptors with these proteoglycans.

In the case of the ACh receptor in skeletal muscle membrane, the signal for the time as well as the site of its localization comes from the innervating nerve of appropriate type (Cohen & Weldon 1980). One is obliged, at least for synaptogenesis during early development, to look for the signal at the nerve terminal. While it remains possible that some diffusible factor released by the nerve terminal could induce the clustering of the receptor, the simplest mechanism for nerve to induce a discrete cluster of ACh receptors would be that the nerve terminal contains membrane-bound factors at its extracellular face that serve as traps for the ACh receptors in the muscle membrane. A large portion of the ACh receptor, including the carbohydrate moiety and the main immunogenic site, is exposed to the extracellular space (Lindstrom 1980), providing favorable conditions for specific interaction with extracellular molecules.

Involvement of Cytoskeleton

Since the early studies on the phenomenon of receptor capping induced by multivalent ligands (Taylor et al 1971), the idea that cytoplasmic filaments are responsible for moving or anchoring plasma membrane proteins has gained increasing popularity. Extensive reviews have appeared on this and related subjects (Edelman 1976, Nicolson 1976, de Petris 1977, Schlessinger 1983). There is yet no conclusive evidence that the filaments provide direct driving force for translocating membrane proteins. It appears more likely that cytoplasmic filaments provide a specific or nonspecific mechanism to restrain lateral motion of membrane proteins (see Koppel et al 1981). Tank et al (1982) showed a dramatic increase in the diffusion coefficients of ACh receptors (from less than 10^{-10} to 3×10^{-9} cm^2/s) in the membrane "blebs" produced by treatment of adult muscle fibers or myoblasts with formaldehyde. From cytochemical staining it appears that the plasma membrane in these blebs is lifted off the underlying cytoskeleton, consistent with the idea that the latter is involved in restricting protein mobility. As myotube ages in culture, the fraction of ACh receptors that is resistant to detergent extraction increases (Prives et al 1982), and there is a correlation between the lateral mobility of ACh receptors and their detergent extractability (Stya & Axelrod 1983b). An "anchoring" effect of cytoskeletal structure was suggested as the cause of receptor immobility in these studies.

Localized cytoplasmic proteins or filaments that attach to the plasma membrane and have specific affinity with the ACh receptors could serve as traps for the formation of ACh receptor clusters. Some cytoskeletal components—vinculin, α-actinin, and filamin—can be detected near the postsynaptic muscle membrane within 2 d after the onset of ACh receptor accumulation at the developing neuromuscular junction (Bloch & Hall 1983). Ultrastructural studies have shown the existence of a cytoplasmic meshwork of filaments that codistribute with the postsynaptic complex or ACh receptor clusters in muscle cells (Gulley & Reese 1981, Peng 1983). All these findings suggest possible involvement of cytoskeletal components in the development or maintenance of the ACh receptor clusters. For any extramembranous component to serve as a "trap" or "anchor" for the ACh receptor, specific affinity of the component with the ACh receptor is required. This is yet to be demonstrated for any of the components that colocalize with the ACh receptor, including those in the extracellular matrix.

Specific binding of an integral membrane protein to cytoskeletal components has been demonstrated in erythrocytes: A transmembrane anion channel, band 3 protein, was shown to bind specifically with a peripheral protein ankyrin, which serves as a link to a cytoskeleton network consisting of spectrin, actin, and band 4.1 protein (Marchesi 1979, Branton et al 1981). The ankyrin-linkage to spectrin appears to anchor the band 3 protein: The lateral diffusion rate of band 3 protein is relatively low in normal erythrocytes, but it increases 50-fold in spectrin-deficient erythrocytes (Koppel et al 1981). Spectrin-like proteins are also found in other cell types, including neurons (Bennett et al 1982), thus suggesting the possibility of similar membrane–cytoskeletal interaction. Such interaction may allow not only the development and maintenance of protein topography, but also the modulation of protein topography. Recent studies on the long-term potentiation in hippocampal neurons (Lynch & Baudry 1983, Siman et al 1984) have suggested that the activity-dependent Ca^{2+} influx in the postsynaptic cell may trigger a modulation of cytoskeletal–membrane interaction that leads to a modulation of either the number or the distribution of functional transmitter receptors in the postsynaptic membrane.

Lateral Interactions within the Lipid Matrix

Specific and nonspecific lipid–protein and protein–protein interactions within the membrane could affect significantly the distribution of membrane proteins. For example, it is known that mixtures of phospholipids can exhibit fluid–fluid immiscibility at physiological temperatures (for references, see Berclaz & McConnell 1981), and many membrane proteins show preference for interaction with specific lipids (see reviews by Gennis & Jonas 1977, Bennett et al 1980); it is thus possible that some proteins would partition preferentially in one lipid phase, resulting in a segregated distribution. The effect of lipids on the distribution of membrane proteins is vividly demonstrated by the effect of cholesterol on the distribution of membrane proteins. Cherry et al (1980) showed that when cholesterol is introduced into a reconstituted lipid bilayer membrane containing bacteriorhodopsin, the protein molecules are induced to segregate into cholesterol-poor domains of the membrane, both above and below the phase transition of the lipid. A direct demonstration of an analogous lipid effect in biological membranes is lacking. A polyene antibiotic filipin, which reacts specifically with membrane cholesterol to form filipin–sterol complexes visible in freeze-fracture replicas, has recently been used to probe the cholesterol distribution in a number of excitable membranes (Andrews & Cohen 1979, Montesano 1979, Nakajima & Bridgman 1981, Pumplin & Bloch 1983). For example, Nakajima & Bridgman (1981) found that the complex is absent from the active zone region of the presynaptic membrane of the frog neuromuscular junction. Such a finding, however, does not necessarily imply that the particle aggregation results from an exclusion of proteins from cholesterol-rich regions of the membrane. Since protein aggregates are densely packed, the absence of cholesterol could be due to its exclusion by the aggregates; alternatively, the aggregation may have prevented access of the filipin to the cholesterol.

Mobile proteins embedded in a fluid lipid bilayer are in many ways similar to colloidal particles in solution. Regardless of the chemical specificity of each protein, the interactions among them must conform to some basic physical forces. Many proteins bear negative charges at the membrane–water interface, and, unless the charges are properly screened, they will exhibit mutual electrostatic repulsion. However, proteins are all composed of similar atoms, and tend to attract one another by van der Waals (electrodynamic) forces. In the absence of extra-membranous constraints, the distribution of membrane proteins will thus be determined by balance of electrostatic repulsion and van der Waals attraction. Factors that reduce the former or increase the latter will favor the formation of protein aggregates. This mechanism of protein aggregation is supported by an experiment on the protein distribution in the erythrocyte membrane. Elgsaeter & Branton (1974) found that the intramembranous particles become strongly aggregated after trypsin and neuraminidase treatments, both of which reduce the cell surface negative charges. Moreover, they observed that the aggregation increased smoothly with increasing ionic strength; this is consistent with the expectation that screening of charges by high ionic strength reduces repulsive force. Ligand binding provides another means for modulating the repulsive and attractive forces. Membrane receptors could be induced to aggregate by the binding of a ligand if the latter reduces their mutual repulsion by screening the charges on the receptors or enhances their mutual van der Waals attraction by changing the conformation of the receptors. No crosslinking of the receptors is necessary in this aggregation mechanism, an interesting feature to consider when accounting for the receptor-aggregating action of many monovalent polypeptides, e.g. insulin, epidermal growth factor, and nerve growth factor (Schlessinger et al 1978, Levi et al 1980). Elegant analyses of colloidal interactions among membrane proteins have been presented by Gingell (1973, 1976). A model for ligand-induced aggregation of membrane proteins based on the thermodynamic consideration of boundary surface tension in a protein aggregate was proposed by Gershon (1978).

With all the attention currently being given to the extracellular matrix and cytoskeleton, one should not overlook the possibility that interaction among membrane proteins themselves could be the primary mechanism for localized aggregation within the membrane. Freeze-fracture replicas that exhibit regular arrays of intramembranous particles, e.g. those observed at the postsynaptic muscle membrane and at the gap junction, vividly remind us that the process of aggregation may not be too different from that of the crystallization or the precipitation of proteins in a supersaturated solution. The behavior of ACh receptors in a "supersensitive" muscle membrane is a good example: Nonspecific surface perturbation, e.g. contacts with a piece of thread (Jones & Vrbova 1974) or a positively charged latex bead (Peng et al 1981), could induce receptor clusters. In any case, it appears that the need for an extramembranous "anchoring" mechanism to localize proteins in the cell membrane is currently overemphasized. By virtue of their size, large protein aggregates may diffuse negligible distances over the lifetime of the protein on the cell surface. In fact, they are more likely to serve as anchoring structures for the extracellular or cytoplasmic components in the aqueous phase, resulting in specific localization of the latter on the two sides of the plasma membrane.

FUNCTIONAL IMPLICATIONS OF PROTEIN MOBILITY

The lateral diffusion of the membrane proteins may merely represent a potentially disruptive force with which the cell must contend in structuring membrane topography during growth, differentiation, and normal functioning. However, a number of studies suggest that lateral diffusion may also provide a structural basis for function and plasticity in the organization of the plasma membrane. In the following, I examine the experimental evidence and some hypothetical models that argue that lateral mobility is functionally important for the interaction between the membrane components, for selective cell–cell recognition/adhesion, for the formation and elimination of nerve connections, and for the reorganization of the excitable membrane.

Lateral Interactions in the Membrane Plane

Lateral mobility of membrane components is advantageous if any significant reaction between them is to occur in the membrane. The idea that cell membranes can provide a platform for chemical reactions was in fact proposed before the general awareness of membrane fluidity. Adam & Delbrück (1968) suggested that chemical reactions can be facilitated if substrates are bound to the cell membrane, because the target-searching time could be reduced by reducing the dimensionality for molecular diffusion. The exact rate advantage expected for reactants confined within the membrane rather than residing in the aqueous phase depends upon a number of factors: the shape and dimension of the diffusion space, the relative diffusion rate of the reactants in aqueous phase vs in membrane, the orientation of reactive groups, and the two-dimensional reaction kinetics (see review McCloskey & Poo 1984). For the present discussion, I consider only examples of reactions in the excitable membrane in which protein mobility may be important.

REACTION IN THE PHOTORECEPTOR DISK MEMBRANES Transduction of photons into a receptor potential in vertebrate photoreceptors may involve lateral motion of rhodopsin in the disk membranes of the photoreceptor. Lateral mobility of rhodopsin is the highest along integral membrane proteins whose diffusion rates have been measured (see section on Protein Mobility). Based on the observation of the fast kinetics of phosphodiesterase (PDE) activity induced by a flash of light, Liebman & Pugh (1979, 1981) proposed that the first step of signal amplification in the phototransduction process involves the collisional activation of hundreds of disk membrane-associated PDE molecules by a single photo-activated rhodopsin, and that the resultant PDE activity is responsible for the light-dependent hydrolysis of cyclic GMP, a putative internal transmitter that regulates the permeability of the plasma membrane. While the exact role of PDE activation and cyclic GMP hydrolysis in phototransduction remains to be elucidated (see Hubbell & Bownds 1979, George & Hagins 1983), rapid lateral diffusion of rhodopsin may be relevant for its function in the disk membrane.

HORMONAL RESPONSE Studies of cAMP-mediated hormonal response have indicated that in many cell types the adenylate cyclase system responsible for catalyzing the synthesis of cAMP consists of three membrane proteins: the hormone (or neurotransmitter) receptor, the guanyl nucleotide-binding (G) protein, and the catalytic unit. Hormone–receptor complexes catalyze the formation of GTP-G protein complexes, which in turn activate the cyclase (Rodbell 1980). The exact topographic relationship between the receptor, the G protein, and the catalytic unit is not known. Many lines of evidence suggest that at least the receptor and catalytic unit are physically separate molecules in the membrane, and the activation of catalytic activity by the hormone–receptor complexes depends upon the diffusion rate of these proteins in the plane of the membrane (Schramm et al 1977, Rimon et al 1978, Hanski et al 1979, see review Ross & Gilman 1980). Rasenick et al (1981) found that incubation of enriched synaptic membrane fractions from rat cerebral cortex with fatty acids that "fluidize" the membrane, or with drugs that presumably disrupt cytoplasmic microtubules, enhanced G protein-mediated activation of cyclase activity. They suggested that the ability of the G protein to diffuse laterally in the membrane is a limiting factor in cyclase activation. So far, no evidence suggests long-range segregation of the three components of the adenylate cyclase system in the plane of membrane. Short-range lateral mobility within mobile domains may thus be sufficient for their functions. A cellular consequence of adenylate cyclase activity is the activation of protein kinases that phosphorylate cytoplasmic and membrane proteins, e.g. ion channels (see Greengard 1979, Levitan et al 1983). Protein kinases and protein phosphatases that regulate the state of phosphorylation of the membrane protein could themselves be membrane-bound. If so, the lateral mobility of these enzymes would also be functionally important.

RECEPTOR CROSS-TALK The hormonal response described above could lead to a modulation of membrane proteins indirectly via intracellular second messengers. There is also evidence of direct interactions between two different transmitter receptors, namely "receptor cross-talk," within the excitable membrane (see review Birdsall 1982). For example, Watanabe et al (1978) found that in isolated membrane preparations of dog myocardium, muscarinic agonists could increase β-adrenergic agonist binding up to threefold. Vasoactive intestinal peptide at physiologically relevant concentrations will increase muscarinic agonist binding by up to 10^5-fold in membrane preparations from cat submandibular gland (Lundberg et al 1982). If such receptor cross-talk involved two physically separate (rather than allosterically linked) receptors in the membrane, the lateral mobility of the receptors would be functionally important.

Binding of a ligand to a membrane receptor can no doubt produce dramatic changes in the receptor's lateral interaction with other membrane proteins, as exemplified by the hormone-induced clustering of receptors and the subsequent association with the hormone–receptor complexes with the structure of "coated pits" (see review Pastan & Willingham 1981). As discussed in a previous section, ligand binding could specifically modify the electrostatic or van der Waals interactions between the ligand–receptor complex and other membrane proteins, leading to a modulation of reactions between them as well as their membrane distribution. With the expanding list of neurotransmitters and neuromodulators currently being identified in the nervous system, the possible direct lateral interactions among various receptors in the plane of cell membrane, especially after agonist and antagonist binding, will certainly be worth investigating.

Enhancement of Selective Cell–cell Adhesion

Selective adhesion between cells is of primary importance for cell–cell interaction and morphogenesis in development. It is probably the underlying basis for neuronal recognition and the stabilization of synaptic connections. Lateral mobility of the surface "recognition/adhesion molecules," at least a short-range mobility, is advantageous for forming adhesive bonding between the cells (see Bell 1978). Long-range lateral mobility would further help to increase the efficiency of bond formation and to recruit (or "trap") more of these recognition/adhesion molecules from non-contacted regions of the cells.

Accumulation of molecules at the cell–cell contact sites has frequently been observed: When ferritin conjugated Con A was used to agglutinate erythrocytes and lymphocytes, Singer (1976) found that ferritin-Con A was concentrated on the contacting zones of the membranes and depleted elsewhere. Since both cell types bind Con A, the tetravalent lectin appeared to act as an intercellular bridging ligand and as a trap for freely diffusing Con A receptors in both membranes. In *Xenopus* nerve-muscle culture, nerve contact induced a marked accumulation of the preexisting muscle surface ACh receptors to the regions of nerve contact, and the effect was nerve-type specific (Anderson et al 1977, Cohen & Weldon 1980). Muscle–muscle contact in the *Xenopus* culture also induces marked accumulation of muscle surface soybean agglutinin receptors, presumably a subpopulation of glycoproteins, to the contact site within 20 min after the contract is made (Chow & Poo 1982). Measurements have shown that the adhesion between these muscle cells increases monotonically during the period of receptor redistribution (M-m. Poo and E. Evans, unpublished observations). With a similar time course, the cells also become electrically coupled (Chow & Poo 1984). The increase in the amplitude of synaptic potentials also correlates with the extent of ACh receptor accumulation at nerve–muscle junctions in *Xenopus* culture (Kidokoro et al 1980). While the causal relationships between various cell surface events is far from clear, the results from experiments on *Xenopus* muscle cells are consistent with the notion that the linkage of complementary molecules on the opposing surfaces, the development of adhesion, and the establishment of cell–cell communication are closely-linked processes, with lateral mobility of membrane components providing an underlying structural basis.

Localization of ACh receptors induced by contact of the proper nerve could also serve to stabilize the appropriate synaptic contact during the early development, if the localization were due to the entrapment of ACh receptors by specific molecules associated with the surface of the nerve terminal. In *Xenopus* embryo ACh receptors appear on the entire embryonic muscle surface before the onset of synaptogenesis (Blackshaw & Warner 1976, Chow & Cohen 1983), and a high density of ACh receptors has long been correlated with the receptiveness of muscle surface to innervation (Katz & Miledi 1964, Jansen et al 1973). The recognition/adhesion of proper nerve terminals could be, in fact, a function served by the carbohydrate moiety of the ACh receptor.

Lateral Signaling in the Membrane Plane

The most interesting implication of *long-range* lateral mobility of membrane proteins in the embryonic cells is the possibility of lateral signaling in the plane of plasma membrane. Consider the following situation in a developing nervous system: A target cell has synthesized a certain amount of recognition/adhesive molecules and incorporated them uniformly on the cell surface. Contact with an appropriate nerve terminal will cause an accumulation with time of a fixed amount of the molecules to the contact site (via lateral diffusion), and at the same time will decrease the concentration of the molecules elsewhere. After a certain number of nerve terminals have made contacts (anywhere on the target cell surface), surface concentration of the molecule at the noncontact region will be depleted to such a level that no more stable adhesions with additional nerves could be made—the target cell becomes refractory to innervation. The lateral mobility of the molecules thus provides a simple mechanism of counting synaptic contacts, through which the target cell can regulate the number of the appropriate innervation sites by metabolic control over the turnover rate of a single molecule, without the need to know where and how many contacts have been made on its surface. Lateral redistribution of the complementary recognition/adhesion molecule in the presynaptic nerve membrane can equally provide a mechanism for the nerve cell to regulate the number of target cells it innervates.

Consider a second situation in which a number of stable contacts have been made initially on the target cell, but the rate of synthesis of the adhesion molecule later drops to a lower level, as a result of active synaptic transmission. The reduction in the amount of the newly synthesized, contact-stabilizing adhesion molecules on the cell surface will lead to a reduction in the number of stable connections, if the molecules initially accumulated at the contact sites undergo turnover and the newly synthesized molecules are incorporated randomly over the entire plasma membrane. Which terminals will remain depends upon an activity-dependent competition between the terminals: The terminals with the stronger "localizing power," e.g. that produce more molecular "traps" or are more electrically active (see next section), will survive and the others will be eliminated. Thus long-range lateral mobility of newly synthesized proteins in the plasma membrane also provide a simple mechanism for the elimination of poly-neuronal innervation in the developing nervous system (Purves & Lichtman 1980).

For the specificity of the recognition/adhesion process and for the economy of the cell metabolism, the best candidates for the hypothesized recognition/

adhesion molecules are various neurotransmitter receptors. These molecules also happen to have the need to be localized at the appropriate nerve contact sites. In addition, if the "localizing power" of the nerve terminal is associated with the level of synaptic activity, it is an inherent positive-feedback mechanism: the more transmitter receptors localized, the more stable and active the synapse will become, and the higher "localizing power" the terminal will possess (see next section).

Modulation of Membrane Protein Distribution

Modulation of the distribution and the number of plasma membrane proteins will no doubt produce functional alteration in the excitable membrane. Long-term modulation of a number of membrane proteins by electrical activity or chemical factors is well-known, as in the case of denervation-induced hypersensitivity to neurotransmitters (see review Thesleff & Sellin 1980) and the "down regulation" of the receptor number induced by the binding of hormone or neurotransmitter molecules (see reviews Raff 1976, Kebabian et al 1977). Redistribution of preexisting membrane proteins as a means of modulating membrane property is a direct consequence of the lateral mobility of membrane protein, and is the subject of the present discussion.

Redistribution of membrane proteins can affect the behavior of the excitable membrane as a whole. For examples, a local reduction in the K^+ channel density or an increase of Na^+ channel density can elevate membrane excitability at that region and lead to repetitive discharge (see also Holden & Yoda 1981). Modulation of channel density in an axonal branch near the point of bifurcation may provide an on-off switch for the impulse transmission into the breach. Redistribution may also change the behavior of the individual protein: Young & Poo (1983b) have shown that electric field induced clustering of preexisting dispersed ACh receptors on isolated *Xenopus* muscle cells can slow down the receptor-channel kinetics significantly, presumably by altering the local membrane environment of the individual ACh receptor. The binding affinity of neurotransmitter and hormone molecules to their membrane receptors could also be affected by the state of receptor aggregation in the membrane.

AGGREGATION FACTORS The topography of plasma membrane proteins could be modulated by extracellular "trophic" substance. This possibility was suggested by the finding that a number of neurally derived factors, when added to a myotube culture, can induce aggregation of muscle surface ACh receptors. Some of these "aggregation factors" increase both the total number of ACh receptors as well as the number of ACh receptor clusters on the muscle surface (Podleski et al 1978, Jessell et al 1979), whereas others only increase the number of clusters (Christian et al 1978, Nitkin et al 1983), apparently by aggregating the preexisting dispersed ACh receptors on the myotube surface.

Although all these factors are polypeptide in nature, they appear in fact to be different molecules. Among them, the protein purified from the extracellular matrix of the electric organ of *Torpedo* (Nitkin et al 1983) showed by far the highest receptor aggregating activity: It is effective at sub-nanomolecular concentrations when added to cultured myotubes. Moreover, antiserum against the purified material from the electric organ also binds to extracellular matrix at frog neuromuscular junction, where natural clustering of ACh receptors occurs.

The main interest in the receptor aggregation factors stems from the expectation that such factors, if locally released by the motor nerve terminal, could be responsible for the clustering of ACh receptors at the muscle end-plate. However, substances released by the nerve terminal, if allowed to diffuse freely in the extracellular space, are unlikely to produce a single discrete aggregate of ACh receptors at the endplate unless other conditions are met, e.g. the aggregating activity has a steep concentration dependence or it triggers a cooperative interaction among the ACh receptors themselves. In this regard, the factor obtained by Nitkin et al (1983) appears most attractive, since it is tightly bound to the insoluble extracellular matrix fraction.

The aggregation of ACh receptors by neuronal factors is currently the only well-documented case of "trophic" modulation of protein distribution. It is not clear whether this type of modulation plays a primary role in the development and maintenance of the nonrandom distribution of many other proteins in excitable membranes. Note that for a trophic factor to convey a localization signal, the simplest mechanism is that the factor itself be localized, e.g. in the cytoplasmic structure or in the extracellular matrix. The problem of localization of membrane proteins then becomes a problem of localization of substances within cytoplasmic or extracellular spaces.

ACTIVITY-DEPENDENT MODULATION Neuronal activity can exert long-term effects on the neuron via numerous cellular mechanisms (see reviews Harris 1981, Zigmond & Bowers 1981). The present discussion is focused on one specific question: Can electrical activity modulate directly and immediately the topography of preexisting proteins in the excitable membranes? It has been suggested recently that the electric field associated with the synaptic current may redistribute membrane receptor/channels or cytoplasmic metabolites toward or away from the postsynaptic zone by an "electrophoretic" mechanism, resulting in a direct modification of synaptic efficiency (Fraser & Poo 1982, Horwitz 1982, 1984). Model calculations (S. H. Young and M-m. Poo, unpublished) based on the decrement of endplate potential recorded at the neuromuscular junction (Fatt & Katz 1951) showed that the intracellular longitudinal electric field within 1 mm of the muscle end-plate is between 0.1 to 1 V/cm during the synaptic current flow. Rapid decrement of synaptic potential along the neck of dendritic spines could generate cytoplasmic field up to 40 V/cm, assuming a model of the spine described by Jack et al (1975). Extracellular fields could be even higher as a result of the restricted intercellular space for current flow. Such electric fields are not negligible, especially when the synaptic activity is high (see also Horwitz 1974).

Direct field effects on synaptic structure has been a recurring notion in the past (Ranck 1964, Elul 1966). Recent experimental demonstration of the effects of externally applied electric fields on the distribution of membrane proteins in cultured nerve and muscle cells (see review Poo 1981, Young et al 1984) has stimulated further inquiry into this possibility. While in vivo tests of the idea are still lacking, a number of *in vitro* findings are suggestive:

1. Extracellular electric fields with an intensity of the order of 1 V/cm are effective in causing lateral migration of cell surface receptors, including ACh receptors in the embryonic muscle cells and lectin receptors in embryonic neurons (see review Poo 1981, Young et al 1984). Pulsatile fields are also effective, but the effect is frequency-dependent (Lin-Liu et al 1984), a property reminiscent of in vivo phenomena.
2. Aggregates of proteins, although diffusionally stable in the membrane, are electrokinetically mobile, suggesting the effectiveness of the field on the established topography (Poo 1981).
3. Electrokinetic mobility of a membrane receptor can, in some cases, be greatly increased by specific binding of an extracellular ligand (McCloskey et al 1984), suggesting a possible in vivo mechanism for controlling the specificity of the electric-field-induced redistribution.

Here it may be worthwhile to consider the relevance of nonspecific forces in the localization of substances. Chemical specificity, which is no doubt the crucial determinant of many cellular processes—from enzymatic reactions to the cell–cell "recognition"—operates nevertheless only within molecular ranges. Nonspecific physical forces are probably responsible for the long-range spatial arrangement of molecules that sets the stage for specific molecular interactions. For example, the electric field associated with the synaptic current could cause a lateral electromigration and local accumulation of the adhesion molecules on the postsynaptic cell surface near the active synapse that favors the formation of specific bonds and the stabilization of the synaptic connection. This provides a natural mechanism for the activity-dependent competition between synapses on the same postsynaptic cell (see the previous section). The powerful modulatory role of nonspecific force is vividly illustrated by the egg rotation experiment originally reported by Penners & Schleip (1928a,b) and recently repeated and confirmed by Jacobson (1982): A twin embryo of *Xenopus laevis*, with two separate neuroaxes, can be produced simply by turning the fertilized egg ventral side up for 1 hr until the second cleavage.

This result suggests that a simple inversion in the direction of gravitational force during a crucial period of development is capable of modulating the localization of specific cytoplasmic components in such a way as to trigger a duplication of a long-chain of specific interactions. In view of the ubiquitous electrical activity within the developing and mature nervous systems, one wonders whether the activity-associated electrical force, which in principle could be much stronger than that of gravity, also plays a powerful modulatory role in the organization of neuronal structure.

CONCLUDING REMARKS

A distinct feature of the neuron is its elaborate morphological differentiation and intricate cellular organization. For the development and maintenance of such a complex structure, macromolecules must be mobile after biosynthesis in order to allow for efficient translocation and they must also attain stable localization when a specific site has been reached. The mobility and localization of membrane proteins is an integral part of this general problem of neuronal organization. It has become increasingly clear that the plasma membrane "proper" exists as a continuum with the cytoplasmic and extracellular matrix. To what extent mobility and localization within the plasma membrane reflects properties of this "extended membrane" is still unknown. One thing is certain, however: Properties of the plasma membrane can no longer be considered in isolation.

There is no doubt that the nervous system is a dynamic structure, and that structural modification, including the possibility of reorganizing the neuronal membrane, provides a natural basis for the functional plasticity of the nervous system. But how dynamic is the neuronal structure? We still do not know whether the topography of transmitter receptors and ionic channels in a mature neuronal membrane is susceptible to modulation under normal physiological conditions. If it is, how susceptible is it? Can neuronal activity lead to an immediate modulation of the protein topography? These questions illustrate the present need for studying the dynamics of neuronal structure in quantitative terms.

ACKNOWLEDGMENTS

I thank M. McCloskey, S. H. Young, and I. Chow for helpful comments on the manuscript. Preparation of this review was supported by grants from the National Institutes of Health (GM 30666, NS 17558) and the National Science Foundation (BNS 8309336).

Literature Cited

Abercrombie, M., Heaysman, J. E. M., Pegram, S. M. 1970. The locomotion of fibroblasts in culture. III. Movements of particles on the dorsal surface of the leading lamella. *Exp. Cell Res.* 62:389–98

Adam, G., Delbrück, M. 1968. Reduction of

dimensionality in biological diffusion processes. In *Structural Chemistry and Molecular Biology*, ed. N. Davidson, A. Rich. pp. 198–215. San Francisco: Freeman

Almers, W., Stanfield, P. R., Stuhmer, W. 1983a. Lateral distribution of sodium and potassium channels in frog skeletal muscle: Measurements with a patch clamp technique. *J. Physiol. London* 336:261–84

Almers, W., Stanfield, P. R., Stuhmer, W. 1983b. Slow changes in currents through sodium channels in frog muscle membrane. *J. Physiol. London* 339:253–71

Andersen, P., Blackstad, T. W., Lømo, T. 1966. Location and identification of excitatory synapses on hippocampal pyramidal cells. *Exp. Brain Res.* 1:236–48

Anderson, D. J. 1983. Acetylcholine receptor biosynthesis: From kinetics to molecular mechanism. *Trends Neurosci.* 6:169–71

Anderson, M. J., Cohen, M. W. 1974. Fluorescent staining of acetylcholine receptors in vertebrate skeletal muscle. *J. Physiol. London* 237:385–96

Anderson, M. J., Cohen, M. W. 1977. Nerve-induced and spontaneous redistribution of acetylcholine receptors on cultured muscle cell. *J. Physiol. London* 268:757–73

Anderson, M. J., Cohen, M. W., Zorychta, E. 1977. Effects of innervation on the distribution of acetylcholine receptors on cultured muscle cells. *J. Physiol. London* 268: 731–56

Anderson, M. J., Fambrough, D. M. 1983. Aggregates of acetylcholine receptors are associated with plaques of a basal lamina heparan sulfate proteoglycan on the surface of skeletal muscle fiber. *J. Cell Biol.* 97: 1396–1411

Andrews, L. D., Cohen, A. I. 1979. Freeze-fracture evidence for the presence of cholesterol in particle-free patches of basal disks and the plasma membrane of retinal rod outer segments of mice and frogs. *J. Cell Biol.* 81:215–18

Axelrod, D. 1980. Crosslinkage and a visualization of acetylcholine receptors on myotubes with biotinylated α-bungarotoxin and fluorescent avidin. *Proc. Natl. Acad. Sci. USA* 77:4823–27

Axelrod, D. 1983. Lateral motion of membrane proteins and biological function. *J. Membr. Biol.* 75:1–10

Axelrod, D., Ravdin, R., Koppel, D. E., Schlessinger, J., Webb, W. W., Elson, E. L., Podleski, T. R. 1976. Lateral motion of fluorescently labeled acetylcholine receptors in membranes of developing muscle cells. *Proc. Natl. Acad. Sci. USA* 73:4594–98

Barnard, E. A. 1979. Visualization and counting of receptors at the light and electron microscope levels. In *The Receptors—A Comprehensive Treatise*, ed. R. D. O'Brien. New York: Plenum

Barrett, E. F., Barrett, J. N. 1976. Separation of two voltage-sensitive potassium currents, and demonstration of a tetrodotoxin-resistant calcium current in frog motoneurons. *J. Physiol. London* 255:737–74

Baylor, D. A., Lamb, T. D., Yau, K.-W. 1979. The membrane current of single rod outer segments. *J. Physiol. London* 288:589–611

Bell, G. I. 1978. Models for the specific adhesion of cells to cells. *Science* 200:618–27

Bennett, J. P., McGill, K. A., Warren, G. B. 1980. The role of lipids in the functioning of a membrane protein: The sarcoplasmic reticulum calcium pump. In *Curr. Top. Membr. Trans.* 14:127–64. New York: Academic

Bennett, V., Davis, J., Fowler, W. E. 1982. Brain spectrin, a membrane-associated protein related in structure and function to erythrocyte spectrin. *Nature* 299:126–31

Berclaz, T., McConnell, H. M. 1981. Phase-equilibria in binary-mixture of dimyristoylphosphatidylcholine and cardiolipin. *Biochemistry* 20:6635–40

Betz, W. J., Caldwell, J. H. 1984. Mapping electric currents around skeletal muscle with a vibrating probe. *J. Gen. Physiol.* 83:143–56

Betz, W. J., Caldwell, J. H., Ribchester, R. R., Robinson, K. R., Stump, R. F. 1980. Endogenous electric field around muscle fibres depends on the Na^+-K^+ pump. *Nature* 287:235–37

Birdsall, N. J. M. 1982. Can different receptors interact directly with each other? *Trends Neurosci.* 5:137–38

Blackshaw, S., Warner, A. 1976. Onset of acetylcholine sensitivity and end-plate activity in developing myotome muscles of *Xenopus*. *Nature* 262:217–218

Bloch, R. 1979. Dispersal and reformation of acetylcholine receptor clusters of cultured rat myotubes treated with inhibitors of energy metabolism. *J. Cell Biol.* 82:620–43

Bloch, R. J., Geiger, G. 1980. The localization of acetylcholine receptor clusters in areas of cell-substrate contact in cultures of rat myotubes. *Cell* 21:25–35

Bloch, R. J., Hall, Z. W. 1983. Cytoskeletal components of the vertebrate neuromuscular junction: Vinculin, α-actinin, and filamin. *J. Cell Biol.* 97:217–23

Bostock, H., Sears, T. A., Sherratt, R. M. 1981. The effects of 3-aminopyridine and tetraethylammonium ions on normal and demyelinated nerve fibers. *J. Physiol. London* 313:301–15

Bostock, H., Sears, T. A. 1978. The internodal axon membrane: Electrical excitability and continuous conduction in segmental demyelination. *J. Physiol. London* 280:273–301

Branton, D., Cohen, C. M., Tyler, J. 1981. Interaction of cytoskeletal proteins on the human erythrocyte membrane. *Cell* 24:24–32

Bray, D. 1970. Surface movements during growth of single explanted neurons. *Proc. Natl. Acad. Sci. USA* 65:905–10

Bretscher, M. S. 1976. Directed lipid flow in cell membranes. *Nature* 260:21–23

Brigant, J. L., Mallart, A. 1982. Presynaptic currents in mouse motor endings. *J. Physiol.* 333:619–32

Brismar, T. 1979. Potential clamp experiments on myelinated nerve fibers from allaxon diabetic rats. *Acta physiol. scand.* 105:384–86

Burden, S. J. 1982. Identification of an intracellular postsynaptic antigen at the frog neuromuscular junction. *J. Cell Biol.* 94:521–30

Burden, S. J., Sargent, P. B., McMahan, U. J. 1979. Acetylcholine receptors in regenerating muscle accumulate at original synaptic site in the absence of the nerve. *J. Cell Biol.* 82:412–25

Catterall, W. 1981. Localization of sodium channels in cultured neuronal cells. *J. Neurosci.* 1:777–83

Chao, N-m., Young, S. H., Poo, M-m. 1981. Localization of cell membrane components by surface diffusion into a "trap." *Biophys. J.* 36:139–53

Cherry, R. J. 1979. Rotational and lateral diffusion of membrane proteins. *Biochim. Biophys. Acta* 559:289–337

Cherry, R. J., Muller, J. U., Holenstein, C., Heyn, M. P. 1980. Lateral segregation of proteins induced by cholesterol in bacteriorhodopsin–phospholipid vesicles. *Biochem. Biophys. Acta* 596:145–51

Chiu, S. Y., Ritchie, J. M. 1980. Potassium channels in nodal and internodal membrane of mammalian myelinated fibres. *Nature* 284:170–71

Chiu, S. Y., Ritchie, J. M. 1981. Evidence for the presence of potassium channels in the internodal region of acutely demyelinated mammalian single nerve fibers. *J. Physiol. London* 313:415–38

Chiu, S. Y., Ritchie, J. M. 1982. Evidence for the presence of potassium channels in the internode of frog myelinated nerve fibres. *J. Physiol. London* 322:485–501

Chiu, S. Y., Ritchie, J. M., Rogart, R. B., Stagg, D. 1979. A quantitative description of membrane currents in rabbit myelinated nerve. *J. Physiol. London* 292:149–66

Chow, I., Cohen, M. W. 1983. Developmental changes in the distribution of acetylcholine receptors on the myotomes of *Xenopus laevis*. *J. Physiol. London* 339:553–71

Chow, I., Poo, M-m. 1982. Redistribution of cell surface receptors induced by cell–cell contact. *J. Cell Biol.* 95:510–18

Chow, I., Poo, M-m. 1984. Formation of electrical coupling between embryonic muscle cells in culture. *J. Physiol. London* 346: 181–94

Christian, C. N., Daniels, M. P., Sugiyama, H., Vogel, Z., Jacques, L., Nelson, P. G. 1978. A factor from neurons increases the number of acetylcholine receptor aggregates on cultured muscle cells. *Proc. Natl. Acad. Sci. USA* 75:4011–15

Chujo, J., Yamada, Y., Yamamoto, C. 1975. Sensitivity of Purkinje cell dendrites to glutamic acid. *Exp. Brain. Res.* 23:293–300

Cohen, M. W., Weldon, P. R. 1980. Localization of acetylcholine receptors and synaptic ultrastructure at nerve–muscle contacts in culture: Dependence on nerve type. *J. Cell Biol.* 86:388–401

Cone, R. A. 1972. Rotational diffusion of rhodopsin in the visual receptor membrane. *Nature New Biol.* 236:39–43

Cragg, B. G., Thomas, P. K. 1964. The conduction velocity of regenerated nerve fibers. *J. Physiol. London* 171:164–75

Cull-Candy, S. G. 1976. Two types of extrajunctional L-glutamate receptors in locust muscle fibers. *J. Physiol. London* 255: 449–64

Dautry-Versat, A., Lodish, H. F. 1983. The Golgi complex and the sorting of membrane and secreted proteins. *Trends Neurosci.* 6:484–90

Dembo, M., Harris, A. K. 1981. Motion of particles adhering to the leading lamella of crawling cells. *J. Cell Biol.* 91:528–36

Del Castillo, J., Katz, B. 1955. On the localization of acetylcholine receptors. *J. Physiol. London* 128:157–81

Dennis, M. J. 1981. Development of the neuromuscular junction: Inductive interactions between cells. *Ann. Rev. Neurosci.* 4:43–68

De Petris, S. 1977. Distribution and mobility of plasma membrane components on lymphocytes. In *Dynamic Aspects of Cell Surface Organization,* ed. G. Poste, G. L. Nicolson, *Cell Surf. Rev.* 3:643–713. Amsterdam: North-Holland

Devreotes, P. N., Gardner, J. M., Fambrough, D. M. 1977. Kinetics of biosynthesis of acetylcholine receptors and subsequent incorporation into the plasma membrane of cultured chick skeletal muscle. *Cell* 10:365–73

Diamond, J. 1965. Variation in the sensitivity to gamma-amino-butyric acid of different regions of the Mauthner neurone. *Nature* 199:773–75

Dichter, M. A., Fischbach, G. D. 1977. The action potential of chick dorsal root ganglion neurons maintained in cell culture. *J. Physiol. London* 267:281–98

Edelman, G. M. 1976. Surface modulation in cell recognition and cell growth. Science[+] 192:218–26

Edidin, M. 1981. Molecular motions and membrane organization and function. In *Membrane Structure,* ed. J. B. Finean, R.

H. Michell, pp. 37–82. Amsterdam: Elsevier/North-Holland Biochemical

Edidin, M., Fambrough, D. M. 1973. Fluidity of the surface of cultured muscle fibers. Rapid lateral diffusion of marked surface antigens. *J. Cell Biol.* 57:27–53

Edidin, M., Zagyansky, Y., Lardner, T. J. 1976. Measurement of membrane protein lateral diffusion in single cells. Na^{+} 191:466–68

Edwards, C., Frisch, H. L. 1976. A model for the localization of acetylcholine receptors at the muscle end-plate. *J. Neurobiol.* 7: 377–81

Elgsaeter, A., Branton, D. 1974. Intramembranous particle aggregation in erythrocyte ghosts. I. The effects of protein removal. *J. Cell Biol.* 63:1018–30

Ellisman, M. H., Levinson, S. R. 1982. Immunocytochemical localization of sodium channel distributions in the excitable membrane of *Electrophorus electricus*. *Proc. Natl. Acad. Sci. USA* 79:6707–11

Elson, E. L., Schlessinger, J. 1979. Long-range motions on cell surfaces. In *The Neurosciences Fourth Study Program,* ed. F. O. Schmitt, F. G. Worden, pp. 691–701. Cambridge, MA: MIT

Elul, R. 1966. Dependence of synaptic transmission on protein metabolism of nerve cells: A possible electrokinetic mechanism of learning? *Nature* 210:1127–31

Ernst, S. A., Mills, J. W. 1977. Basolateral plasma membrane localization of ouabain-sensitive sodium transport sites in the secretory epithelium of the avian salt gland. *J. Cell Biol.* 75:74–94

Evans, W. H. 1980. A biochemical dissection of the functional polarity of the plasma membrane of the hepatocyte. *Biochim. Biophys. Acta* 604:27–64

Fain, G. L., Lisman, J. E. 1982. Membrane conductances of photoreceptors. *Prog. Biophys. Mol. Biol.* 37:91–147

Fambrough, D. M. 1974. Acetylcholine receptors. Revised estimates on extrajunctional receptor density in denervated rat diaphragm. *J. Gen. Physiol.* 64:468–72

Fambrough, D. M. 1979. Control of acetylcholine receptors in skeletal muscle. *Physiol. Rev.* 59:165–216

Fambrough, D. M., Bayne, E. K. 1983. Multiple forms of (Na^{+}-K^{+})-ATPase in the chicken. Selective detection of the major nerve, skeletal muscle, and kidney form by a monoclonal antibody. *J. Biol. Chem.* 258:3926–35

Fambrough, D. M., Bayne, E. K., Gardener, J. M., Anderson, M. J., Wakshull, E., Rotundo, R. L. 1982. Monoclonal antibodies to skeletal muscle cell surface. In *Neuroimmunology,* ed. J. Brockes, pp. 49–90. New York: Plenum

Fatt, P., Katz, B. 1951. An analysis of the end-plate potential recorded with an intracellular electrode. *J. Physiol. London* 117:109–28

Fertuck, H. C., Salpeter, M. M. 1974. Localization of acetylcholine receptor by ^{125}I-labeled α-bungarotoxin binding at mouse motor endplates.*Proc. Natl. Acad. Sci. USA* 71:1376–81

Fertuck, H. C., Salpeter, M. M. 1976. Quantitation of junctional and extrajunctional acetyl-choline receptors by electron microscope autoradiography after ^{125}I-α-bungarotoxin binding at mouse neuromuscular junctions. *J. Cell Biol.* 69:144–58

Fischbach, G. D., Cohen, S. A. 1973. The distribution of acetylcholine sensitivity over uninnervated and innervated muscle fibers grown in cell culture. *Dev. Biol.* 31:147–58

Frank, E., Fischbach, G. D. 1979. Early events in neuromuscular junction formation in vitro: Induction of acetylcholine receptors in the postsynaptic membrane and morphology of newly formed synapses. *J. Cell Biol.* 83: 143–58

Fraser, S. E., Poo, M-m. 1982. Development, maintenance, and modulation of patterned membrane topography: Models based on the acetylcholine receptors. *Curr. Top. Dev. Biol.* 17:77–100

Freeman, J. A., Snipes, G. J., Mayes, B., Weiss, J., Norden, J. J. 1982. Cultured retinal neurite growth cones generate steady currents that might play a role in development. *Int. Soc. Dev. Neurosci. Abstr.*

Freeman, J. A., Weiss, J. M., Snipes, G. J., Mayes, B., Norden, J. J. 1981. Growth cones of goldfish retinal neurites generate DC currents and orient in an electric field. *Soc. Neurosci. Abstr.* 7:550

Fritz, L. C., Brockes, J. P. 1983. Immunochemical properties and cytochemical localization of the voltage-sensitive sodium channel from the electroplax of the eel (*Electrophoresus electricus*). *J. Neurosci.* 3:2300–9

Froehner, S. C., Gulbrandsen, V., Hyman, C., Jeng, A. Y., Neubig, R. R., Cohen, J. B. 1981. Immunofluorescence localization at the mammalian neuromuscular junction of the M_r 43,000 protein of Torpedo postsynaptic membrane. *Proc. Natl. Acad. Sci. USA* 78:5530–34

Frye, L. D., Edidin, M. 1970. The rapid intermixing of cell surface antigens after formation of mouse–human heterokaryons. *J. Cell Sci.* 7:319–35

Gennis, R. B., Jonas, A. 1977. Protein–lipid interactions. *Ann. Rev. Biophys. Bioeng.* 6:195–238

George, J. S., Hagins, W. A. 1983. Control of Ca^{+2} in rod outer segment disks by light and cyclic GMP. *Nature* 303:344–48

Gershon, N. D. 1978. A model for capping of membrane receptors based on boundary surface effect. *Proc. Natl. Acad. Sci. USA* 75:1357–60

Gingell, D. 1973. Membrane permeability change by aggregation of mobile glycoprotein units. *J. Theoret. Biol.* 38:677–79

Gingell, D. 1976. Electrostatic control of membrane permeability via intramembranous particle aggregation. In *Mammalian Cell Membranes,* ed. G. A. Jamieson, D. M. Robinson, pp. 198–223. London: Butterworths

Gingell, D., Ginsberg, L. 1978. Problems in the physical interpretation of membrane interaction and fusion. In *Membrane Fusion,* ed. G. Poste, G. L. Nicolson, pp. 791–833. Amsterdam: Elsevier/North-Holland

Greengard, P. 1979. Cyclic nucleotides, phosphorylated proteins, and the nervous system. *Fed. Proc.* 38:2208–17

Grinvald, A., Farber, I. 1981. Optical recording of calcium action potentials from growth cones of cultured neurons with a laser microbeam. *Science* 212:1164–67

Gulley, R. L., Reese, T. S. 1981. Cytoskeletal organization at the postsynaptic complex. *J. Cell Biol.* 91:298–302

Hagins, W. A., Penn, R. D., Yoshikami, S. 1970. Dark current and photocurrent in retinal rods. *Biophys. J.* 10:380–412

Hall, Z. W. 1973. Multiple forms of acetylcholinesterase and their distribution in endplate and non-endplate regions of rat diaphragm muscle. *J. Neurobiol.* 4:343–52

Hall, Z. W., Lubit, B. W., Schwartz, J. H. 1981. Cytoplasmic actin in postsynaptic structures at the neuromuscular junction. *J. Cell Biol.* 90:789–92

Hanski, E., Rimon, G., Levitzki, A. 1979. Adenylate cyclase activation by the β-adrenergic receptors as a diffusion controlled process. *Biochemistry* 18:846–53

Harris, E. J., Kuffler, S. W., Dennis, M. J. 1971. Differential chemosensitivity of synaptic and extrasynaptic areas on the neuronal surface membrane in parasympathetic neurons of the frog, tested by microapplication of acetylcholine. *Proc. R. Soc. London Ser. B* 177:541–53

Harris, W. A. 1981. Neural activity and development. *Ann. Rev. Physiol.* 43:689–710

Heinemann, S., Bevan, S., Kullberg, R., Lindstrom, J., Rice, J. 1977. Modulation of the acetylcholine receptor by anti-receptor antibody. *Proc. Natl. Acad. Sci. USA* 74:3090–94

Heuser, J. E., Reese, T. 1977. Structure of the synapse. In *Handbook of Physiology. The Nervous System,* Vol. 1, *Cellular Biology of Neurons,* ed. E. R. Kandel, pp. 261–94, Baltimore: Williams & Wilkins

Hild, W., Tasaki, I. 1962. Morphological and physiological properties of neurons and glial cells in tissue culture. *J. Neurophysiol.* 25:277–304

Holden, A. V., Yoda, M. 1981. The effects of ionic channel density on neuronal function. *J. Theoret. Neurobiol.* 1:60–81

Horwitz, B. 1982. Neuronal plasticity: Its relation to electrophoretic migration in response to postsynaptic potential gradients. *Soc. Neurosci. Abstr.* 8:709

Horwitz, B. 1984. Electrophoretic migration due to postsynaptic potential gradients: Theory and application to autonomic ganglion neurons and to dendritic spines. *Neuroscience* 12:887–906

Hubbell, W. L., McConnell, H. M. 1968. Spin-label studies of the excitable membranes of nerve and muscle. *Proc. Natl. Acad. Sci. USA* 61:12–16

Hubbell, W. L., Bownds, M. D. 1979. Visual transduction in vertebrate photoreceptors. *Ann. Rev. Neurosci.* 2:17–34

Hume, R. I., Role, L. W., Fischbach, G. D. 1983. Acetylcholine release from growth cones detected with patches of acetylcholine receptor-rich membranes. *Nature* 305: 632–34

Jack, J. J. B., Noble, D., Tsien, R. W. 1975. *Electric Current Flow in Excitable Cells.* Oxford: Clarendon

Jacobson, K. 1980. Fluorescence recovery after photobleaching: Lateral mobility of lipids and proteins in model membrane and on single cell surfaces. In *Lasers in Biology and Medicine,* ed. F. Hillenkamp, R. Pratesi, C. A. Sacchi, pp. 271–88. New York: Plenum

Jacobson, K., Wu, E., Poste, G. 1976. Measurement of translational mobility of concanavalin A in glycerol-saline solutions and on the cell surface by fluorescence recovery after photo-bleaching. *Biochim. Biophys. Acta* 433:215–22

Jacobson, M. 1982. Origins of the nervous system in amphibians. In *Neuronal Development,* ed. N. C. Spitzer, pp. 45–93, New York: Plenum

Jaffe, L. F. 1979. Control of development by ionic currents. In *Membrane Transduction Mechanisms,* ed. R. A. Cone, J. E. Dowling, pp. 199–231. New York: Raven

Jan, L. Y., Jan. Y. N. 1976. L-glutamate as an excitatory transmitter at the Drosophila larval neuromuscular junction. *J. Physiol. London* 262:215–36

Jansen, J. K. S., Lomø, T., Nicolaysen, K., Westgaard, R. H. 1973. Hyperinnervation of skeletal muscle fibers: Dependence on muscle activity. *Science* 181:559–61

Jessell, T. M., Siegel, R. E., Fischbach, G. D. 1979. Induction of acetylcholine receptors on cultured skeletal muscle by a factor extracted from brain and spinal cord. *Proc. Natl. Acad. Sci. USA* 76:5397–5401

Jones, R., Vrbova, G. 1974. Two factors responsible for the development of denervation hypersensitivity. *J. Physiol. London* 236:517–38

Kado, R. T. 1973. Aplysia giant cell: Soma-axon voltage clamp current differences. *Science* 182:843–45

Kandel, E. R. 1981. Calcium and the control of synaptic strength by learning. *Nature* 293:459–87

Katz, B., Miledi, R. 1964. The development of acetylcholine sensitivity in nerve-free segments of skeletal muscle. *J. Physiol. London* 170:389–96

Katz, B., Miledi, R. 1969. Tetrodotoxin-resistant electrical activity in presynaptic terminals. *J. Physiol. London* 203:459–87

Kebabian, J. W., Zatz, M., Kocsis, J. D. 1977. Modulation of receptor sensitivity in the pineal: The role of cyclic nucleotides. *Soc. Neurosci. Symp.* 2:376–98

Kelly, P. T., Cotman, C. W., Gentry, C., Nicolson, G. L. 1976. Distribution and mobility of lectin receptors on synaptic membranes of identified neurons in the central nervous system. *J. Cell Biol.* 71:487–96

Kelly, R. G., Hall, Z. W. 1982. Immunology of the neuromuscular junction. In *Neuroimmunology,* ed. J. Brockes, pp. 1–39, New York: Plenum

Kidokoro, Y., Anderson, M. J., Gruener, R. 1980. Changes in synaptic potential properties during acetylcholine receptor accumulation and neurospecific interactions in *Xenopus* nerve-muscle cell culture. *Dev. Biol.* 78:78–85

Ko, P. K., Anderson, M. J., Cohen, M. W. 1977. Denervated skeletal muscle fibers develop patches of high acetylcholine receptor density. *Science* 196:540–42

Kocsis, J. D., Waxman, S. G. 1980. Absence of potassium conductance in central myelinated axons. *Nature* 287:348–49

Koda, L. Y., Partlow, L. M. 1976. Membrane marker movement on sympathetic axons in tissue culture. *J. Neurobiol.* 7:157–72

Koppel, D. E., Sheetz, M. P., Schindler, M. 1981. Matrix control of protein diffusion in biological membranes. *Proc. Natl. Acad. Sci. USA* 78:3576–80

Kyte, J. 1976. Immunoferritin determination of the distribution of ($Na^+ + K^+$) ATPase over the plasma membranes of renal convoluted tubules. I. Distal segment. *J. Cell Biol.* 68:304–18

Lawson, D., Raff, M. C., Gomperts, B., Fewtrell, C., Gilula, N. B. 1977. Molecular events during membrane fusion. A study of exocytosis in rat peritoneal mast cells. *J. Cell Biol.* 72:242–59

Lehouelleur, J., Schmidt, H. 1980. Extracellular recording of localized activity in denervated frog slow muscle fibers. *Proc. R. Soc. London Ser. B* 209:403–13

Lentz, T. L., Chester, J. 1977. Localization of acetylcholine receptors in central synapses. *J. Cell Biol.* 75:258–67

Levi, A., Shechter, Y., Neufeld, E. J., Schlessinger, J. 1980. Mobility, clustering, and transport of nerve growth factor in embryonal sensory cells and in a sympathetic neuronal cell line. *Proc. Natl. Acad. Sci. USA* 77: 3469–73

Levitan, I. B., Lemos, J. R., Novak-Hofer, I. 1983. Protein phosphorylation and the regulation of ion channels. *Trends Neurosci.* 6:495–99

Liebman, P. A., Entine, G. 1974. Lateral diffusion of visual pigment in photoreceptor disk membrane. *Science* 185:457–59

Liebman, P. A., Pugh, E. N. Jr. 1979. The control of phosphodiesterase in rod disk membranes: Kinetics, possible mechanisms, and significance for vision. *Vision Res.* 19:375–80

Liebman, P. A., Pugh, E. N. Jr. 1981. Control of rod disk membrane phosphodiesterase and a model for visual transduction. *Curr. Top. Membr. Trans.* 15:157–70

Lin-Liu, S., Adey, R., Poo, M-m. 1984. Migration of cell surface concanavalin A receptors induced by pulsed electric fields. *Biophys. J.* 45:1211–18

Lindstrom, J. 1980. Biology of myasthenia gravis. *Ann. Rev. Pharmacol.* 20:337–58

Llinás, R., Hess, R. 1976. Tetrodotoxin-resistant dendritic spikes in avian Purkinje cells. *Proc. Natl. Acad. Sci. USA* 73:2520–23

Llinás, R., Nicolson, C. 1971. Electrophysiological properties of dendrites and somata in alligator Purkinje cells. *J. Neurophysiol.* 34:532–51

Llinás, R., Sugimori, M. 1979. Calcium conductance in Purkinje cell dendrites: Their role in development and integration. *Prog. Brain Res.* 51:323–32

Llinás, R., Steinberg, I. Z., Walton, K. 1976. Presynaptic calcium currents and their relation to synaptic transmission: Voltage clamp study in squid giant synapse and theoretical model for the calcium gate. *Proc. Natl. Acad. Sci. USA* 73:2918–22

Llinás, R. 1979. The role of calcium in neuronal function. In *The Neurosciences. Fourth Study Program,* ed. F. O. Schmitt, F. G. Worden, pp. 555–72. Cambridge: MIT

Lundberg, J. M., Hedlund, B., Bartfai, T. 1982. Vasoactive intestinal peptide enhances muscarinic ligand binding in cat submandibular salivary gland. *Nature* 295:147–49

Lynch, G., Baudry, M. 1983. Origins and manifestations of neuronal plasticity in the hippocampus. In *Clinical Neurosciences,* ed. W. Willis. New York: Churchill-Linvinstone

Marchesi, V. T. 1979. Spectrin: Present status of a putative cyto-skeletal protein of the red cell membrane. *J. Membr. Biol.* 51:101–31

Matus, A., DePetris, S., Raff, M. C. 1973. Mobility of concanavalin A receptors in myelin and synaptic membranes. *Nature New Biol.* 244:278–80

Mazurkiewicz, J. E., Hossler, F. E., Barrnett,

R. J. 1978. Cytochemical demonstration of sodium, potassium-adenosine phosphatase by a heme peptide derivative of ouabain. *J. Histochem. Cytochem.* 26:1042–52

McCloskey, M., Liu, Z.-y., Poo, M-m.1984. Lateral electromigration and diffusion of Fc receptors in rat basophilic leukemia cells: Effects of ligand binding. *J. Cell Biol.* In press

McCloskey, M., Poo, M-m. 1984. Protein diffusion in cell membranes: Some biological implications. *Int. Rev. Cytol.* 87:19–81

McMahan, U. J., Sargent, P. B., Rubin, L. A., Burden, S. J. 1980. Factors that influence the organization of acetylcholine receptors in regenerating muscle are associated with the basal lamina at the neuromuscular junction. In *Ontogenesis and Functional Mechanisms of Peripheral Synapses,* ed. J. Taxi, pp. 345–54. Amsterdam: Elsevier/North Holland

Miledi, R., Parker, I. 1981. Calcium transients recorded with arsenazo III in the presynaptic terminal of the squid giant synapse. *Proc. R. Soc. London Ser. B* 212:197–211

Montesano, R. 1979. Inhomogeneous distribution of filipin-sterol complexes in smooth muscle cell plasma membrane. *Nature* 280:328–29

Moody-Corbett, F., Cohen, M. W. 1982. Influence of nerve on the formation and survival of acetylcholine receptor and acetylcholinesterase patches on embryonic *Xenopus* muscle cells in culture. *J. Neurosci.* 2:633–46

Nakajima, Y., Bridgman, P. C. 1981. Absence of filipin-sterol complexes from the membranes of active zones of acetylcholine receptor aggregates at frog neuromuscular junction. *J. Cell Biol.* 88:453–58

Nicolson, G. L. 1976. Transmembrane control of the receptors on normal and tumor cells. I. Cytoplasmic influence over cell surface components. *Biochim. Biophys. Acta* 457:57–108

Nitkin, R. M., Godfrey, E. W., Wallace, B. G., McMahan, U. J. 1983. Characterization of the ACHR aggregating molecules in extracellular matrix fractions from electric organ and muscle. *Soc. Neurosci. Abstr.* 9:1179

Okun, L. M. 1972. Isolated dorsal root ganglion neurons in culture: Cytological maturation and extension of electrically active processes. *J. Neurobiol.* 3:111–51

Orci, L., Perrelet, A., Friend, D. S. 1977. Freeze-fracture of membrane fusions during exocytosis in pancreatic B-cells. *J. Cell Biol.* 75:23–30

Orida, N., Poo, M-m. 1980. On the developmental regulation of acetylcholine receptor mobility in the *Xenopus* muscle membrane. *Exp. Cell Res.* 130:281–90

Palade, G. 1975. Intracellular aspects of the process of protein synthesis. *Science* 189:347–58

Pastan, I. N., Willingham, M. C. 1981. Journey to the center of the cell: Role of the receptosome. *Science* 214:504–9

Patel, N. 1984. *A study of growth and orientation of embryonic neuron in culture.* PhD dissertation, Univ. Calif., Irvine

Patel, N., Poo, M-m. 1982. Orientation of neurite growth by extracellular electric field. *J. Neurosci.* 2:483–96

Patrick, J., Heinemann, S., Schubert, D. 1978. Biology of cultured nerve and muscle cells. *Ann. Rev. Neurosci.* 1:417–44

Peng, H. B., Cheng, P.-C., Luther, P. W. 1981. Formation of ACh receptor clusters induced by positively charged latex beads. *Nature* 292:831–34

Peng, H. B. 1983. Cytoskeletal organization of the presynaptic nerve terminal and the acetylcholine receptor cluster in cell culture. *J. Cell Biol.* 97:489–98

Penners, A., Schleip, W. 1928a. Die Entwicklung der Schultzeschen Doppelbildungen aus dem Ei von *rana fusca,* Teil I–IV. *Z. Wiss. Zool. Abt. A* 130:305

Penners, A., Schleip, W. 1928b. Die Entwicklung der Schultzeschen Doppelbildungen aus dem Ei von *rana fusca,* Teil V und VI. *Z. Wiss. Zool. Abt. A* 131:1

Peters, R. 1981. Translational diffusion in the plasma membrane of single cells as studied by fluorescence microphotolysis. *Cell Biol. Int. Rep.* 5:733–60

Peters, R., Peters, J., Tews, K., Bahr, W. 1974. A microfluorimetric study of translational diffusion in erythrocyte membranes. *Biochim. Biophys. Acta* 367:282–94

Pfenninger, K. H. 1978. Organization of neuronal membranes. *Ann. Rev. Neurosci.* 1:445–71

Pfenninger, K. H., Bunge, R. P. 1974. Freeze-fracturing of nerve growth cones and young fibers. A study of developing plasma membrane. *J. Cell Biol.* 63:180–96

Podleski, T. R., Axelrod, D., Ravdin, P., Greenberg, I., Johnson, M. M., Salpeter, M. M. 1978. Nerve extract induces increase and redistribution of acetylcholine receptors on cloned muscle cells. *Proc. Natl. Acad. Sci. USA* 75:2035–39

Poo, M-m. 1981. *In situ* electrophoresis of membrane components. *Ann Rev. Biophys. Bioeng.* 10:245–76

Poo, M-m. 1982. Rapid lateral diffusion offunctional acetylcholine receptors in embryonic muscle membrane. *Nature* 295332–34

Poo, M-m., Cone, R. A. 1974. Lateral diffusion of rhodopsin in the photoreceptor membrane. *Nature* 247:438–41

Poo, M-m., Lam, J. W., Orida, N. K., Chao, A. W. 1979. Electrophoresis and diffusion in the plane of cell membrane. *Biophys. J.* 26:1–22

Prives, H., Fulton, A. B., Penman, S., Daniels, M. P., Cristian, C. N. 1982. Interaction of the cytoskeletal framework with acetylcholine receptors on the surface of embryonic muscle cells in culture. *J. Cell Biol.* 92:231–36

Pumplin, D. W., Bloch, R. J. 1983. Lipid domains of acetylcholine receptor clusters detected with saponin and filipin. *J. Cell. Biol.* 97:1043–54

Pumplin, D. W., Fambrough, D. M. 1983. (Na^+ + K^+)-ATPase correlated with a major group of intramembrane particles in freeze-fracture replicas of cultured chick myotubes. *J. Cell Biol.* 97:1214–25

Purves, D., Lichtman, J. W. 1980. Elimination of synapses in the developing nervous system. Na^+210:153–57

Purves, D., Sackman, B. 1974. Membrane properties underlying spontaneous activity of denervated muscle fibers. *J. Physiol. London* 239:125–53

Raff, M. 1976. Self-regulation of membrane receptors. *Nature* 259:255–66

Ranck, J. B. 1964. Synaptic"learning" due to electroosmosis: A theory. *Science* 144:187–89

Ransom, B. R., Bullock, P. N., Nelson, P. G. 1977. Mouse spinal cord in cell culture. III. Neuronal chemosensitivity and its relationship to synaptic activity. *J. Neurophysiol.* 40:1163–77

Rasenick, M. M., Stein, P. J., Bitensky, M. W. 1981. The regulatory subunit of adenylate cyclase interacts with cytoskeletal components *Nature* 294:560–62

Rimon, G., Hanski, E., Braun, S., Levitzki, A. 1978. Mode of coupling between hormone receptors and adenylate cyclase elucidated by modulation of membrane fluidity. *Nature* 276:394–96

Ritchie, J. M. 1982. Sodium and potassium channels in regenerating and developing mammalian myelinated nerves. *Proc. R. Soc. London Ser. B* 215:273–87

Ritchie, J. M., Rang, H. P., Pellegrino, R. 1981. Sodium and potassium channels in de-myelinated and remyelinated mammalian nerve. *Nature* 294:257–59

Ritchie, J. M., Rogart, R. B. 1977. Density of sodium channels in mammalian myelinated nerve fiber and the nature of the axonal membrane under the myelin sheath. *Proc. Natl. Acad. Sci. USA* 74:211–15

Rodbell, M. 1980. The role of hormone receptors and GTP-regulatory proteins in membrane transduction. *Nature* 284:17–22

Rogart, R. 1981. Sodium channels in nerve and muscle membrane. *Ann Rev. Physiol.* 43:711–25

Ross, E. M., Gilman, A. G. 1980. Biochemical properties of hormone-sensitive adenylate cyclase. *Ann. Rev. Biochem.* 49:533–64

Ross, W. N., Stuart, A. E. 1978. Voltage sensitive calcium channels in the presynaptic terminals of a decrementally-conducting photoreceptor. *J. Physiol. London* 274:173–91

Rotundo, R. L., Fambrough, D. M. 1980a. Synthesis, transport and fate of acetylcholinesterase in cultured chick embryo muscle cells. *Cell* 22:583–94

Rotundo, R. L., Fambrough, D. M. 1980b. Secretion of acetylcholinesterase: Relation to acetylcholine receptor metabolism. *Cell* 22:595–602

Saffman, P. G., Delbrück, M. 1975. Brownian motion in biological membranes. *Proc. Natl. Acad. Sci. USA* 72:3111–15

Sanes, J. R., Hall, Z. W. 1979. Antibodies that bind specifically to synaptic sites on muscle fiber basal lamina. *J. Cell Biol.* 83:357–70

Schelessinger, J. 1983. Mobilities of cell-membrane proteins: How are they modulated by the cytoskeleton? *Trends Neurosci.* 6:360–63

Schinitzky, M., Henkart, P. 1979. Fluidity of cell membranes—current concepts and trends. *Int. Rev. Cytol.* 60:121–47

Schlessinger, J., Schechter, Y., Cuatrecasas, P., Willingham, M. C., Pastan, I. 1978. Quantitative determination of the lateral diffusion coefficients of the hormone-receptor complexes of insulin and epidermal growth factor on the plasma membrane of cultured fibroblasts. *Proc. Natl. Acad. Sci. USA* 75:5353–57

Schramm, M., Orly, J., Eimerl, S., Korner, M. 1977. Coupling of hormone receptors to adenylate cyclase of different cells by cell fusion. *Nature* 268:310–13

Schwartzkroin, P. A., Andersen, P. 1975. Glutamic acid sensitivity of dendrites in hippocampal slices *in vitro*. *Adv. Neurol.* 12:45–51

Shepherd, G. M. 1979. *The Synaptic Organization of the Brain.* Oxford: Oxford Univ. Press

Siman, R., Baudry, M., Lynch, G. 1984. Brain fodrin: A substrate of calpain I, an endogenous calcium-activated protease. *Proc. Natl. Acad. Sci. USA* 81:3572–76

Singer, S J. 1976. The fluid mosaic model of membrane structure. Some applications to ligand–receptor and cell–cell interactions. In *Surface Membrane Receptors*, ed. R. A. Bradshaw, W. A. Frazier, R. C. Merrell, D. I. Gottlieb, R. A. Hogue-Angeletti, pp. 1–24. New York: Plenum

Singer, S. J., Nicolson, G. L. 1972. The fluid mosaic model of the structure of cell membranes. *Science* 175:720–31

Smith, K. J., Bostock, H., Hall, S. M. 1982. Saltatory conduction precedes remyelination in axons demyelinated with lysophosphatidylcholine. *J. Neurol. Sci.* 54:13–31

Spitzer, N. C. 1979. Ionic channels in development. *Ann Rev. Neurosci.* 2:263–97

Stockbridge, N., Ross, W. N. 1984. Localized Ca^{2+} and calcium-activated potassium conductances in terminals of a barnacle photoreceptor. *Nature* 309:266–68

Strichartz, G. R. 1977. The composition and structure of excitable nerve membrane. In *Mammalian Cell Membranes,* ed. G. A. Jamieson, D. M. Robinson, 3:172–205. London: Butterworths

Stuhmer, W., Almers, W. 1982. Photobleaching through glass micropipettes: Sodium channels without lateral mobility in the sarcolemma of frog skeletal muscle. *Proc. Natl. Acad. Sci. USA* 79:946–50

Stya, M., Axelrod, D. 1983a. Diffusely distributed acetylcholine receptors can participate in cluster formation on cultured rat myotubes. *Proc. Natl. Acad. Sci. USA* 80:449–53

Stya, M., Axelrod, D. 1983b. Mobility and detergent extractibility of acetylcholine receptors on cultured rat myotubes: A correlation. *J. Cell Biol.* 97:49–51

Sytkowski, A. J., Vogel, Z., Nirenberg, M. W. 1973. Development of acetylcholine receptor clusters on cultured muscle cells. *Proc. Natl. Acad. Sci. USA* 70:270–74

Takeuchi, A., Takeuchi, N. 1964. The effect on crayfish muscle of iontophoretically applied glutamate. *J. Physiol. London* 170:296–317

Takeuchi, A., Takeuchi, N. 1965. Localized action of gamma-aminobutyric acid on the crayfish muscle. *J. Physiol. London* 177:225–38

Tank, D. W., Wu, E-S., Webb, W. W. 1982. Enhanced molecular diffusibility in muscle membrane blebs: Release of lateral constraints. *J. Cell Biol.* 92:207–12

Taylor, R. B., Duffus, W. P. H., Raff, M. C., dePetris, S. 1971. Redistribution and pinocytosis of lymphocyte surface immunoglobulin molecules induced by anti-immunoglobulin antibody. *Nature New Biol.* 233:225–29

Taylor, R. E., Moore, J. W., Cole, K. S. 1960. Analysis of certain errors in squid axon voltage clamp measurements. *Biophys. J.* 1:161–202

Thesleff, S., Sellin, L. C. 1980. Denervation supersensitivity. *Trends Neurosci.* 3:122–26

von Wedel, R. J., Carlson, S. S., Kelly, R. B. 1981. Transfer of synaptic vesicle antigens to the presynaptic plasma membrane during exocytosis. *Proc. Natl. Acad. Sci. USA* 78:1014–19

Vigny, M., Keonig, J., Rieger, F. 1976. The motor end-plate specific form of acetylcholine: Appearance during embryogenesis and reinnervation of rat muscle. *J. Neurochem.* 27:1347–52

Watanabe, A. M., McConnaughey, M. M., Strawbridge, R. A., Fleming, J. W., Jones, L. R., Besch, H. R. 1978. Muscarinic cholinergic receptor modulation of β-adrenergic receptor affinity for catecholamines. *J. Biol. Chem.* 253:4833–36

Weaver, D. L. 1983. Diffusion-mediated localization on membrane surfaces. *Biophys. J.* 41:81–86

Weinberg, C. B., Reiness, C. G., Hall, Z. W. 1981a. Topographical segregation of old and new acetylcholine receptors at developing ectopic endplates in adult rat muscle. *J. Cell Biol.* 88:215–23

Weinberg, C. B., Sanes, J. R., Hall, Z. W. 1981b. Formation of neuromuscular junctions in adult rats: Accumulation of acetylcholine receptors, acetylcholinesterase and components of synaptic basal lamina. *Dev. Biol.* 84:255–68

Willard, A. L. 1980. Electrical excitability of outgrowing neurites of embryonic neurones in cultures of dissociated neural plate of *Xenopus laevis. J. Physiol.* 301:115–28

Wolf, D. E., Ziomek, C. A. 1983. Regionalization and lateral diffusion of membrane proteins in unfertilized and fertilized mouse eggs. *J. Cell Biol.* 96:1786–90

Wong, R. K. S., Prince, D. A., Basbaum, A. I. 1979. Intradendritic recordings from hippocampal neurons. *Proc. Natl. Acad. Sci. USA* 76:986–90

Young. S. H., McCloskey, M., Poo, M-m. 1984. Migration of cell surface receptors induced by extracellular electric field: Theory and applications. In *Receptors,* ed. M. Conn. New York: Academic

Young, S. H., Poo, M-m. 1983a. Rapid lateral diffusion of extrajunctional acetylcholine receptors in *Xenopus* tadpole myotomes. *J. Neurosci.* 3:1888–99

Young. S. H., Poo, M-m. 1983b. Topographical rearrangement of ACh receptors alters channel kinetics. *Nature* 304:161–63

Young, S. H., Poo, M-m. 1983c. Spontaneous release of transmitter from growth cones of embryonic neurons. *Nature* 305:634–37

Zieglgansberger, W., Champagnat, J. 1979. Cat spinal motoneurons exhibit topographic sensitivity to glutamate and glycine. *Brain Res.* 160:95–104

Zigmond, R. E., Bowers, C. W. 1981. Influence of nerve activity on the macromolecular content of neurons and their effector organs. *Ann Rev. Physiol.* 43:673–87

Ann. Rev. Neurosci. 1985. 8:407–30

STIMULUS SPECIFIC RESPONSES FROM BEYOND THE CLASSICAL RECEPTIVE FIELD:

Neurophysiological Mechanisms for Local–Global Comparisons in Visual Neurons

John Allman, Francis Miezin, EveLynn McGuinness

Division of Biology, California Institute of Technology, Pasadena, California 91125

INTRODUCTION

We perceive the visual world as a unitary whole, yet one of the guiding principles of nearly a half century of neurophysiological research since the early recordings by Hartline (1938) has been that the visual system consists of neurons that are driven by stimulation within small discrete portions of the total visual field. These classical receptive fields (CRFs) have been mapped with the excitatory responses evoked by a flashed or moving stimulus, usually a spot or bar of light. Most of the visual neurons, in turn, are organized in a series of maps of the visual field, at least 10 of which exist in the visual cortex in primates as well as additional topographic representations in the lateral geniculate body, pulvinar and optic tectum (Allman 1977, Newsome & Allman 1980, Allman & Kaas 1984). It has been widely assumed that perceptual functions that require the integration of inputs over large portions of the visual field must be either collective properties of arrays of neurons representing the visual field, or features of those neurons at the highest processing levels in the visual system, such as the cells in inferotemporal or posterior parietal cortex that typically possess very large receptive fields and do not appear to be organized in visuotopic maps. These assumptions have been based on the results of the many studies in which receptive fields were mapped with con-

0147–006X/85/0301–0407$02.00

ventional stimuli, presented one at a time, against a featureless background. However, unlike the neurophysiologist's tangent screen, the natural visual scene is rich in features, and there is a growing body of evidence that in many visual neurons stimuli presented outside the CRF strongly and selectively influence neural responses to stimuli presented within the CRF. These results suggest obvious mechanisms for local–global comparisons within visuotopically organized structures. Such broad and specific surround mechanisms could participate in many functions that require the integration of inputs over wide regions of the visual space such as the perceptual constancies, the segregation of figure from ground, and depth perception through motion parallax. In the first section of this paper, we trace the historical development of the evidence of response selectivity for visual stimuli presented beyond the CRF; in the second, examine the anatomical pathways that subserve these far-reaching surround mechanisms; and in the third, explore the possible relationships between these mechanisms and perception.

STIMULUS SELECTIVITY BEYOND THE CLASSICAL RECEPTIVE FIELD

Retina and Lateral Geniculate Nucleus

Barlow (1953) was the first to probe the apparently silent regions beyond the CRF. While recording from retinal ganglion cells in the frog, he simultaneously turned on a spot of light in the receptive field and a second spot outside the field, and noted that the second spot suppressed the response to stimulation within the receptive field. Kuffler (1953) reported that receptive fields of retinal ganglion cells in the cat were divided into a center and an antagonistic surround. If the cell responded to turning on a spot of light in the center, it responded to turning off a spot in a concentric surrounding region; he also recorded off-centers with on-surrounds. Since stimulation of both the center and the surround gave excitatory responses, both would be parts of the CRF. For the same reason, the CRF would include the "on" and "off" regions for neurons in the lateral geniculate nucleus (Hubel & Wiesel 1961).

McIlwain (1964) activated optic tract axons and lateral geniculate neurons by presenting a flashing spot within their CRFs and found that stimulation with a moving dark spot up to 90° away influenced (usually facilitated) the response. McIlwain termed this phenomenon the "periphery effect" and discovered that it was produced by mechanisms within the retina, since bilateral optic tract section central to the optic tract recording site did not reduce the effect. The effect did not exhibit much selectivity for particular stimuli so long as the movement had sufficient speed and excursion. He also noted that a flashing spot in the classical receptive field was not required for some neurons. Subsequently, Levick et al (1965) and McIlwain (1966) found that the flashing spot was not necessary in general to observe the increased discharge rate due to movement in the surround. McIlwain (1966) recognized that this observation created a problem as to what to include in the CRF, a quandary that has never been resolved for the periphery effect. Cleland et al (1971) and Ikeda & Wright (1972) found the periphery effect in cat retinal ganglion cells with transient responses and faster axonal conduction velocities but not in cells with sustained responses with slower conduction velocities. Ikeda & Wright (1972) discovered that the periphery effect was suppressed by the presence of stationary contrast within the CRF. It was then determined that the effect could be greatly strengthened in both retinal ganglion cells and lateral geniculate neurons in cats by abruptly shifting a grating that occupied most of the visual field except for the region near the CRF (Kruger & Fischer 1973, Fischer & Kruger 1974). The effect was independent of the direction of displacement of the grating. In the lateral geniculate nucleus in macaque monkeys, Kruger (1977) found the effect in 90% of the neurons in the magnocellular laminae, which respond transiently and with short latencies. The effect was strong and excitatory in magnocellular neurons and was suppressed by steady illumination of the CRF. For neurons in the lateral geniculate parvocellular laminae, which respond in a sustained fashion and at longer latencies, the effect was present in only 30% of the cells and was weak and inhibitory. Also recording from the lateral geniculate nucleus in macaque monkeys, Marrocco et al (1982) demonstrated responses from beyond the CRF for neurons that did not exhibit the periphery effect. By turning on and off a spot in the receptive field and measuring the effect of rotating a radial grating beyond the classical surround, they observed both facilitatory and inhibitory effects that could be abolished by cooling the visuotopically corresponding portions of the striate cortex and thus inactivating the striate-geniculate feedback.

Optic Tectum

Suppressive regions surrounding the CRF for neurons in the optic tectum have been reported in the frog (Grusser-Cornehls et al 1963), the ground squirrel (Michael 1972), the cat (McIlwain & Buser 1968), and the macaque monkey (Wurtz et al 1980). The first evidence for directional selectivity for responses to stimuli presented beyond the CRF was discovered by Sterling & Wickelgren (1969) in the cat optic tectum. They presented a bar moving in the preferred direction in the center and a second bar moving in the surround. When the bar in the surround moved in the same direction as that preferred for the center, the cell was more suppressed than when the bar in the surround moved in the opposite direction. Also recording from the cat optic tectum, Rizzolatti et al (1974) stimulated the CRF and presented a second moving 10° spot at distances as much as 120° away. They found that the suppressive surround was present in 90% of the cells and extended throughout much of the visual field, although it tended to be stronger in the hemifield ipsilateral to the CRF. Rizzolatti et al also found that the suppressive effect was much stronger if the center and surround stimuli were synchronous. They did not find evidence for a directionally selective influence from the surround and suggested that they were studying a second, broader surround mechanism rather than that described by Sterling & Wickelgren (1969). It is also possible that a single spot stimulus presented in a remote part of the visual field was inadequate to demonstrate selectivity in the distant surround and that a moving pattern subtending a larger area would have been more effective. In the intermediate and deep layers of the optic tectum in pigeons, Frost et al (1981) stimulated the CRF with a moving spot and tested the effect of presenting different directions of movement of random dot patterns in the surround. They found that the surrounds were directionally selective and were more than 100° in diameter. Many of the cells were facilitated by background movement opposite to the direction of spot movement in the center. The responses from the CRF with the *random dot background stationary* were very broadly tuned for direction. Frost & Nakayama (1983) found that the surround effect in the pigeon depended on the direction of movement of the center stimulus, such that the direction of greatest suppression by surround movement was the same direction as the movement of the center stimulus.

Visual Cortex

Hubel & Wiesel (1962) described a class of neurons in the first visual area (V–I) in the cat, the simple cells, that like the Kufflerian cells possessed "on" and "off" subregions within their CRFs. Recording from the second (V–II) and third (V–III) visual areas in the cat, Hubel & Wiesel (1965) found a class of neurons, the lower order hypercomplex cells, in which they were able to map out antagonistic regions flanking the CRF. They were unable to obtain a response when they stimulated these regions alone, but when they simultaneously presented a slit stimulus in the CRF and another slit of the same orientation in the antagonistic flank, the response from the first slit was suppressed. Hubel & Wiesel (1965) concluded, "A cell responds when an appropriately oriented line (slit, bar, or edge) is shone or moved anywhere within the activating area, and the response is prevented with similar and simultaneous stimulation of the antagonistic region." They were thus the first to suggest that stimulation of regions beyond the CRF had more than merely a suppressive influence, that it could have a *selective* antagonistic effect. They estimated that the antagonistic flanks extended 2–3° beyond the CRF.

Recording in V–I in the cat, Jones (1970) presented an edge moving in the CRF and measured the effect of moving a second edge at different distances. In 65% of his cells, the second edge exerted an influence from locations beyond the CRF. He observed both inhibitory and facilitatory surround effects, which in some cases extended 20° from the receptive field although most extended less than 8°. Jones also noted that the more peripheral the location of the CRF, the greater the distance over which a second edge could exert an inhibitory influence. Blakemore & Tobin (1972) described a complex cell with an orientation-selective CRF and an antagonistic orientation-selective surround. They found that the response produced by an optimally oriented bar was abolished by presenting a grating of the same orientation in the surround and was slightly facilitated by an orthogonal grating in the surround.

Bishop et al (1973) studied simple cells in V–I in cats by continuously stimulating the CRF and measuring the effect of test stimuli presented at different locations. Through these techniques they were able to map out powerful but nonorientation selective inhibitory regions that extended 2 to 6° beyond the CRF. Recording intracellularly, Creutzfeldt et al (1974) mapped inhibitory areas 3–4° in diameter that overlapped excitatory areas that were about 2° in diameter. They found that the inhibitory influences were directional and to some extent orientation-selective. Maffei & Fiorentini (1976) found facilitatory and inhibitory regions in the surrounds of both simple and complex cells. They found that the facilitatory surround regions were orientation-selective, but reported that the inhibitory regions were not very selective for orientation. Maffei & Fiorentini also found that the facilitatory and inhibitory surround regions were tuned for the spatial frequency of sine-wave gratings. The inhibition was strongest for the spatial frequency for which the CRF was maximally sensitive. They estimated that the facilitatory and inhibitory surround regions together were three times the size of the CRFs. Fries et al (1977) and Nelson & Frost (1978) found that simple and hypercomplex cells have antagonistic orientation-selective surrounds beyond their CRFs and that these surrounds are also sometimes directionally selective as well. Hammond & McKay (1981) tested the influence of moving random dot patterns on the responses of simple cells stimulated with optimally oriented moving bars. They found antagonistic direction-selective regions that extended beyond the CRFs for the simple cells and were comparable in size to the CRFs of nearby complex cells.

Using a two-stimulus paradigm in V–I in cats that was similar to their experiments in the optic tectum, Rizzolatti & Camarda (1977) found only very weak effects resulting from surround stimulation more than 30° from the CRF. However, in lateral suprasylvian visual cortex, their findings were similar to their previous results in the optic tectum, namely broad inhibitory surrounds that were stronger in the hemifield ipsilateral to the CRF and were not directionally selective. Most neurons in lateral suprasylvian visual cortex are directionally selective (Hubel & Wiesel 1969, Spear & Baumann 1975). Recently, von Grunau & Frost (1983) have tested surround mechanisms in this region with the same methods used in the pigeon optic tectum experiments. In 9 of

11 cells tested quantitatively, they found directionally selective surrounds; the preferred direction for the center was the direction of greatest suppression by the surround, and the opposite direction was either less inhibitory or facilitatory. It is probable in this case, as in the optic tectum, that a moving array of dots subtending a large portion of the visual field is a more effective test for directional selectivity in the surround than a single spot. The moving array of random dots also more closely approximates natural viewing conditions in which the visual image is commonly a texture-filled scene.

In primates, the middle temporal visual area contains a high concentration of directionally selective neurons (Baker et al 1981, Maunsell & Van Essen 1983a). We have investigated surround mechanisms in the middle temporal area in the owl monkey (Miezin et al 1982, Allman et al 1984a). Middle temporal neurons are very responsive to moving random dot patterns. We stimulated the CRF with random dots moving in the preferred direction and at the preferred velocity and tested the effect of different directions of random dot movement in the surround. Forty-four percent of our middle temporal cells had antagonistic directionally selective surrounds and half of these cells were facilitated by background movement in the direction opposite to that preferred by the center (see Figure 1). An additional 8% were maximally suppressed by background movement in the preferred direction for the center but were powerfully and specifically facilitated by a shearing background movement in one direction 90° to the preferred direction for the center. This specific facilitation indicates that center–surround relations are not limited to simple antagonistic interactions. Another 30% of our middle temporal cells were suppressed by all directions of movement. Finally, 8% showed no effect from surround stimulation; these cells were often recorded immediately adjacent to cells with directionally selective surrounds and shared with these adjacent cells the same CRFs and the same directional preferences. It is possible that at a higher stage of neural processing the outputs of these two types of cells are compared. Such a comparison would enable the system to determine whether a particular stimulus movement was an isolated occurrence or part of a larger pattern of stimuli moving in one direction.

The histogram in the top half of Figure 2 illustrates the combined normalized response of 42 middle temporal neurons of all three types to random dots moving in the preferred direction within their CRFs with the background stationary. The responses from the CRFs began abruptly in the third bin following the onset of movement and ceased abruptly in the third bin following the offset of movement. Each bin represents 40 msec. There was a transient response lasting about 600 msec, followed by a response sustained throughout the remainder of the stimulus presentation in the CRF. The lower histogram in Figure 2 indicates that the inhibitory responses from the region beyond the CRF began abruptly in the fourth bin following the onset of background

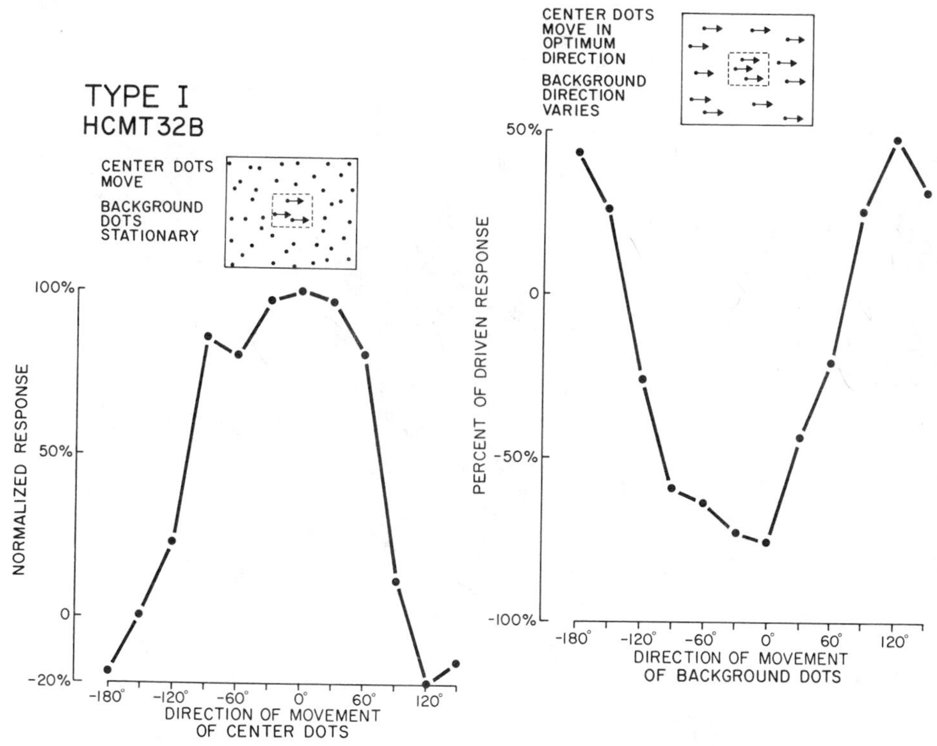

Figure 1 Direction-selective neuron with an antagonistic direction-selective surround recorded from middle temporal area in the owl monkey. The *left graph* depicts the response of the cell to 12 directions of movement of an array of random dots within an area coextensive with its CRF. The response is normalized so that 0% is equal to the average level of spontaneous activity sampled for 2-sec periods before each presentation. Negative percentages in the left graph indicate inhibition relative to the level of spontaneous activity. In the left graph, the response to the optimum direction is 100%. The *right graph* depicts the response of the cell to different directions of background movement while the CRF was simultaneously stimulated with an array of dots moving in the cell's preferred direction. In the right graph, the CRF was stimulated by the array moving in the optimum direction during the 2-sec sample periods preceding background movement, and thus a response of 100% in the left graph is equivalent to 0% in the right graph. The stimulus conditions are depicted schematically above each graph. In the experiment, the dots were 50% dark–50% light and the background was much larger relative to the center than is depicted schematically.

movement and ceased abruptly in the fourth bin following the offset of background movement. Thus the response from beyond the CRF required somewhat less than 40 msec additional processing time beyond that required for the CRF. The lower histogram also indicates that there was a transient rebound in the response from the CRF following the offset of background movement.

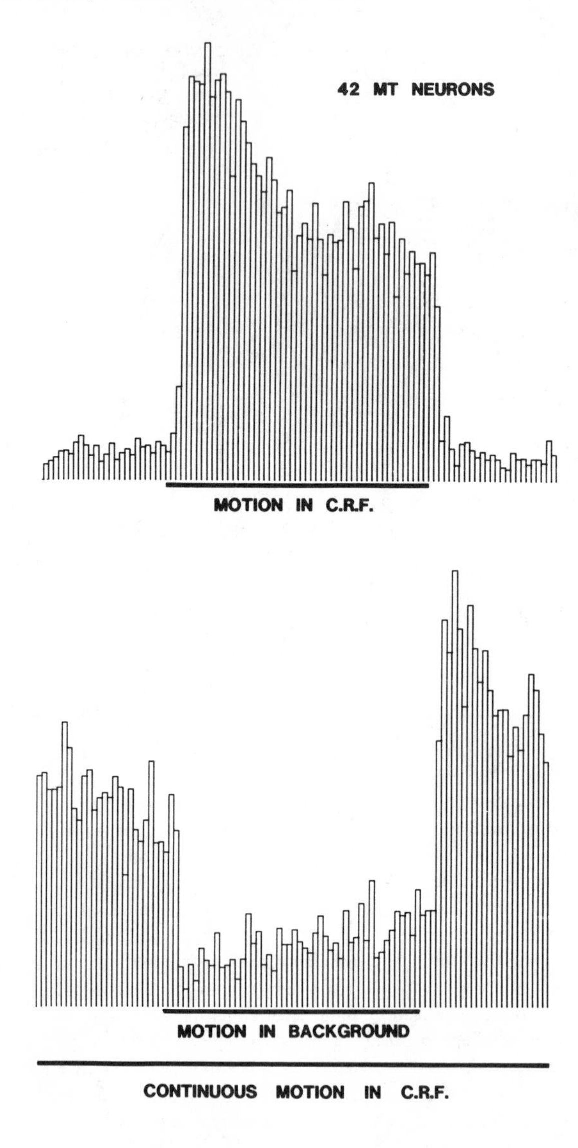

Figure 2 The *top histogram* illustrates the combined responses of 42 middle temporal neurons to random dots moving for a 2-sec period in the preferred direction within their CRFs, with the background stationary. In the *lower histogram,* the same middle temporal neurons were stimulated continuously with random dots moving in the preferred direction within their CRFs and then tested for a 2-sec period in which the random dots in the surrounding regions also moved in the same direction. Each bin represents 40 msec. The histograms were constructed by normalizing with the largest 40 msec bin in the histogram for each cell and then combining the histograms.

Background movement also has a profound influence on responses of middle temporal neurons to moving bar stimuli. Figure 3 illustrates that background movement in the preferred direction (180°) can totally abolish the response to bar stimuli, whereas background movement in the anti-preferred direction (0°) can powerfully facilitate such responses. Background movement in the preferred direction inhibited nearly all middle temporal neurons; background movement in the anti-preferred direction facilitated 44% of middle temporal neurons. Figure 3 also shows that the background movement can influence the *sharpness of tuning* of the responses to bar stimuli moving in different directions, because the responses to the nonoptimal directions of 120°, 270°, and 300° that were present when the background was stationary were virtually abolished when the background dots moved at 0°.

In the foregoing experiments, the random dots in the surround moved at the same speed as in the center. We discovered that the velocity of the random dots moving in the preferred direction in the background also influenced the response to bar stimuli. In Figure 4 the left graph illustrates the velocity tuning curve for a neuron tested with a bar stimulus of optimum length, width, and contrast moving in the cell's preferred direction against a background of stationary random dots. The cell's preferred velocity was 16° per second. In the right graph in Figure 4, the bar moved at 16° per second and the velocity of the random dot background was varied. The result was profound inhibition produced by background stimulation at the preferred velocity of 16° per second. About one-third of middle temporal cells exhibited this V-shaped pattern; the maximum inhibition resulted when the background dots moved at the preferred velocity for the bar stimulus; in the remaining middle temporal cells the inhibition tended to decrease with background velocity. The V-shaped velocity for the CRF is characteristic of the majority of V–II neurons in the owl monkey (Allman et al 1984b).

We mapped the extent of the surround in ten middle temporal cells by masking off parts of the screen while stimulating both the center and surround with random dots moving in the preferred direction at the preferred velocity. Figure 5 illustrates the results. The CRF was surrounded by a masking annulus of variable outside diameter. In only one cell in this series, ANMT22D–I, were we able to create an annulus sufficiently large to eliminate the suppressive effect of background movement, and this was with an annulus eight times the diameter of the CRF. The data suggest that the surrounds are seven to ten times the diameter of the CRFs of middle temporal neurons. The areas of the CRFs for these cells increased with eccentricity and ranged from 25 to 700 deg^2. The smallest surround in this sample of middle temporal neurons would thus be about 1200 deg^2 and the others would range upward, with the largest including virtually the whole visual field.

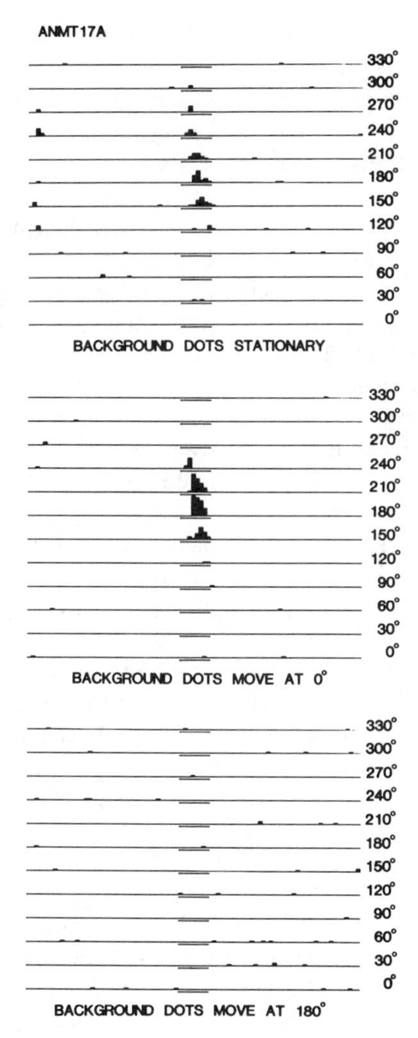

Figure 3 The responses of a middle temporal neuron to a bar moving in different directions superimposed on a background of random dots. The bar was oriented orthogonally to the direction of movement. The results of each of the 12 directions (0° through 330°) are shown in histograms consisting of a fore period, an underscored stimulus presentation period, and an after period. The largest histogram bin contains 26 spikes.

Very recently Tanaka et al (1984) recorded from the middle temporal area in macaque monkeys. The response to a moving bar was suppressed by background movement in the same direction in one half of their middle temporal neurons; one-quarter were facilitated by background movement in the opposite direction. In these neurons the effective zone extended well beyond the CRF.

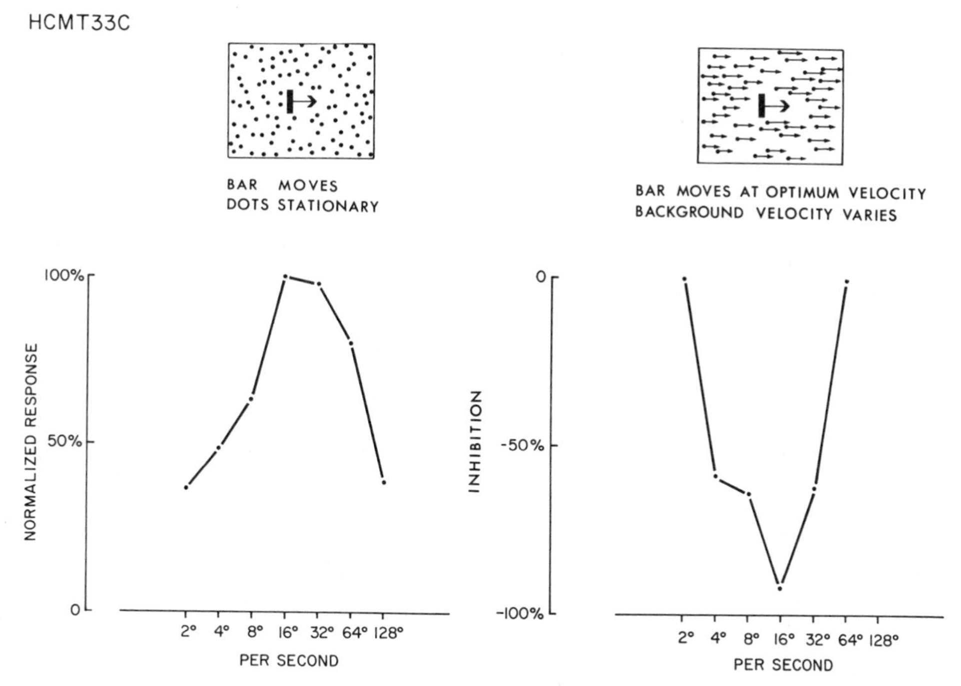

Figure 4 The effect of center and background velocity on a neuron recorded from the middle temporal area in the owl monkey. The *left graph* is a velocity-tuning curve for a bar moving in the preferred direction with the background stationary. The *right graph* is a velocity-tuning curve for background movement while simultaneously presenting the bar moving at the optimum velocity (16° per second). In the background velocity experiments, the background dots stimulated both the CRF and the surround; however, in these experiments covering the surround eliminated or greatly reduced the inhibitory effect.

There is a strong quantitative similarity in other aspects of the functional properties of middle temporal neurons recorded from the macaque and the owl monkeys (Baker et al 1981, Maunsell & Van Essen 1983a, Albright 1984). It remains to be seen whether the lower incidence of cells influenced by moving background in the macaque middle temporal area reflects a genuine functional difference or amthodological difference between the studies conducted in the two species.

Recent recordings from neurons in the V4 complex in macaque monkeys have revealed broad surround regions tuned for orientation, spatial frequency, and color (Desimone et al 1984, R. Desimone and S. J. Schein, personal communication). Moran et al (1983) found that while the CRFs for V4 cells located near the vertical meridian extended an average of only 0.6° into the ipsilateral hemifield, the inhibitory surrounds extended at least 16° into the ipsilateral hemifield. This suppression by stimulation in the ipsilateral hemi-

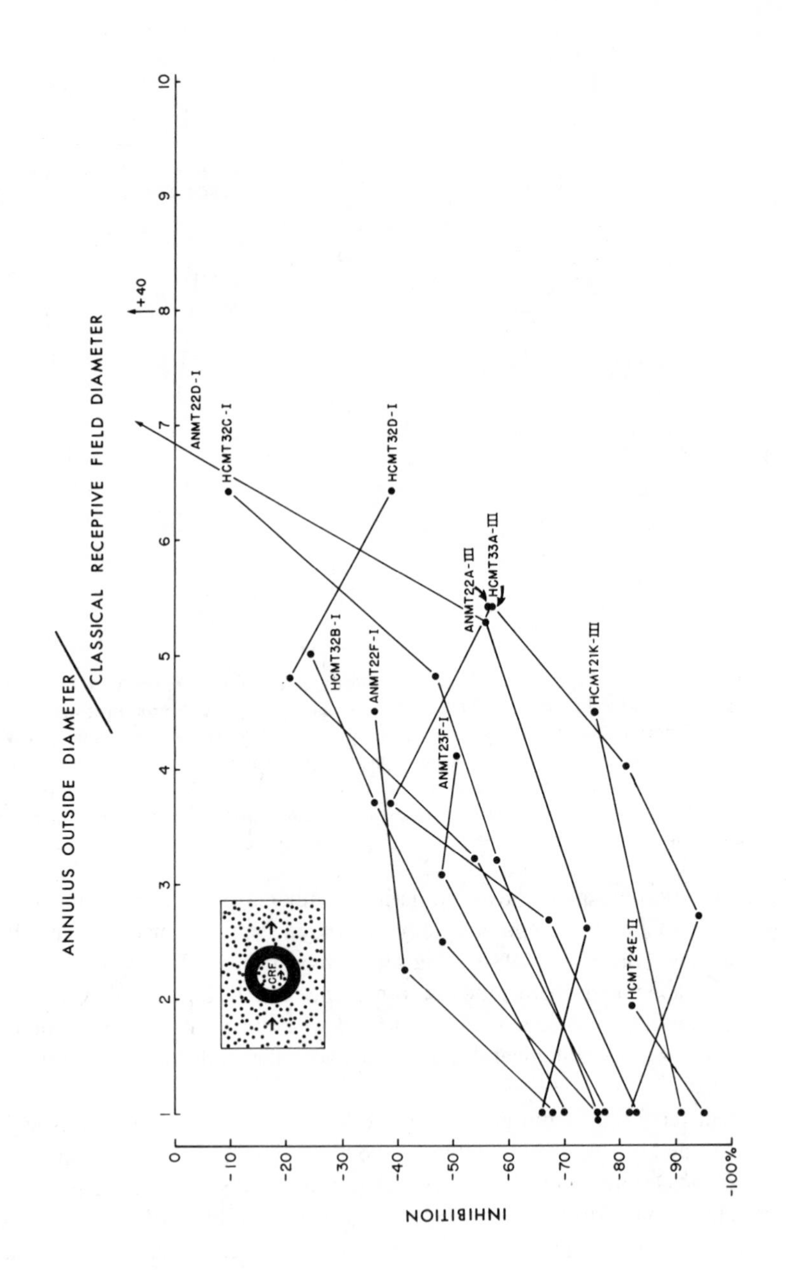
ANNULUS OUTSIDE DIAMETER / CLASSICAL RECEPTIVE FIELD DIAMETER
1
2
3
4
5
6
7
8
9
10
+40
INHIBITION
0
-10
-20
-30
-40
-50
-60
-70
-80
-90
-100%
CRF
ANMT22D-I
HCMT32C-I
HCMT32D-I
ANMT22A-III
HCMT33A-III
HCMT32B-I
ANMT22F-I
ANMT23F-I
HCMT21K-III
HCMT24E-II

field was greatly reduced by section of the corpus callosum. Zeki (1983a,b) reported two types of neurons in macaque V4: wave-length selective and color-coded. The responses of the first type were highly dependent on the wavelength of light illuminating the CRF. The responses of the color-coded cells depended on the natural color of the object in the CRF under normal illumination, but not on the spectral content of the light reflected from the object. Thus, the color-coded cells exhibited color constancy over a certain range of illumination conditions. The response of the color coded cells depended on the color of objects located outside the CRF, which indicated that they possess surround mechanisms with complex and as yet undetermined properties (Zeki 1983b). The influence of the surround on the color of an object was limited to the ipsilateral hemifield in a corpus-callosum sectioned human (Land et al 1983), which restricts the locus of the effect to the cortex, but regions other than the V4 complex, particularly V–II and the ventral posterior area (VP), may be involved as well (Burkhalter & Van Essen 1982).

Finally, von der Heydt et al (1984) discovered that about one third of the neurons in V–II in awake macaque monkeys respond to illusory contours when the real contours evoking the response were located entirely outside the CRF. V–I neurons were unresponsive under the same conditions. The perception of illusory contours might be considered as a type of constancy, since the visual system is interpolating a continuous contour from an interrupted contour, which under natural conditions would be produced by a partially occluding surface. The tropical forest environment, in which primates evolved, abounds with occluding foliage and branches, and the ability to reconstruct surfaces that are partially hidden from view would be very adaptive.

Extensive surrounds beyond the CRF are not limited to the visual system. Barn owls, which hunt in darkness using sound localization, possess auditory neurons with sharply defined spatial receptive fields that are organized into an orderly representation of auditory space in the midbrain nucleus mesencephalicus lateralis dorsalis (MLD) (Knudsen & Konishi 1978a). These receptive fields are mapped in the owl's auditory space by moving a sound source in an

Figure 5 The effect of varying the outside diameter of masking annuli of surround inhibition of the response from the CRF in 10 neurons recorded from middle temporal area in the owl monkey. The stimulus conditions are depicted schematically, but in the experiments, the dots were 50% dark–50% light and the surround was much larger. The dots in the center and surround moved in the optimum direction for the CRF. The inside diameter of the masking annulus corresponds approximately to the diameter of the CRF. The abscissa corresponds to the ratio of the outside diameter of the masking annulus to the diameter of the CRF. A value of 1 is equivalent to stimulation without the masking annulus. The areas of the CRFs were: ANMT22A–III = 25 deg^2; ANMT22D–I = 35 deg^2; ANMT22F–I = 48 deg^2; HCMT21K–III = 100 deg^2; HCMT33A–III = 133 deg^2; ANMT23F–I = 175 deg^2; HCMT32C–I = 343 deg^2; HCMT32D–I = 343 deg^2; HCMT24E–II = 560 deg^2; HCMT32B–I = 700 deg^2.

anechoic chamber. Knudsen & Konishi (1978b) probed the regions beyond these receptive fields by stimulating the MLD neurons with a sound source located in the CRF and measuring the effect of moving a second sound source through the remainder of the auditory field in a manner analogous to our experiments. By using this technique they demonstrated that the second sound source had an inhibitory effect throughout most of auditory space beyond the CRF. Thus neurons in the owl's auditory space mapped MLD would be capable of making the same sort of local–global comparisons within a representation of space that exists in the middle temporal area and other visual structures.

Summary of Responses from Beyond the Classical Receptive Field

Most of the known response properties described in the studies carried out within the CRF of neurons at various levels in the visual system (selectivity for direction and velocity of movement, orientation, spatial frequency, and wavelength) are matched in some neurons by antagonistic tuned mechanisms in the surrounding parts of the visual field from which no direct response can be obtained but which nonetheless exert a strong influence on the responses obtained from stimuli presented within the CRF. One significant parameter, relative depth between the center and the surround, has yet to be investigated, but we predict the existence of neurons with antagonistic surround mechanisms tuned for depth. The "periphery effect" and the responses to illusory contours indicate that under some conditions, excitatory responses can be obtained from apparently silent surrounds without any direct stimulation of what would normally be considered the CRF. Thus, the concept of the CRF, as useful as it is, does have some vexing limitations.

ANATOMICAL CONNECTIONS AND SURROUND SPECIFICITY

Dowling & Boycott (1966) suggested that McIlwain's (1964) periphery effect was mediated by amacrine cells, since there are serial amacrine–amacrine and amacrine–ganglion cell synapses that could transmit the effect across the retina. Werblin's (1972) intracellular recordings demonstrated that amacrine cells were depolarized by remote retinal stimulation. These results suggest that responses from beyond the CRF also might be mediated by serial chains of interneurons at higher levels in the visual system. Although this remains a valid possibility, there is an increasing body of anatomical data indicating that cortical neurons extend their axons horizontally beyond the dimensions that would correspond to their CRFs. Recently, Rockland et al (1982) injected the neural tracer, horseradish peroxidase, into V–I in tree shrews and found horizontal connections arrayed in periodic stripes in layers I–III throughout a large portion of this area. Rockland et al and Mitchison & Crick (1982) have proposed several schemes for the elaboration of receptive field mechanisms on the basis of this intrinsic connectivity. Since the CRFs in V–I in tree shrews are small and in precise visuotopic order (Kaas et al 1972), these periodic intrinsic connections are probably related to surround mechanisms beyond the CRF. The striking similarity of this striped pattern of intrinsic connectivity with the topographic distribution of orientation preferences revealed by single neuron recording and metabolic uptake of 2-deoxyglucose in V–I in tree shrews (Humphrey & Norton 1980, Humphrey et al 1980) suggests that this intrinsic connectivity is related to orientation preferences. Since most surround mechanisms are antagonistic, it is likely that this intrinsic connectivity participates in elaboration of antagonistic orientation-selective surrounds, possibly mediated by inhibitory interneurons. Such antagonistic orientation-selective surrounds have often been found for visual cortex neurons in cats. Gilbert & Wiesel (1983) injected single neurons in V–I in cats with horseradish peroxidase and found that the axons extended horizontally over distances of up to 4 mm and that collaterals within the axonal fields were distributed in repeating clusters with an average periodicity of about 1 mm. This pattern of clustered axonal collateralizations may also be related to the elaboration of stimulus-specific surround mechanisms.

In V–I in monkeys, horizontal intrinsic fibers extend well beyond the distance of about 2 mm that would be expected for connecting a cell with adjacent cells sharing portions of its CRF (Hubel & Wiesel 1974). Fisken et al (1975) made small lesions in V–I and traced degenerating fiber 1.5 to 2 mm horizontally in most layers, except in the stria of Gennari, where fibers extended horizontally for up to 4 mm. More recently Rockland & Lund (1983) injected horseradish peroxidase in V–I in macaque and squirrel monkeys and found a periodic lattice-like pattern of dense connections extending 2 mm from the margin of the injection in layers II–III and 3 mm in the stria of Gennari. Similar lattice-like patterns of intrinsic connections have been found to extend 2 to 3 mm in V2 and 3.5 to 4 mm in V4 in monkeys (Rockland 1983). These patterns of connectivity may participate in the elaboration of stimulus-specific surround mechanisms. It is particularly interesting that the horizontal connections are most extensive for V–I in the stria of Gennari, as this layer receives an input from the magnocellular laminae of the lateral geniculate nucleus via layer IV–C–α (Hubel & Wiesel 1972, Lund et al 1975), contains a high proportion of directionally selective neurons (Dow 1974, Livingstone & Hubel 1984), and projects to the middle temporal area (Spatz 1975, Lund et al 1975, Tigges et al 1981, Maunsell & Van Essen 1983b). Montero (1980) found by using two separate tracers that the input from V–I to the middle temporal area terminates in a series of bands within partially overlapping projections from adjacent sites in V–I. These partially overlapping striate projections may contribute, possibly via interneurons, to the large direction selective surrounds in the middle temporal area. Small injections of tritiated proline in the middle temporal area have revealed extensive connections within the area that also terminate in a series of bands, and thus could also contribute to large direction-selective surrounds (Maunsell & Van Essen 1983b, Weller et al 1984). Although the transcallosal connections of the middle temporal area are much heavier near the representation of the vertical meridian, they extend throughout most of the area (Newsome & Allman 1980) and thus may contribute to the portion of the surround extending into the opposite half of the visual field as do transcallosal connections in V4 (Moran et al 1983). Large portions of the V4 complex are richly supplied with callosal connections (Van Essen et al 1982).

Another source of stimulus-specific input to surrounds is from noncortical structures. Sherk & LeVay (1983) found that unilateral destruction of the claustrum in cats resulted in a 50% reduction in the number of end-stopped cells in the ipsilateral V–I with other visual response properties apparently unaffected. End-stopping is produced by inhibitory regions flanking the CRF (Hubel & Wiesel 1965). Neurons in the visual claustrum of the cat are arranged in a representation of the visual field (LeVay & Sherk 1981) and respond best to long bars or slits with a definite preferred orientation (Sherk & LeVay 1981). Thus, neurons in the visual claustrum seem well suited to provide input, possibly through cortical interneurons, to antagonistic orientation-specific surrounds in the visual cortex (Sherk & LeVay 1983). Comparable experiments have not been conducted in primates; however, the claustrum has been found to project to the visual cortex in galagos (Carey et al 1979) and baboons (Riche & Lanoir 1978). Another possible source of stimulus-specific input to visual cortical surrounds is from the pulvinar. The pulvinar contains several topographic representations of the visual field (Allman et al 1972, Gattass et al 1978, Bender 1981) and many direction- and orientation-selective neurons (Gattass et al 1979, Bender 1982).

The visual system contains many descending pathways (Tigges et al 1981, Maunsell & Van Essen 1983b), such as from the middle temporal area to V–I and V–II, which could contribute to surround mechanisms in the recipient structure. The CRFs in the higher area typically are larger than those at a comparable eccentricity in the lower area, and thus the CRFs in the higher area might match the dimensions of the true receptive field including the surround in the lower area (F. Crick, personal communication). Marrocco et al (1982) has demonstrated that interruption of the striate–geniculate pathway by cooling striate cortex eliminates surround responses from regions beyond the CRF in many lateral geniculate neurons. Small injection sites in the superior temporal visual area project to the entire extent of the ipsilateral middle temporal area in the owl monkey (Weller et al 1984) and may be another source of the large surrounds present for neurons in this structure.

In summary, surround mechanisms have been demonstrated to be mediated by (*a*) transcallosal connections with the representation of the opposite hemifield, (*b*) input from the claustrum, (*c*) descending input. In addition, it is very likely that intrinsic connections with adjacent parts of the visuotopic map that extend beyond the neurons's CRF contribute to surround mechanisms. All of these connections subserving surround mechanisms may be further mediated by interneurons. Finally, it should be pointed out that although we emphasize in the chapter anatomical connections that transcend the topographic organization of the CRFs, a great many connections do reflect the visuotopic organization of CRFs. This, of course, has been the subject of considerable investigation (for recent research in primates see Weller & Kaas 1983).

SURROUNDS AND PERCEPTION

Stimulus Selectivity Beyond the Classical Receptive Field and Figure–Ground Discrimination

Stimulus-specific responses from beyond the CRF seem ideally suited for discriminating figure from ground and preattentive vision (Treisman & Gelade 1980, Julesz 1981). Julesz's elementary units of figure–ground discrimination, the "textons," are based on differences in motion, color, orientation, etc. that are strikingly similar to the tuned antagonistic interactions between the CRF and the background in visual cortical and tectal neurons. This preattentive system is capable of guiding focal attention with a latency of about 50 msec (Julesz 1984), which is slightly longer than the time required for the response from the regions beyond the CRF to influence the response from within the CRF in middle temporal neurons (see Figure 2).

Stimulus Selectivity Beyond the Classical Receptive Field and Perceptual Constancies

The function of the visual system is to extract behaviorally significant features embedded in a complex optical array over a very broad range of environmental conditions. Its first task is to discriminate discontinuities in the optic array. Local antagonistic center–surround mechanisms clearly have this role. A second and more difficult task is to make good estimates of the qualities of objects in the visual field, their color and motion for example, on the basis of rather imperfect optical information imaged on the photoreceptor layer of the retina. Thus, the wavelength composition of the retinal image will depend on environmental lighting conditions that may vary enormously, yet the behaviorally significant task may involve judging the ripeness of fruit based on its color. Retinal image motion can be produced by movement of the eye, movement of the animal, or movement in the environment, yet the system's task is to determine the motion of objects relative to other objects and to the observer.

In both cases more than just information restricted to a small locality on the retinal surface is required to make veridical judgments. Land's (1959a,b, 1983) experiments indicate that the system compares the wavelength composition of the light reflected by the object with that of other objects in the surrounding visual field and is able to extract color constancy over a broad range of lighting conditions. The determination of the motion of objects in the environment similarly requires the integration of motion information over a large portion of the visual field. To determine object motion relative to the observer requires further input concerning eye and head position. These position inputs are usually thought to be derived from motor commands to the eye muscles (von Helmholtz 1925), from the vestibular system, and perhaps from proprioceptors in the eye muscles; however, another parallel source of position information could be derived from the visual image itself (Gibson 1966, Koenderink 1984) and implemented through surround mechanisms.

The neurons in several of the extrastriate cortical visual areas have properties that seem well suited for performing the local–global comparisons that serve as the basis of perceptual constancy functions and the discrimination of figure from background. They possess large stimulus-specific surrounds that extend well beyond their CRFs and are capable of comparing stimuli present within their CRFs and other stimuli present throughout large portions of the visual field. Their CRFs are embedded in representations of the visual field that may provide the spatial structure to the perceptual image. Recently Ingle & Shook (1984) found that area 18 lesions in gerbils produced a selective deficit in the animal's ability to make local–global comparisons. It is remarkable that more than 40 years ago Kluver (1942) proposed on the basis of behavioral deficits in monkeys following occipital lobectomy that the visual cortex is necessary for spatial discrimination and perceptual constancies. He made the analogy that just as mechanisms have evolved to maintain a constant internal physiological environment in higher organisms, which create for them an independence from external environmental conditions, similarly the perceptual constancy functions of the visual cortex "guarantee varying degrees of freedom from the visual milieu." Such constancy mechanisms are of enormous adaptive value but are energetically and computationally very expensive, and yield only a partial independence from the physical environment. For example, common experience reminds us that objects may in some instances appear differently colored under different lighting conditions.

Surround Mechanisms and the Perception of Depth Through Motion

Von Helmholtz (1925) observed: "Suppose, for instance, that a person is standing still in a thick woods, where it is impossible for him to distinguish, except vaguely and roughly, in the mass of foliage and branches all around him what belongs to one tree and what to another, or how far apart the separate trees are, etc. But the moment he begins to move forward, everything disentangles itself, and immediately he gets an apperception of the material contents of the woods and their relations to each other in space, just as if he were looking at a good stereoscopic view of it." Nakayama & Loomis (1974) have suggested a division of labor between *stereopsis* and *kineopsis:* "Retinal disparity, based on a relatively small interpupillary distance, probably controls behavior which is directed at the near environment; whereas optical velocity information (kineopsis), based on much greater displacements of a single eye, controls more distantly directed behavior." Von Helmholtz (1925) concluded: "The apparent angular velocities of objects in the field of view will be inversely proportional to their real distances away; and, consequently, safe conclusions can be drawn as to the real distance of the body from its apparent angular velocity."

Nakayama & Loomis (1974) postulated a simple neural mechanism that could serve as the basis for the analysis of optical flow patterns that occur as a viewer moved through its environment, with the images of objects located at different distances from the viewer moving at different velocities across the retina. They hypothesized the existence of a class of neurons possessing a velocity-selective center with an antagonistic velocity-selective surround. Such neurons would be suppressed by an optical flow field of uniform velocity but would detect differential velocities such as would result from sweeping past objects at different distances from the viewer. In our studies of neurons in the middle temporal area (Allman et al 1984a) and V–II (Allman 1984b), we have provided the first experimental confirmation of this hypothesis in the discovery of neurons sensitive to background movement. It is easy to imagine how an antagonistic velocity-selective center-surround mechanism, as first proposed by Nakayama & Loomis and demonstrated in our study, could subserve the spatial, velocity-discriminating function required for depth perception through motion parallax or optical flow patterns. It should be pointed out that the characteristic symmetrical V-shaped background velocity tuning curves obtained from the majority of V–II and one third of middle temporal neurons do not discriminate between the condition in which the background movement is faster than the preferred velocity and the condition in which the background movement is slower than the preferred velocity. Thus, these neurons could register the relative magnitude of the depth difference between center and surround but not whether the center was nearer or farther than the background. The middle temporal cells in which inhibition simply increases with background velocity might help to resolve this ambiguity.

The velocity–distance relationship postulated by von Helmholtz (1925) obtains only when the observer fixates on very distant objects. If the observer fixates on an object at a given depth while he or she is in motion, objects beyond the fixation plane will move in the same direction as the observer while objects nearer than the fixation plane will move in the opposite direction (Gordon 1965). Rogers & Graham (1979) produced compelling, unambiguous depth illusions based on motion parallax with random dot patterns displayed on an oscilloscope screen that confirmed the depth relationships predicted by this optical configuration. There exists additional motion-related depth information in the visual scene described so graphically by von Helmholtz (1925). As Gibson (1979) has emphasized, the disappearance or emergence of background from behind an occluding surface is a strong cue for depth. The depth percept elicited by kinetic occlusion is very powerful and can override conflicting stereoscopic cues (Royden et al 1984). The antagonistic direction-selective center-surround mechanism may serve the computations for depth perception through kinetic occlusion by helping to identify which surfaces in an array are in motion with respect to other surfaces.

CONCLUSIONS

The function of the visual system is not merely to create a set of precise neural analogues of the optic image on the photoreceptors, but beyond this, to reconstruct behaviorally significant features of the visual environment on the basis of imperfect and unconstant optical stimuli. Gibson (1950, 1966, 1979), Land (1959a,b, 1983), and Ramachandran & Anstis (1983) have emphasized the influence of the context of the whole visual field on perception at any one locality within the field. The brain contains many maps of the visual field as revealed by the topographic organization of CRFs, but the total receptive fields (TRFs) for many neurons in these maps may be much larger and even extend throughout much of the visual field. The TRFs provide mechanisms for local–global comparisons embedded in visuotopic matrices that may serve as the basis for many functions in vision, such as the perceptual constancies, figure–ground discrimination, and depth perception through motion. The surrounds explored thus far usually exert selective *antagonistic* influences on their CRFs, but the existence of more complex surround mechanisms is indicated by middle temporal neurons that are strongly facilitated by background motion shearing at 90° to the preferred direction for the CRF, by the responses to illusory contours in V–II (von der Heydt et al 1984), and by the influence of background color patches on the properties of color-coded neurons in the V4 complex (Zeki 1983b). The successful exploration of these complex surround mechanisms calls for collaboration among psychophysicists, mathematical modelers, and neurophysiologists. There exist some very promising beginnings for this endeavor (Horn 1974, Nakayama & Loomis 1974, Ballard et al 1983, Land 1983, Reichardt et al 1983). The exploration of surround mechanisms will be vital to our understanding of the role that each cortical visual area plays in the perceptual processes.

ACKNOWLEDGMENTS

We thank Drs. Francis Crick, Robert Desimone, John Maunsell, and Terrence Sejnowski for many helpful discussions and Leslie Wolcott for drawing the illustrations. This work was supported by grants from the National Institutes of Health (EY–03851), the Pew Memorial Trust and the L. S. B. Leakey Foundation.

Literature Cited

Albright, T. D. 1984. Direction and orientation selectivity of neurons in visual area MT of the macaque. *J. Neurophysiol.* In press

Allman, J. M. 1977. Evolution of the visual system in the early primates. In *Progress in Psychobiology and Physiological Psychology*, ed. J. M. Sprague, A. N. Epstein, 7:1–53. New York: Academic

Allman, J. M., Kaas, J. H. 1984. Superior temporal visual area (ST) in the owl monkey. In preparation

Allman, J. M., Kaas, J. H., Lane, R. H., Miezin, F. M. 1972. Representation of the visual field in the inferior nucleus of the pulvinar in the owl monkey (Aotus trivirgatus). *Brain Res.* 40:291–302

Allman, J., Miezin, F., McGuinness, E. 1984a. Direction and velocity specific responses from beyond the classical receptive field in cortical visual area MT. *Perception*. In press

Allman, J., Miezin, F., McGuinness, E. 1984b. Direction and velocity specific responses from beyond the classical receptive field in the first and second cortical visual areas. In preparation

Baker, J. F., Petersen, S. E., Newsome, W. T., Allman, J. 1981. Visual response properties of neurons in four extrastriate visual areas of the owl monkey *(Aotus trivirgatus):* A quantitative comparison of medial, dorsomedial, dorsolateral, and middle temporal areas. *J. Neurophysiol.* 45:397–416

Ballard, D. H., Hinton, G. E., Sejnowski, T. J. 1983. Parallel visual computation. *Nature* 306:21–26

Barlow, H. B. 1953. Summation and inhibition in the frog's retina. *J. Physiol.* 119:69–88

Bender, D. E. 1981. Retinotopic organization of macaque pulvinar. *J. Neurophysiol.* 46:672–93

Bender, D. E. 1982. Receptive-field properties of neurons in the macaque inferior pulvinar. *J. Neurophysiol.* 48:1–17

Bishop, P. O., Coombs, J. S., Henry, G. H. 1973. Receptive fields of simple cells in the cat striate cortex. *J. Physiol.* 231:31–60

Blakemore, C., Tobin, E. A. 1972. Lateral inhibition between orientation detectors in the cat's visual cortex. *Exp. Brain Res.* 15:439–40

Burkhalter, A., Van Essen, D. C. 1982. Processing of color, form and disparity in visual areas V2 and VP of ventral extrastriate cortex in the macaque. *Neurosci. Abstr.* 8:811

Carey, R. G., Fitzpatrick, D., Diamond, I. T. 1979. Layer I of striate cortex of *Tupaia glis* and *Galago senegalensis:* Projections from thalamus and claustrum revealed by retrograde transport of horseradish peroxidase. *J. Comp. Neurol.* 186:393–438

Cleland, B. G., Dubin, M. W., Levick, W. R. 1971. Sustained and transient neurones in the cat's retina and lateral geniculate nucleus. *J. Physiol.* 217:473–96

Creutzfeldt, O. D., Kuhnt, U., Benevento, L. A. 1974. An intracellular analysis of visual cortical neurones to moving stimuli: Responses in a cooperative neuronal network. *Exp. Brain Res.* 21:251–74

Desimone, R., Schein, S. J., Albright, T. D. 1984. Form, color, and motion analysis in prestriate cortex of macaque monkey. In *Study Group on Pattern Recognition Mechanisms*, ed. C. Chagas. Vatican City: Pontifical Acad. Sci.

Dow, B. M. 1974. Functional classes of cells and their laminar distribution in monkey visual cortex. *J. Neurophysiol.* 37:927–46

Dowling, J. E., Boycott, B. B. 1966. Organization of the primate retina: Electron microscopy. *Proc. R. Soc. London Ser. B* 166:80–111

Fischer, B., Kruger, J. 1974. The shift effect in the cat's lateral geniculate nucleus. *Exp. Brain Res.* 21:225–27

Fisken, R. A., Garey, L. J., Powell, T. P. S. 1975. The intrinsic, association and commissural connections of area 17 of the visual cortex. *Philos. Trans. R. Soc. London Ser. B* 272:487–536

Fries, W., Albus, K., Creutzfeldt, O. D. 1977. Effects of interacting visual patterns on single cell responses in cat's striate cortex. *Vision Res.* 17:1001–8

Frost, B. J., Nakayama, K. 1983. Single visual neurons code opposing motion independent of direction. *Science* 220:744–45

Frost, B. J., Scilley, P. L., Wong, S. C. P. 1981. Moving background patterns reveal double-opponency of directionally specific pigeon tectal neurons. *Exp. Brain Res.* 43:173–85

Gattass, R., Oswaldo-Cruz, E., Sousa, A. P. B. 1979. Visual receptive fields of units in the pulvinar of cebus monkey. *Brain Res.* 160:413–30

Gattass, R., Sousa, A. P. B., Oswaldo-Cruz, E. 1978. Single unit response types in the pulvinar of the cebus monkey to multisensory stimulation. *Brain Res.* 158:75–87

Gibson, J. J. 1950. *The Perception of the Visual World.* Boston: Houghton Mifflin

Gibson, J. J. 1966. *The Senses Considered as Perceptual Systems.* Boston: Houghton Mifflin

Gibson, J. J. 1979. *The Ecological Approach to Visual Perception.* Boston: Houghton Mifflin

Gilbert, C. D., Wiesel, T. N. 1983. Clustered intrinsic connections in cat visual cortex. *J. Neurosci.* 3:1116–33

Gordon, D. A. 1965. Static and dynamic visual fields in human space perception. *J. Opt. Soc. Am.* 55:1296–1303

Grusser-Cornehls, U., Grusser, O. J., Bullock, T. H. 1963. Unit responses in the frog's tectum to moving and non-moving visual stimuli. *Science* 141:820–22

Hammond, P., MacKay, D. M. 1981. Modulatory influences of moving textured backgrounds on responsiveness of simple cells in feline striate cortex. *J. Physiol.* 319:431–42

Hartline, H. K. 1938. The response of single optic nerve fibers of the vertebrate eye to illumination of the retina. *Am. J. Physiol.* 121:400–15

Horn, B. K. P. 1974. Determining lightness from an image. *Comput. Graph. Image Process.* 3:277–99

Hubel, D. H., Wiesel, T. N. 1961. Integrative action in the cat's lateral geniculate body. *J. Physiol.* 155:385–98

Hubel, D. H., Wiesel, T. N. 1962. Receptive fields, binocular interaction and functional architecture in the cat's visual cortex. *J. Physiol.* 160:106–54

Hubel, D. H., Wiesel, T. N. 1965. Receptive fields and functional architecture in two nonstriate visual areas (18 and 19) of the cat. *J. Neurophysiol.* 28:229–89

Hubel, D. H., Wiesel, T. N. 1969. Visual area of the lateral suprasylvian gyrus (Clare–Bishop area) of the cat. *J. Physiol.* 202:251–60

Hubel, D. H., Wiesel, T. N. 1972. Laminar and columnar distribution of geniculocortical fibers in the macaque monkey. *J. Comp. Neurol.* 146:421–50

Hubel, D. H., Wiesel, T. N. 1974. Uniformity of monkey striate cortex: A parallel relationship between field size, scatter, and magnification factor. *J. Comp. Neurol.* 158:295–306

Humphrey, A. L., Norton, T. 1980. Topographic organization of the orientation column system in the striate cortex of the tree shrew *(Tupaia glis).* I. Microelectrode recording. *J. Comp. Neurol.* 192:531–47

Humphrey, A. L., Skeen, L. C., Norton, T. 1980. Topographic organization of the orientation column system in the striate cortex of the tree shrew *(Tupaia glis).* II. Deoxyglucose mapping. *J. Comp. Neurol.* 192:549–66

Ikeda, H., Wright, M. J. 1972. Functional organization of the periphery effect in retinal ganglion cells. *Vision Res.* 12:1857–79

Ingle, D. J., Shook, B. L. 1984. Action-oriented approaches to visuo-spatial brain functions. In *Brain Mechanisms and Spatial Vision,* ed. D. Ingle, D. Lee, M. Jeannerod. The Hague: Nijhot

Jones, B. H. 1970. Responses of single neurons in cat visual cortex to a simple and more complex stimulus. *Am. J. Physiol.* 218:1102–7

Julesz, B. 1981. Textons, the elements of texture perception, and their interactions. *Nature* 290:91–97

Julesz, B. 1984. Toward an axiomatic theory of preattentive vision. In *Dynamic Aspects of Neocortical Function,* ed. G. Edelman, W. M. Cowan. In press

Kaas, J. H., Hall, W. C., Killackey, H., Diamond, I. T. 1972. Visual cortex of the tree shrew *(Tupaia glis):* Architectonic subdivisions and representation of the visual field. *Brain Res.* 42:491–96

Kluver, H. 1942. Functional significance of the geniculostriate system. *Biol. Symp.* 7:253–300

Knudsen, E. I., Konishi, M. 1978a. Space and frequency are represented separately in auditory midbrain of the owl. *J. Neurophysiol.* 41:870–84

Knudsen, E. I., Konishi, M. 1978b. Center-surround organization of auditory receptive fields in the owl. *Science* 202:778–80

Koenderink, J. J. 1984. Space, form and optical deformations. In *Brain Mechanisms and Spatial Vision,* ed. D. Ingle, D. Lee, M. Jeannerod. The Hague: Nijhot

Kruger, J. 1977. Stimulus dependent colour specificity of monkey lateral geniculate neurones. *Exp. Brain Res.* 30:297–311

Kruger, J., Fischer, B. 1973. Strong periphery effect in cat retinal ganglion cells: Excitatory responses in on- and off-centre neurones to single grid displacements. *Exp. Brain Res.* 18:316–18

Kuffler, S. W. 1953. Discharge patterns and functional organization of mammalian retina. *J. Neurophysiol.* 28:37–68

Land, E. H. 1959a. Color vision and the natural image. Part I. *Proc. Natl. Acad. Sci. USA* 45:115–29

Land, E. H. 1959b. Color vision and the natural image. Part II. *Proc. Natl. Acad. Sci. USA* 45:636–44

Land, E. H. 1983. Recent advances in retinex theory and some implication for cortical

computations: Color vision and the natural image. *Proc. Natl. Acad. Sci. USA* 80:5163–69
Land, E. H., Hubel, D. H., Livingstone, M. S., Perry, S. H., Burns, M. M. 1983. Colour-generating interactions across the corpus callosum. *Nature* 303:616–18
LeVay, S., Sherk, H. 1981. The visual claustrum of the cat II. The visual field map. *J. Neurosci.* 9:981–92
Levick, W., Oyster, C., Davis, D. 1965. Evidence that McIlwain's periphery effect is not a stray light artifact. *J. Neurophysiol.* 28:555–59
Livingstone, M. S., Hubel, D. H. 1984. Anatomy and physiology of a color system in the primate visual cortex. *J. Neurosci.* 4:309–56
Lund, J. S., Lund, R. D., Hendrickson, A. E., Bunt, A. H., Fuchs, A. F. 1975. The origin of efferent pathways from the primary visual cortex, area 17, of the macaque monkey as shown by retrograde transport of horseradish peroxidase. *J. Comp. Neurol.* 164:287–304
Maffei, L., Fiorentini, A. 1976. The unresponsive regions of visual cortical receptive fields. *Vision Res.* 16:1131–39
Marrocco, R. T., McClurkin, J. W., Young, R. A. 1982. Modulation of lateral geniculate nucleus cell responsiveness by visual activation of the cortico-geniculate pathway. *J. Neurosci.* 2:256–63
Maunsell, J. H. R., Van Essen, D. C. 1983a. Functional properties of neurons in middle temporal visual area of the macaque monkey. I. Selectivity for stimulus direction, speed, and orientation. *J. Neurophysiol.* 49:1127–47
Maunsell, J. H. R., Van Essen, D. C. 1983b. The connections of the middle temporal visual area (MT) and their relationship to a cortical hierarchy in the macaque monkey. *J. Neurosci.* 3:2563–86
McIlwain, J. T. 1964. Receptive fields of optic tract axons and lateral geniculate cells: Peripheral extent and barbiturate sensitivity. *J. Neurophysiol.* 27:1154–73
McIlwain, J. T. 1966. Some evidence concerning the physiological basis of the periphery effect in the cat's retina. *Exp. Brain Res.* 1:265–71
McIlwain, J. T., Buser, P. 1968. Receptive fields of single cells in the cat's superior colliculus. *Exp. Brain Res.* 5:314–25
Michael, C. R. 1972. Functional organization of cells in superior colliculus of the ground squirrel. *J. Neurophysiol.* 35:833
Miezin, F., McGuinness, E., Allman, J. 1982. Antagonistic direction specific mechanisms in area MT in the owl monkey. *Neurosci. Abstr.* 8:681
Mitchison, G., Crick, F. 1982. Long axons within the striate cortex: Their distribution, orientation, and patterns of connection. *Proc. Natl. Acad. Sci. USA* 79:3661–65
Montero, V. M. 1980. Patterns of connections from the striate cortex to cortical visual areas in superior temporal sulcus of macaque and middle temporal gyrus of owl monkey. *J. Comp. Neurol.* 189:45–59
Moran, J., Desimone, R., Schein, S. J., Mishkin, M. 1983. Suppression from ipsilateral visual field in area V4 of the macaque. *Neurosci. Abstr.* 9:957
✓Nakayama, K., Loomis, J. M. 1974. Optical velocity patterns, velocity-sensitive neurons, and space perception: A hypothesis. *Perception* 3:63–80
Nelson, J. I., Frost, B. 1978. Orientation selective inhibition from beyond the classic visual receptive field. *Brain Res.* 139:359–65
Newsome, W. T., Allman, J. M. 1980. Interhemispheric connections of visual cortex in the owl monkey, *Aotus trivirgatus*, and the Bushbaby, *Galago senegalensis*. *J. Comp. Neurol.* 194:209–33
Ramachandran, V. S., Anstis, S. M. 1983. Perceptual organization in moving patterns. *Nature* 304:529–31
Reichardt, W., Poggio, T., Hausen, K. 1983. Figure–ground discrimination by relative movement in the visual system of the fly. Part II: Towards the neural circuitry 1. *Biol. Cybernet.* 46:1–30
Riche, D., Lanoir, J. 1978. Some claustro-cortical connections in the cat and baboon as studied by retrograde horseradish peroxidase transport. *J. Comp. Neurol.* 177:435–44
Rizzolatti, G., Camarda, R. 1977. Influence of the presentation of remote visual stimuli on visual responses of cat area 17 and lateral suprasylvian area. *Exp. Brain Res.* 29:107–22
Rizzolatti, G., Camarda, R., Grupp, L. A., Pisa, M. 1974. Inhibitory effect of remote visual stimuli on the visual responses of the cat superior colliculus: Spatial and temporal factors. *J. Neurophysiol.* 37:1262–75
Rockland, K. S. 1983. Lattice-like intrinsic neural connections in macaque prestriate visual cortex. *Neurosci. Abstr.* 9:476
Rockland, K. S., Lund, J. S. 1983. Intrinsic laminar lattice connections in primate visual cortex. *J. Comp. Neurol.* 216:303–18
Rockland, K. S., Lund, J. S., Humphrey, A. L. 1982. Anatomical banding of intrinsic connections in striate cortex of tree shrews *(Tupaia glis)*. *J. Comp. Neurol.* 209:41–58
Rogers, B., Graham, M. 1979. Motion parallax as an independent cue for depth perception. *Perception* 8:125–34
Royden, C., Baker, J., Allman, J. 1984. Illusions of depth produced by moving random dots. In preparation

Sherk, H., LeVay, S. 1981. The visual claustrum of the cat III. Receptive field properties. *J. Neurosci.* 1:993–1002

Sherk, H., LeVay, S. 1983. Contribution of the cortico-claustral loop to receptive field properties in area 17 of the cat. *J. Neurosci.* 3:2121–27

Spatz, W. B. 1975. An efferent connection of the solitary cells of Meynert. A study with horseradish peroxidase in the marmoset *Callithrix. Brain Res.* 92:450–55

Spear, P. D., Baumann, T. P. 1975. Receptive field characteristics of single neurons in lateral suprasylvian visual area of the cat. *J. Neurophysiol.* 38:1403–20

Sterling, P., Wickelgren, B. G. 1969. Visual receptive fields in the superior colliculus of the cat. *J. Neurophysiol.* 32:1–15

Tanaka, K., Saito, H., Fukada, Y., Hikosaka, K., Yukie, M., Iwai, E. 1984. Two groups of neurons responding to local and whole field movements in the macaque MT area. *Neurosci. Abstr.* 10:474

Tigges, J., Tigges, M., Anschel, S., Cross, N. A., Letbetter, W. D., McBride, R. L. 1981. Areal and laminal distribution of neurons interconnecting the central visual cortical area 17, 18, 19, and MT in squirrel monkey *(Saimiri). J. Comp. Neurol.* 202:539–60

Triesman, A. M., Gelade, G. 1980. A feature-integration theory of attention. *Cognitive Psychol.* 12:97–136

Van Essen, D. C., Newsome, W. T., Bixby, J. L. 1982. The pattern of interhemispheric connections and its relationship to extrastriate visual areas in the macaque monkey. *J. Neurosci.* 2:265–83

Von Grunau, M., Frost, B. J. 1983. Double-opponent-process mechanism underlying RF-structure of directionally specific cells of cat lateral suprasylvian visual area. *Exp. Brain Res.* 49:84–92

von Helmholtz, H. 1925. *Treatise on Physiological Optics.* Optical Soc. Am.

von der Heydt, R., Peterhans, E., Baumgartner, G. 1984. Illusory contours and cortical neuron responses. *Science* 224:1260–62

Weller, R. E., Kaas, J. H. 1983. Retinotopic patterns of connections of area 17 with visual areas V–II and MT in macaque monkeys. *J. Comp. Neurol.* 220:253–79

Weller, R. E., Wall, J. T., Kaas, J. H. 1984. Cortical connections of the middle temporal visual area (MT) and the superior temporal cortex in owl monkeys. *J. Comp. Neurol.* 228:81–104

Werblin, F. S. 1972. Lateral interaction at inner plexiform layer of vertebrate retina: antagonistic responses to change. *Science* 175:1008–10

Wurtz, R. H., Richmond, B. J., Judge, S. J. 1980. Vision during saccadic eye movements. III. Visual interactions in monkey superior colliculus. *J. Neurophysiol.* 43:1168–81

Zeki, S. 1983a. Colour coding in the cerebral cortex: The reaction of cells in monkey visual cortex to wavelengths and colours. *Neuroscience* 9:741–765

Zeki, S. 1983b. Colour coding in the cerebral cortex: The responses of wavelength-selective and colour-coded cells in monkey visual cortex to changes in wavelength composition. *Perception* 9:767–81

Ann. Rev. Neurosci. 1985. 8:431–55

NEUROPEPTIDES IN IDENTIFIED *APLYSIA* NEURONS

Rashad-Rudolf J. Kaldany, John R. Nambu, and Richard H. Scheller

Department of Biological Sciences, Stanford University, Stanford, California 94305

INTRODUCTION

Biologically active peptides are frequently used as extracellular chemical messengers in the central and peripheral nervous systems of both invertebrates and vertebrates. In this review we discuss many of the current issues concerning neuropeptide gene organization, biosynthesis, processing, and physiological activities. The long-range goal of this research is to elucidate the precise contributions of neuroactive peptides to the governing of animal behavior. The gastropod mollusc *Aplysia californica* serves as an appropriate model system for addressing these issues.

The Aplysia Nervous System

The central nervous system of *Aplysia* consists of approximately 20,000 neurons that are organized into four pairs of symmetric ganglia in the head, and a single asymmetric abdominal ganglion. Several criteria, including location, size, color, and physiological activity, have been applied to identify individual neurons (Kandel 1976). These neurons can attain sizes of up to 1 mm in diameter and contain correspondingly large amounts of DNA, RNA, and protein (Coggeshall et al 1971, Aswad 1978). Furthermore, many of the neurons exhibit distinguishing patterns of electrical activity. These characteristics have facilitated a correlation of the physiological and biochemical activities in several identified nerve cells with specific behaviors. The best-characterized group of *Aplysia* central neurons is the abdominal ganglion. These cells govern a number of reflex and fixed action patterns, including withdrawal of the gill and mantle organs as well as inking and egg laying (Kandel 1976). The abdominal ganglion also regulates a host of visceral functions such as the excretion

0147-006X/85/0301-0431$02.00

of waste products, cardiac output, and certain aspects of respiration. Many *Aplysia* central neurons utilize peptides as extracellular messengers, and are ideal for studying the roles of these molecules in behavioral processes.

Some peptidergic neurons in *Aplysia* are particularly easy to identify because their cell bodies contain large numbers of dense core vesicles. These cells thus take on a whitish appearance, rather than the orange carotenoid pigmentation characteristic of most other neurons. Some of the abdominal ganglion peptidergic neurons are highlighted in Figure 1. Preliminary investigations of the proteins synthesized in these "white cells" indicates that in all cases a single prominent product dominates protein biosynthesis (Wilson 1971, Gainer & Wollberg 1974, Berry 1976a, Loh & Gainer 1975). Pulse chase studies showed that these proteins are cleaved into smaller peptide products that are, in many instances, localized to a vesicle fraction (Aswad 1978, Loh et al 1977). The conclusions of these studies advanced the hypothesis that the "white cells" use peptide messengers.

Insight into the physiological activities of the peptides was initially gained by injecting cellular extracts into the animal and observing the resultant behavioral and physiological change. This approach was particularly fruitful with the bag cells and abdominal ganglion neuron R15. Upon injection, extracts of the bag cell neurons or the atrial gland (an exocrine gland located on the hermaphroditic duct) elicit egg laying and the associated behaviors (Kupfermann 1967, Arch et al 1978). R15 extracts contain protease sensitive material that upon injection causes the animal to take up water (Kupfermann & Weiss 1976). The peptide(s) of this cell may thus function in an osmoregulatory capacity. Further evidence for this hypothesis comes from the fact that R15 receives input from the osphradium, a structure near the gill that is sensitive to salinity changes.

Characterized Neuropeptides

Although it has long been thought that biologically active peptides are common in the *Aplysia* central nervous system, isolation and characterization of these molecules has proven difficult. The greatest success in this area has been with the egg-laying system. Strumwasser and colleagues made use of the "egg-laying" bioassay to isolate a 36-amino-acid peptide (the egg-laying hormone, ELH) from the bag cell neurons (Chiu et al 1979) and two 34-amino-acid peptides (peptides A and B) from the atrial gland (Heller et al 1980). Microsequencing techniques were used to determine the primary amino acid sequences of these molecules. More recently, the primary sequences of a second bag cell messenger, α-bag cell peptide (Rothman et al 1983a), and the small cardioactive peptide-B (Morris et al 1982) have been determined. The existence of numerous other *Aplysia* neuropeptides has been suggested by studies using a myriad of antibodies to both vertebrate and invertebrate peptides. For exam-

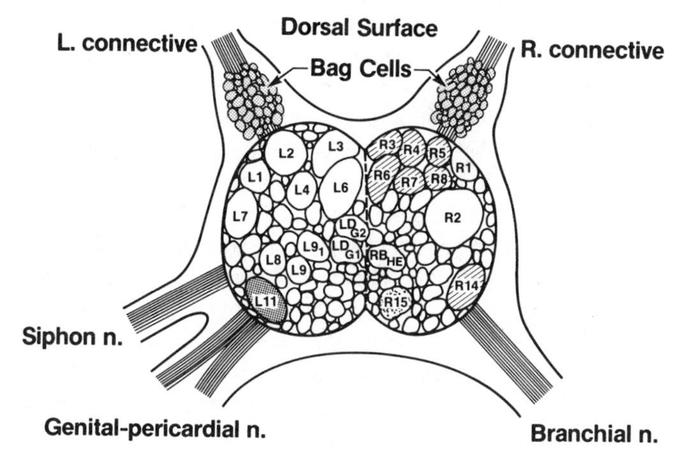

Figure 1 Schematic representation of the dorsal surface of the abdominal ganglion. Representative cells are labeled L or R (designating left or right hemiganglion) and have an identifying number. Peptidergic neurons are indicated with various shadings. The bag cells are grouped in two large clusters of about 400 neurons each on the rostral portion of the ganglion. Cells R3–8 and R14 (striped area) each send a single axon down the branchial nerve that terminates on the efferent vein of the gill at the base of the heart. R15 (speckled area) is a neurosecretory cell thought to be involved in controlling salt and water balance. L11 (dotted area) is a cholinergic cell that also uses one or more neuropeptides in intercellular communication.

ple, antibodies to FMRFamide and leucine-enkephalin have been shown to stain specifically several cells of the abdominal ganglion (Hopkins et al 1982). Until these molecules or the corresponding genes are isolated, however, one must be cautious of this type of observation because of the potential for non-specific cross-reactivity.

MOLECULAR GENETIC APPROACH TO NEUROPEPTIDE CHARACTERIZATION

A more recent approach to understanding the roles of neuropeptides has been to use the techniques of molecular biology to clone the genes encoding these products. The identification and characterization of neuropeptide genes can be of value for several reasons. Application of recombinant DNA methodology can facilitate the determination of the primary sequence of nervous system proteins without requiring their biochemical isolation and purification. Analysis of DNA coding sequences not only permits precise assignment of the amino acid structure of known peptides, but can demonstrate the existence of previously undescribed products, as occurred with γ-melanocyte-stimulating hormone (Nakanishi et al 1979) and the calcitonin-gene-related-peptide (Rosenfeld et al 1983). As it has become apparent that neuropeptides are synthesized as portions of larger precursor proteins (Herbert et al 1981), this type of analysis is useful in predicting whether specific peptides are encoded by the same genes or the same mRNA transcripts.

At least part of the information necessary to control gene expression may be directly specified by the primary sequence of nucleotides, through various regulatory elements such as promoters, terminators, or enhancers. Analysis of these regions may prove valuable in elucidating the basis for specific spatial and temporal patterns of neuropeptide gene expression. Two mechanisms already known to regulate the production of nervous system peptides by acting at the level of nucleic acid structure include alternate splicing of mRNAs (Nawa et al 1983) and differential use of poly A+ addition sites (Amara et al 1982). Only through a detailed analysis of the corresponding DNA and RNA molecules can a complete understanding of neuropeptide biosynthesis be obtained.

Methodology

Although a number of methods are available to isolate specific clones of interest, the strategy of differential (+/−) screening has been particularly useful for isolating DNA sequences that encode various biologically active peptides in *Aplysia*. This procedure relies upon the premise that a particular gene is transcribed in a tissue-specific fashion. Thus, poly A+ mRNA is extracted from both expressive and nonexpressive tissues and converted into radiolabeled cDNA via the enzyme reverse transcriptase. These cDNAs are then used as probes to screen recombinant DNA libraries containing either whole genomic (chromosomal) restriction enzyme fragments or cDNAs derived from tissue-specific mRNA populations. Clones hybridizing selectively to probes made from the expressive tissue more likely contain the sequences of interest.

When coupled with the high levels of peptide synthesis exhibited by certain *Aplysia* neurons, application of this methodology has facilitated the isolation of coding sequences expressed in collections of neurons and in some cases single identified neurons. Thus, neuropeptide genes were identified from the atrial gland and bag cells as well as abdominal ganglion neurons R3–14, R15, L11, and buccal ganglion cells B1 and B2. This methodology is attractive not only because it allows for the identification of cell-specific products, but also because no previous structural characterization of the product itself is necessary. In contrast, other cloning procedures rely upon prior amino acid sequence data to permit the synthesis of corresponding oligonucleotide probes, or the availability of specific antibodies (Noda et al 1982, Young & Davis 1983).

APLYSIA NEUROPEPTIDE GENES

The ELH Gene Family

A small multigene family has been identified and shown to encode a battery of related peptides involved in mediating the egg-laying behavior of *Aplysia*

(Scheller et al 1982). In adult animals, the ELH precursor comprises up to 50% of the total protein synthesized by the bag cells (Arch 1972). Along with results from in vitro translation experiments, this observation suggests that a correspondingly large amount of ELH mRNA is present in these cells. High levels of this mRNA are not thought to be present in other tissues (McAllister et al 1983). Thus, the strategy of differential screening, as described above, was selected to isolate ELH-encoding genes.

A total of six distinct clones were shown to have ELH homology, thus demonstrating that ELH is encoded in a small gene family (Figure 2). Results from genomic Southern blot analysis using radiolabeled ELH DNA to probe *Aplysia* sperm DNA further support this notion, as several different restriction enzyme fragments were observed to have ELH homology (Scheller et al 1983a). The exact gene copy number is somewhat difficult to assess due to a high degree of polymorphism that is maintained in these animals. Nonetheless, studies to date suggest that the ELH gene family consists of five to nine discrete members.

Structural characterization of the ELH clones was accomplished through restriction enzyme mapping, electron microscopy heteroduplex analysis, and DNA sequencing. In order to determine where these genes are expressed, DNA from each clone was digested to completion with the restriction enzyme Eco RI and probed with ^{32}P-labeled cDNA from the bag cells or atrial gland via Southern blot hybridization (Southern 1975). At low hybridization stringency (i.e.: where lower homology duplexes are stable), all major Eco RI fragments hybridized to both bag cell and atrial gland probes. However, as the stringency was increased, the fragments hybridized only to bag cell cDNA or atrial gland cDNA, but not to both. Furthermore, among the three clones having more than one hybridizing Eco RI restriction enzyme fragment (ELH 1,3, and 8), one was found to hybridize to the bag cell probe and the other to the atrial gland probe. This result implies that these clones each possess two distinct transcriptional regions. These data also suggest that several structurally related genes are contained in the six ELH clones and that tissue-specific transcription of these genes occurs within the bag cell and atrial gland tissues.

Subsequent DNA sequence analysis of these regions clearly demonstrated the validity of this assessment. ELH clones 1, 3, and 8 were each found to contain two distinct genes, one encoding an atrial gland product, peptide B, and the other coding for ELH, produced in the bag cells. Separated by roughly 9 kb of genomic DNA, these genes are closely linked on the chromosome and are transcribed from opposite strands of DNA. Among the other clones, ELH 7 contains a single copy of an ELH gene while ELH 13 and 18 both contain a single peptide A gene. These six clones contain a total of nine related genes, at least three of which (A, B, and ELH) are non-allelic.

Comparison of the restriction enzyme maps and nucleotide sequence data implies that these genes are roughly 90% homologous. That this homology

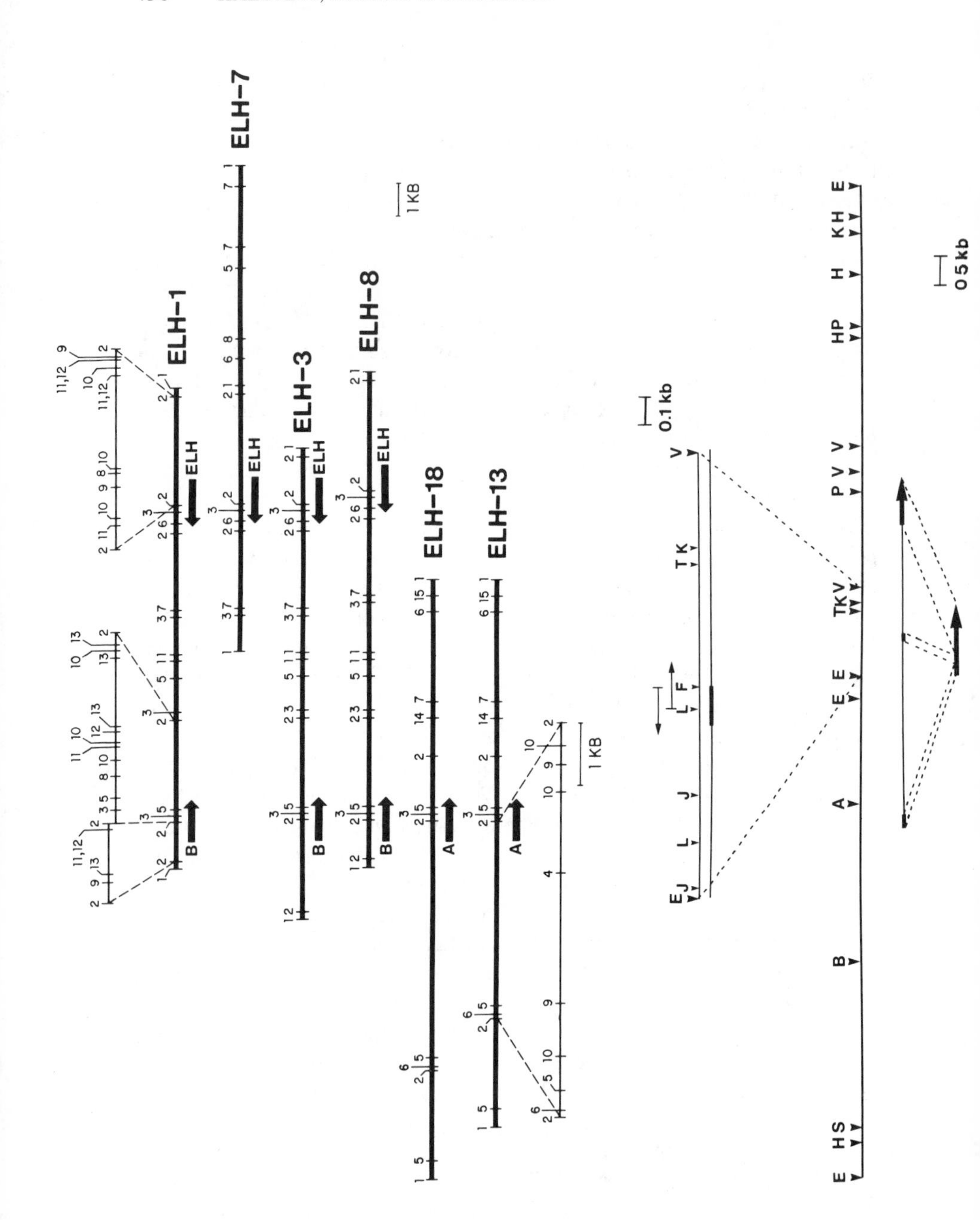
ELH-1
ELH-7
ELH-3
ELH-8
ELH-18
ELH-13
ELH
B
A
1 KB
0.1 kb
0 5 kb

extends past the coding regions of these genes is evidenced by electron microscopy studies. When DNA from clone 8, containing an ELH gene, is allowed to anneal to DNA from clone 18, which contains a single peptide A gene, a heteroduplex structure is formed that extends through the coding sequences and into neighboring flanking regions. Similarly, when clone-1 DNA is allowed to self-anneal, a large stem-loop structure, diagnostic of inverted repeats, results. The clone-1 repeat unit is roughly 6 kb, much longer than either the ELH or B peptide genes themselves.

Initially, examination of ELH gene expression was undertaken via northern blot hybridization, with ELH encoding DNA used to probe poly A+ mRNA from both bag cell and atrial gland tissue (Scheller et al 1982). Three discrete bag cell transcripts of 1600, 1450, and 1200 nucleotides were found to be homologous to ELH DNA. In addition, two homologous transcripts of 1400 and 1150 nucleotides were observed to be present among atrial gland RNA. No hybridization was found to occur with RNA from several other tissues.

More recently, further analysis of ELH gene expression has been performed through the characterization of corresponding mRNA sequences (A.C. Mahon et al, *J. Neurosci.*, submitted). This was accomplished by isolating ELH cDNA clones from either abdominal ganglion/bag cell or atrial gland cDNA libraries. Nucleotide sequence analysis of these clones suggests that at least two distinct ELH encoding mRNAs are produced within the bag cells. Atrial gland cDNAs, which code for peptides A and B, have been similarly characterized. These transcripts not only contain sequences from the coding regions of their respective genes, but in addition, 5′ untranslated regions that are not contiguous with the transcribed regions of the previously described ELH genomic clones. This finding requires the presence of at least one intervening sequence within these genes. Preliminary restriction mapping and blotting data on newly isolated genomic clones suggest that transcription is initiated several kb upstream of the precursor coding regions (A. C. Mahon et al, *J. Neurosci.*, submitted). Characterization of the precise organization of these sequences is currently being pursued. Such information should eventually be useful in elucidating the nature of regulation used to specifically express genes encoding ELH in

Figure 2 *Aplysia* neuropeptide genes. A. The ELH gene family. Restriction enzyme maps of the six isolated ELH recombinant clones. Gene identity and direction of transcription are indicated by the labeled *arrows*. Numbers above the clones indicate various restriction enzyme cleavage sites: 1. Eco RI; 2. PstI; 3. XhoI; 4. StuI; 5. PvuII; 6. HindIII; 7. BglI; 8. XbaI; 9. AvaII; 10. HincII; 11. HaeIII; 12. HhaI; 13. HpaII; 14. BamHI; 15. SalI.

B. The R3-14 peptide gene. Restriction enzyme map of the R3-14 genomic clone and the organizaiton of the coding region. Positions of exons and introns and the direction of transcription are indicated below by *arrows*. DNA sequencing strategy used to define intron/exon boundaries is shown above. Letters refer to restriction enzyme cleavage sites: A. SalI; B. BamHI; E. EcoRI; F. HinfI; H. HindIII; J. AvaII; K. KpnI; L. HaeIII; P. PstI; S. SmaI; T. SstI; PvuII.

the bag cells and those for peptides A and B in the atrial gland. The high degree of tissue specific expresion exhibited by these closely related genes promises to provide an attractive model system for studying gene regulation in the nervous system.

Highly homologous members of the ELH gene family almost certainly arose from a common ancestor. This "proto-ELH" gene probably gave rise to the present day A, B, and ELH precursor genes through a series of discrete duplication events, occurring first within a single transcription unit (i.e.: intragenically) to produce a polyprotein gene, and then via duplications of large regions of the chromosome to generate a family of these genes. The existence of multiple gene copies allows for increased diversification of coding potentials, as variants can evolve without a concomittant loss of the original gene products. These variants, freed from the specific selective constraints imposed on their progenitor, could then diverge to generate new products, exhibit new patterns of expression, or both.

Although the precise number and identity of transcriptionally active ELH genes is as yet uncertain, distinct patterns of expression clearly are exhibited by the various family members and, through a series of point mutations and insertion/deletion events, the protein coding potentials of the A, B, and ELH genes have diverged.

IN SITU HYBRIDIZATION In situ hybridization has been used to detect mRNA in tissue sections. Individual ganglia or other tissues are fixed, embedded in paraffin, and cut into 5μ sections. Cloned DNA fragments that have been radioactively labeled by nick translation incorporating ^{125}I-dCTP are denatured and hybridized to the tissue sections; the tissue is then coated with emulsion and exposed. After developing, the tissue is stained with geimsa and examined in the light microscope. The in situ technique is potentially useful in detecting the transcription of genes that are not translated and is a technique that can be used when looking for the expression of sequences for which the protein sequence is not known. The most severe disadvantage of this technique, particularly for the nervous system, is the fact that only nerve cell bodies, the axon hillock, and the proximal axons are made visible; more distal neuronal processes cannot be traced with the in situ technique. The sensitivity and specificity of the immunocythochemical technique (see below) may turn out to be better suited for these kinds of studies than the in situ technique; however, these issues await more detailed comparative studies.

The developmental expression of the ELH gene family has been studied using in situ hybridization (McAllister et al 1983). Cells that hybridize to the ELH gene probes were observed along the inner lining of the body wall and on tissue strands leading from the body wall to the ganglion. These observations gave rise to the hypothesis that the bag cell neurons arise in proliferative zones in the caudal ectoderm and then migrate to their eventual positions in the central nervous system. In contrast, the atrial gland, which expresses other members of the ELH gene family, seems to develop as a specialization of the basement membrane in the white hermaphroditic duct. Proliferation at this region results in a highly infolded region of tissue that forms a continuous lumen with the duct. No cells are seen outside the area of the duct, thus suggesting that migration is not involved in the maturation of the ELH-expressing cells in the atrial gland.

Two interesting conclusions arise from these observations. First, the data suggest that the bag cells are committed to their eventual neuronal fate while still along the body wall and before they are fully differentiated as neurons. Second, the closely related genes are expressed in cells following different developmental paths and must have evolved specific mechanisms to accomplish this task. Presumably the specificity of expression arises at the level of transcription; however, this remains to be determined.

The R3-14 Peptide-Precursor Gene

In order to examine the peptidergic nature of abdominal ganglion neurons R3–14, the identification and characterization of cell-specific mRNA transcripts was undertaken. Poly A+ mRNA was extracted from 50 individually dissected R2, R15, RUQ, LUQ, and R14 cells. This RNA was then used to direct the synthesis of ^{32}P-labeled cDNA probes, and through a differential screening analysis, a 1.2 kb cDNA clone specific for cells R3–14 was isolated from an abdominal ganglion cDNA library (Nambu et al 1983). Structural characterization was accomplished through restriction enzyme mapping and DNA sequence analysis. Identification of the poly-A tail facilitated assignment of the 3′-end of the message. The cDNA clone contains a 324 bp open reading frame that is flanked by a 173 bp untranslated 5′ region and a 724 bp 3′ noncoding end. The coding region predicts a 108-amino-acid protein, the presumptive neuropeptide precursor synthesized in these cells. Genomic Southern blot analysis revealed that the 3′ untranslated region contains a repetitive sequence element present in many copies along the genome. Further analysis of the genomic representation was accomplished through characterization of an R3–14 peptide genomic clone that was isolated from a genomic DNA library.

Unlike the gene family arrangement for ELH, the R3–14 neuropeptide precursor is encoded by a gene that is present in only a single copy per haploid genome. This gene is contained within approximately 7 kb of genomic DNA and is comprised of three small exons (I, II, III) interrupted by two introns (Figure 2). The direction of transcription has been assigned. Northern blot analysis using the coding region of the cDNA as a probe reveals a 1250 nucleotide mRNA transcript in the R3–14 cells of the abdominal ganglion; homologous transcripts are not detectable in other neural or non-neural tissues.

The position of the introns has been determined via DNA sequence analysis of the region flanking the middle exon (II) (Taussig et al 1984b). Exon I encodes the complete 5′ untranslated portion of the mRNA transcript. Exon II begins at a position two nucleotides upstream of the initiator methionine codon and contains sequences specifying up to amino acid 42. Exon III then includes the remainder of the protein-coding region and the complete 3′ untranslated end.

The organization of exons and introns within the R3–14 peptide gene may furnish some insight into the evolutionary origins of the gene and its products. Exons have been proposed to code for discrete functional domains within eukaryotic proteins, whereas introns promote recombination between these domains (Gilbert 1978). This notion may be best supported by work done on the immunoglobulin heavy chain genes, where individual exons code for each constant region segment (Sakano et al 1980). The R3–14 peptide gene contains three exons that may specify discrete functional domains of the protein precursor. Exon II encodes the complete hydrophobic signal sequence as well as most of the negatively charged portion of the precursor. Exon III encodes the remainder of the precursor, which is largely positively charged. These oppositely charged portions are thought to be liberated from the precursor as independent peptide products (see below). This organization may have arisen by the joining of segments once separated in the genome which then came under common transcriptional control. The joining event could have occurred through recombination within intervening sequences to give rise to the current gene. This mechanism may be important for bringing together previously unrelated products to produce novel proteins with unique functional capacities. Within the nervous system, new combinations of such peptides might allow for enhanced degrees of cellular interaction.

The L11 Peptide-Precursor Gene

A mRNa transcript produced selectively in the peptidergic neuron L11 was identified via isolation in an L11 specific cDNA clone from the abdominal ganglion cDNA library (Taussig et al 1984a). This clone was isolated by use of the differential screening procedure described above. The 1.1 kb cDNA clone contains both a 286 bp 5′ and a 390 bp 3′ noncoding portion. The coding region is 453 bp in length and specifies a 151-amino-acid protein. This protein is thought to be the precursor of one or more neuroactive peptides.

Northern blot analysis using the L11 cDNA to probe RNA from a variety of tissues suggests that homologous transcripts may be present at low levels in other ganglia. This prospect is currently being addressed by screening for similar sequences in head ganglia cDNA libraries. The precise structural organization of the corresponding gene(s) is also presently under investigation. Results from genomic Southern blot analyses suggest the existence of a single large gene extensively interrupted by intervening sequences (R.-R. J. Kaldany and R. H. Scheller, unpublished observations).

NEUROPEPTIDES: DISTRIBUTION, PRECURSOR STRUCTURE, AND PROCESSING

An important goal in these studies is to understand the physiological role of neuropeptides. The amino acid sequence of the appropriate reading frame on the strand containing poly A+ can be deduced from the nucleotide sequences of the cell-specific genes that have been characterized. In general, only one long open reading frame is found, and the resulting protein has a molecular weight consistent with the size of the presumptive precursor obtained by pulse/chase experiments. Moreover, the deduced sequence contains stretches of hydrophobic amino acids at the NH_2 terminus that probably serve as signal sequences (see below). These criteria serve to identify the correct reading frame. We can then use the deduced amino acid sequence to identify potential cleavage sites and the resulting peptides. Once actual cleavage sites are identified, the predicted peptides can either be synthesized if they are simple enough, or isolated from the various tissues. The isolation of larger peptides would be facilitated by prior knowledge of the sequence. For instance, antibodies raised against synthetic peptides whose sequence is analogous to portions of the large peptides can be used as affinity reagents. The purified endogenous peptides can then be used to determine their physiological roles.

Precursors and Signal Sequences

Secreted proteins have been shown to be synthesized as parts of larger proteins. One of the most ubiquitous characteristics of these proteins is a hydrophobic stretch of 15–30 amino acids at the NH_2 terminal end. This hydrophilic stretch, called a *signal* or *leader sequence,* is essential for directing the preproprotein through the rough endoplasmic reticulum and into the cisternae (Loh et al 1984). In all the *Aplysia* neuropeptide genes, the initiator methionine is followed by a largely hydrophobic stretch of amino acids. The hydrophobic region after the initiator methionine is 4 residues long for ELH, and 10 residues long for peptide A and B precursors. The presence of this hydrophilic region at the amino terminal end is a common feature of signal sequences (Loh & Gainer 1983, Docherty & Steiner 1982). Of the next 14 amino acids in all three gene products, 8 are hydrophobic. Cell-free translations from total poly A+ RNA templates can be used to determine the size of the primary translation products. Since a large percentage (up to 50%) of the mRNA codes for the precursors, the analysis using a mixture of poly A+ RNA from single cells or homogeneous tissues such as atrial glands or bag cells is relatively simple. Cell-free translations of poly A+ RNA from the bag cells, atrial glands (Scheller et al 1982), and the R3–14 cells (Nambu et al 1983) yielded a prevalent product whose size was consistent with the molecular weight of the precursor deduced from the nucleotide sequence. mRNA from the bag cells and atrial glands was translated in the presence of microsomes from dog pancreas, a system known to support several of the post-translational events that occur in vivo, including cleavage of the signal sequence. A reduction of about 2 kDa in the molecular weight of the translated products was observed (Scheller et al 1983a). Hence, we conclude that a signal sequence is part of the initial translation products of these genes.

To analyze the processing of these precursor proteins, pulse/chase experiments have been most fruitful. The characteristic electrophysiological activity of various individual neurons in isolated ganglia has been observed for up to 45 days and has been found to remain constant. The biochemical characteristics of these neurons also appear to remain intact, and this property has been exploited to study the synthesis and processing of peptides. Abdominal ganglia can be bathed in artificial seawater containing glucose and trace amounts of tritiated amino acids. The labeled amino acids are taken up into the cells and incorporated into proteins. This pulse/chase technique has been used to study precursor synthesis and processing in other systems, most notably the pro-opiomelanocortin system (Eipper & Mains 1980, Herbert et al 1981). As peptidergic cells devote a large portion of their synthetic machinery to the synthesis of peptides, the number of proteins that incorporate tritiated amino acids is small. In *Aplysia,* cells can be dissected free after appropriate pulses of radioactively labeled amino acids and individual cell proteins can then be analyzed (Loh & Gainer 1975, Aswad 1978, Berry 1976a). Alternately, cells can be filled directly with labeled amino acids by pressure injections through a micro electrode; we have found that this technique yields a higher degree of incorporation of label. The pressure injection of isotope is advantageous because the electrical activity of the cell can also be monitored. Using the techniques described above, we observed a prevalent porduct of 11 kDa (R14 and R15) and 13 kDa (L11) after a brief pulse of ^{3}H-leucine (Loh & Gainer 1975, Aswad 1978, R.-R. J. Kaldany et al, unpublished). This represents a reduction of about 2 kDa in the size of the proteins deduced from the DNA sequence and from cell-free translation experiments in the case of R3–14.

The R14 and L11 pre-proteins each have a highly hydrophobic stretch, of 22 and 24 amino acids respectively, before reaching a sequence of hydrophilic residues. The signal sequence cleavage site for the R14 precursor has been determined. A 2-hr pulse using ^{3}H-valine was used to label the precursor. Proteins were extracted and separated by high-pressure liquid chromatography. Peaks were sized by SDS-polyacrylamide gel electrophoresis. The peak coinciding with the correct molecular weight for the precursor was subjected to amino-terminal sequence analysis. A peak of radioactivity was observed at the third, ninth, and twentieth cycles. This result is consistent with the cleavage occurring at the carboxy terminal end of the serine at position 23 of the precursor. The resulting precursor would have a molecular weight of 10.5 kDa, which is consistent with the size of the precursor found in vivo. (Scheller et al 1984).

Precursor Processing

In the last few years, the primary structure of several neuropeptide precursors has been determined. By aligning the sequence of isolated and characterized peptides with the amino acid sequence of the precursors, the sites of precursor cleavage can be deduced. Pairs of basic residues are the most common sites of precursor cleavage; of these, the most common is the Lys-Arg sequence (Loh & Gainer 1983). The sequence Arg-Arg is also used, though less frequently and Lys-Lys has been observed in one case. Cleavage at single Arg residues also occurs. It should be noted, moreover, that not all paired basic residues are used as cleavage recognition sites; this implies that the tertiary structure of the protein or peptide is also of importance.

PROCESSING OF THE ELH GENE FAMILY PRECURSORS Four peptides have been isolated and characterized from bag cells and are found to be encoded in the ELH precursor (Figure 3). Several other peptides are predicted from the DNA sequence of the ELH precursor gene. Reading from the amino to the carboxy terminal end of the precursor, four small peptides, called β, γ, δ, and α-bag-cell peptides (BCPs) are bounded by paired basic residues except for the amino terminal end of α-BCP, which is flanked by a single arginine. A peptide has been isolated that has an amino acid composition identical to β-BCP. Three structurally related peptides, called α-BCPs (1–7, 1–8, 1–9), have been isolated and their amino acid sequence determined (Rothman et al 1983a). Rothman et al speculate that the 9-amino-acid peptide is released from the precursor; a carboxypeptidase-like enzyme would then sequentially produce the 8- and 7-amino-acid peptides.

The next identified sequence is that of ELH. The amino terminal isoleucine of ELH is flanked by a Lys-Arg sequence, whereas the carboxy terminal Lys is followed by the sequence Gly-Lys-Arg. The acidic peptide sequence is found at the carboxy terminal end of the ELH precursor immediately after the Gly-Lys-Arg following the ELH sequence. The stop codon defines the end of the acidic peptide. Approximately equimolar amounts of ELH, acidic peptide, and α-BCP are recovered from bag cell extracts (Rothman et al 1983a).

Two peptides from the atrial gland, peptides A and B, have been characterized; the peptides are derived from two independent genes. The A and B

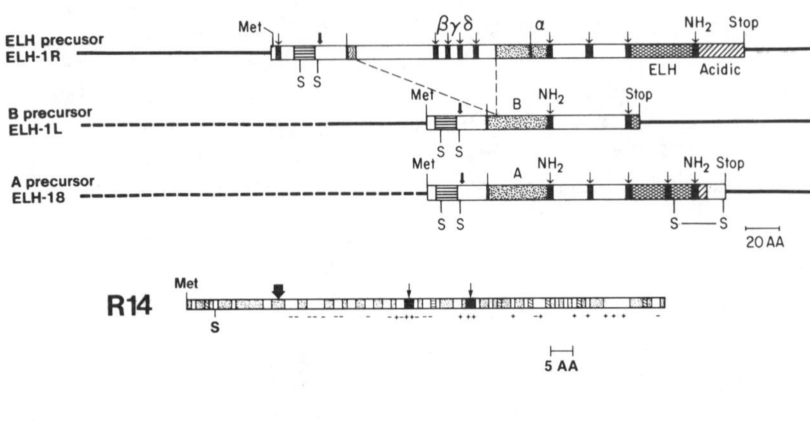

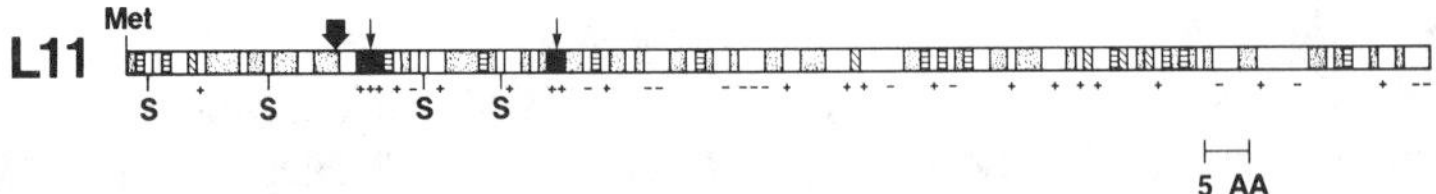

Figure 3 Aplysia neuropeptide precursors. A. The ELH, Peptide A and B precursor proteins. In each of the three proteins the initiator methionine is followed by a hydrophobic region (*horizontal stripes*). An **S** indicates the location of a cysteine residue. *Thick arrows* indicate the putative cleavage site of the signal sequence. A *line* above the sequence represents a potential cleavage at a single arginine residue, while *small arrows* represent potential or known cleavage sites at sequences of 2, 3 or 4 basic residues. If carboxyterminal amidation is believed to occur, an NH_2 is written above the arrow. The A/B peptide homology is represented by *stippled boxes*. The ELH homology is represented by *crosshatched boxes* (▒). The acidic peptide hemology is designated by *angled parallel lines* (▨).
B. The R3–14 peptide precursor. Potential cleavage sites are indicated by *small arrows*, the cleavage position of the signal sequence by a *large arrow*, and the charged amino acids by +/− signs. Hydrophobic regions are designated by the *stippled pattern*, histidine residues by *angled parallel lines*, and proline residues by horizontal lines. **S** indicates cysteine residues.
C. The L11 peptide precursor. Potential cleavage sites, putative position of signal sequence cleavage, and various amino acid types are indicated as in B.

genes and the peptide precursors share a large degree of sequence homology. In both precursors, a single arginine is found 34 amino acids after the initiator methionine. The 34 amino acids of the peptides A and B follow this arginine. The carboxyl terminus of these peptides is flanked by the sequence Gly-Lys-Arg, the same carboxy terminal sequence as for ELH in the ELH precursor. ELH and peptides A and B have amidated carboxytermini, consistent with the finding that carboxy terminal glycines are enzymatically degraded to yield amidated carboxy terminal ends. Amidation is believed to confer stability to peptides against proteolytic degradation by carboxypeptidases. This enhanced stability is probably most important for peptides that must migrate long distances to reach their targets. ELH and peptides A and B are all believed to be released into the circulation and to diffuse to distant sites of action.

Several other potential cleavage sites are present in the precursors derived from the ELH gene family sequences. Which of these are used, and what other prossible peptides are coproduced? One good candidate, an Arg-Arg sequence following the Lys-Arg at the carboxy terminal end of β-BCP, would release a pentapeptide, γ-BCP whose sequence is identical to β-BCP except for the substitution of the acidic residue Asp for the basic residue His at the carboxy terminal end. Many single arginines may also be cleavage-recognition sites.

PROCESSING OF THE R3–14 AND THE L11 PRECURSORS Peptides from the R3–14 or L11 cells have not yet been fully characterized. Hence, there is only indirect evidence as to the cleavage sites. By chasing cells labeled with ^{3}H-leucine for various periods of time, two peptides of molecular weight 4.6 and 6.2 kDa were suggested to be derived from the R14-specific precursor (Aswad 1978, Berry 1976b). The 6.2 kDa protein was presumed to be the final product. This conclusion was based on the persistence of this protein after a 3-hr chase and because it is a prevalent product of this cell as determined by Coomassie brilliant blue staining. By a similar analysis, L11 was shown to synthesize a major product of 9–11 kDa.

Knowledge of the amino acid sequence will greatly facilitate the delineation of the cleavage sites of the precursors from the R3–14, L11, bag cells, and atrial glands. The R3–14 and L11 precursors both contain two sets of paired basic residues. It has already been suggested that cleavage at the paired basic residues of the R3–14 precursor would release a 12-amino-acid peptide. The flanking peptides would have molecular weights of 3.3 kDa and 5.0 kDa (Nambu et al 1983), in reasonable agreement with the in vivo pulse chase experiments. Similarly, in the L11 precursor, cleavage at the paired basic residues would release a 20-amino-acid peptide and a protein of about 9 kDa.

Two approaches, both based on the incorporation of labeled amino acids, have been used to test these kinds of predictions. SDS-poly-acrylamide gel electrophoresis is used to separate proteins of greater than 3 kDa. By a judicious choice of amino acids, we can tell whether a certain region of the precursor is found in a specific molecular weight band. For instance, in the R3–14 cells, the amino acid histidine is found only in the carboxy terminal half of the R3–14 precursor. Using tritiated amino acids, we have observed two major products of molecular weight 6.2 kDa and 3.2 kDa in the R3–14 cells. The 3.2 kDa protein, however, does not incorporate any ^{3}H-histidine. We conclude from these results that the Arg-Arg sequence is cleaved to yield the 3.2 kDa protein. this protein would correspond to the negatively charged region at the NH_2 terminal end of the precursor.

A second approach is to subject isolated peptides to NH_2 terminal sequence analysis (see above). We assume that cleavage occurs at single arginines or at paired basic residues. The cycle at which peaks of radioactivity are released corresponds to the location of the amino acid after a particular cleavage site. The use of at most two amino acids should yield an unambiguous result. Of course, the carboxy terminus cannot be determined in this way. We have obtained evidence that the Lys-Arg sequence and one of two Gly-Arg sequences are utilized as cleavage recognition sites. These two cleavages would yield a 21-amino-acid peptide with an amidated carboxy terminus (R: R. J. Kaldany et al, unpublished observations).

The processing of the R3–14 or L11 precursors is not as simple as originally proposed by Aswad nor does it occur solely at paired basic residues. Pulse chase experiments of either set of cells has yielded several products ranging in molecular weight from 10 kDa to 3 kDa. Moreover, several small molecular weight (less than 3 kDa) peptides have been detected. ^{3}H-amino acids were injected into the R14 cell body. After a 6-hr incubation, labeled proteins were extracted from the whole ganglion. The proteins were fractionated by HPLC. The HPLC profile was similar to that which is observed when the whole ganglion is incubated with the ^{3}H-amino acids and the R14 cell is dissected out. However, the relative incorporation of the label changes dramatically. When the amino acids are injected and proteins from the whole ganglion are extracted, a far larger proportion of the label is observed in the small molecular weight peptides instead of the larger proteins. One interpretation of these results is that processing of precursors is a slow process; several cleavages may be necessary to yield final products. The initial cleavages may occur in the cell body; however, the final, relatively small products may not accumulate until the vesicles are well outside the cell body, thus implying that the processing enzymes are copackaged in the vesicles.

Immunocytochemistry

It is important to determine the distribution of cells that express the different peptide precursors in the organism and to obtain a complete picture of the terminal field distribution. In regard to the egg-laying hormone system, this has been accomplished using antibodies generated against ELH and peptide A, both of which had been isolated. These affinity-purified antibodies along with in situ hybridization studies revealed a network of ELH-containing interneurons dispersed throughout the major ganglia of the *Aplysia* central nervous system. The function of these cells remains unclear and their possible involvement in the mediation of the egg-laying behavior is being considered (see below). It is also possible that these neurons may simply use ELH and associated peptides as transmitters for functions unrelated to egg laying.

If the peptides of interest are present in small amounts it becomes very difficult to isolate enough material to raise antibodies. To circumvent these problems we synthesized peptides using information from the nucleotide sequence of the cDNA clone. These peptides were then coupled to protein carriers (either BSA or BGG) and used to raise both rabbit and guinea pig antibodies. Indirect immunofluorescence was used to locate the positions of cell bodies and processes containing the products of the gene. Kreiner et al (1984), using antibodies against three domains of the precursor deduced from the cDNA clone specific to R14 and R3–8, found only 12 cells (R3–14) to contain immunoreactivity. No other cells in the central nervous system or peripheri were immunoreactive, a finding consistent with the lack of hybridization of mRNA from other central ganglia with this clone in the northern blot analysis. Furthermore, antibodies generated from all three regions of the precursor react with the same cells. These data are reassuring in that they provide additional evidence that we have determined the correct nucleotide sequence and selected the proper reading frame to determine the amino acid sequence of the precursor. Even a 99.9% level of accuracy in nucleotide sequence determination may be inadequate, because a simple deletion or insertion would give rise to a frame shift error that could define the wrong precursor protein.

Numerous immunoreactive processes and varicosities are seen in the sheath surrounding the abdominal ganglion. In addition, each cell sends an axon out the branchial nerve that terminates on the efferent vein at the base of the heart. R14 sends additional immunoreactive processes to the abdominal ganglion artery and the anterior aorta. Similar findings were observed using autoradiographic techniques to detect the unusually elevated uptake of glycine in these neurons and processes (Price & McAdoo 1979, 1981).

The distribution of immunoreactive bag cell and R3–14 processes suggests that the extracellular messengers used by these neurons may have central and peripheral targets. Innervation of the vascularized connective tissue sheath that surrounds the ganglia is an effective means for delivering the secreted products to both the neuronal targets and the circulation. If the peptides are stable enough to survive the journey through the circulation, a host of peripheral actions may be elicited in concert with central actions.

Processing Enzymes

One of the most intriguing properties of neuropeptide biosynthesis is the possibility for differential precursor processing as a mechanism of regulating information flow within and between neurons. This regulation could be a consequence of the stimulation or inhibition of neuronal activity in fully differentiated cells or it could occur as a function of development. Differential processing implies regulation of enzymatic activity and this may occur at many levels, from genetic regulation to post-translational alterations of the substrate. To determine the site(s) of regulation, purification of the enzymes is essential. Recently, the initial characterization of an enzyme from the atrial gland of *Aplysia* that cleaves at single arginine residues was reported (Wallace et al

1984). The assay consisted of incubating an extract of atrial gland tissue with dynorphin A, a 17-amino-acid peptide containing one Arg-Arg sequence as the sixth and seventh residues from the NH_2 terminus and a single Arg at the nineth position. A 2-hr incubation of dynorphin A with the enzyme extract converted the substrate, in a 90–100% yield, to dynorphin 1–8. Dynorphin 1–9, which includes a carboxyterminal arginine, was a much poorer substrate. The authors concluded that the peptidase may cleave dynorphin A directly at the amino side of the arginine residue at position 9. In this manner, the atrial gland of *Aplysia* may be a useful tissue for studying and isolating enzymes involved in post-translational processing of secretory peptides.

RECEPTORS

The activity of chemical messengers is mediated by receptors. The characterization of receptors for peptides, and in particular neuropeptides, is just beginning. The receptors for insulin (Fujita-Yamaguchi et al 1983), nerve growth factor (Puma et al 1983), leutinizing hormone-releasing gonadotropin (Dattatreyamurty et al 1983), and epidermal growth factor (Cohen et al 1982) have been purified. The isolation of the opiate receptors has been more difficult to achieve. The existence of four types of opiate receptors has been postulated on the basis of binding studies using natural and synthetic opiates and their antagonists (Akil et al 1984). Are these receptors structurally related? Are they encoded by family of genes as are the opiates themselves? Are the mechanisms by which they mediate the effects of the opiates related?

The egg-laying hormone system presents a similar situation. A set of genes generates a battery of neuropeptides, each with a distinct spectrum of biological activities. How many receptors mediate the effects of these factors? For example, how related are the receptors for the structurally similar peptides A and B, and α-BCP? A preliminary report of the action of ELH on the in vitro system (Dudek & Tobe 1978) augurs well for attempts to isolate and characterize the ELH receptor. By quantitating the release of eggs from a section of ovotestis as a function of ELH concentration, a concentration of 1 nM was found to be sufficient to release eggs (Rothman et al 1983b). ELH was the only component of bag cell extracts that had this activity. The ovotestis is a large tissue and may represent a rich source of receptor for this neuropeptide. The purification of the ELH receptor would represent the first step in a molecular analysis of such molecules.

PHYSIOLOGY AND BEHAVIOR

Egg-laying is a seasonal and episodic behavior. Normally, the bag cells are quiescent (Kupfermann & Kandel 1970); however, they fire long-lasting bursts just before egglaying in vivo. When injected into an animal, an extract of bag cells (Kupfermann 1967, Kupfermann & Kandel 1970) or atrial glands (Arch et al 1978, Heller et al 1980) will elicit the array of behavioral responses normally associated with egg-laying. The complex fixed-action pattern associated with egg-laying includes an inhibition of feeding; cessation of locomotion; enhanced contraction of the siphon and gill, known as respiratory pumping, which is thought to result in greater blood oxygenation; and a waving of the head, which helps wind the egg-string as it is being extruded. Bag cells prelabeled with radioactive amino acids were shown to release several peptides during the afterdischarge, among which are ELH and acidic peptide (Stuart et al 1980). These experiments suggest that the components responsible for triggering egg-laying behavior are contained within the bag cells and/or the atrial gland. Injections of purified ELH will elicit egg-laying but not all the other behaviors associated with this fixed action. In this section, we review the evidence to support the notion that some, if not all, the components that elicit egg-laying are encoded by the ELH gene family.

Action potentials in the bag cells can be elicited by electrical stimulation of the pleuro-visceral connective; the electrical activity is synchronous in all the cells of the cluster. After a stimulation, the bag cells will continue to discharge independently synchronous action potentials (Kaczmarek et al 1978). The biological trigger for bag cell afterdischarge is unknown. Peptides A and B can initiate afterdischarge of bag cells in in vitro experiments. Strumwasser suggests that these peptides are released from the atrial gland during copulation and act as the natural trigger of egg-laying (Strumwasser et al 1980, Strumwasser 1983). Animals whose abdominal ganglia have been surgically removed will lay eggs in response to ELH injections but not in response to peptide B injection. One can conclude, in accordance with the working hypothesis of Strumwasser, that the abdominal ganglion is a necessary mediator of the action of peptide B in eliciting egg-laying. (The experiments have only been done using peptide B, but the same results would presumably be observed with peptide A.) Two observations are, however, inconsistent with this hypothesis:

1. Copulation alone is insufficient to trigger egg-laying.
2. The atrial gland is an exocrine gland; since the duct into which peptides are believed to be released leads directly to the gonopore, it is unclear how peptides A and B would reach the bag cells.

Discharge of the bag cells (frequently called afterdischarge) produces several types of effects on identified abdominal ganglia neurons (Mayeri et al 1979a,b). The responses have been divided into four categories:

1. burst augmentation, an effect on the bursting pacemaker activity of cell R15;

2. prolonged excitation, in which normally silent cells of the left lower quadrant fire tonically;
3. slow inhibition, in which left upper quadrant cells L2, L3, L5, and L6 are hyperpolarized;
4. transient excitation, manifest by depolarization of cells L1 and R1, which sometimes fire a short train of spikes, and of siphon motor neurons, whose activity increases for a period of several minutes (Brownell & Schaefer 1982).

A correlation of these effects with specific peptides has been attempted using peptides either purified from the ganglion or synthesized chemically. These peptides are perfused into the abdominal ganglion through the vasculature leading to the ganglion, while simultaneously recording intracellularly from identified neurons. ELH appears to be responsible for the burst augmentation of R15 and the prolonged excitation of the left lower quadrant cells. These effects may last for periods of up to several hours. α-BCP inhibits the left upper quadrant cells (Rothman et al 1983a). A peptide with the amino acid composition of β-BCP may mimic the transient excitation of L1 and R1. The effects of α-BCP and β-BCP are of short duration and are abolished if protease inhibitors are not concomitantly added or if the peptide is not applied over a desheathed ganglion. Residues 2–7 in α-BCPs are identical to a six-residue sequence in the carboxy terminal region of peptides A and B. This region of homology may represent the active site of peptides A and B, since α-BCPs also have excitatory effects on bag cells (Scheller et al 1983b). Release of α-BCPs have been suggested to play an autoexcitatory role in bag-cell afterdischarge. In documenting the neuronal and behavioral effects of the peptides released during bag cell afterdischarge, one must correlate afterdischarge with effects on neurons controlling a specific component of egg-laying behavior. This has been done for one component, respiratory pumping, in which afterdischarge, increased activity of siphon motor neurons, and increased frequency and amplitude of siphon contraction have been described (Brownell & Schaefer 1982). At present, it is unclear how effects of afterdischarge on other neurons in the abdominal ganglion relate to the physiological and behavioral components of egg-laying.

The buccal ganglia control the motor output that underlies certain components of feeding behavior. ELH applied to buccal ganglia in vitro activates a pair of neurons in these ganglia (Stuart & Strumwasser 1980) and this effect may be related to the observation that feeding is inhibited during egg-laying. Each of these neurons has an ipsilateral axon in buccal nerve 4. The functions of the neurons that were studied are as yet unknown; however, purified ELH applied to isolated ganglia caused activation of at least one axon in each buccal nerve 4 and one axon in each cerebrobuccal connective nerve. The buccal nerve 4 axon, which can be stimulated by ELH application, excites the radula extractor muscles (Ram 1982). The recording of the activity of this muscle during egg-laying may clarify the connection, if any, of these observations with the inhibition of feeding during egg-laying.

The central program for locomotion is located in the head ganglia (cerebral, pleural, pedal, and buccal). The majority of foot and body wall motor neurons are located in the pedal and pleural ganglia (Hening et al 1979, Fredman & Jahan-Parwar 1980). When injected into the bloodstream of nonlocomoting animals, serotonin triggers locomotion (Mackey & Carew 1983). Animals whose pedal and pleural ganglia have been surgically isolated also respond to serotonin in a qualitatively similar manner. Bag cell extracts have an antagonistic effect in this assay, an effect which can be abolished by prior treatment of the extract with pronase. Thus, the inhibition of locomotion during egg-laying may be mediated by peptides released from bag cells. If this is indeed the case, how do peptides from the bag cells reach the head ganglia?

By releasing neurotransmitter substances at their synapses, neurons are able to communicate rapidly through point-to-point contact. Like conventional neurotransmitters, neuroactive peptides may act locally. ELH and α-BCPs appear to fulfill such a role, as they affect individual neurons in the abdominal ganglion. Neuropeptides can also act as hormones, diffusing to distant sites. ELH, which is a 36-amino-acid peptide with an amidated carboxyl terminus, may be stable enough to diffuse to the ovotestis causing the contraction of smooth muscle follicles and resulting in the release of eggs. Similarly, ELH may diffuse, or be transported via the circulation, to the head ganglia and modulate the activity of neurons that affect behaviors associated with egg-laying. Alternatively, the extensive neural circuitry observed immunocytochemically using anti-ELH antibodies may be responsible for affecting head ganglia neurons (see above and McAllister et al 1983).

The complex fixed-action pattern resulting in egg-laying consists of a coordinated control of the activity of several ganglia and non-neuronal tissue. How is this coordinate control achieved? The neural circuitry responsible for the individual behavioral conponents and the point of action of the peptides remains to be determined. The demonstration of a causal relationship is never an easy task; clearly, however, more electrophysiological experiments are needed to assess the potential role of the cells and the molecules mediating the egg-laying behavior.

The organization of multiple functional products into a single polyprotein precursor is certainly of general importance in the biosynthesis of neuroactive peptides. The biological success of this strategy is evidenced by its implementation in entities ranging in structural complexity from viruses (Rancaniello & Baltimore 1981) to the human brain (Comb et al 1982). This arrangement does have several critical implications:

1. It may insure the stoichiometric synthesis and release of products that could then be used in related functional capacities. With respect to the *Aplysia* neuropeptides discussed in this paper, the ELH precursor seems likely to serve in this fashion.
2. By utilizing alternate processing pathways, cells in distinct tissues may liberate very different sets of end products from the same initial precursor. Differential processing of the pro-opiomelanocortin precursor in the anterior and intermediate lobes of the pituitary gland presents an especially lucid example of this capability (Herbert et al 1981).
3. It is possible to generate temporarily distinct sets of products from a given precursor without requiring changes in transcription if processing pathways can be altered in response to endogeneous cellular activities or external stimuli.

The capacity to generate highly diverse combinations of active peptides is of particular relevance within the nervous system, where the numerical complexity of cell types and interconnections reaches a zenith. The polyprotein stratagem clearly affords this capability with a minimal expenditure of genetic material. All of the *Aplysia* neuropeptide precursors reviewed appear to be polyproteins—each containing several products that may be biologically active. Presumably, this arrangement satisfies specific functional requirements of both neural and endocrine systems in the mediation of cellular communication.

SUMMARY AND CONCLUSIONS

Extensive electrophysiological experiments on *Aplysia* neurons have resulted in an understanding of simple behaviors in terms of the activities of a single identified neurons. Beginning with the work of Kupfermann & Kandel, neuropeptides in *Aplysia* have become increasingly implicated as chemical agents that control or affect behavior. Several neuropeptides have been isolated and characterized; recently, the genes that code for several of these neuropeptides have been isolated. Studies of neuropeptide gene expression and the behaviors affected thereby have been bridged in the egg-laying hormone neuroendocrine system. The role of polyproteins in coordinating complex, fixed-action patterns is beginning to emerge. The continued investigation of this neuroendocrine system, and the other cell-specific polyproteins that have been characterized more recently, promises to yield further insights into the roles of neuropeptides in governing behavior.

ACKNOWLEDGMENTS

The authors would like to thank Paul Taghert and Mark Schaefer for critically reviewing the manuscript. Unpublished work was supported by grants to R. H. S. from the NIH, McKnight Foundation, and March of Dimes.

Literature Cited

Akil, H., Watson, S. J., Young, E., Lewis, M. E., Khachaturian, H., Walker, J. M. 1984. Endogenous opioids: Biology and function. *Ann. Rev. Neurosci.* 7:223–55

Amara, S. G., Jonas, V., Rosenfeld, M. G., Ong, E. J., Evans, R. M. 1982. Alternative RNA processing in calcitonin gene expression generates mRNAs encoding different polypeptide products. *Nature* 298:240–44

Arch, S. 1972. Biosynthesis of the egg-laying hormone (ELH) in the bag cell neurons of *Aplysia californica*. *J. Gen. Physiol.* 60: 1263–78

Arch, S., Smock, R. T., Gurvis, R., McCarthy, C. 1978. Atrial gland induction of the egg-laying response in *Aplysia californica*. *J. Comp. Physiol.* 128:67–70

Aswad, D. 1978. Biosynthesis and processing of presumed neurosecretory proteins in single identified neurons of *Aplysia californica*. *J. Neurobiol. 9:267–84*

Berry, R. W. 1976a. A comparison of the 12,000 dalton proteins synthesized by *Aplysia* neurons L11 and R15. *Brain Res.* 115:456–66

Berry, R. W. 1976b. Processing of low molecular weight proteins by identified neurons of *Aplysia*. *J. Neurochem.* 26:229–31

Brownell, P. H., Schaefer, M. E. 1982. Activation of a long lasting motor program by bag cell neurons in *Aplysia*. *Soc. Neurosci. Abstr.* 8:736

Chiu, A. Y., Hunkapiller, M., Heller, E., Stuart, D. K., Hood, L. E., Strumwasser, F. 1979. Neuropeptide egg-laying hormone of *Aplysia californica:* Purification and primary structure. *Proc. Natl. Acad. Sci. USA* 76:6656–60

Coggeshall, R. E., Yaksta, B. A., Swartz, F. J. 1971. A cytophotometric analysis of the DNA in the nucleus of the giant cell, R-2, in *Aplysia*. *Chromosoma* 32:205–12

Cohen, S., Java, R. A., Sawyer, S. T. 1982. Purification and characterization of epidermal growth factor receptor/protein kinase from normal mouse liver. *Proc. Natl. Acad. Sci. USA* 79:6237-41

Comb, M., Seeburg, P. H., Adelman, J., Eiden, L., Herbert, E. 1982. Primary structure of the human met- and leu-enkephalin precursor and its mRNA. *Nature* 295:663–71

Dattatreyamurty, B., Rothman, P., Saxena, B. B. 1983. Isolation of luteinizing hormone-chorionic gonadotropin receptor in high yield from bovine corpora lutea. *J. Biol. Chem.* 258:3140–58

Docherty, K., Steiner, D. F. 1982. Post-translational proteolysis in polypeptide hormone biosynthesis. *Ann. Rev. Physiol.* 44:625–38

Dudek, F. E., Tobe, S. S. 1978. Bag cell peptide acts directly on ovotestis of *Aplysia californica:* Basis for an *in vitro* bioassay. *Gen. Comp. Endocrinol.* 36:618–27

Eipper, B. A., Mains, R. E. 1980. Structure and biosynthesis of pro-adrenocorticotropin/endorphin related peptides. *Endocrin. Rev.* 1:1–27

Fredman, S. M., Jahan-Parwar, B. 1980. Role of pedal ganglia motor neurons in pedal wave generation in *Aplysia*. *Brain Res. Bull.* 5:179–93

Fujita-Yamaguchi, Y., Choi, S., Sakamoto, Y., Itakura, K. 1983. Purification of insulin receptor with full binding activity. *J. Biol. Chem.* 258:5045–49

Gainer, H., Wollberg, Z. 1974. Specific protein metabolism in identifiable neurons of *Aplysia californica*. *J. Neurobiol.* 5:243–61

Gilbert, W. 1978. Why genes in pieces? *Nature* 271:501

Heller, E., Kaczmarek, L. K., Hunkapiller, M. W., Hood, L. E., Strumwasser, F. 1980. Purification and primary sequence of two neuroactive peptides that cause bag cell afterdischarge and egg-laying in *Aplysia*. *Proc. Natl. Acad. Sci. USA* 77:2328–32

Hening, W. A., Walters, E. T., Carew, T. J., Kandel, E. R. 1979. Motor neuronal control of locomotion in *Aplysia*. *Brain Res.* 179:231–53

Herbert, E., Burnberg, N., Lissitsky, J. C., Civelli, O., Uhler, M. 1981. Pro-opiomelanocortin: A model for the regulation of expression of neuropeptides in pituitary and brain. *Neurosci. Commun.* 1:16–27

Hopkins, W. E., Stone, L. S., Rothman, B. S., Basbaum, A. F., Mayeri, E. 1982. Egg-laying hormone, leucine-enkephalin, and serotonin immunoreactivity in the abdominal ganglion of *Aplysia:* A light microscope study. *Soc. Neurosci. Abstr.* 8:857

Kaczmarek, L. K., Jenning, K., Strumwasser, F. 1978. Neurotransmitter modulation, phosphodiesterase inhibitor effects, and cyclic AMP correlates of afterdischarge in peptidergic neurites. *Proc. Natl. Acad. Sci. USA* 75:5200–4

Kandel, E. R. 1976. *The Cellular Basis of Behavior.* San Francisco: Freeman

Kreiner, T., Rothbard, J., Schoolnick, G. K., Scheller, R. H. 1984. Antibodies to synthetic peptides defined by cDNA cloning reveal a network of peptidergic neurons in *Aplysia*. *J. Neurosci.* 4(10):2581–89

Kupfermann, I. 1967. Stimulation of egg-laying. Possible neuroendocrine function of bag cells of abdominal ganglion of *Aplysia californica*. *Nature* 216:814–15

Kupfermann, I., Kandel, E. R. 1970. Stimulation of egg-laying by extracts of neuroendocrine cells (bag cells) of abdominal ganglion of *Aplysia*. *J. Neurobiol.* 33:877–81

Kupfermann, I., Weiss, K. R. 1976. Water regulation by a presumptive hormone contained in identified neurosecretory cell R15 of

Aplysia. J. Gen. Physiol. 67:113–23

Loh, Y. P., Brownstein, M. J., Gainer, H. 1984. Proteolysis in neuropeptide processing and other neural functions. *Ann. Rev. Neurosci.* 7:189–222

Loh, Y. P., Gainer, H. 1975. Low molecular weight specific proteins in identified molluscan neurons. 1. Synthesis and storage. *Brain Res.* 92:181–92

Loh, Y. P., Gainer, H. 1983. Biosynthesis and processing of neuropeptides. In *Brain Peptides,* ed. D. T. Kreiger, M. J. Brownstein, J. B. Martin, pp. 79–116. New York: Wiley

Loh, Y. P., Sarne, Y., Daniels, M. P., Gainer, H. 1977. Subcellular fractionation studies related to the processing of neurosecretory proteins in *Aplysia* neurons. *J. Neurochem.* 29:135–39

Mackey, S., Carew, T. J. 1983. Locomotion in *Aplysia:* Triggering by serotonin and modulation by bag cell extract. *J. Neurosci.* 3:1469–77

Mayeri, E., Brownell, P., Branton, W. D., Simon, S. B. 1979a. Multiple, prolonged actions of neuroendocrine bag cells in *Aplysia*. I. Effects on bursting pacemaker neurons. *J. Neurophysiol.* 42:1165–84

Mayeri, E., Brownell, P., Branton, W. D. 1979b. Multiple, prolonged actions of neuroendocrine bag cells on neurons in *Aplysia*. II. Effects on beating pacemaker and silent neurons. *J. Neurophysiol.* 42:1185–97

McAllister, L. B., Scheller, R. H., Kandel, E. R., Axel, R. 1983. *In situ* hybridization to study the origin and fate of identified neurons. *Science* 222:800–8

Morris, H. W., Panico, M., Karplus, A., Lloyd, P. E., Riniker, B. 1982. Elucidation by FAB-MS of the structure of a new candidate peptide from *Aplysia*. *Nature* 300:643–45

Nakanishi, S., Inoue, A., Kita, T., Nakamura, M., Chang, A. C. Y., Cohen, S. N., Numa, S. 1979. Nucleotide sequence of cloned cDNA for bovine corticotropin-β-lipotropin precursor. *Nature* 278:423–27

Nambu, J. R., Taussig, R., Mahon, A. C., Scheller, R. H. 1983. Gene isolation with cDNA probes from identified *Aplysia* neurons: Neuropeptide modulators of cardiovascular physiology. *Cell* 35:47–56

Nawa, H., Hirose, T., Takashima, H., Inayama, S., Nakanishi, S. 1983. Nucleotide sequence of cloned cDNAs for two types of bovine brain substance P precursor. *Nature* 306:32–35

Noda, M., Takahashi, H., Tanabe, T., Toyosato, M., Furutani, Y. S., Hirose, T., Asai, M., Inayama, S., Miyata, T., Numa, S. 1982. Primary structure of α-subunit precursor of *Torpedo californica* acetylcholine receptor deduced from cDNA sequences. *Nature* 299:793–97

Price, C. H., McAdoo, D. J. 1979. Anatomy and ultrastructure of the axons and terminals of neurons R3-14 in *Aplysia*. *J. Comp. Neurol.* 188:647–77

Price, C. H., McAdoo, D. J. 1981. Localization of axonally transported [^{3}H]glycine in vesicles of identified neurons. *Brain Res.* 219:307–15

Puma, P., Buxser, S. E., Watson, L., Kelleher, D. J., Johnson, G. L. 1983. Purification of the receptor for nerve growh factor from A875 melanoma cells by affinity chromatography. *J. Biol. Chem.* 258:3370–75

Racaniello, V. R., Baltimore, D. 1981. Molecular cloning of poliovirus cDNA and determination of the complete nucleotide sequence of the viral genome. *Proc. Natl. Acad. Sci. USA* 8:4887–91

Ram, J. L. 1982. *Aplysia* egg-laying hormone increases excitatory input into a retractor muscle of the buccal mass. *Brain Res.* 236:505–10

Rosenfeld, M. G., Mermod, J.-J., Amara, S. G., Swanson, L. W., Sawchenko, P. E., Rivier, J., Vale, W. W., Evans, R. M. 1983. Production of a novel neuropeptide encoded by the calcitonin gene via tissue specific RNA splicing. *Nature* 304:129

Rothman, B. S., Mayeri, E., Brown, R. O., Ylan, P.-M., Shively, E. 1983a. Primary structure and neuronal effects of α-bag cell peptide. A second candidate neurotransmitter encoded by a single gene in bag cell neurons of *Aplysia*. *Proc. Natl. Acad. Sci. USA* 80:5753–57

Rothman, B. S., Weir, G., Dudek, F. E. 1983b. Egg-laying hormone: Direct action on the ovotestis of *Aplysia*. *Gen. Comp. Endocrinol.* 52:134–41

Sakano, H., Maki, R., Kurosawa, Y., Roeder, W., Tonegawa, S. 1980. Two types of somatic recombination are necessary for the generation of complete immunoglobulin heavy-chain genes. *Nature* 286:676–83

Scheller, R. H., Jackson, J. F., McAllister, L. B., Schwartz, J. H., Kandel, E. R., Axel, R. 1982. A family of genes that codes for ELH, a neuropeptide eliciting a stereotyped pattern of behavior in *Aplysia*. *Cell* 28:707–19

Scheller, R. H., Jackson, J. F., McAllister, L. B., Rothman, B. S., Mayeri, E., Axel, R. 1983a. A single gene encodes multiple neuropeptides mediating a stereotyped behavior. *Cell* 35:7–22

Scheller, R. H., Kaldany, R.-R., Kreiner, T., Mahon, A. C., Nambu, J. R., Schaefer, M., Taussig, R. 1984. Neuropeptides: Mediators of behavior in *Aplysia*. *Science* 225(4668):1300–8

Scheller, R. H., Rothman, B. S., Mayeri, E. 1983b. A single gene encodes multiple peptide transmitter candidates involved in a stereotyped behavior. *Trends Neurosci.*6:340–45

Southern, E. M. 1975. Detection of specific sequences among DNA fragments separated by gel electrophoresis. *J. Mol. Biol.* 98:503–17

Stuart, D. K., Chiu, A. Y., Strumwasser, F. 1980. Neurosecretion of egg-laying hormone and other peptides from electrically active bag cell neurons of *Aplysia*. *J. Neurophysiol.* 43:488–98

Stuart, D. K., Strumwasser, F. 1980. Neuronal sites of action of a neurosecretory peptide, egg-laying hormone, in *Aplysia californica*. *J. Neurophysiol.* 43:499–519

Strumwasser, F., Kaczmarek, L. K., Chiu, A. Y., Heller, E., Jenning, K., Viele, D. 1980. Peptides controlling behavior in *Aplysia*. In *Peptides: Integrators of Cell and Tissue Function,* ed. F. Bloom, pp. 197–218. New York: Raven

Strumwasser, F. 1983. Peptidergic neurons and neuroactive peptides in molluscs: From behavior to genes. In *Brain Peptides,* ed. D. T. Kreiger, M. J. Brownstein, J. B. Martin, pp. 183–216. New York: Wiley

Taussig, R., Kaldany, R. R., Scheller, R. H. 1984a. A cDNA clone encoding neuropeptides isolated from *Aplysia* neuron L11. *Proc. Natl. Acad. Sci. USA* 81:4988–92

Taussig, R., Picciotto, M. R., Scheller, R. H. 1984b. Two introns define functional domains of a neuropeptide precursor in *Aplysia*. *UCLA Symp*. In press

Wallace, E. F., Webber, E., Barchas, J. D., Evans, C. J. 1984. A putative processing enzyme from *Aplysia* that cleaves dynorphin A at the single argining residue. *Biochem. Biophys. Res. Commun.* 119:415–22

Wilson, W. L. 1971. Molecular weight distribution of proteins synthesized in single identified neurons of *Aplysia*. *J. Gen. Physiol.* 57:26–40

Young, R. A., Davis, R. W. 1983. Yeast polymerase II genes: Isolation with antibody probes. *Science* 222:778–81

Ann. Rev. Neurosci. 1985. 8:457–94

VERTEBRATE NEUROETHOLOGY: DEFINITIONS AND PARADIGMS

David Ingle

The Rowland Institute, 100 Cambridge Parkway, Cambridge, Massachusetts 02142

David Crews

Institute of Reproductive Biology, Department of Zoology,
University of Texas at Austin, Austin, Texas 78712

INTRODUCTION

Neuroethology has its historical origins in three scientific disciplines: comparative neuroanatomy, comparative and physiological psychology, and comparative ethology. Neuroethology, however, avoids the emphasis on structure characteristic of comparative neuroanatomy and on task-oriented problems popular with psychologists, while drawing heavily on behavioral adaptations, the hallmark of comparative ethological investigations. Neuroethological studies measure an animal's competence, not its capabilities, by testing the organism under naturalistic conditions rather than under artificial conditions that may give rise to misleading results. Although the boundaries of neuroethology are arbitrary, its focus is on complex behaviors and not on simple reflexes such as coughing and eye-wiping. Instead of determining how simple reflexes become organized, integrated, and synthesized to form the foundation of complex behaviors (cf Teitelbaum et al 1983), neuroethological studies begin with the behavioral sequence itself, concentrating on motivated or goal-directed behaviors that have adaptive value.

Behavior is best viewed as the final common pathway in a dynamic process involving the organism's external and internal milieux. One can arbitrarily divide this complex interaction into four aspects:

0147-006X/85/0301-0457$02.00

1. the way in which the organism has become sensitive to certain physical, biotic, and social factors as key stimuli;
2. the mechanisms by which these stimuli are presented and integrated within the brain;
3. how this representation of the environment results in changes in the organism's internal state;
4. how those physiological changes in the internal milieu influence the manner in which the organism interacts with its environment.

Superimposed on these physiological issues is the ethologist's traditional concern for development, ecological fitness, and evolutionary history.

Because one's definition of "neuroethology" depends upon one's definition of ethology, it seems useful to begin this review by identifying the questions raised by ethologists concerning the mechanisms of behavior and discussing their methods for description and analysis of behavior. Certainly, ethologists focus their attention upon the behavior of animals in nature or within laboratory environments where natural behaviors easily emerge. Ethological studies are typically concerned with (a) natural stimuli that elicit biologically important behaviors such as feeding, fleeing, courtship, and fighting; (b) the spatiotemporal structure of ensuing action pattens (pursuit, biting, threatening, calling, etc), and (c) the motivational, developmental, and physiological conditions that determine which alternative response is likely to be elicited by a given set of external stimuli. The ethologist is typically uninterested in knowing the range of stimuli that an animal might be trained to detect in the laboratory, or which arbitrary responses might be conditioned (such as pressing a lever). The ethologist ignores questions regarding the durability of memory for man-made geometric stimuli but is fascinated by a bird's ability to remember a large number of places where they have previously hidden seeds (Shettleworth 1983).

Although ethologists seldom share the enthusiasm of psychologists for measuring detection thresholds or limits of discrimination, they emphasize that a given species will spontaneously attend to certain key stimuli (originally called "sign stimuli") that signify food, enemy, mate, or parent. Because many key stimuli that elicit social behavior in the lower vertebrates (fishes, amphibians, or reptiles) become effective without any explicit learning, early ethologists postulated "Innate Releasing Mechanisms" (IRMs) for the translation of key stimuli into behavioral responses.[1]

[1]Abbreviations used: AH-POA, anterior hypothalamus–preoptic area; AVT, arginine-8 vasotocin; CF, contant frequency; DTAM, dorsal tegmental area of the medulla; DVR, dorsal ventricular ridge; FCR, flexor carpi radialis; FM, frequency-modulated; IRMs, innate releasing mechanisms; LHRH, luteinizing hormone-releasing hormone; n. pre-V, pre-trigeminal nucleus; PDVR, posterior division of the dorsal ventricular ridge; POA, preoptic area; SR, sterno-radialis; VIN, ventral infundibular nucleus; VMN, ventromedial nucleus.

Although a large number of neuroethologists have focused upon stimulus-specific IRMs and allied sensorimotor guidance mechanisms, an ethological description of behavior equally embraces details of the social and motivational context of behavior, and of the final sequence of motor patterns through which the animal achieves its goal. Those neuroethologists concerned with motivation or with behavioral endocrinology tend to view the behavioral sequence as a whole, and to consider the multiplicity of external and internal conditions that bias the animals' response to relevant stimuli. The price for this sophisticated view of causality is that anatomical and physiological tracing of relevant neural pathways proves much more difficult. Although we know many important facts abut the coding of visual and acoustic stimuli, it has been difficult to trace the flow of information through areas of limbic system and hypothalamus related to eating, fighting, or mating. It is typically even harder to guess the spatial and temporal order of brain events that lead to the programming of motor patterns in vertebrates. Therefore, gradual understanding of motivational systems has been won by focusing upon relatively simple stimulus-response linkages (e.g. triggering of lordosis in females by the male's tactile input) or by plotting the key loci in the brain where active substances such as hormones are bound by sensitive neurons. Because single neurons recorded in freely moving animals may show good correlations with sensory events or with consequent motor patterns, it is likely that single neuron recording will become more effective in detailing "motivational processes" as the physiologist moves forward stepwise from the sensory side, or backwards from the motor event. We emphasize that it is the ethologist's documentation of relatively consistent relationships among motivation, stimulus, and response which engenders optimism that recording in the freely moving animal will allow a more refined means of tracing the flow of influence toward the final common pathway than will current uses of focal lesions, electrical stimulation, and chemical mapping of the brain.

Studies of motivated behavior in various animals sometimes reveal common themes despite the specialization of input and output organization. For example, the feeding behavior of the frog reflects an interaction between direct activation of the optic tectum by retinal input and central modulation of the tectum by diencephalic structures (see Case Study 4.) Studies of aggression in cats show that diencephalic events modify sensorimotor reflexes involving midbrain and brainstem (Bandler & Flynn 1971); electrical activation of hypothalamus facilitates the cat's tendency to *orient* toward a moving object, to respond to whisker pressure by biting, and to respond to touching the paws by *swiping*. The sexual response of a female rat to the grasping behavior of the male can also be regarded as a "patterned reflex" facilitated by the action of estrogen on key neurons within the diencephalon or brainstem (see Case Study 5). Whereas one finds striking commonalities among vertebrate classes such as the evocation of feeding or aggression after stimulation of the hypothalamus in fish, bird, and cat, the stimuli effective for a common behavior (mating) and the hormonal mechanisms involved can differ dramatically within a class or even within a genus. Thus the neuroethologist must combine an awareness of common operational principles in brain function with his particular knowledge of the special adaptations of each group of animals.

One assumes the mechanisms underlying behavior as well as the natural selection have evolved in response to specific environmental conditions. Although we recognize that this assumption may prove overly narrow in the future (Gould & Vrba 1982), such an approach is useful for understanding the evolution and adaptation of brain-behavior relations. Thus, consideration of the physical and social environments in which animals live can give valuable clues as to which stimuli are biologically relevant and hence can provide candidates for neuroethological studies. In a very real sense, then, ethological investigators regard the laboratory and the field as complementary to one another. The laboratory is the only possible arena for determining the physiological bases of species-typical behaviors, whereas the field is a valuable testing ground for determining the adaptive function of the behavior in question.

The hypothesis of IRMs has led some neuroethologists to probe the sensory projection systems for neurons attuned to particular events such as the call of a potential mate or the visual configuration of a prey. Neurons that fulfilled these expectations were called "mate detectors" or "prey detectors." Our review catalogs some successes in the search for neural detectors of key stimuli. On the other hand, psychologists have demonstrated unequivocally that internal stimuli, such as hormones and biological rhythms, are also critical determinants of the kind or the intensity of responses that key stimuli may elicit. While emphasizing natural stimuli, ethologists use simplified artificial stimuli (resembling the natural stimuli in a limited fashion) to evoke species-typical behaviors. In many cases a combination of simple stimuli (e.g. color plus form to simulate the visual image of a mate or the addition of two tonal frequencies to simulate a call) does surprisingly well in evoking a natural behavioral response. Although these results do not mean that the natural stimulus gestalt has been fully characterized, they are enormously useful for probing the nervous system in search for "detector" neurons.

Many vertebrate neuroethologists are now engaged in study of the physiological bases of so-called IRMs. They typically choose animals with specialized sensory capacities that adapt well to the laboratory. Once they achieve a first-order description of sensory neurons with interesting selective responses, a common next step is to make comparisons with species of the same group (e.g. among frogs). These comparative studies typically reveal that each species has a distinctive signalling pattern and possesses the matching sensory filter to detect signals of its own species (Capranica 1976, 1983). The comparative method often confirms the suggestion that the brain of each animal attends to relatively few biologically significant stimuli within a potentially enormous range of sensory inputs.

In the following sections, we review a few successful lines of work exemplary of the goals and methods of vertebrate neuroethology. We omit one notably successful area of work, because it has been recently well-reviewed: the behavior of electrical fishes (see Heiligenberg & Bastian 1983). It is hoped that this "case study" approach conveys the flavor and the excitement of this growing scientific field, although it cannot pretend to cover the rich variety of problems now under laboratory investigation. At the end of our review we attempt to do justice to the future potential of this field by pointing out some promising new approaches and opportunities for collaboration with workers in other established branches of neurobiology. Thus our intention is to establish first the "mainline" approach to neuroethology, but to conclude with an awareness of expanding borders and new horizons.

THE SENSORY BASES OF LOCALIZATION AND DISCRIMINATION

Case Study 1: Mate Selection by Frogs

An important chapter in neuroethology was initiated by Frishkopf et al (1968) when they demonstrated that the dominant frequencies in the mating call of the bullfrog (*Rana catesbeiana*) are well-matched by the spectral sensitivities of fibers within the bullfrog's auditory nerve. As with other species, an effective call contains frequencies mainly in the lower and upper frequency ranges (about 200 and 1500 Hz, respectively). In fact, adding tones in the middle range (about 800 Hz) will actually inhibit the calling response of other males or the approach of females. This effect appears to be a practical device for insuring that gravid females ignore the calls of sexually immature males, whose frequency falls within the adult's middle frequency range of spectral sensitivities of fibers within the bula. When one examines the variety of auditory nerve unit responses to pure tones, it becomes clear that axons discharge maximally to only one of three frequency ranges: 200–300 Hz, 600–8800 Hz, or 1200–1500 Hz, i.e. just those frequency bands which provide behaviorally relevant excitation or inhibition. Capranica (1976, 1983) went on to test this generalization for other groups of frogs and toads. For example, in the green tree frog (*Hyla cinerea*), the male's call has energies at 900 Hz and 3100 Hz, but these correspond to middle- and high-frequency auditory nerve fibers, in contrast to the bullfrog. In some species, such as the cricket frog, the male's call has only a high-frequency peak. Elegant anatomical studies by Capranica & Moffat (1980) have further established that the low and midfrequency nerve responses are derived from a specialized peripheral organ, the amphibian papilla, while the high frequency response is derived from a separate organ, the basilar papilla. This fact indicates that evolution of specialized hearing mechanisms began with a unique design of peripheral transduction mechanisms, as well as differentiation of central neural mechanisms.

Capranica and others naturally suspected that "mating call detector" neurons in the bullfrog should have excitatory inputs from both high and low frequency afferent fibers, and probably inhibitory inputs from middle frequency fibers. However, careful examination of responses of auditory neurons within both medulla and midbrain (the torus semicircularis) were disappointing in that all neurons studies were still attuned to only a single frequency range. More recent studies by Mudry et al (1977) with the green tree frog seem to indicate that the next level of auditory processing (the central nucleus of the dorsal thalamus) may be the site of convergence of the dual auditory channels. She found that gross potentials evoked by pure tones were maximal when both high and middle frequency tones were presented simultaneously; the amplitude of the potential was then facilitated in a nonlinear manner, suggesting that many intrinsic soma discharged only in response to high-plus-low frequency inputs from the torus. An interesting by-product of this experiment was Mudry's finding that cooling the tree frogs shifted the maximal response of thalamus from tones of 900 Hz down to a frequency of only 500 Hz, without changing the optimal high frequency input sensitivity. Gerhardt & Mudry (1980) then confirmed with behavioral studies the prediction that gravid female tree frogs would preferentially approach speaker-generated calls (phonotaxis) in cold weather when the low-frequency component was reduced to 500 Hz. This experimental sequence provides an instructive example of the two-way extrapolation between physiology and behavior characteristic of successful neuroethology.

Another important chapter in the Anuran auditory story comes from the work of Narins, who studied the call of the tropical frog, *Eleutherodactylis coqui,* commonly called the Coqui. This dual-note call (Co + Qui) appears to have a dual function in communicating different signals to males and females simultaneously (Narins & Capranica 1978). Females respond mainly to high frequency "Qui" sound by orienting toward the sound source, while males respond to the lower-frequency "Co" sound by emitting a response call which heralds male competition. Remarkably, the auditory systems exhibit sexual dimorphism, such that the female auditory nerve contains a majority of "Qui"-sensitive units and the male auditory nerve contains a preponderance of "Co"-sensitive units (Narins & Capranica 1980). It is thus possible for males and females to hear quite different elements of the same social conversation.

In man, the localization of sound in the horizontal plane is accomplished mainly by attention to the interaural differences in the time-of-arrival of a particular sound at the two ears. A similar neural mechanism, sensitive to dichotic time differences, appears to operate in the frog. Feng & Capranica (1976a,b) have shown that single auditory neurons within the medulla may respond selectively to sounds arriving from either the left or right sides. Such directional discrimination is lost for all neurons after severing one auditory nerve. This mechanism is sufficient for determining the accuracy of horizontal (ground level) localization during phototaxis, but would not aid vertical sound localization. However, Gerhardt & Rheinlander (1982) with green frogs and Passmore et al (1984) with the painted reed frog (*Hyperolius marmoratus*) have demonstrated localization by female frogs of elevated speakers broadcasting the male's mating call. Passmore et al note that the virbrational phase of the upper eardrum is opposite that of the lower portion and that this asymmetry might lead to vertical directionality in sensitivity of the eardrum.

Case Study 2: Sound Localization in the Barn Owl

The owl, by contrast, has acute ability to localize prey by sound in both the vertical and horizontal dimensions: He must orient the head accurately toward a brief rustling sound in order to program a flight directly toward the small animal whose movements generated the salient sound. Knudsen & Konishi (1979) have shown that dichotic cues are used by the barn owl (*Tyto alba*) for sound localization: Directed attacks can be elicited toward a loudspeaker in a dark room, but accuracy of head-turning is lost after occlusion of a single ear. The auditory cues for discrimination of target elevation are also binaural, but depend upon intensity differences between the two sides instead of time-of-arrival differences. Since one ear is situated a bit higher than the other, the intensity of an elevated stimulus is slightly greater for the ear pointed upward. Knudsen & Konishi (1980) have used physiological techniques to describe how these two stimulus dimensions can be encoded within the owl's brain so as to collaborate in specifying the target direction.

These authors were able to locate a region within the owl's midbrain (the MLD nucleus) in which a two-dimensional representation of auditory space could be found (Konishi 1983). Within this map, a single neuron responded to a sound source at the appropriate asymmetry within one domain (e.g. intensity) only when input from the other domain (time difference) was also within the range of disiparities to which that cell was tuned. Thus, activation of the locus-specific neuron demanded the summation of appropriate inputs from each domain. A comparable problem has been defined by the recent report of Fuzessory & Pollak (1984) based on recording within the auditory system of the moustache bat (*Pteronotus parnellii rubiginosus*). Their external ear generates binaural spectral cues, which contain two-dimensional spatial information. Single neurons of the inferior colliculus have spatial receptive fields that reflect the interaural intensity disparities for a particular harmonic component of the bat's echo. This use of tonotopic organization for generation of a spatial map contrasts with the temporal disparity used by the owl.

Since recording in other species has revealed separate encoding of sound intensity or temporal disparities at lower brain levels, it appears that the midbrain spatial map of the owl is formed by a highly coincidental pattern of convergence of inputs from different peripheral coding systems. Such a map might be formed automatically if inputs from the peripheral systems are already topographically arranged (as is the visual system) before they arrive in the midbrain. Alternatively, the pattern of coincidences established via early experience could play a role in the formation of the map. Knudsen et al (1984) have recently demonstrated that experience does play such a role: For owlets reared with one ear plugged, the midbrain map forms so as to normalize the perceived targets within its spatial coordinates. Consequently, when the ear is later unplugged, the receptive field locations for a given region of the MLD are skewed in the expected direction. What began as a neuroethological venture has now led to an exciting problem for the developmental physiologist.

Case Study 3: Echolocation by Bats

Several laboratories in the US and in Germany have focused their efforts on analysis of the behavioral uses and physiological correlates of acoustic signals in different bat species. These animals are a functionally diverse group, with varying dependence upon acoustic or visual information and various modes of feeding, so that ethologists have a rich field for investigating adaptations of the brain to either unimodal or multimodal perception as well as to diverse sensorimotor tasks. The remarkable specialization for processing feedback from emitted acoustic signals enables many bats to perform spatial and pattern analyses typically achieved only by vision. The lessons to be learned from the study of bats are not only of ethological value but are of broad general interest for theory of information processing. A substantial summary of the acoustic physiology and behavior of bats appears in a recent volume edited by Ewert et al (1983a).

The bats' sonar signals are particularly stereotyped and accurately formed with the purpose of providing several classes of information once the reflected sound patterns (echos) are compared with the pattern, intensity, and timing of emitted sound. Bats using echos emit short sounds consisting of two distinct parts: a long constant-frequency (CF) component, followed by a short frequency-modulated (FM) component. The echo returns while the pulse is still being emitted. Simmons et al (1983a,b) argue that the remarkable accuracy of bats in estimating target distance (1 cm at a 3-m distance) or horizontal location ($\pm$ 1.5 deg) demands that they cross-correlate the phase of echo with the emitted signal. This can be done best using the wide-band component. Simmons infers from acuity experiments with trained bats that they must be able to determine the time-of-occurrence of an echo with an accuracy of about 0.5 μsec. He predicts that with such an acute sensitivity, dichotic integration should provide the measured acuity for directional hearing. Vertical localization (Lawrence 1981) is also surprisingly good (±3 deg) and this must reflect a discrimination among complex sounds or secondary echoes produced by the external ear's anatomy (see Fuzessory & Pollack 1984).

The constant-frequency component of the echo provides unique information for detection of relative motion between the flying bat and his target (the rate at which he is gaining on a moth, or approaching an obstacle). In fact, bats are sensitive to small changes in frequency of the echo (the so-called Doppler shift) due to changes in target distance (Simmons et al 1983b, Schnitzler & Henson 1980, Schnitzler & Ostwald 1983). The constant-frequency component provides still further information since the variations in the target distance due to wing motion of the flying prey will register in the bat's brain as "flutter" of the acoustical image. The wing frequency provides a useful cue as to the nature of the prey, even when sizes are equal (Schnitzler & Ostwald 1983). Variations in echo frequency from the constant-frequency component (Doppler shift information) is of such special significance to many bats that Pollak et al (1983) explored the tonotopic map in the inferior colliculus of the mustache bat (*Pteronotus parnelli rubiginosus*) with this issue in mind. In fact, one-half of the structure was devoted to representation of frequencies very close to the constant frequency of 60 kHz. This dramatic hypertrophy of this portion of the cochlear map is analogous to the magnification of the primate fovea representation upon the visual cortex.

The elegant physiological studies of Suga and co-workers indicate that these sources of auditory information are indeed encoded within single neurons within the hypertrophied auditory cortex of bats (reviewed in Suga 1982, Suga et al 1983). For example, in the moustached bat (*Pteronotus parnelli rubiginosus*) one area is attuned to shifts in the echo from the typical 62 kHz signal: the best frequencies of these neurons are between 61–63 kHz. A shift off the maximal locus of excitation along one axis of this cortical map is therefore a measure of the change in distance of the moving target, or impeding surface (Suga & O'Neill 1979). Such neurons seem useful for detection of fluttering insects as shown by Schnitzler & Ostwald (1983). In another cortical zone (the FM-FM area) an important stimulus parameter is pulse-echo delay (Suga et al 1979, O'Neill & Suga 1979). Along one axis, cells are arranged according to their optimal echo delay (from 0.4–18 msec). Some cells are also specialized for encoding echo amplitude and provide information for coding target angle. Obviously a combination of target distance and target angle is required to judge the real size of the target. Finally, an area termed "CF-CF" is described by Suga et al (1983) that contains neurons specialized to detect the magnitude of the Doppler shift, i.e. the velocity in the radial direction. These neurons do not encode small variations in echo delays. Suga notes that the precision of echolation is likely to be compromised by jamming signals from other bats in a large colony, but that individual variations in emitted constant-frequency values appear to minimize this risk. He provides some evidence that cortical regions of individual bats have "best frequency" ranges, which vary in relation to that individual's typical emitted frequency, a fact consistent with the report of Pollak et al (1983) for cells of the bat's inferior colliculus. Is it possible that young bats tune their own brains to this selected constant-frequency range during a plastic period during development?

Given the expansion of our knowledge of bat behavior and associated cortical mechanisms, it behooves comparative anatomists to document the pattern of connections within the auditory system and between auditory and "association" areas. An interesting first step has been made by Kobler et al (1983), who placed HRP within the so-called frontal association cortex of bats (the projection target of the thalamic mediodorsal nucleus) and discovered a strong auditory-frontal projection. In addition, there is a unique projection to frontal cortex from a special division of the thalamic medial geniculate nucleus (Zook & Casseday 1982). This information should delight neuropsychologists who suspect that the frontal cortex plays a role in representation of spatial surroundings and should stimulate ethologists to examine the spatial abilities of bats in relation to their familiarity with the testing environment.

Case Study 4: Role of the Anuran Optic Tectum in Prey-catching

Functional studies of the vertebrate tectum began with the experiments of Hess et al (1946) using electrical point stimulation of various brain sites in freely behaving cats. They found that the focal stimulation of the anterior tectum elicited head and eye orientation toward the rostral lateral visual field, while activation of the caudal tectum produced large-amplitude turns as if to fixate imagined objects in the caudal field. Hess called this stereotyped sequence the "visual grasp reflex" and suggested that its natural function was to explore novel or biologically significant events by aiming the fovea of the eye at the target. His student, Akert (1949), demonstrated a similar visuomotor mapping via electrical activation of the trout's optic tectum: Stimulation of the lateral tectum produced conjugate eye movements and body flexion toward the opposite visual field, but stimulation of the rostral tectum produced ocular convergence and a snapping movement, as if toward prey. This version of the visual-grasp reflex could be taken literally! Studies with a variety of animals confirm the universal existence of a tectal efferent "motor map" that is in register with the retinotectal projection map (Ingle & Sprague 1975).

Lesion studies provide complementary evidence that the tectum subserves orientation toward important stimuli. In fishes and frogs, tectum ablation eliminates all overt responses toward moving food stimuli (Ingle 1973, 1977, Springer et al 1977) as well as all tendencies to avoid looming "threat" stimuli. A compelling demonstration that the frog's optic tectum is sufficient to direct both kinds of orienting behavior may be found in frogs in which one optic tract has regenerated to the ipsilateral tectum some months following a unilateral tectum ablation. Using the eye with "wrong-way" projections to the tectum, Ingle (1973) found that prey-strikes were directed away from the prey to mirror-symmetrical locations, whereas looming black disks elicited jumps toward the threat. The selective use of rewiring in fishes and frogs (where good regeneration is the rule) may be used in the future to produce new kinds of "rewiring" of sensorimotor connections that will allow more definitive tests of hypotheses than can be made by lesions alone.

The description of frog retinal ganglion cell terminals as recorded in the optic tectum was a landmark study in neurophysiology (Lettvin et al 1959, 1961). For the first time, single unit responses were classified using approximations of "natural" stimuli, rather than by turning on or off various patterns of light. Lettvin discovered that the most superficial axon terminals in the frog's tectum (classes 1 and 2) responded to moving objects below a certain angular size (optimally about 4–5 deg) but not to the turning of lights on or off. "On/off" axons (class 3) were recorded at the next deeper level and sustained "off" units (class 4) at the deepest retinal fiber lamina. Lettvin was of course aware that small moving objects are optimal stimuli for triggering feeding behavior of frogs and that large briskly moving objects (such as would excite the class-3 and class-4 fibers) tend to elicit avoidance jumps by alert frogs. Lettvin regarded these separate retinal channels as sending to the brain "useful parcels of biological information," but he did not imagine (as some critics have suggested) that the IRM was simply determined within the retina. He was in fact the first to point out the dynamic and complex properties of many intrinsic neurons within the frog's tectum (Lettvin et al 1961).

The next step in decoding the retinotectal system was to correlate the kinds of tectal cell response with the selectivity shown by the whole animal. Ewert (1968) performed initial studies on toads (*Bufo bufo*) and Ingle (1968) studied frogs (*Rana pipiens*) using floor-level moving dummy stimuli at fixed distances from the stationary subject. Both frogs and toads tended to prefer objects in the 4–8 deg size range when moved at a distance of about 8 cm. However, Ingle found that the optimal angular size of the dummy prey objects changed appreciably with distance, approximating "size distance constancy," i.e. that preferences were for the same "real size." This rule was more rigorously established by size-constancy experiments with toads (Ewert & Gebauer 1973) and with frogs (Ingle & Cook 1977). It is important to realize that the tendency of these animals to snap preferentially at very large angle (20 deg) stimuli when very close cannot be explained by a selective response of tectal cells to superficial "small spot detector" units. As viewing distance is reduced, there must be a switch in either the response of a single type of neuron from class-1+2 retinal inputs to dominance by class-3 inputs, or a switch in activity between two classes of size-specific tectal neurons.

A remarkable observation of Ewert (1968, 1970; Ewert et al 1983b) led to a new line of experimentation which may have resolved these questions. Toads with knife-cuts that disconnected the tectum from the more medial pretectal cell groups produced animals that would turn and strike at any moving object, irrespective of size and without showing the typical course of habituation during repeated stimulation. The toad without a pretectum became an automaton in regard to his feeding behavior! Ewert & von Weitersheim (1974) then found that a major class of tectal cells that normally respond best to small worm-like moving objects could not be found in the disinhibited toad's tectum. In order to decide whether this class of "prey sensitive" neurons lose their size-selectivity after pretectal damage, Ewert (personal communication) injected a toxin (kainic acid) into the pretectum while recording size-sensitive tectal cells, and found that soon after the injection the cells began to respond to any large moving object. Ingle (1983a) used these facts to construct a neural model of size-constancy in which it is postulated that changes in the eye's accommodation state that relate to depth of the prey (Collett 1977) are accompanied by changes in pretectal inhibition of the tectum, allowing single tectal cells to alter their size preferences among moving objects. Extensive projections from pretectum to tectum were first demonstrated in frogs (before their demonstration in mammals) by Trachtenberg & Ingle (1974) as a follow-up of Ewert's neurobehavioral discoveries.

Recently, some related facts concerning the morphology and destination of tectal efferent neurons have added details to the model of prey-selection. Ingle (1983b) confirmed earlier observations that descending tectal efferent fibers terminate in both ipsilateral and contralateral brainstem. A surprising behavioral result emerged after severing the second pathway as it crosses the ventral tegmental midline: Split-tegmentum frogs retained the ability to avoid threat and to snap in response to nearby prey, but totally lost the ability to turn toward the prey. Thus the taxic and consummatory components of prey-catching originate from different tectal output cells. This dual-response control model is further supported by the effects of severing the ipsilateral tectal efferents at the level of the isthmus (or by cutting the lateral branch of this efferent bundle in the anterior medulla). Frogs without this group of tectal efferents are unable to initiate snapping at nearby prey in the opposite lateral visual field (they remain passive), but they usually do turn to fixate more distant objects along the same line of sight. Thus the fixation response alone (when not integrated with a snap) can be programmed via the contralateral efferent projection.

Studies with retrograde filling of tectal cells from implants in the medulla reveal a striking difference between cells projecting to opposite sides of the brainstem. Those emitting ipsilateral axons have dendrites that terminate at the class-3 level (and sometimes additionally above or below), whereas cells with ipsilateral descending axons have visible dendritic ramifications only at the class-2 level. If this inference from the light microscope is confirmed by electron microscopy, the data suggest that turns toward distant prey are triggered via class-2 fibers alone, while the initiation of snapping can be achieved via inputs from either class-3 or class-2 fibers. When stimuli are close, it is suggested (Ingle 1983a) that pretectal inhibition is low and class-3 fibers are sufficient to elicit snapping. Recent studies (D. Ingle, unpublished observations) make this plausible by showing that surgical deletion of class-2 fibers from a large area of dorsal tectum still allows the frog to snap at overhead stimuli. Whether such frogs could still orient to objects in the lateral field in the absence of class-2 fibers is a critical question, as yet unanswered. Further details of feeding behavior of frogs with either class-2 or class-3 fibers deleted may help to understand why some fiber-cell connections are highly specific filters and why other cells integrate multiple fiber types.

Frogs with a complete hemisection of the isthmus attempt to avoid looming black disks by jumping forward, but lack the taxic component that normally directs the frog away from the approaching threat. This deficit also results from smaller lesions that interrupt the medial portion of the ipsilateral projection of the medulla. In fact, the medial ipsilateral efferent fibers abut the contralateral tectal efferent route, suggesting that both avenues converge in a posterior medulla "orienting center" that turns the frog toward the ipsilateral field (toward one eye's field and away from the other's). In fact, the body-turn sequence appears identical for prey-catching and avoidance movements. The idea of a general-purpose motor system for orientation agrees with the conclusions of Comer & Grobstein (1977) that prey-turns or snaps elicited by light touch of the limbs are not dependent upon the optic tectum, but upon a tactile-sensitive region of tegmentum just ventral to the tectum. The ability of frogs to jump around barriers is also programmed independently of the tectum (Ingle 1977, 1980, 1983b); this implies still another independent avenue from the sensorium to the final common pathway for body orientation.

There is good evidence that the mammalian tectum works according to the same general plan inferred from studies of fish and amphibians. Ablation of the tectum abolishes orienting to small moving targets (Schneider 1969, Casagrande & Diamond 1974, Goodale & Milner 1982) and eliminates avoidance of sudden overhead threat (Merker 1980). It is interesting that Merker achieved a high level of threat avoidance in hamsters by keeping them in a "wild" state: They lived in tunnels and foraged for seeds in an open place with no view of the experimenter. Without such precautions many laboratory animals readily habituate to the variety of events around them and will not show the full response to "predator-like" stimuli. One may wonder about the naturalness of hamsters or gerbils orienting to moving seeds at the end of a wire. However, while seeds are not normally so animated, insects are: Both hamsters (Finlay et al 1980) and gerbils (D. Ingle, unpublished data) readily chase and eat crickets moving within a large arena. From this observation the question arose as to whether gerbils (unlike frogs) are able to anticipate the trajectory of a fast-moving target and orient ahead of the stimulus in order to intercept it. Ingle et al (1979) documented this anticipatory turning ability via film analysis. Later studies (Ingle & Shook 1984) showed that this ability was dependent upon an intact striate cortex. After an occipital ablation, the destriate gerbil, like the frog, oriented only on the basis of immediate position, and did not take motion direction into account. The point of this comparative exercise is that the neuroethologist may profitably compare species rather distant from one another in phylogeny. Provided that some basic features of their behavior are quite similar, the complexities acquired by one group (mammals) may be accounted for by the evolution of new pieces of brain (striate cortex) rather than simply by remodeling the old mechanisms (tectum).

NEUROETHOLOGICAL ANALYSIS OF HORMONE-DEPENDENT BEHAVIORS

Behavioral biologists have long noted that the behaviors associated with reproduction tend to be characteristic of each species and each sex. These behaviors are crucial to both reproductive function at a physiological level as well as at a social level. A large body of research has shown how successful reproduction depends on the precise coordination and synchronization of myriad factors: The individual's reproductive capacity is influenced by the physical, biotic, and social cues in its immediate environment as well as by its developmental and evolutionary history. Further, it is useful to think not only of the neural and endocrine processes occurring within the individual but also of the individual's interaction with other individuals. Species-typical behaviors often are embedded in a chain of complementary stimulus-response interactions (Beach 1978), each stage of which depends upon the preceding events and, at the same time, sets the stage for those to follow. The context of a situation also can play as important a role as hormones in influencing perception (Beach 1983, Goldfoot & Neff 1984). Given that the individual's behavior both influences, and is influenced by, its physiological state as well by the context in which it finds itself, is it feasible to take a holistic approach in which the environment, behavior, and physiology are considered simultaneously (Crews 1980, 1983b)? The following case studies indicate that this question can be answered positively provided that the animals and the behaviors chosen for study take into account that the causal mechanisms and the functional outcomes of behavior were naturally selected to be coordinated under specific sets of social and physiological conditions.

Case Study 5: Neuroendocrine Control of Rat Lordosis Behavior

Perhaps the most thoroughly analyzed species-typical behavioral response is the lordosis behavior of the female laboratory rat. The sequence of events underlying this hormone-dependent behavioral response, including their neurophysiological correlates as well as the manner in which behavior interacts with the female's hormonal milieu, has been the subject of intense investigation (see reviews by Komisaruk 1978, Pfaff 1980). Experiments have demonstrated that if a female rat is in the appropriate hormonal state, whether in natural estrus or primed with exogenous estrogen and progesterone, the courtship and copulatory behavior of the male will elicit certain postural changes by the female, ultimately resulting in the female elevating her rump by a dorsiflexion of her spine. This lordosis behavior can be viewed profitably in terms of a reflex arc.

Stimulation of specific areas of the female's body, particularly the flanks, tail base, and perineum, are crucial for the occurrence of lordosis. Sensory input from the female's internal genital tract generated by the male's intromissions and ejaculation are also important. These stimuli together activate lordosis behavior via separate afferent neural pathways. The sensory afferents supplying the perineal region travel in the pudendal nerve and enter the spinal cord. Transection of different regions of the spinal cord demonstrates that the anterolateral columns are the critical ascending neural pathway, projecting into the central gray of the midbrain. At the level of the spinal gray many units respond to cutaneous stimulation such as would be provided by the male's grasping action, including responses to both hair movement and skin deformation. Estrogen has profound effects on this peripheral afferent system, as well as within the brain. The sensitivity of the peripheral mechanoreceptors and the receptive field size of the pudendal nerve increase with estrogen treatment in ovariectomized females.

Estrogen is taken up and bound in specific regions of the brain, including the medial preoptic area (POA), the medial anterior and basal hypothalamus, the limbic system, and the mesencephalon. The hypothalamus is a major integrative area, and it is here that we see the pivotal role played by estrogen in modulating the neural mechanisms subserving lordosis behavior. For example, estrogen increases biosynthetic and secretory activity of cells in the ventromedial hypothalamus and increases electrical excitability of medial and basal hypothalamic neurons. Estrogen also regulates the release of luteinizing hormone-releasing hormone (LHRH). However, whereas both estradiol-concentrating neurons and the hypothalamic neurons producing LHRH overlap in their distribution in the brain, recent studies indicate that the genomic effects of estrogen mediated by nuclear estradiol concentration do not include a direct effect on LHRH-containing neurons (Shivers et al 1983). This surprising finding suggests that the genomic regulatory effects of estrogen must be mediated by another, as yet unidentified, class of neurons.

Lesions in the medial anterior hypothalamus and ventromedial nucleus (VMN) of the hypothalamus result in severe deficits in lordosis behavior. This effect is transitory, however, thus indicating that these cells are not part of a direct reflex arc but influence the descending brainstem pathways that facilitate lordosis. In complementary studies, electrical stimulation of the VMN facilitates lordosis in estrogen-primed ovariectomized female rats. Similarly, sexual receptivity can be induced by implantation of minute amounts of steroid hormones into the medial anterior hypothalamus and VMN (Barfield et al 1983). Interestingly, systemic estrogen treatment decreases electrical activity of POA neurons, yet mating behavior is activated by implants of estradiol into the medial preoptic area, a major site of estrogen-concentrating cells. Direct application of anisomycin, a protein synthesis inhibitor, into the VMN inhibits lordosis, yet if placed in the preoptic area, lordosis is facilitated (Meisel & Pfaff 1984). The finding that behavioral responsiveness to estrogen is increased following lesions of the preoptic area but is inhibited when the preoptic area is stimulated also suggests that estrogen acts at the level of the preoptic area by inhibiting neural activity. Thus, these opposing actions of the VMN and the preoptic area may serve to modulate female sexual receptivity.

The neuronal projections from the VMN and the preoptic area can be traced to the mesencephalic central gray, with the axons descending through the hypothalamus and medial forebrain bundle in layers. Electrical stimulation of this area facilitates lordosis behavior, whereas lesions abolish it. Microinjection of fluorescent dyes reveal descending fibers in the central gray that project to the reticular formation of the lower brainstem (Morrell & Pfaff 1982, Morrell et al 1984). The hormone-dependent hypothalamic output signals that are transformed and relayed in the central gray are carried in the lateral vestibulospinal and reticulospinal tracts, terminating in the spinal cord. Lesions of these descending pathways disrupt lordosis behavior, whereas stimulation of the lateral vestibular nucleus, the source of the lateral vestibulospinal tract, facilitates it. The final common pathway, the motoneurons controlling the muscles responsible for elevating the rump, the most crucial component in the lordosis behavior, are located in the medial and ventral borders of the ventral horn. The deep-back muscles used during lordosis consist of both fast-twitch and slow-twitch fibers, and recent histochemical and EMG studies indicate the elicitation of the lordotic posture involves a "ballistic" response upon adequate sensory input (Schwartz-Giblin et al 1983).

The behavior of the copulating male not only activates the neuroendocrine circuitry underlying the lordotic response, but also sets the stage for more long-lasting physiological changes. The vaginocervical stimulation a female receives during mating has profound effects on her reproductive physiology and subsequent receptive behavior. The number of intromissions a female receives prior to the ejaculation of sperm into the vagina has been shown to be crucial to the induction of pregnancy in the species (see reviews by Adler 1978, 1983). In many rodent species, only females that have received the species-typical series of mounts and intromissions become pregnant (see review by Dewsbury 1978).

The mechanism by which the multiple intromissions and sperm plug trigger pregnancy has been detailed. Using the 2-deoxyglucose (2-DG) method, Allen et al (1981) have shown that vaginocervical stimulation increases neural activity in the medial preoptic area and reticular formation. This is the beginning of a neuroendocrine reflex that culminates in the release of pituitary hormones (LH and prolactin) crucial to establishing the progestational state necessary for a successful pregnancy. Further, the vaginal plug that is deposited during mating is important for the transport of sperm from the female rat's vagina into the uterus. Indeed, the pre-ejaculatory intromissions act in concert with the deposition of the vaginal plug to ensure the transport of sperm. Females receiving the appropriate number of intromissions necessary to prime the neuroendocrine system but receiving sperm only (and no plug) will not become pregnant. If the plug is dislodged shortly after its deposition, sperm transport is disrupted. Thus, the post-ejaculatory interval that characterizes the mating sequence in the rat appears to serve to permit the female adequate time without copulatory stimulation to facilitate sperm transport from the vagina to the uterus. The reader should keep in mind that equally detailed information is available on the neural and endocrine mechanisms and their functional outcomes for the male rat (see reviews by Sachs & Barfield 1976, Hart 1978). Furthermore, the reader should realize that the female rat's mating sequence involves active solicitation of the male (McClintock 1983) as well as her reflexive lordosis response. This goal-directed behavior has thus far not received due attention of the neuroethologist or behavioral endocrinologist.

Case Study 6: Neuroendocrine Events Underlying Amphibian Sexual Behavior

Another outstanding example of neuroethological research concerns the neuroendocrine mechanisms underlying the sexual behavior of *Anuran* amphibians. Many frogs during the breeding season emit three distinct call-types: mate calls, territorial calls, and release calls. Each call-type is performed by specific individuals in a specific context. For example, territorial males produce one call-type to attract mates and another call-type to establish territorial boundaries. The release call is usually the only call-type produced by unreceptive females and is elicited when they are clasped by a breeding male. These calls are characteristic to the species and can serve as premating reproductive isolating mechanisms in areas inhabited by closely related species.

The hormonal control of reproductive behavior in *Ranid* frogs and salamanders has until recently confounded investigators. Although removal of either the testes or the pituitary abolishes courtship behavior in these groups, systemic replacement with testicular steroids will not maintain or restore sexual behavior (see review by Crews & Silver 1984). The only known exceptions have been the South African clawed frog, *Xenopus laevis* (Kelley & Pfaff 1978, Wetzel & Kelley 1983), and the rough-skinned newt, *Taricha granulosa* (Moore 1978). It is surprising that male sex hormones are thus ineffective since the preoptic areas of both *Rana* and *Xenopus* are known to concentrate them (reviewed in Kelley & Pfaff 1978), and in *R. pipiens* both direct application of testosterone into the preoptic area or electrical stimulation of this region will induce mating behavior (reviewed in Crews & Silver 1984). The paradox may be clarified by the demonstration of Moore and his colleagues (reviewed in Moore 1983) that the neurohypophseal polypeptide, arginine-8 vasotocin (AVT), synergizes with androgen to induce sexual behavior in the rough-skinned newt, *T. granulosa,* by acting directly on cells in the brain (Moore & Miller 1983) as well as increasing the influx of testosterone into the brain by changing the permeability of the blood-brain barrier (Moore et al 1981).

Kelley and her colleagues have made considerable progress in establishing the neural and endocrine bases of two male reproductive behaviors (calling and clasping) in *X. laevis* (Kelley 1980). Male mate-calling in this species consists of a series of rapidly emitted clicks, with two click rates (fast and slow) predominating (Wetzel & Kelley 1983). The clicks often occur in trains (trills), with alternating fast and slow trills constituting the mate call. The clicking sound is produced by movements of the arytenoid discs at the anterior end of the larynx and is effected by the paired bipennate muscles that are innervated by neurons of the cranial nerve nucleus IX–X (Kelley 1980).

The neural circuit for calling has been traced using the HRP technique (Wetzel & Kelley 1981). These studies have revealed that the primary source of input to the laryngeal motoneurons is from the dorsal tegmental area of the medulla (DTAM) (homologous to Schmidt's pre-V nucleus, see below), which, in turn, receives diencephalic input from the thalamus, striatum, and preoptic area. Significantly, all levels of the neural circuit for mate calling contain steroid-concentrating cells, including the larynx and the motoneurons that innervate it (Kelley 1980, Kelley et al 1981, Wetzel & Kelley 1983). Hormonal modulation of auditory sensitivity is also apparent in the gravid female *Xenopus* that responds to mate calls. Studies using the 2-DG technique (Kelley 1980, Paton et al 1982) indicate that the superior olive, lateral lemniscus, torus semicircularis, and posterocentral thalamus are all functionally active during acoustic stimulation. The laminar nucleus of the torus contains estradiol and dihydrotestosterone-concentrating cells and may play a role in localization of a calling male by conspecifics (Morrell et al 1975, Kelley 1980, 1982).

The neural mechanisms subserving mate calling behavior in *Ranid* frogs have also received attention. Patterned calling such as the species-typical mate calls emitted by breeding male frogs can be elicited by stimulation of the preoptic area and the ventral infundibular nucleus (VIN) (Schmidt 1966, 1968). Bilateral lesions of the preoptic area and VIN abolish mate calling. Electrical stimulation of a small medullary nucleus, the pre-trigeminal nucleus (n. pre-V; homologous to DTAM in *Xenopus*), and nucleus IX–X will reliably elicit vocalizations in response to tapes of conspecific calls that are similar acoustically to spontaneous mate calls of breeding males. The demonstration that the preoptic area is a major sex-hormone-concentrating site in the amphibian brain gives support to Schmidt's earlier suggestion that the preoptic area acts as an "androgen-sensitive activator" of a more posterior mate calling center including the nucleus pre-V.

Schmidt (1971, 1973, 1974) has proposed a model of the neural mechanisms involved in *Anuran* mate calling. This model postulates the existence of a motor coordination center ("efferent vocal center") in the region of the hypoglossal and vagus nuclei that is responsible for the generation of patterned calling. The pretrigeminal nucleus is a sensory correlation center ("afferent vocal center") that receives and integrates sensory input and transmits it to the more posterior motor coordination site. It is this latter center that the androgen-concentrating neurons in the preoptic area activate to trigger mate calling during the breeding season.

If a sexually active male comes in contact with another frog, he will immediately clasp it with his forelimbs. Kelley has also examined the neural and endocrine bases of clasping behavior in *Xenopus*. It appears that clasping is primarily a spinal reflex; the two most important muscles are the flexor carpi radialis (FCR) and the sterno-radialis (SR) of the arm. The motoneurons that innervate the FCR and SR are found throughout the brachial plexus, with SR neurons tending to be more medial than FCR neurons (Erulkar et al 1981). Clasping behavior is under sex hormone control, and autoradiographic studies have indicated that the distributions of hormone-concentrating neurons differ depending upon the hormone (Kelley & Pfaff 1978). Motoneurons within the brachial enlargement of the spinal cord are labeled only after testosterone or dihydrotestosterone administration. The metabolic capabilities of the spinal cord have also been examined and the brachial enlargement found to have substantial 5 α-reductase activity (Erulkar et al 1981). Finally, recent studies indicate that administration of androgens alters the electrophysiological activity of the neurons involved in clasping (Erulkar et al 1981).

The release call is another species-typical vocal behavior in amphibians that has received the attention of neuroethologists. Receptivity in many frogs is characterized by the female's silence. If a nongravid female or a male is clasped in the pectoral region, it will emit a release call in an effort to disengage the male. Gravid females do not emit the release call until after they have oviposited. The release call is produced by air being passed through apposed vocal cords as the glottis is rhythmically opened and closed (see review by Diakow & Raimondi 1981). This and the accompanying contractions of the trunk muscles stimulate the male to release his clasp.

Noble & Aronson (1942) reported 40 years ago that complete removal of the forebrain, diencephalon, optic tectum, cerebellum, and anterior tegmentum did not prevent the elicitation of the release call in male *R. pipiens*. Localized lesions in the torus semicircularis (homologous to the mammalian inferior colliculus) only temporarily abolish the behavior, but recovery does not occur if the trigeminoisthmic tegmentum is destroyed. Interestingly, electrical stimulation of this area, in particular the pre-trigeminal nucleus (n. pre-V), will evoke normal release calls (Schmidt 1974).

Diakow and her colleagues have examined the stimuli-modulating the release call in *R. pipiens*. Because denervation of the skin of the trunk abolishes the call in response to manual stimulation, Diakow suggested that the trunk stimulation received during clasping is responsible for the initiation of the release call in nongravid females. Observing that distension of the abdomen by intraperitoneal injection of physiological saline caused unreceptive females to allow males to mount and clasp, Noble & Aronson (1942) hypothesized that intra-abdominal pressure from the ovulation egg mass inhibits the release call in gravid females. This was elegantly confirmed by Diakow (1977), who showed that artificial distension of both males and females reduced the frequency of release calls in response to manual stimulation or after clasping by sexually active males. Distended females are also observed to remain in amplexus for longer periods than control females. This increase in girth, and its accompanying increase in intra-abdominal pressure and distension, is effective in silencing females whether the increase is accompanied by fluid accumulation or not (Diakow 1978). Although Diakow's observations are in accord with Noble & Aronson's hypothesis, she points out (Diakow & Nemiroff 1981) that in order to test this hypothesis adequately, it is necessary to find and to quantify a normally occurring increase in intra-abdominal pressure associated with inhibition of the call.

The ecological relevance of this crucial mechanism in amphibian reproduction becomes clear when the dependence on water is considered. To this end the discovery (Diakow 1978, Diakow & Raimondi 1981) that administration of AVT inhibits the release call was the first evidence that AVT plays a pivotal role in potentiating amphibian reproductive behavior. It is possible that AVT inhibits the release call by stimulating water accumulation or consequent pressure within the body. Both prostaglandin E_2 and prostaglandin $F_2\alpha$ silence gonadectomized females. This and the observation that the action of AVT on receptivity can be retarded by indomethacin, a prostaglandin synthesis inhibitor, suggests that the effect of AVT is mediated by prostaglandin (Diakow & Nemiroff 1981). Although inhibition of the release call normally occurs around the time of ovulation, the relationship between AVT function and gonadal hormone activity in females is not known. Arginine vasotocin will inhibit the call in ovariectomized females, whereas administration of estrogen or progesterone without AVT fails to inhibit the release call (Diakow et al 1978, Diakow & Raimondi 1981). There is a sexual dimorphism in responsiveness to AVT; females are more responsive than males to the hormone (Raimondi & Diakow 1981).

Case Study 7: Neuroendocrine Control of Reptilian Sexual Behavior

A great deal is known about the interaction of hormones, behavior, and the environment in the control of reproductive behavioral displays in the green anole lizard, *Anolis carolinensis* (see reviews by Crews 1980, Crews & Greenberg 1981). The significant feature of these investigations is that by studying these lizards in nature and in laboratory environments that simulate natural conditions, it has been possible to reveal the ecological function of the neuroendocrine mechanisms that underlie behavior. Research on the green anole has not only identified the environmental cues important in regulating the timing of the annual reproductive cycle, but it has also been possible to trace the effects of these biologically relevant stimuli from their perception and integration by the brain to the alterations they cause in physiological state and to the resulting changes in the organism's interaction with its environment.

Temperature is the major environmental cue initiating the transition from reproductive inactivity to breeding in this lizard. Rising temperatures in the spring induce the animal's emergence from winter dormancy. When the testes mature in the male and androgen concentrations in the systemic circulation are high, breeding territories are established. Male agonistic displays (assertion and challenge) vary between individuals yet are characteristic of the species. Following the establishment of a breeding territory, the male begins to court females using another species-typical display (courtship).

In both the male and the female green anole there is an association between the peak phase in gonadal activity (gamete maturation and maximal sex steroid hormone secretion) and mating behavior. As in many animals, especially laboratory and domesticated mammals and birds exhibiting this associated pattern of reproductive activity, courtship behavior in the male green anole depends on the functional integrity of the hypothalamic-pituitary-gonadal axis.

The hypothalamus and preoptic area are major integrative areas for the internal and external stimuli that regulate reproduction. Removal of the gonads abolishes sexual behavior, whereas sex hormone replacement therapy reinstates the complete repertoire of sexual displays in both males and females.

The distribution of sex-steroid-hormone-concentrating neurons in the green anole resembles that found in other vertebrates (Martinez-Vargas et al 1978, Morrell et al 1979). In addition to being selectively bound in the accessory sex structures, testosterone, dihydrotestosterone, and estradiol are concentrated in the anterior hypothalamus–preoptic area (AH-POA). Both areas are involved in the regulation of challenge and courtship displays in the male as well as in the hormone feedback regulation of pituitary gonadotropin secretion (see review by Crews 1979). Destruction of the AH-POA results in an immediate and rapid decline in both sexual and aggressive androgen-treated male green anoles. Lesions rostral to this area also reduce reproductive behavior in intact males, but apparently this effect is indirect through destruction of cell bodies producing gonadotropin-releasing hormone. Sexual behavior can be restored in castrated lizards by implantation of testosterone into the AH-POA. Intracranial implantation of estrogen or dihydrotestosterone also stimulates reproductive behavior in castrated lizards. Interestingly, the sequence in which reproductive behaviors are reinstated after testosterone implantation—aggressive behavior followed by courtship behavior—is the same as that shown by reproductively inactive males receiving environmental stimulation and by castrated males given systemic androgen treatment.

As in mammals, the basal hypothalamus is a major site of steroid hormone feedback regulation of pituitary gonadotropin release. When a lesion is made in the anterior basal hypothalamus, the testes collapse and atrophy. Lesions of the posterior basal hypothalamus have no detectable influence on testicular activity of intact males. The basal hypothalamus also appears to be a major behavioral integrative area as lesions in this area result in a rapid and permanent decline in aggressive and courtship behaviors.

Forebrain structures mediate the display of both aggressive and sexual displays in the green anole lizard. By taking advantage of the almost complete decussation of the optic tracts, the function of different nuclear groups can be examined by making unilateral lesions and then testing the animal when the eye projecting to the damaged (versus the undamaged) hemisphere is covered. Using this technique, Greenberg and his colleagues found that challenge displays, but not assertion displays, were abolished in lizards sustaining lesions in the paleostriatum and the major descending pathway from the paleostriatum, the lateral forebrain bundle (Greenberg 1978, Greenberg et al 1979). The posterior division of the dorsal ventricular ridge (PDVR) in reptiles is homologous as a field to the mammalian amygdala (Northcutt 1981). Lesions in the ventromedial nucleus of the PDVR do not affect the exhibition of challenge or assertion displays but courtship displays were impaired or abolished (Greenberg et al 1984).

There is some evidence that midbrain structures play an important role in the control of the courtship display. A prominent component of this display in *Anolis* lizards is the extension of the hyoid process, thereby exposing a patch of brightly colored skin called the dewlap. Androgens are concentrated in areas of the torus semicircularis that are homologous to those regions of the nucleus intercollicularis that exhibit hormone uptake in songbirds (see chapter by Konishi, this volume). In several bird species, stimulation of this area elicits species-typical vocal displays, whereas lesions abolish vocalizations (Konishi & Gurney 1982). Stimulation of areas immediately peripheral to the central nucleus of the torus semicircularis elicits vocalizations in the Tokay gekko (*Gekko gekko*) and gular extension of the green iguana (*Iguana iguana*) and the western collared lizard (*Crotaphytus collaris*), behaviors that are prominent components in the reproductive displays of these species. It is possible that these two sexual signals, one auditory (singing in birds) and the other visual (gular extension in lizards), are subserved by homologous hormone-dependent neural mechanisms.

An interest in the autonomic mechanisms that underlie behavior that may evolve or be "ritualized" into display behavior has kindled an examination of the color change phenomenon shown in many lizards (Crews & Greenberg 1981, Greenberg & Crews 1983). In the green anole, body color changes rapidly during agonistic interactions and also as a result of long-term dominant-subordinate relationships between males. The pattern of color change and the hormones detectable in the blood of such animals indicates a profound involvement of acute and chronic stress endocrinology in their social organization (Greenberg et al 1984). Because the dermal chromatophores of green anoles are devoid of direct innervation, this species may be utilized as a model for stress-related neuroendocrine responses to various forms of social stimulation.

A similar reciprocal interaction between the internal and external milieu exists in the female green anole. Warm temperatures are necessary, but not sufficient, for stimulating ovarian activity. Ovarian growth is initiated by pituitary gonadotropin secretion, but the rate of ovarian growth is facilitated by the courtship behavior of the male. In *Anolis* lizards, as in higher primates, the ovaries alternate in the production of a single ovum (Jones 1978). The nature and pattern of sex steroid hormone secretion during successive follicular cycles of the breeding season has been examined (Jones et al 1983). In the first follicular cycle of the season, estrogen concentrations increase as the largest ovarian follicle matures, falling after ovulation; at the time of ovulation progesterone levels rise. In subsequent cycles during the breeding season, estrogen concentrations remain low although there continues to be a periovulatory surge in progesterone.

These regular follicular cycles are precisely correlated with periods of sexual receptivity, which are characterized by the female performing a neckbend in response to the courting male. This behavior can be induced in ovariectomized females by the administration of appropriate dosages of estrogen and progesterone; the dosage of estrogen, the interval between estrogen and progesterone administration, and the time of year are critical in the control of female sexual receptivity (Wu et al 1985).

Estrogen, like testosterone and dihydrotestosterone, is concentrated in specific areas of the brain of the female green anole. Although receptors for these three sex steroid hormones exist in many of the same neural structures in both the male and the female, some areas are sensitive only to estrogens or to androgens. Although we know that steroid receptors in the female green anole have characteristics similar to those described in mammals (Tokarz et al 1981), the site(s) of estrogen action have yet to be determined.

Females of many species show an abrupt decline in sexual receptivity following mating and the green anole is no exception. This rapid change in the female green anole's behavior is due to the intromission by the male. Mating appears to initiate a neuroendocrine reflex involving prostaglandins; administration of prostaglandin $F_{2\alpha}$ quickly inhibits sexual receptivity in ovariectomized, estrogen-treated females (Tokarz & Crews 1981). Although transitory, the nonreceptive behavior of such females is indistinguishable from that shown after mating by intact, preovulatory females or by ovariectomized females. While it has been established that prostaglandin exerts its action at the level of the brain, the specific area(s) have not been ascertained.

Because representative species of every vertebrate class have been found to exhibit sex hormone mediation of mating behavior, this has been taken as evidence for a phylogenetic conservatism of neuroendocrine mechanisms controlling sexual behavior in vertebrates. However, with a few exceptions (see below), all of the species studied in this regard have shown an *associated* reproductive tactic in which gonadal activity is maximal at the time of mating. Many vertebrates, however, exhibit a different reproductive tactic in which mating occurs at a time when the gonads are not producing gametes and circulating levels of sex hormones are low; this annual reproductive pattern has been termed a *dissociated* reproductive tactic (Crews 1984a). Of considerable interest, then, are the neuroendocrine mechanisms controlling sexual behavior in animals that mate when hormones are at their lowest level.

Although only two species exhibiting a dissociated reproductive tactic have been studied experimentally, there is reason to believe the neuroendocrine mechanisms controlling sexual behavior in such species differ in important ways from those of species exhibiting an associated reproductive tactic. For example, in the Asian musk shrew (*Suncus murinus*), sexual receptivity in the female is independent of the presence of the ovaries (Dryden & Anderson

1977). More information is available on the neuroendocrine mechanisms controlling sexual behavior in the red-sided garter snake. *Thamnophis sirtalis parietalis*.

The red-sided garter snake spends half of the year underneath the ground in limestone caverns that serve as subterranean hibernacula. The snakes emerge in the spring, at which time the majority of matings occur. The breeding season is brief, lasting usually less than three weeks. After mating, the snakes disperse to feeding grounds; in the fall, they return to the hibernacula from which they had emerged the previous spring.

Unlike green anole lizards, which depend primarily upon visual cues, the garter snake depends primarily upon the chemical senses, having two parallel, but nonoverlapping chemical sense systems, the vomeronasal system and the main olfactory system. In a series of elegant studies, Halpern and co-workers (reviewed in Halpern 1983) have examined the sensory projections and functions of these systems in the control of behavior of garter snakes. She first traced the projections of the vomeronasal system and the olfactory system. The primary projection of the vomeronasal organ is to the glomerular layer of the accessory olfactory bulb; secondary projections are to the nucleus sphericus via the accessory olfactory tract and third-order axons project via the stria terminalis to the preoptic area and the ventromedial nucleus of the hypothalamus. The axons of neurons in the olfactory epithelium, on the other hand, project to the glomerular layer of the main olfactory bulb and then to the olfactory cortex via the lateral olfactory tract.

The snake's flicking tongue serves to transport odorants to the ducts leading to the vomeronasal organ lying above the roof of the mouth. The odorants come into direct contact with the dendritic tips of the vomeronasal bipolar neurons. Halpern and her colleagues have demonstrated experimentally that the perception of biologically relevant chemical cues depends on a functioning vomeronasal system. Those behaviors involving the detection of prey odors and social and sexual pheromones are abolished when the vomeronasal nerves are sectioned; lesions of the main olfactory nerves have no effect on these stimulus-specific behaviors.

Whereas courtship behavior in the adult garter snake is dependent on the vomeronasal system, it is independent of sex hormone control (see reviews by Crews 1983a, Crews et al 1984). Courtship behavior in male garter snakes consists of a stereotyped movement in which the male presses his chin against the female's back and moves rapidly up-and-down the length of her body. This behavior is elicited by a pheromone released from the dorsal skin surface of the female; the attractant pheromone is related to vitellogenin, a blood-borne precursor of yolk (Garstka & Crews 1981). Male garter snakes will not exhibit courtship behavior unless they first experience a prolonged period of constant low temperatures. On emerging from hibernation, the male will exhibit intense courtship behavior for a short period; males will not show intense sexual behavior again unless they are hibernated again. Male red-sided garter snakes castrated before entering into hibernation, or even on emergence the preceding year, continue to court females on emergence from hibernation; indeed, so long as males have gone through a period of low-temperature dormancy, castration or androgen replacement will neither inhibit nor prolong the display of sexual behavior. Courtship behavior does not appear to depend on other elements of the hypothalamo-pituitary-gonadal system. For example, the pituitary can be removed on emergence from hibernation, or even before the animal enters hibernation, and the males will still exhibit intense courtship behavior on emergence. Even implantation of testosterone directly into the AH-POA will not induce sexual behavior in males maintained under summer-like conditions (Friedman & Crews 1985a). Indices of gonadal activity were altered following intrahypothalamic hormone administration, however, thus indicating that the sex steroid concentrating sites in the AH-POA (Halpern et al 1982) are involved in the feedback control mechanisms regulating testicular growth.

Attempts to induce courtship behavior in noncourting male red-sided garter snakes maintained under summer-like conditions have been unsuccessful (Crews et al 1984, Garstka et al 1982). Administration of a variety of sex hormones in various concentrations and in various combinations have only occasionally, and then inconsistently, induced sexual behavior. Further, administration of a variety of hypothalamic hormones and neurotransmitters has also failed to elicit courtship behavior. When taken together, these data suggest that courtship behavior in the male red-sided garter snake is under inhibitory control by neural mechanisms sensitive to temperature but not to sex hormone concentrations in the blood.

Although courtship behavior in the adult male red-sided garter snake is independent of sex hormone control, these hormones do appear to play a role in the development of the behavior. Treatments of neonate or yearling castrated males with testosterone will induce courtship behavior (Crews unpublished data); however, only those male snakes treated with androgen courted females on emergence from hibernation. Because male red-sided garter snakes do not reach sexual maturity until the second summer after birth, it is possible that the androgens produced during the first testicular recrudescence program the brain to respond to the shifts in temperature that precede emergence. Other evidence that would support this interpretation is the finding that lesions in the AH-POA of males as they emerge from hibernation will terminate the male's sexual behavior (Friedman & Crews 1985b). It is notable that the AH-POA is a major integrative area of temperature-sensitive neurons. It would be of interest to determine whether decline in courtship behavior in these males is accompanied by a disruption of their perception of temperature.

The neuroendocrine mechanisms in females exhibiting the dissociated reproductive tactic probably differ from those of females that exhibit the associated reproductive tactic. The long- and short-term consequences of mating are especially striking in such females. Copulation stimulates follicle maturation as well as ovulation in the female musk shrew (Dryden & Anderson 1977). In the garter snake, copulation or the presence of sperm in oviducal storage tubules are necessary but not sufficient stimuli for pregnancy; mating initiates a neuroendocrine reflex that is characterized by a transient surge in circulating levels of estradiol ending with ovulation six weeks later (Garstka et al 1985); this in some respects parallels endocrine events characterizing silent estrus in sheep (Foster & Ryan 1984).

CONCLUDING REMARKS: LESSONS FROM NEUROETHOLOGY

By now the reader should be aware of the neuroethologist's prejudices and strategies as well as of some productive results of considering brain function from this particular vantage point. Although the ways in which these lessons can apply to other established disciplines within the neurosciences will depend upon which brain systems and which questions are under consideration, we have some specific recommendations to make based upon present knowledge. These recommendations stem from what we regard as the virtues of neuroethology: it is naturalistic, comparative, and multidisciplinary.

Advantages of the Naturalistic Approach

Two sister disciplines to neuroethology (physiological psychology and clinical neuropsychology) also attempt to relate anatomy, physiology, and function of particular brain components or systems. Within these areas, research on cortical mechanisms of visual pathways is particularly vigorous and well-supported. The majority of vertebrate neurophysiologists who study inbred rats, domesticated cats, and rhesus monkeys do not favor natural stimuli for their studies of single unit response properties. For example, the visual selectivities of units in occipital neocortex are typically evaluated through the use of geometrical patterns unlikely to be confronted in nature. Such experiments are often pursued in an attempt to make comparative research relevant to the understanding of human perception, with the assumption that what is important are visual filters useful for recognition of general features of the environment common to most animals. The extent to which special cortical mechanisms have evolved for specialized functions is not really known among common laboratory species.

How can knowledge about "natural stimuli" prove useful in guiding physiological studies of mammalian cortex? The study of infant visual perception has helped to give an "ethological flavor" to studies of neocortex in the monkey. Adapting a simple method formerly used to demonstrate innate pattern preferences in newly hatched chicks, Fantz (1967) showed that human infants exhibit reliable preferences for fixating or scanning certain visual patterns. Among the more salient patterns attracting infant gaze was a schematic of the human face. Recently, physiologists have demonstrated that a significant subset of neurons in the monkey's association cortex (superior temporal gyrus) are sensitive to face-like schema (Perrett & Rolls 1983). These responses are often specific to particular features subsumed under the class "face" rather than simply neural correlates of emotional responses elicited by faces; some neurons respond to eyes alone, some to the lower part of the face, and some to the hairline. However, further studies would have to be done in infant monkeys or in socially deprived adults to confirm the hypothesis that this cortical area harbors an IRM sensitive to faces. Studies of clinical syndromes resulting from damage to association cortices of man can be fascinating when viewed as instructive anecdotes, yet difficult to analyze systematically using standard clinical tests. Some neuropsychologists have begun to examine natural behaviors of patients, including their reactions to the emotional content of other faces, their own facial display of affect, or their use of gestures (Kolb & Milner 1981a,b). Following the popularity of "neurolinguistics," there appears to be a bright future for both clinical and comparative studies of cortical mechanisms underlying social communication.

A recent symposium (see Ingle & Shook 1984) on the interfacing of neurophysiology and spatial perception included studies of natural skills of man and animals for running, jumping, and catching objects as well as anticipation of collision. For example, it was shown that one could perform the equivalent of a psychophysical study of "collision perception" by examining films to assess the timing of a jump by a volleyball player or the timing of wing-folding by plummeting gannets (Lee & Reddish 1981, Lee & Young 1984). In studying the use of velocity-sensitive or loom-detector units in the cortex, it may in fact be much easier to analyze effects of an extra-striate ablation in a dog playing a ball-catching game or avoiding human chasers than it will be to train the standard laboratory cat to compare moving spots on an oscilloscope. It is quite possible that students of the limbic system would also profit from switching attention from cats to a more naturalistic analysis of social behavior of dogs, where the richness of interaction compares with that of primates (Beach 1970, Beach et al 1982). While dozens of neuroanatomists will throw up their hands at this suggestion, it may be that the time spent extrapolating anatomical findings from the cat to the dog will be more than justified by the ability to focus on meaningful behavioral categories for later lesion or recording studies.

Finally, let us mention that studies of memory may be enriched by the naturalistic approach of ethologists. An outstanding example is the potency of imprinting in birds (Lorenz 1935, Hess 1959), in which an important kind of memory is laid down quickly in an otherwise naive animal. The review of

Horn (1983) indicates that the neural changes accompanying such massive channeling of behavior in chicks are large enough to be detected by ultrastructural (synapse number) and chemical (protein synthesis) measurements in the forebrain. An important new direction in the physiology of memory concerns the proposition that some key facets of spatial memory are represented by the activity of single neurons of the rat's hippocampus (O'Keefe & Nadel 1978). There is evidence that animals quickly and efficiently establish spatial memories within minutes of merely exploring a new environment. For example, gerbils that explore a small object only a few times will spontaneously discriminate the location of that object within an arena one week later (Cheal 1978). There is a dearth of information on the spatial memory efficiency of lower vertebrates, such as fish, which show good evidence of route-finding and homing within natural lakes. As yet, we lack good information on whether they have evolved a "hippocampus-like" system that is necessary for such learning.

Advantages of the Comparative Approach

One concern of the comparative approach is the evolution of homologous systems. Homology is a primary focus for both ethologists and neuroanatomists (Northcutt 1984, Wilczynski 1984). We have earlier suggested that the optic tectum (a homologous structure among vertebrates) has similar behavioral roles in feeding and avoidance behavior from fish to primates. If this hypothesis is valid, then functional extrapolations from definite knowledge of tectum in a fish or frog can at least suggest critical experiments for students of mammalian tectum: e. g. cutting the crossed tectal projections to the brainstem (the pre-dorsal bundle) should produce a loss of orienting toward prey but not the ability to detect prey. In this way, the frog provides a predictive model insofar as homologies are valid.

A second example of the comparative approach by ethologists is the analysis of differences in morphology or in behavior of closely-related species. Species differences in components of behavior represent natural experiments that reflect the end results of ecological and evolutionary constraints. Where two animals differ in only a few major respects, the neural basis of these differences may be detectable against a background of similarities. Grüsser & Grüsser-Cornhels (1976) noted that *Ranid* frogs have more "stationary spot detectors" (class 1 fibers of Lettvin) at the tectal surface than does the toad, *Bufo bufo,* but less class 1 fibers than does the tree frog, *Hyla septrionalis*. The appropriate behavioral comparisons remain to be carried out, but it seems highly likely that the ecological demands of a moving hunter of ground prey such as the toad are significantly different from those of frogs, which are largely sit-and-wait predators. Capranica et al (1973) have shown that where the calls of geographically separate cricket frogs *Acris* are distinct, they respond selectively to their own geographical neighbors on the basis of dialect. If their auditory system has shifted its sensitivity just enough to account for this difference, it would be interesting to know whether this small step is realized by changes in the peripheral auditory organ or within central circuits. In applying this kind of comparison to the mammalian nervous system, the phylogenetic distance of the species chosen for comparison will vary with the system studied. Thus, one might expect differences in vocalization detectors among groups of new world monkeys, but probably no significant differences in visual cortical neurons among *Felidae*.

Finally, a challenging kind of comparative approach selects particular animals for their extreme specialization in order to see whether new rules for perception or action have emerged, rather than to find a model for simplification of a known problem. The auditory skills of bats are particularly interesting because acoustic information is used to perform tasks (pattern discrimination, localization in depth) that usually require the use of vision. It turns out that bats also represent spatial relationships in the form of "maps" in their auditory cortex, but these central maps cannot be derived from the primal topography of the receptor surface as is the case for vision. Yet the logical basis of receptive field organization may provide similar to that in the visual system, e.g. cat striate cells are sensitive to relative motion of figure and ground, as certain cortical neurons in the bat are sensitive to magnitude of the Doppler shift in echos, rather than encoding absolute frequencies of the echo. Knudsen et al (1984) also point out that the owl's auditory map in the midbrain must arise by developmental principles somewhat different from the spatial map found in the mammalian visual thalamus. Yet the role of experience in shaping the owl's map (or those in the bat's cortex) may work by principles analogous to those which fine-tune properties of neurons in the visual cortex, such as their binocularity.

Advantages of the Multidisciplinary Approach

Finally, we believe that the most revealing neuroethological studies are multidisciplinary. First, investigations should be conducted in both the laboratory and in the field if possible (Schneirla 1950): Although the laboratory allows dissection of a behavior into its causal mechanisms, it often does not illuminate the "meaning" of the behavior to the organism. It is only when studies are conducted in nature or in laboratory habitats that simulate nature that the adaptive function of the behavior becomes evident. Classical examples of this cross-fertilization of the laboratory and the field are the studies of bat-moth interaction of Roeder (1974) and the studies on the species-specificity of frog calls reviewed above. Hopkins' studies (1983) of communication in electric fish in the field uncovered differences between both the species and the sexes that proved to be powerful tools in understanding the behavior of these animals.

A second feature of a multidisciplinary approach is the incorporation of techniques that probe the different levels of biological organization, making evident the relations between the levels. We have presented some examples of how a change in the physical environment can be traced ultimately to both the molecular level (gene expression) as well as to the populational level (gene flow). Thus, the idea that signals and receptors co-evolve can be extended to neural systems that coopt external (Thiessen 1983) and internal (Crews 1984b) cues to fine-tune complex processes such as thermoregulation or reproductive behavior.

Structures and function of the brain are a consequence of gene expression. The genetic basis of vertebrate behavior has long been the subject of intensive investigation among psychologists, but it is only recently that the neuroendocrine mechanism has been examined in this regard. For example, the development of androgen-insensitive and hypogonadal strains of laboratory rodents has been particularly exciting. In the latter instance, immunohistochemical studies have demonstrated that the sterility is due to a lack of LHRH. Krieger and her colleagues have found that if the preoptic area from a fetus of a heterozygous fertile strain is transplanted into the preoptic area of an adult homozygous sterile animal, their gonads will produce gametes (Krieger et al 1982, Gibson et al 1984a); immunohistochemical and electrophysiological studies indicate that the transplanted preoptic area establishes connections and produces LHRH necessary for normal pituitary gonadotropin secretion. Although the behavior of hypogonadal mice receiving such transplant has not been documented, one assumes these new neural connections are also involved in the display of mating behavior (Gibson et al 1984b). The eyeless axolotl (*Ambystoma mexicanum*) has also proven useful for studies of neural development (Harris 1983).

The third aspect of multidisciplinary research was first articulated by Niko Tinbergen, who pointed out four distinct, but related, questions when studying behavior: What is the causal basis of the behavior? How does the behavior develop during the individual's lifetime? How does the behavior function to adapt the animal to the environment in which it lives? And how did the behavior evolve? Even if an investigator concentrates on only one of these facets, a complete understanding of the behavior under study will not emerge unless the other facets are at least considered. For example, studies of the development and causal mechanisms of sexual behavior have long been mutually illuminating. We know that steroid hormones act early in development to influence the pattern of growth and death of neurons; this effect is more marked in certain areas (e.g. limbic system) than in other areas (see review by Arnold & Gorski 1984). We know also that in many adult laboratory and domesticated animals, gonadal steroid hormones and hypothalamic peptide hormones activate sexual behavior. In this regard it is significant that a major site of LHRH-concentrating neurons in vertebrates is the accessory olfactory bulb (Demski 1984) through which pheromones exert potent effects on both physiology and behavior (Vandenbergh 1983).

Consideration of the biology of the animal can also greatly facilitate neuroethological investigations. For example, the harsh conditions in which garter snakes live in Canada would lead one to predict corresponding adaptations in the neuroendocrine mechanism underlying male courtship behavior and ovarian growth in the female. Or, it has long been known that in some passerine birds both the male and the female sing, producing "duets." How do the neural systems subserving song in species in which both sexes sing differ from those species in which only the male sings (Brenowitz & Arnold 1983, Brenowitz et al 1984)? Ecological studies can also shed light on the development of neuroendocrine controlling mechanisms. For example, environmental constraints, social organization, and energetic requirements can completely alter the timing of many physiological and behavioral events (Bronson 1979, 1984).

In conclusion, the naturalistic, comparative, multidisciplinary approach of the neuroethologist provides an alternative to the normative approach that dominates most other neuroscience disciplines concerned with the neurobiology of behavior. On the one hand, the neuroethological approach may reveal extreme cases invaluable for the development of animal models, while on the other hand, it will emphasize the range and diversity of the alternative solutions to adaptive problems all vertebrates share. The authors heartily agree with T. H. Bullock's (1984) recent admonition, "Neuroscience is part of biology, more specifically zoology, and it suffers from tunnel vision unless continuous with ethology, ecology, and evolution."

ACKNOWLEDGMENTS

We wish to thank the following individuals for helpful comments on the manuscript: C. Diakow, N. Greenberg, M. Halpern, D. Kelley, F. Moore, D. Pfaff, and J. Simmons. We owe a special debt to the Neurosciences Institute of the Neurosciences Research Program, which provided facilities and assistance in completing the manuscript. Research support was provided to David Ingle by NIH NS-13592 and NSF BNS 83-03946 and to David Crews by NIH HD-16687 and NIMH Research Scientist Development Award MH-00135.

Literature Cited

Adler, N. T. 1978. On the mechanisms of sexual behavior and their evolutionary constraints. In *Biological Determinants of Sexual Behavior* ed. J. B. Hutchison, pp. 657–95. Chichester: Wiley. 822 pp.

Adler, N. T. 1983. The neuroethology of reproduction. In *Advances in Vertebrate Neuroethology*, ed. J.-P. Ewert, R. R. Capranica, D. J. Ingle, pp 1033–65. New York: Plenum. 1238 pp.

Akert, K. 1949. Der visuelle Greifreflex. *Helv. Physiol. Pharmacol. Acta* 7:112–34

Allen, T. O., Adler, N. T., Greenberg, J. H., Reivich, M. 1981. Vaginocervical stimulation selectively increases metabolic activity in the rat brain. *Science* 211:1070–72

Arnold, A. P., Gorski, R. A. 1984. Gonadal steroid induction of structural sex differences in the central nervous system. *Ann. Rev. Neurosci.* 7:413–42

Barfield, R. J., Rubin, B. S., Glaser, J. H., Davis, P. G. 1983. Sites of action of ovarian hormones in the regulation of oestrus behavior in rats. In *Hormones and Behavior in Higher Vertebrates*, ed. J. Balthazart, E. Pröve, R. Gilles, pp. 2–17. Berlin: Springer-Verlag. 489 pp.

Beach, F. A. 1978. Animal models for human sexuality. In *Sex, Hormones and Behavior, Ciba Found. Symp.* 62:113–43. Amsterdam: Excerpta Medica. 382 pp.

Beach, F. A. 1970. Coital behavior in dogs, VIII. Social affinity, dominance and sexual preference in the bitch. *Behavior* 36:131–48

Beach, F. A. 1983. Hormones and psychological processes. *Can. J. Psychol.* 37:193–210

Beach, F. A., Buehler, M. G., Dunbar, I. F. 1982. Competitive behavior in male, female, and pseudohermaphroditic female dogs. *J. Comp. Physiol. Psychol.* 96:855–74

Bandler, R. J., Flynn, J. P. 1971. Visual patterned reflex during hypothalamically elicited attack. *Science* 171:703–6

Brenowitz, E. A., Arnold, A. P. 1983. Neural correlates of avians song duetting. *Soc. Neurosci. Abstr.* 9:538

Brenowitz, A. E., Arnold, A. P., Levin, R. N. 1984. Comparison of male and female vocal control regions in duetting birds. *Soc. Neurosci. Abstr.* 10:402

Bronson, F. H. 1979. The reproductive ecology of the house mouse. *Q. Rev. Biol.* 54:265–99

Bronson, F. H. 1984. Mammalian reproduction: An ecological perspective. *Biol. Reprod.* In press

Bullock, T. H. 1984. Comparative neuroscience holds promise for quiet revolutions. *Science* 225:473–78

Capranica, R. R. 1976. Morphology and physiology of the auditory system. In *Handbook of Frog Neurobiology*, ed. R. Llinas, W. Precht, pp. 551–75. Berlin/Heidelberg/New York: Springer-Verlag

Capranica, R. R. 1983. Neurobehavioral correlates of sound communication in anurans. See Ewert et al 1983a, pp. 701–30

Capranica, R. R., Frishkopf, L. S., Nevo, E. 1973. Encoding of geographical dialects in the auditory system of the cricket frog. *Science* 182:1272–75

Capranica, R. R., Moffat, A.J.M. 1980. Nonlinear properties of the peripheral auditory system of anurans. In *Comparative Studies of Auditory Processing in Vertebrates*, ed. A. Popper, R. Fay, pp. 139–66. Berlin/Heidelberg/New York: Springer-Verlag

Casagrande, V. A., Diamond, I.T. 1974. Ablation study of the superior colliculus in the tree shrew (*Tupaia glis*). *J. Comp. Neurol.* 156:207–38

Cheal, M. 1978. Stimulus-elicited investigation in the Mongolian gerbil *Meriones unguiculatus*). *Worm Runner's Digest* 20:26–32

Collett, T. 1977. Stereopsis in toads. *Nature* 267:349–51

Comer, C., Grobstein, P. 1978. Prey acquisition in atectal frogs. *Brain Res.* 153:217–21

Crews, D. 1979. Neuroendocrinology of lizard reproduction. *Biol. Reprod.* 20:51–73

Crews, D. 1980. Interrelationships among ecological, behavioral, and neuroendocrine processes in the reproductive cycle of *Anolis carolinensis* and other reptiles. In *Advances in the Study of Behavior*, ed. J. S. Rosenblatt, C. G. Beer, R. A. Hinde, M. C. Busnell, 2:1–74. New York: Academic

Crews, D. 1983a. Control of male sexual behaviour in the Canadian red-sided garter snake. See Barfield et al 1983, pp. 398–406

Crews, D. 1983b. Regulation of reptilian reproductive behavior. See Ewert et al 1983a, pp. 997–1032

Crews, D. 1984a. Gamete production, sex steroid secretion, and mating behavior uncoupled. *Horm. Behav.* 18:21–28

Crews, D. 1984b. Functional associations in behavioral endocrinology. In *Masculinity/Femininity*,ed. J. Reinisch. Bloomington, IN: Indiana Univ. Press. In press

Crews, D., Camazine, B., Diamond, M., Mason, R., Tokarz, R. R., Garstka, W. R. 1984. Hormonal independence of courtship behavior in the male garter snake. *Horm. Behav.* 18:29–41

Crews, D., Greenberg, N. 1981. Function and causation of social signals in lizards. *Am. Zool.* 21:273–94

Crews, D., Silver, R. 1984. Reproductive physiology-behavior interactions in nonmammalian vertebrates. In *Handbook of Behavioral Neurobiology*,ed. N. T. Adler, D. W. Pfaff. New York: Plenum. In press

Demski, L. 1984. The evolution of neuroanatomical substitutes of reproductive behavior: Sex steroid and LHRH-specific pathways including the terminal nerve. *Am. Zool.* In press

Dewsbury, D. A. 1978. The comparative method in studies of reproductive behavior. In *Sex and Behavior: Status and Prospectus*,ed. T. E. McGill, D. A. Dewsbury, B. D. Sachs, pp. 83–112. New York: Plenum

Diakow, C. 1978. A hormonal basis for breeding behavior in female frogs. *Science* 199:1456–57

Diakow, C. 1977. Initiation and inhibition of the release croak of *Rana pipiens*. *Physiol. Behav.* 19:607–10

Diakow, C., Nemiroff, A. 1981. Vasotocin, prostaglandin and female behavior in the frog, *Rana pipiens*. *Horm. Behav.* 15:86–93

Diakow, C., Raimondi, D. 1981. Physiology of *Rana pipiens* reproductive behavior: A proposed mechanism for inhibition of the release call. *Am. Zool.* 21:295–304

Diakow, C., Wilcox, J. N., Woltmann, R. 1978. Female frog reproductive behavior elicited in the absence of the ovaries. *Horm. Behav.* 11:183–89

Dryden, D. L., Anderson, J. N. 1977. Ovarian hormone: Lack of effect on reproductive structures of female Asian musk shrews. *Science* 197:782–84

Erulkar, S. D., Kelley, D. B., Jurman, M. E., Zemlan, F. P., Schneider, G. T., Krieger, N. R. 1981. Modulation of the neural control of the clasp reflex in male *Xenopus laevis* by androgens: A multidisciplinary study. *Proc. Natl. Acad. Sci. USA* 78:5876–80

Ewert, J.-P. 1968. Der Einflussvon Zwischenhirndefekten auf die Visuomotorik im Beute- und Fluchtverhalten der Erdkröte (*Bufo bufo L.*). *Z. Vergl. Physiol.* 61:41–70

Ewert, J.-P. 1970. Neural mechanisms of prey-catching and avoidance behavior in the toad (*Bufo bufo L.*). *Brain Behav. Evol.* 3:36–56

Ewert, J.-P., Capranica, R. R., Ingle, D. J., eds. 1983a. *Advances in Vertebrate Neuroethology.* New York/London: Plenum. 1238 pp.

Ewert, J.-P., Burghagen, H., Schürg-Pfeiffer. 1983b. Neuroethological analysis of the innate releasing mechanism for prey-catching behavior in toads. See Ewert et al 1983a, pp. 413–76

Ewert, J.-P., Gebauer, L. 1973. Grössenkonstanz phänomene im Beutefangverhalten der Erdkröte (*Bufo bufo L.*) *J. Comp. Physiol.* 85:303–15

Ewert, J.-P., von Wietersheim, A. 1974. Der Einflussvon Thalamus/Praetectum- Defekten auf die Antwort von Tectum-Neuronen gegenüber visuellen Mustern bei der Kröte (*Bubo bufo L.*) *J. Comp. Physiol.* 92:149–60

Fantz, R. L. 1967. Visual perception and experience in infancy. In *Early Behavior*, pp. 181–224. New York: Wiley

Feng, A. S., Capranica, R. R. 1976a. Sound localization in anurans: I. Evidence of binaural interaction in the dorsal medullary nucleus of bullfrogs (*Rana catesbeiana*). *J. Neurophysiol.* 39:871–81

Feng, A. S., Capranica, R. R. 1976b. Binaural interaction in the superior olivary nucleus of the green tree frog (*Hyla cinerea*). *J. Neurophysiol.* 41:43–54

Finlay, B. L., Sengelaub, D. R., Berg, A. T., Cairns, S. J. 1980. A neuroethological approach to hamster vision. *Behav. Brain Res.* 1:479–96

Foster, D. L., Ryan, K. D. 1984. Puberty in the lamb: Sexual maturation of a seasonal breeder in a changing environment. In *Control of the Onset of Puberty II*, ed. P. C. Sizonenko, M. L. Aubert, M. Grumbach. Baltimore: Williams & Wilkins. In press

Friedman, D., Crews, D. 1985a. Role of the anterior hypothalamus–preoptic area in the regulation of courtship behavior in the Canadian red-sided garter snake (*Thamnophis sirtalis parietalis*): Implantation studies. *Horm. Behav.* In press

Friedman, D., Crews, D. 1985b. Role of the anterior hypothalamus–preoptic area in the regulation of courtship behavior in the Canadian red-sided garter snake (*Thamnophis sirtalis parietalis*): Lesion studies. *Behav. Neurosci.* In press

Frishkopf, L. S., Capranica, R. R., Goldstein, M. H., Jr. 1968. Neural coding in the bullfrog's auditory system—a teleological appraoch. *Proc. IEEEE* 56:969–80

Fuzessory, Z. M., Pollak, G. D. 1984. Neural mechanisms of sound localization in an echolocating bat. *Science* 225:725–27

Garstka, W. R., Camazine, B., Crews, D. 1982. Interactions of behavior and physiology during the annual reproductive cycle of the red-sided garter snake, *Thamnophis sirtalis parietalis. Herpetologica* 38:104–23

Garstka, W., Crews, D. 1981. Female sex pheromones in the skin and circulation of a snake. *Science* 214:681–83

Garstka, W., Tokarz, R. R., Halpert, A., Diamond, M., Crews, D. 1985. Social and hormonal control of yolk synthesis and deposition in the red-sided garter snake *Thamnophis sirtalis parietalis. Horm. Behav.* In press

Gerhardt, H. C., Mudry, K. M. 1980. Temperature effects on frequency preference and mating call frequencies in the green tree frog *Hyla cinerea* (Anura: Hylidae). *J. Comp. Physiol.* 137:1–6

Gerhardt, H. C., Rheinlander, J. 1982. Localization of elevated sound sources by the green tree frog. *Science* 217:663–64

Gibson, M. J., Charlton, H. M., Perlow, M. J., Zimmerman, E. A., Davis, T. F., Krieger, D. T. 1984a. Preoptic area brain grafts in hypogonadal (hpg) female mice abolish effects on congenital hypothalamic gonadotropin-releasing hormone (GnRH) deficiency. *Endocrinology* 114:1938–40

Gibson, M. J., Krieger, D. T., Charlton, H. M., Zimmerman, E. A., Silverman, A. J., Perlow, M. J. 1984b. Mating and pregnancy can occur in genetically hypogonadal mice with preoptic area brain grafts. *Science* 225:949–51

Goldfoot, D. A., Neff, D. A. 1984. On measuring behavioral sex differences in social contexts. In *The Handbook of Behavioral Neurobiology*, Vol. 8: *Neurobiology of Reproduction*, ed. N. Adler, D. Pfaff. New York: Plenum. In press

Goodale, M. A., Milner, A. D. 1982. Fractionating orientation behavior in rodents. In *Analysis of Visual Behavior*, ed. D. J. Ingle, M. A. Goodale, R. J. W. Mansfield, pp. 267–299. Cambridge, MA: MIT Press. 834 pp.

Gould, S. J., Vrba, E. S. 1982. Exaptation—

a missing term in the science form. *Paleobiology* 8:4–15
Greenberg, N. 1978. Ethological considerations in the experimental study of lizard behavior. In *Behavior and Neurology of Lizards*, ed. N. Greenberg, P. D. MacLean. US Dept H.E.W. Publ. No. (ADM) 77-491
Greenberg, N., Crews, D. 1983. Physiological ethology of aggression in amphibians and reptiles. In *Hormones and Aggression*, ed. B. Svare, pp. 469–506. New York: Plenum
Greenberg, N., MacLean, P. D., Ferguson, J. L. 1979. Role of the paleostriatum in species-typical display behavior of the lizard (*Anolis carolinensis*). *Brain Res.* 172:217–28
Greenberg, N., Scott, M., Crews, D. 1984. Role of the amygdala in the reproductive and aggressive behavior of the lizard, *Anolis carolinensis*. *Physiol. Behav.* 32:147–51
Grüsser, G.-J., Grüsser-Cornhels, U. 1976. Neurophysiology of the anuran visual system. In *Frog Neurobiology*, ed., R. Llinas, W. Precht, pp. 297–385. Berlin: Springer-Verlag
Halpern, M. 1983. Nasal chemical senses in snakes. See Ewert et al 1983a, pp. 141–76
Halpern, M., Morrell, J., Pfaff, D. 1982. Cellular ^{3}H-estradiol and ^{3}H-testosterone localization in the brains of garter snakes: An autoradographic study. *Gen. Comp. Endocrinol.* 46:211–24
Harris, W. A. 1983. The eyeless axolotl: Experimental embryogenetics and the development of the nervous system. *Trends Neurosci.* 6:505–10
Hart, B. L. 1978. Hormone, spinal reflexes, and sexual behavior. In *Biological Determinants of Sexual Behaviour*, ed. J. Hutchison, pp. 319–48. Chichester: Wiley. 822 pp.
Hess, E. H. 1959. Imprinting: An effect of early experience. *Sci. Am.* 130:133–41
Hess, W. R., Burgs, S., Bucher, V. 1946. Motorische Funktion des Tektal- und Tegmentalgebeites. *Monatsschr. Psychiatr. Neurol.* 112:1–52
Heiligenberg, W., Bastian, J. 1984. The electrical sense of weakly electric fish. *Ann. Rev. Physiol.* 46:561–84
Hopkins, C. D. 1983. Neuroethology of species recognition in electroreception. See Ewert et al 1983a, pp. 871–82
Horn, G. 1983. Information storage in the brain: A study of imprinting in the domestic chick. See Ewert et al 1983a, pp. 511–41
Ingle, D. J. 1968. Visual releasers of prey catching behavior in frogs and toads. *Brain Behav. Evol.* 1:500–18
Ingle, D. J. 1973. Two visual systems in the frog. *Science* 181:1053–55
Ingle, D. J. 1977. Detection of stationary objects by frogs following optic tectum ablation. *J Comp. Physiol. Psychol.* 91:1359–64
Ingle, D. J. 1980. Some effects of pretectum lesions on the frog's detection of stationary objects. *Behav. Brain Res.* 1:139–63
Ingle, D. J. 1983a. Brain mechanisms of localization in frogs and toads. See Ewert et al 1983a, pp. 177–226
Ingle, D. J. 1983b. Prey selection in frogs and toads: A neuro-ethological model. In *Handbook of Behavioral Neurobiology*, ed. E. Satinoff, P. Teitelbaum, pp. 235–61. New York: Plenum
Ingle, D. J., Cheal, M., Dizio, P. 1979. Cine analysis of visual orientation and pursuit by the Mongolian gerbil. *J. Comp. Physiol. Psychol.* 93:919–28
Ingle, D. J., Cook, J. 1977. The effect of viewing distance upon size preference of frogs for prey. *Vision Res.* 17: 1009–14
Ingle, D. J., Shook, B. 1984. Action-oriented approaches to visuospatial brain functions. In *Brain Mechanisms and Spatial Vision*, ed. D. J. Ingle, D. Lee, M. Jeannerod. The Hague: Nijhoff. In press
Ingle, D. J., Sprague, J. M. 1975. Sensorimotor functions of the midbrain tectum. *Neurosci. Res. Program Bull.* 13:169–288
Jones, R. E. 1978. Ovarian cycles in nonmammalian vertebrates. In *The Vertebrate Ovary*, ed. R. E. Jones, pp. 731–62. New York: Plenum
Jones, R. E., Guillette, L. J., Summers, C. H., Tokarz, R. R., Crews, D. 1983. Ovarian follicular and luteal condition, and related steroid hormone levels, during the estrous cycle of the lizard, *Anolis carolinensis*. *J. Exp. Zool.* 227:145–54
Kelley, D. B. 1980. Auditory and vocal nuclei in the frog brain concentrate sex hormones. *Science* 107:553–55
Kelley, D. B. 1982. Female sex behaviors in the South African clawed frog, *Xenopus laevis:* Gonadotropin-releasing, gonadotropic, and steroid hormones. *Horm. Behav.* 16:158–74
Kelley, D. B., Pfaff, D. W. 1978. Generalizations from comparative studies on neuroanatomical and endocrine mechanisms of sexual behavior. See Hart 1978, pp. 225–54
Kelley, D. B., Wetzel, D., Hannigan, P. 1981. Mate calling in the South African clawed frog, *Xenopus leavis:* Hormonal modulation of neuroeffectors. *XVIIth Int. Congr.* (Abstr.)
Knudsen, E. I., Esterly, S. D., Knudson, P. F. 1984. Monaural occulation alters sound localization during a sensitive period in the barn owl. *J. Neuroscience* 4:1001–11
Knudsen, E. I., Konishi, M. 1979. Mechanisms of sound localization in the barn owl (*Tyto alba*). *J. Comp. Physiol.* 133:13–21
Knudson, E. I., Konishi, M. 1980. Monaural occlusion shifts the receptive field locations

of auditory midbrain units in the owl. *J. Neurophysiol.* 44:687–95

Kobler, J. B., Isbey, S. F., Casseday, J. H. 1983. Evidence for connections between auditory cortex and frontal cortex of the bat, *Pternotus parnellii. Neurosci. Soc. Abstr.* 9:956

Kolb, B., Milner, B. 1981a. Performances of complex arm and facial movements after focal brain lesions. *Neuropsychologia* 19:491–503

Kolb, B., Milner, B. 1981b. Observations on spontaneous facial expressions after focal cerebral excisions and after intra-carotid injections. *Neuropsychologia* 19:505–14

Komisaruk, B. K. 1978. The nature of the neural substrate of female sexual behaviour in mammals and its hormonal specificity: Review and speculations. In *Biological Determinants of Sexual Behavior*, ed. J. Hutchison, pp. 349–93. Chichester: Wiley

Konishi, M. 1983. Localization of acoustic signals in the owl. See Ewert et al 1983a, pp. 227–45

Konishi, M., Gurney, M. E. 1982. Sexual differentiation of brain and behavior. *Trends Neurosci.* 5(1):20–23

Krieger, D. T., Perlow, M. J., Gibson, M. J. Davis, T. F., Zimmerman, E. A., Ferin, M., Charlton, H. M. 1982. Brain grafts reverse hypogonadism in gonadotropin-releasing hormone deficiency. *Nature* 298:468–70

Lawrence, B. D. 1981. *Minimum discriminable vertical for the echolocating bat (Eptesicus fuscus).* M. A. thesis, Washington Univ., St. Louis, Mo.

Lee, D. N., Reddish, P. E. 1981. Plummeting gannets: A paradigm of ecological optics. *Nature* 293:293–94

Lee, D. N., Young, D. C. 1984. Visual timing of interceptive action. In *Brain Mechanisms and Spatial Vision*, ed. D. Ingle, M. Jeannerod, D. Lee. The Hague: Martinus Nijhoff. In press

Lettvin, J. Y., Maturana, H. R., McCulloch, W. S., Pitts. W. H. 1959. What the frog's eye tells the frog's brain. *Proc. IRE* 47:1940–51

Lettvin, J. Y., Maturana, H. R. McCulloch, W. S., Pitts, W. H. 1961. Two remarks on the visual system of the frog. In *Sensory Communication*, ed. W. A. Rosenblith. Cambridge, MA: MIT Press

Lorenz, K. 1935. Der Kumpan in der Umwelt des Vogels. *J. Ornitho.* 83:137–413

Martinez-Vargas, M. C., Keefer, D. A., Stumpf, W. E. 1978. Estrogen localization in the brain of the lizard, *Anolis carolinensis. J. Exp. Zool.* 205:141–47

McClintock, M. K. 1983. The behavioral endocrinology of rodents: A functional analysis. *Bioscience* 33:573–77

Meisel, R. L., Pfaff, D. W. 1984. RNA and protein synthesis inhibitors: Effects on sexual behaviors in female rats. *Brain Res. Bull.* 12:187–93

Merker, B. H. 1980. The sentinel hypothesis: A role for the mammalian superior colliculus. PhD dissertation. Department of Psychol., Mass. Instit. Technol., Cambridge, Mass.

Moore, F. L. 1978. Differential effects of testosterone plus dihydrotestosterole on male courtship behavior of castrated newts. *Taricha granulosa. Horm. Behav.* 11:202-8

Moore, F. L. 1983. Behavioral endocrinology of amphibian reproduction. *BioScience* 33:557–61

Moore, F. L., Miller, L. J. 1983. Arginine vasotocin induces sexual behavior of newts by acting on cells in the brain. *Peptides* 4:97–102

Moore, F. L., Zoeller, R T., Spielvogel, S. P., Baum, M. J., Han, S.-J., Crews, D., Tokarz, R. R. 1981. Arginine vasotocin enhances influx of testosterone into the newt brain. *Comp. Biochem. Physiol.* 70A:115–17

Morrell, J. I., Crews, D., Ballin, A., Morgentaler, A., Pfaff, D. W. 1979. ^{3}H-estradiol, ^{3}H-testosterone, ^{3}H-dihydrotestosterone in the brain of the lizard. *Anolis carolinensis:* An autoradographic study. *J. Comp. Neurol.* 188:201–24

Morrell, J. I., Kelley, D. B., Pfaff, D. W. 1975. Autoradiographic localization of hormone-concentrating cells in the brain of an amphibian, *Xenopus laevis*. II. Estradiol. *J. Comp. Neurol.* 164:63–78

Morrell, J. I., Pfaff, D. W. 1982. Characterizations of estrogen-concentrating hypothalamic neurons by their axonal projections. *Science* 217:1273–76

Morrell, J. I., Schwanzel-Fukuda, M., Fahrbach, S. E., Pfaff, D. W. 1984. Axonal projections and peptide content of steroid hormone concentraing neurons. *Peptides*. In press

Mudry, K. M., Constantine-Paton, M., Capranica, R. R. 1977. Auditory sensitivity of the diencephalon of the leopard frog (*Rana p. pipiens*). *J. Comp. Physiol.* 144:1–13

Narins, P. M., Capranica, R. R. 1978. Communicative significance of the two-note call of the tree frog (*Eleutherodactylus coqui*). *J. Comp. Physiol.* 127:1–9

Narins, P. M., Capranica, R. R. 1980. Neural adaptations for processing the two-note call of the Puerto Rican tree frog (*Eleutherodactylus coqui*). *Brain Behav. Evol.* 17:48–66

Noble, G. K., Aronson, L. R. 1942. The sexual behavior of Anura. 1. The normal mating pattern of *Rana pipiens*. *Bull. Am. Mus. Nat. Hist.* 80:127–42

Northcutt, R. B. 1981. Evolution of the telencephalon in nonmammals. *Ann. Rev. Neurosci.* 4:301–50

Northcutt, R. G. 1984. Evolution of the vertebrate central nervous system: Patterns and processes. *Am. Zool.* In press

O'Keefe, J., Nadel, L. 1978. *The Hippocampus as a Cognitive Map*. Oxford: Clarendon. 570 pp.

O'Neill, W. E., Suga, N. 1979. Target range-sensitive neurons in the auditory cortex of the mustache bat. *Science* 203:69–73

Passmore, M. I., Capranica, R. R., Telford, S. R., Bishop. P. J. 1984. Phonotaxis in the painted reed frog (*Hyperloius marmoratus*.) *J. Comp. Physiol. A* 154:189–97

Paton, J. A., Kelley, D. B., Sejnowski, T. J., Yoklowski, M. L. 1982. Mapping the auditory central nervous system of *Xenopus laevis* with 2-deoxyglucose autoradiography. *Brain Res*. 249:15–22

Perrett, D. I., Rolls, E. T. 1983. Neural mechanisms underlying the visual analysis of faces. See Ewert et al 1983a, pp. 543–66

Pfaff, D. W. 1980. *Estrogens and Brain Function*. New York: Spring-Verlag. 282 pp.

Pollak, G. D., Bodenhamer, R. D., Zook, J. M. 1983. Cochleotophic organization of the mustache bat's inferior collicluus. See Ewert 1983a, pp. 925–35

Raimondi, D., Diakow, C. 1981. Sex dimorphism in responsiveness to hormonal induction of female behavior in frogs. *Physiol. Behav.* 27:167–70

Roeder, K. D. 1974. Some neuronal mechanisms of simple behavior. *Adv. Study Behav.* 5:1–46

Sachs, B. D., Barfield, R. J. 1976. Functional analysis of masculine behavior in the rat. *Adv. Study Behav.* 7:92–154

Schmidt, R. S. 1966. Central mechanisms of frog calling. *Behavior* 26:251–85

Schmidt, R. S. 1968. Preoptic activation of frog mating behavior. *Behavior* 30:239–57

Schmidt, R. S. 1971. A model of the central mechanisms of male anuran acoustic behavior. *Behavior* 39:288–317

Schmidt, R. S. 1973. Central mechanisms of frog calling. *Am. Zool.* 13:1169–77

Schmidt, R. S. 1974. Neural correlates of frog calling. Trigeminal tegmentum. *J. Comp. Physiol.* 92:229–54

Schmidt, R. S. 1968. Preoptic activation of frog mating behavior. *Behavior* 30:239–57

Schneider, G. E. 1969. Two visual systems: Brain mechanisms for localization and discrimination are dissociated by tectal and cortical lesions. *Science* 163:895–902

Schneirla, T. C. 1950. The relationship between observation and experimentation in the field study of behavior. *Ann. NY Acad. Sci.* 51:1022–44

Schnitzler, H.-U., Henson, O. W. Jr. 1980. Performance of animal airborne sonar systems. I. Microchiroptera. In *Animal Sonar Systems*, ed. R. G. Bushnel, J. F. Fish, pp. 109–81.New York: Plenum

Schnitzler, H.-U., Ostwald, J. 1983. Adaptations for the detection of fluttering insects by echolocation in horseshoe bats. See Ewert 1983a, pp. 801–27

Schwartz-Giblin, S., Rosello, L., Pfaff, D. W. 1983. A histochemical study of lateral longissimus muscle in the rat. *Exp. Neurol.* 79:497–518

Shettleworth, S. J. 1983. Memory in food-hoarding birds. *Sci. Am.* 248:102–9

Shivers, B. D., Harlan, R. E., Morrell, J. I., Pfaff, D. W. 1983. Absence of oestradiol concentration in cell nuclei of LHRH-communoreactive neurones. *Nature* 304:345–47

Simmons, J. A., Kick, S. A., Lawrence, B. D., Hale, C., Bard, C., Escudie, B. 1983a. Acuity of horizontal angle discrimination by the echolocating bat (*Eptesicus tuscus*). *J. Comp. Physiol.* 153:321–30

Simmons, J. A., Kick, S. A., Lawrence, B. D. 1983b. Localization with biosonar signals in bats. See Ewert et al 1983a, pp. 247–60

Springer, A. D., Easter, S. S., Agranoff, B. W. 1977. The role of the optic tectum in various visually mediated behaviors of goldfish. *Brain. Res.* 128:393–404

Suga, N. 1982. Functional organization of the auditory cortex: Representation beyond tonotopy in the bat. In *Cortical Sensory Organization*, ed. C. N. Woolsey, 3:157–218. New Jersey: Humana

Suga, N., Niwa, H., Taniguchi, I. 1983. Representation of bisonar information in the auditory cortex of the mustached bat, with emphasis on representation of target velocity information. See Ewert et al 1983a, pp. 829–67

Suga, N., O'Neill, W. E. 1979. Neural axis representing target range in the auditory cortex of the mustached bat. *Science* 206:351–53

Suga, N., O'Neill, W. E., Manabe, T. 1979. Harmonic-sensitive neurons in the auditory cortex of the mustache bat. *Science* 203:270–74

Teitelbaum, P., Schallert, T., Whishaw, I.Q. 1983. Sources of spontaneity in motivated behavior. In *Handbook of Behavioral Neurobiology*, ed. E. Satinoff, P. Teitelbaum, 6:23–65. New York: Plenum

Thiessen, D. D. 1983. Thermal constraints and influences on communication. *Adv. Study Behav.* 13:147–89

Tokarz, R. R., Crews, D. 1981. Effect of prostaglandins on sexual receptivity in the female lizard, *Anolis carolinensis*. *Endocrinology* 109:451–58

Tokarz, R. R., Crews, D., McEwen, B.S., 1981. Estrogen-sensitive progestion binding in the brain and oviduct of the lizard, *Anolis carolinensis*. *Brain Res*. 220:95–

Trachtenberg, M. C., Ingle, D. J. 1974. Thalamo-tectal projections in the frog. *Brain Res*. 79:419–30

Vandenbergh, J. G. (ed.) 1983. *Pheromones and Reproduction in Mammals*. New York: Academic. 297 pp.

Wetzel, D., Kelley, D. 1981. Neural circuitry for mate calling in male South African clawed frogs. *Xenopus laevi*. *Soc. Neurosci. Abstr.* 7:269

Wetzel, D., Kelley, D. 1983. Androgen and gonadotropin effects of male mate calls in South African clawed frogs, *Xenopus laevis*. *Horm. Behav.* 17:388–404

Wilczynski, W. 1984. Central nervous systems subserving a homoplasous periphery. *Am. Zool*. In press

Wu, J., Whittier, J. M., Crews, D. 1985. Role of progesterone in the control of sexual receptivity in *Anolis carolinensis*. *Gen. Comp. Endocrinol*. In press

Zook, J. M., Casseday, J. H. 1982. Origin of ascending projections to inferior colliculus in the moustache bat, *Pteronotus parnellii*. *J. Comp. Neurol*. 207:14–28

Ann. Rev. Neurosci. 1985. 8:495–545

POSTNATAL DEVELOPMENT OF VISION IN HUMAN AND NONHUMAN PRIMATES

Ronald G. Boothe, Velma Dobson, and Davida Y. Teller*

Departments of Psychology, Ophthalmology, and Physiology/Biophysics, Regional Primate Research Center and Child Development and Mental Retardation Center, University of Washington, Seattle, Washington 98195

INTRODUCTION

In this chapter we summarize the growing literature on the optical, anatomical, physiological, and behavioral development of vision in human and nonhuman primates. The literature in this area is new and diverse, and to our knowledge it has not been previously reviewed as a whole. Emphasis is on postnatal development, in both its normal and its most common abnormal forms. All monkeys in the studies reviewed were Old World monkeys.

References to the large body of developmental studies on nonprimate species have been omitted (for reviews see Grobstein & Chow 1976, Tees 1976, Movshon & Van Sluyters 1981, Sherman & Spear 1982, Mitchell & Timney 1984). We have also omitted a large body of literature on embryology and prenatal anatomical development (see Rakic & Goldman-Rakic 1982, van Hof-van Duin & Mohn 1984). Of necessity, extended discussions of theoretical issues are precluded. Extended works on the topic of visual development can be found in Banks & Salapatek (1983), Aslin et al (1981), and the October 1983 issue of *Behavioural Brain Research*.

A major reason for studying the development of vision in monkeys is to establish and use the monkey as a model for studying normal and abnormal human visual development. Many of the behavioral and anatomical studies reviewed below suggest major similarities of development, with rough parity at birth, and development proceeding about four times faster in Old World monkey infants than in human infants. To emphasize this 4:1 conversion factor

*Current address: Yerkes Primate Center, and Departments of Psychology and Ophthalmology, Emory University, Atlanta, Georgia 30322.

0147-006X/85/0301-0495$02.00

we specify human age in months and monkey age in weeks throughout this review. Ideally such a conversion factor will eventually allow highly informed guesses to be made from invasive monkey studies to the time course of normal human anatomical, physiological, and behavioral development, and from monkey deprivation studies to human sensitive periods for naturally occurring visual deprivation and for clinical treatment.

The recent availability of a variety of new techniques and methodologies has contributed heavily to the growth of this field. These include methods in physiological optics such as near retinoscopy (Mohindra 1977) and photorefraction (Howland & Howland 1974, Kaakinen 1979, Howland et al 1983), modern variants of behavioral methods such as preferential looking (Gwiazda et al 1978, Teller 1979), and operant techniques for use with both human toddlers (Mayer & Dobson 1980, 1982, Atkinson et al 1981, Birch et al 1983c, Abramov et al 1985) and young monkeys (Boothe & Sackett 1975, Boothe 1981). Finally, studies of underlying neural mechanisms have been greatly facilitated by the combined use of several new anatomical methods in conjunction with physiological recordings from single units (e.g. Hubel 1982, Wiesel 1982).

NORMAL POSTNATAL DEVELOPMENT

Optics and Eye Movements

OPTICS

Schematic eye Simple schematic eyes have been proposed for normal eyes (Lotmar 1976, Enoch & Hamer 1983) and for aphakic eyes (Enoch & Hamer 1983).

Optical quality Optical line spread functions obtained from infant monkey eyes demonstrate that optical quality for well-focused retinal images is very good at birth and improves rapidly to adult levels by about nine weeks after birth (Williams & Boothe 1981). The quality of human infant optics, evaluated by ophthalmoscopy, is also reported to be very good within days after birth (R. Kalina 1979 and personal communication).

Optic media It has been argued, especially on the basis of visual evoked potential (VEP) studies of infant spectral sensitivity, that the optic media are clearer in infants than in young adults (e.g. Werner 1982). However, behavioral studies of spectral sensitivity tend to contradict this conclusion, and the matter remains unresolved. These studies of spectral sensitivity are reviewed below.

Refractive error Banks (1980) has provided a comprehensive recent review in this area. Studies of the spherical component of human infants' refractive errors have usually been conducted using cycloplegic retinoscopy. Full-term newborns tend to be hyperopic (mean = +2 D) but highly variable (s.d. = 2 D). Preterm newborns tend to be one to two diopters less hyperopic than full terms and even more variable (s.d. = 2.5 D). The amount of hyperopia tends to reach a maximum within the first six postnatal months and to decrease thereafter. Infant monkey eyes are similarly hyperopic, and move toward emmetropia during the first few postnatal months (M. Neuringer et al 1984 and personal communication; F. Young personal communication). Part but not all of the hyperopia seen in young infants is probably due to the small eye artifact (Glickstein & Millodot 1970).

Astigmatism Early estimates of the cylindrical component of human infant optics were highly varied (see Banks 1980). The results of more recent studies using cycloplegic retinoscopy (Ingram & Barr 1979, Fulton et al 1980), near retinoscopy (Mohindra et al 1978, Gwiazda et al 1984b), and photorefraction (Howland et al 1978, Braddick et al 1979, Howland & Sayles 1984) have consistently indicated that the incidence of astigmatism is significantly higher in infants than adults, and remains high during most of the preschool years (Atkinson et al 1980, Mohindra & Held 1981, Gwiazda et al 1984b, Howland & Sayles 1984). M. Neuringer (personal communication) also reports a high incidence of astigmatism in infant monkeys. Much of the astigmatism is against-the-rule in both species (Fulton et al 1980, Dobson et al 1981, 1984, Gwiazda et al 1984b, Howland & Sayles 1984, M. Neuringer, personal communication).

It has been suggested that part of the high incidence of against-the-rule astigmatism seen in infants might be due to the fact that refractive error is measured along the optic rather than the visual axis of the eye (Banks 1980, London & Wick 1982). Nevertheless, the variability in the incidence of against-the-rule astigmatism among different populations (Atkinson & Braddick 1982c, 1983, Thornetal 1984) and the fact that there is no rapid decrease in the incidence of astigmatism that parallels the rapid decrease in the angle between the optic and visual axes during the first postnatal year (London & Wick 1982) suggest that some other factor must be responsible. Indeed, recent photokeratometric measurements by Howland (1982) have shown that an infant's astigmatism is highly correlated with the curvature of the infant's cornea.

Accommodation Recent reviews of the literature on accommodation in infants can be found in Banks & Salapatek (1983) and Aslin (1984). Although one-month-old human infants' accommodation changes from sleep to wakefulness (Haynes et al 1965), one-month-olds often do not accommodate appropriately to targets at different distances. Infants' accommodation is at least roughly appropriate to target distance by three to four months postnatal (Haynes et al 1965, Banks 1980, Brookman 1983, Braddick et al 1979). Similarly, infant monkeys accommodate poorly during the early postnatal weeks, but accommodative slopes near one are seen by five weeks postnatal (Howland et al 1982). The relative lack of accommodative change shown by very young infants probably arises from a large depth of field and an inability to detect blur rather than from a motor deficit (Owens & Held 1978, Braddick et al 1979, Green et al 1980).

It has been hypothesized that the location of the young infant's fixed accommodative distance is at the infant's position of accommodative rest, or dark focus (Owens & Held 1978, Braddick et al 1979, Green et al 1980). However, there are no measurements available for the location of the dark focus in very young infants of either species. For 3- to 12-month-old human infants, the dark focus has been reported to be at 70 cm (Aslin & Dobson 1983), a value similar to the 80 cm dark focus reported for adults (Leibowitz & Owens 1978), but quite different from the 19 cm fixed focal distance reported for one-month-olds (Haynes et al 1965).

Pupil Neonatal monkeys and humans show a pupillary response to light, including a consensual response (Hines 1942, Salapatek & Banks 1978, Mendelson 1982a), but few quantitative studies of the magnitude or time course of the pupillary response have been reported. In human infants, the pupil size of one-month-olds is 1 to 2 mm smaller than that of adults at all luminances, whereas the pupil size of two-month-olds is similar to that of adults at luminances above 7 cd/m^2 but smaller than that of adults at lower luminances (Salapatek & Banks 1978). Binocular summation is reported to be absent from the pupillary response of human infants until about four months postnatal (Birch & Held 1983, cf Shea et al 1985).

EYE MOVEMENTS Ultrasound technology has now advanced to the point where studies of fetal eye movements are possible (Birnholz 1981, Bots et al 1981, Prechtl & Nijhuis 1983). Aslin (1984) has recently reviewed the development of eye movements during the first postnatal year.

Saccades The ability to execute a saccade in the correct direction to a stimulus in the near periphery is present in the very young human infant (Tronick & Clanton 1971, Tronick 1972, Harris & MacFarlane 1974, Aslin & Salapatek 1975, Schneck et al 1982, 1985). However, eye movement recordings in one- to three-month-old infants have shown that localization of a peripheral target is accomplished by a series of hypometric saccades (Aslin & Salapatek 1975, Salapatek et al 1980), even when the infants are allowed to make head movements to accompany the eye movements (Goodkin 1980, Roucoux et al 1982, 1983, Regal et al 1983). The latency of the initial saccade following target presentation is significantly longer than that of adults (Aslin & Salapatek 1975, Regal et al 1983). In contrast, when infants view more complex stimuli, the frequency of hypometric saccades is very low, and infants' saccadic amplitudes and velocities are equivalent to those of adults (L. Hainline et al 1984b).

Smooth pursuit Although Kremenitzer et al (1979) found evidence of an occasional smooth pursuit movement in human newborns when target velocity was low, other researchers have found that infants do not show smooth pursuit until they reach six to eight weeks postnatal, but instead track moving objects with a series of saccades (reviewed by Aslin 1981, 1985). As infants get older, smooth pursuit movements become interspersed among saccades during tracking of target movement and, although the accuracy of pursuit depends on target velocity, the proportion of smooth pursuit eye movements increases with age (Aslin 1981, Roucoux et al 1982). Infant monkeys will visually track a moving object within the first few days after birth (Boothe et al 1982b, Mendelson 1982a), but eye movements are sometimes reported to be jerky rather than smooth (Hines 1942, Mowbray & Cadell 1962).

Optokinetic nystagmus (OKN)[a] A visually induced nystagmus can be elicited in both human and monkey infants soon after birth by moving large stripes through the visual field (Ordy et al 1962, 1965, Atkinson 1979, Boothe et al 1982b), and this response has been used by a number of researchers to study various aspects of pattern perception (reviewed by Dobson & Teller 1978, Boothe 1981, Aslin 1985). The response shown by very young infants is not identical to an adult OKN, however, in that the slow-phase velocity is more restricted than that of adults (Kremenitzer et al 1979) and the form of the response to vertical vs horizontal movement is somewhat different (Hainline et al 1984a).

Another immature feature of the OKN system is that monocular OKN can be elicited only in the temporal-to-nasal direction in human infants less than two months of age (Atkinson 1979, Atkinson & Braddick 1981), and in monkey infants less than two weeks of age (Atkinson 1979). Temporal-to-nasal OKN is reported to be more vigorous than nasal-to-temporal OKN in full-term human infants less than five months of age (Naegele & Held 1982) and in preterm infants less than six months corrected age (age from due date) (van

[a]Technical symbols and abbreviations: °, degrees of visual angle; ′, minutes of visual angle; cd/m^2, candles per square meter; CFF, critical flicker frequency; CSF, contrast sensitivity function; cy/deg, cycles per degree; dLGN, dorsal lateral geniculate nucleus; ERG, electroretinogram; OKN, optokinetic nystagmus; VEP, visual evoked potential; V′λ, spectral luminous efficiency function for scotopic vision.

Hof-van Duin & Mohn 1984). Furthermore, monocular optokinetic afternystagmus cannot be elicited by nasal-to-temporal movement until infants reach four to five months postnatal (Schor et al 1983).

Vergence Studies of the development of fusional and accommodative vergence have been reviewed by Aslin (1984). Fusional vergence in human infants less than four months of age is slower and of lesser magnitude than that of adults (Slater & Findlay 1975, Aslin 1977). Accommodative vergence, as indicated by appropriate vergence movements under monocular viewing conditions, is present by two months (Aslin & Jackson 1979). There are no reports on the development of vergence movements in monkeys.

Scan paths The large literature on infant scan paths in response to patterned stimuli has been reviewed recently by Banks & Salapatek (1983) and Haith (1980); see also Bronson (1982) and Hainline & Lemerise (1982). Mendelson (1982b) has conducted studies of visual scanning in infant monkeys.

Visuomotor A variety of visuomotor tasks such as visual tracking of large and small objects, visually guided reaching and placing, and responses to avoid an impending collision have been studied in both human (Greenman 1963, White 1971, Bower 1975, Peterson et al 1980, Yonas 1981) and monkey infants (Boothe et al 1982b, Mendelson 1982a,b). Neonates do not respond consistently on these tasks. Tasks that involve only orienting motor responses and coarse sensory capacities (e.g. tracking large objects) generally emerge earliest. Tasks requiring finer sensory processing (e.g. tracking small objects) or more complicated motor responses (e.g. visually guided reaching) emerge at progressively later ages. By six weeks in monkeys, and six months in humans, normal infants will ordinarily respond appropriately on a wide range of visuomotor tasks.

Anatomy and Physiology

ANATOMY

Retina The major anatomical development of retinal neurons and synapses takes place prenatally in both humans and monkeys (Mann 1964, Smelser et al 1974, Hollenberg & Spira 1973, Spira & Hollenberg 1973). The primary postnatal changes observed in the primate retina are associated with differentiation of the macular region. In the monkey, ganglion cell and bipolar layers have already thinned in the macular area at the time of birth, leaving a characteristic foveal depression similar in gross histological appearance to that of an adult (Samorajski et al 1965, Hendrickson & Kupfer 1976). However, the outer segments of foveal cone photoreceptors continue to increase in length during the early postnatal weeks, and receptor packing density increases by at least a factor of two in linear dimensions between birth and adulthood (Hendrickson & Kupfer 1976).

The human macula is even more immature at birth, with incomplete migration of the ganglion cell and inner nuclear layers out of the foveal region, the continued presence of the transient layer of Chievitz, and only a single layer of short, thick immature cones (Mann 1964, Abramov et al 1982, Hendrickson & Yuodelis 1984). The fovea remains immature until after 15 months, when cone outer segments are still less than half the length of adults' and the transient layer of Chievitz is still present. Even at 45 months, when the fovea is nearly adult-like, cone outer segments remain shorter than those of adults. Cone density increases, from 168/mm at 45 days postpartum to the adult level of 418/mm by 45 months (Hendrickson & C. Yuodelis 1984).

Dorsal lateral geniculate nucleus (dLGN) In humans, neurons of the parvocellular layers of the dLGN grow rapidly over the early postnatal months and reach adult size near the end of the first year. The neurons of the magnocellular layers show continued rapid growth throughout the first year, and adult size is not reached until at least 24 months postnatal (Hickey 1977). In monkeys, no significant growth has been reported in the parvocellular laminae after eight days, but significant growth occurs between eight days and adulthood in the magnocellular layers (Headon et al 1981b).

The most striking difference between mature and immature dLGN neurons is the presence of many spines and filopodia on the somas and dendrites of the immature neurons. The number of spines reaches a maximum at birth in the monkey and at four months postnatal in the human, and declines to adult levels by twelve weeks in monkey and nine months in man (Garey & Saini 1981, de Courten & Garey 1982, Garey & de Courten 1983). Few growth cones are seen postnatally in monkeys, but some dLGN neurons nevertheless continue to show postnatal increases in the numbers of secondary and tertiary branches (Garey & Saini 1981).

Striate cortex Striate cortex of a newborn monkey exhibits an adult-like pattern of lamination. The total number of neurons shows a postnatal decrease of 16% (O'Kusky & Colonnier 1982b). Dendritic branching patterns in monkey cortex appear adult-like for the most part at birth (Boothe et al 1979). In human cortex some differentiation of neurons and growth of dendrites occurs during the first few postnatal months (Takashima et al 1980).

The most notable feature of the infant primate visual cortex in both humans and monkeys is the presence of many more spines and synapses than are seen in the adult (Lund et al 1977, Boothe et al 1979, Mates & Lund 1983a–c, Takashima et al 1980, Huttenlocher et al 1982, Garey & de Courten 1983).

In monkeys a dense population of spines is found at early postnatal ages even on cell populations that do not have many spines in the adult (Lund et al 1977). Quantitative counts reveal that spine frequency in monkey cortex increases until eight weeks postnatal and then decreases to adult levels (Boothe et al 1979). This pruning back of spines follows different time courses for different neuronal populations. The number of spines on cortical cells in layers that receive primarily a parvocellular geniculate input decreases to adult levels sooner than does the number of spines on cells in layers receiving primarily a magnocellular input. Some neuronal populations are still above adult levels as late as 40 weeks after birth in monkeys (Boothe et al 1979).

O'Kusky & Colonnier (1982b) conducted quantitative counts of synapses in monkey striate cortex. They estimate that there are approximately 389 billion synapses at birth, rising to 741 billion at 24 weeks and then decreasing to 381 billion in the adult. Huttenlocher et al (1982) have shown a similar trend in humans, although with a somewhat different time course. They found a rapid increase in synaptic density occurring through eight months postnatal. Synaptic density then decreases, with the adult level, which is 60% of the density at eight months, being reached at about eleven years.

Electron microscopy of normal and Golgi-impregnated tissue has been used to show the relationship between developmental changes in spine numbers and synapses. Type 1 synaptic contacts are initially formed on dendritic shafts and are then carried out on spine outgrowths. During maturation some spines shrink and detach along with their synapses. Thus, changes in numbers of spines reflect changes in Type 1 synapses (Mates & Lund 1983a,b). Type 2 synapses form on cell somas, and they, too, first increase and then decrease in number during normal postnatal development (Mates & Lund 1983c).

Myelination Magoon & Robb (1981) examined development of myelin in human optic nerve and tract. They concluded that some fibers in the optic nerve are myelinated by term, but the amount of myelin about individual nerves increases dramatically up to two years of age and continues to increase less rapidly thereafter.

O'Kusky & Colonnier (1982a) examined the development of glial cells in monkey striate cortex. Oligodendrocytes show a tenfold increase from birth to maturity, with a corresponding increase in the density of myelinated axons. Total numbers of astrocytes and microglia do not change significantly postnatally.

Development of binocular pathways The cells of the dLGN have already segregated out into left- and right-eye laminae prior to birth in both monkeys (Rakic 1977a,b) and humans (Hitchcock & Hickey 1980). In monkey striate cortex the terminals from left- and right-eye dLGN laminae completely overlap prenatally until about three weeks prior to birth (Rakic 1976). These cortical inputs then begin to segregate out into ocular dominance columns. At the time of birth in monkeys ocular dominance columns can be demonstrated physiologically (Hubel et al 1977, LeVay et al 1980), anatomically (Rakic 1976, Hebel et al 1977, LeVay et al 1980), and with 2-deoxyglucose methods (Des Rosiers et al 1978). Segregation into mature columns having little or no overlap is complete by about six weeks after birth.

SINGLE UNIT PHYSIOLOGY

Receptive fields Little information is available about the development of the receptive field properties of single neurons in primates. No studies are available regarding the development of receptive fields of primate retinal ganglion cells, and studies of dLGN cells have been limited to studies of spatial resolution, reviewed below.

Despite the postnatal anatomical changes described above, there are reports that the general receptive field properties of cortical cells appear to be quite mature at birth. Wiesel & Hubel (1974) reported that the response characteristics of most cells in newborn macaques are normal by adult standards. Simple, complex, and hypercomplex receptive fields, with about the same range of orientation specificity as expected from adults, were found in newborn monkeys. Ocular dominance and sequences of orientation shifts were also reported to be orderly and adult-like. However, Blakemore & Vital-Durand (1981b) reported that cortical cells in newborn Patos monkeys were considerably less mature. They reported that many cells were not orientation selective and showed little low-spatial frequency falloff.

Spatial resolution Larger deficits have been found in studies of the development of spatial resolution in monkey dLGN cells and cortical cells in the region of the foveal projection (Blakemore & Vital-Durand 1981a,b, 1982). In the newborn, there is little apparent variation across the visual field in the spatial resolution characteristics of dLGN cells. There is little difference in spatial resolution between newborn and adult dLGN cells at eccentricities greater than 10 degrees. Thus, at the physiological level the development of acuity appears to involve primarily improvements in resolution of neurons in the central visual field. Between birth and adulthood there is a sevenfold improvement in the resolution of central dLGN neurons, from less than 5 cy/deg at birth to greater than 35 cy/deg in the adult. The time course of development extends over at least the first 30 postnatal weeks.

Blakemore & Vital-Durand (1981a,b, 1982) showed that contrast sensitivity functions (CSFs) for some cortical cells in adult monkeys show contrast sensitivity values greater than 70 and high frequency cutoffs greater than 30 cy/deg. In newborns, peak sensitivity is only about 10 and high frequency cutoffs for the best units are less than 5 cy/deg. There is thus a large developmental improvement in the CSFs of neurons in the dLGN and cortex that respond to the central visual field, with both sensitivity and acuity improving by factors of more than six. Development of spatial resolution in these dLGN and cortical neurons closely parallels the development of behaviorally measured acuity (described below) (Blakemore & Vital-Durand 1981a).

ELECTRORETINOGRAM (ERG) AND VISUAL EVOKED POTENTIAL (VEP) ERG and VEP techniques have been used to study a variety of aspects of visual development. These studies are reviewed below, with the behavioral studies of similar visual functions.

Visual Function

ABSOLUTE AND INCREMENT THRESHOLDS

Absolute Threshold The average absolute thresholds of one- and three-month-old humans are elevated by about 1.7 and 1.0 log unit, respectively, above those of adults when all groups are tested with large (17°) test fields. Calculations based on these data show that the isomerization of between 10,000 and 100,000 photopigment molecules in a 17° area is sufficient to generate a visual response in one-month-olds (Powers et al 1981). The dark-adapted spectral sensitivities of one- and three-month-old infants, tested behaviorally, agree well with $V'\lambda$ (Powers et al 1981). By VEP measures, there is good agreement with $V'\lambda$ between about 430 and 600 nm (Werner 1982, cf Teller 1982). Thus, there can be little doubt that a rhodopsin-based receptor system is active in young infants, and that it is the most sensitive system at absolute threshold over most or all of the visible spectrum.

Spatial summation At absolute threshold one- and three-month-olds exhibit complete spatial summation over areas as large as 9° and 5.5°, respectively, while adult controls demonstrate complete summation over only about 2.5° (Hamer & Schneck 1984). Summation areas have also been studied at controlled eccentricities between 9° and 36°. One-month-olds show complete summation over 17° or more at all eccentricities. The fact that summation areas are larger in infants than in adults implies that the difference between infant and adult absolute thresholds will vary with field size; being, for example, about 2.5 log units for 3° fields and 1.5 log units for 17° fields in one-month-olds (Schneck et al 1982, 1985).

Adaptation Photopic background fields elevate the thresholds of young infants (Peeples & Teller 1978, Pulos et al 1980). Systematic increases of background luminance over the scotopic and mesopic range lead to systematic elevations of increment thresholds. When test fields of 6° to 10° are used, the slopes of infant Weber functions increase from about 0.6 in young infants to 0.95 in three- to four-month-olds (Hansen & Fulton 1981, Dannemiller & Banks 1983). Since young infants have large summation areas, it seems likely that 6° to 10° stimuli are effectively "small" for young infants, and that the age-related changes in slope seen with these fields can be attributed to age-related changes in spatial summation (cf Barlow 1972). Larger fields lead to slopes closer to one, at least in two-month-olds under rod isolation conditions (Brown 1984).

The ERG has also been used to study the development of adaptation processes in human infants (Fulton & Hansen 1982).

COLOR VISION Infant color vision has been reviewed recently by Teller & Bornstein (1985).

Photopic spectral sensitivity VEP studies of photopic spectral sensitivity in human adults and infants show reasonable agreement between the two ages over the mid- and long-wavelength spectral regions. The sensitivities of young infants in the short-wavelength region tend to be elevated relative to those of adults, but this difference disappears by four months postnatal (Dobson 1976, Moskowitz-Cook 1979). Behavioral testing indicates good agreement between two- to three-month-old infants and adults throughout the spectrum (Peeples & Teller 1978). The absolute shapes of photopic spectral sensitivity curves vary between studies, in reasonable accord with variations in stimulus parameters.

Chromatic adaptation The shapes of two-month-old infants' spectral sensitivity curves between 460 and 560 nm vary with chromatic adaptation, indicating the presence of at least two receptor types active in this spectral region. However, most two-month-olds tested with short wavelength test lights against a broadband "yellow" background of about 1.5 log cd/m^2 exhibited greater sensitivity at 500 than at 460 nm; thus, by this test, they failed to demonstrate the presence of short wavelength sensitive cones. Short wavelength sensitive cones were evident in the response of most three-month-olds and adults (Pulos et al 1980).

Wavelength discrimination Two-month-old infants can discriminate lights of many but not all wavelength compositions from white light (Peeples & Teller 1975, Teller et al 1978c). According to classical color theory (e.g. Boynton 1979), two-month-olds must therefore be at least dichromats, but not necessarily trichomats. Infant macaque monkeys discriminate all wavelengths of light from white light at the earliest ages tested to date, six to eight weeks, and must therefore be trichromats by that age (Boothe et al 1975).

Adaptations of the classical tests for dichromacy have also been used on human infants. Two-month-olds can discriminate between members of a tritan pair, 547 vs 420 nm (Varner et al 1984). Virtually all three-month-olds and many two-month-olds can make Rayleigh discriminations (Bornstein 1976, Hamer et al 1982). There is therefore strong evidence that two- to three-month-olds have all three of the adult cone types, although the possibilities of rod involvement or anomalous juvenile pigments have not been ruled out.

Test field size influences both luminance and chromatic (Rayleigh) discriminations in infants (Packer et al 1985). This and the overall pattern of infant color vision data have been used to argue that the discrimination failures seen in very young infants are probably due to immaturities of spatial processing or postreceptoral chromatic mechanisms rather than to any absence or anomaly of receptor types (Hamer et al 1982, Packer et al 1985). However, no studies specifically designed to probe postreceptoral chromatic processing in infant subjects have yet been reported.

SPATIAL RESOLUTION

Spatial summation Studies of spatial summation at absolute threshold (Hamer & Schneck 1984, Schneck et al 1985) and the effects of field size on chromatic discrimination (Packer et al 1985) are described above.

Acuity: behavioral measures In the developmental literature, grating acuity is by far the most intensively studied of all visual functions. The extensive early literature on human infants is reviewed by Dobson & Teller (1978), and references to the early monkey literature may be found in Teller et al (1978b) and Boothe (1981).

Behavioral studies of both humans and monkeys consistently show that grating acuity is about 0.5 to 1 cy/deg in very young infants—more than six octaves, or two orders of magnitude, below the 30 to 60 cy/deg shown by adults of both species. Grating acuity improves slowly and monotonically over the early postnatal months in both species (Fantz et al 1962, Ordy et al 1964, 1965, Teller et al 1974, 1978b, Gwiazda et al 1978, Allen 1979, Goldblatt et al 1980, Lee & Boothe 1981, Neuringer et al 1984), and reaches adult levels at three to five years postnatal in children (Mayer & Dobson 1980, 1982, Birch et al 1983c) and about one year in monkeys (Boothe 1981, 1984). The time courses of development of grating acuity in humans and monkeys can be made to superimpose approximately by multiplying the monkey age axis by a factor of four (cf Teller 1981). Thus, the grating acuity data provide the strongest available evidence for the 4:1 rule in the visual literature. As a convenient mnemonic device it can be stated that infant acuity, expressed in cy/deg, is approximately equal to age, in months for human infants and in weeks for monkey infants.

The finding that grating acuity does not reach adult levels until children reach about five years of age is in agreement with studies in which Allen cards, the "E" game, or Snellen letters have been used to measure acuity in preschool children (reviewed in Friendly 1978, Mayer & Dobson 1982). When eye charts with "crowded" optotypes are used to test acuity, acuity does not become adult-like until about age ten (Atkinson & Braddick 1982a, Hohmann & Haase 1982). Optotypes have not been used to study the development of acuity in infant monkeys.

Excellent agreement of average grating acuities at a given age is usually obtained, across laboratories and variations of technique, when adequate sample sizes are used. Population standard deviations of about 0.5 octave have been reported (Allen 1979, Dobson 1980, Gwiazda et al 1980b, Mayer & Dobson 1982). These values may, however, be overestimates, as they include the effects of measurement error as well as true individual differences. Age norms solid enough to be useful in the evaluation of human infants at risk for vision disorders are rapidly becoming available.

Several attempts have been made to develop rapid techniques for clinical assessment of infant acuity (Dobson et al 1978a, Gwiazda et al 1980b, Mayer et al 1982). Controversy has arisen in this field concerning the detailed effects of the shapes of infant psychometric functions, number of trials, stimulus presentation strategies, and scoring techniques on the bias and variability of acuity estimates (Held et al 1979, Teller et al 1982, Banks et al 1982, Wolfe et al 1983, Teller 1983b, S. McKee et al, personal communication).

Acuity: visual evoked potential (VEP) measures In VEP studies of infant acuity, acuity has often been estimated by measuring VEP amplitude as a function of either grating frequency or grating contrast, and extrapolating to zero VEP amplitude. Such studies have indicated that human infants' acuity improves rapidly in the early postnatal months, and that acuity within one octave of adult levels is reached by six months to one year (Marg et al 1976, Sokol & Dobson 76, Pirchio et al 1978, Sokol 1978, T. Norcia and C. Tyler, personal communication). Possible causes of the discrepancy between behavioral and VEP acuity estimates have been discussed by Dobson & Teller (1978) and Sokol (1978), but the discrepancy is not fully resolved.

Several aspects of the pattern-elicited VEP do not mature until early childhood. There is a negative peak that appears in the human pattern VEP at five months, and does not reach adult amplitude or latency until six years (Spekreijse 1978, 1983, De Vries-Khoe & Spekreijse 1982). Also, VEP latencies to high spatial frequency stimuli do not decrease to adult values until after five years (Sokok & Jones 1979, Moskowitz & Sokol 1980, 1983).

Acuity: electroretinogram (ERG) measures Although there have been a number of studies of the development of flash-elicited ERGs in human infants (reviewed by Maurer 1975), it is only recently that ERG studies of human infants' spatial vision have been carried out, and preliminary reports are inconsistent. Odom et al (1983a,b) found that 3.5-month-olds showed ERG resolution thresholds (30 cy/deg) as good as those of adults, whereas their VEP resolution thresholds (8.5 cy/deg) were poorer than those of adults. Fiorentini et al (1983a), on the other hand, found that the infant's ERG acuity was poorer than that of adults, and ERG acuity development paralleled VEP acuity development between two and six months.

Acuity: optical vs neural limits It has been argued that the poor acuity seen in young infants cannot be attributed primarily to either poor optical quality or receptor sampling frequency (Boothe 1983). The excellent correspondence between acuity and the cutoff frequencies of dLGN neurons serving the central visual field suggests that much of the normal development of acuity depends on the maturation of neural elements that lie in the retina or dLGN and receive projections from the central visual field (Blakemore & Vital-Durand 1981a,b, 1982).

Contrast sensitivity Under the best conditions, newborns respond to contrasts of about 10% (Adams 1982), as do preterm infants tested at 37 weeks gestational age (Shepard 1985). Infants two to three months old respond to contrasts of 5% to 8% (Peeples & Teller 1975, Atkinson et al 1977a,b, Banks & Salapatek 1978).

Contrast sensitivity functions (CSFs) With the exception of Harris et al's (1976) study of one six-month-old, behavioral studies of CSFs in human infants have been limited to one- to three-month-olds. One-month-olds fail to show a low frequency roll-off, and their overall sensitivity is greatly reduced. The shape of the CSF in two- and three-month-olds is similar to that of adults, but the overall sensitivity remains greatly reduced and the entire function is shifted toward lower spatial frequencies (Atkinson et al 1974, 1977b, Banks & Salapatek 1976, 1978, 1981), with cutoff frequencies consistent with the grating acuity norms cited above. VEP results indicate that newborns do not show a low-frequency roll-off (Atkinson et al 1979). VEP studies in which contrast sensitivity has been estimated by measuring VEP amplitude as a function of grating contrast and extrapolating to zero VEP amplitude have shown that infants' CSFs are nearly adult-like by seven months postnatal (Harris et al 1976, Pirchio et al 1978).

The development of the CSF has also been studied in preschool- and school-age children. Although the shape of the CSF is adult-like at these ages, the overall sensitivity is less than that of adults until children reach six to 12 years of age, perhaps due to differences in the response strategies of children vs adults (Derefeldt et al 1979, Beazley et al 1980, Atkinson et al 1981, Bradley & Freeman 1982, Abramov et al 1985).

CSFs have been obtained longitudinally from infant monkeys from five to 60 weeks postnatal (Boothe et al 1980, Boothe 1984, and R. Boothe, personal communication). Monotonic increases in sensitivity are seen at all spatial frequencies, but sensitivities to different frequencies develop at different rates, with lower frequencies asymptoting to adult levels sooner than do higher frequencies. For example, sensitivity to 1–5 cy/deg gratings has reached adult levels by about 20 weeks, whereas sensitivity to gratings finer than 15 cy/deg is still improving at 40 weeks. This fact results in developmental changes in the shape of the CSF. Qualitatively, these changes can be described as shifts of the CSF upward (increased sensitivity to contrast) and to the right (responding to higher spatial frequencies). CSF peak frequency increases from an average of 1 to 2 cy/deg at ten weeks to an adult level of about 5 cy/deg by 20 weeks. Cutoff frequency increases from about 10 cy/deg at ten weeks to adult values of greater than 30 cy/deg by 30 to 50 weeks.

If the 4:1 rule is used, the data from one- to three-month-old human infants, six-week to adult monkeys, and young children fit together in a reasonably consistent sequence (cf Teller & Boothe 1980) descriptive of the entire range of development from birth through adulthood.

Infant CSFs can be used to predict infants' responses to a variety of patterns, including defocused faces (Atkinson et al 1977c), and gratings of varying regularity and duty cycle (Banks & Salapatek 1981, 1983, Banks & Stephens 1982).

Single vs multiple channels The question of single vs multiple spatial frequency channels has been addressed in three recent studies by means of both masking (Fiorentini et al 1983b, Banks et al 1985) and compound grating (Banks et al 1985) paradigms. The data agree that three-month-olds demonstrate the presence of more than a single channel. Six-week-olds demonstrate only a single channel behaviorally with both paradigms (Banks et al 1985), but a six-week-old tested with VEP measures and a masking paradigm demonstrated multiple channels (Fiorentini et al 1983b). The finding that three-month-olds show contrast constancy but six-week-olds do not supports the hypothesis that three-month-olds have multiple spatial frequency channels but six-week-olds do not (Stephens & Banks 1985).

Stimulus orientation Although human infants can discriminate among stimulus orientations during the first two postnatal months (McGurk 1972, Leehey et al 1975, Slater & Sykes 1977, Maurer & Martello 1980), there is no evidence for variation in acuity with orientation in non-astigmatic infants less than six months of age (Teller et al 1974, Gwiazda et al 1978, 1980a, Dobson et al 1978b). Many older infants (Gwiazda et al 1978, 1980a, Birch et al 1983c) and young children (Mayer 1977, 1983a, Beazley et al 1982) show slightly better acuity for vertical and horizontal gratings than for oblique gratings, as do adults; however, there is considerable variability among both children and adults (Birch et al 1983c, Mayer 1983a, Gwiazda et al 1984a). These variations in acuity with orientation may be the result of differential amounts of exposure to different stimulus orientations (Mayer 1983b) or the individual's past or present refractive error (see section on astigmatism, below).

Stimulus luminance For luminances above 10 cd/m^2, human infant acuity does not vary with changes in luminance (Dobson et al 1983). Reductions in luminance below 10 cd/m^2 result in reduced acuity (Dobson 1983, Dobson et al 1983, Dobson & Maier 1984) and depressed CSFs (Fiorentini et al 1980, cf Banks & Salapatek 1981). The reduction in contrast sensitivity shown by four- and six-month-old infants from photopic to mesopic levels is less than that shown by adults (Fiorentini et al 1980).

Target size One- and two-month-old humans show better acuity for 19° than 10° targets, but this difference disappears by three months (Atkinson et al 1983, cf Dobson & Teller 1978). Spinelli et al (1983) reported that VEP amplitude increased with target size in a three-month-old infant tested with mid to low spatial frequencies, but there was no amplitude change when the target size of a stimulus in the high spatial frequency range for the infant was increased from 2° to 14°. They argued from this finding that sensitivity to higher spatial frequencies is confined to a differentiated macular region at this age.

Stimulus distance One- to five-month-old humans show less fixation of a far target than a near target (McKenzie & Day 1972, de Schonen et al 1978). However, there is no evidence that acuity or contrast sensitivity varies as a function of distance in young infants (Fantz et al 1962, Salapatek et al 1976, Atkinson et al 1977b), presumably because of the young infant's large depth of field (Salapatek et al 1976, Braddick et al 1979, Banks 1980, Green et al 1980).

Stimulus movement Although human infants show a fixation preference for moving stimuli over stationary stimuli (Fantz & Nevis 1967, Carpenter 1974, Volkmann & Dobson 1976), there is no evidence that acuity or contrast sensitivity is better for moving than for stationary gratings. Atkinson et al (1977a) found no difference in one- to three-month-olds' contrast sensitivity for drifting vs stationary gratings, and Dobson et al (1978b) found no difference in two-month-olds' acuity for stationary gratings vs phase-alternated checkerboards. However, it is noteworthy that VEP amplitude vs spatial frequency functions, which are often used as the basis of VEP acuity measurements, do vary as a function of the rate at which the stimuli are phase-alternated (Moskowitz & Sokol 1980).

Monocular vs binocular testing Monocular acuity measured with preferential looking procedures tends to be one-half to one octave poorer than acuity measured binocularly (Atkinson et al 1982b, Mayer et al 1982, Dobson 1983).

Gestational age Data on acuity development in infants born prior to term were reviewed recently by van Hof-van Duin & Mohn (1984). In general, preterm infants' acuity is similar to, although perhaps slightly poorer than, that of full-term infants of the same gestational age; however, it is clearly poorer than that of full-term infants of the same postnatal age (Miranda 1970, Fantz et al 1975, Dobson et al 1980, Dubowitz et al 1980, 1983, Baraldi et al 1981, van Hof-van Duin et al 1983, van Hof-van Duin & Mohn 1984). A similar result has been shown in infant monkeys of known gestational ages (Lee & Boothe 1981).

Vernier acuity Two studies of human infant vernier acuity have been reported recently. Using moving vernier targets and a sound synchronized with the motion of the target, Shimojo et al (1984) found a three-octave increase in vernier resolution, from about 16′ at two months to about 2′ at eight months. Vernier and grating acuities were similar in absolute values at two months but vernier acuity exceeded grating acuity by more than an octave by eight months. In contrast, Manny & Klein (1984), using stationary targets, found considerably lower values for both vernier and grating acuity. Both forms of acuity improved by about three octaves between birth and six months, with vernier acuity exceeding grating acuity by roughly one octave at all ages. The discrepancies between the studies remain to be resolved. Nonetheless, it is clear that vernier acuities improve considerably in the early months, and that they exceed grating acuity at a few months postnatal if not at birth.

TEMPORAL RESOLUTION Critical flicker frequencies (CFFs) of one-, two-, and three-month-old human infants have been shown to be about 40, 50 and 52 Hz, respectively; CFFs of adult controls were 53–54 Hz in the same study. A reduction in luminance to 3.4 cd/m^2 reduced CFF by 10 to 15 Hz for both two-month-olds and adults (Regal 1981). Thus, unlike visual acuity, temporal resolution appears to mature rapidly and to be adult-like within about three months postnatal. Recent data on infant temporal CSFs indicate that six- and twelve-week-old infants are less sensitive than are adults at low and middle temporal frequencies, and six-week-olds do not show the low-frequency fall-off in sensitivity that is characteristic of the functions of adults and twelve-week-olds (Hartmann & Banks 1984).

BINOCULARITY A recent review of infant binocularity has been presented by Aslin & Dumais (1980).

Binocular disparity Three techniques have been developed recently to study infants' responsiveness to binocular disparity: behavioral responses to line stereograms (Held et al 1980) and to random dot stereograms (Fox et al 1980, Shea et al 1980) and VEPs to random dot stereograms (Petrig et al 1981). Data from the three techniques agree well (cf Teller 1982) in showing that few infants respond to disparity cues at two months, but virtually all infants respond by five to seven months. Extremely rapid onsets of stereopsis in longitudinally tested infants have been reported, with a decrease in the disparity threshold from 58′ to 1′ occurring within a very few weeks' time. Responsiveness to crossed disparity generally precedes responsiveness to uncrossed disparity by two to six weeks (Held et al 1980, Birch et al 1982).

The onset of responsiveness to disparity in repeating bar patterns is the same for small (9°) as for large (21°) fields (Birch et al 1983b). The large fields allow multiple opportunities for fusion even in the presence of vergence errors up to about 25 prism diopters. This and other arguments are used to support the conclusion that the onset of responsiveness to disparity should be attributed to the development of sensory mechanisms (e.g. disparity-tuned cortical cells) rather than to the refinement of vergence control (Birch et al 1983b, Held 1984).

Binocular function The onset of VEPs to the initiation and termination of binocular correlation (Braddick et al 1980), binocular summation of the pupillary response (Birch & Held 1983), aversion to rivalrous stimuli (Birch et al 1983a), and binocular symmetry of the OKN response (Atkinson 1979, Naegele & Held 1982) also occur during the third to sixth month postnatal (Birch 1983). Held (1984) has compared the developmental time courses of these binocular functions to the segregation of cortical ocular dominance columns seen in monkeys (LeVay et al 1980). He suggests that the development of binocular function may occur at the time that left-eye and right-eye inputs become segregated in cortical layer IV.

Summary of Normal Development

One of the most striking aspects of normal visual development to emerge from this review is the level of immaturity of infant primate visual systems. Development is more than just growth, and infants are not just little adults. Different visual functions emerge in different age ranges, and with different time courses. On the human time scale, flicker sensitivity appears to be adult-like by three months postnatal, stereopsis and other binocular functions emerge between three and seven months, and acuity and contrast sensitivity undergo a long, slow development that is not complete until several years after birth.

These behavioral changes are accompanied by anatomical and physiological changes that are just as complex and impressive. A massive postnatal remodeling of synaptic connections occurs between cortical neurons. Physiological properties such as the spatial resolution of single units in the dLGN improve by a factor of six or more. The challenge of this field is to continue the exploration of the dependence of visual function on visual substrate in the developmental context.

VISUAL DEPRIVATION AND AMBLYOPIA

Overview

Given the level of immaturity of the visual system at birth, the amount of postnatal development that is required to reach adult levels of functioning, and the ample evidence of the importance of the visual environment to visual development (Wiesel 1982), it is remarkable that vision develops normally in most cases. However, there are a variety of human clinical conditions in which vision does not develop normally. The most prevalent outcome of these conditions is amblyopia, commonly referred to as "lazy eye." There are several forms of amblyopia, but they all have in common the symptoms of abnormal binocular vision, poor spatial vision (acuity and contrast sensitivity), or both. Von Noorden (1980) estimates that over 2% of the human population is afflicted with amblyopia, and that amblyopia is responsible for loss of vision in more people than all ocular diseases and trauma put together.

Most forms of amblyopia are associated with ocular problems that produce a disruption of the normal visual input. From the most to the least severe, these problems include dense cataracts, ptosis, and aphakia, which can lead to a complete loss of spatial pattern in the retinal image; optical defocus, which removes high spatial frequencies from the retinal image; and strabismus, which deprives the visual system of simultaneous, corresponding images from the two eyes. Within each of these categories, the severity of the disruption is also correlated with the degree of imbalance of the sensory input to the two eyes. Conditions of severe imbalance in which only one eye is deprived typically cause the most severe effects, whereas if both eyes are equally deprived the resulting deficits are less severe.

All of these forms of deprivation have now been mimicked, more or less closely, by deprivation experiments in primates, and all produce visual losses that mimic, more or less exactly, the spatial vision and binocularity deficits seen in human amblyopes. Sensitive periods for many of these deprivation conditions have been explored in at least a preliminary fashion. Furthermore, modern developmental studies of infants of both species allow one to trace the onsets of visual deficits in naturally occurring or artificially imposed forms of visual deprivation. A large literature in support of the hypothesis that visual deprivation during an early sensitive period is one of the primary factors responsible for producing many types of amblyopia has now accumulated from such sources (von Noorden 1980, Vaegan et al 1981, Wiesel 1982, Odom 1983, Teller 1983a).

The various forms of visual deprivation are reviewed below, in the order from most to least severe. For each form of deprivation, the clinical, animal, and developmental literatures are reviewed together, as are the effects of deprivation upon visual function and upon visual substrate. In reviewing visual function, we have attempted to distinguish clearly between the two separate visual functions—spatial vision and binocularity—commonly lumped together under the title of amblyopia, in order to emphasize the possibility that different functional losses may be caused by different distortions of the underlying anatomical and physiological processing.

Due to space limitations we include only the studies of human and monkey deprivation that seem directly related to the deficits in spatial vision and binocularity found in human amblyopia. Several other important topics in abnormal development have been omitted. These include induced myopia in monkeys (e.g. Young 1970, Wiesel & Raviola 1977, 1979, von Noorden & Crawford 1978a, Thorn et al 1982) and humans (Young 1970, Robb 1977a, O'Leary & Millodot 1979, Hoyt et al 1981, Johnson et al 1982), the effects of early lesions to the visual system (e.g. Dineen & Hendrickson 1981, Dineen et al 1982a,b, Hendrickson & Dineen 1982), and the effects of teratogens (Merigan & Weiss 1980), nutritional deficits (e.g. Malinow et al 1980, Neuringer et al 1984), and light damage (see December 1980 issue of *Vision Research*). We have also neglected the effects of deprivation on spectral sensitivity (Harwerth et al 1981), temporal modulation sensitivity (Harwerth et al 1983a), and the accommodative control system (Kiorpes & Boothe 1984, Smith & Harwerth 1984; see also Otto & Safra 1978, Hokoda & Ciuffreda 1982), as well as the extensive clinical literature on the effects of various treatments on amblyopia in humans (reviewed by Birnbaum et al 1977, Flynn & Cassady 1978).

Pattern Deprivation

In human patients, pattern deprivation can come about from conditions such as ptosis, uncorrected aphakia following cataract extraction, or the use of patching for therapeutic reasons. Pattern deprivation has been produced experimentally in monkeys by dark rearing, surgically removing the lens from the eye, or suturing the eyelids closed. Infant humans or monkeys who experience pattern deprivation to only one eye often show a profound loss of spatial vision in the deprived eye. Binocular deprivation is usually less severe.

MONOCULAR PATTERN DEPRIVATION

Effects on spatial vision In humans, Anderson & Baumgartner (1980) have shown that histories of congenital and early unilateral ptosis can be found in some cases of otherwise unexplained monocular visual acuity deficits. However, most cases of ptosis are accompanied by strabismus, anisometropia, or ocular disease, so that it is unclear to what extent the reduced acuity is attributable to the ptosis itself.

Awaya et al (1973, 1976, 1979, 1980) have shown that long-lasting visual acuity deficits sometimes occur in an eye subjected to only one week of patching following entropion surgery in infancy. When both eyes are treated in sequence, the second-treated eye suffers the more. When monocular patching lasts longer than one week, the proportion of infants and young children who show acuity deficits increases and the prognosis for good acuity in the deprived eye decreases (Awaya et al 1979, 1980, von Noorden 1981b). Von Noorden (1981b) has reported that several months of monocular occlusion occurring as late as the age of five years can cause uncorrectable acuity loss.

In monkeys, the effects of monocular pattern deprivation, and the sensitive periods for its effects upon spatial vision, have been investigated in several laboratories. Monkeys reared with monocular lid-suture or aphakia instituted during the first twelve weeks after birth often have extremely poor acuity and contrast sensitivity when the deprived eye is tested (von Noorden et al 1970, von Noorden 1973a, Hendrickson et al 1977, Harwerth et al 1981, 1983a, Boothe et al 1984). Even periods of monocular lid-suture as short as two weeks instituted during the period of maximum sensitivity are sufficient to cause deficits in spatial vision (von Noorden 1973a, Harwerth et al 1983a). However, Wiesel (1982) reports one case of a monkey that had one eye closed for nine days beginning at three weeks of age that showed no acuity loss. Lid-suture instituted at ages older than twelve weeks does not produce acuity deficits (von Noorden 1973a, Hubel et al 1977).

Reports of recovery from the original deprivation effects are mixed. Von Noorden (1973a) reports little or no recovery of pattern vision in the deprived eye of a lid-sutured monkey even after extended periods of reverse suture, but Hendrickson et al (1977) found some recovery following several months of reverse suture. Hendrickson et al (1977) also found that placing a central lesion in the nondeprived retina speeded up recovery of acuity in the deprived eye. But Harwerth et al (1984) found no recovery of spatial vision in lid-sutured monkeys following enucleation of the non-deprived eye.

Effects on binocularity Awaya et al (1973, 1976, 1979) reported that all patients who showed acuity deficits following short-term patching also showed a lack of binocularity, with eccentric fixation and no simultaneous perception.

Harwerth et al (1983a) report that binocular summation of CSFs is absent in monkeys raised with monocular lid-suture.

Anatomical and physiological studies Several studies of anatomical and physiological changes in the visual systems of monocularly lid-sutured monkeys have been conducted. No effects have been found in overall numbers of synapses in monkey retina (von Noorden 1973a), dLGN (Wilson & Hendrickson 1981), or cortex (von Noorden 1973a, O'Kusky & Colonnier 1982b). However, other clear effects have been found.

Retina It has been reported that monocular lid-suture lasting 24 months, but not shorter periods, produces statistically significant decreases in size and density of parafoveal retinal ganglion cells (von Noorden et al 1977).

dLGN anatomy Cells in the layers of the dLGN that receive input from the deprived eye are smaller than cells in layers receiving input from the non-deprived eye. Cells in both parvocellular and magnocellular subdivisions are affected. This result has been found following monocular lid-suture (Headon & Powell 1973, Headon et al 1982, LeVay et al 1980, Vital-Durand et al 1978, von Noorden 1973b, von Noorden & Middleditch 1975, von Noorden & Crawford 1978b) and monocular occlusion that blocked out all light (von Noorden & Crawford 1981a), and in monocularly aphakic monkeys (von Noorden & Crawford 1977).

The interlaminar cell size differences in the dLGN are apparent following deprivation instituted during the first 12 weeks after birth (von Noorden & Crawford 1978b, Hubel et al 1977,LeVay et al 1980, Vital-Durand et al 1978). Long-term monocular lid-suture instituted at ages older than 12 weeks leads to shrinkage in all dLGN layers (Headon et al 1981a). This shrinkage is not apparent when making within-animal comparisons because there are no interlaminar differences.

Reverse suture during the first eight postnatal weeks can reverse the differences in dLGN cell size (Vital-Durand et al 1978, Garey & Vital-Durand 1981). Simply reopening the deprived eye has little or no effect.

dLGN physiology Hubel et al (1977) reported that the receptive fields in dLGN cells appear normal following monocular lid-suture, and Blakemore & Vital-Durand (1981a) found that lid-suture for as long as seven months had no obvious effects on the spatial resolution of dLGN neurons.

Cortex In striate cortex most of the studies performed to date of anatomical and physiological changes following monocular lid-suture have been concentrated on the topic of ocular dominance and binocularity. Changes in the ocular dominance system can be demonstrated most clearly in layer IV of the striate cortex. Following monocular lid-suture, deprived eye columns decrease in width while those from the nondeprived eye increase (Hubel et al 1977, Blakemore et al 1978, LeVay et al 1980, Swindale et al 1981).

Recordings from single units in monocularly lid-sutured animals reveal that there are few binocularly driven cells, and that most of the monocular neurons can be driven only by stimuli delivered to the non-deprived eye (Baker et al 1974, Crawford et al 1975, Hubel et al 1977, Blakemore et al 1978). Von Noorden & Crawford (1981a) found the same results following monocular occlusion. There are no reports of the spatial resolution properties of cortical cells in monocularly lid-sutured monkeys. Blakemore & Vital-Durand (1981a) report that spatial resolution could not be measured in cortical cells from the deprived eyes of their animals because of the scarcity and poor response properties of these cells.

Baker et al (1974) found changes in ocular dominance in prestriate cortex similar to the changes reported for striate cortex.

The sensitive period during which layer IV ocular dominance columns are susceptible to monocular lid-suture is relatively short, spanning the first ten weeks, with greater sensitivity during the first six weeks. Physiological ocular dominance of single cortical neurons outside of layer IV can be affected during a much longer period, extending to at least one year, but not to adulthood (LeVay et al 1980).

Reverse suture at three or four weeks will cause true reversal of column sizes in layer IV of the cortex, such that the originally deprived eye columns are now larger (LeVay et al 1980, Swindale et al 1981). Reverse suture at six weeks of age will produce a recovery of the columns to normal widths, but not a complete reversal. Reverse suture at one year of age does not lead to any recovery. No change in the area occupied by the deprived eye is found at any age if the deprived eye is simply opened rather than reverse suturing the eyes. LeVay et al (1980) report that the parvocellular inputs to the cortex remain susceptible to reverse suture longer than the magnocellular inputs. Swindle et al (1981) failed to find such differences between sublayers of layer IV that receive primarily parvocellular or magnocellular input.

Binocular competition A number of lines of evidence suggest that the changes in anatomical and physiological ocular dominance columns caused by monocular deprivation are related in some way to competition between pathways or signals from the two eyes. Von Noorden et al (1976) showed that the adverse effect of unilateral visual deprivation on dLGN cell growth was absent in those portions of deprived dLGN laminae that were located opposite a degenerated patch of nondeprived dLGN. The binocular segment of the monkey dLGN is more susceptible to the effects of monocular deprivation than is the monocular segment (von Noorden & Middleditch 1975), although long-term deprivation may obliterate the difference. Rakic (1981) found that if one eye is removed prenatally, the dLGN fails to segregate into left-eye and right-eye laminae, and cortical ocular dominance columns fail to develop. Hubel et al (1977) note that it is only within the regions of cortex receiving binocular input that input from the deprived eye is reduced.

The first place in the visual system where binocularity competitive interactions are likely to occur is layer IV of visual cortex, where the left- and right-eye dLGN terminals converge on common postsynaptic cells (Hubel et al 1977). There is a close relationship between the extent of cortical layer IVC innervated by the dLGN and cell soma size in the dLGN (Vital-Durand et al 1978, LeVay et al 1980, Garey & Vital-Durand 1981, Swindale et al 1981). For this reason the interlaminar differences in dLGN cell soma size, reviewed above, are generally considered to be secondary to the relative loss of terminal fields in the cortex by cells in the deprived layers of dLGN.

The sensitive period for changing physiological ocular dominance outside of layer IV in the cortex extends longer than the sensitive period for changing ocular dominance columns in layer IV, or for producing laminar differences in dLGN cell size (von Noorden & Crawford 1978b, 1979, Blakemore et al 1978, 1981, LeVay et al 1980). This result suggests that two stages of competitive processes may be needed to account for the effects of monocular lid-suture on cortical development. dLGN afferents to the cortex first compete for cortical cells in layer IV, possibly by rules along the lines suggested by Hubel et al (1977), LeVay et al (1980), or Swindale (1980). Axons from these first order cortical cells then compete for synaptic connections to higher order cortical cells, possibly by rules along the lines suggested by Hebb (1949) and Changeaux & Danchin (1976).

Summary The effects of monocular pattern deprivation on behavioral function, anatomy, and physiology are severe. Behaviorally, spatial resolution is so poor that in many cases it cannot even be tested, and binocular functions such as stereopsis and binocular summation are often completely absent. Anatomical changes are seen in the retina and dLGN, and in layer IV of the cortex. These changes are seen in both parvocellular and magnocellular central pathways. Physiological responses seem little affected up to the level of the dLGN, but at the cortex few neurons can be driven effectively by the deprived eye.

The fact that acuity and binocularity are disrupted behaviorally and that anatomical and physiological ocular dominance pathways are both affected has led to a common assumption that these effects are causally related. However, exceptions have been noted that suggest caution in drawing strong inferences from physiology to behavior. For example, Hubel (1979) reports the case of a monocularly deprived monkey that was reverse sutured and showed behavioral recovery of acuity to near normal levels in the formerly deprived eye. Physiological recordings, however, revealed that cortical ocular dominance was almost completely shifted away from the formerly deprived eye. If this animal had not been tested behaviorally, it would probably have been assumed that the monkey's formerly deprived eye was severely amblyopic.

BINOCULAR PATTERN DEPRIVATION

Effects on spatial vision Several anecdotal reports in the clinical literal describe persons who had lost vision early in life and were not corrected of their abnormalities until adulthood (e.g. Gregory & Wallace 1963, Valvo 1968, Ackroyd et al 1974). It is generally reported that these persons never attain useful vision, but few detailed studies of specific visual functions have been reported.

Binocular pattern deprivation has been produced in monkeys by rearing them in darkness and by bilateral lid-suture. Behavioral studies of dark-reared monkeys indicate that binocular pattern deprivation is more harmful to visual responsiveness than to spatial resolution. Regal et al (1976) report that when monkeys emerged from 24 weeks of dark-rearing they appeared on the basis of informal observations to be blind. For example, they would not blink or try to avoid an impending collision. However, their visual acuity was only moderately depressed, to one to two octaves below normal. Riesen et al (1964) obtained similar results from monkeys deprived of pattern vision for 20 or 60 days by a combination of dark-rearing and diffusing contact lenses, and Hyvarinen et al (1981a,b) report similar results for bilateral lid-suture.

Hoyt (1980) reports that binocular occlusion of human newborns for less than two weeks during phototherapy does not affect acuity at age five.

Effects on binocularity No studies of the effects of binocular pattern deprivation on binocularity have been reported.

Anatomical and physiological studies Several parts of the visual system from binocularly deprived monkeys have been examined anatomically and physiologically.

Retina Hendrickson & Boothe (1976) found that dark-rearing of up to six months' duration produced no qualitative changes in the number, size, or staining characteristics of retinal ganglion cells. This is in contrast to the dramatic effects reported in dark-reared chimpanzees, in whom large numbers of retinal ganglion cells degenerate (Chow et al 1957). However, it has been suggested (Hendrickson & Boothe 1976) that the chimpanzee results might have been a secondary deficit caused by self-imposed retinal ischemia, since Chow et al reported that the chimpanzees adopted the stereotypy of pressing on their eyeballs while in the dark.

dLGN Long-term binocular lid-suture produces abnormally small cells in all layers of the dLGN (Headon & Powell 1978, Vital-Durand et al 1978). Hendrickson & Boothe (1976) found no qualitative changes in cell size or staining characteristics of dLGN cells in monkeys dark-reared for up to six months.

Cortex A normal anatomical pattern of ocular dominance columns developed in layer IV of striate cortex in a single monkey dark-reared until seven weeks of age (LeVay et al 1980). Physiological recordings from single units in the cortex of binocularly lid-sutured monkeys find large numbers of visually unresponsive units (Crawford et al 1975). Approximately equal numbers of cortical cells can be driven by each eye following bilateral lid-suture, but very few cells respond to both eyes (Wiesel & Hubel 1974, Crawford et al 1975).

Blakemore & Vital-Durand (1981a,b) studied the spatial tuning properties of cortical cells in binocularly lid-sutured monkeys. They found that the development of spatial resolution of cortical cells is retarded in binocularly deprived monkeys. A monkey deprived from birth for 18 weeks showed spatial resolution equivalent to that of a newborn, and therefore considerably poorer than that of a normal 18-week-old monkey. Blakemore & Vital-Durand (1981a) comment on the differences they found between the spatial resolution characteristics of cortical neurons in binocularly deprived monkeys and the characteristics of dLGN neurons in monocularly deprived monkeys. Noting that the spatial tuning properties of the former were affected and those of the latter were not, they speculate that there might be two neural filters for spatial information, one peripheral (perhaps in the retina or dLGN) and one in the striate cortex; and that visual deprivation may disrupt the maturation of the cortical filter but not the maturation of the peripheral filter.

Hyvarinen et al (1981a,b) studied physiological responses from single neurons in area 19 and in parietal cortex following bilateral lid-suture. They reported that multi-modal neurons show a decreased responsiveness to visual stimuli and an increased responsiveness to somatic stimulation.

Summary There have been relatively few studies of binocularly pattern-deprived humans or monkeys. Behaviorally, the spatial resolution deficits caused by binocular deprivation are moderate compared to the spatial resolution deficits seen in the deprived eye in monocular deprivation. Physiological results suggest that spatial resolution of individual neurons is unaffected until the cortical level. There have been no behavioral studies of binocular function. Physiological results showing that most binocular neurons are missing from striate cortex suggest that binocular functioning may be disrupted.

The most striking behavioral result of binocular deprivation is a lack of visual responsiveness. This may be related to the finding that other sensory modalities increase their influences on multi-modal neurons, at the expense of visual influence.

CATARACTS

Effects on spatial vision Cataracts, or opacities of the lens, scatter the light and interfere with the formation of a clear, high contrast retinal image. They can vary from tiny noncentral irregularities to large, dense, central opacities that scatter all of the incoming light and deprive the eye of virtually all pattern vision.

When cataracts are removed surgically, the defocused aphakic eyes continue to be almost totally deprived of pattern vision unless the patient is fitted with an appropriate optical correction. Even when corrected, aphakes are focused only at a fixed distance and therefore continue to be exposed to optical defocus much of the time. Thus, the effects of cataracts and aphakia would be expected to range from total pattern deprivation, as described in the previous section, to moderate optical defocus, described in the following section.

Studies of CSFs in individual patients support this view. Some monocular aphakes show no evidence of pattern vision, a result similar to that of patients who were raised with complete monocular pattern deprivation, whereas other monocular aphakes show deficits primarily in the mid to high spatial frequency range, as do many patients raised under conditions of optical defocus (Hess et al 1981, Maurer et al 1983a).

Surgical removal of congenital cataracts has traditionally failed to restore acuity to normal values, particularly in the unilateral case. When a unilateral cataract is dense and present from birth, the prognosis for visual acuity better than 20/200 is usually very poor (see reviews by Hiles & Wallar 1977, Vaegan & Taylor 1979, Helveston et al 1980). In general, partial cataracts and cataracts with onset after infancy have much less serious consequences, especially if surgery and optical correction are done promptly (Hiles & Wallar 1977, Frey et al 1973, Frank & France 1977, Vaegan & Taylor 1979, Helveston et al 1980, von Noorden 1981b). Patients who acquire bilateral cataracts after six months of age have a good prognosis for acuity of 20/60 or better, even when the duration between the onset of the cataract and post-surgical fitting of optical correction is more than six months (Hiles & Wallar 1977, Taylor et al 1979).

Several recent reports have indicated that if surgery is performed and optical correction completed before two to three months postnatal, good visual acuity can be preserved. Patients with bilateral congenital cataracts who are operated on and optically corrected by three months of age have a good chance of developing acuity of 20/60 or better, but delay of surgery decreases the probability that acuity will be better than 20/200 (Taylor et al 1979, Jacobson et al 1981b, Rogers et al 1981, Gelbart et al 1982, Mohindra et al 1983a). Even in the unilateral case, acuity in the aphakic eye can be maintained within the normal range, at least until two years of age (Enoch & Rabinowicz 1976, Enoch et al 1979, Odom et al 1981, Lewis et al 1983b, Beller et al 1981, Jacobson et al 1981b, 1983).

Studies of infants born with congenital cataracts who received early surgery and optical correction indicate that acuity development proceeds normally until eighteen months, but then appears to be arrested, so that acuity falls one to two octaves below the normal range by the time these children are two to three years of age (Atkinson & Braddick 1982a, Lewis et al 1983a).

Infant monkeys have been subjected to lensectomy in one eye soon after birth to model the effects of cataract removal in human infants (Gammon et al 1984). One of these aphakic monkeys, reared with no optical correction until 24 weeks of age, could make brightness discriminations but could not do an acuity task if all brightness cues were faded out. Moderate levels of acuity, near 5 cy/deg, were maintained in a second aphakic monkey that was fitted with a +40 D continuous wear contact lens throughout the rearing period (Boothe et al 1984).

Effects on binocularity Binocularity and stereopsis have been reported to be severely reduced or eliminated in patients with a history of unilateral (Frank & France 1977, Beller et al 1981, Hess et al 1981) or bilateral (Gelbart et al 1982) cataracts. Asymmetrical optokinetic nystagmus (OKN) has been reported to occur in both the aphakic and normal eyes of children treated for unilateral congenital cataracts, but not in either eye of children treated for unilateral traumatic cataracts incurred after three years of age (Maurer et al 1983b).

Anatomical and physiological studies There are no direct studies of the effects of cataracts on anatomy and physiology. The effects of pattern deprivation and aphakia on anatomy and physiology are discussed in the previous sections (monocular and binocular pattern deprivation). The effects of optical defocus on anatomy and physiology are discussed in the following section.

Optical Defocus

Less drastic forms of pattern deprivation are produced by optical defocus. In this case, high spatial frequencies in the retinal image are most attenuated, while lower spatial frequencies are relatively less affected. For spherical errors (myopia and hyperopia), the loss affects contours of all orientations in the retinal image; for astigmatism, the loss varies with contour orientation.

ANISOMETROPIA Anisometropia is defined as a difference in refractive error between the two eyes.

Effects on spatial vision Anisometropia often occurs in conjunction with monocular deficits in spatial vision (see review by Tanlamai & Goss 1979).

In general, hyperopic anisometropes are more likely to have reduced acuity in one eye than are myopic anisometropes, perhaps because hyperopes tend to use the eye that requires the lesser accommodative effort for focusing, while for myopic anisometropes there exists a distance for each eye that will produce a clear image without accommodation (Jampolsky et al 1955, Schapero 1971). In addition, the prevalence of amblyopia increases with increasing amounts of anisometropia. Tanlamai & Goss (1979), for example, found the prevalance of amblyopia to be 100% among hyperopes with at least 4.5 D and myopes with at least 6.5 D of anisometropia, and 50% among hyperopes with at least 2.5 D and myopes with at least 4.5 D of myopia. There is also evidence that the depth of amblyopia is correlated with the degree of anisometropia (reviewed recently by Kivlin & Flynn 1981).

The CSFs of the amblyopic eyes of anisometropes are depressed, and there is some evidence that there is more attenuation at high than at low spatial frequencies (e.g. Gstalder & Green 1971, Sjöstrand 1978, Bradley & Freeman 1981). Tests of two anisometropic amblyopes indicated that acuity at eccentricities of 20° and beyond were similar to acuity values of normals at the same retinal eccentricities (Sireteanu & Fronius 1981). Levi & Klein (1982a,b) report that vernier acuity, Snellen acuity, and grating acuity are depressed by proportionate amounts in anisometropic amblyopes. However, Gstalder & Green (1971) and Howell et al (1983) report that Snellen acuity is substantially lower than grating acuity, and Freeman & Bradley (1980) report that vernier acuity is reduced proportionately more than grating acuity in anisometropic amblyopes. Suprathreshold distortions of spatial vision are also seen in some anisometropic amblyopes (Hess et al 1978).

Anisometropia has been simulated in monkeys by securing a high powered lens (− 10D) in front of one eye (Smith et al 1983). Monkeys reared under these conditions showed deficits in their CSFs at high spatial frequencies when tested in adulthood. Differences in refractive error between the two eyes have also been simulated in monkeys by rearing infants with chronic atropinization of one eye during early life. The atropinized eye in infant monkeys raised under these conditions sometimes shows deficits in contrast sensitivity to high spatial frequencies, and these deficits can be permanent (Boothe et al 1982a, Harwerth et al 1983a).

Early correction is thought to be important in preventing anisometropic amblyopia (e.g. Jampolsky et al 1955). In a recent developmental study, Mayer et al (1982) found that the acuity difference between eyes shown by three anisometropic children less than five years of age was eliminated by optical correction, whereas the acuity difference shown by a five-year-old was not.

Von Noorden (1981a) has reported that the use of atropinization of one eye of two- to four-year-old children as a treatment for other kinds amblyopia can itself produce amblyopia of the originally non-amblyopic, atropinized eye.

Effects on binocularity Many, but ot all, anisometropic amblyopes show deficits in binocularity, as evidenced by abnormal fixation patterns (Sen 1980), lack of stereopsis (Cooper & Feldman 1978, Kani 1978), poor interocular transfer (Sireteanu et al 1981, van Hoff-van Duin & Mohn 1981), lack of binocular summation (Lennerstrand 1978, Sireteanu et al 1981, cf Levi et al 1979), or poorer acuity in the amblyopic eye under binocular than monocular conditions (Awaya & von Noorden 1972). Monkeys raised with monocular atropinization (Harwerth et al 1983a) or with a -10 D lens in front of one eye (Smith et al 1983) fail to demonstrate binocular summation in spatial contrast sensitivity.

Anatomy and physiology: dLGN Von Noorden et al (1983) found interlaminar differences in a postmortem examination of the dLGN of a human anisometrope. Similar interlaminar cell size differences have been found in monkey dLGN following chronic atropinization of one eye (Hendrickson et al 1982). In both human and monkey studies, statistically significant differences were found only in the parvocellular layers.

Cortex Deoxyglucose methods have been used by Hendrickson et al (1982) to examine cortical ocular dominance columns in the monocularly atropinized monkeys tested behaviorally by Boothe et al (1982a). Columns demonstrated by stimulating the untreated eye were wider than columns produced by stimulating the atropinized eye. This change in cortical ocular dominance was confirmed with physiological recordings, but here an interaction was found between spatial tuning and binocularity. Monocular cortical cells driven by the untreated eye responded to higher spatial frequencies than did cortical cells driven by the atropinized eye. The few binocular cells that were encountered responded to higher frequencies through their untreated than through their atropinized eye. For these reasons, when low spatial frequencies were used as stimuli, many cells could be found that responded to the deprived eye. On the other hand, when high spatial frequency gratings were used, then almost all cortical neurons responding to the grating were driven only through the non-deprived eye (Hendrickson et al 1982). This result suggests caution in speculating about changes in physiological properties or visual functions on the basis of changes in ocular dominance columns.

Summary The effects of anisometropia on spatial vision are quite different than the effects of monocular pattern deprivation. Acuity is only moderately depressed and measurements of CSFs have shown that it is primarily the high spatial frequencies that are affected. This difference is reflected in the anatomical and physiological deficits. Only the parvocellular central pathways seem to be affected anatomically. Physiologically, the deprived eye is able to drive cortical cells that respond to low spatial frequencies, and only loses its connections to cells tuned to higher spatial frequencies.

Monocular optical defocus disrupts binocular function behaviorally, and also leads to loss of most binocular cortical cells. However, the ocular dominance shifts produced by monocular defocus appear to occur predominantly in cells that respond to high spatial frequencies.

AMETROPIA

Effects on spatial vision Adults who are high myopes or high hyperopes often show acuity deficits even when they are tested with optical correction (Borish 1970).

There have been two recent developmental studies in this area. Fiorentini & Maffei (1976) found that nine young adults with myopia greater than 5 D who had been optically corrected since age six, and one who had been corrected since age two, showed depressed CSFs at all spatial frequencies even when wearing optical correction. On the other hand, Mohindra et al (1983a) showed that the acuity deficits shown by four uncorrected high hyperopes 30 weeks to three years of age were eliminated when they were tested with optical correction.

The effects of ametropia have not been studied in monkeys.

ASTIGMATISM

Effects on spatial vision Human adult astigmats who were not optically corrected as children often show acuity and contrast sensitivity differences for stimuli of different orientations (meridional amblyopia) that are predictable from the axis and form of their astigmatism, even when they are tested with optical correction (Freeman et al 1972, Mitchell et al 1973, Freeman & Thibos 1973, 1975a,b, Mitchell & Wilkinson 1974, Freeman 1975, Cobb & MacDonald 1978). Similar results have been seen in monkeys (Harwerth et al 1980, 1983b).

Several recent studies support the likelihood of a causal relationship between the presence of astigmatism during early childhood and the presence of meridional amblyopia, as well as the possible importance of early correction. Monkeys reared with experimentally imposed astigmatism show axis-appropriate variations in acuity and contrast sensitivity (Boothe & Teller 1982). A single naturally astigmatic infant monkey tested longitudinally showed no meridional amblyopia until about five or six weeks postnatal (Teller et al 1978a). Meridional amblyopia has been looked for and not found in a small number of human astigmats who were optically corrected by age seven (Mitchell et al 1973, Cobb & MacDonald 1978). A small number of astigmatic children less than three years of age, tested while wearing corrective lenses, did not exhibit meridional amblyopia (Mohindra et al 1978, Teller et al 1978a, Held 1978, Mohindra et al 1983a).

Gwiazda et al (1984a) followed six children with documented infantile astigmatism, and tested them at age six years, after the astigmatism had disappeared. Meridional amblyopia consistent with the orientation of their infantile astigmatism was found in the four children who had a myopic focus at six months, but not in the two children who were compound hyperopic astigmats at six months. This result suggests that the unusual patterns of meridional amblyopia seen in some non-astigmatic adult humans (Daugman 1983, Mayer 1983a) and monkeys (Williams et al 1981) may be the result of infantile astigmatism.

Anatomy and physiology Kiorpes et al (1979) recorded from cortical cells in two monkeys that had been reared with artificially imposed astigmatism. These monkeys had been tested behaviorally and had been shown to have better acuity and contrast sensitivity to vertical than to horizontal gratings (Boothe & Teller 1982). No significant differences were found in the numbers of cells encountered as a function of central axis orientation. However, horizontally oriented cells had poorer contrast sensitivities and lower average peak spatial frequencies than normal. Wiesel (1982) reports a case of a monkey raised viewing only vertical stripes with one eye, while the other eye was sutured closed. Cortical cells that preferred horizontal orientations were driven by both eyes. Cells that preferred vertical tended to be monocular and driven by the exposed eye.

Strabismus

Strabismus is a lack of good binocular alignment. Strabismus often occurs in conjunction with amblyopia, but the direction of causality has been a much-debated issue, and indeed, both directions may well occur (von Noorden 1980). It is not obvious what form of visual deprivation is produced by strabismus. It has been suggested that the double images produced by misalignment lead to binocular rivalry and suppression of one image (von Noorden 1980), with a consequent deprivation possibly introduced at central visual levels. A hypothesis of a more direct form of deprivation has been articulated more recently. Ikeda (1979) argues that strabismus deprives the eye of its normal binocular fixation patterns, and that this may lead to a lack of well-focused images in the fovea of the nonfixating eye. If the patient adopts the strategy of alternating fixation, then both eyes probably receive, alternately, well-focused foveal images. However, if the patient adopts a fixation preference for one eye, then the image falling upon the other fovea will often be out of focus, thus depriving that fovea of high spatial frequencies much of the time.

Effects on spatial vision Patients with strabismus show CSFs that are depressed across some or all spatial frequencies (Levi & Harwerth 1977, Hess & Howell 1977, Sjöstrand 1978, Thomas 1978). In addition, they show perceptual distortions for suprathreshold stimuli (Hess et al 1978, Rentschler & Hilz 1979, Bedell & Flom 1981, 1983) and suffer the effects of "crowding," which results in their optotype line acuity (Rentschler et al 1980, Levi & Klein 1982a,b, Howell et al 1983) and vernier acuity (Levi & Klein 1982a,b) being much worse than would be predicted from their grating acuity. On the other hand, at retinal eccentricities greater than 10°, the amblyopic eyes of strabismic amblyopes show acuity values equal to those found for normals (Kirschen & Flom 1978, Avetisov 1979, Sireteanu & Fronius 1981, Bedell 1982, cf Hess & Jacobs 1979).

Recent deprivation and developmental studies support the hypothesis that strabismus can be a primary causal factor leading to acuity deficits. Surgically induced esotropia initiated in monkeys during the first 12 weeks after birth typically leads to spatial vision deficits including acuity and contrast sensitivity losses (von Noorden & Dowling 1970, von Noorden 1973a, Kiorpes & Boothe 1980, Kiorpes et al 1984a, Harwerth et al 1983a). As in human patients, the spatial vision deficits do not develop if the monkey adopts an alternating fixation pattern (von Noorden & Dowling 1970, Kiorpes et al 1984c). This may account for reports (e.g. Wiesel 1982) that monkeys with surgical strabismus often fail to show acuity deficits.

The development of spatial vision deficits in congenitally and surgically esotropic human and monkey infants has been the subject of several recent studies. Preferential looking studies of human infant patients suggest that some congenital esotropes maintain equal acuity in the two eyes until they reach four to six months postnatal, after which they begin to show acuity differences (Jacobson et al 1981a, 1982, Birch 1983, Birch & Stager 1984). Similarly, developmental studies in monkeys demonstrate that a surgically induced esotropia initiated near birth disrupts the normal pattern of acuity development in the deviated eye, and the onset of an interocular acuity difference is sometimes delayed for several weeks after the esotropia is induced (Kiorpes & Boothe 1980, Kiorpes et al 1984a). Preferential looking and VEP studies show that the relative acuity difference between eyes of strabismic human infants and young children can sometimes be manipulated rapidly and dramatically with occlusion therapy (Thomas et al 1979, Mohindra et al 1979, 1983b, Katsumi et al 1981, Odom et al 1981, 1982, Jacobson et al 1981a, 1982, 1983, Lennerstrand et al 1982).

The question of the optimum time for eye alignment surgery in strabismic patients remains a much debated issue (cf Jampolsky 1977, Parks 1977, Robb 1977b). There is some evidence that the incidence of poor spatial resolution is lower in congenital esotropes whose eyes are aligned surgically before one to two years of age than in those who undergo surgery after age two (Zak & Morin 1982, but cf Taylor 1974, Ing 1981, 1983), but the likelihood that different kinds of patients have surgery at different times makes interpretation difficult.

Jampolski (1978) has made compelling arguments that the paralytic strabismus produced surgically in monkeys does not model very adequately the comitant strabismus commonly found in humans. Recently, two monkey models of strabismus that are more appropriate have been studied. It has been documented that various forms of strabismus occur naturally in infant monkeys, and that amblyopia sometimes occurs in monkeys with a naturally occurring comitant strabismus (Kiorpes & Boothe 1981, Kiorpes et al 1984b,c). The presence or absence of acuity loss in these monkeys is correlated predictably with the fixation pattern adopted by the monkey, providing further justification that these monkeys are good models of the common kinds of human strabismic amblyopia.

A second monkey model of comitant strabismus has been produced by rearing young monkeys with prisms over both eyes (Smith et al 1979, Crawford & von Noorden 1980a,b). The prisms serve to dissociate binocular input and mimic in some ways the sensory input of comitant strabismus. Harwerth et al (1983a) tested these monkeys behaviorally and found that their CSFs were depressed at mid and high spatial frequencies.

Effects on binocularity Older children and adults who have experienced strabismus frequently show a lack of stereopsis (e.g. Cooper & Feldman 1978, Kani 1978), poor binocular summation and interocular transfer (e.g. Ware & Mitchell 1974, Mitchell et al 1975, Wade 1976, Lema & Blake 1977, Lennerstrand 1978, Blake & Cormack 1979, Anderson et al 1980, Sireteanu et al 1981, van Hof-van Duin & Mohn 1981, Keck & Price 1982, cf Levi et al 1979), and poorer acuity in the amblyopic eye under binocular than monocular conditions (Awaya & von Noorden 1972). Data from Hohmann & Creutzfeldt (1975) and Banks et al (1975) indicate that the earlier the surgical correction of strabismus and the later the onset of strabismus, the better are interocular transfer and binocular interaction. Binocular summation and interocular transfer have been shown to occur in the peripheral retina of strabismic amblyopes (Sireteanu et al 1981).

Very few studies of binocularity in infants with abnormal binocular experience have been reported. Preliminary reports show a lack of stereopsis and other forms of binocularity in strabismic infants (Bechtoldt & Hutz 1979, Braddick et al 1980, Shea & Aslin 1982, Birch 1983, Leguire et al 1983). Monocular OKN testing has revealed that asymmetrical monocular OKN is common among infants and young children with esotropia but rare among those with exotropia (Atkinson & Braddick 1981, Naegele & Held 1983).

Harwerth et al (1983a) failed to find binocular summation in surgically strabismic monkeys, and Wiesel (1982) reports that his surgically strabismic monkeys showed no fusion. Harwerth et al (1983a) found deficits in binocular summation in monkeys that wore prisms during early life, and Crawford et al (1983b) report that prism-reared monkeys were stereoblind to random dot stereograms.

Anatomy and physiology: dLGN Reports of interlaminar differences in dLGN cell size have been found following surgically induced strabismus (von Noorden 1973b, von Noorden & Middleditch 1975, Crawford & von Noorden 1979). Parvocellular layers are affected more strongly than magnocellular layers. However, Crawford & von Noorden (1980a,b) report that prism-rearing did not produce any cell size differences between dLGN layers, presumably because these animals alternate fixation.

Cortex In the striate cortex there are few binocular cells following surgically induced strabismus (Crawford & von Noorden 1979) or prism-rearing (Crawford & von Noorden 1980a,b, Crawford et al 1983a). This loss of binocular cells occurs even when the strabismic monkeys are reared in a repetitive unidirectional environment consisting exclusively of vertical stripes (von Noorden & Crawford 1981b). Sometimes, in addition to the loss of binocular cells, there are shifts in ocular dominance such that most monocular cells tend to be driven by one eye. In other cases, approximately equal numbers of cells remain driven by either eye. It has been suggested that the loss of binocular cells is associated with the functional loss of binocularity and that the shift of ocular dominance toward one eye is associated with acuity loss (Crawford & von Noorden 1979, 1980a,b, Crawford et al 1983a,b).

Von Noorden & Crawford (1979) report that the sensitive period during which strabismus can produce its effects on dLGN cell size is over by 12 weeks. However, physiological ocular dominance of cortical neurons can be influenced at later ages.

There are no reports of the spatial resolution properties of single neurons in strabismic monkeys.

Summary of Visual Deprivation and Amblyopia

A common theme beginning to emerge from the studies of abnormal visual development is that amblyopia is not a general amorphous loss of acuity and binocularity. Careful psychophysical testing has revealed that amblyopes have specific deficits that are different when tested in the central visual field than in the periphery, different when tested with low spatial frequencies than with high spatial frequencies, and affected by factors such as optotype crowding.

Furthermore, different kinds of amblyopia (e.g. stimulus deprivation, anisometropic, strabismic) are associated with different specific deficits.

Similarly, a common theme beginning to emerge from the animal deprivation studies is that specific kinds of deprivation produce changes or deficits in specific subsystems of the visual pathways. For example, specific kinds of deprivation affect magnocellular pathways differently than parvocellular pathways, cortical layer IV differently than supragranular layers, and neurons tuned to low spatial frequencies differently than neurons tuned to high spatial frequencies. A major task for the next generation of experiments in this field is to relate specific kinds of deprivation to specific behavioral deficits and to changes in specific subsystems of the visual machinery.

OVERALL SUMMARY

A major justification for carrying out developmental studies and deprivation studies in monkeys is to attack the detailed ontological and causal questions that must usually remain formally unexplored in retrospective human clinical cases. Neither nature nor the ophthalmic surgeon assigns infants randomly to disease entities nor treatment groups, but prospective developmental studies in humans and laboratory studies in monkeys can come closer to doing so. Much initial progress has been made in these areas. Conditions that are thought to cause specific kinds of amblyopia in humans, such as strabismus, anisometropia, and stimulus deprivation, have been produced experimentally in monkeys. Developmental studies, conducted in parallel in humans and monkeys, have been used to document the time of onset of visual deficits, and to monitor the short- and long-term effects of optical or surgical treatment. The field is ready for a more detailed exploration of the significant parameters of deprivation (both clinical and experimental), and of the sensitive periods for both deprivation and treatment; in short, for systematic studies of the necessary and sufficient conditions for normal visual development in primates.

In this review we have tried to emphasize interdisciplinary parallels wherever such parallels were present. Yet in reviewing the literature on primate visual development, it has become obvious that there is much isolation of the subject matter of human from monkey studies, of behavioral from anatomical and physiological studies, and of studies of normal development from studies of the effects of deprivation. For example, amblyopia is by definition a deficit in visual acuity as well as in binocularity, and models in monkeys of the losses of spatial vision seen in human amblyopia are becoming established; yet only a handful of studies deal with the effects of deprivation on the spatial resolution properties of cells in the primate visual system. The absence of studies of the neural bases of normal development is also particularly striking. It is hoped that the gaps pointed out by the organization of this chapter will motivate researchers to study areas in which interdisciplinary efforts could yield rich returns.

ACKNOWLEDGMENTS

We thank the many colleagues who provided us with preprints and unpublished information, Dr. John Flynn for comments on the manuscript, Lawson Sebris for technical support, and Marjorie Zachow for secretarial assistance. Supported by NIH grants EY 03956 (R. G. B.), EY 02581 (V. D.), EY 02920, and EY 04470 (D. Y. T.).

Literature Cited

Abramov, I., Gordon, J., Hendrickson, A., Hainline, L., Dobson, V., LaBossiere, E. 1982. The retina of the newborn human infant *Science* 217:265–67

Abramov, I., Hainline, L., Turkel, J., Lemerise, E., Smith, H., Gordon, J., Petry, S. 1985. Rocket-ship psychophysics: Assessing visual functioning in young children. *Invest. Ophthalmol. Visual Sci.* In press

Ackroyd, C., Humphrey, N. K., Warrington, E. K. 1974. Lasting effects of early blindness. *Q. J. Exp. Psychol.* 26:114–24

Adams, R. J. 1982. Newborn's detection of visual contrast. *Infant Behav. Dev. (Special ICIS Issue)* 5:2

Allen, J. L. 1979. *The development of visual acuity in human infants during the early postnatal weeks.* Phd Dissertation, Univ. Washington, Seattle.

Anderson, P., Mitchell, D. E., Timney, B. 1980. Residual binocular interaction in stereoblind humans. *Vision Res. 20*:603–11

Anderson, R. L., Baumgartner, S. A. 1980. Amblyopia in ptosis. *Arch. Ophthalmol.* 98:1068–69

Aslin, R. N. 1977. Development of binocular fixation in human infants. *J. Exp. Child Psychol.* 23:133–50

Aslin, R. N. 1981. Development of smooth pursuit in human infants. In *Eye Movements: Cognition and Visual Perception,* ed. D. F. Fisher, R. A. Monty, J. W. Senders. Hillsdale, NJ: Erlbaum

Aslin, R. N. 1985. Motor aspects of visual development in infancy. In *Handbook of Infant Perception,* ed. P. Salapatek, L. B. Cohen. New York: Academic. In Press

Aslin, R. N., Alberts, J. R., Petersen, M.R., eds. 1981. *Development of Perception, Vol. 2, The Visual System.* New York: Academic

Aslin, R. N., Dobson, V. 1983. Dark vergence and dark accommodation in human infants. *Vision Res.* 23:1671–78

Aslin, R. N., Dumais, S. T. 1980. Binocular vision in infants: A review and a theoretical framework. *Adv. Child Dev. Behav.* 15: 53–94

Aslin, R. N., Jackson, R. W. 1979. Accommodative-convergence in young infants: Development of a synergistic sensory-motor system. *Can. J. Psychol. 33*:222–31

Aslin, R. N., Salapatek, P. 1975. Saccadic localization of visual targets by the very young human infant. *Percept. Psychophys.* 17: 293–302

Atkinson, J. 1979. Development of optokinetic nystagmus in the human infant and monkey infant: An analogue to development in kittens. In *Developmental Neurobiology of Vision,* ed. R. D. Freeman, pp. 277–87. (NATO Adv. Study Inst. Ser., Ser. A, Life Sci., Vol. 27). New York: Plenum

Atkinson, J., Braddick, O. 1981. Development of optokinetic nystagmus in infants: An indicator of cortical binocularity? In *Eye Movements: Cognition and Visual Perception,* ed. D. F. Fisher, R. A. Monty, J. W. Senders, pp. 53–64. Hillsdale, NJ: Erlbaum

Atkinson, J., Braddick, O. 1982a. Assessment of visual acuity in infancy and early childhood. *Acta Opthamol.* 157:18–26

Atkinson, J., Braddick, O. 1982c. The use of isotropic photorefraction for vision screening in infants. *Acta Ophthalmol.* 157:36–45

Atkinson, J., Braddick, O. 1983. Vision screening and photorefraction—The relation of refractive errors to strabismus and amblyopia. *Behav. Brain Res.* 10:71–80

Atkinson, J., Braddick, O., Braddick, F. 1974. Acuity and contrast sensitivity of infant vision. *Nature* 247:403-4

Atkinson, J., Braddick, O., French, J. 1979. Contrast sensitivity of the human neonate measured by the visual evoked potential. *Invest. Ophthalmol. Visual Sci.* 18:210–13

Atkinson, J., Braddick, O., French, J. 1980. Infant astigmatism: Its disappearance with age. *Vision Res.* 20:891–93

Atkinson, J., Braddick, O., Moar, K. 1977a. Contrast sensitivity of the human infant for moving and static patterns. *Vision Res.* 17:1045–47

Atkinson, J., Braddick, O., Moar, K. 1977b. Development of contrast sensitivity over the first 3 months of life in the human infant. *Vision Res*. 17:1037–44

Atkinson, J., Braddick, O., Moar, K. 1977c. Infants' detection of image defocus. *Vision Res*. 17:1125–26

Atkinson, J., Braddick, O., Pimm-Smith, E. 1982b. 'Preferential looking' for monocular and binocular acuity testing of infants. *Br. J. Ophthalmol*. 66:264–68

Atkinson, J., French, J., Braddick, O. 1981. Contrast sensitivity function of preschool children. *Br. J. Ophthalmol*. 65:525–29

Atkinson, J., Pimm-Smith, E., Evans, C., Braddick, O. J. 1983. The effects of screen size and eccentricity on acuity estimates in infants using preferential looking. *Vision Res*. 23:1479–83

Avetisov, E. S. 1979. Visual acuity and contrast sensitivity of the amblyopic eye as a function of the stimulated region of the retina. (Trans. D. G. Kirschen, E. D. Shlyakhov, M. C. Flom.) *Am. J. Optomet. Physiol. Opt*. 56:465–69

Awaya, S., Miyake, Y., Imaizumi, Y., Shiose, Y., Kanda, T., Komuro, K. 1973. Amblyopia in man, suggestive of stimulus deprivation amblyopia. *Jpn. J. Ophthalmol*. 17:69–82

Awaya, S., Miyake, Y., Shiose, Y., Kanda, T., Kawase, Y. 1976. Stimulus deprivation amblyopia in man. In *Transactions Second International Strabismological Meeting*, pp. 62–73. Paris-Marseille: Diffusion Generale de Libraire

Awaya, S., Sugawara, M., Miyake, S. 1979. Observations in patients with occlusion amblyopia. Results of treatment. *Trans. Ophthalmol. Soc. UK* 99:447–54

Awaya, S., Sugawara, M., Miyake, S., Isomura, Y. 1980. Form vision deprivation amblyopia and the results of its treatment—With special reference to the critical period. *Jpn. J. Ophthalmol*. 24:241–50

Awaya, S., von Noorden, G. K. 1972. Visual acuity of amblyopic eyes under monocular and binocular conditions: Further observations. *J. Pediat. Ophthalmol*. 9:8–13

Baker, F. A., Grigg, P., von Noorden, G. K. 1974. The effects of visual deprivation and strabismus on the response of neurons in the visual cortex of the monkey including studies on the striate and prestriate cortex in the normal animal. *Brain Res*. 66:185–208

Banks, M. S., 1980. The development of visual accommodation during early infancy. *Child Dev*. 51:646–66

Banks, M. S., 1980. The development of visual accommodation during early infancy. *Child Dev*. 51:646–66

Banks, M. S., Aslin, R. N., Letson, R. D. 1975. Sensitive period for the development of human binocular vision. *Science* 190: 675–77

Banks, M. S., Salapatek, P. 1976. Contrast sensitivity function of the infant visual system. *Vision Res*. 16:867–69

Banks, M. S., Salapatek, P. 1978. Acuity and contrast sensitivity in 1-, 2-, and 3-month-old human infants. *Invest. Ophthalmol. Visual Sci*. 17:361–65

Banks, M. S., Salapatek, P. 1981. Infant pattern vision: A new approach based on the contrast sensitivity function. *J. Exp. Child Psychol*. 31:1–45

Banks, M. S., Salapatek, P. 1983. Infant visual perception. In *Biology and Infancy*, ed. M. Haith, J. Campos, pp. 435–571. *Handb. Child Psychol*. New York: Wiley

Banks, M. S., Stephens, B. R. 1982. The contrast sensitivity of human infants to gratings differing in duty cycle. *Vision Res*. 22: 739–44

Banks, M. S., Stephens, B. R., Dannemiller, J. L. 1982. A failure to observe negative preference in infant acuity testing. *Vision Res*. 22:1025–31

Banks, M. S., Stephens, B. R., Hartman, E. E. 1985. The development of basic mechanisms of pattern vision. I. Spatial frequency channels. *J. Exp. Psychol*. In press.

Baraldi, P., Ferrari, F., Fonda, S., Penne, A. 1981. Vision in the neonate (full-term and premature): Preliminary result of the application of some testing methods. *Documenta Ophthalmol*. *51*:101–12

Barlow, H. B. 1972. Dark and light adaptation: Psychophysics. In *Handbook of Sensory Physiology*, Vol. 7 (Part 4), *Visual Psychophysics*, pp. 1–28. New York: Springer-Verlag

Beazley, L. D., Illingworth, D. J., Jahn, A., Greer, D. V. 1980. Contrast sensitivity in children and adults. *Br. J. Ophthalmol*. 64:863–66

Beazley, L. D., O'Connor, W. M., Illingworth, D. J. 1982. Adult levels of meridional anisotropy and contrast threshold in 5-year olds. *Vision Res*. 22:135–38

Bechtoldt, H. P., Hutz, C. S. 1979. Stereopsis in young infants and stereopsis in an infant with congenital esotropia. *J. Pediat. Ophthalmol. Strabismus* 16:49–54

Bedell, H. E. 1982. Symmetry of acuity profiles in esotropic amblyopic eyes. *Human Neurobiol*. 1:221–24

Bedell, H. E., Flom, M. C. 1981. Monocular spatial distortion in strabismic amblyopia. *Invest. Ophthalmol. Visual Sci*. 20:263–68

Bedell, H. E., Flom, M. C. 1983. Normal and abnormal space perception *Am. J. Optomet. Physiol. Opt*. 60:426–35

Beller, R., Hoyt, C. S., Marg, E., Odom, J. V. 1981. Good visual function after neonatal surgery for congenital monocular cataracts. *Am. J. Ophthalmol*. 91:559–65

Birch, E. E. 1983. Assessment of binocular function during infancy. *Opthal. Paediat.*

Genet. 2:43–50

Birch, E. E., Gwiazda, J., Bauer, J. A. Jr., Naegele, J., Held, R. 1983c. Visual acuity and its meridional variations in children aged 7 to 60 months. *Vision Res.* 23:1019–24

Birch, E., Gwiazda, J., Held, R. 1982. Stereoacuity development for crossed and uncrossed disparities in human infants. *Vision Res.* 22:507–13

Birch, E. E., Gwiazda, J., Held, R. 1983b. The development of vergence does not account for the onset of ster.opsis. *Perception* 12:331–36

Birch, E. E., Held, R. 1983. The development of binocular summation in human infants. *Invest. Ophthalmol. Visual Sci.* 24:1103–7

Birch, E. E., Shimojo, S., Held, R. 1983a. The development of aversion to rivalrous stimuli in human infants *Invest. Ophthalmol. Visual Sci. Suppl.* 24:92

Birch, E. E., Stager, D. R. 1984. Monocular acuity and stereopsis in infantile esotropia. *Invest. Ophthalmol. Visual Sci. Suppl.* 25:217

Birnbaum, M. H., Koslowe, K., Sanet, R. 1977. Success in amblyopia therapy as a function of age: A literature survey. *Am. J. Optomet. Physiol. Opt.* 54:269–75

Birnholz, J. C. 1981. The development of human fetal eye movement patterns. *Science* 213:679–81

Blake, R., Cormack, R. H. 1979. Psychophysical evidence for a monocular visual cortex in stereoblind humans. *Science* 203:274–75

Blakemore, C., Garey, L. J., Vital-Durand, F. 1978. The physiological effects of monocular deprivation and their reversal in the monkey's visual cortex. *J. Physiol.* 283: 223–62

Blakemore, C., Vital-Durand, F. 1981a. Postnatal development of the monkey's visual system. *The Fetus and Independent Life. CIBA* 86:152–71

Blakemore, C., Vital-Durand, F. 1981b. Development of spatial resolution and contrast sensitivity in monkey visual cortex. *Neurosci. Abstr.* 7:140

Blakemore, C., Vital-Durand, F. 1982. Development of contrast sensitivity by neurones in monkey striate cortex. *J. Physiol.* 334: 18–19

Blakemore, C., Vital-Durand, F., Garey, L. J. 1981. Recovery from monocular deprivation in the monkey. 1. Reversal of physiological effects in the visual cortex. *Proc. R. Soc. London Ser. B* 213:399–423

Boothe, R. G. 1981. Development of spatial vision in infant macaque monkeys under conditions of normal and abnormal visual experience. See Aslin et al 1981, pp. 217–41

Boothe, R. G. 1983. Optical and neural factors limiting acuity development: Evidence obtained from a monkey model. *Curr. Eye Res.* 2:211–15

Boothe, R. G. 1984. Development of contrast sensitivity in infant macaque monkeys. *Neurosci. Abstr.* 10:1158

Boothe, R. G., Gammon, J. A., Tigges, M., Wilson, J. R. 1984. Behavioral measurements of acuity obtained from aphakic monkeys raised with extended wear soft contact lenses. *Invest. Ophthalmol. Visual Sci. Suppl.* 25:216

Boothe, R. G., Greenough, W. T., Lund, J. S., Wrege, K. 1979. A quantitative investigation of spine and dendrite development of neurons in visual cortex (area 17) of *Macaca nemestrina* monkeys. *J. Comp. Neurol.* 186:473–90

Boothe, R. G., Kiorpes, L., Hendrickson, A. 1982a. Anisometropic amblyopia in Macaca nemestrina monkeys produced by atropinization of one eye during development. *Invest. Ophthalmol. Visual Sci.* 22:228–33

Boothe, R. G., Kiorpes, L., Regal, D. M., Lee, C. P. 1982b. Development of visual responsiveness in *Macaca nemestrina* monkeys. *Dev. Psychol.* 18:665–70

Boothe, R. G., Sackett, G. P. 1975. Perception and learning in infant rhesus monkeys. In *The Rhesus Monkey* ed. G. H. Bourne, 1:343–63. New York: Academic

Boothe, R. G., Teller, D. Y. 1982. Meridional variations in acuity and CSFs in monkeys reared with externally applied astigmatism. *Vision Res.* 22:801–10

Boothe, R. G., Teller, D. Y., Sackett, G. P. 1975. Trichromacy in normally reared and light deprived infant monkeys *(Macaca nemestrina)*. *Vision Res.* 15:1187–91

Boothe, R. G., Williams, R., Kiorpes, L., Teller, D. Y. 1980. Development of contrast sensitivity in infant *Macaca nemestrina* monkeys. *Science* 208:1290–92

Borish, I. M. 1970. *Clinical Refraction.* Chicago: Professional Press. 3rd ed.

Bornstein, M. H. 1976. Infants are trichromats. *J. Exp. Child Psychol.* 21:425–45

Bots, R. S. G. M., Nijhuis, J. G., Martin, C. B., Prechtl, H. F. R. 1981. Human fetal eye movements: Detection in utero by ultrasonography *Early Human Dev.* 5:87–94

Bower, T. G. R. 1975. Infant perception of the third dimension and object concept devel opment. In *Infant Perception: From Sensation to Cognition,* Vol. 2, ed. L. B. Cohen, P. Salapatek. New York: Academic

Boynton, R. M. 1979. *Human Color Vision.* New York: Holt, Rinehart & Winston

Braddick, O., Atkinson, J., French, J., Howland, H. C. 1979. A photorefractive study of infant accommodation. *Vision Res.* 19:1319–30

Braddick, O., Atkinson, J., Julesz, B., Kropfl, W., Bodis-Wolner, I., et al. 1980. Cortical binocularity in infants. *Nature* 288:363–65

Bradley, A., Freeman, R. D. 1981. Contrast sensitivity in anisometropic amblyopia. *Invest. Ophthalmol. Visual Sci*. 21:467–76

Bradley, A., Freeman, R. D. 1982. Contrast sensitivity in children. *Vision Res*. 22: 953–59

Bronson, G. W. 1982. *The Scanning Patterns of Human Infants: Implications for Visual Learning*. Norwood, NJ: Ablex

Brookman, K. E. 1983. Ocular accommodation in human infants. *Am. J. Optomet. Physiol. Opt*. 60:91–99

Brown, A. M. 1984. Scotopic sensitivity in 2-month-old infants. *Invest. Ophthalmol. Visual Sci. Suppl*. 25:161

Carpenter, G. C. 1974. Visual regard of moving and stationary faces in early infancy. *Merrill-Palmer Q. Behav. Dev*. 20:181–94

Changeux, J. -P., Danchin, A. 1976. Selective stabilisation of developing synapses as a mechanism for the specification of neuronal networks. *Nature* 264:705–12

Chow, K. L., Risen, A. H., Newell, F. W. 1957. Degeneration of retinal ganglion cells in infant chimpanzees reared in darkness. *J. Comp. Neurol*. 107:27–42

Cobb, S. R., MacDonald, C. F. 1978. Resolution acuity in astigmats: Evidence for a critical period in the human visual system. *Br. J. Physiol. Opt*. 32:38–49

Cooper, J., Feldman, J. 1978. Random-dot-stereogram performance by strabismic, amblyopic, and ocular-pathology patients in an operant-discrimination task. *Am. J. Optomet. Physiol. Opt*. 55:599–609

Crawford, M. L. J., Blake, R., Cool, S. J., von Noorden, G. K. 1975. Physiological consequences of unilateral and bilateral eye closure in macaque monkeys: Some further observations. *Brain Res*. 84:150–54

Crawford, M. L. J., Smith, E. L., Harwerth, R. S., von Noorden, G. K. 1983a. Stereoblind monkeys have few binocular neurons. *Neurosci. Abstr*. 9:1218

Crawford, M. L. J., von Noorden, G. K. 1979. The effects of short-term experimental strabismus on the visual system in *Macaca mulatta*. *Invest. Ophthalmol. Visual Sci*. 18:496–505

Crawford, M. L. J., von Noorden, G. K. 1980a. Optically induced comitant strabismus in monkeys. *Invest. Ophthalmol. Visual Sci*. 19:1105–9

Crawford, M. L. J., von Noorden, G. K. 1980b. Concomitant strabismus and cortical eye dominance in young rhesus monkeys. *Trans. Ophthalmol. Soc. UK* 99:369–74

Crawford, M. L. J., von Noorden, G. K., Meharg, L. S., Rhodes, J. W., Harwerth, R. S., et al. 1983b. Binocular neurons and binocular function in monkeys and children. *Invest. Ophthalmol. Visual Sci*. 24:491–95

Dannemiller, J. L., Banks, M. S. 1983. The development of light adaptation in the human infant. *Vision Res*. 23:599–609

Daugman, J. G. 1983. Visual plasticity as revealed in the two dimensional modulation transfer function of a meridional amblyope. *Human Neurobiol*. 2:71–76

De Courten, C., Garey, L. J. 1982. Morphology of the neurons in the human lateral geniculate nucleus and their normal development. *Exp. Brain Res*. 47:159–71

Derefeldt, G., Lennerstrand, G., Lundh, B. 1979. Age variations in normal human contrast sensitivity. *Acta Ophthalmol*. 57: 679–90

Des Rosiers, M. H., Sakurada, O., Jehle, J., Shinohara. M., Kennedy. C., Sokoloff, L. 1978. Functional plasticity in the immature striate cortex of the monkey shown by the [14c]deoxyglucose method. *Science* 200: 447–49

De Schonen, S., McKenzie, B., Maury, L, Bresson, F. 1978. Central and peripheral object distances as determinants of the effective visual field in early infancy. *Perception* 7:499–506

De Vries-Khoe, L., Spekreijse, H. 1982. Maturation of luminance and pattern EPs in man. *Documenta Ophthalmol. Proc. Ser*. 31: 461–75

Dineen, J., Hendrickson, A. 1981. Age-correlated differences in the amount of retinal degeneration after striate cortex lesions in monkeys. *Invest. Ophthalmol. Visual Sci*. 21:749–52

Dineen, J., Hendrickson, A., Keating, E. G. 1982a. Alterations of retinal inputs following striate cortex removal in adult monkey *Exp. Brain Res*. 47:446–56

Dineen, J., Vermeire, B., Boothe, R. 1982b. Contrast sensitivity changes in an infant monkey with extensive transsynaptic ganglion cell loss following striate cortex lesions. *Neurosci. Abstr*. 8:295.

Dobson, V. 1976. Spectral sensitivity of the two-month infant as measured by the visually evoked cortical potential. *Vision Res*. 16:367–74

Dobson, V. 1980. Behavioral tests of visual acuity in infants. *Int. Ophthal. Clinics* 20:233–50

Dobson, V. 1983. Clinical applications of preferential looking measures of visual acuity. *Behav. Brain Res*. 10:25–38

Dobson, V., Fulton, A. B., Manning, K., Salem, D., Peterson, R. A. 1981. Cycloplegic refractions of premature infants. *Am. J. Ophthalmol*. 91:490–95

Dobson, V., Fulton, A. B., Sebris, S. L. 1984. Cycloplegic refractions of infants and young children: The axis of astigmatism. *Invest. Ophthalmol. Visual Sci*. 25:83–87

Dobson, V., Maier, J. S. 1984. Infants' visual acuity at scotopic and mesopic luminances.

Invest. Ophthalmol. Visual Sci. Suppl. 26:219
Dobson, V., Mayer, D. L., Lee, C. P. 1980. Visual acuity screening of preterm infants. *Invest. Ophthalmol. Visual Sci.* 19: 1498–1505
Dobson, V., Salem, D., Carson, J. B. 1983. Visual acuity in infants—The effect of variations in stimulus luminance within the photopic range. *Invest. Ophthalmol. Visual Sci.* 24:519–22
Dobson, V.,Teller, D. Y. 1978. Visual acuity in human infants: A review and comparison of behavioral and electrophysiological studies. *Vision Res.* 18:1469–83
Dobson, V., Teller, D. Y., Belgum, J. 1978b. Visual acuity in human infants assessed with stationary stripes and phase-alternated checkerboards. *Vision Res.* 18:1233–38
Dobson, V., Teller, D. Y., Lee, C. P., Wade, B. 1978a. A behavioral method for efficient screening of visual acuity in young infants. *Invest. Ophthalmol. Visual Sci.* 17:1142–50
Dubowitz, L., Dubowitz, V., Morante, A. 1980. Visual function in the newborn: A study of preterm and full-term infants. *Brain Dev.* 2:15–27
Dubowitz, L. M. S., Mushin, J., Morante, A., Placzek, M. 1983. The maturation of visual acuity in neurologically normal and abnormal newborn infants. *Behav. Brain Res.* 10:39–45
Enoch, J. M., Hamer, R. D. 1983. Image size correction of the unilateral aphakic infant. *Ophthal. Pediat. Genet.* 2:153–65
Enoch, J. M., Rabinowicz, I. M. 1976. Early surgery and visual correction of an infant born with unilateral eye lens opacity *Documenta Ophthalmol* 41:371–82
Enoch, J. M., Rabinowicz, I. M., Campos, E. C. 1979. Post surgical contact lens correction of infants with sensory deprivation amblyopia associated wtih unilateral congenital cataract. *J. Jpn. Contact Lens Soc.* 21:95–104
Fantz, R. L., Fagan, J. F. III, Miranda, S. B. 1975. Early visual selectivity. In *Infant Perception: From Sensation to Cognition,* ed. L. B. Cohen, P. Salapatek, pp. 249–345. New York: Academic
Fantz, R. L., Nevis, S. 1967. Pattern preferences and perceptual-cognitive development in early infancy. *Merrill-Palmer Q.* 13: 77–108
Fantz, R. L., Ordy, J. M., Udelf, M. S. 1962. Maturation of pattern vision in infants during the first six months. *J. Comp. Physiol. Psychol.* 55:907–17
Fiorentini, A., Maffei, L. 1976. Spatial contrast sensitivity of myopic subjects. *Vision Res.* 16:437–38
Fiorentini, A., Pirchio, M., Spinelli, D. 1980. Scotopic contrast sensitivity in infants evaluated by evoked potentials. *Invest. Ophthalmol. Visual Sci.* 19:950–55
Fiorentini, A., Pirchio, M., Spinelli, D. 1983a. Development of retinal and cortical responses to pattern reversal in infants: A selective review. *Behav. Brain Res.* 10:99–106
Fiorentini, A., Pirchio, M., Spinelli, D. 1983b. Electrophysiological evidence for spatial frequency selective mechanisms in adults and infants. *Vision Res.* 23:119–27
Flynn, J. T., Cassady, J. C. 1978. Current trends in amblyopia therapy. *Ophthalmology (Trans. Am. Acad. Ophthalmol.)* 85:428–50
Fox, R., Aslin, R. N., Shea, S. L., Dumais, S. T. 1980. Stereopsis in human infants. *Science* 207:323–24
Frank, J. W., France, T. D. 1977. Visual acuity and binocularity in children with unilateral acquired aphakia. *J. Pediat. Ophthalmol.* 14:200–4
Freeman, R. D. 1975. Contrast sensitivity in meridional amblyopia. *Invest. Ophthalmol.* 14:78–81
Freeman, R. D., Bradley, A. 1980. Monocularly deprived humans: Nondeprived eye has supernormal vernier acuity. *J. Neurophysiol.* 43:1645–53
Freeman, R. D., Mitchell, D. E., Millodot, M. 1972. A neural effect of partial visual deprivation in humans. *Science* 175:1384–86
Freeman, R. D., Thibos, L. N. 1973. Electrophysiological evidence that abnormal early visual experience can modify the human brain. *Science* 180:876–78
Freeman, R. D., Thibos, L. N. 1975a. Contrast sensitivity in humans with abnormal visual experience. *J. Physiol.* 247:687–710
Freeman, R. D., Thibos, L. N. 1975b. Visual evoked responses in humans with abnormal visual experience. *J. Physiol.* 247:711–24
Frey, T., Friendly, D., Wyatt, D. 1973. Reevaluation of monocular cataracts in children. *Am. J. Ophthalmol.* 76:381–88
Friendly, D. 1978. Preschool visual acuity screening tests. *Trans. Am. Opthalmol. Soc.* 76:383–480
Fulton, A. B., Dobson, V., Salem, D., Mar, C., Petersen, R. A., Hansen, R. M. 1980. Cyclopegic refractions in infants and young children. *Am. J. Ophthalmol.* 90:239–47
Fulton, A. B., Hansen, R. M. 1982. Background adaptation in human infants. *Documenta Opthalmol. Proc. Ser.* 31:191–97.
Gammon, J. A., Wilson, J. R., Tigges, M. 1984. Production of anisometropia in infant monkeys as a model to study visual development after cataract removal. *Invest. Ophthalmol. Visual Sci. Suppl.* 25:215
Garey, L. J., De Courten, C. 1983. Structual development of the lateral geniculate nucleus and visual cortex in monkey and man. *Behav. Brain Res.* 10:3–13
Garey, L. J., Saini, K. D. 1981. Golgi studies

of the normal development of neurons in the lateral geniculate nucleus of the monkey. *Exp. Brain Res.* 44:117–28

Garey, L. J., Vital-Durand, F. 1981. Recovery from monocular deprivation in the monkey. II. Reversal of morphological effects in the lateral geniculate nucleus. *Proc. Royal Soc. London Ser. B* 213:425–33

Gelbart, S. S., Hoyt, C. S., Jastrebski, G., Marg, E. 1982. Long-term visual results in bilateral congenital cataracts. *Am. J. Ophthalmol.* 93:615–21

Glickstein, M., Millodot, M. 1970. Retinoscopy and eye size. *Science* 168:605–6

Goldblatt, A., Strauss, S., Hess, P. 1980. A replication and extension of findings about the development of visual acuity in infants. *Infant Behav. Dev.* 3:179–82

Goodkin, F. 1980. The development of mature patterns of head-eye coordination in the human infant. *Early Human Dev.* 4:373–86

Green, D. G., Powers, M. K., Banks, M. S. 1980. Depth of focus, eye size and visual acuity. *Vision Res.* 20:827–35

Greenman, G. W. 1963. Visual behavior of newborn infants. In *Modern Perspectives in Child Development,* ed. A. J. Solnit, S. A. Provence. New York: Hallmark

Gregory, R. L., Wallace, J. G. 1963. Recovery from early blindness: A case study. *Exp. Psychol. Soc. Monogr.* 2

Grobstein, P., Chow, K. L. 1976. Receptive field organization in mamalian cortex: The role of individual experience in development. In *Development of Behavior and the Nervous System,* ed. G. Gottlieb, 3:155–95. New York: Academic

Gstalder, R. J., Green, D. G. 1971. Laser interferometric acuity in amblyopia. *J. Pediatric Ophthalmol. Strabismus* 8:251–56

Gwiazda, J., Brill, S., Mohindra, I., Held, R. 1978. Infant visual acuity and its meridional variation. *Vision Res.* 18:1557–64

Gwiazda, J., Brill, S., Mohindra, I., Held, R. 1980a. Preferential looking acuity in infants from two to fifty-eight weeks of age. *Am. J. Optomet. Physiol. Opt.* 57:428–32

Gwiazda, J., Scheiman, M., Held, R. 1984a. Anisotropic resolution in children's vision. *Vision Res.* 24:527–31

Gwiazda, J., Scheiman, M., Mohindra, I., Held, R. 1984b. Astigmatism in children: Changes in axis and amount from birth to six years. *Invest. Ophthalmol. Visual Sci.* 25:88-92

Gwiazda, J., Wolfe, J. M., Brill, S., Mahindra, I., Held, R. 1980b. Quick assessment of preferential looking acuity in infants *Am. J. Optomet. Physiol. Opt.* 57:420–27

Hainline, L., Lemerise, E. 1982. Infants' scanning of geometric forms varying in size. *J. Exp. Child Psychol.* 33:235–56

Hainline, L., Lemerise, E., Abramov, I., Turkel, J. 1984a. Orientational asymmetries in small-field optokinetic nystagmus in human infants. *Behav. Brain Res.* In press

Hainline, L., Turkel, J., Abramov, I., Lemerise, E., Harris, C. M. 1984b. Characteristics of saccades in human infants. *Vision Res.* In press

Haith, M. M. 1980. *Rules that Babies Look By.* Hillsdale, NJ: Erlbaum

Hamer, R. D., Alexander, K., Teller, D. Y. 1982. Raleigh discriminations in young human infants. *Vision Res.* 22:575–87

Hamer, R. D., Schneck, M. E. 1984. Spatial summation in dark-adapted human infants. *Vision Res.* 24:77–85

Hansen, R. M., Fulton, A. B. 1981. Behavioral measurement of background adaptation in infants. *Invest. Ophthalmol. Visual Sci.* 21:625–29

Harris, L., Atkinson, J., Braddick, O. 1976. Visual contrast sensitivity of a 6-month-old infant measured by the evoked potential. *Nature* 264:570–71

Harris, P., MacFarlane, A. 1974. The growth of the effective visual field from birth to seven weeks. *J. Exp. Child Psychol.* 18:340–48

Hartmann, E. E., Banks, M. S. 1984. Development of temporal contrast sensitivity in human infants. *Invest. Ophthalmol. Visual Sci. Suppl.* 25:220

Harwerth, R. S., Crawford, M. L. J., Smith, E. L., Boltz, R. L. 1981. Behavioral studies of stimulus deprivation amblyopia in monkeys. *Vision Res.* 21:779–89

Harwerth, R. S., Smith, E. L., Boltz, R. L. 1980. Meridional amblyopia in monkeys. *Exp. Brain Res.* 39:351–56

Harwerth, R. S., Smith, E. L., Boltz, R. L., Crawford, M. L. J., von Noorden, G. K. 1983a. Behavioral studies on the effect of abnormal early visual experience in monkeys: Spatial modulation sensitivity. *Vision Res. 23*:1501–10

Harwerth, R. S., Smith, E. L., Boltz, R. L., Crawford, M. L. J., von Noorden, G. K. 1983a. Behavioral studies on the effect of abnormal early visual experience in monkeys: Temporal modulation sensitivity. *Vision Res.* 23:1511–18

Harwerth, R. S., Smith, E. L., Crawford, M. L. J., von Noorden, G. K. 1984. Effects of enucleation of the nondeprived eye on stimulus deprivation amblyopia in monkeys. *Invest. Ophthalmol. Visual Sci.* 25:10–18

Harwerth, R. S., Smith, E. L., Okundaye, O. J. 1983b. Oblique effects, vertical effects and meridional amblyopia in monkeys. *Exp. Brain Res.* 53:142–50

Haynes, H., White, B. L., Held, R. 1965. Visual accommodation in human infants. *Science* 148:528–30

Headon, M. P., Powell, T. P. S. 1973. Cellular changes in the lateral geniculate nucleus of infant monkeys after suture of the eyelids. *J.*

Anat. 116:135–45
Headon, M. P., Powell, T. P. S. 1978. The effect of bilateral eye closure upon the lateral geniculate nucleus in infant monkeys. *Brain Res.* 143:147–54
Headon, M. P., Sloper, J. J., Hiorns, R. W., Powell, T. P. S. 1981a. Shrinkage of cells in undeprived laminae of the monkey lateral geniculate nucleus following late closure of one eye. *Brain Res.* 229:187–92
Headon, M. P., Sloper, J. J. Hiorns, R. W., Powell, T. P. S. 1981b. Cell sizes in the lateral geniculate nucleus of normal infant and adult rhesus monkeys. *Brain Res.* 229: 183–86
Headon, M. P., Sloper, J. J., Powell, T. P. S. 1982. Initial hypertrophy of cells in undeprived laminae of the lateral geniculate nucleus of the monkey following early monocular visual deprivation. *Brain Res.* 238:439–44
Hebb, D. O. 1949. *The Organization of Behavior.* New York: Wiley
Held, R. 1978. Development of visual acuity in normal and astigmatic infants. In *Frontiers in Visual Science,* ed. S. J. Cool, E. L. Smith, III. New York: Springer-Verlag
Held, R. 1984. Binocular vision . . . behavioral and neuronal development. In *Neonate Cognition: Beyond the Blooming, Buzzing Confusion,* ed. J. Mehler, R. Fox. Hillsdale, NJ: Erlbaum. In press.
Held, R., Birch, E., Gwiazda, J. 1980. Stereoacuity of human infants. *Proc. Natl. Acad. Sci. USA* 77:5572–74
Held, R., Gwiazda, J., Brill, S., Mohindra, I., Wolfe, J. 1979. Infant visual acuity is underestimated because near threshold gratings are not preferentially fixated. *Vision Res.* 19:1377–79
Helveston, E. M., Saunders, R. A., Ellis, F. D. 1980. Unilateral cataracts in children. *Ophthalmic Surg.* 11:102–8
Hendrickson, A. E., Boles, J., McLean, E. 1977. Visual acuity and behavior of monocularly deprived monkeys after retinal lesions. *Invest. Ophthalmol. Visual Sci.* 16:469–73
Hendrickson, A. E., Boothe, R. 1976. Morphology of the retina and dorsal lateral geniculate nucleus in dark-reared monkeys *(Macaca nemestrina). Visual Res.* 16: 517–21
Hendrickson, A. E., Dineen, J. T. 1982. Hypertrophy of neurons in dorsal lateral geniculate nucleus following striate cortex lesions in infant monkeys. *Neurosci. Lett.* 30:217–22
Hendrickson, A. E., Kupfer, C. 1976. The histogenesis of the fovea in the macaque monkey. *Invest. Ophthalmol. Visual Sci.* 15:746–56
Hendrickson, A. E., Movshon, J., Eggers, H., Gizzi, M., Kiorpes, L., Boothe, R. 1982. Anatomical and physiological effects of early unilateral blur. *Invest. Ophthalmol. Visual Sci. Suppl.* 22:237
Hendrickson, A. E., Youdelis, C. 1984. The morphological development of the human fovea. *Ophthalmology* 91:603–12
Hess, R. F., Campbell, F. W., Greenhalgh, T. 1978. On the nature of the neural abnormality in human amblyopia; Neural aberrations and neural sensitivity loss. *Pflügers Archiv.* 377:201–7
Hess, R. F., France, T. D., Tulunay-Keesey, U. 1981. Residual vision in humans who have been monocularly deprived of pattern stimulation in early life. *Exp. Brain Res.* 44: 295–311
Hess, R. F., Howell, E. R. 1977. The threshold contrast sensitivity function in strabismic amblyopia: Evidence for a two type classification. *Vision Res.* 17:1049–55
Hess, R. F., Jacobs, R. J. 1979. A preliminary report of acuity and contour interactions across the amblyopes visual field. *Vision Res.* 19:1403–8
Hickey, T. L. 1977. Postnatal development of the human lateral geniculate nucleus: Relationship to a critical period for the visual system. *Science* 198:836–38
Hiles, D. A., Wallar, P. H. 1977. Visual results following infantile cataract surgery. *Int. Ophthalmol. Clinics* 17:265–82
Hines, M. 1942. The development and regression of reflexes, postures, and progression in the young macaque. *Contribut. Embryol. Carnegie Inst.* 30:153–95
Hitchcock, P. F., Hickey, T. L. 1980. Prenatal development of the human lateral geniculate nucleus. *J. Comp. Neurol.* 194:395–411
Hohmann, A., Creutzfeldt, O. D. 1975. Squint and the development of binocularity in humans. *Nature* 254:613–14
Hohmann, A., Haase, W. 1982. Development of visual line acuity in humans. *Ophthalmic Res.* 14:107–12
Hokoda, S. C., Ciuffreda, K. J. 1982. Measurement of accommodative amplitude in amblyopia. *Ophthalmol. Physiol. Optics* 2:205–12
Hollenberg, M. J., Spira, A. W. 1973. Human retinal development: Ultrastructure of the outer retina. *Am. J. Anat.* 137:357–86
Howell, E. R., Mitchell, D. E., Keith, C. G. 1983. Contrast thresholds for sine gratings of children with amblyopia. *Invest. Ophthalmol. Visual Sci.* 24:782–87
Howland, H. C. 1982. Infant eyes: Optics and accommodation. *Current Eye Res.* 2:217–24
Howland, H. C., Atkinson, J., Braddick, O., French, J. 1978. Infant astigmatism measured by photorefraction. *Science* 202:331–32
Howland, H. C., Boothe, R., Kiorpes, L. 1982. Accommodative defocus does not limit

development of acuity in infant *Macaca nemestrina* monkeys. *Science* 215:1409–11

Howland, H. C., Braddick, O., Atkinson, J., Howland, B. 1983. Optics of photorefraction: Orthogonal and isotropic methods. *J. Opt. Soc. Am.* 73:1701–8

Howland, H. C., Howland, B. 1974. Photorefraction: A technique for the study of refractive state at a distance. *J. Opt. Soc. Am.* 64:240–49

Howland, H. C., Sayles, N. 1984. Photorefractive measurements of astigmatism in infants and young children. *Invest. Ophthalmol. Visual Sci.* 25:93–102

Hoyt, C. S. 1980. The long-term visual effects of short-term binocular occlusion of at-risk neonates. *Arch. Ophthalmol.* 98:1967–70

Hoyt, C. S., Stone, R. D., Fromer, C., Billson, F. A. 1981. Monocular axial myopia associated with neonatal eyelid closure in human infants. *Ophthalmology* 91:197–200

Hubel, D. H. 1979. The visual cortex of normal and deprived monkeys. *Am. Sci.* 67:532–43

Hubel, D. H. 1982. Exploration of the primary visual cortex. *Nature* 299:515–24

Hubel, D. H., Wiesel, T. N., LeVay, S. 1977. Plasticity of ocular dominance columns in monkey striate cortex. *Philos. Trans. R. Soc. London Ser. B* 278:377–409

Huttenlocher, P. R., de Courten, C., Garey, L. J., van der Loos, H. 1982. Synaptogenesis in human visual cortex—Evidence for synapse elimination during normal development. *Neurosci. Lett.* 33:247–52

Hyvarinen, J., Carlson, S., Hyvarinen, L. 1981a. Early visual deprivation alters modality of neuronal responses in area 19 of monkey cortex. *Neurosci. Lett.* 26:239–44

Hyvarinen, J., Hyvarinen, L., Linnankoski, I. 1981b. Modification of parietal association cortex and functional blindness after binocular deprivation in young monkeys. *Exp. Brain Res.* 42:1–8

Ikeda, H. 1979. Is amblyopia a peripheral defect? *Trans. Ophthalmol. Soc. UK 99*: 347–52

Ing, M. R. 1981. Early surgical alignment for congenital esotropia. *Trans. Am. Ophthalmol. Soc.* 79:625–63

Ing. M. R. 1983. Early surgical alignment for congenital esotropia. *J. Pediat. Ophthalmol. Strabismus* 20: 11–18.

Ingram, R. M., Barr, A. 1979. Changes in refraction between the ages of 1 and 3½ years. *Br. J. Ophthalmol.* 63:339–42

Jacobson, S. G., Mohindra, I., Held, R. 1981a. Age of onset of amblyopia in infants with esotropia. *Documenta Ophthalmol. Proc. Ser.* 30:210–16

Jacobson, S. G., Mohindra, I., Held, R. 1981b. Development of visual acuity in infants with congenital cataracts. *Br. J. Ophthalmol.* 65:727–35

Jacobson, S. G., Mohindra, I., Held, R. 1982. Visual acuity of infants with ocular diseases. *Am. J. Ophthalmol.* 93:198–209

Jacobson, S. G., Mohindra, I., Held, R. 1983. Monocular visual form deprivation in human infants. *Documenta Ophthalmol.* 55:199–211

Jampolsky, A. 1977. When should one operate for congenital strabismus? In *Controversy in Ophthalmology*, ed. R. J. Brockhurst, S. A. Boruchoff, B. T. Hutchinson, S. Lessell, pp. 416–22. Philadelphia: Saunders

Jampolsky, A. 1978. Unequal visual inputs and strabismus management: A comparison of human and animal strabismus. In *Symposium on Strabismus: Transactions of the New Orleans Academy of Ophthalmology*, pp. 358–492. St. Louis: Mosby

Jampolsky, A., Flom, B. C., Weymouth, F. W., Moses, L. E. 1955. Unequal corrected visual acuity as related to anisometropia. *Arch. Ophthalmol.* 54:893–905

Johnson, C. A., Post, R. B., Chalupa, L. M., Lee, T. J. 1982. Monocular deprivation in humans: A study of identical twins. *Invest. Ophthalmol. Visual Sci.* 23:135–38

Kaakinen, K. 1979. A simple method for screening of children with strabismus, anisometropia or ametropia by simultaneous photography of the corneal and the fundus reflexes. *Acta Ophthalmol.* 57:161–71

Kalina, R. E. 1979. Examination of the premature infant. *Ophthalmology* 86:1690–94

Kani, W. 1978. Stereopsis and spatial perception in amblyopes and uncorrected ametropes. *Br. J. Ophthalmol.* 62:756–62

Katsumi, O., Oguchi, Y., Uemura, Y. 1981. Assessment of visual ability in infantile esotropia using preferential looking method. *Jpn. J. Ophthalmol.* 25:457–63

Keck, M. J., Price, R. L. 1982. Interocular transfer of the motion aftereffect in strabismus. *Vision Res.* 22:55–60

Kiorpes, L., Boothe, R. G. 1980. The time course for the development of strabismic amblyopia in infant monkeys *(Macaca nemestrina)*. *Invest. Ophthalmol. Visual Sci.* 19:841–45

Kiorpes, L., Boothe, R. G. 1981. Naturally occurring strabismus in monkeys. *Invest. Ophthalmol. Visual Sci.* 20:257–63

Kiorpes, L., Boothe, R. G. 1984. Accommodative range in amblyopic monkeys *(Macaca nemestrina)*. *Vision Res.* In press

Kiorpes, L., Boothe, R. G., Carlson, M. R. 1984a. Acuity development in surgically strabismic monkeys. *Invest. Ophthalmol. Visual Sci. Suppl.* 25:216

Kiorpes, L., Boothe, R. G., Carlson, M. R., Alfi, D. 1984b. Frequency of naturally occurring strabismus in monkeys. *J. Pediatr.*

Ophthalmol. Strabismus. In press

Kiorpes, L., Carlson, M. R., Boothe, R. G. 1984c. *Studies of strabismus and amblyopia in infant monkeys*. Presented at Am. Assoc. Pediat. Ophthalmol. Strabismus, Vail, Colo.

Kiorpes, L., Thorell, L. G., Boothe, R. G. 1979. Response properties of cortical cells in monkeys with experimentally produced meridional amblyopia. *Invest. Ophthalmol. Visual Sci. Suppl.* 18:194

Kirschen, D. G., Flom, M. C. 1978. Visual acuity at different retinal loci of eccentrically fixating functional amblyopes. *Am. J. Optomet. Physiol. Optics* 55:144–50

Kivlin, J. D., Flynn, J. T. 1981. Therapy of anisometropic amblyopia. *J. Pediat. Ophthalmol. Strabismus* 18:47–56

Kremenitzer, J. P., Vaughan, H. G. Jr., Kurtzberg, D., Dowling, K. 1979. Smooth-pursuit eye movements in the newborn infant. *Child Dev.* 50:442–48

Lee, C. P., Boothe, R. G. 1981. Visual acuity development in infant monkeys *(Macaca nemestrina)* having known gestational ages. *Vision Res.* 21:805–9

Leehey, S. C., Moskowitz-Cook, A., Brill, S., Held, R. 1975. Orientational anisotropy in infant vision. *Science* 190:900–2

Leguire, L. E., Rogers, G. L., Fellows, R. R. 1983. Toward a clinical test for stereopsis in human infants. *Invest. Ophthalmol. Visual Sci. Suppl.* 24:34

Leibowitz, H. W., Owens, D. A. 1978. New evidence for the intermediate position of relaxed accommodation. *Documenta Ophthalmol.* 46:133–47

Lema, S. A., Blake, R. 1977. Binocular summation in normal and stereoblind humans. *Vision Res.* 17:691–95

Lennerstrand, G. 1978. Binocular interaction studied with visual evoked responses (VER) in humans with normal or impaired binocular vision. *Acta Ophthalmol.* 56:628–37

Lennerstrand, G., Andersson, G., Axelsson, A. 1982. Clinical assessment of visual functions in infants and young children. *Acta Ophthalmol.* 157:63–67

LeVay, S., Wiesel, T. N., Hubel, D. H. 1980. The development of ocular dominance columns in normal and visually deprived monkeys. *J. Comp. Neurol.* 191:1–51

Levi, D. M., Harwerth, R. S. 1977. Spatio-temporal interactions in anisometropic and strabismic amblyopia. *Invest. Ophthalmol. Visual Sci.* 16:90–95

Levi, D. M., Harwerth, R. S., Smith, E. L. 1979. Humans deprived of normal binocular vision have binocular interactions tuned to size and orientation. *Science* 206:852–54

Levi, D. M., Klein, S. 1982a. Differences in vernier discrimination for gratings between stribismic and anisometropic amblyopes. *Invest. Ophthalmol. Visual Sci.* 23:398–407

Levi, D. M., Klein, S. 1982b. Hyperacuity and amblyopia. *Nature* 298:268–70

Lewis, T. L., Maurer, D., Brent, H. P. 1983a. The development of visual resolution following treatment for bilateral congenital cataracts. *Invest. Ophthalmol. Visual Sci. Suppl.* 24:133

Lewis, T. L., Maurer, D., Brent, H. P. 1983b. *An OKN test of resolutions*. Presented at Soc. Res. Child Dev. Meet., Detroit, Mich.

London, R., Wick, B. C. 1982. Changes in angle lambda during growth: Theory and clinical applications. *Am. J. Optomet. Physiol. Opt.* 59:568–72

Lotmar, W. 1976. A theoretical model for the eye of new-born infants. *Albrecht von Graefes Arch. Klin. Exp. Ophthal.* 198:179–85

Lund, J. S., Boothe, R. G., Lund, R. D. 1977. Development of neurons in the visual cortex (area 17) of the monkey *(Macaca nemestrina):* A golgi study from fetal day 127 to postnatal immaturity. *J. Comp. Neurol.* 176:149–88

Magoon, E. H., Robb, R. M. 1981. Development of myelin in human optic nerve and tract. *Arch. Ophthalmol.* 99:655–59

Malinow, M. R., Feeneyburns, L., Peterson, L. H., Klein, M. L., Neuringer, M. 1980. Diet-related macular anomalies in monkeys. *Invest. Ophthalmol. Visual Sci.* 19:857–63

Manni, I. 1964. *The Development of the Human Eye*. London: British Med. Assoc.

Manny, R., Klein, S. 1984. The development of vernier acuity in infants. *Current Eye Res.* 3:453–62

Marg, E., Freeman, D. N., Peltzman, P., Goldstein, P. J. 1976. Visual acuity development in human infants: Evoked potential measurements. *Invest. Ophthalmol.* 15: 150–53

Mates, S. L., Lund, J. S. 1983a. Neuronal composition and development in lamina 4C of monkey striate cortex. *J. Comp. Neurol.* 221:60–90

Mates, S. L., Lund, J. S. 1983b. Spine formation and maturation of Type 1 synapses on spiny stellate neurons in primate visual cortex. *J. Comp. Neurol.* 221:91–97

Mates, S. L., Lund, J. S. 1983c. Developmental changes in the relationship between type 2 synapses and spiny neurons in the monkey visual cortex. *J. Comp. Neurol.* 221:98–105

Maurer, D. 1975. Infant visual perception: Methods of study. In *Infant Perception: From Sensation to Cognition*, ed. L. B. Cohen, P. Salapatek, 2:1–76. New York: Academic

Maurer, D., Lewis, T. L., Brent, H. P. 1983b. Peripheral vision and optokinetic nystagmus in children with unilateral congenital cataract. *Behav. Brain Res.* 10:151–62

Maurer, D., Lewis, T. L., Tytla, M. E. 1983a. Contrast sensitivity in cases of unilateral congenital cataract. *Invest. Ophthalmol. Visual Sci. Suppl.* 24:21

Maurer, D., Martello, M. 1980. The discrimination of orientation by young infants. *Vision Res.* 20:201–4

Mayer, D. L., Dobson, M. V. 1980. Assessment of vision in young children: A new operant approach yields estimates of acuity. *Invest. Ophthalmol. Visual Sci.* 19:566–70

Mayer, D. L., Dobson, V. 1982. Visual acuity development in infants and young children, as assessed by operant preferential looking. *Vision Res.* 22:1141–52

Mayer, D. L., Fulton, A. B., Hansen, R. M. 1982. Preferential looking acuity obtained with a staircase procedure in pediatric patients. *Invest. Ophthalmol. Visual Sci.* 23:538–43

Mayer, M. J. 1977. Development of anisotropy in late childhood. *Vision Res.* 17:703–10

Mayer, M. J. 1983a. Non-astigmatic children's contrast sensitivities differ from anisotropic patterns of adults. *Vision Res.* 23:551–59

Mayer, M. J. 1983b. Practice improves adults' sensitivity to diagonals. *Vision Res.* 23: 547–50

McGurk, H. 1972. Infant discrimination of orientation. *J. Exp. Child Psychol.* 14:151–64

McKenzie, B. E., Day, R. H. 1972. Object distance as a determinant of visual fixation in early infancy. *Science* 178:1108–10

Mendelson, M. J. 1982a. Clinical examination of visual and social responses in infant rhesus monkeys. *Dev. Psychol.* 18:658–64

Mendelson, M. J. 1982b. Visual and social responses in infant rhesus monkeys. *Am. J. Primatol.* 3:333–40

Merigan, W. H., Weiss, B. 1980. *Neurotoxicity of the Visual System*. New York: Raven

Miranda, S. B. 1970. Visual abilities and pattern preferences of premature infants and full-term neonates. *J. Exp. Child Psychol.* 10:189–205

Mitchell, D. E., Freeman, R. D., Millodot, M., Haegerstrom, G. 1973. Meridional amblyopia: Evidence for modification of the human visual system by early visual experience. *Vision Res.* 13:535–58

Mitchell, D. E., Reardon, J., Muir, D. W. 1975. Interocular transfer of the motion after-effect in normal and stereoblind observers. *Exp. Brain Res.* 22:163–73

Mitchell, D. E., Timney, B. 1984. Postnatal development of function in the mammalian visual system. *Handb. Physiol.* (Sect. 1)3:507–55

Mitchell, D. E., Wilkinson, F. 1974. The effect of early astigmatism on the visual resolution of gratings. *J. Physiol.* 243:739–56

Mohindra, I. 1977. A non-cycloplegic refraction technique for infants and young children. *J. Am. Optomet. Assoc.* 48:518–23

Mohindra, I., Held, R. 1981. Refraction in humans from birth to five years. *Documenta Ophthalmol. Proc. Ser.* 28:19–27

Mohindra, I., Held, R., Gwiazda, J., Brill, S. 1978. Astigmatism in infants. *Science* 202:329–31

Mohindra, I., Jacobson, S. G., Held, R. 1983a. Binocular visual form deprivation in human infants. *Documenta Ophthalmol.* 55:237–49

Mohindra, I., Jacobson, S. G., Thomas, J., Held, R. 1979. Development of amblyopia in infants. *Trans. Ophthalmol. Soc. UK* 99:344–46

Mohindra, I., Jacobson, S. G., Zwann, J., Held, R. 1983b. Psychophysical assessment of visual acuity in infants with visual disorders. *Behav. Brain Res.* 10:51–58

Moskowitz-Cook, A. 1979. The development of photopic spectral sensitivity in human infants. *Vision Res.* 19:1133–42

Moskowitz, A., Sokol, S. 1980. Spatial and temporal interaction of pattern-evoked cortical potentials in human infants. *Vision Res.* 20:699–708

Moskowitz, A., Sokol, S. 1983. Developmental changes in the human visual system as reflected by the latency of the pattern reversal VEP. *Electroencephalogr. Clin. Neurophysiol.* 56:1–15

Movshon, J. A., Van Sluyters, R. C. 1981. Visual neural development. *Ann. Rev. Psychol.* 32:477–522

Mowbray, J. B., Cadell, T. E. 1962. Early behavior patterns in rhesus monkeys. *J. Comp. Physiol. Psychol.* 55:350–57

Naegele, J. R., Held, R. 1982. The postnatal development of monocular optokinetic nystagmus in infants. *Vision Res.* 22:341–46

Naegele, J. R., Held, R. 1983. Development of optokinetic nystagmus and effects of abnormal visual experience during infancy. In *Spatially Oriented Behavior*, ed. A. Heim, M. Jeannerod, pp. 155–74. New York: Springer-Verlag

Neuringer, M., Conner, W. E., Petten, C. V., Barstad, L. 1984. Dietary omega-3 fatty acid deficiency and visual loss in infant rhesus monkeys. *J. Clin. Invest.* 73:272–76

Odom, J. V. 1983. Effects of visual deprivation on monocular acuities of humans and animals. *Am. J. Optomet. Physiol. Opt.* 60: 472–80

Odom, J. V., Dawson, W. W., Romano, P. E., Maida, T. M. 1983a. Human pattern evoked retinal response (PERR): Spatial tuning and development. *Documenta Ophthalmol. Proc. Ser.* 37:265–71

Odom, J. V., Hoyt, C. S., Marg, E. 1981. Effect of natural deprivation and unilateral eye patching on visual acuity of infants and children. *Arch. Ophthalmol.* 99:1412–16

Odom, J. V., Hoyt, C. S., Marg, E. 1982. Eye patching and visual evoked potential acuity in children four months to eight years old. *Am. J. Optomet. Physiol. Opt.* 59:706–17

Odom, J. V., Maida, T. M., Dawson, W. W., Romano, P. E. 1983b. Retinal and cortical pattern responses: A comparison of infants and adults. *Am. J. Optomet. Physiol. Opt.* 60:369–75

O'Kusky, J., Colonnier, M. 1982a. Postnatal changes in the number of astrocytes, oligodendrocytes, and microglia in the visual cortex (Area 17) of the macaque monkey: A stereological analysis in normal and monocularly deprived animals. *J. Comp. Neurol.* 210:307–15

O'Kusky, J., Colonnier, M. 1982b. Postnatal changes in number of neurons and synapses in visual cortex (Area 17) of macaque monkey: A stereological analysis in normal and monocularly deprived animals. *J. Comp. Neurol.* 210:291–306

O'Leary, D. J., Millodot, M. 1979. Eyelid closure causes myopia in humans. *Experientia* 35:1478–79

Ordy, J. M., Latanick A., Samorajlki T., Massopust L. C. 1964. Visual acuity in newborn primate infants. *Proc. Soc. Exp. Biol. Med.* 115:677–80

Ordy, J. M., Massopust, L. C., Wolin, L. R. 1962. Postnatal development of the retina, electroretinogram, and acuity in the rhesus monkey. *Exp. Neurol.* 5:364–82

Ordy, J. M., Samorajlke, T. S., Collins, R. L., Nagy, A. R. 1965. Postnatal development of vision in a subhuman primate *(Macaca mulatta). Arch. Ophthalmol.* 73:674–86

Otto, J., Safra, D. 1978. Accommodation in amblyiopic eyes. *Metabolic Ophthalmol.* 2:139–42

Owens, D. A., Held, R. 1978. The development of ocular accommodation. In *Cahiers de l'Orthoptie*, ed. R. Pigassou-Albuoy, 3:10–23. Toulouse: A. F. I. M.

Packer, O., Hartmann, E. E., Teller, D. Y. 1985. Infant color vision: The effect of test field size on Rayleigh discriminations. *Vision Res.* In press

Parks, M. M. 1977. Operate early for congenital strabismus. In *Controversy in Ophthalmology*, ed. R. J. Brockhurst, S. A. Boruchoff, B. T. Hutchinson, S. Lessell, pp. 423–30. Philadelphia: Saunders

Peeples, D. R., Teller, D. Y. 1975. Color vision and brightness discrimination in two-month-old human infants. *Science* 189:1102–3

Peeples, D. R., Teller, D. Y. 1978. White-adapted photopic spectral sensitivity in human infants. *Vision Res.* 18:49–53

Peterson, L., Yonas, A., Fisch, R. O. 1980. The development of blinking in response to impending collision in preterm, fullterm, and postterm infants. *Infant Behav. Dev.* 3: 155–65

Petrig, B., Julesz, B., Kropfl, W., Baumgartner, G., Anliker, M. 1981. Development of stereopsis and cortical binocularity in human infants: Electrophysiological evidence. *Science* 213:1402–5

Pirchio, M., Spinelli, D., Fiorentini, A., Maffei, L. 1978. Infant contrast sensitivity evaluated by evoked potentials. *Brain Res.* 141:179–84

Powers, M. K., Schneck, M., Teller, D. Y. 1981. Spectral sensitivity of human infants at absolute visual threshold. *Vision Res.* 21:1005–16.

Prechtl, H. F. R., Nijhuis, J. G. 1983. Eye movements in the human fetus and newborn. *Behav. Brain Res.* 10:119–24

Pulos, E., Teller, D. Y., Buck, S. L. 1980. Infant color vision: A search for short wavelength-sensitive mechanisms by means of chromatic adaptation. *Vision Res.* 20: 485–93

Rakic, P. 1976. Prenatal genesis of connections subserving ocular dominance in the rhesus monkey. *Nature* 261:467–71

Rakic, P. 1977a. Genesis of the dLGN in the rhesus monkey: Site of origin, kinetics of proliferation, routes of migration and pattern of distribution of neurons. *J. Comp. Neurol.* 176:23–52

Rakic, P. 1977b. Prenatal development of the visual system in rhesus monkey. *Philos. Trans. R. Soc. London Ser. B* 278:245–60

Rakic, P. 1981. Development of visual centers in the primate brain depends on binocular competition before birth. *Science* 214: 928–31

Rakic, P., Goldman-Rakic, P. S. 1982. Development and modifiability of the cerebral cortex. *Neurosci. Res. Program Bull.* 20: 429–611

Regal, D. M. 1981. Development of critical flicker frequency in human infants. *Vision Res.* 21:549–55

Regal, D. M., Ashmead, D. H., Salapatek, P. 1983. The coordination of eye and head movements during early infancy: A selective review. *Behav. Brain Res.* 10:125–32

Regal, D. M., Boothe, R., Teller, D. Y., Sackett, G. P. 1976. Visual acuity and visual responsiveness in dark-reared monkeys *(macaca nemestrina) Vision Res.* 16: 523–30

Rentschler, I., Hilz, R. 1979. Abnormal orientation selectivity in both eyes of strabismic amblyopes. *Exp. Brain Res.* 37:187–91

Rentschler, I., Hilz, R., Brettel, H. 1980. Spatial tuning properties in human amblyopia cannot explain the loss of optotype acuity. *Behav. Brain Res.* 1:433–43

Riesen, A. H., Ramsey, R. L., Wilson, P. 1964.

Development of visual acuity in rhesus monkeys deprived of patterned light during early infancy. *Psychonomic Sci.* 1:33–34

Robb, R. M. 1977a. Refractive errors associated with hemangiomas of the eyelids and orbit in infancy. *Am. J. Ophthalmol.* 83: 52–58

Robb, R. M. 1977b. When should one operate for congenital strabismus? In *Controversy in Ophthalmology,* ed. R. J. Brockhurst, S. A. Boruchoff, B. T. Hutchinson, S. Lessell, pp. 431–33. Philadelphia: Saunders

Rogers, G. L., Tischler, C. L., Tsou, B. H., Hertle, R. W., Fellows, R. R. 1981. Visual acuities in infants with congenital cataracts operated on prior to 6 months of age. *Arch. Ophthalmol.* 99:999–1003.

Roucoux, A., Culee, C., Roucoux, M. 1982. Gaze fixation and pursuit in head free human infants. In *Physiological and Pathological Aspects of Eye Movements,* ed. A. Roucoux, M. Crommelinck, pp. 23–31. The Hague: Junk Publ.

Roucoux, A., Culee, C., Roucoux, M. 1983. Development of fixation and pursuit eye movements in human infants. *Behav. Brain Res.* 10:133–39

Salapatek, P., Aslin, R. N., Simonson, J., Pulos, E. 1980. Infant saccadic eye movements to visible and previously visible targets. *Child Dev.* 51:1090–94

Salapatek, P., Banks, M. S. 1978. Infant sensory assessment: Vision. In *Communicative and Cognitive Abilities—Early Behavioral Development,* ed. F. D. Minifie, L. L. Lloyd, pp. 61–106. Baltimore: University Park

Salapatek, P., Bechtold, A. G., Bushnell, E. W. 1976. Infant visual acuity as a function of viewing distance. *Child Dev.* 47:860–63

Samorajski, T., Keefe, J. R., Ordy, J. M. 1965. Morphogenesis of photoreceptor and retinal ultrastructure in a sub-human primate. *Vision Res.* 5:639–48

Schapero, M. 1971. *Amblyopia.* Radnor, Penna.: Chilton

Schneck, M. E., Hamer, R. D., Packer, O. S., Teller, D. Y. 1985. Area-threshold relations at controlled retinal locations in 1-month-old infants. *Vision Res.* In press

Schneck, M., Packer, O., Teller, D. Y. 1982. Spatial summation at controlled retinal locations in one-month-old human infants. *Invest. Ophthalmol. Visual Sci. Suppl.* 22:44

Schor, C. M., Narayan, V., Westall, C. 1983. Postnatal development of optokinetic after nystagmus in human infants. *Vision Res.* 23:1643–47

Sen, D. K. 1980. Anisometropic amblyopia. *J. Pediat. Ophthalmol. Strabismus* 17: 180–84

Shea, S. L., Aslin, R. N. 1982. Stereopsis in strabismic and potentially strabismic children. *Infant Behav. Dev. (Special ICIS Issue)* 5:212

Shea, S. L., Doussard-Roosevelt, J. A., Aslin, R. N. 1985. Pupillary measures of binocular luminance summation in infants and stereoblind adults. *Invest. Ophthalmol. Visual Sci.* In press

Shea, S. L., Fox, R., Aslin, R. N., Dumais, S. T. 1980. Assessment of stereopsis in human infants. *Invest. Ophthalmol. Visual Sci.* 19:1400–4

Shepherd, P. A., Fagan, J. F. III, Kleiner, K. A. 1985. Visual pattern detection in preterm neonates. *Infant. Behav. Dev.* In press

Sherman, S. M., Spear, P. D. 1982. Organization of visual pathways in normal and visually deprived cats. *Physiol. Rev.* 62: 738–855

Shimojo, S., Birch, E. E., Gwiazda, J., Held, R. 1984. Development of vernier acuity in infants. *Vision Res.* 24:721–28

Sireteanu, R., Fronius, M. 1981. Naso-temporal asymmetries in human amblyopia: Consequence of long-term interocular suppression. *Vision Res.* 21:1055–63

Sireteanu, R., Fronius, M., Singer, W. 1981. Binocular interaction in the peripheral visual field of humans with strabismic and anisometropic amblyopia. *Vision Res.* 21: 1065–74

Sjöstrand, J. 1978. Contrast sensitivity in amblyopia: A preliminary report. *Metabolic Ophthalmol.* 2:135–37

Slater, A. M., Findlay, J. M. 1975. Binocular fixation in the newborn baby. *J. Exp. Child Psychol.* 20:248–73

Slater, A. M., Sykes, M. 1977. Newborn infants' visual responses to square wave gratings. *Child Dev.* 48:545–54

Smelser, G. K., Ozanics, V., Rayborn, M., Sagun, D. 1974. Retinal synaptogenesis in the primate. *Invest. Ophthalmol.* 13:340–61.

Smith, E. L., Bennet, M. J., Harwerth, R. S., Crawford, M. L. J. 1979. Binocularity in kittens reared with optically induced squint. *Science* 204:875–77

Smith, E. L., Harwerth, R. S. 1984. Behavioral measurements of accommodative amplitude in rheses monkeys. *Vision Res.* In press

Smith, E. L., Harwerth, R. S., Crawford, M. L. J. 1983. Optically-induced anisometropia in monkeys: Behavioral studies. *Invest. Ophthalmol. Visual Sci. Suppl.* 24:22

Sokol, S. 1978. Measurement of infant visual acuity from pattern reversal evoked potentials. *Vision Res.* 18:33–41

Sokol, S., Dobson, V. 1976. Pattern reversal visually evoked potentials in infants. *Invest. Ophthalmol.* 15:58–62

Sokol, S., Jones, K. 1979. Implicit time of pattern evoked potentials in infants: An index of maturation of spatial vision. *Vision Res.* 19:747–55

Spekreijse, H. 1978. Maturation of contrast EPs and development of visual resolution. *Arch. Ital. Biol.* 116:358–69

Spekreijse, H. 1983. Comparison of acuity tests and pattern evoked potential criteria: Two mechanisms underly acuity maturation in man. *Behav. Brain Res.* 10:107–17

Spinelli, D., Pirchio, M., Sandini, G. 1983. Visual acuity in the young infant is highest in a small retinal area. *Vision Res.* 23: 1133–36

Spira, A. W., Hollenberg, M. J. 1973. Human retinal development: Ultrastructure of the inner retinal layers. *Dev. Biol.* 31:1–21

Stephens, B. R., Banks, M. S. 1985. The development of basic mechanisms of pattern vision. II. Contrast constancy. *J. Exp. Child Psychol.* In press

Swindale, N. V. 1980. A model for the formation of ocular dominance stripes. *Proc. R. Soc. London Ser. B* 208:243–64

Swindale, N. V., Vital-Durand, F., Blakemore, C. 1981. Recovery from monocular deprivation in the monkey. III. Reversal of anatomical effects in the visual cortex. *Proc. R. Soc. London Ser. B* 213:435–50

Takashima, S., Chan, F., Becker, L. E., Armstrong, D. L. 1980. Morphology of the developing visual cortex of the human infant. *J. Neuropathol. Exp. Neurol.* 39:487–501

Tanlamai, T., Goss, D. A. 1979. Prevalence of monocular amblyopia among anisometropes. *Am. J. Optomet. Physiol. Opt.* 56:704–15

Taylor, D. M. 1974. Is congenital esotropia functionally curable? *J. Pedat. Ophthalmol.* 11:3–35

Taylor, D. M., Vaegan, Morris, J. A., Rogers, J. E., Warland, J. 1979. Amblyopia in bilateral infantile and juvenile cataract. *Trans. Ophthalmol. Soc. UK* 99:170–75

Tees, R. C. 1976. Mammalian perceptual development. In *Studies on Development of Behavior and the Nervous System,* Vol 3, ed. G. Gottlieb. New York: Academic

Teller, D. Y. 1979. The forced-choice preferential looking procedure: A psychophysical technique for use with human infants. *Infant Behav. Dev.* 2:135–53

Teller, D. Y. 1981. The development of visual acuity in human and monkey infants. *Trends Neurosci.* 4:21–24

Teller, D. Y. 1982. Scotopic vision, color vision, and stereopsis in infants. *Current Eye Res.* 2:199–210

Teller, D. Y. 1983a. The development of visual acuity in human and monkey infants: Basic and clinical studies. *J. Japan. Assoc. Strabismus Amblyopia* 77:104–12

Teller, D. Y. 1983b. Measurement of visual acuity in human and monkey infants: The interface between laboratory and clinic. *Behav. Brain Res.* 10:15–23

Teller, D. Y., Allen, J. L., Regal, D. M., Mayer, D. L. 1978a. Astigmatism and acuity in two primate infants. *Invest. Ophthalmol. Visual Sci.* 17:344–49

Teller, D. Y., Boothe, R. G. 1980. The development of vision in infant primates. *Trans. Ophthalmol. Soc. UK* 99:333–37

Teller, D. Y., Bornstein, M. H. 1985. Infant color vision. In *Handbook of Infant Perception,* ed. P. Salapatek, L. B. Cohen. In press

Teller, D. Y., Mayer, D. L., Makous, W. L., Allen, J. L. 1982. Do preferential looking techniques underestimate infant visual acuity? *Vision Res.* 22:1017–24

Teller, D. Y., Morse, R., Borton, R., Regal, D. 1974. Visual acuity for vertical and diagonal gratings in human infants. *Vision Res.* 14:1433–39

Teller, D. Y., Peeples, D. R., Sekel, M. 1978c. Discrimination of chromatic from white light by two-month-old human infants. *Vision Res.* 18:41–48

Teller, D. Y., Regal, D. M., Videen, T. O., Pulos, E. 1978b. Development of visual acuity in infant monkeys *(Macaca nemestrina)* during early postnatal weeks. *Vision Res.* 18:561–66

Thomas, J. 1978. Normal and amblyopic contrast sensitivity functions in central and peripheral retinas. *Invest. Ophthalmol. Visual Sci.* 17:746–53

Thomas, J., Mohindra, I., Held, R. 1979. Strabismic amblyopia in infants. *Am. J. Optomet. Physiol. Opt.* 56:197–201

Thorn, F., Doty, R. W., Gramiak, R. 1982. Effect of eyelid suture on development of ocular dimensions in macaques. *Current Eye Res.* 1:727–33

Thorn, F., Fang, L., Held, R. 1984. Orthogonal astigmatic axes in Chinese and caucasian infants. *Invest. Ophthalmol. Visual Sci. Suppl.* 25:221

Tronick, E. 1972. Stimulus control and the growth of the infant's effective visual field. *Percept. Psychophys.* 11:373–75

Tronick, E., Clanton, C. 1971. Infant looking patterns. *Vision Res.* 11:1479–86

Vaegan, Arden, G., Fells, P. 1981. Amblyopia: Some possible relations between experimental models and clinical experience. In *Paediatric Ophthalmology,* ed. D. Taylor, K. Wylson. New York: Dekker

Vaegan, Taylor, D. 1979. Critical period for deprivation amblyopia in children. *Trans. Ophthalmol. Soc. UK* 99:432–39

Valvo, A. 1968. Behavior patterns and visual rehabilitation after early and long lasting blindness. *Am. J. Ophthalmol.* 65:19–24

Van Hof-van Duin, J., Mohn, G., Fetter,

W. P. F., Mettau, J. W., Baerts, W. 1983. Preferential looking acuity in preterm infants. *Behav. Brain Res.* 10:47–50
Van Hof-van Duin, J., Mohn, G. 1984. Vision in the preterm infant. In *Continuity of Neural Functions from Pre- to Postnatal Life,* ed. H. F. R. Prechtl. Oxford: Blackwells. In press
Varner, D., Cook, J. E., McDonald, M., Teller, D. Y. 1984. Discrimination of a tritonopic parir by one- and two-month-old infants. *Invest. Ophthalmol. Visual Sci. Suppl.* 25:162
Vital-Durand, F., Garey, L. J., Blakemore, C. 1978. Monocular and binocular deprivation in the monkey: Morphological effects and their reversibility. *Brain Res.* 158:45–64
Volkmann, F. C., Dobson, M. V. 1976. Infant responses of ocular fixation to moving visual stimuli. *J. Exp. Child Psychol.* 22:86–99
Von Noorden, G. K. 1973a. Experimental amblyopia in monkeys. Further behavioral observations and clinical correlations. *Invest. Ophthalmol.* 12:721–26
Von Noorden, G. K. 1973b. Histological studies of the visual system in monkeys with experimental amblyopia. *Invest. Ophthalmol.* 12:727–38
Von Noorden, G. K. 1980. *Burian and Von Noorden's Binocular Vision and Ocular Motility: Theory and Management of Strabismus.* St. Louis, Mo.: Mosby
Von Noorden, G. K. 1981a. Amblyopia caused by unilateral atropinization. *Ophthalmology* 88:131–33
Von Noorden, G. K. 1981b. New clinical aspects of stimulus deprivation amblyopia. *Am. J. Ophthalmol.* 92:416–21
Von Noorden, G. K., Crawford, M. L. J. 1977. Form deprivation without light deprivation produces the visual deprivation syndrome in Macaca mulatta. *Brain Res.* 129:37–44
Von Noorden, G. K., Crawford, M. L. J. 1978a. Lid closure and refractive error in macaque monkeys. *Nature* 272:53–54
Von Noorden, G. K., Crawford, M. L. J. 1978b. Morphological and physiological changes in the monkey visual system after short-term lid suture. *Invest. Ophthalmol. Visual Sci.* 17:762–68
Von Noorden, G. K., Crawford, M. L. J. 1979. The sensitive period. *Trans. Ophthalmol. Soc. UK* 99:442–46
Von Noorden, G. K., Crawford, M. L. J. 1981a. The effects of total unilateral occlusion vs lid suture on the visual system of infant monkeys. *Invest. Ophthalmol. Visual Sci.* 21:142–46
Von Noorden, G. K., Crawford, M. L. J. 1981b. Failure to preserve cortical binocularity in strabismic monkeys raised in a unidirectional visual environment. *Invest. Ophthalmol. Visual Sci.* 20:665–70
Von Noorden, G. K., Crawford, M. L. J., Levacy, R. A. 1983. The lateral geniculate nucleus in human anisometropic amblyopia. *Invest. Ophthalmol. Visual Sci.* 24:788–90
Von Noorden, G. K., Crawford, M. L. J., Middleditch, P. R. 1976. The effects of monocular visual deprivation: Disuse or binocular interaction. *Brain Res.* 111: 277–85
Von Noorden, G. K., Crawford, M. L. J., Middleditch, P. R. 1977. Effect of lid suture on retinal ganglion cells in Macaca mulatta. *Brain Res.* 122:437–44
Von Noorden, G. K., Dowling, J. E. 1970. Experimental amblyopia in monkeys. II. Behavioral studies of strabismic amblyopia. *Arch. Ophthalmol.* 84:215–20
Von Noorden, G. K., Dowling, J. E., Ferguson, D. C. 1970. Experimental amblyopia in monkeys. I. Behavioral studies of stimulus deprivation amblyopia. *Arch. Ophthalmol.* 84:206–14
Von Noorden, G. K., Middleditch, P. R. 1975. Histology of the monkey lateral geniculate nucleus after unilateral lid closure and experimental strabismus: Further observations. *Invest. Ophthalmol. Visual Sci.* 14:674–83.
Wade, N. J. 1976. On interocular transfer of the movement aftereffect in individuals with and without normal binocular vision. *Perception* 5:113–18
Ware, C., Mitchell, D. E. 1974. On interocular transfer of various visual aftereffects in normal and stereoblind observers. *Vision Res.* 14:731–34
Werner, J. S. 1982. Development of scotopic sensitivity and the absorption spectrum of the human ocular media. *J. Opt. Soc. Am.* 72:247–58
White, B. L. 1971. *Human Infants: Experience and Psychological Development.* Englewood Cliffs, N.J.: Prentice-Hall
Wiesel, T. N. 1982. Postnatal developmental of the visual cortex and the influence of environment. *Nature* 299:583–91
Wiesel, T. N., Hubel, D. H. 1974. Ordered arrangement of orientation columns in monkeys lacking visual experience. *J. Comp. Neurol.* 158:307–18
Wiesel, T. N., Raviola, E. 1977. Myopia and eye enlargement after neonatal lid fusion in monkeys. *Nature* 266:66–68
Wiesel, T. N., Raviola, E. 1979. Increase in axial length of the macaque monkey eye after corneal opacification. *Invest. Ophthalmol. Visual Sci.* 18:1232–36
Williams, R., Boothe, R. 1981. Development of optical quality in the infant monkey *(Macaca nemestrina)* eye. *Invest. Ophthalmol. Visual Sci.* 21:728–36
Williams, R., Boothe, R., Kiorpes, L., Teller,

D. 1981. Oblique effects in normally reared monkeys *(Macaca nemestrina)*: Meridional variations in contrast sensitivity measured with operant techniques. *Vision Res*. 21:1253–66

Wilson, J. R., Hendrickson, A. E. 1981. Neuronal and synaptic structure of the dorsal lateral geniculate nucleus in normal and monocularly deprived macaca monkeys. *J. Comp. Neurol*. 197:517–39

Wolfe, J. M., Gwiazda, J., Held, R. 1983. The meaning of non-monotonic psychometric functions in the assessment of infant preferential looking acuity. A reply to Banks et al (1982) and Teller et al (1982). *Vision Res*. 23:917–20

Yonas, A. 1981. Infants' responses to optical information for collision. See Aslin et al 1981, pp. 313–34

Young, F. A. 1970. Development of optical characteristics for seeing. In *Early Experience and Visual Information Processing in Perceptual and Reading Disorders*, pp. 35–61. Washington, DC: Natl. Acad. Sci.

Zak, T. A., Morin, J. D. 1982. Early surgery for infantile esotropia: Results and influence of age upon results. *Can. J. Ophthalmol*. 17:213–18

Ann. Rev. Neurosci. 1985. 8:547–83

SPATIAL FREQUENCY ANALYSIS IN THE VISUAL SYSTEM

Robert Shapley

Rockefeller University, New York, New York 10021

Peter Lennie

Center for Visual Science, University of Rochester, Rochester, New York 14627

Within the last 15 years, a method called "spatial frequency analysis" has been applied widely to the study of receptive fields of neurons in the visual pathway. Out of this work have emerged new concepts of how the brain analyzes and recognizes visual images. The aim of this paper is to explain why "spatial frequency analysis" is useful, and to review the insights into visual function that have resulted from its application. It is important to note at the outset that while spatial frequency analysis can provide a comprehensive description of the behavior of neurons in which signals are summed linearly (see below), it has much more limited application to the behavior of neurons that combine signals nonlinearly. Because of this, the greatest insights into visual information processing have come and probably will continue to come from a combined use of space, time, spatial frequency, and temporal frequency measurements.

The earliest applications of spatial frequency analysis were motivated by the idea that visual physiology and psychophysics could be more closely related by a uniform approach to the problems of spatial vision. This approach has continued to bind together the psychophysics and physiology of spatial vision, and the interested reader will find relevant psychophysical work discussed in reviews by Braddick et al (1978), De Valois & De Valois (1980), Robson (1980), Westheimer (1984), and Kelly & Burbeck (1984).

0147-006X/85/0301-0547$02.00

WHAT IS SPATIAL FREQUENCY ANALYSIS?

A filter is a stimulus-response machine that responds better to certain stimuli than to others. Visual neurons may be viewed as neural filters of visual signals. The natural method for studying filters is systems analysis, a procedure that allows one to characterize a filter by means of its response to a simple set of inputs (stimuli). Most scientists are familiar with filters that operate on time varying signals. For example, a "low pass" filter is used to remove the high-frequency components from a wave form such as a recording of neuronal electrical activity. Less familiar, perhaps, are filters for wave forms that vary in space rather than in time. A microscope objective, for instance, is a spatial filter that, in addition to magnifying, removes some high frequency spatial information and thereby limits the spatial resolution of the microscope. Spatial frequency analysis is a specialization of systems analysis designed for the study of the spatial filtering properties of optical imaging devices. It has been extended to the study of the imaging capacity of the eye (Campbell & Green 1965) and the study of single visual neurons (e.g. Enroth-Cugell & Robson 1966). Using this approach, one determines the spatial filtering characteristics of a cell by measuring its response to a set of sinusoidal gratings.

Figure 1 shows the luminance profile of a sinusoidal grating. Along one axis (perpendicular to the bars of the grating), the luminance is a sinusoidal function of position:

$$\mathrm{L(x)} = \mathrm{L_0} + \mathrm{L_1}\sin(2\pi kx + \phi) = L_0\,[1 + c\sin(2\pi kx + \phi)]. \qquad (1)$$

Here $L(x)$ is the luminance as a function of position, L_0 the mean luminance, L_1 the amplitude modulation of the luminance in the grating, c (equal to L_1/L_0) the contrast, ϕ the spatial phase, and k is the spatial frequency, in cycles or periods of the grating per unit of visual angle (usually degrees). Along the other axis of the grating (parallel to the bars), the luminance is fixed; thus,

$$L(x,y) = L(x), \qquad \text{for all } y. \qquad (2)$$

This is why we say a grating is one-dimensional. The contrast c determines how dark or light the bars of the grating are, and is independent of the mean level, L_0. At low values of c, the grating is almost indistinguishable from its gray background, whereas at high values the bars of the grating appear much lighter (or darker) than the background.

SPATIAL FILTERING BY NEURONS

Spatial filtering properties may be determined by measuring the neural response to sine gratings at several spatial frequencies. Contrast may be either fixed or variable. If spatial frequency is varied while contrast is fixed, one measures

PERIODIC STIMULUS

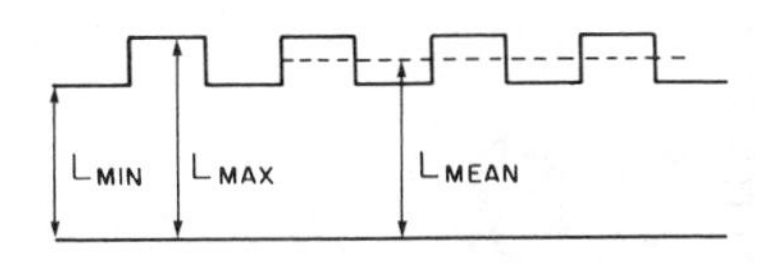

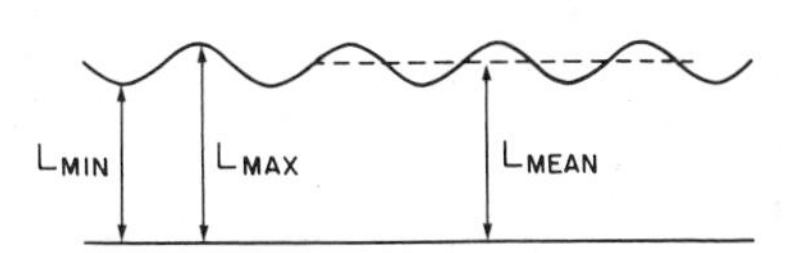

Figure 1 Luminance profiles of periodic spatial patterns. Top, a square wave grating: bottom, a sine wave grating. The contrast of such patterns is $(L_{max} - L_{min})/(L_{max} + L_{min})$, where the luminances L_{mean}, and L_{max}, and L_{min} are as indicated in the figure.

a *spatial frequency response function*. If the spatial frequency is varied and the contrast is adjusted to produce a criterion response, one can then determine the *spatial frequency sensitivity function,* known also for historical reasons as the *contrast sensitivity function* (Enroth-Cugell & Robson 1966, Campbell & Robson 1968). The contrast sensitivity is the reciprocal of the contrast required to produce a criterion response. If the response of the neuron is strictly proportional to contrast, then the spatial frequency response and sensitivity functions are identical. A typical spatial frequency sensitivity function for a retinal ganglion cell is shown in Figure 2. It illustrates a general finding: visual neurons are selectively sensitive (tuned) to a particular range of spatial frequencies. Above and below the optimal spatial frequency, the contrast sensitivity falls. For the cell that provided the results of Figure 2, and other ganglion cells like it, the tuning is rather broad. For cells in the visual cortex, the tuning may be quite sharp (see below). Thus, we can treat visual neurons as tuned filters for spatial frequencies in the same way that auditory physiologists treat auditory neurons as tuned filters for sound frequencies (cf Campbell 1974, Robson 1975, De Valois & De Valois 1980)

In the typical visual neurophysiological experiment, the eyes are motionless, or almost so, and responses are evoked by temporal modulation of contrast or by motion of the sine-wave grating pattern across the visual field. In the following discussion of spatial filtering and spatial summation, we assume that stimuli are modulated at a fixed rate. The rate of modulation (or the rate of drift) used in a particular experiment does influence spatial frequency response functions as described in the section *Spatiotemporal Separability and Coupling in Receptive Fields*.

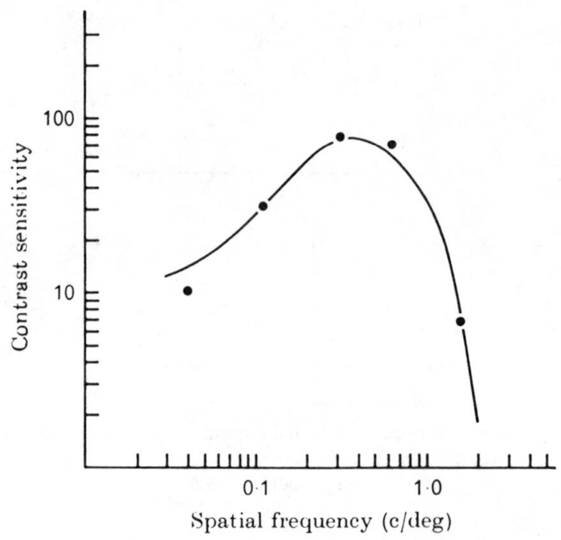

Figure 2 The spatial frequency sensitivity function for an X-ganglion cell in the cat's retina. *Filled circles* show the reciprocal of the contrast required to produce a criterion response of 10 imp/sec. The *smooth curve* is from the Difference of Gaussians model. From Enroth-Cugell & Robson (1966).

Fourier Analysis and Synthesis

LINEARITY The concept of linearity is important because, if a neuron acts as a linear spatial filter, measurement of its spatial frequency response function allows one to predict its response to any arbitrary spatial pattern. Linearity means that the response to a stimulus that is the sum of two simpler stimuli is the sum of the responses to each of the simpler stimuli presented alone. For example, one might temporally modulate the illumination of a small spot in the center of the receptive field (stimulus 1) or modulate an annulus around the spot (stimulus 2). If the cell were combining signals in a linear manner, the response to simultaneous modulation of stimulus 1 and stimulus 2 (stimulus 1 + 2) would be simply the sum of the responses to stimulus 1 and stimulus 2 given separately. Thus in a linear filter there is no interaction between stimuli; their effects are just added.

SINE WAVES The reader might wonder why sinusoidal gratings are used for spatial frequency analysis. It is because sine functions are uniquely suited for analyzing linear filters. The sine function is the one spatially extended function that passes through a linear filter unchanged in form. A linear filter may change the amplitude of the sinusoid, or displace it relative to the input (that is, shift its phase) by an amount specified by the amplitude and phase of the spatial frequency response, but it never changes its shape or frequency. If a sine wave goes in, a sine wave of the same frequency comes out. This property is not possessed by other waveforms. For example, the cornea and lens of the eye form a linear optical imaging system, a linear spatial filter. For most spatial stimuli—like spots, lines, or bars—passage through the eye's spatial filter produces a change in the spatial distribution of light because of blur and scatter. However, a sine grating is imaged as a sine grating on the retina; blur and scatter cause only a loss of contrast.

THE FOURIER REPRESENTATION Any waveform may be represented as the *sum* of sinusoidal waveforms of different frequencies, amplitudes, and phases. The coefficients of the terms in the sum are obtained by calculating the *Fourier transform* of the original waveform. The Fourier transform is a complex-valued function of frequency, meaning that at each frequency it gives the amplitude and phase of the sine at that frequency (see Bracewell 1978). The amplitude of the Fourier transform as a function of frequency is often called the *amplitude spectrum*. The phase of the transform as a function of frequency is called the *phase spectrum*. The Fourier representation of waveforms as a sum of sines is a mathematical property of functions and is independent of whether one is dealing with linear or nonlinear systems. Thus, for the analysis of spatial vision, any pattern can be represented as a sum of sine gratings of the appropriate amplitudes and phases.

The three facts just presented allow one to predict the response of a linear spatial filter to any spatial pattern. Let us review them:

1. A linear spatial filter simply adds the responses to sums of stimuli.
2. Sine waves are unchanged in form when passed through a linear filter.
3. Any input waveform (or output waveform) may be represented as a sum of sinusoids where the coefficients in the sum are given by the Fourier transform.

The procedure for predicting the response of the filter to any stimulus is then as follows: Construct the Fourier representation of the stimulus. Each sinusoidal component in the sum passes through the linear filter by having its amplitude multiplied by the amplitude of the *spatial frequency response* at the frequency of the sine, and its phase added to the phase of the spatial frequency response. The Fourier representation of the output waveform is the sum of all the sinusoids that have been acted upon by the filter in this way. This means that the Fourier transform of the output of the filter is simply the *product* of the Fourier transform of the input waveform multiplied by the spatial frequency response. The output can then be calculated from its Fourier representation by inverse Fourier transformation (see Bracewell 1978 again). By this procedure, called Fourier synthesis, one can predict the response of a linear system to any arbitrary input. A clear example is the prediction of the response of a retinal ganglion cell to a step of luminance on its receptive field from the measured spatial frequency response function (Enroth-Cugell & Robson 1984).

The behavior of many visual neurons is linear enough that spatial frequency analysis can be used to characterize their visual function completely. What if the neurons are nonlinear? Although the procedures of Fourier synthesis do not work for nonlinear neurons, spatial frequency analysis has been used to dissect linear from nonlinear components, and to gain some insight into the nature of the nonlinearity. Since conventional receptive field maps cannot be used to predict the response of a nonlinear neuron to arbitrary stimuli, what little we do understand about the nature of visual processing in these cells has come mainly from the application of spatial frequency analysis.

SPATIAL FREQUENCY ANALYSIS AND RECEPTIVE FIELDS

The value of spatial frequency analysis can be seen in dealing with the problem of how to describe the receptive field of a retinal ganglion cell. We use this as an example, and then go on to discuss other major insights that have been provided by this approach.

It is often useful to know the distribution of sensitivity within the receptive field: We may want to know how the properties of a receptive field depend upon the morphology of the ganglion cell's dendritic field and the density of its synaptic contacts, or we may want to establish how the ganglion cell limits spatial frequency resolution and contrast sensitivity measured psychophysically.

Four experimental methods have been widely used to obtain the spatial sensitivity distribution of a ganglion cell's receptive field. The *point weighting function* is defined as the sensitivity for a point stimulus as a function of the position of the stimulus in the visual field. It can be measured directly with a small spot of light placed in all positions in the receptive field. If, instead of using a small spot, one measures sensitivity with a line at positions along a diameter of the receptive field, the corresponding sensitivity profile is called the *line-weighting function*. A third method is to measure the *area-sensitivity curve*. This involves centering a spot of light on the receptive field, and measuring sensitivity as a function of spot area. The fourth method is by *Fourier transformation of the spatial frequency sensitivity function*. If one can show that a ganglion cell approximates a linear system, one can calculate the (one-dimensional) spatial distribution of sensitivity, or line-weighting function, from the spatial frequency response (or sensitivity) function. It is just the Fourier transform of the spatial frequency response.

Although all these methods prove useful, and for a linear cell can provide equivalent descriptions of the receptive field, there are practical drawbacks: eye movements, prolonged measurement times, and local saturation. The point weighting function is tedious to obtain, and is vulnerable to errors introduced by eye-movements. Moreover, in regions of the receptive field where the sensitivity is low, very intense spots are required to measure sensitivity and these may saturate pre-ganglionic elements in the retina. The line weighting function suffers from the same problems, although it is more easily and quickly obtained. The area-sensitivity method avoids local saturation, but is rather insensitive to the fine structure of the receptive field, and cannot tell one about the presence or absence of radial symmetry. The measurement of the spatial frequency response or sensitivity has fewer drawbacks. The stimulus is spatially extended, and in general we do not depend upon it being precisely positioned on the receptive field—a great advantage when dealing with very small receptive fields of the kind found in the monkey's fovea. The grating can be of rather low contrast to avoid saturation, and most applications require the eye to be stable for only short periods. Useful information can therefore be obtained even in the presence of eye movements, as long as they are not too frequent.

Evaluation of Receptive Field Models

Rodieck (1965) was the first to represent the receptive field of the cat's ganglion cell as overlapping center and surround mechanisms. Each mechanism may be thought of as a group of receptors and interneurons whose signals are pooled together. Light-evoked signals generated within each pool are summed, and the resulting signals from center and surround are summed at the ganglion cell. The center mechanism has a narrow spatial distribution of sensitivity, and the surround a rather broader one. Rodieck proposed that these two spatial distributions could be approximated by Gaussian surfaces with different extents of spread, as in Figure 3B. Formally, he proposed that the point weighting function, $S(r)$, could be written as

$$S(r) = \mathrm{k_c} \exp[-(r/r_c)^2] - k_s \exp[-r/r_s)^2] \tag{3}$$

where k_c is the peak sensitivity of the center, at $r=0$, and k_s is the peak local sensitivity of the surround, also at $r=0$. The spatial spread of the center is r_c; at $r=r_c$ the sensitivity has declined from the peak by a factor of e^{-1}, or 1/2.718. The spatial spread of the surround is r_s.

Certain properties of the spatial frequency sensitivity function can be understood in terms of Rodieck's model. The spatial frequency resolution limit (the highest spatial frequency that can be resolved by the cell) is due to the finite size of the center, and is roughly proportional to the reciprocal of the center's spread, r_c (Cleland et al 1979, So & Shapley 1979, 1981, Linsenmeier et al 1982). The grating of optimal spatial frequency is one to which the center is

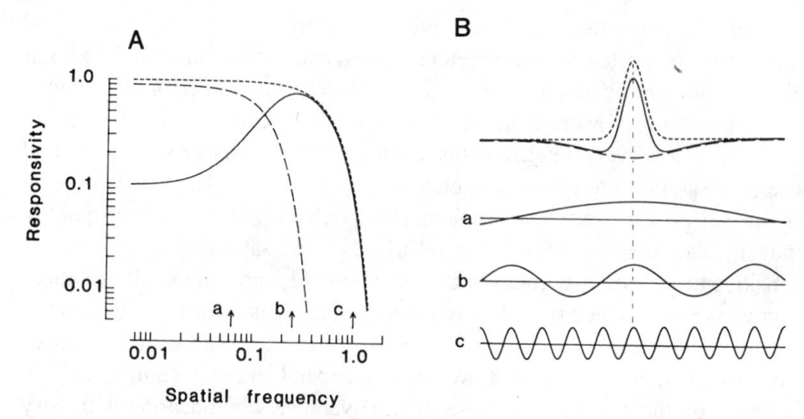

Figure 3 Difference of Gaussians model in (A) *spatial frequency* and (B) *space*. In *A*, the unbroken curve is the spatial frequency response of Rodieck's Difference of Gaussians model. The *dashed curves* show the spatial frequency responses of the center *(fine dashes)* and surround *(coarse dashes)*, each of which is a Gaussian function of spatial frequency. In this example, the ration $k_s r_s^2/k_c r_c^2$ was 0.9, and the ratio r_s/r_c was 4. These are typical values for X-cells in the cat's retina. In *B*, the *upper set of curves* represents the Difference of Gaussians model; the *solid curve* is the sum of the narrower center Gaussian *(fine dashes)* and the broader surround Gaussian *(coarse dashes)*. These three curves are the inverse Fourier transforms of the corresponding spatial frequency responses in *A*. The three lower curves in *B* are profiles of sinusoidal gratings presented to the receptive field with even symmetry, a position of maximal response for a contrast reversal grating. The spatial frequencies of these three sinusoids, in the units in which the receptive field dimensions are expressed, are labeled a,b,c in *A*. From Enroth-Cugell & Robson (1984).

quite sensitive but the surround is very insensitive (see Figure 3B). The loss of sensitivity seen at low spatial frequencies reflects the increasingly effective antagonism from the surround.

We can go beyond this qualitative description to estimate precisely the center's spread, r_c, and peak sensitivity, k_c, and the surround's spread, r_s, and sensitivity, k_s, from measurements of spatial frequency sensitivity. Enroth-Cugell & Robson (1966) showed that if the distribution of sensitivity within the receptive field of a linearly behaving ganglion cell is a difference of Gaussians *in space*, then the spatial frequency sensitivity function will be a difference of Gaussians *in spatial frequency* (as long as the system can be treated as a linear one). The smooth curve in Figure 3A, which is the spatial frequency sensitivity function of Rodieck's model, is a difference of Gaussian functions in spatial frequency. At high spatial frequency, the sensitivity falls as a Gaussian function of spatial frequency. The spread of this function is the reciprocal of the product of π (3.1416) times the center's spread ($1/\pi r_c$), so the size of the center can be estimated with great accuracy from the high-frequency roll-off of the spatial frequency sensitivity curve. Cleland et al (1979) and So & Shapley (1981) showed that this estimate of center size based on spatial frequency measurements is exactly the same as that based on line weighting or area-sensitivity measurements, but is more easily obtained. Thus, spatial frequency analysis can be used to provide a comprehensive, precise description of the receptive field mechanisms of neurons that act as linear spatial filters.

NEW INSIGHTS INTO RETINAL MECHANISMS

Having demonstrated how receptive fields can be studied with spatial frequency analysis, we proceed to review what has been learned from the application of these new techniques.

Analysis of Receptive Field Types

THE EXISTENCE OF Y-CELLS Enroth-Cugell & Robson's (1966) description of two previously unrecognized classes of ganglion cells that they called X and Y provided the first clear evidence for parallel processing in the retina. Their discovery resulted from the application of techniques of systems analysis and is one of the early successes of the approach. Although it had been known for some time that the conduction velocities of retinal afferents fell into distinct groups, the relation between visual functional properties and conduction velocity, which was established by Cleland et al (1971), had to wait for the insight that there were functionally distinct types of cell (see Rodieck 1979, Lennie 1980, Stone 1983). We shall discuss these results critically because the methods used to obtain them have since been applied to the investigation of receptive fields of geniculate and cortical cells.

Two differences between X- and Y-cells were revealed by their responses to gratings:

1. When stimulated by drifting gratings, X-cells responded with a modulation of their impulse discharge rate in synchrony with the passage of grating cycles across their receptive fields, but with a negligible change in their mean impulse rate. Y-cells responded to drifting gratings with both a modulation of the impulse rate and an increase in the average impulse rate. The difference was especially clear for gratings of high spatial frequencies near the resolution limit of the cell. Under these conditions, an X-cell continued to produce only a modulated response with no change in mean rate; the discharge of a Y-cell was modulated very little but mean rate was clearly elevated (Enroth-Cugell & Robson 1966). In these experiments, the average illumination falling on the ganglion cells' receptive fields was constant. If the average input to a linear system is constant, then the average output must remain constant, so the elevation of average impulse rate caused by drifting gratings is evidence for a nonlinearity in the retinal network leading to the Y-cell (Enroth-Cugell & Robson 1966, Victor & Shapley 1979b).

2. X- and Y-cells were distinguished by their responses to temporal modulation of the contrast of standing sinusoidal gratings. The stimulus used by Enroth-Cugell & Robson can be represented as

$$L(x,t) = L_0[1 + c \sin(2\pi kx + \phi)\, M(t)] \tag{5}$$

where c is contrast, k is spatial frequency, and ϕ is spatial phase or position in the field. The temporal modulation $M(t)$ was a square wave that went from 0 to 1 repetitively. When $M(t)$ was 0, the stimulus was a spatially uniform screen at the mean luminance L_0; when $M(t)$ was 1, the stimulus was the sine grating $L_0[1 + c \sin(2\pi kx + \phi)]$. The spatial frequency chosen was near the optimum for the cell, a point to which we return below. The spatial phase, ϕ, was adjusted to evoke a maximal response. (Note that by "response" we mean a modulation in the rate of impulse firing synchronized with the stimulus. For a cell with an even symmetric receptive field, the maximum response occurs when the peak or trough of the grating is located on the axis of symmetry; see Figure 3B.) Then the spatial phase was changed until a phase was found at which the cell gave the least response. For X-cells this minimum was no response at all, and the spatial phase or position at which it was obtained was called the "null position." The existence of this "null position" is consistent with the hypothesis that the pathways that lead to the X cell are linear. No null position could be found for Y-cells. At the position of least response, introduction of the grating elicited a burst of impulses, and withdrawal of the grating produced an (almost) identical burst. The absence of a null position, and the peculiar excitatory "on-off" character of the Y-cell's response, reveal a nonlinearity in the retinal network that leads to the Y-cell.

The implications of the "null test" experiment became clearer after later work by Hochstein & Shapley (1976a,b). They investigated, for ganglion cells in cat, how the amplitude of responses to gratings undergoing contrast reversal depended upon spatial phase. The contrast-reversal grating can be represented formally by Eq. 5, with the temporal modulation signal $M(t)$, equal to $\sin(2\pi ft)$, where f is the temporal frequency of contrast reversal. The grating was presented at a series of spatial phases, and for each position the modulated discharge rate was analyzed into its Fourier components: the fundamental component at the temporal frequency of the stimulus modulation, and harmonic components at frequencies that were twice, three times, and higher multiples of the stimulus temporal frequency. Hochstein & Shapley (1976a) found that the response of an X-cell was mainly at the fundamental frequency of the stimulus, and they observed that the amplitude of the fundamental response component varied sinusoidally with spatial phase, as shown in Figure 4. That is, the X-cell response could be written:

$$R_x(k,f,c,\phi) = S_x(k,f,c)\sin(\phi) \tag{6}$$

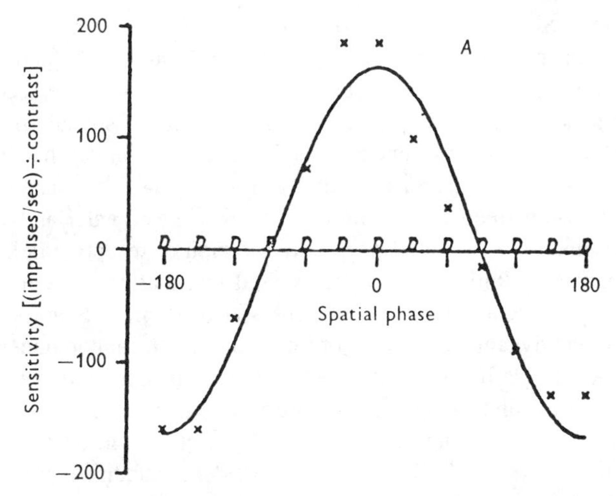

Figure 4 Variation of response amplitude with spatial phase, for an X-cell. Sensitivity for the fundamental response of an on-center X-ganglion cell excited by contrast reversal gratings was determined at several spatial phases, and is plotted as X's. Negative sensitivity is when the phase of the response changed by 180 deg. Points marked *D* are the amplitudes of the second harmonic, which were in the noise. The *smooth curve* is a sine function. From Hochstein & Shapley (1976a).

where $S_x(k,f,c)$ expresses the fact that the response depended on spatial frequency, k, temporal frequency, f, and contrast, c. The "null positions" of Enroth-Cugell & Robson (1966) are just those places where the sinusoidal function of spatial phase in Eq. 6 equals zero. Three properties of the receptive field are sufficient to produce the sinusoidal variation of response amplitude observed in X-cells: linearity of local responses in the photoreceptors, linear spatial summation of signals from different local regions, and homogeneity of temporal properties across the receptive field. When the contrast of a grating is modulated, the amplitude of modulation at each point in the receptive field is a sinusoidal function of the spatial phase, as can be seen from Eq. 5. If the local responses in photoreceptors are linear, then the local response will have the same sinusoidal dependence on spatial phase as the stimulus. If the ganglion cell simply adds these local responses, the grand sum will be a sum of sinusoids of spatial phase, all of which have the same argument, and therefore the sum will also be a sinusoid of phase. This reasoning is correct if the local responses are synchronous, i.e. if the spatial and temporal aspects of the summed response are separable (see section on Spatiotemporal Separability, below). However, if the receptive field is not spatiotemporally separable, the amplitude of the response may not vary sinusoidally with spatial phase even though local responses are summed linearly (Movshon et al 1978a).

Hochstein & Shapley (1976a,b) found in the responses of Y-cells a fundamental component whose amplitude varied sinusoidally with spatial phase, but (especially at high spatial frequencies) there was also a large second harmonic component, and its amplitude was independent of spatial phase. This remarkable behavior of the second harmonic component implies that the receptive field of the Y-cell contains rectifying mechanisms (or "subunits") that have higher spatial frequency resolution than the linear center and surround mechanisms and are distributed over a wide region of receptive field (Hochstein & Shapley 1976b). The different spatial distributions of sensitivity of fundamental and second harmonic responses (Hochstein & Shapley 1976b), and the different dynamic properties of linear and nonlinear components (Victor & Shapley 1979a,b), led to the idea that subunits exist independently of the center and surround, as shown in Figure 5. This idea has been corroborated by Frishman & Linsenmeier (1982), who used pictrotoxin, an antagonist to gamma-aminobutyric acid (GABA), to suppress the subunit input to the Y-cell without effect on the center or surround. A drifting grating will excite a large population of subunits at different temporal phases, and will therefore give rise to the elevation of mean impulse rate observed by Enroth-Cugell & Robson (1966). The subunits resolve grating patterns about as well as X-cells from the same retinal locus, suggesting that a single retinal interneuron could be the source of the X-cell center and the Y-cell subunit (cf Victor & Shapley 1979b).

Nonlinear subunits seem to be distributed far beyond the boundaries of the classical center and surround of the receptive field. Krüger & Fischer (1973) were able to evoke second harmonic responses from ganglion cells by using contrast reversal gratings that covered a large part of the visual field, excluding

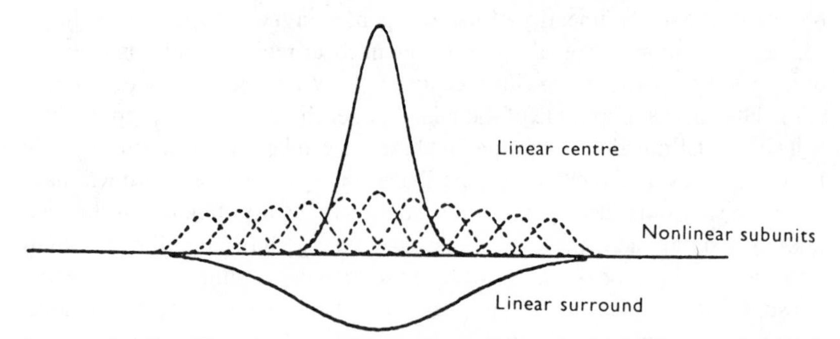

Figure 5 The nonlinear subunit model for cat Y-retinal ganglion cells. The classical center-surround organization of the Y-cell is indicated by the *solid curves*. The *dashed curves* are approximations to the sensitivity profiles of the many nonlinear subunits. Measurements of the spatial frequency sensitivities indicate that subunits are about 1/3 as wide as the Y-cell center mechanism, and about as wide as the centers of neighboring X-cells. Subunits extend farther from the center of the receptive field than is indicated here. From Hochstein & Shapley (1976b).

the classical receptive field and a substantial region around it. Barlow et al (1977) later found that this "shift effect" was pronounced in Y- but not in X-cells, and Derrington et al (1979) showed that its spatial and temporal properties were like those of the subunits within the classical receptive field.

The results of the null test and its modifications are sometimes interpreted incorrectly to mean that *center-surround signal summation* is linear in X-cells and nonlinear in Y-cells. The following argument demonstrates why such as interpretation is a mistake. The most useful spatial gratings for identifying neurons as X- or Y-cells are those of high spatial frequency, near the spatial frequency resolution limit of the cell. These patterns necessarily must be stimulating the neural elements of highest spatial frequency resolution, and therefore spatially the smallest, in the receptive field. The large, low-resolution receptive field surround mechanism will not be able to distinguish such high spatial frequency patterns from uniform illumination and therefore will not respond to them. Thus, the usual spatial summation experiment measures the linearity or nonlinearity *of the smallest receptive field mechanism(s)*. In the X-cell, the smallest receptive field mechanism is the linear center. In the Y-cell, the smallest receptive field mechanism is the nonlinear subunit. The linearity of center-surround summation is a separate issue. In fact, Hochstein & Shapley (1976a) found some indications that center-surround summation was nonlinear in some X-cells. Therefore, they concluded that one should not call X-cells "linear" and Y-cells "nonlinear," but rather one ought to realize that the tests of spatial summation revealed the presence of the small nonlinear subunits in Y-cells but not in X-cells.

Spatial frequency analysis also revealed a functional visual difference between X- and Y-cells: they were optimally tuned to different spatial frequencies and their fundamental spatial frequency resolution was markedly different. X-cells could follow drifting sine gratings with temporally modulated firing up to a spatial frequency three times higher, on the average, than in neighboring Y-cells (Hochstein & Shapley 1976b, Cleland et al 1979). That is, the spatial frequency resolution for a fundamental response was three times higher in X-cells than in Y-cells. Cleland et al (1979) demonstrated the decrease of spatial frequency resolution with increasing retinal eccentricity in both X- and Y-cells, with the X/Y ratio in fundamental spatial frequency resolution staying about 3. Furthermore, the *optimal spatial frequency,* the spatial frequency of the grating that produced the biggest response, was higher for X-cells than for Y-cells at a given retinal locus (Victor & Shapley 1979a). X- and Y-cells are tuned to different parts of the spatial frequency range: the Y-cells for coarse patterns, the X-cells for fine patterns. These results on spatial frequency resolution and tuning are consistent with previously presented concepts about X- and Y-cell receptive fields. The lower spatial frequency resolution of Y-cells implies that they have larger centers than nearby X-cells. The X-cell center is about the same size as each nonlinear subunit, that is, about one third the diameter of the Y-cell center.

Properties of Receptive Field Mechanisms in X-cells

Besides being used to sort cells into functional categories, spatial frequency analysis has been employed to investigate the properties of retinal receptive field mechanisms, as the following examples illustrate.

ASYMMETRY OF RECEPTIVE FIELD Rodieck's (1965) model represents the ganglion cell's receptive field as the difference of two Gaussian functions. The usual analysis of spatial contrast sensitivity curves (Enroth-Cugell & Robson 1966) includes the assumption that these are concentric and therefore the sensitivity profile has even symmetry around an axis through the middle of the receptive field. This assumption can be examined by establishing whether the spatial frequency response function obtained with a moving grating pattern depends upon the direction in which the grating moves. Following Dawis et al (1984), let us consider gratings drifting from right to left as having positive spatial frequency, and those drifting from left to right as having negative spatial frequency. Then one can consider the symmetry around the axis of zero spatial frequency of the function that relates phase of response and spatial frequency. For a receptive field that is even-symmetric, the phase of responses obtained for opposite directions of motion will be an even-symmetric function of spatial frequency (Dawis et al 1984). By applying this analysis to the responses of X-cells driven by moving gratings, Dawis et al found that the receptive fields of most X-cells were somewhat asymmetric.

SPATIOTEMPORAL SEPARABILITY AND COUPLING IN RECEPTIVE FIELDS If the spatial frequency response or sensitivity function were independent of the temporal frequency of drift or contrast modulation, then the receptive field would be spatiotemporally separable. Spatiotemporal separability means that the temporal response properties at all points of the receptive field are the same, so that the point weighting function can be factored into temporal and spatial components; i.e. the spatial distribution of sensitivity would not depend on the temporal frequency at which the measurements were made. For example, Rodieck's (1965) model, discussed above, describes a receptive field that is spatiotemporally separable. Although this is a good approximation when stimulus modulation is in the low temporal frequency range, there is considerable evidence against spatiotemporal separability and for spatiotemporal coupling in cat X-cell receptive fields (Victor & Shapley 1979a, Derrington & Lennie 1982, Enroth-Cugell et al 1983, Dawis et al 1984). All this work implies that the surround mechanism of the receptive field either has a slightly longer delay than the center mechanism, or that signals from the surround undergo somewhat more temporal filtering than those from the center.

INHOMOGENEITY OF SURROUND The simplest interpretation of the observation that the receptive field is spatiotemporally coupled is that center and surround are each spatiotemporally separable, but that center and surround have different temporal response properties (Derrington & Lennie 1982, Enroth-Cugell et al 1983, Dawis et al 1984). The center mechanism does appear to be spatiotemporally separable because values of the center's spatial spread, r_c, estimated from contrast sensitivity functions, do not depend upon the temporal frequency at which the measurement is made. However, estimates of the surround spatial spread, r_s, obtained by Dawis et al (1984), from their analysis of the amplitude and phase functions of the responses to moving gratings, suggest that for most X-cells the extent of the surround depends upon temporal frequency. The surround itself therefore seems to be spatiotemporally coupled, implying that different regions of the surround have different temporal response properties.

ORIENTATION SELECTIVITY Cleland & Enroth-Cugell (1968) and Hammond (1974) showed, by mapping receptive fields of cat ganglion cells with spots of light, that many receptive fields were in fact elliptical. Spatial frequency analysis provides a rapid and powerful method for determining the ellipticity. Levick & Thibos (1982) have used it to show that the major axis of a receptive field center tends to lie along a line connecting the receptive field to the area centralis. The orientation of major and minor axes of the receptive field center can be determined by choosing a grating of spatial frequency near the spatial frequency resolution limit of the cell (which gives a pure central response), and observing how the sensitivity (or amplitude of response) varies with orientation. A complete description of the receptive field can be obtained from spatial frequency sensitivity functions (or corresponding response functions), at many different orientations, for one can then establish whether center or surround, or both, are elliptical. Levick & Thibos' (1982) observations show that, for most X-cells, orientation biases diminish as spatial frequency is reduced and the surround contributes to the response; this suggests that in such units only the center is elliptical.

EFFECT OF ADAPTATION LEVEL ON RECEPTIVE FIELD ORGANIZATION It is well-established that at very low levels of illumination the surround of the receptive field of a cat's ganglion cell becomes relatively less effective (Barlow et al 1957), but there has been no agreement on whether or not the distribution of sensitivity within the receptive field changes with the transition from rod to cone signals, and whether the size of the center increases as the level of light adaptation is reduced. Conventional spot maps of the receptive field have not provided clear answers because they cannot easily show what is happening to the surround, but spatial frequency sensitivity functions obtained at different levels of illumination provide a simple and precise way to characterize the changes in receptive field organization. For the cat these measurements show (Derrington & Lennie 1982) that the center of the receptive field is enlarged somewhat (r_c may increase by up to 30%) as the adaptation level falls from photopic to low scotopic. Derrington & Lennie found no systematic change in r_s, the spread of the surround.

SPATIAL FREQUENCY ANALYSIS APPLIED TO THE LATERAL GENICULATE NUCLEUS

The lateral geniculate nucleus (LGN) in the thalamus is in most mammals the major relay to the visual cortex from the retina. The visual function of the lateral geniculate nucleus has remained somewhat mysterious. Is the LGN simply a relay nucleus, or is substantial visual information processing done there? Spatial frequency response measurements have provided new insights into the function of the lateral geniculate nucleus.

Distinct Classes of Cell

CAT In the cat's retina, X- and Y-cells (and some other types) can be distinguished by a number of behaviors, including, for example, the time-courses of their responses to standing contrast in the receptive field (Cleland et al 1971). In the LGN, however, several of the criteria used in the retina do not clearly reveal distinct classes of cells whose properties can be easily related to those of the ganglion cells that drive them. Spatial frequency analysis is one of the more robust methods for revealing classes of LGN neuron and providing information about their retinal inputs. By the application of tests of linearity of spatial summation used in the retina (see above) the great majority of relay cells in the dorsal A and A_1 layers of the cat's LGN have been shown to have linear spatial summation (like X-cells) or nonlinear spatial summation of the type characteristic of Y-cells (Shapley & Hochstein 1975, Derrington & Fuchs 1979, So & Shapley 1979, Lehmkuhle et al 1980). This is not a trivial finding, for it implies that the X-cells of the LGN receive no inputs from Y- cells in the retina.

Spatial frequency sensitivity functions provide further evidence that these two groups of LGN cells are driven only by X- and Y- ganglion-cells, respectively. The spatial frequency sensitivity or response curves obtained from the two types in the LGN (So & Shapley 1981, Troy 1983a) are like those of their counterparts in the retina (Enroth-Cugell & Robson 1966, Derrington & Lennie 1982) and can be described by the difference-of-Gaussians function, with center and surround having spatial spreads and sensitivities rather like those of retinal cells. The fundamental spatial frequency resolutions and the optimal spatial frequencies of X- and Y-geniculate cells differ by about a factor of 3, following the behavior of their retinal inputs (Derrington & Fuchs 1979, So &

Shapley 1979, Lehmkuhle et al 1980, Troy 1983a). This means that the visual functional differences noted for X- and Y-cells in the retina persist in the LGN: the X-cells are most sensitive to patterns that contain relatively high spatial frequencies, while the Y-cells are most sensitive to patterns of low spatial frequency. The similarities between Y-cells in the LGN and their retinal counterparts extend to the spatial frequency sensitivities of the geniculate Y-cell's fundamental and second harmonic responses: The second harmonic component has lower peak sensitivity but better spatial-frequency resolution than its fundamental component and, as in the retina, the spatial frequency response of the second harmonic closely resembles that of the fundamental of neighboring X-cells (So & Shapley 1981). A few cells in the dorsal layers of the LGN appear to receive inputs from both X- and Y- ganglion-cells. These had been identified from measurements of conduction-velocity by Cleland et al (1971); their distinctive visual properties arise from the fact that when a contrast-modulated grating is used to find the highest resolvable spatial frequency, the fundamental and second harmonic components of response share the same limit (So & Shapley 1979).

In the more ventral C laminae of the cat's LGN, many of the neurons are driven by slow-conducting axons. By applying the test for linearity of spatial summation, Sur & Sherman (1982) were able to distinguish two groups of cells in the C laminae. One showed linear spatial summation, and the other showed a distinctive nonlinearity (second harmonic response) that dominated the response at all spatial frequencies. Sur & Sherman measured the spatial and temporal frequency sensitivity functions of these linear and nonlinear cells in the C laminae. Peak sensitivities of both types were about fourfold less than those of X-cells found in the A or A1 laminae. For the linearly summating cells (and the second harmonic component of response of nonlinear ones), the highest resolvable spatial frequency was close to that of the fundamental component of response of a Y-cell.

MONKEY The use of a spatial frequency analysis has shed new light on the organization of the lateral geniculate nucleus of the monkey. The neurons in the monkey's LGN are organized into six layers. The four most dorsal layers are composed of small neurons and are called parvocellular layers. The two ventralmost layers, containing larger neurons, are named the magnocellular layers. The most widely used scheme for the classification of cells in the LGN of macaque was devised by Wiesel & Hubel (1966), who distinguished three classes of parvocellular neuron by the chromatic and spatial organization of their receptive fields. Their largest class, type I, had concentrically organized receptive fields with centers and surrounds that had different spectral sensitivities, i.e. the receptive fields were chromatically and spatially opponent. Type III cells had spatially but not chromatically opponent receptive fields,

Type II cells had chromatically but not spatially opponent receptive fields. Spatial frequency analysis, exploiting both chromatic and achromatic gratings, suggests that type I and type III cells form a single group. Almost all parvocellular units show linear spatial summation like X-cells in cat (Shapley et al 1981, Blakemore & Vital-Durand 1981). The spatio-temporal frequency sensitivities of type I and type III cells are indistinguishable when obtained with achromatic gratings (Derrington & Lennie 1984). When coupled with the evidence that type III cells do in fact have chromatically opponent receptive fields (Padmos & Van Norren 1975, Derrington et al 1984), these results suggest that type I cells and type III cells in the parvocellular laminae belong to the same population.

Wiesel & Hubel (1966) identified two classes of neuron in the magnocellular layers of the LGN: type III with spatially opponent and chromatically non-opponent fields like parvocellular type III cells; and type IV, whose receptive field contained a central region excited by a broad range of wavelengths, enclosed by a suppressive surround that appeared to be driven mainly by long wavelength cones. Magnocellular and parvocellular type III neurons clearly differ in visual contrast sensitivity and in the conduction velocity of their inputs, so they should not be considered as a single type. Measurements of spatial contrast sensitivity to achromatic gratings and of linearity of spatial summation do not differentiate magnocellular type III and IV cells (Kaplan & Shapley 1982, Derrington & Lennie 1984). Moreover, an analysis of responses to chromatic gratings of different spatial frequencies (Derrington et al 1984) shows that all magnocellular neurons have weak chromatic opponency in their receptive fields, so the type III/IV distinction probably does not represent a qualitative division in the magnocellular layers.

On a number of indices that help distinguish X- and Y-cells in the cat's retina (speed of conduction of axons, responses to standing contrast, responses to fast-moving objects), magnocellular neurons have been thought to be more like cat Y-cells than X-cells (Dreher et al 1976). Parvocellular neurons, by the same tests, behave more like X-cells. Several groups (Dreher et al 1976, Sherman et al 1976, Schiller & Malpeli 1978) have suggested that the parvocellular and magnocellular cells are the primate's counterparts to X- and Y-cells in the cat. Spatial frequency analysis has revealed that in important respects this parallel is misleading. First, most cells in the magnocellular division of the macaque's LGN show linear spatial summation, like that of X-cells in cat (Kaplan & Shapley 1982, Derrington & Lennie 1984); 15–25% show a pronounced nonlinearity of spatial summation, like that seen in the cat's Y-cells. However, there is no compelling evidence from the distribution of an index of nonlinearity that magnocellular cells fall into two distinct groups like X- and Y-cells in cat (Derrington & Lennie 1984). Second, the contrast sensitivities of magnocellular cells are much higher than those of parvocellular ones (Kaplan & Shapley 1982, Hicks et al 1983, Derrington & Lennie 1984) and are similar to those of X- and Y-cells in the A and A1 laminae of the lateral geniculate nucleus of the cat. Thus, in terms of visual contrast sensitivity, monkey parvocellular neurons are unlike cat X-cells, and in terms of spatial summation, most magnocellular neurons are unlike cat Y-cells. By showing that neurons in the LGN fall into fewer classes than had previously been discerned, and by drawing attention to the differences between the contrast sensitivities of parvocellular and magnocellular neurons, spatial frequency analysis has thrown into sharp relief the very substantial differences between the properties of cells in the two divisions of the LGN.

Transfer Function of LGN Cells

Receptive fields of neurons in the LGN of both cat and monkey bear a strong superficial resemblance to the receptive fields of the ganglion cells that drive them. The visual physiologist is therefore challenged to discover what, if any, transformation of the visual signal is undertaken by the LGN. Spatial frequency analysis provides a powerful tool for doing this, because if the LGN neuron behaves as a linear or quasilinear filter, its transfer properties (the transformation imposed on the signals reaching it from ganglion cells) can be directly established by exploiting a property of transfer functions: dividing the overall spatial frequency response function obtained from the LGN cell (this would reflect both LGN and ganglion cell properties) by that of the ganglion cell that drives it gives one the transfer function of the LGN cell alone. A further advantage of spatial frequency analysis is that it permits a uniform treatment of the spatial and the temporal transfer properties.

Some evidence indicates that in cat the surround of the LGN receptive field is relatively stronger than in the receptive field of a ganglion cell (Hubel & Wiesel 1961), and that most LGN cells give more transient responses to standing contrast than do retinal ganglion cells (Cleland et al 1971). These differences between retina and LGN can be very precisely characterized by comparing spatial and temporal contrast sensitivity functions of LGN cells and the ganglion cells that drive them. So & Shapley (1981) measured the spatial contrast sensitivities of X- and Y-cells in the LGN and also the contrast sensitivities of an associated S-potential, the extracellularly recorded synaptic potential that represents the retinal input to the cell (Kaplan & Shapley 1984). One surprising finding to emerge from this work is that the spatial frequency responses of relay cells in the LGN were, with one exception out of ten, identical to those of the S-potential. Thus, at least in the lightly anesthetized animal, the LGN does little filtering of signals evoked by stimuli of moderate contrast. Coenen & Vendrik (1972) found that the level of arousal had a marked effect upon the capacity of an optic tract fiber to drive an LGN cell, so the LGN may well change its transfer properties in the more deeply anaesthetized animal. The exceptional cell studied by So & Shapley showed, by comparison with its S-potential, a substantial loss of sensitivity to low spatial frequencies, indicating increased antagonism from the periphery of the receptive field. The work of Troy (1983b) and Dawis et al (1984) shows that spatiotemporal coupling in LGN cells resembles that observed in retinal ganglion cells.

SPATIAL FREQUENCY ANALYSIS APPLIED TO VISUAL CORTEX

Visual cortical neurons are more selective for visual stimuli than are neurons in the retina or LGN. In particular, visual cortical cells are much more highly tuned on the dimensions of spatial frequency and orientation than are their geniculate inputs. Spatial frequency analysis therefore provides a powerful method for characterizing the visual properties of cortical cells.

Implications for Classification

Most recent work on classification has extended or revised Hubel & Wiesel's division of cat cortical cells into "simple," "complex," and "hypercomplex" categories according to their receptive field properties. Simple and complex cells were originally distinguished on the basis of linearity of spatial summation, assessed qualitatively rather than quantitatively. By this we mean that Hubel & Wiesel (1962) called cortical cells "simple" if the map of the receptive field, obtained with stationary, flashing, small spots or bars, could be used to predict responses to wider or longer bars moved across the field. Simple receptive fields usually had separate "on" and "off" areas. "Complex" cells failed this qualitative test of linearity. Their receptive fields, if they could be mapped with spots at all, usually had overlapping "on" and "off" regions. Moreover, even though a complex cell had a wide receptive field, its response to wide bars might be poorer than to narrower bars.

Although the distinction between simple and complex cells has been progressively refined (cf Hubel & Wiesel 1977, Gilbert 1977), its validity and significance have been repeatedly challenged. One question is whether the cells fall naturally into categories or lie on a continuum of receptive field properties. A related question is whether the simple-complex distinction reflects different underlying cortical connectivity. Hubel & Wiesel (1962, 1977) argued from an examination of receptive field properties that information was processed serially—from geniculate afferent to simple cell to complex cell—but that conclusion has been challenged by the view that simple and complex cells are driven by parallel inputs (cf Stone 1983). Both the "simple-complex" distinction and the issue of serial vs parallel processing are susceptible to quantitative analysis. Spatial frequency analysis, a particularly useful method for studying spatial summation, has been employed in many of the recent studies dealing with these issues.

Unless stated otherwise, the work discussed in the following sections refers to observations on striate cortex (area 17).

DISTINCT CELL CLASSES Several papers have described the responses of simple and complex cells to moving gratings. In the earliest, Maffei & Fiorentini (1973) showed that cells that were "simple" by Hubel & Wiesel's criteria responded to drifting sine gratings mainly with a modulated impulse rate synchronous with the passage of bars across the receptive field. Complex cells produced mainly an elevated mean discharge rate that became modulated only when the spatial frequency was very low. This different between simple and complex cells has been confirmed by several investigators (Movshon et al 1978a, De Valois et al 1978, 1982, Glezer et al 1980, Dean & Tolhurst 1983).

Dean & Tolhurst (1983) asked whether the modulated or unmodulated nature of the response to a moving grating allows the investigator to sort cells in the cat's cortex into discrete categories, or merely to arrange them along a continuum of modulated vs unmodulated response. These workers found a continuous distribution of a quantity they called "relative modulation," the ratio of the amplitude of the fundamental response component to the mean impulse rate, in the response of the cell to a drifting grating of optimal spatial frequency and orientation. Relative modulation was close to zero in many complex cells and was greater than 1 in many simple cells (classified according to their behavior on Hubel & Wiesel's tests). Although the average relative modulation of simple cells was much higher than that of complex cells, the distributions of relative modulation for simple and complex cells overlapped, so that Dean & Tolhurst could not make a reliable classification of cells into two types with this response measure alone. In the monkey, however, De Valois et al (1982) found that the distribution of relative modulation was bimodal; the peak below a relative modulation of 1 was occupied by the complex cells, while the peak around 1.5 was occupied by simple cells. Whether this difference between species is real remains to be investigated. The question of the discreteness of the simple and complex categories therefore remains troubling, and requires further investigation with quantitative techniques. As described below, the use of contrast reversal gratings enables a more reliable classification of cells as simple or complex.

PARALLEL INPUTS The simple cell in cat responds to a moving grating predominantly with a discharge modulated at the temporal frequency at which cycles of the grating move across the receptive field. In this respect the simple cell resembles an X-cell in the retina or lateral geniculate nucleus. The complex cell responds mainly with an elevation of its impulse rate, and thus resembles a retinal or geniculate Y-cell. Since X- and Y- afferents from the lateral geniculate nucleus project in parallel to the cortex, and since these different fibers make monosynaptic connections with different types of cortical neurons (Hoffmann & Stone 1971, Stone 1972; other papers reviewed in Stone 1983), X- and Y-cells have been suggested to provide the principal monosynaptic drive to simple and complex cells, respectively. This is the antithesis of Hubel & Wiesel's (1962, 1977) suggestion that geniculate cells drive simple cells, which in turn drive complex cells. The use of sine gratings to examine the spatial summation properties of simple and complex cortical neurons has helped to clarify some aspects of the disagreement.

SPATIAL SUMMATION Movshon et al (1978a,b) and De Valois et al (1982) used Hochstein & Shapley's modification of Enroth-Cugell & Robson's null test (see above) to investigate spatial summation in the receptive fields of cortical cells. De Valois et al (1982) found that the responses of all simple cells in macaque (like those of retinal X-cells) depended strongly on spatial phase, with clear null positions. Movshon et al (1978a,b) found a somewhat more complicated picture in area 17 of the cat. Many simple cells, called "linear simple cells," behaved like retinal X-cells in having a sinusoidal dependence of response on spatial phase. However, Movshon et al (1978a) also found a population of simple cells that had no "null positions" yet still seemed to sum signals linearly because their responses contained no harmonic distortion. These cells responded at the modulation frequency at all spatial phases, but the amplitude of response varied rather little with spatial phase, although its temporal phase varied continuously. Maffei & Fiorentini (1973) described similar behavior in all ten simple cells they studied in this way. As we discussed above, three conditions are sufficient for the amplitude of response to vary sinusoidally with spatial phase: (*a*) linear local responses; (*b*) linear spatial summation; (*c*) homogeneity of temporal response characteristics across the receptive field. The non-nulling simple cells fail to meet the third condition. Movshon et al (1978a) noted that such cells always had odd-symmetric receptive fields. They conjectured that the two main receptive field subregions had somewhat different temporal frequency responses. Movshon et al (1978a) also found a small population of so-called "nonlinear simple cells" that were simple by Hubel & Wiesel's (1962) criteria, but nevertheless produced second harmonic responses at certain spatial phases.

Complex cells in cat (Movshon et al 1978b) and monkey (De Valois et al 1982) respond very nonlinearly to contrast reversing gratings, giving second harmonic responses at all spatial phases (cf Kulikowski & Bishop 1981, Pollen & Ronner 1982b). However, the receptive field organization of complex cells is quite unlike that of retinal and geniculate Y-cells. In Y-cells of the retina, the amplitude of the nonlinear response does not vary with spatial phase; indeed, this remarkable constancy allowed Victor & Shapley (1979b) to estimate, at around 100, the number of nonlinear subunits in the Y-cell's receptive field. However, the amplitudes of the nonlinear responses of complex cells show a pronounced variation with spatial phase (Movshon et al 1978b, De Valois et al 1982). This result implies that the number of subunits in their receptive fields is rather small (Glezer et al 1980). Hochstein & Spitzer (1985) found in all the complex cells they studied a marked variation with phase in the amplitude of response at all spatial frequencies. They estimate the number of functional subunits in the receptive field of the complex cell at two to ten.

Evidence from spatial frequency analysis of spatial summation thus suggests that simple and complex cortical cells differ quantitatively in the linearity of their spatial summation. The properties of simple cells appear to reflect those of X-cell inputs. The *spatial summation* properties of complex cells reflect not the properties of the Y-class of geniculate inputs but rather seem to reflect inputs from a small number of X-geniculate-cells and/or simple cells. The overlap in properties of simple and some complex cells (Dean & Tolhurst 1983) could result from a continuous distribution in the number of inputs a complex cell receives. Perhaps a complex cell with only a small number of subunits resembles a simple cell to some extent, while a complex cell with many subunits is more distinctive. The evidence from spatial summation experiments with gratings suggests that simple and complex cells receive some direct input from the LGN, although for both types this seems to arise predominantly from X-afferents. However, there is also indirect evidence (which we review below) about spatial frequency bandwidths that indicates X-Y convergence on a sub-class of complex cells (Tolhurst & Thompson 1981). More direct evidence on geniculo-cortical functional connectivity from cross-correlation experiments (Tanaka 1983) supports these notions that simple and complex cells receive X-afferent input from the LGN in parallel, and that some complex cells are excited by both X- and Y-geniculate cells.

Hawken & Parker (1984) provide further evidence for parallel processing in monkey striate cortex. They measured the contrast sensitivity of cortical cells in layer IV, and found that cells in layer IVcα had higher contrast sensitivities than those in IVcβ. The different sensitivities probably reflect differences between the sensitivities of the magnocellular afferents, which project to layer IVCα, and parvocellular afferents, which project to layer IVCβ (see above).

SPATIAL TUNING AND SPATIAL INTERACTIONS IN CORTEX

Neurophysiological work on the spatial frequency tuning of cortical neurons has had particular significance because of its relevance to the idea that the human visual system contains highly tuned "channels" or spatial frequency

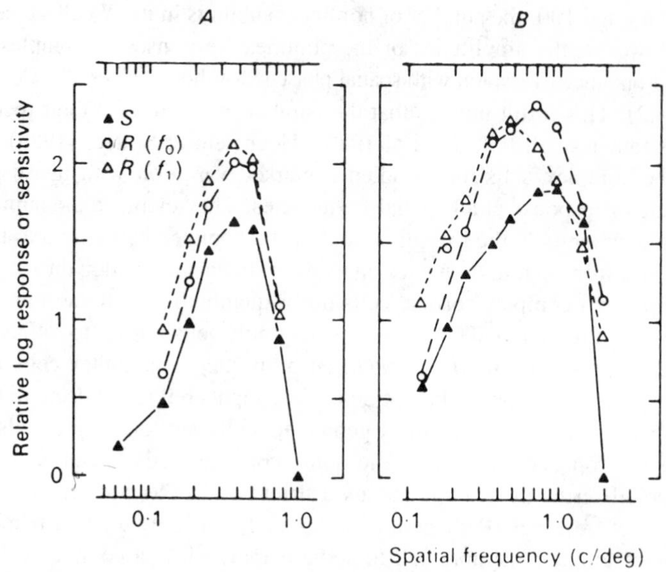

Figure 6 Spatial frequency response and sensitivity curves for simple and complex cells in the cat's striate cortex. In these experiments gratings of high contrast (0.5) moved steadily across the receptive field. *Open circles* show response measured as the mean discharge rate; *open triangles* show response measured as the amplitude of the fundamental Fourier component of discharge. *Filled triangles* show sensitivities. The graph in *A* was from a simple cell; that in *B* from a complex cell. From Movshon et al (1978c).

filters (cf Campbell & Robson 1968, Blakemore & Campbell 1969, Graham 1977).

Bandwidth and Best Frequency

Cooper & Robson (1968) and Campbell et al (1969) first showed, and Maffei & Fiorentini (1973) later confirmed, that many cortical neurons in cat are highly tuned for spatial frequency. Ikeda & Wright (1975), Tolhurst & Movshon (1975), and Movshon et al (1978c) found narrow spatial frequency tuning curves for both simple and complex cells in the cat. Narrowly tuned cells in the monkey's striate cortex have been studied by De Valois et al (1977, 1978, 1982). Figure 6 (from Movshon et al 1978c) shows typical examples of simple and complex cells narrowly tuned for spatial frequency.

The degree of spatial tuning may be quantified in terms of the width of the tuning curve, or "bandwidth," which is usually defined as the ratio of the higher to the lower frequency at which the sensitivity has declined to one half the peak sensitivity. The bandwidth is usually expressed in octaves, the logarithm of the bandwidth to the base 2. The distributions of bandwidth for simple and complex cells are very similar, both in cats and in monkeys. The average bandwidth of visual cortical cells in both cats and monkeys is around 1.5 octaves, i.e. the ratio of the higher to lower spatial frequencies at half-peak sensitivities is on average $2^{1.5}$ or 2.8 (Movshon et al 1978c, Kulikowski & Bishop 1981, Tolhurst & Thompson 1981, De Valois et al 1977, 1982). The distribution of bandwidths for a population of cells is, however, quite broad. The monkey's foveal cortex contains about as many simple and complex cells with a bandwidth of 2.5 octaves as cells with a bandwidth of 0.7 octaves (De Valois et al 1982). In the cat, the distributions appear to be somewhat tighter, but still there are about as many simple and complex cells with a bandwidth of one octave as two (Movshon et al 1978c, Kulikowski & Bishop 1981). By comparison, the average bandwidth for an X-cell in the LGN of the cat is from three to four octaves (So & Shapley 1981, Troy 1983a,b, cf Thibos & Levick 1983) and in the monkey may exceed five octaves (Kaplan & Shapley 1982, Hicks et al 1983, Derrington & Lennie 1984). The narrower bandwidth of cortical neurons must be caused by intracortical interactions.

Tolhurst & Thompson (1981) introduced the quantity, *normalized bandwidth,* which they defined as the difference between the higher and lower spatial frequencies at half sensitivity, divided by the spatial frequency of the peak sensitivity. Using this measure, they discovered that cells near the junction of layers III and IV in the cat striate cortex tended to have broader bandwidths than cells in other cortical laminae. Y-geniculate-afferents terminate in this region (Ferster & LeVay 1978, Gilbert & Wiesel 1979). Wider bandwidths might be associated with convergence of X- and Y-afferents onto single cortical neurons (Tanaka 1983), because Y-cells are tuned to lower spatial frequencies than are X-cells. In a region of the visual field where, for example, LGN X-cells respond best at 1 c/deg and LGN Y-cells respond best around 0.33 c/deg, a cortical cell that received inputs from both types would have a wider ranger of responsiveness than one that received input from X-cells only or Y-cells only. The relatively rare wider-bandwidth neurons were interspersed among the more numerous narrower-band cells that presumably receive only X-cell input.

Even at one locus in the cortex, representing a single site on the retina, there is a broad distribution of optimal spatial frequencies in the population of cortical neurons. In the region of the cat's cortex that represents the area centralis, neurons are tuned to spatial frequencies from 0.3 up to 3 c/deg (Movshon et al 1978c, Tolhurst & Thompson 1981). De Valois et al (1982) reported a somewhat larger range of spatial frequency (centered near 3 c/deg) for monkey visual cortex that represents the fovea.

Fourier Analysis

The broad distribution of spatial frequency optima and the narrow bandwidths of individual neurons has led many to propose that the visual cortex performs some kind of Fourier analysis of the image into spatial frequency components (Campbell & Robson 1968, Cooper & Robson 1968, Campbell et al 1969, Blakemore & Campbell 1969, Pollen et al 1971, Maffei & Fiorentini 1973, Glezer et al 1973, Robson 1975, Graham 1977, Glezer & Cooperman 1977, Maffei 1978, Movshon et al 1978c, Maffei et al 1979, Robson 1980, Tolhurst & Thompson 1981, Kulikowski & Bishop 1981, Kulikowski et al 1982, De Valois et al 1982, Ginsburg 1982, Pollen & Ronner 1983).

The requirements of a rigorous spatial Fourier analyzer include linearity, spatial delocalization, narrow bandwidth, spatial homogeneity, and encoding of amplitude and phase. Linearity is required because otherwise activity of an element in the cortical lattice would not signify unambiguously the presence of a particular spatial frequency in the stimulus image. Nonlinearities would create spurious "cross talk." Spatial delocalization is a consequence of linearity and narrow bandwidth, since the spatial distribution of sensitivity of a neuron that is both narrowly tuned and also linear must have many peaks and troughs of sensitivity spread across space. Spatial homogeneity is required in order that sine waves be undistorted after filtering; a change of the distance scale with position would cause the response to a spatial sinusoid to be non-sinusoidal. Encoding of amplitude and phase are required because both are needed to identify an image unambiguously.

Most of the requirements for strict Fourier analysis are not met by the retino-cortical pathway. The behavior of many cortical cells is very nonlinear (see below). The requirement of spatial delocalization together with narrow spatial tuning also is not satisfied [cf Westheimer (1984) and our section on Nonlinearities below]. Moreover, psychophysical experiments (cf Kelly & Burbeck 1984), physiological experiments (e.g. Cleland et al 1979, So & Shapley 1979, Linsenmeier et al 1982), and neuroanatomy (reviewed in Stone 1983) all reveal that the receptive field sizes of retinal ganglion cells and cortical cells are not uniform (homogeneous) across the visual field. To a first approximation the center's spread, r_c, is proportional to the distance of the center of the receptive field from the fovea or area centralis. If r_c's vary systematically with retinal position, then so also do the spatial filtering properties of the receptive fields. The existence of such retinal inhomogeneity has caused advocates of spatial Fourier analysis to propose that the spatial image may be analyzed into spatial Fourier components over small patches of visual field (Robson 1975). Within each patch, the spatial filtering properties would be homogeneous. This idea of "patch-wise" Fourier analysis is somewhat intermediate between the concept of rigorous Fourier analysis and strictly local feature detection. The idea is an attractive one because it is consistent with the narrow spatial frequency tuning and spread of best frequencies of cortical neurons, but it is weakened to the extent that the neurons behave nonlinearly.

Marr (1981) offered an alternative view of the narrow spatial tuning of cortical cells. He suggested that the principal task of early visual processing was to identify the borders of objects. These borders may be obscured by noise or camouflage if they are viewed with only a single broad-band spatial filter (or edge detector). Marr hypothesized that coincidence of responses to a border in two or more spatial channels might be a less error-prone way of identifying the presence of visual contours. He suggested that visual cortical cells might be responsive specifically to zero-crossings in the spatially filtered visual image, and that these zero-crossings correspond to the edges or borders in the visual scene. The hypothesis may not be correct in detail, but has the attractive feature of being consistent with the pervasive nonlinearity of cortical signal processing, and in particular with the existence of neurons that are narrowly tuned for spatial frequency yet have localized line-weighting functions.

Linearity has been the most important issue in the debate about Fourier analysis by the cortex. We review the evidence in the following sections.

SYNTHESIS EXPERIMENTS Several experiments have shown that cortical cells behave in some respects like linear filters, so that one can use knowledge of their responses to sinusoidal gratings to predict their responses to arbitrary stimuli. For example, De Valois et al (1979) used checkerboard patterns to stimulate cells in area 17 of cat and monkey cortices. A checkerboard may be represented as a sum of sine gratings at different spatial frequencies and orientations, i.e. it may be analyzed into Fourier components in two dimensions, spatial frequency and orientation. The lowest spatial frequency gratings in this sum are aligned along the 45° diagonals of the checkerboard, and have as their period the length of the check diagonals. The diagonal components are also the components of highest amplitude in the checkerboard. Thus, if cells act like narrow band spatial filters, one might expect them to respond to these diagonal components if presented with a checkerboard stimulus pattern in which only the fundamental components at the diagonals fell within their tuning curves. This is exactly what De Valois et al (1979) observed. Furthermore, when the size of the checkerboard was increased, the cells responded to higher harmonics in the checkerboard pattern at other orientations away from 45°.

An experiment by Albrecht et al (1980), in which an attempt was made to predict responses to bars from the spatial frequency tuning curve, provides another example of Fourier synthesis. The spatial frequency tuning curves and the bar width tuning curves for two neurons are plotted in Figure 7. It is immediately obvious that the cells are much more sharply tuned for spatial frequency than they are for width of bar. This becomes easy to understand when we consider that, because of their sharp edges, bars are broad-band stimuli in the domain of spatial frequency. As width is increased, the fre-

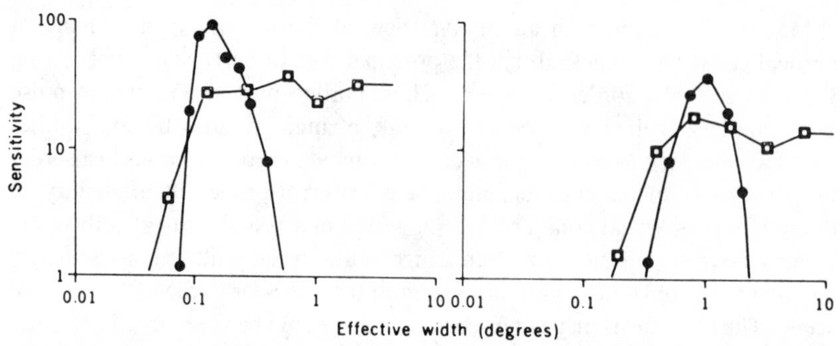

Figure 7 Selectivity for spatial frequency and selectivity for bar width in monkey and cat visual cortical cells. Sensitivities for bars are plotted as *squares;* sensitivities for gratings as *closed circles*. In both cases sensitivity was the reciprocal of the contrast required to evoke a response of criterion amplitude. The "effective width" for a bar was just its width; for a grating it was taken to be one half cycle of the sine waveform. *A* shows tuning curves for a simple cell in monkey, *B* those for a complex cell in cat. Bar sensitivities should be multiplied by a factor of two, to correct for a difference between the way in which contrast is defined for bars and gratings. From Albrecht, et al (1980).

quencies in the Fourier representation of the bar continue to lie in the passband of the narrowly tuned (in spatial frequency) cortical cells. Similar reasoning explains the broad tuning curves of cortical cells for square wave gratings (Schiller et al 1976, Maffei et al 1979), whose Fourier representations are rich in high harmonics of the fundamental frequency. Even when the fundamental frequency is too low to stimulate the cells, the high harmonics in the square wave will fall within the cells' tuning curves and act as effective stimuli.

Other experiments to examine the Missing Fundamental Illusion (Maffei et al 1979, Albrecht & De Valois 1981), or responses to compound stimuli and/or moving stimuli (Maffei et al 1979, Pollen & Ronner 1982b, Movshon et al 1984), have shown that the spatial frequency tuning curve often provides enough information to predict the response to a compound stimulus.

To what extent spatial phase and symmetry are important for the analysis of visual images remains an open question. This problem is now especially relevant in view of Pollen & Ronner's (1981) report that adjacent simple cells respond to the same drifting sine grating with a 90° phase shift. Their result suggests that simple cells could be arranged in even symmetric and odd symmetric pairs, and could thereby be used for accurate encoding of position from phase. Although more work is needed to prove the existence of even-odd pairs of simple cells, such an arrangement would provide a powerful mechanism for spatial localization (Marcelja 1980, Kulikowski et al 1982).

Nonlinearities

Despite the above evidence on the successes of Fourier synthesis, the nonlinearity of visual signal processing in many cortical cells provides a reason for doubting the importance of spatial Fourier analysis in vision. For many cells the line-weighting function cannot be predicted from the spatial frequency tuning curve under the assumption of linear signal summation (Glezer et al 1980, Kulikowski & Bishop 1981), and nonlinearities such as adaptation or habituation are prevalent. However, the nature of some of these nonlinearities suggests that they are precisely what make the cells highly tuned spatial frequency filters.

LINE-WEIGHTING AND SPATIAL FREQUENCY TUNING Let us consider the relation between the spatial frequency tuning curves and the line weighting functions of individual simple cells (as in Movshon et al 1978a, Glezer et al 1980, Kulikowski & Bishop 1981, Kulikowski et al 1982). If the neurons summed signals linearly across their receptive fields, one would be able to predict the line-weighting function from the spatial frequency response, or vice versa, by Fourier transforming one member of the pair into the other.

A linear spatial filter is characterized not only by the magnitude of its response to different spatial frequencies but also by the phase of its response. The phase of response is directly linked to the symmetry of the spatial sensitivity profile (Dawis et al 1984), which for a cortical neuron may be even symmetric (as are most retinal ganglion cells and LGN cells to a first approximation), or odd symmetric, like that of proposed "edge detectors" (Hubel & Wiesel 1962, Tolhurst 1977). Or the profile may be neither even nor odd symmetric, but could be the sum of even and odd symmetric functions. One therefore needs to know the phase of the response in order to predict the line-weighting function from the spatial tuning curve. Response phase has been measured rarely (cf Glezer et al 1980), but by assuming even or odd symmetry the line-weighting function can sometimes be reconstructed (e.g. Movshon et al 1978a, Kulikowski & Bishop 1981). Since many cells have neither clearly even nor clearly odd symmetric receptive fields (Kulikowski & Bishop 1981, Glezer et al 1980), presumably the line-weighting functiongs could be predicted more accurately were the phases of responses also measured. Glezer et al (1980) measured the phase as well as the amplitude of the response to drifting gratings. In general, their predicted line-weighting functions did not agree with measurements. This discrepancy suggests a role for inhibitory interactions between cortical cells.

Neurons that have narrow bandwiths in spatial frequency should have many oscillations in their line-weighting functions, from positive to negative sensitivity. On the assumption of linearity, narrow bandwidth of the Fourier representation mathematically implies such "ringing" of the line-weighting function. For example, if the Fourier representation were infinitesimally narrow—an impulse in the frequency domain—then the line weighting function would be a sine wave. However, as Westheimer (1984) has remarked, the measured line-weighting functions rarely exhibit the spatial oscillation demanded of a linear spatial filter with the observed spatial frequency tuning. Line-weighting functions of simple cells reveal too few antagonistic sub-regions and these are too insensitive to be consistent with the spatial frequency responses, under the assumption that the cells act as linear spatial filters. In other words, the line-weighting functions are too localized in space. This suggests some sort of nonlinearity in simple cells, and the experiments of De Valois & Tootell (1983) indicate what kind it might be. They found that gratings of spatial frequencies to which a simple cell was "silent" (i.e. gratings that had no direct effect on the cell) could inhibit the response to a simultaneously presented grating of a spatial frequency that fell within the pass-band of the cell. This failure of superposition is clear evidence of a nonlinearity. Spatial-frequency-specific inhibition was mainly from high frequencies to lower frequencies and served to sharpen the spatial tuning curve on the high frequency side; it could very easily make the spatial frequency tuning curve narrower than would be expected from the line-weighting function. Thus, the cortex seems to have adopted a nonlinear path to reconcile conflicting demands: minimizing simultaneously positional and spatial frequency uncertainty (Marcelja 1980, Kulikowski et al 1982, Pollen & Ronner 1982b).

The line-weighting function and the spatial frequency tuning curve of a complex cell cannot be used to predict one another (Movshon et al 1978b, Pollen & Ronner 1982a). However, this discrepancy is not surprising, since nonlinear interactions in the receptive field themselves define a complex cell. The subunit models of Movshon et al (1978b) and Hochstein & Spitzer (1984) imply that the spatial frequency tuning should be characteristic of one of the subunits in the complex cell field, or should arise from interaction between subunits. As a result, the size of the receptive field and the line-weighting function of a complex cell give no clue to its spatial frequency resolution or its optimal spatial frequency.

PATTERN ADAPTATION AND THE CORTICAL CONTRAST GAIN CONTROL Pattern adaptation caused by exposure to high contrasts reflects the operation of highly nonlinear mechanisms. Maffei et al (1973) and Movshon & Lennie (1979) found that after exposure to adapting patterns of high contrast, the responsiveness of simple and complex cells was much reduced. The effects of pattern adaptation on spatial frequency selectivity are complicated: for some cells the whole tuning curve is depressed, but for others the loss of sensitivity is confined to a narrow band of spatial frequencies (Movshon & Lennie 1979). In many cells, pattern adaptation shifts the contrast-response curve to a range of lower-sensitivity (Ohzawa et al 1982, Albrecht et al 1984). This cortical mechanism (in some ways analogous to the retinal contrast gain control described by Shapley & Victor 1978) serves to extend the range over which contrast may be discriminated, and prevents saturation of the response of a cortical neuron. However, such a strong nonlinearity in cortical signal processing of patterns is inconsistent with the concept that the visual system is designed to carry out spatial Fourier analysis: the response to a compound stimulus will certainly not be simply the superposition of the responses to each stimulus presented alone.

ORIENTATION-DEPENDENT INHIBITION Orientation-dependent inhibition is another substantial nonlinearity: gratings at nonoptimal orientations, which produce no directly measurable modulation of a simple cell's discharge, can inhibit the response to a stimulus at the optimal orientation (Burr et al 1981, Morrone et al 1982). It would be interesting to know whether this orientation-dependent inhibition, which is found mainly in simple cells, is related to the spatial frequency inhibition described by De Valois & Tootell (1983). Morrone et al (1982) point out that orientation-dependent inhibition makes simple cells rather selective for one-dimensional patterns. Another consequence is that a cell may not respond to a sinusoidal grating to which it is tuned if that grating is presented together with gratings at other orientations or spatial frequencies, as is typical of natural visual patterns. This would be a serious violation of superposition.

DIRECTIONAL SELECTIVITY Bishop and his colleagues (Bishop et al 1971a,b, 1973) have emphasized that many simple cells respond preferentially to stimuli moving in a particular direction. Although it is possible to construct a directionally selective neuron with only linear interactions, the evidence of Bishop et al implies a nonlinear mechanism. We know too little about the mechanism to speculate about its possible relationship to the other nonlinear ones mentioned above. Directional selectivity in complex cells is also common, but is perhaps to be expected in neurons defined by their rich nonlinear properties.

Direction-selective units in the striate cortex of macaque respond only to the component of object motion perpendicular to their preferred orientation. By contrast, some neurons in the middle temporal area of the macaque monkey respond to the direction of motion of a two-dimensional pattern (Movshon et al 1984). These neurons thus seem to have a specific nonlinearity that renders them responsive to the direction of motion of compound patterns comprised of gratings that have different orientations. Movshon et al argue that such a nonlinearity is required to support a corresponding human visual capacity.

Cytoarchitectonics: Spatial Frequency Rows or Columns?

In the visual cortex, significant stimulus attributes appear to be segregated into columns parallel to the radial direction, while neurons with different projection targets are segregated into different layers. Three methods have been used to determine how cells with similar spatial frequency tuning are arranged in the cortical lattice: *(a)* making long electrode tracks at a shallow angle with respect to the surface of the cortex; *(b)* making electrode tracks at several angles and reconstructing them with accurate histological marking; *(c)* labeling impulse activity in cortical cells with radioactive 2-deoxyglucose (Sokoloff et al 1977). Sadly, these three methods yield three different answers. By making approximately tangential electrode tracks, Maffei & Fiorentini (1977) found similarly tuned cells arranged in rows or layers. Using the second technique of accurate electrode track marking, Tolhurst & Thompson (1982) found clusters of cells of similar spatial frequency tuning with no obvious relation to either layers or columns. Berardi et al (1982) report that changes of spatial frequency and changes of orientation are negatively correlated in tracks through the cortex. Thus, when orientation changes rapidly, as on a tangential track, spatial frequency tends to remain fixed. When orientation tends to remain fixed, as on a radial track down a column, spatial frequency tends to change. None of the electrophysiological results are consistent with the interpretation of experiments in which cortex has been labeled with deoxyglucose: Tootell et al (1981) found in cat that stimulation by gratings of a single spatial frequency led to the presence of label in columns.

In the monkey, Tootell et al (1983) found that stimulation by low spatial frequencies caused deoxyglucose labeling of the cytochrome oxidase "blobs" in striate and cytochrome oxidase strips in peristriate cortex. A stimulus grating of 7 c/deg presented at all orientations labeled all the cortex not marked by cytochrome oxidase. This is not particularly compelling evidence for spatial frequency columns, though it does imply a great degree of cortical specificity. If the cortex were really subdivided into spatial frequency columns, one would not expect a grating of a single spatial frequency to label all the cortex unoccupied by "blobs."

The real mystery about spatial frequency columns is why the deoxyglucose label should indicate a columnar organization that is not present in the electrophysiology. Perhaps the presence of concentrations of deoxyglucose marks not only excited cells, but also active endings that produce inhibitory effects on the (presumed) unlabeled cells at the same site. This possibility bears on another mystery, namely the real nature of orientation columns. As Bauer et al have indicated (1983, Bauer 1982) electrode tracks that are accurately radial through area 17 of monkey and cat appear to reveal a discontinuity of preferred orientation (of almost 90°) at the layer IV–V border. Such a discontinuity has never been indicated in the deoxyglucose labeling following stimulation at a single orientation (reviewed in Hubel & Wiesel 1977). Schoppmann & Stryker (1981) showed that deoxyglucose labeling was associated with increased excitation in area 17 of cat, but presented results only for layer IV. The crucial issue seems to be whether deoxyglucose labeling is associated with increased excitation of cells in layers V and VI, or with increased inhibition. Until this question is answered, the existence of spatial frequency columns and even of orientation columns must be in doubt. It is possible that, for both spatial frequency and orientation, the positions of tuning curves for cells in layers I–IV are highly correlated along a radial track, but are uncorrelated or even negatively correlated with the tuning curves of cells in layers V and VI along the same radial track.

ACKNOWLEDGMENTS

This work was supported by grants from the National Eye Institute of the National Institute of Health: grant R01 EY-1472 to RMS, grant R01 EY-04440 to PL, and grant EY-5P30 01319 to the Center for Visual Science.

Literature Cited

Albrecht, D. G., De Valois, R. L. 1981. Striate cortex responses to periodic patterns with and without the fundamental harmonics. *J. Physiol.* 319:497–514

Albrecht, D. G., De Valois, R. L., Thorell, L. G. 1980. Visual cortical neurons: Are bars or gratings the optimal stimuli? *Science* 207:88–90

Albrecht, D. G., Farrar, S. B., Hamilton, D. B. 1984. Spatial contrast adaptation characteristics of neurones recorded in the cat's visual cortex. *J. Physiol.* 347:713–39

Barlow, H. B., Levick, W. R. 1969. Changes in the maintained discharge with adaptation level in the cat retina. *J. Physiol.* 202:699–718

Barlow, H. B., Fitzhugh, R., Kuffler, S. W. 1957. Change of organization in the receptive fields of the cat's retina during dark adaptation. *J. Physiol.* 137:338–54

Barlow, H. B., Derrington, A. M., Harris, L. R., Lennie, P. 1977. The effects of remote retinal stimulation on the responses of cat retinal ganglion cells. *J. Physiol.* 269:177–94

Bauer, R. 1982. A high probability of an orientation shift between layers 4 and 5 in central parts of the cat striate cortex. *Exp. Brain Res.* 48:245–55

Bauer, R., Dow, B. M., Snyder, A. Z., Vautin, R. 1983. Orientation shift between upper and lower layers in monkey visual cortex. *Exp. Brain Res.* 50:133–45

Berardi, N., Bisti, S., Cattaneo, A., Fiorentini, A., Maffei, L. 1982. Correlation between the preferred orientation and spatial frequency of neurones in visual areas 17 and 18 of the cat. *J. Physiol.* 323:603–18

Bishop, P. O., Coombs, J. S., Henry, G. W. 1971a. Responses to visual contours: Spatiotemporal aspects of excitation in the receptive fields of simple cells *J. Physiol.* 219:625–57

Bishop, P. O., Coombs, J. S., Henry, G. H. 1971b. Interaction effects of visual contours on the discharge frequency of simple striate neurons. *J. Physiol.* 219:659–87

Bishop, P. O., Coombs, J. S., Henry, G. H. 1973. Receptive fields of simple cells in the cat striate cortex. *J. Physiol.* 231:31–60

Blakemore, C. B., Campbell, F. W. 1969. On the existence of neurones in the human visual system selectively sensitive to the orientation and size of retinal images. *J. Physiol.* 203:237–60

Blakemore, C., Vital-Durand, F. 1981. Distribution of X- and Y-cells in the monkey's lateral geniculate nucleus. *J. Physiol.* 320:17–18P

Bracewell, R. N. 1978. *The Fourier Transform and Its Applications* New York: McGraw-Hill. 2nd ed.

Braddick, O., Campbell, F. W., Atkinson, J. 1978. Channels in vision: Basic aspects. *Handb. Sensory Physiol.* 7:3–38

Burr, D., Morrone, C., Maffei, L. 1981. Intracortical inhibition prevents simple cells from responding to textured patterns. *Exp. Brain Res.* 43:455–58

Campbell, F. W. 1974. The transmission of spatial information through the visual system. In *The Neurosciences: Third Study Program,* ed. F. O. Schmitt, F. G. Worden. Cambridge, Mass.: MIT Press

Campbell, F. W., Cooper, G. F., Enroth-Cugell, C. 1969. The spatial selectivity of the visual cells of the cat. *J. Physiol.* 203:223–35

Campbell, F. W., and Green, D. G. 1965. Optical and retinal factors affecting visual resolution. *J. Physiol.* 181:576–93

Campbell, F. W., Robson, J. G. 1968. Application of Fourier analysis to the visibility of gratings. *J. Physiol.* 197:551–56

Cleland, B.G., Enroth-Cagell, C. 1968, Quantitative aspects of sensitivity and summation in the cat retina.*J. Physiol,* 198:17-38

Cleland, B. G., Levick, W. R. 1974. Brisk and sluggish concentrically organized ganglion cells in the cat's retina. *J. Physiol.* 240:421–56

Cleland, B., Dubin, M. W., Levick, W. R. 1971. Sustained and transient neurones in the cat's retina and lateral geniculate nucleus. *J. Physiol.* 228:649–80

Cleland, B. G., Harding, T. H., Tulunay-Keesey, U. 1979. Visual resolution and receptive field size: Examination of two kinds of cat retinal ganglion cell. *Science* 205:1015–17

Coenen, A. M. L., Vendrik, A. J. H. 1972. Determination of the transfer ratio of cat's geniculate neurons through quasi-intracellular recordings and the relationships with the level of alertness. *Exp. Brain Res.* 14:227–42

Cooper, G. F., Robson, J. G. 1968. Successive transformations of spatial information in the visual system. *IEE Nat. Phys. Lab. Conf. Proc.* 42:134–43

Dawis, S., Shapley, R., Kaplan, E., Tranchina, D. 1984. The receptive field organization of X cells in the cat: Spatiotemporal coupling and asymmetry. *Vis. Res.* 24:549–64

Dean, A. F., Tolhurst, D. J. 1983. On the distinctness of simple and complex cells in the visual cortex of the cat. *J. Physiol.* 344:305–25

Derrington, A. M., Fuchs, A. F. 1979. Spatial and temporal properties of X and Y cells in the cat lateral geniculate nucleus. *J. Physiol.* 293:347–64

Derrington, A. M., Lennie, P. 1982. The influence of temporal frequency and adaptation level on receptive field organization of retinal ganglion cells in the cat. *J. Physiol.* 333:343–66

Derrington, A. M., Lennie, P. 1984. Spatial and temporal contrast sensitivities of neurons in lateral geniculate nucleus of macaque. *J. Physiol.* 357: In press

Derrington, A. M., Lennie, P., Wright, M. J. 1979. The mechanism of peripherally evoked responses in retinal ganglion cells. *J. Physiol.* 289:299–310

Derrington, S. M., Krauskopf, J., Lennie, P. 1984. Chromatic mechanisms in lateral geniculate nucleus of macaque. *J. Physiol.* 357: In press

De Valois, K. K., Tootell, R. B. H. 1983. Spatial-frequency-specific inhibition in cat striate cortex cells. *J. Physiol.* 336:359–76

De Valois, K. K., De Valois, R. L., Yund, E. W. 1979. Responses of striate cortex cells to grating and checkerboard patterns. *J. Physiol.* 291:483–505

De Valois, R. L., De Valois, K. K. 1980. Spatial Vision. *Ann. Rev. Psychol.* 31:309–41

De Valois, R. L., Albrecht, D. G., Thorell, L. G. 1977. Spatial tuning of LGN and cortical cells in monkey visual system. In *Spatial Contrast*, ed. H. Spekreijse, L. H., van der Tweel, pp. 60–63. Amsterdam: North Holland

De Valois, R. L., Albrecht, D. G., Thorell, L. G. 1978. Cortical cells: Bar detectors or spatial frequency filters? In *Frontiers in Visual Science*, ed. S. J. Cool, E. L. Smith, pp. 544–56. New York: Springer Verlag

De Valois, R. L., Albrecht, D. G., Thorell, L. G. 1982. Spatial frequency selectivity of cells in macaque visual cortex. *Vis. Res.* 22:545–59

Dreher, B., Fukada, Y., Rodieck, R. W. 1976. Identification, classification, and anatomical segregation of cells with X-like and Y-like properties in the lateral geniculate nucleus of old-world primates. *J. Physiol.* 258:433–52

Enroth-Cugell, C., Robson, J. G. 1966. The contrast sensitivity of retinal ganglion cells of the cat. *J. Physiol.* 187:517–52

Enroth-Cugell, C., Robson, J. G. 1984. The Friedenwald Lecture. Functional characteristics and diversity of cat retinal ganglion cells. *Invest. Ophthalmol Vis. Sci.* 25:250–67

Enroth-Cugell, C., Robson, J. G., Schweitzer-Tong, D. E., Watson, A. B. 1983. Spatiotemporal interactions in cat retinal ganglion cells showing linear spatial summation. *J. Physiol.* 341:279–307

Ferster, D., LeVay, S. 1978. Axonal arborizations of lateral geniculate neurons in striate cortex of cat. *J. Comp. Neurol.* 182:923–44

Frishman, L. J., Linsenmeier, R. A. 1982. Effects of picrotoxin and strychnine on nonlinear responses of Y-type cat retinal ganglion cells. *J. Physiol.* 324:347–63

Gilbert, C. D. 1977. Laminar differences in receptive field properties of cells in cat primary visual cortex. *J. Physiol.* 268:391–421

Gilbert, C. D., Wiesel, T. N. 1979. Morphology and intracortical projections of functionally characterised neurones in the cat visual cortex. *Nature* 280:120–25

Ginsburg, A. P. 1982. On a filter approach to understanding the perception of visual form. In *Recognition of Pattern and Form*, ed. D. G. Albrecht. Berlin: Springer Verlag

Glezer, V. D., Ivanoff, V. A., Tscherbach, T. A. 1973. Investigation of complex and hypercomplex receptive fields of visual cortex of the cat as spatial frequency filters. *Vis. Res.* 13:1875–1904

Glezer, V. D., Cooperman, A. M. 1977. Local spectral analysis in the visual cortex. *Biol. Cybern.* 28:101–8

Glezer, V. D., Tscherbach, T. A., Gauselman, V. E., Bondarko, V. M. 1980. Linear and nonlinear properties of simple and complex receptive fields in area 17 of the cat visual cortex. *Biol. Cybern.* 37:195–208

Graham, N. 1977. Visual detection of aperiodic spatial stimuli by probability summation among narrowband channels. *Vision Res.* 17:637–52

Hammond, P. 1974. Cat retinal ganglion cells: Size and shape of receptive field centres. *J. Physiol.* 242:99–118

Hawken, M. J., Parker, A. J. 1984. Contrast sensitivity and orientation selectivity in lamina IV of the striate cortex of Old World monkeys. *Exp. Brain Res.* 54:367–72

Hicks, T. P., Lee, B. B., Vidyasagar, T. R. 1983. The responses of cells in macaque lateral geniculate nucleus to sinusoidal gratings. *J. Physiol.* 337:183–200

Hochstein, S., Shapley, R. M. 1976a. Quantitative analysis of retinal ganglion cell classifications. *J. Physiol.* 262:237–64

Hochstein, S., Shapley, R. M. 1976b. Linear and nonlinear subunits in Y cat retinal ganglion cells. *J. Physiol.* 262:265–84

Hochstein, S., Spitzer, H. 1985. One, few, infinity. Linear and nonlinear processing in visual cortex. In *Models of the Visual Cortex*, ed. D. Rose, V. G. Dobson. New York: Wiley

Hoffmann, K.-P., Stone, J. 1971. Conduction velocity of afferents to cat visual cortex: A correlation with cortical receptive field properties. *Brain Res.* 32:460–66

Hubel, D. H., Wiesel, T. N. 1961. Integrative action in the cat's lateral geniculate body. *J. Physiol.* 155:385–98

Hubel, D. H., Wiesel, T. N. 1962. Receptive fields, binocular interaction, and function architecture in the cat's visual cortex. *J. Physiol.* 160:106–54

Hubel, D. H., Wiesel, T. N. 1977. Ferrier Lecture. Functional architecture of macaque monkey visual cortex. *Proc. Ry. Soc. London Ser. B.* 198:1–59

Ikeda, H., Wright, M. J. 1975. Spatial and temporal properties of "sustained" and "transient" neurones in area 17 of the cat's visual cortex. *Exp. Brain Res.* 22:362–82

Kaplan, E., Shapley, R. M. 1982. X and Y cells in the lateral geniculate nucleus of macaque monkeys. *J. Physiol.* 330:125–43

Kaplan, E., Shapley, R. M. 1984. The origin of the S(Slow) potential in the mammalian lateral geniculate nucleus. *Exp. Brain Res.* 55:111–16

Kelly, D. H., Burbeck, C. A. 1984. Critical problems in spatial vision. *CRC Crit. Rev. Biomed. Eng.* 10:125–77

Krüger, J., Fischer, B. 1973. Strong periphery effect in cat retinal ganglion cells. Excitatory responses in on- and off-center neurones to single grid displacements. *Exp. Brain Res.* 18:316–18

Kulikowski, J. J., Bishop, P. O. 1981. Linear analysis of the responses of simple cells in the cat visual cortex. *Exp. Brain Res.* 44:386–400

Kulikowski, J. J., Marcelja, S., Bishop, P. O. 1982. Theory of spatial position and spatial frequency relations in the receptive fields of simple cells in the visual cortex. *Biol. Cybern.* 43:187–98

Lehmkuhle, S., Kratz, K. E., Mangel, S. C., Sherman, S. M. 1980. Spatial and temporal sensitivity of X- and Y- cells in dorsal lateral geniculate nucleus of the cat. *J. Neurophysiol.* 43:520–41

Lennie, P. 1980. Parallel visual pathways: A review. *Vis. Res.* 20:561–94

Levick, W. R., Thibos, L. N. 1982. Analysis of orientation bias in cat retina. *J. Physiol.* 329:243–61

Linsenmeier, R. A., Frishman, L. J., Jakiela, H. G., Enroth-Cugell, C. 1982. Receptive field properties of X and Y cells in the cat retina derived from contrast sensitivity measurements. *Vis. Res.* 22:1173–83

Maffei, L. 1978. Spatial frequency channels: Neural mechanisms. *Handb. Sensory Physiol.* 8:39–66

Maffei, L., Fiorentini, A. 1973. The visual cortex as a spatial frequency analyzer. *Vision Res.* 13:1255–67

Maffei, L., Fiorentini, A. 1977. Spatial frequency rows in the striate visual cortex. *Vision Res.* 17:257–64

Maffei, L., Fiorentini, A., Bisti, S. 1973. Neural correlate of perceptual adaptation to gratings. *Science* 182:1036–38

Maffei, L., Morrone, M. C., Pirchio, M., Sandini, G. 1979. Responses of visual cortical cells to periodic and non-periodic stimuli. *J. Physiol.* 296:27–47

Marcelja, S. 1980. Mathematical description of the responses of simple cortical cells. *J. Opt. Soc. Am.* 70:1297–1300

Marr, D. 1981. *Vision*. San Francisco: Freeman

Morrone, M. C., Burr, D. C., Maffei, L. 1982. Functional implications of cross-orientation inhibition of cortical visual cells. I. Neurophysiological evidence. *Proc. R. Soc. London Ser. B* 216:335–54

Movshon, J. A., Lennie, P. 1979. Pattern selective adaptation in visual cortical neurones. *Nature* 278:850–51

Movshon, J. A., Thompson, I. D., Tolhurst, D. J. 1978a. Spatial summation in the receptive fields of simple cells in the cat's striate cortex. *J. Physiol.* 283:53–77

Movshon, J. A., Thompson, I. D., Tolhurst, D. J. 1978b. The receptive field organization of complex cells in the cat's striate cortex. *J. Physiol.* 283:79–99

Movshon, J. A., Thompson, I. D., Tolhurst, D. J. 1978c. Spatial and temporal contrast sensitivity of neurones in areas 17 and 18 of the cat's visual cortex. *J. Physiol.* 283:101–20

Movshon, J. A., Adelson, E. H., Gizzi, M. S., Newsome, W. T. 1984. The analysis of moving visual patterns. In *Pattern Recognition Mechanisms*, ed. C. Chagas, R. Gat-

tass, C. G. Gross. Rome: Vatican Press
Ohzawa, I., Sclar, G., Freeman, R. D. 1982. Contrast gain control in the cat visual cortex. *Nature* 298:266–68
Padmos, P., Van Norren, D. V. 1975. Cone systems interaction in single neurons of the lateral geniculate nucleus of the macaque. *Vis. Res.* 15:617–19
Pollen, D. A., Lee, J. R., Taylor, J. H. 1971. How does the striate cortex begin the reconstruction of the visual world? *Science?* *173:74–77*
Pollen, D. A., Ronner, S. F. 1981. Phase relationship between adjacent simple cells in the visual cortex. *Science* 212:1409–11
Pollen, D. A., Ronner, S. F. 1982a. Spatial frequency selectivity of periodic complex cells in the visual cortex of the cat. *Vis. Res.* 18: 665–82.
Pollen, D. A., Ronner, S. F. 1982b. Spatial computation performed by simple and complex cells in the visual cortex of the cat. *Vis. Res.* 22:101–18
Pollen, D. A., Ronner, S. F. 1983. Visual cortical neurons as localized spatial frequency filters. *IEEE Trans. Syst. Man. Cybern.* SMC–13:907–16
Robson, J. G. 1975. Receptive fields: Spatial and intensive representations of the visual image. *Handb. Percept.* 5:81–115
Robson, J. G. 1980. Neural images: The physiological basis of spatial vision. In *Visual Coding and Adaptability,* ed. C. Harris, pp. 177–214. Hillsdale, NJ: Erlbaum
Rodieck, R. W. 1965. Quantitative analysis of cat retinal ganglion cell response to visual stimuli. *Vis. Res.* 5:583–601
Rodieck, R. W. 1979. Visual pathways. *Ann. Rev. Neurosci.* 2:193–225
Schiller, P. H., Finlay, B. L., Volman, S. F. 1976. Quantitative studies of single-cell properties in monkey striate cortex. III. Spatial frequency. *J. Neurophysiol.* 39:1334–51
Schiller, P. H., Malpeli, J. G. 1978. Functional specificity of lateral geniculate nucleus laminae of the rhesus monkey. *J. Neurophysiol.* 41: 788–97
Schoppmann, A., Stryker, M. P. 1981. Physiological evidence that the 2-deoxyglucose method reveals orientation columns in cat visual cortex. *Nature* 293:574–75
Shapley, R. M., Hochstein, S. 1975. Visual spatial summation in two classes of geniculate cells. *Nature* 256:411–13
Shapley, R. M., Victor, J. D. 1978. The effect of contrast on the transfer properties of cat retinal ganglion cells, *J. Physiol.* 285:275–98
Shapley, R., Kaplan, E., Soodak, R. 1981. Spatial summation and contrast sensitivity of X and Y cells in the lateral geniculate nucleus of the macaque. *Nature* 292:543–45
Sherman, S. M., Wilson, J. R., Kaas, J. H., Webb, S. V. 1976. X and Y cells in the dorsal lateral geniculate nucleus of the owl monkey. *Science* 192:475–77
So, Y. T., Shapley, R. M. 1979. Spatial properties of X and Y cells in the lateral geniculate nucleus of the cat and conduction velocities of their inputs. *Exp. Brain Res.* 36:533–50
So, Y. T., Shapley, R. M. 1981. Spatial tuning of cells in and around lateral geniculate nucleus of the cat: X and Y relay cells and perigeniculate interneurons. *J. Neurophysiol.* 45:107–20
Sokoloff, L., Reivich, M., Kennedy, C., Des Rosiers, M. H., Patlak, C. S., Pettigrew, K. D., Sakurada, O., Shinohara, M. 1977. The [^{14}C]deoxyglucose method for the measurement of local cerebral glucose utilization: Theory, procedure, and normal values in the conscious and anesthetized albino rat. *J. Neurochem.* 28:897–916
Stone, J. 1972. Morphology and physiology of the geniculo-cortical synapse in the cat: The question of parallel input to the striate cortex. *Invest. Ophthalmol.* 11:338–46
Stone, J. 1983. *Parallel Processing in the Visual System: The Classification of Retinal Ganglion Cells and Its Impact on the Neurobiology of Vision.* New York: Plenum
Stone, J., Fukuda, Y. 1974. Properties of cat retinal ganglion cells: A comparison of W-cells with X- and Y-cells. *J. Neurophysiol.* 37:722–48
Sur, M., Sherman, S. M. 1982. Linear and nonlinear W-cells in the C-laminae in the cat lateral geniculate nucleus. *J. Neurophysiol.* 47:869–84
Tanaka, K. 1983. Cross-correlation analysis of geniculostriate neuronal relationships in cats. *J. Neurophysiol.* 49:1303–18
Thibos, L. N., Levick, W. R. 1983. Spatial frequency characteristics of brisk and sluggish ganglion cells of the cat's retina. *Exp. Brain Res.* 51:16–22
Tolhurst, D. J. 1977. Symmetry and receptive fields. In *Spatial Contrast,* ed. H. Spekreijse, L. H. van der Tweel, pp. 36–38. Amsterdam: North Holland
Tolhurst, D. J., Movshon, J. A. 1975. Spatial and temporal contrast sensitivity of striate cortical neurones. *Nature* 257:674–75
Tolhurst, D. J., Thompson, I. D. 1981. On the variety of spatial frequency selectivities shown by neurons in area 17 of the cat. *Proc. R. Soc. London Ser. B* 213:183–99
Tolhurst, D. J., Thompson, I. D. 1982. Organization of neurones preferring similar spatial frequencies in cat striate cortex. *Exp. Brain Res.* 48:217–27
Tootell, R. B. H., Silverman, M. S., De Valois, R. L. 1981. Spatial frequency columns in primary visual cortex. *Science* 214:813–15
Tootell, R. B. H., Silverman, M. S., De Valois,

R. L., Jacobs, G. H. 1983. Functional organization of the second cortical visual area in primates. *Science* 220:737–39

Troy, J. B. 1983a. Spatial contrast sensitivities of X and Y type neurones in the cat's dorsal lateral geniculate nucleus. *J. Physiol.* 344:399–417

Troy, J. B. 1983b. Spatio-temporal interaction in neurones of the cat's dorsal lateral geniculate nucleus. *J. Physiol.* 344:419–32

Victor, J. D., Shapley, R. M. 1979a. Receptive field mechanisms of cat X and Y retinal ganglion cells. *J. Gen. Physiol.* 74:275–98

Victor, J. D., Shapley, R. M. 1979b. The nonlinear pathway of Y ganglion cells in the cat retina. *J. Gen. Physiol.* 74:671–89

Westheimer, G. 1984. Spatial vision. *Ann. Rev. Psychol.* 35:201–26

Wiesel, T. N., Hubel, D. H. 1966. Spatial and chromatic interactions in the lateral geniculate body of the rhesus monkey. *J. Neurophysiol.* 29:1115–56

SUBJECT INDEX

T

V

CUMULATIVE INDEXES

CONTRIBUTING AUTHORS, VOLUMES 4–8

CHAPTER TITLES, VOLUMES 4–8